Modern Biocatalysis
Advances Towards Synthetic Biological Systems

Catalysis Series

Series editors:
Bert Klein Gebbink, *Utrecht University, The Netherlands*
Jose Rodriguez, *Brookhaven National Laboratory, USA*

Titles in the series:
1: Carbons and Carbon Supported Catalysts in Hydroprocessing
2: Chiral Sulfur Ligands: Asymmetric Catalysis
3: Recent Developments in Asymmetric Organocatalysis
4: Catalysis in the Refining of Fischer–Tropsch Syncrude
5: Organocatalytic Enantioselective Conjugate Addition Reactions: A Powerful Tool for the Stereocontrolled Synthesis of Complex Molecules
6: N-Heterocyclic Carbenes: From Laboratory Curiosities to Efficient Synthetic Tools
7: *P*-Stereogenic Ligands in Enantioselective Catalysis
8: Chemistry of the Morita–Baylis–Hillman Reaction
9: Proton-Coupled Electron Transfer: A Carrefour of Chemical Reactivity Traditions
10: Asymmetric Domino Reactions
11: C–H and C–X Bond Functionalization: Transition Metal Mediation
12: Metal Organic Frameworks as Heterogeneous Catalysts
13: Environmental Catalysis Over Gold-Based Materials
14: Computational Catalysis
15: Catalysis in Ionic Liquids: From Catalyst Synthesis to Application
16: Economic Synthesis of Heterocycles: Zinc, Iron, Copper, Cobalt, Manganese and Nickel Catalysts
17: Metal Nanoparticles for Catalysis: Advances and Applications
18: Heterogeneous Gold Catalysts and Catalysis
19: Conjugated Linoleic Acids and Conjugated Vegetable Oils
20: Enantioselective Multicatalysed Tandem Reactions
21: New Trends in Cross-Coupling: Theory and Applications
22: Atomically-Precise Methods for Synthesis of Solid Catalysts
23: Nanostructured Carbon Materials for Catalysis
24: Heterocycles from Double-Functionalized Arenes: Transition Metal Catalyzed Coupling Reactions
25: Asymmetric Functionalization of C–H Bonds
26: Enantioselective Nickel-catalysed Transformations
27: N-Heterocyclic Carbenes: From Laboratory Curiosities to Efficient Synthetic Tools, 2nd edition
28: Zeolites in Catalysis: Properties and Applications
29: Biocatalysis: An Industrial Perspective

30: Dienamine Catalysis for Organic Synthesis
31: Metal-free Functionalized Carbons in Catalysis: Synthesis, Characterization and Applications
32: Modern Biocatalysis: Advances Towards Synthetic Biological Systems

How to obtain future titles on publication:
A standing order plan is available for this series. A standing order will bring delivery of each new volume immediately on publication.

For further information please contact:
Book Sales Department, Royal Society of Chemistry, Thomas Graham House, Science Park, Milton Road, Cambridge, CB4 0WF, UK
Telephone: +44 (0)1223 420066, Fax: +44 (0)1223 420247
Email: booksales@rsc.org
Visit our website at www.rsc.org/books

Modern Biocatalysis
Advances Towards Synthetic Biological Systems

Edited by

Gavin Williams
North Carolina State University, USA
Email: gjwillia@ncsu.edu

and

Mélanie Hall
University of Graz, Austria
Email: melanie.hall@uni-graz.at

ROYAL SOCIETY
OF CHEMISTRY

Catalysis Series No. 32

Print ISBN: 978-1-78262-726-5
PDF ISBN: 978-1-78801-045-0
EPUB ISBN: 978-1-78801-453-3
ISSN: 1757-6725

A catalogue record for this book is available from the British Library

The Royal Society of Chemistry is a charity, registered in England and Wales, Number 207890, and a company incorporated in England by Royal Charter (Registered No. RC000524), registered office: Burlington House, Piccadilly, London W1J 0BA, UK, Telephone: +44 (0) 207 4378 6556.

For further information see our web site at www.rsc.org

Printed in the United Kingdom by CPI Group (UK) Ltd, Croydon, CR0 4YY, UK

Preface

Biocatalysis—the use of enzymes for chemical transformations—has a long history in providing mankind with all sorts of molecules. While early processes almost exclusively relied on naturally occurring whole-cell systems, the implementation of molecular biology-based methods has rendered the manipulation of isolated enzymes routine work. This in turn has led to improved knowledge about enzymatic mechanisms and boosted enzyme-orientated research in multiple areas. The past years have seen impressive advances leading to sophisticated tools and innovative techniques for the design and development of bio-based processes for the production of (fine) chemicals. The synergy between synthetic biology and biocatalysis is now strongly emerging as an important trend for future sustainable processes and we felt the need to merge these two complementary branches, which have been evolving mainly concurrently.

Excellent books are available for chemists wishing to implement natural catalysts in synthetic processes, while recent books on synthetic biology focus on bottom-up creation of new modular parts, circuit design, and chassis engineering but are not enzyme centric. Specialized and focused reviews in prominent journals on the development of robust and efficient biosynthetic routes are becoming more frequent, but a unifying platform was still missing. This book includes a number of contributions to document the current merging of traditional biocatalysis with more synthetic biology-based approaches, and keeps enzymes as the central protagonists.

The book is organized into five sections. In Section I, *Accessing New Enzymes*, several contributions exemplify the technical diversity at hand to identify new enzymes, which largely benefits from ever-improving computational power. In Section II, *Understanding and Engineering Enzymes*, we gathered experts to highlight how modifying protein sequence and structure

Catalysis Series No. 32
Modern Biocatalysis: Advances Towards Synthetic Biological Systems
Edited by Gavin Williams and Mélanie Hall

Published by the Royal Society of Chemistry, www.rsc.org

in silico and *in vivo* is fundamental to obtaining crucial catalytic insights used to tailor enzyme properties. In Section III, *Enzymes from Secondary Metabolism*, several contributions highlight the remarkable ability of enzymes in secondary metabolism to construct complex natural products from simple small molecule building blocks. Approaches to engineer and optimize these pathways are also described. Section IV, *Biocatalysis for Modern Synthesis*, focuses on elaborate enzyme-based processes developed for the synthesis of fine chemicals. Importantly, multi-step reactions now combine various types of (bio)catalysts and are particularly well suited for the cost-effective generation of enantiopure molecules of high value. Finally, Section V, *Applied Biocatalysis*, reflects the technological input required to implement natural or engineered enzymes in industrial settings, and finishes with an opening on the promising use of enzymes in remediation, according to which biodegradation processes might witness a restored interest.

We hope that readers will find this book helpful to connect all areas that biocatalysis—and synthetic biology—as progressive field now encompasses, and to identify current and emerging trends towards the development of efficient synthetic biological systems in a broad sense.

We thereby hope to trigger mutual inspiration for the currently unfolding generation of hybrid chemists/biologists who are committed to render current and upcoming synthetic processes more sustainable.

Tools exist, let's use them!

Mélanie Hall (Graz, Austria)
Gavin Williams (Raleigh, USA)

Contents

Catalysis Series No. 32
Modern Biocatalysis: Advances Towards Synthetic Biological Systems
Edited by Gavin Williams and Mélanie Hall

Published by the Royal Society of Chemistry, www.rsc.org

Section I: Accessing New Enzymes

CHAPTER 1

Genome Mining for Enzyme Discovery

ANNE ZAPARUCHA,* VÉRONIQUE DE BERARDINIS AND CARINE VAXELAIRE-VERGNE

Génomique Métabolique, Genoscope, Institut François Jacob, CEA CNRS, Univ Evry, Univ Paris-Saclay, 91057 Evry, France
*Email: azaparuc@genoscope.cns.fr

1.1 Introduction

Nature appears as the veteran protein engineer since it began its bioengineering 'experiments' billions of years ago.[1] A number of strategies have been developed to exploit the extraordinary large source of enzymes contained in genome and metagenome sequences and to discover novel biocatalysts. Historically, strategies are based on *in vivo* selection on individual or collections of strains, and for a decade, on metagenomes (consortium of uncultivated microorganisms). In brief, DNA was extracted from mixed microbial populations and size-selected inserts were cloned into suitable expression vectors.[2] Screening for enzymatic activities is generally performed *in situ* and based on an indicator medium. Positive clones are then sequenced to identify the genes of interest. Such an approach is very effective but restricted to enzymes for which a generic assay, most of the time a colorimetric one, can monitor the activities (*e.g.* lipases, amylases, oxidases, *etc.*).[3] Moreover, the screening is performed without overexpression of the protein, resulting in a limited sensitivity. It is estimated that biotechnology has missed up to 99% of existing microbial resources by using traditional

Catalysis Series No. 32
Modern Biocatalysis: Advances Towards Synthetic Biological Systems
Edited by Gavin Williams and Mélanie Hall

Published by the Royal Society of Chemistry, www.rsc.org

screening techniques, and when the desired transformation has been identified, the isolation of enzymes from wild-type strains is usually time-consuming and, most of the time, the protein loses much of its catalytic activity.[4]

In contrast, genome sequence information enables direct cloning of the targeted genes using the polymerase chain reaction (PCR) and thus an efficient expression in a proper heterologous host strain, even if PCR errors and/or expression drawbacks may be encountered. The emergence of high-throughput sequencing (or Next Generation Sequencing, NGS) in the mid-2000s has generated an incredible amount of sequences deposited in databases (in October 2016, around 67 000 000 of protein sequences from 509 000 species have been deposited in the TrEMBL database, http://www.uniprot.org/uniprot/ TrEMBL source) and more sequences are expected in the future. In addition, the deep sequencing of metagenomes from diverse environments offers a huge reservoir of unexploited enzymes which reflect specific metabolic requirements for a defined process [*e.g.* waste water treatment, bioremediation (see Chapter 18)], or a particular ecological niche.[5]

The exploration of the extraordinary amount of available genomic resources can be rationalized to optimize the experimental effort by computational methods that try to reveal the sequence/function relationships of proteins. In addition, the ability of an enzyme to not only transform its metabolic substrate, but also catalyze the same chemical transformation for a range of different substrates, expands the field of conversion possibilities. This property, inherent to many enzymes and outlined as a potential for biocatalyst discovery for dozens of years, expands the chemical capability of enzymes and the chemistry performed by living cells.[6–9] The search for new biocatalysts is then mostly based on the hypothesis of substrate promiscuity, since unnatural substrates are often targeted in organic synthesis.

From a genome sequence, a plethora of information is available, from the function reflected in the name of an enzyme to conserved patterns/signatures and even the (predicted) structure; features that can be parsed to search for new biocatalysts.[10] Most of the time, to handle the huge amount of data, the information is retrieved through comprehensive organized databases or using computational approaches. For example, the database BRENDA (BRaunschweig ENzyme DAtabase; www.brenda-enzymes.org) contains extensive details on a full suite of known enzyme substrates, thus providing comprehensive indications about the biocatalytic potential of enzymes.[11]

All these data offer a great potential to identify new enzymes for biotechnological applications: discovery of novel enzymes with new properties, enhanced or inverted (chemo-, regio-, stereo-) selectivity, altered pH- or temperature profile, improved stability (temperature, solvent *etc.*), substrate or product inhibition, enhanced catalytic efficiency.

In this chapter, we will present the different genome mining approaches for enzyme discovery. Case studies relevant to synthetic applications will be described.

1.2 Text-based Searches Using Enzyme Name

Protein sequences obtained from sequencing of single gene, entire genome or microorganism consortia are available through public databases such as UniprotKB or NCBI databases. Functional prediction of proteins (name annotation) is mainly performed automatically by sequence comparison with already annotated enzymes. *In silico* screening of public databases for a specific enzyme using the name of the enzymatic function as query has long been one of the easiest way to find new enzymes. However, this approach suffers from two main drawbacks: (1) the experimentally established functions only concern a tiny fraction of the enzymes, since function is mainly extrapolated from a small number of characterized proteins (partially inventoried in the Swissprot section of Uniprot); (2) it is limited by the lack of novelty in the features of the newly identified enzymes.[12]

The explosion of the amount of data produced by the NGS generates a growing number of sequences without reliable annotation. Today, nearly 40% of the sequences stored in the most comprehensive protein database UniProtKB are labeled as "uncharacterized protein." In addition, at least 20% of assigned functions are estimated to be wrong in databases.[12] Consequently, the annotation query approach does not make the most out of the data potential, and this will be even truer in the future. Nonetheless, it has been successfully applied in many projects regarding various enzyme families including, but not limited to, nitrilases, cytochromes P450, glycosylhydrolases and halogenases, from fungal and bacterial origin.[13–19] It may also be used as first-line search then sharpened with additional criteria, *e.g.*, sequence alignment, conserved motifs/key residues, phylogeny or 3D model analysis, to search for particular enzyme features, substrate specificity, thermophily, or new homologs of known enzyme with broad biotechnological applications.[20–25]

Thus, by combining an annotation query with analysis of the genomic context of the putative gene of interest, Zhu *et al.* discovered a nitrilase highly active towards the targeted substrate mandelonitrile (Figure 1.1).[26] In prokaryotic genomes, the genes encoding enzymes involved in the same biosynthetic pathway are generally co-localized in gene clusters; therefore analysis of the genomic organization (genes in their genomic neighborhood) of one gene within the chromosome provides clues on the natural function of the encoded enzymes, especially regarding their substrates. Firstly, a query using "nitrilase" as an identifier to the NCBI Gene database (http://www.ncbi.nlm.nih.gov/entrez/query.fcgi?db=gene) gave hits which were sorted to keep only bacterial hits containing a carbon–nitrogen hydrolase domain and 300–385 amino acid length. Among the remaining 16 putative nitrilase genes, the genomic context of the bll6402 gene from *B. japonicum*

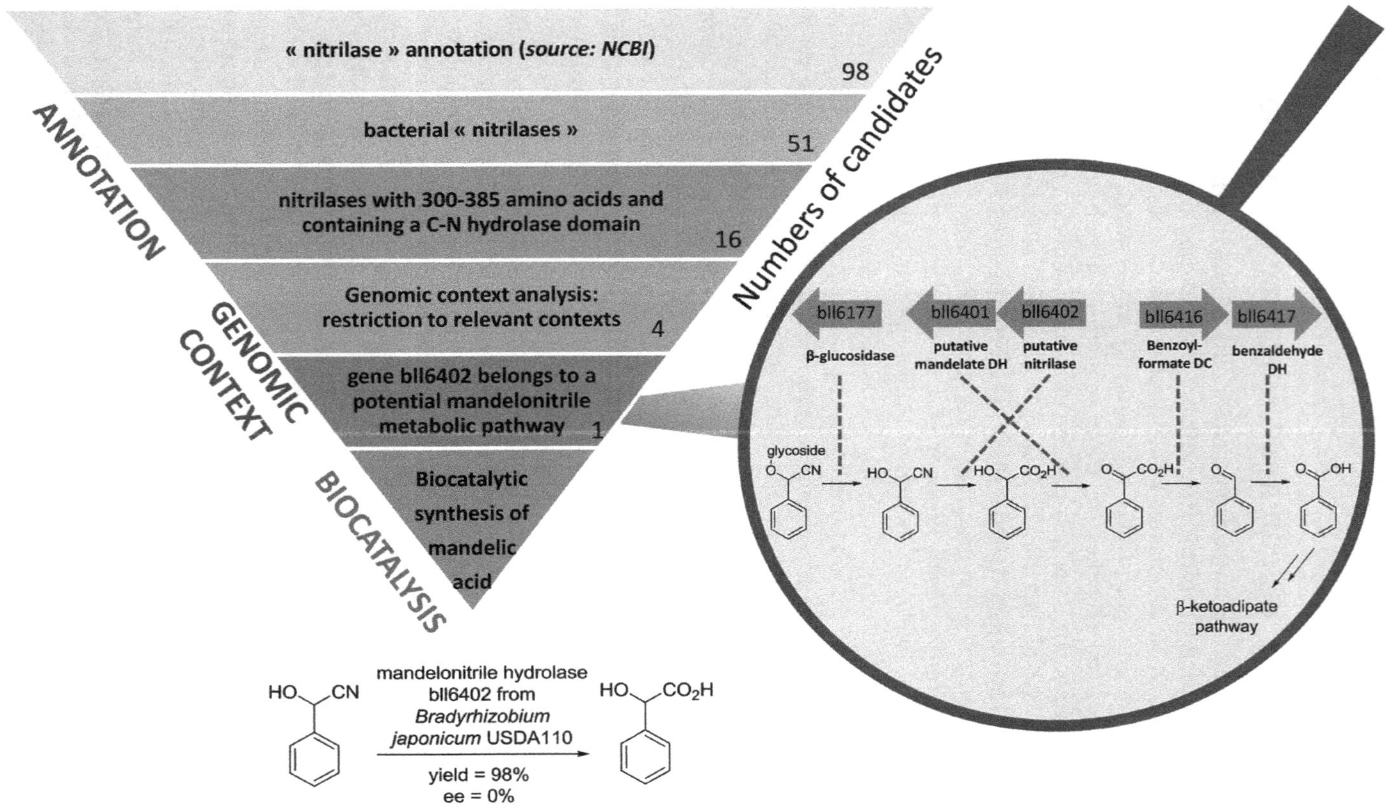

Figure 1.1 Discovery of a nitrilase active towards mandelonitrile by combining annotation query with analysis of the genomic context.

USDA110 suggested a mandelonitrile metabolic pathway. The bll6402 nitrilase was indeed found to be a mandelonitrile hydrolase, which effectively catalyzed the hydrolysis of mandelonitrile and derivatives to the corresponding carboxylic acids.

Given that many open reading frames have no predicted function or are incorrectly annotated, the text-based approach appears to be less attractive in the context of broad research projects. To continue to be viable, this method should be accompanied by an effort to accurately annotate enzyme families. There are some curated databases in which it can be of interest to perform text-based search, examples include the Carbohydrate-Active enZYmes Database (CAZy: http://www.cazy.org/) or the Lipase Engineering Database (http://www.led.uni-stuttgart.de/).

1.3 Sequence-driven Approaches

This section presents the different ways to access new enzymes by using at least one described protein or nucleotide sequence. These approaches certainly belong to the most common ones to discover new enzymes. It is based on the analysis of the primary sequence of proteins as a whole or in specific portions.

With a characterized enzyme and related gene as a starting point, new enzymes performing the same or similar reactions can be identified. The gene sequence encoding the known protein can be used as template to target unsequenced genes by designing primers for their amplification or to target already sequenced genes by pairwise protein sequence alignment tools. These approaches explore the steadily expanding protein-sequence data space and open the way for the efficient discovery of novel biocatalysts.

1.3.1 Probe Technology Based on PCR Primer Design

Sequence homology-based screening, involving polymerase chain reaction (PCR)-based approaches targeting novel genes similar to already known ones, is very fruitful. It also enables exploration of complex DNA mixtures from metagenomes for which no genomic data is available. This approach is based on the design of PCR primers using as the template a parental gene encoding for an enzyme with similar catalytic activity to that of the targeted enzyme. These primers are then degenerated, allowing a set of primers with all combinations of alternative triplets of nucleotides encoding each codon at all positions. These primers are used for PCR amplification of close homologous genes.

Since most of the time these primers are designed in inner conserved regions, the new genes are often only partially amplified and the gene sequences are usually completed by inverse PCR to obtain the flanking sequences. The full gene is later cloned into an expression vector. For example, new type I BVMO genes were discovered by using highly degenerate oligonucleotides. Those primers were used for amplification of internal conserved region of type I BVMO genes present in total DNA isolated from strains able to grow on alicyclic compounds.[27] As mentioned above, this method is largely applicable for metagenome exploration, as illustrated by the discovery from marine metagenomes of new laccases with alkalescence-dependent activity, by using highly degenerate primers designed to target the conserved region in copper-binding sites of the laccases.[28] Other enzymes from different families were discovered through this approach, such as alcohol dehydrogenases,[29] lipases,[30] cytochromes P450,[31] 2,5-diketo-D-gluconic acid reductases[32] or alpha-amino acid pyruvate transaminases.[33]

By combining this probe technology with genomic context analysis, new enzymes have been identified in strains producing novel natural products (Figure 1.2). The paradigm used in this approach was that products with identical or similar structural elements are produced by biosynthetic pathways that contain highly homologous enzymes.[34] For example, halogenases were used as suitable targets to predict the biosynthetic potential of different *Actinomycetes* strains. It was predicted that *Actinomycetes* that harbor putative halogenase sequences in a particular genetic context have the potential to synthesize compounds that belong to the respective natural product class.

Nevertheless, it should be noted that the PCR probe-based screening approach is inherently conservative since the primers reflect conserved amino acid sequence motifs to match the targeted genes with reasonable likelihood.

1.3.2 Pairwise Sequence Alignment-based Strategy

To identify new enzymes, a "sequence identity" strategy can be applied among the complete nucleotide and protein sequences indexed in databases. Unlike the probe technology described earlier in this chapter, in the "sequence identity" strategy, the search is conducted through available genomic data. One practical advantage is that the genes can easily be cloned from the original source organism or synthesized. Enzymes with experimentally validated catalytic activity are usually used as queries for pairwise protein sequence alignment (as BLAST algorithm, https://blast.ncbi.nlm.nih.gov/Blast.cgi), mainly against protein databases such as UniprotKB or GenBank. The modulation of BLAST parameters, for example the percentage of identity, allows retrieval of sequences more or less similar to the query enzymes.

When many sequences are then retrieved, one can reduce their number to proteins representative of the functional diversity of the enzyme family. This can be guided by a partition of the family into putative iso-functional subfamilies. The simplest method is a clustering based on protein sequence identity. Other methods can be used, such as phylogenetic analyses, genomic contexts or structural classification (Figure 1.3). Candidate enzymes are then selected from each hypothesized iso-functional group in order to clone their corresponding genes.

In addition, some research groups used refined criteria to select enzymes with very little similarity to the biocatalysts used as parental sequences, like

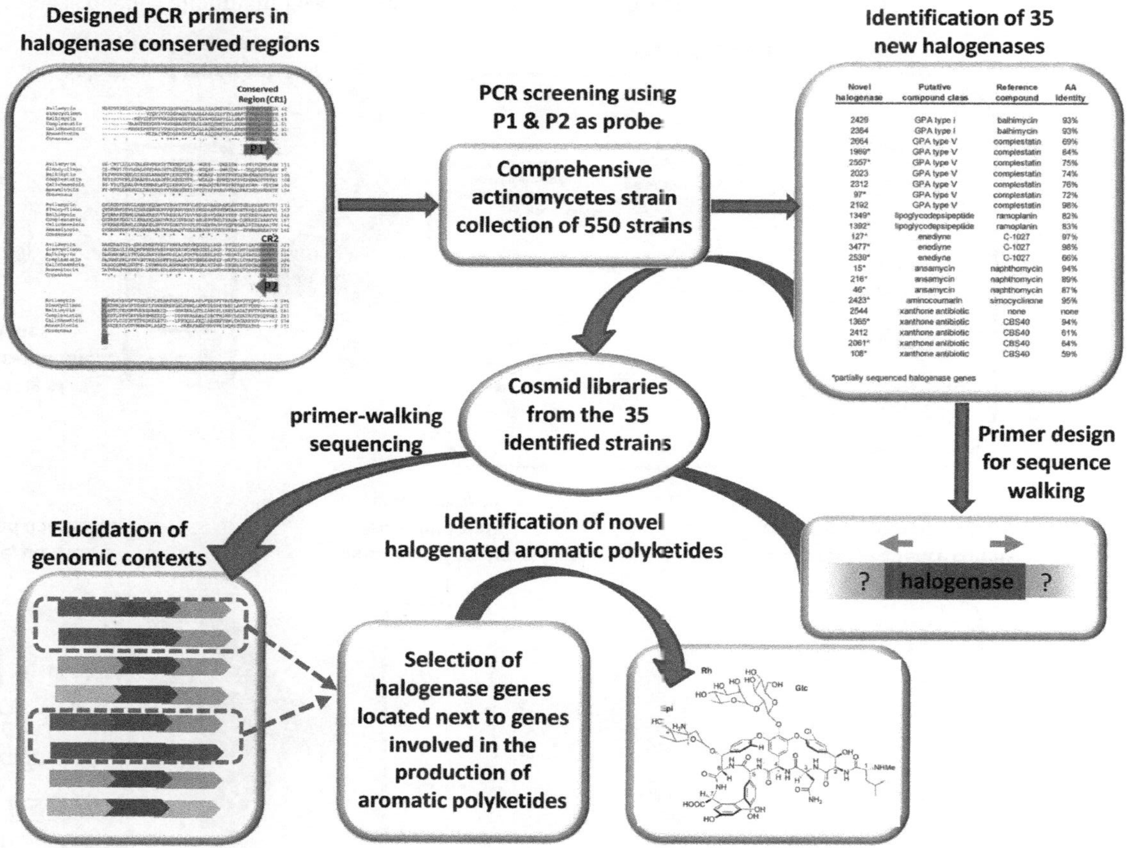

Figure 1.2 Combined strategy to identify new natural products (inspired by the Pelzer study[34]).

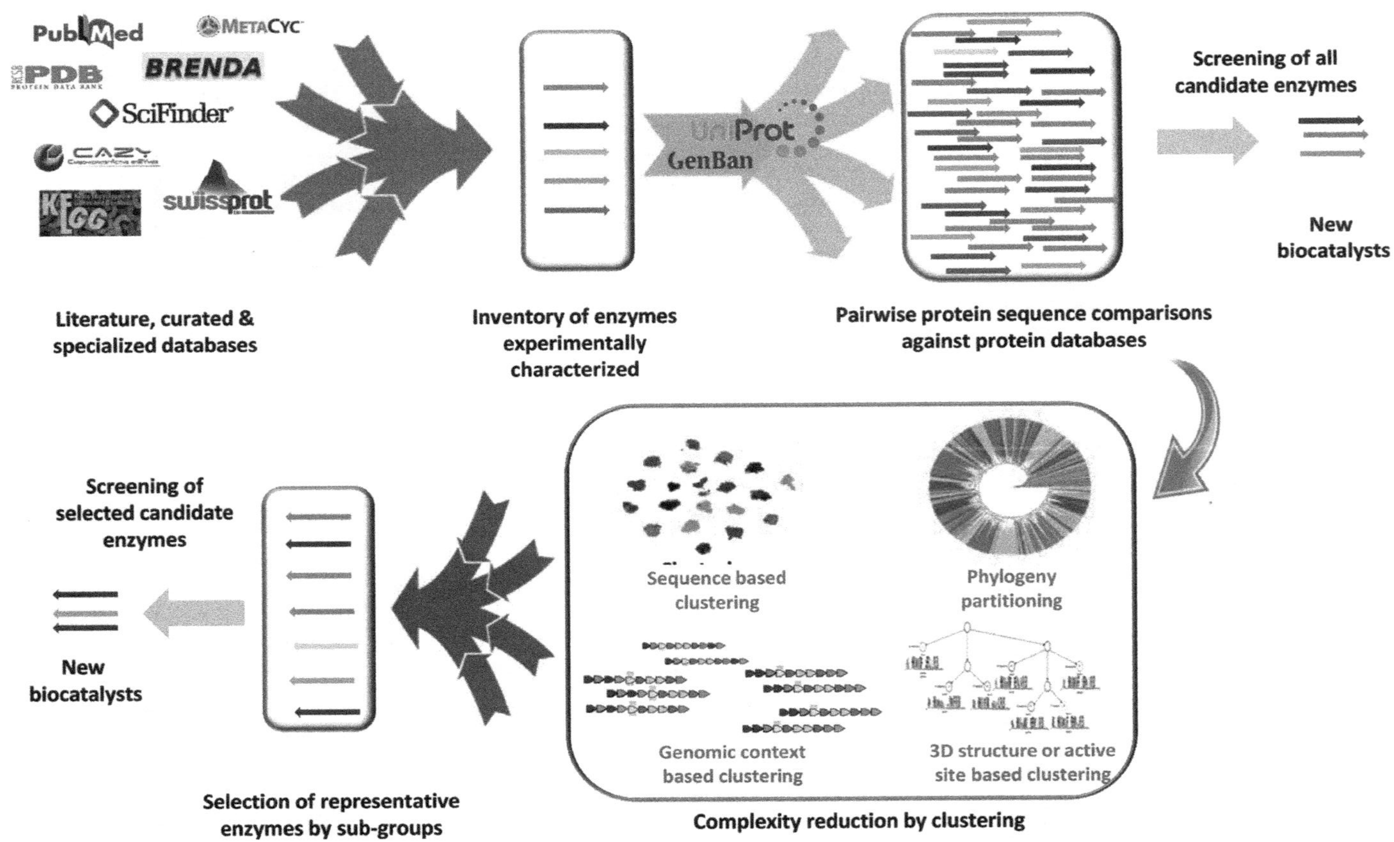

Figure 1.3 Pairwise sequence alignment-based strategy for new enzyme discovery.

the construction of a dendogram or phylogenetic tree of the candidate proteins, to catch variabilities in substrate ranges.[35] This selection can also be restricted to a subgroup of strains with particular features such as thermophily to find thermostable/thermoactive enzymes.[10,24]

Selected genes are then amplified from genomic DNA by PCR amplification and cloned into expression vectors. The number of genes to be cloned depends on the availability of the original DNA strain from research group or institution strain collections, or from commercial suppliers. If the DNA is not available, the gene can be artificially synthesized. Efficient production of heterologous proteins in the host organism can be limited however by the rarity of certain tRNAs that are abundant in the organisms from which the heterologous proteins are derived. Forced high-level expression of heterologous proteins can deplete the pool of rare tRNAs of the host organism and stall translation. Gene sequence optimization with synthetic genes, by reassigning codon usage to the one of the host organism, can allow increase in the over-production of heterologous enzymes. Nevertheless, currently, this is not economically viable for providing a large number of genes and the use of optimized expression host organisms for rare codons can be an issue. A number of technical improvements in molecular biology, including Ligation Independent Cloning (LIC), have permitted the development of efficient universal cloning well adapted to automated cloning processed in 96 microwell plates.[36] The development of cell-free protein synthesis systems avoiding cloning, cell transformation, induction and culture steps, allows direct expression of enzymes in microtiter plates. This high throughput method was used for the discovery of new omega-transaminases among a microbial community.[37]

Among the many enzymes discovered thanks to this approach, we can highlight the work done on the nitrilase family.[38] Database searches using the BLASTP (protein query against protein database) programs and the multiple alignment of amino acid sequences COBALT (Constraint-based Multiple Alignment Tool) enabled the discovery of new nitrilases in filamentous fungi.[14] Similar approach gave access to a nitrilase toolbox for the organic chemist.[39,40] Recently, Guérard-Hélaine *et al.* have screened a large collection of aldolases. A liquid chromatography–mass spectrometry assay led to the discovery of new dihydroxyacetone (DHA) aldolases, wrongly annotated as transaldolases (Figure 1.4).[41]

By using the *Vibrio fluvialis* ω-transaminase (ω-TA) sequence, Kaulmann *et al.* recruited 15 new ω-TAs with low sequence identities (31–38%). Among them, the ω-TA from *Chromobacterium violaceum* was found useful for the stereoselective amination of a large substrate range, in particular for ketodiol conversion.[42] Lavandera *et al.* found an exceptionally solvent-tolerant alcohol dehydrogenase from *Paracoccus pantotrophus*.[43] First, they selected clones of *Paracoccus pantotrophus* DSM11072 for their oxido-reductive behavior. From the sequenced genomes, they identified a short-chain alcohol dehydrogenase gene. The corresponding enzyme stereoselectively reduced various ketones in organic co-solvents.

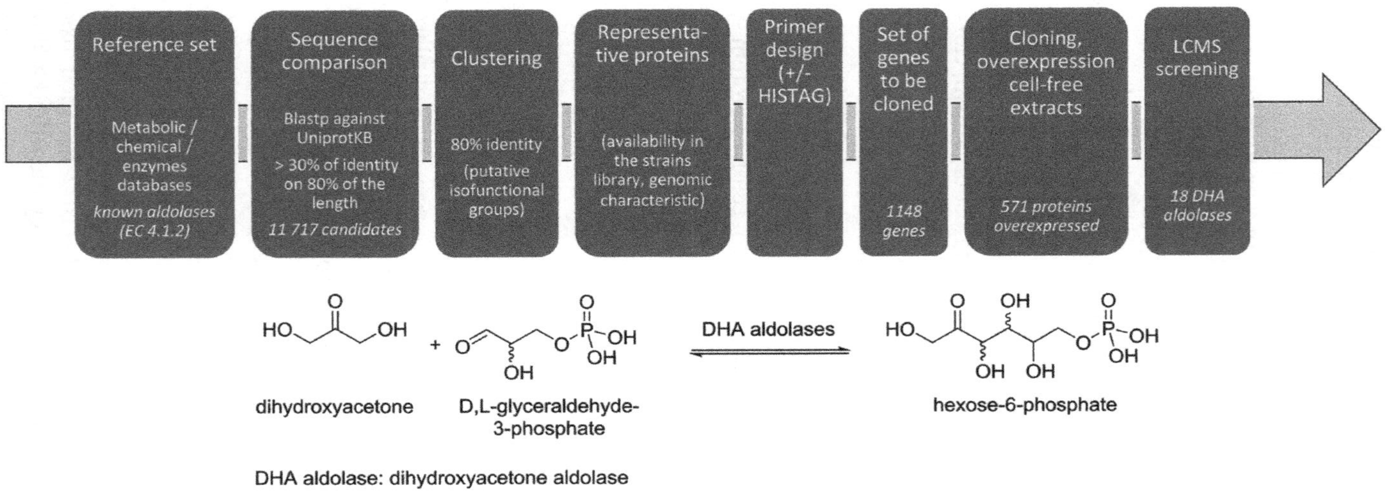

Figure 1.4 New DHA aldolases discovered by genome mining.

It should be noted that the sequence identity searches, depending on the requirements, can be done in different ways: queries with moderate to high sequence identity (>50%) will rather provide enzymes with similar substrate specificity subtypes but different regio/stereoselectivities or catalytic turnovers. Recent reports for transaminases with 95% sequence identity and for nitrilases are good examples.[44,45] As reported by Veselá *et al.*, genes coding for putative nitrilases with moderate similarities (52–69%) to known nitrilases were selected by mining the GenBank database, synthesized artificially and expressed in *Escherichia coli.* Their substrate specificities were determined, which allow classification of the enzymes according to their subtypes (aromatic nitrilase, arylacetonitrilase, aliphatic nitrilase, cyanide hydratase). Those substrate profiles were largely in accordance with those predicted from bioinformatic analysis (Table 1.1).[45]

On the other hand, biocatalysts with different substrate scopes are targeted in lower sequence identity queries, as shown by Furuya and Kino in their paper on the discovery of novel cytochromes P450.[35]

For a focus on a particular catalytic activity, some specific databases have been created. An internet resource dedicated to imine reductases (Imine Reductase Engineering Database, https://ired.biocatnet.de) was recently established by BLAST search with the amino acid sequence of the first characterized IRED reported from *Streptomyces* in 2011.[46,47] In conjunction with selectivity and structural data, a more sophisticated sequence similarity network analysis helps to predict (*R*)- or (*S*)-selectivities from sequence alone.[48] This database is now part of a bigger database – BioCatNet – (https://www.biocatnet.de) combining sequences, structures and experimental data on various protein families with the aim of facilitating protein engineering.[49]

In the preceding examples, queries were conducted with characterized enzymes and their related genes. When looking for enzymes catalyzing an uncommon reaction, with no identified gene, mining the metabolic databases using the searching module based on reactant and product substructures could allow identification of targeted enzymes. Thanks to this strategy, wild amine dehydrogenases were very recently discovered; it is worthy to note that the preceding examples of such enzymatic reductive activity only dealt with engineered amino acid dehydrogenases.[50]

Additionally, industrially relevant biocatalysts were found. This is particularly the case for arylesterases found by Wang *et al.*[51] From 74 proteins screened for esterase activity, they identified three enzymes (RpEST-1, RpEST-2 and PpEST-3) active towards *p*-nitrophenyl esters, displaying the best combination of catalytic activity, thermal stability and solvent stability. Another example is natural reductase from *Candida glabrata* discovered by Ma *et al.*[52] From six enzymes screened for their reductive activity, CgKR1 from *Candida glabrata* exhibited very high activity towards methyl *o*-chlorobenzoylformate. This allowed the preparative synthesis of methyl (*R*)-*o*-chloromandelate (CMM), precursor of the widely used platelet aggregation inhibitor clopidogrel, at 300 $g L^{-1}$ scale (Scheme 1.1).

Table 1.1 Substrate profile of nitrilases and preparative hydrolysis of dinitriles.

Enzyme	Predicted substrate specificity subtype	Catalytic activity (U mg^{-1} of protein)				Products (ratio, yield%)
		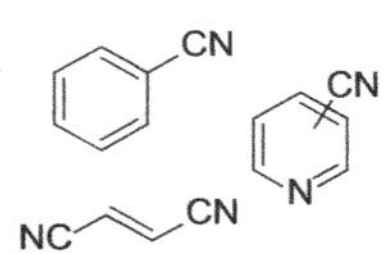	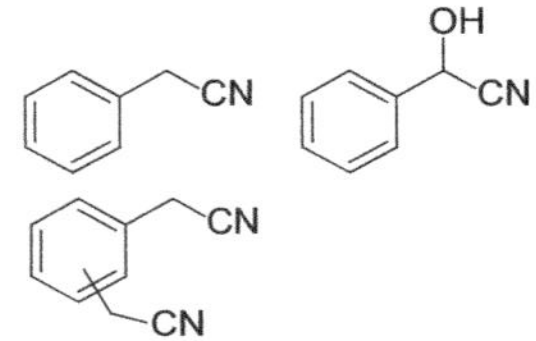	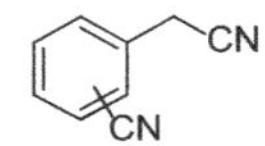	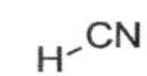	
NitAk1	aromatic nitrilase	0.3–39	0–2.6	0.4–0.7	0	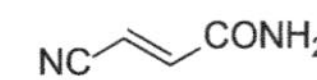 87%
NitAd	arylacetonitrilase	0.03–2.9	2.6–64	16–43	0	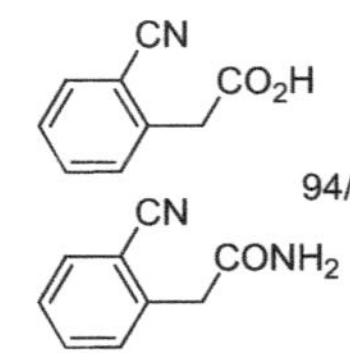 94/6, 85%
NitAk2	arylacetonitrilase	0.02–3.9	1.6–40	7.1–34	0	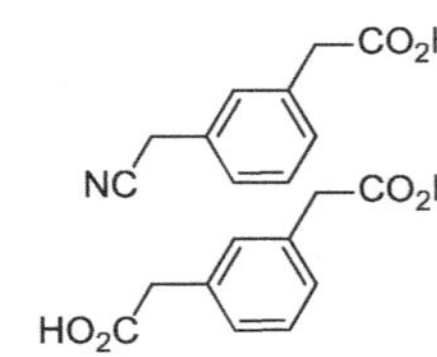 93/7, 73%
NitMp	arylacetonitrilase	0.08–9.9	9.4–241	59–159	3.5	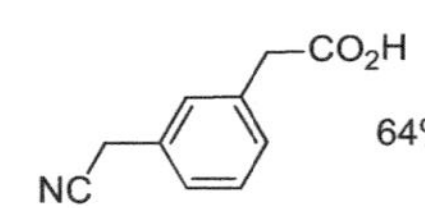 64%

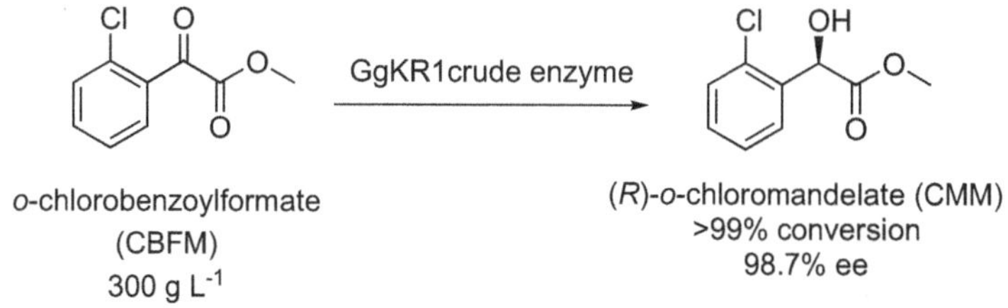

Scheme 1.1 Preparative synthesis of (*R*)-*o*-chloromandelate. CgKR1: carbonyl reductase from *Candida glabrata.*

This sequence-comparison approach is a way of getting around the constraint of annotation, as proteins with false annotation or without predicted function are retrieved.[41] This aspect was stressed by Zhu *et al.* in their report on the novel γ-lactamase from *Bradyrhizobium japonicum* USDA 6, previously unnamed, showing 49% identity to the query protein.[53]

The strength of this approach was illustrated by a bioinformatic strategy integrating the different clustering systems and set up to investigate a Pfam family. In the present case, it turned out that the discovered enzymes were not relevant for biotechnological purposes, but this integrated strategy could be applied to evaluate the biocatalytic potential of any family.[54,55]

1.3.3 Signature-/Key Motif-based Strategy

Rather than conducting the search for new enzymes using the primary sequence of a protein as a whole, one can focus on specific portions of this sequence. Proteins are classified into families based on the presence of important domains or conserved sequence features. These signatures are built using different computational approaches that usually use as a starting point a multiple sequence alignment of proteins sharing a set of characteristics. In InterPro, a database providing functional analysis of proteins, patterns, profiles, fingerprints and hidden Markov models (HMMs) from a number of different databases, are brought together into a single searchable resource, offering convenient access to their predictive capabilities (Figure 1.5).[56]

Moreover, specific sequence motifs/key residues can allow the unambiguous discrimination of the targeted family from the vast number of other sequences among a superfamily, as illustrated for halohydrin dehalogenases belonging to the short-chain dehydrogenase/reductase (SDR) superfamily.[57] Therefore, protein signatures can be very relevant for the discovery of new enzymes. A few groups have employed a sequence pattern search which proved quite fruitful. In a seminal paper of 2005, Fraaije *et al.* identified novel Baeyer–Villiger monooxygenases (BVMOs) thanks to the protein sequence motif [FXGXXXHXXXW(DP)] described earlier by the same group. PAMO, a thermostable monooxygenase from *Thermobifida fusca*, was found to be highly active towards phenylacetone ($k_{cat}/K_m = 32\,000\ M^{-1}\ s^{-1}$).

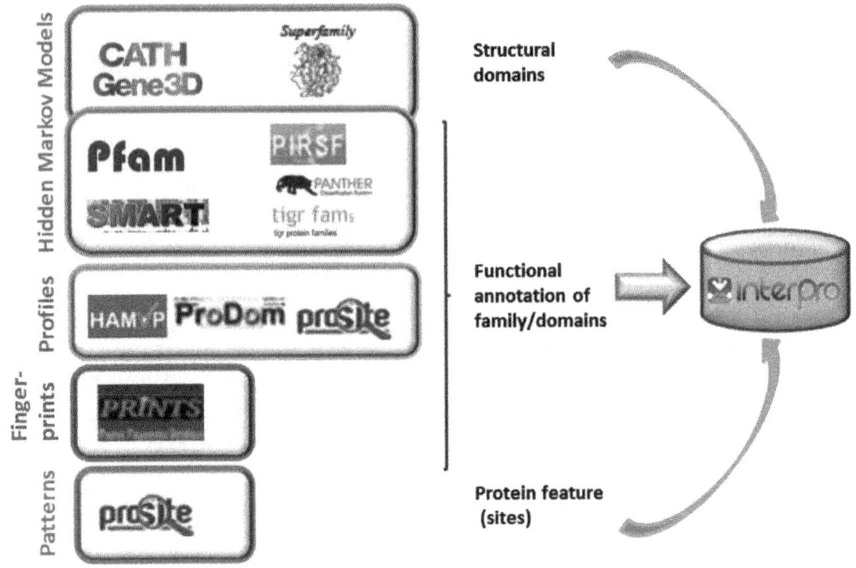

Figure 1.5 InterPro member databases grouped by signature construction method.

The authors pointed to the difficulty to predict the enzyme substrate specificity on the basis of its sequence; indeed, despite its high sequence identity (53%) with steroid monooxygenase, PAMO has no catalytic activity towards progesterone.[58] More recently, Wetzl *et al.* used HMM of the C-terminal domain of six known IREDs to find new ones of bacterial origin. After clustering of the protein sequences matching this HMM hypothesized to be responsible for the catalytic properties, enzymes representative of the IRED sequence space were selected and tested. Interestingly, relationships between stereochemistry, substrate structure and clustering have been observed, as illustrated with IR-10/12 and 14 in Figure 1.6.[59] Two very recently identified IRED-specific motifs, the cofactor binding motif GLGxMGx$_5$[ATS]x$_4$Gx$_4$[VIL]WNR[TS]x$_2$[KR] and the active site motif Gx[DE]x[GDA]x[APS]x$_3${K}x[ASL]x[LMVIAG], should help to discover many more IREDs.[60]

Similarly, the first discovery of Fe-type nitrile hydratase was recently achieved using a conserved motif in alpha-subunit as probe. This motif – KNVIVCSLCSCTAWPILGLPPTWYKSFEYRARVVREPR – containing the iron-binding motif CSLCSC, was selected after sequence alignment of all characterized Fe-type NHases. The nitrile hydratase from *Pseudomonas putida* F1 showed efficient catalytic properties on small aliphatic nitriles but also on some aromatic nitriles.[61]

Arylmalonate decarboxylases (AMDases, EC 4.1.1.76) are very rare and their ability to decarboxylate α-disubstituted malonic acid derivatives to optically pure products without cofactors makes them attractive and promising candidates for the use as biocatalysts in industrial processes. In a

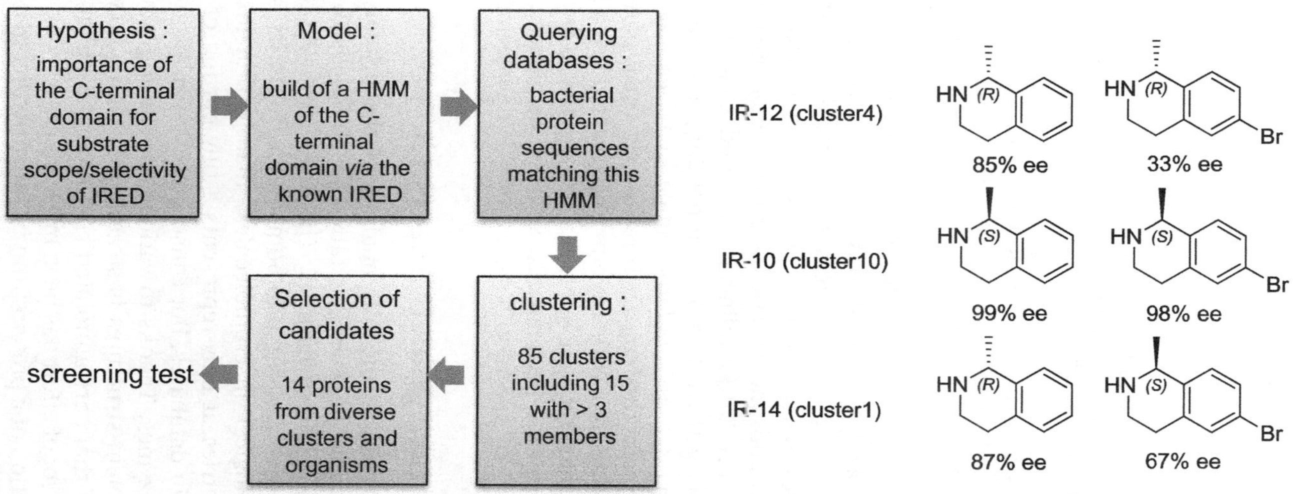

Figure 1.6 Examples of IRED discovered by consensus C-terminal domain search.

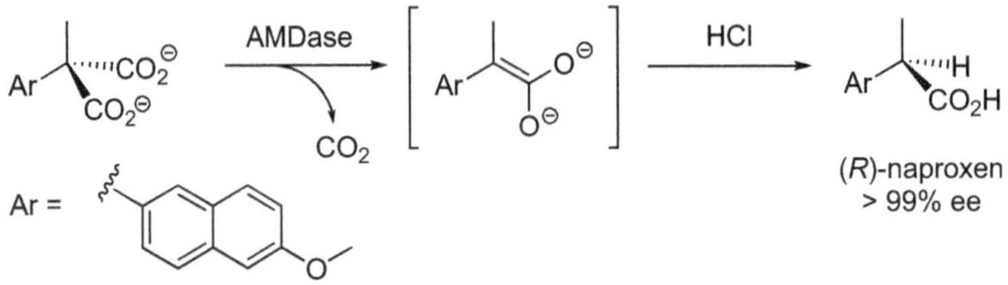

Scheme 1.2 Arylmalonate decarboxylase reaction. Conversion of 2-aryl-2-methyl malonate to (*R*)-arylpropionic acid, illustrated by naproxen. AMDase: arylmalonate decarboxylase.

comprehensive survey, through the development of a search algorithm, Maimanakos *et al.* identified sequence patterns in AMDases, allowing them to discover 58 new AMDases from genomes and metagenomes (Scheme 1.2).[62]

An interesting example in the field of enzymes with industrial potential is the carboxylic acid reductase (CAR) from *Mycobacterium marinum* discovered by Akhtar *et al.* This protein was selected because it holds three consensus sequences characteristic of a previously characterized CAR enzyme: (i) ATP domain, (ii) phosphopantetheine attachments site (LGGXSXXA) and (iii) Rossman fold for NADPH binding. It was found to convert a wide range of aliphatic fatty acids (C6–C18) into corresponding aldehydes, making it a useful catalyst for the synthesis of fatty acid-derived chemical commodities.[63]

In the next few years, this approach relying on small portions of the sequence directly associated with the enzymatic activity (co-factor binding sites, binding residues, catalytic sites, *etc.*) is expected to develop thanks to the growing number of experimentally validated sequences, which should lead to the discovery of new motifs/signatures inside enzyme families.

1.4 3D Structure-guided Approach

Although very substantial, the sole analysis of the primary sequence for predicting the functions of genes generally leads to the discovery of enzymes belonging to already known families. The Brookhaven Protein Data Bank (RSCB PDB http://www.rcsb.org/pdb/home/home.do) indexes the 3D structural data of biological macromolecules, mainly proteins. With nearly 115 000 protein entries, it only represents a tiny portion of the sequences listed in the protein databases, but constitutes a great resource for the discovery of novel enzymes. Thanks to various structural genomic initiatives, the number of protein structures is steadily increasing; however, a significant proportion of them are annotated as proteins of unknown function.[64] For two decades, a lot of effort has been made to develop bioinformatic tools to predict the function of a protein from its 3D structure. In their pioneering work, the Thornton group proposed a methodology based on the recognition of 3D active site templates.[65–67] The Catalytic Site Atlas (CSA), a database providing catalytic residue annotation for enzymes in the Protein Data Bank, was built.[68,69] Other approaches, also based on the conservation of residues in space associated with the catalytic function of the protein, have been developed by different groups.[70–72]

1.4.1 Exploring 3D Structures of Proteins

Besides the common sequence-based approach, a structure-guided approach can be undertaken for enzymes belonging to families or sub-families for which at least one member has a solved structure. In this way, four structures with unknown function belonging to the cluster of "ornithine-aminotransferase (OAT)-like proteins" were identified as amine transaminases (ATA) and biochemically characterized by Steffen-Munsberg *et al.*[73] Two of the four enzymes were found to be rather promiscuous with moderate to excellent activities towards structurally various substrates associated with good to total stereoselectivity (Scheme 1.3).

Exploration of both structural and sequence databases allowed the same group to report the identification of so-far unknown (*R*)-stereoselective transaminases.[74] Careful analysis of the 3D structures of enzymes belonging to fold IV PLP-enzymes led to the identification and prediction of key motifs for stereoselectivity. A sequence-based algorithm was developed and used to search protein databases for enzymes carrying these key motifs. Seventeen enzymes were found to have the desired features.

1.4.2 Active Site Topology/Constellation-guided Strategy

Focusing on highly conserved regions associated with catalysis has led to the development of protein function annotation algorithms that specifically focus on matching catalytic residue geometries.[75,76] As in *de novo* computational enzyme design (see Chapter 4), the first step consists of identifying the catalytic residues, cofactor and substrate binding residues and their relative spatial positions forming the minimal catalytic active site constellation.[77] This template can then be applied in searches in structural databases (Figure 1.7).

This method allowed Steinkellner *et al.* to identify two promiscuous ene-reductases (Old Yellow Enzyme, OYE). These two enzymes, *Ph*ENR from *Pyrococcus horikoshii* and *Tt*ENR from *Thermus thermophilus*, have completely different sequence and fold compared to typical OYEs. Remarkably, an inverted stereopreference was predicted and experimentally confirmed (Scheme 1.4). As outlined by the authors, one of the benefits of the approach lies in the fact that all the hits are proteins stable enough to have their structure determined.[78]

Also based on active site topology, another approach focuses on mechanistic and structural characteristics. Partial reaction, reaction intermediate or transition state, are elements to identify enzymes sharing common mechanistic attribute within a protein family.[79–81] Thus, by modeling high-energy intermediates that mimic the transition state, Hermann *et al.*

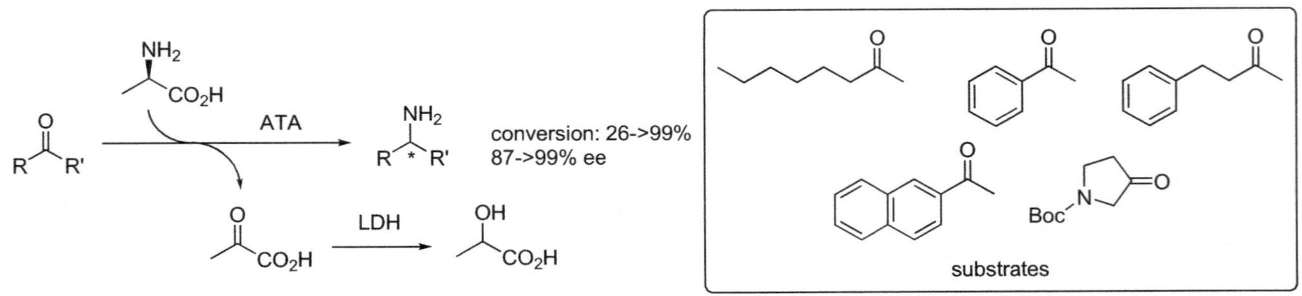

Scheme 1.3 Conversion and enantiomeric excess values of the amines obtained by asymmetric synthesis from the corresponding ketones. ATA: amine transaminase, LDH: lactate dehydrogenase.

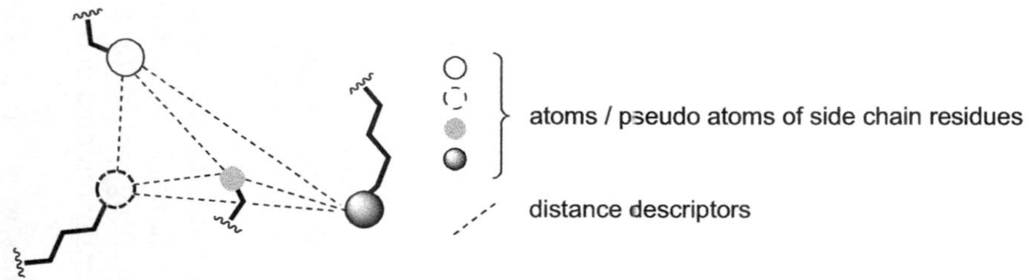

Figure 1.7 Catalytic active site constellation.

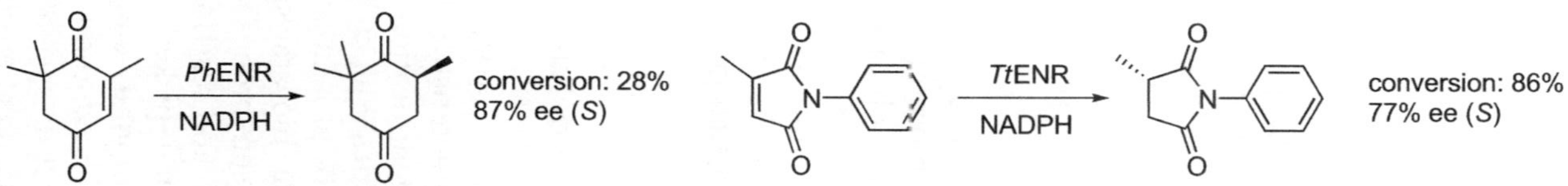

Scheme 1.4 Conversion and enantiomeric excess values of the compounds obtained by asymmetric reduction from the corresponding enones. *Ph*ENR: ene-reductase from *Pyrococcus horikoshii*, *Tt*ENR: ene-reductase from *Thermus thermophilus*.

established the function of Tm0936, an enzyme of unknown function from *Thermotoga maritima* belonging to the amidohydrolase superfamily, from its structure.[82] Homologs of the Tm0936 were later identified by docking putative substrates to modeled enzyme structures.[83] Recently, by such a mechanism-guided approach focusing on stabilization of the transition state, Kürten *et al.* showed that an esterase with the α/β-fold displayed reaction promiscuity and exhibited amidase activity.[84]

All these methodologies for the functional assignment of proteins of unknown functions have been developed for a better understanding of metabolism and enzymology from a fundamental perspective. Until now, they have hardly been used in programs for the identification of enzymes with biotechnological applications (see Chapter 2), but without doubt they present a high potential.

1.5 Conclusion

Owing to the tremendous microbial biodiversity, the various genome mining approaches described in this chapter emphasize the great potential of nature as a reservoir of biocatalysts. The large number of available sequences can be seen as a technical obstacle for the discovery of the desired catalytic activity, but rational genome mining (*i.e.* combining traditional genome sequence comparison or annotation work with other approaches, such as identification of key motifs, functional analysis of the genetic organization of the putative targeted gene, 3D structure-guided strategy) can optimize the screening effort.

Besides the powerful protein engineering methods based on modification of already characterized enzymes (see Chapter 7), mining genomes still constitutes a powerful and complementary way to access new biocatalysts. In protein engineering studies, parental sequences generally govern the overall accessible sequence space, while novel sequences can offer high variability and exhibit characteristics that are difficult to access by laboratory evolution. In addition, identification of new enzymes with at least some marginal activities provides quickly the most efficient starting point for enzyme improvements by protein engineering, which generally remains necessary to obtain industrial biocatalysts.[85]

In 1976, Jensen proposed that ancient enzymes were characterized by broad substrate and reaction scope ("generalist enzymes") and that natural enzyme evolution picked up and fine-tuned these different activities to generate contemporary enzymes with specific catalytic functions.[86] In addition, to re-specialize the function of an enzyme, it seems that specialized enzymes are first de-specialized by natural mutations to obtain a generalist enzyme (promiscuous enzyme), before being re-specialized to a new function.[8,87] Nowadays, inspired by natural enzyme evolution, the identification of all-rounder frequent hit enzymes as generalist enzymes is sought as a framework for protein engineering.[88–90]

It is highly probable that many valuable enzymatic activities have still to be found among wild-type enzymes resulting from billions of years of natural evolution. To explore this reservoir of uncharacterized enzymes, new ways to make use of nature's richness are required. Within this framework, modern bioinformatic tools, barely utilized in the frame of biocatalysis, will certainly be of great benefit in the near future for the discovery of new biocatalysts. We can mention antiSMASH for the assignment of enzyme functions among biosynthetic gene clusters identified by automatic genomic programs, or the conservation of types of chemical transformations among metabolic networks, which enables capture of relevant metabolic contexts and thus assignment of potential enzymatic function.[91,92] Moreover, the profiling of orphan enzymes will provide more activity-attributed sequences, thereby multiplying the reference sequences available for the discovery of novel biocatalysts by sequence comparison approaches.

References

1. L. Fernández-Arrojo, M.-E. Guazzaroni, N. López-Cortés, A. Beloqui and M. Ferrer, *Curr. Opin. Biotechnol.*, 2010, **21**, 725.
2. K. N. Lam, J. Cheng, K. Engel, J. D. Neufeld and T. C. Charles, *Front. Microbiol.*, 2015, **6**, 1196.
3. D. Reyes-Duarte, M. Ferrer and H. García-Arellano, in *Lipases and Phospholipases: Methods and Protocols*, ed. G. Sandoval, Humana Press, Totowa, NJ, 2012, pp. 101–113.
4. P. Lorenz, K. Liebeton, F. Niehaus and J. Eck, *Curr. Opin. Biotechnol.*, 2002, **13**, 572.
5. M. Ferrer, F. Martínez-Abarca and P. N. Golyshin, *Curr. Opin. Biotechnol.*, 2005, **16**, 588.
6. A. Babtie, N. Tokuriki and F. Hollfelder, *Curr. Opin. Chem. Biol.*, 2010, **14**, 200.
7. K. Hult and P. Berglund, *Trends Biotechnol.*, 2007, **25**, 231–238.
8. I. Nobeli, A. D. Favia and J. M. Thornton, *Nat. Biotechnol.*, 2009, **27**, 157.
9. U. T. Bornscheuer and R. J. Kazlauskas, *Angew. Chem., Int. Ed.*, 2004, **43**, 6032.
10. G. A. Behrens, A. Hummel, S. K. Padhi, S. Schätzle and U. T. Bornscheuer, *Adv. Synth. Catal.*, 2011, **353**, 2191.
11. A. Chang, I. Schomburg, S. Placzek, L. Jeske, M. Ulbrich, M. Xiao, C. W. Sensen and D. Schomburg, *Nucleic Acids Res.*, 2015, **43**, D439.
12. A. M. Schnoes, S. D. Brown, I. Dodevski and P. C. Babbitt, *PLoS Comput. Biol.*, 2009, **5**, e1000605.
13. L. J. Basile, R. C. Willson, B. T. Sewell and M. J. Benedik, *Appl. Microbiol. Biotechnol.*, 2008, **80**, 427.
14. O. Kaplan, K. Bezouška, A. Malandra, A. B. Veselá, A. Petříčková, J. Felsberg, A. Rinágelová, V. Křen and L. Martínková, *Biotechnol. Lett.*, 2011, **33**, 309.

15. O. Kaplan, A. B. Veselá, A. Petříčková, F. Pasquarelli, M. Pičmanová, A. Rinágelová, T. C. Bhalla, M. Pátek and L. Martínková, *Mol. Biotechnol.*, 2013, **54**, 996.
16. T. Furuya and K. Kino, *ChemSusChem*, 2009, **2**, 645.
17. H. Suzuki, F. Okazaki, A. Kondo and K.-I. Yoshida, *Appl. Microbiol. Biotechnol.*, 2013, **97**, 2929.
18. F. Zhang, X.-M. Zhang, Y.-R. Yin and W.-J. Li, *J. Biosci. Bioeng.*, 2015, **120**, 644.
19. H. Deng, L. Ma, N. Bandaranayaka, Z. Qin, G. Mann, K. Kyeremeh, Y. Yu, T. Shepherd, J. H. Naismith and D. O'Hagan, *ChemBioChem*, 2014, **15**, 364.
20. G. A. Aleku, H. Man, S. P. France, F. Leipold, S. Hussain, L. Toca-Gonzalez, R. Marchington, S. Hart, J. P. Turkenburg, G. Grogan and N. J. Turner, *ACS Catal.*, 2016, **6**, 3880.
21. Y. Fang, S. Wang, S. Liu and Y. Jiao, *Int. J. Biol. Macromol.*, 2015, **80**, 334.
22. A. Riedel, M. Mehnert, C. E. Paul, A. H. Westphal, W. J. Van Berkel and D. Tischler, *Front. Microbiol.*, 2015, **6**, 1073.
23. M. E. Vaquero, L. I. de Eugenio, M. J. Martínez and J. Barriuso, *PLoS ONE*, 2015, **10**, e0124882.
24. J. Abdoul-Zabar, I. Sorel, V. Hélaine, F. Charmantray, T. Devamani, D. Yi, V. de Berardinis, D. Louis, P. Marlière, W.-D. Fessner and L. Hecquet, *Adv. Synth. Catal.*, 2013, **355**, 116.
25. S. Lutz, J. Lichter and L. Liu, *J. Am. Chem. Soc.*, 2007, **129**, 8714.
26. D. Zhu, C. Mukherjee, E. R. Biehl and L. Hua, *J. Biotechnol.*, 2007, **129**, 645.
27. J. B. Van Beilen, Z. Li, W. A. Duetz, T. H. M. Smits and B. Witholt, *Oil Gas Sci. Technol.*, 2003, **58**, 427.
28. Z. Fang, T. Li, Q. Wang, X. Zhang, H. Peng, W. Fang, Y. Hong, H. Ge and Y. Xiao, *Appl. Microbiol. Biotechnol.*, 2011, **89**, 1103.
29. N. Itoh, K. Isotani, Y. Makino, M. Kato, K. Kitayama and T. Ishimota, *Enzyme Microb. Technol.*, 2014, **55**, 140.
30. Q. Wang, H. Wu, A. Wang, P. Du, X. Pei, H. Li, X. Yin, L. Huang and X. Xiong, *J. Biol. Chem.*, 2010, **285**, 41509.
31. M. Kubota, M. Nodate, M. Yasumoto-Hirose, T. Uchiyama, O. Kagami, Y. Shizuri and N. Misawa, *Biosci., Biotechnol., Biochem.*, 2005, **69**, 2421.
32. W. H. Eschenfeldt, L. Stols, H. Rosenbaum, Z. S. Khambatta, E. Quaite-Randall, S. Wu, D. C. Kilgore, J. D. Trent and M. I. Donnelly, *Appl. Environ. Microbiol.*, 2001, **67**, 4206.
33. H. Yun, S. Lim, B.-K. Cho and B.-G. Kim, *Appl. Environ. Microbiol.*, 2004, **70**, 2529.
34. A. Hornung, M. Bertazzo, A. Dziarnowski, K. Schneider, K. Welzel, S. E. Wohlert, M. Holzenkampfer, G. J. Nicholson, A. Bechthold, R. D. Sussmuth, A. Vente and S. Pelzer, *ChemBioChem*, 2007, **8**, 757.
35. T. Furuya and K. Kino, *Appl. Microbiol. Biotechnol.*, 2010, **86**, 991.
36. C. Aslanidis and P. J. de Jong, *Nucleic Acids Res.*, 1990, **18**, 6069.

37. Y.-C. Kwon, K.-H. Lee, H.-C. Kim, K. Han, J.-H. Seo, B.-G. Kim and D.-M. Kim, *Appl. Environ. Microbiol.*, 2010, **76**, 6295.
38. C. O'Reilly and P. D. Turner, *J. Appl. Microbiol.*, 2003, **95**, 1161.
39. C. Vergne-Vaxelaire, F. Bordier, A. Fossey, M. Besnard-Gonnet, A. Debard, A. Mariage, V. Pellouin, A. Perret, J.-L. Petit, M. Stam, M. Salanoubat, J. Weissenbach, V. De Berardinis and A. Zaparucha, *Adv. Synth. Catal.*, 2013, **355**, 1763.
40. F. Bordier, M. Stam, E. Darii, S. Tricot, A. Fossey, J. Rohault, A. Debard, A. Mariage, V. Pellouin, J.-L. Petit, A. Perret, D. Vallenet, M. Salanoubat, J. Weissenbach, C. Vergne-Vaxelaire, V. de Berardinis and A. Zaparucha, *J. Mol. Catal. B: Enzym.*, 2014, **107**, 79.
41. C. Guérard-Hélaine, V. de Berardinis, M. Besnard-Gonnet, E. Darii, M. Debacker, A. Debard, C. Fernandes, V. Hélaine, A. Mariage, V. Pellouin, A. Perret, J.-L. Petit, M. Sancelme, M. Lemaire and M. Salanoubat, *ChemCatChem*, 2015, 7, 1871.
42. U. Kaulmann, K. Smithies, M. E. B. Smith, H. C. Hailes and J. M. Ward, *Enzyme Microb. Technol.*, 2007, **41**, 628.
43. I. Lavandera, A. Kern, M. Schaffenberger, J. Gross, A. Glieder, S. de Wildeman and W. Kroutil, *ChemSusChem*, 2008, **1**, 431.
44. N. van Oosterwijk, S. C. Willies, J. Hekelaar, A. C. Terwisscha van Scheltinga, N. J. Turner and B. W. Dijkstra, *Biochemistry*, 2016, **55**, 4422.
45. A. B. Veselá, L. Rucká, O. Kaplan, H. Pelantová, J. Nešvera, M. Pátek and L. Martínková, *Appl. Microbiol. Biotechnol.*, 2016, **100**, 2193.
46. K. Mitsukura, M. Suzuki, S. Shinoda, T. Kuramoto, T. Yoshida and T. Nagasawa, *Biosci., Biotechnol., Biochem.*, 2011, **75**, 1778.
47. P. N. Scheller, S. Fademrecht, S. Hofelzer, J. Pleiss, F. Leipold, N. J. Turner, B. M. Nestl and B. Hauer, *ChemBiochem*, 2014, **15**, 2201.
48. G. Grogan and N. J. Turner, *Chem. - Eur. J.*, 2016, **22**, 1900.
49. P. C. F. Buchholz, C. Vogel, W. Reusch, M. Pohl, D. Rother, A. C. Spieß and J. Pleiss, *ChemBiochem*, 2016, **17**, 2093.
50. O. Mayol, S. David, E. Darii, A. Debard, A. Mariage, V. Pellouin, J.-L. Petit, M. Salanoubat, V. de Berardinis, A. Zaparucha and C. Vergne-Vaxelaire, *Catal. Sci. Technol.*, 2016, **6**, 7421.
51. L. Wang, V. Mavisakalyan, E. R. M. Tillier, G. W. Clark, A. V. Savchenko, A. F. Yakunin and E. R. Master, *Microb. Biotechnol.*, 2010, **3**, 677.
52. H. Ma, L. Yang, Y. Ni, J. Zhang, C.-X. Li, G.-W. Zheng, H. Yang and J.-H. Xu, *Adv. Synth. Catal.*, 2012, **354**, 1765.
53. S. Zhu, C. Gong, D. Song, S. Gao and G. Zheng, *Appl. Environ. Microbiol.*, 2012, **78**, 7492.
54. R. C. de Melo-Minardi, K. Bastard and F. Artiguenave, *Bioinformatics*, 2010, **26**, 3075.
55. K. Bastard, A. A. T. Smith, C. Vergne-Vaxelaire, A. Perret, A. Zaparucha, R. De Melo-Minardi, A. Mariage, M. Boutard, A. Debard, C. Lechaplais, C. Pelle, V. Pellouin, N. Perchat, J.-L. Petit, A. Kreimeyer, C. Medigue, J. Weissenbach, F. Artiguenave, V. De Berardinis, D. Vallenet and M. Salanoubat, *Nat. Chem. Biol.*, 2014, **10**, 42.

56. T. Vorapreeda, C. Thammarongtham, S. Cheevadhanarak and K. Laoteng, *Microbiology*, 2015, **161**, 1613.
57. M. Schallmey, J. Koopmeiners, E. Wells, R. Wardenga and A. Schallmey, *Appl. Environ. Microbiol.*, 2014, **80**, 7303.
58. M. W. Fraaije, J. Wu, D. P. H. M. Heuts, E. W. van Hellemond, J. H. L. Spelberg and D. B. Janssen, *Appl. Microbiol. Biotechnol.*, 2005, **66**, 393.
59. D. Wetzl, M. Berrera, N. Sandon, D. Fishlock, M. Ebeling, M. Müller, S. Hanlon, B. Wirz and H. Iding, *ChemBioChem*, 2015, **16**, 1749.
60. S. Fademrecht, P. N. Scheller, B. M. Nestl, B. Hauer and J. Pleiss, *Proteins: Struct., Funct., Bioinf.*, 2016, **84**, 600.
61. X. Pei, L. Yang, G. Xu, Q. Wang and J. Wu, *J. Mol. Catal. B: Enzym.*, 2014, **99**, 26.
62. J. Maimanakos, J. Chow, S. K. Gaßmeyer, S. Güllert, F. Busch, R. Kourist and W. R. Streit, *Front. Microbiol.*, 2016, 7, 1332.
63. M. K. Akhtar, N. J. Turner and P. R. Jones, *Proc. Natl. Acad. Sci. U. S. A.*, 2013, **110**, 87.
64. N. Nadzirin and M. Firdaus-Raih, *Int. J. Mol. Sci.*, 2012, **13**, 12761.
65. A. C. Wallace, N. Borkakoti and J. M. Thornton, *Protein Sci.*, 1997, **6**, 2308.
66. J. A. Barker and J. M. Thornton, *Bioinformatics*, 2003, **19**, 1644.
67. R. A. Laskowski, J. D. Watson and J. M. Thornton, *J. Mol. Biol.*, 2005, **351**, 614.
68. C. T. Porter, G. J. Bartlett and J. M. Thornton, *Nucleic Acids Res.*, 2004, **32**, D129.
69. N. Furnham, G. L. Holliday, T. A. P. de Beer, J. O. B. Jacobsen, W. R. Pearson and J. M. Thornton, *Nucleic Acids Res.*, 2014, **42**, D485.
70. M. von Grotthuss, D. Plewczynski, K. Ginalski, L. Rychlewski and E. I. Shakhnovich, *BMC Bioinf.*, 2006, 7, 1.
71. T. McKay, K. Hart, A. Horn, H. Kessler, G. Dodge, K. Bardhi, K. Bardhi, J. L. Mills, H. J. Bernstein and P. A. Craig, *J. Struct. Funct. Genomics*, 2015, **16**, 43.
72. N. Nadzirin, E. J. Gardiner, P. Willett, P. J. Artymiuk and M. Firdaus-Raih, *Nucleic Acids Res.*, 2012, **40**, W380.
73. F. Steffen-Munsberg, C. Vickers, A. Thontowi, S. Schätzle, T. Tumlirsch, M. Svedendahl Humble, H. Land, P. Berglund, U. T. Bornscheuer and M. Höhne, *ChemCatChem*, 2013, **5**, 150.
74. M. Hohne, S. Schatzle, H. Jochens, K. Robins and U. T. Bornscheuer, *Nat. Chem. Biol.*, 2010, **6**, 807.
75. J. P. Nilmeier, D. A. Kirshner, S. E. Wong and F. C. Lightstone, *PLoS ONE*, 2013, **8**, e62535.
76. S. C. Izidoro, R. C. de Melo-Minardi and G. L. Pappa, *Bioinformatics*, 2015, **31**, 864.
77. A. Zanghellini, *Curr. Opin. Biotechnol.*, 2014, **29**, 132.
78. G. Steinkellner, C. C. Gruber, T. Pavkov-Keller, A. Binter, K. Steiner, C. Winkler, A. Łyskowski, O. Schwamberger, M. Oberer, H. Schwab, K. Faber, P. Macheroux and K. Gruber, *Nat. Commun.*, 2014, **5**, 4150.

79. J. A. Gerlt and P. C. Babbitt, *Annu. Rev. Biochem.*, 2001, **70**, 209.
80. S. C. H. Pegg, S. D. Brown, S. Ojha, J. Seffernick, E. C. Meng, J. H. Morris, P. J. Chang, C. C. Huang, T. E. Ferrin and P. C. Babbitt, *Biochemistry*, 2006, **45**, 2545.
81. J. A. Gerlt, P. C. Babbitt, M. P. Jacobson and S. C. Almo, *J. Biol. Chem.*, 2012, **287**, 29.
82. J. C. Hermann, R. Marti-Arbona, A. A. Fedorov, E. Fedorov, S. C. Almo, B. K. Shoichet and F. M. Raushel, *Nature*, 2007, **448**, 775.
83. D. S. Hitchcock, H. Fan, J. Kim, M. Vetting, B. Hillerich, R. D. Seidel, S. C. Almo, B. K. Shoichet, A. Sali and F. M. Raushel, *J. Am. Chem. Soc.*, 2013, **135**, 13927.
84. C. Kürten, B. Carlberg and P.-O. Syrén, *Catalysts*, 2016, **6**, 90.
85. N. J. Turner, *Nat. Chem. Biol.*, 2009, **5**, 567.
86. R. A. Jensen, *Annu. Rev. Microbiol.*, 1976, **30**, 409.
87. O. Khersonsky and D. S. Tawfik, *Annu. Rev. Biochem.*, 2010, **79**, 471.
88. T. Devamani, A. M. Rauwerdink, M. Lunzer, B. J. Jones, J. L. Mooney, M. A. O. Tan, Z.-J. Zhang, J.-H. Xu, A. M. Dean and R. J. Kazlauskas, *J. Am. Chem. Soc.*, 2016, **138**, 1046.
89. D. G. Guzzetti, A. Lebrun, M. Subileau, E. Grousseau, E. Dubreucq and J. Drone, *ACS Catal.*, 2016, **6**, 5345.
90. P. A. Romero and F. H. Arnold, *Nat. Rev. Mol. Cell Biol.*, 2009, **10**, 866.
91. T. Weber, K. Blin, S. Duddela, D. Krug, H. U. Kim, R. Bruccoleri, S. Y. Lee, M. A. Fischbach, R. Müller, W. Wohlleben, R. Breitling, E. Takano and M. H. Medema, *Nucleic Acids Res.*, 2015, **43**, W237.
92. M. Sorokina, C. Medigue and D. Vallenet, *BMC Bioinf.*, 2015, **16**, 1.

CHAPTER 2

Exploiting Natural Diversity for Industrial Enzymatic Applications

YASUHISA ASANO*[a,b] AND RICHARD METZNER[a,b]

[a] Biotechnology Research Center and Department of Biotechnology, Toyama Prefectural University, 5180 Kurokawa, Imizu, Toyama 939-0398, Japan; [b] Asano Active Enzyme Molecule Project, ERATO, JST, 5180 Kurokawa, Imizu, Toyama 939-0398, Japan
*Email: asano@pu-toyama.ac.jp

2.1 Introduction

Our screening strategies for finding promising industrial enzymes can be categorized into three main methodologies, which can be combined to obtain the most effective results depending on the constraints of time and resources. The first involves detection of the desired enzyme activities directly in biological sources. The second strategy follows a genetic approach based on similarities of their DNA sequences to those of known activity (see Chapter 1). This sequence-based approach is extremely sensitive as a result of modern bioinformatic tools, so it is an indispensable technique and suitable for isolating clones from deposited DNA sequences. Since the relationship between the primary structure and the corresponding enzyme activity cannot always be established with confidence, we use software like INTMSAlign to reduce the number of false-positives. The third strategy is

Catalysis Series No. 32
Modern Biocatalysis: Advances Towards Synthetic Biological Systems
Edited by Gavin Williams and Mélanie Hall

Published by the Royal Society of Chemistry, www.rsc.org

highlighted by describing an example of creating a new enzyme activity based on a combined structural and directed evolution approach (see Chapter 7).

From a practical point of view, it is easier to handle microorganisms such as bacteria, yeasts and moulds as likely sources of enzymes, as compared with utilizing higher organisms. Microorganisms are often kept in laboratories for applied microbiology as stock cultures and can be obtained from major public institutions which collect and distribute the microbial cultures. Microbial screening for enzymes is carried out by cultivating the microorganisms, followed by biochemical detection of the enzyme activities. Microorganisms are also often selectively isolated from nature by the "enrichment culture technique," sometimes associated with the "acclimation technique", because it is often hard to measure all the enzyme activities among many microbial cultures. The activities of some enzymes can be reflected in their growth, so microorganisms having the desired activity for a chemical transformation can be selected simply by colony formation from the vast numbers of microorganisms in nature.[1,2] The direct isolation of microorganisms with new enzyme activities from nature, and the confirmation and further search for wider distribution of the new activities among the stock cultures is straightforward. This method is even more effective when combined with searching and expressing DNA according to known genetic information in heterologous hosts such as *Escherichia coli*. Such an example is seen in the case of the discovery and the distribution studies of the new lyase "aldoxime dehydratase" (Section 2.5) as a result of ongoing research for nitrile-synthesizing enzymes following the discovery of nitrile hydratase, an important industrial enzyme.[3]

Direct isolation of microorganisms from nature is often used in screening for enzymes involved in the initial stage of the metabolism of synthetic chemicals, the assimilation pathways of which can be sought by forming colonies as indicators, when the target compounds to be transformed are used as sole sources of nutrition, such as carbon, nitrogen, sulfur, phosphorus, *etc.* Depending on the target compounds, a wide variety of the cultures are obtained from any natural samples. Our recent examples are described below.

The initial steps of direct isolation of microorganisms and measurement of the activity from the natural organisms are usually followed by enzyme purification, analysis of their structure and improvement of their properties by a genetic approach such as cDNA cloning and expression in various hosts, high-throughput screening and mutagenesis based on the structural information of proteins.

2.2 Screening Enzymes from Soil Microbes, Plants, and Animals (Millipede) by Activity Measurement

Below are some examples of isolation of enzymes from microorganisms having properties that are suitable for industrial applications such as biotransformation of chemicals and the enzymatic analyses of biological samples.

2.2.1 Hydrolases

We studied the kinetic resolution of (±)-3,3,3-trifluoro-2-hydroxy-2-methylpropanamide (**1**) to obtain (*R*)- or (*S*)-3,3,3-trifluoro-2-hydroxy-2-methylpropanoic acid (**2**) with efficient enantioselective preparation of both enantiomers of these valuable molecules (Scheme 2.1). We isolated *Shinella* sp. R-6 and *Arthrobacter* sp. S-2 from soil samples showing enantioselective hydrolytic activity on the racemic amide.[4] The strains utilized **1** as the sole nitrogen source. *Shinella* sp. R-6 was capable of hydrolyzing the target amide *R*-selectively, while *Arthrobacter* sp. S-2 showed *S*-selective activity. Both enantiomers of **1** were successfully prepared with these two strains *via* a two-step whole-cell reaction. From *Arthrobacter* sp. S-2, a novel *S*-enantioselective amidase acting on **1** was purified[5] and successfully used in the kinetic resolution of **1** to obtain (*S*)-**2** and (*R*)-**1** (Scheme 2.1). Shaw *et al.* also described the kinetic resolution of (±)-**1** to obtain (*R*)-**2** and (*S*)-**1**, employing an enantioselective amidase from *Klebsiella oxytoca*.[6]

2.2.2 Oxidoreductases

Enzymes catalyzing the N^{α}-oxidation of L-arginine have been reported for *Pseudomonas putida* P2 and the cyanobacterium *Synechococcus elongatus*.[7,8] However, these enzymes have not been purified nor characterized with respect to details including substrate specificity. We screened for micro-organisms producing amino acid oxidases specific to L-arginine from soil samples. FAD-dependent L-arginine oxidase (EC 1.4.3.-) specific to L-arginine was discovered in newly isolated *Pseudomonas* sp. TPU 7192 and we developed a simple enzymatic method for the determination of L-arginine using this enzyme (Scheme 2.2).[9] Among the L- and D-amino acids tested, the enzyme activity toward L-arginine was the highest, followed by L-lysine with around 10%, and L-canavanine as well as L-homoarginine with below 1% activity. The K_M values for L-arginine and L-lysine were 149 μM and 5.25 mM, respectively, indicating that the enzyme has high affinity towards L-arginine and is specific for L-arginine at low concentration. It was surprizing that it is still possible to isolate an enzyme useful for micro-determination of L-arginine from soil samples.

Focusing on the micro-determination of L-tryptophan, we could show that some enzymes responsible for biosynthesis of secondary metabolites display narrower substrate specificity for the allocation of the micro-determination of biological samples, compared to those evolved for biodegradation. Evidently, metabolism-oriented screening can provide enzymes highly selective for individual biomolecules. Two kinds of tryptophan oxidases involved in bis-indole antibiotic biosynthesis were used to develop assays for selective L-tryptophan determination. A putative L-amino acid oxidase from *Streptomyces* sp. TP-A0274 – StaO involved in staurosporine biosynthesis – was heterologously expressed, biochemically characterized, and identified as a selective tryptophan oxidase.[10] In a similar fashion, L-amino acid oxidase

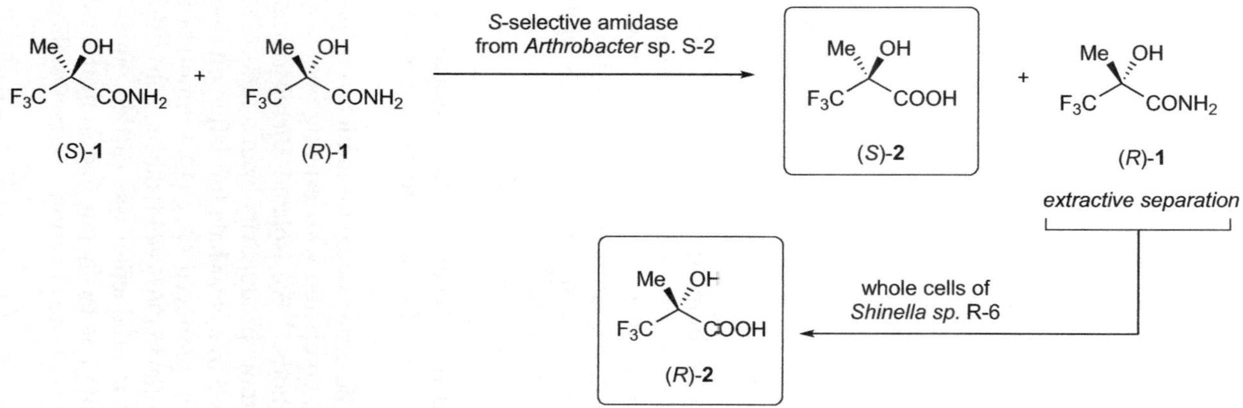

Scheme 2.1 Stepwise enantioselective hydrolysis of *rac*-3,3,3-trifluoro-2-hydroxy-2-methylpropanamide (**1**) by *S*-amidase from *Arthrobacter* sp. S-2 and whole-cells of *Shinella* sp. R-6.

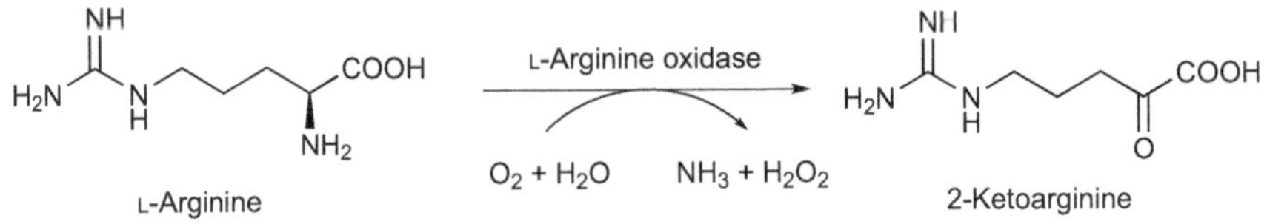

Scheme 2.2 Oxidative deamination of L-arginine catalyzed by L-arginine oxidase from *Pseudomonas* sp.

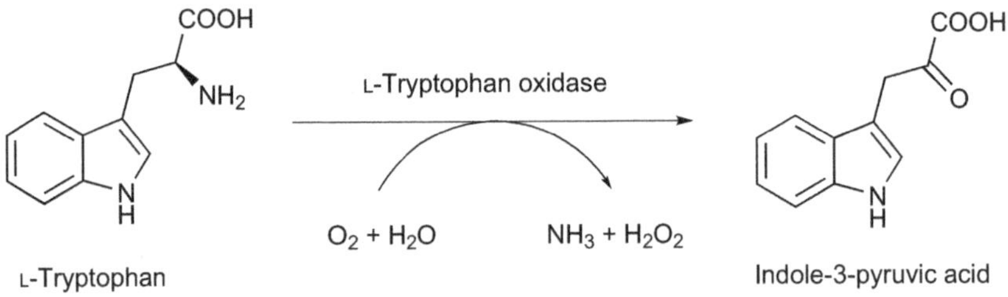

Scheme 2.3 Oxidative deamination of L-tryptophan (catalyzed by L-tryptophan oxidases StaO and VioA, applied in L-tryptophan micro-determination in human plasma samples.

VioA, involved in violacein biosynthesis in *Chromobacterium violaceum*,[11] was obtained. StaO and VioA share similar narrow substrate specificity and high affinity for L-tryptophan (Scheme 2.3). VioA and StaO, which share very low sequence identity (<20%), can both be used for selective quantification of L-tryptophan in human plasma as accurately as HPLC determination following AccQ-tag derivatization.[12]

Screening of natural enzymes is not limited to analytical applications on micromolar levels, but can also be used to obtain efficient catalysts for the production of bulk and fine chemicals. As an example, we developed an enzymatic production method for lumichrome, a photodegradation product of riboflavin used industrially as photosensitizer and fluorescent dye.[13] This biotransformation may be better suited for lumichrome production in terms of simplicity and yield, compared with existing chemical, photolytic, and other biochemical methods.[14] We isolated *Microbacterium* sp. strain TPU 3598 with high lumichrome productivity from a total of eighteen soil samples, which was identified by a clear halo formation on riboflavin-containing plate cultures. The strain produced 4.7 g (19.4 mmol) of lumichrome from 7.6 g (20.2 mmol) of riboflavin (96% yield) during cultivation in 500 mL scale (Figure 2.1). Lumichrome production was carried out in suspension, and the resulting lumichrome was easily purified from the cultivation medium or the reaction mixture *via* centrifugation and recrystallization.[15]

2.2.3 Lyases

Based on studies on the enzymes responsible for plant cyanogenesis, we discovered several new hydroxynitrile lyases (HNL) by direct enzyme activity

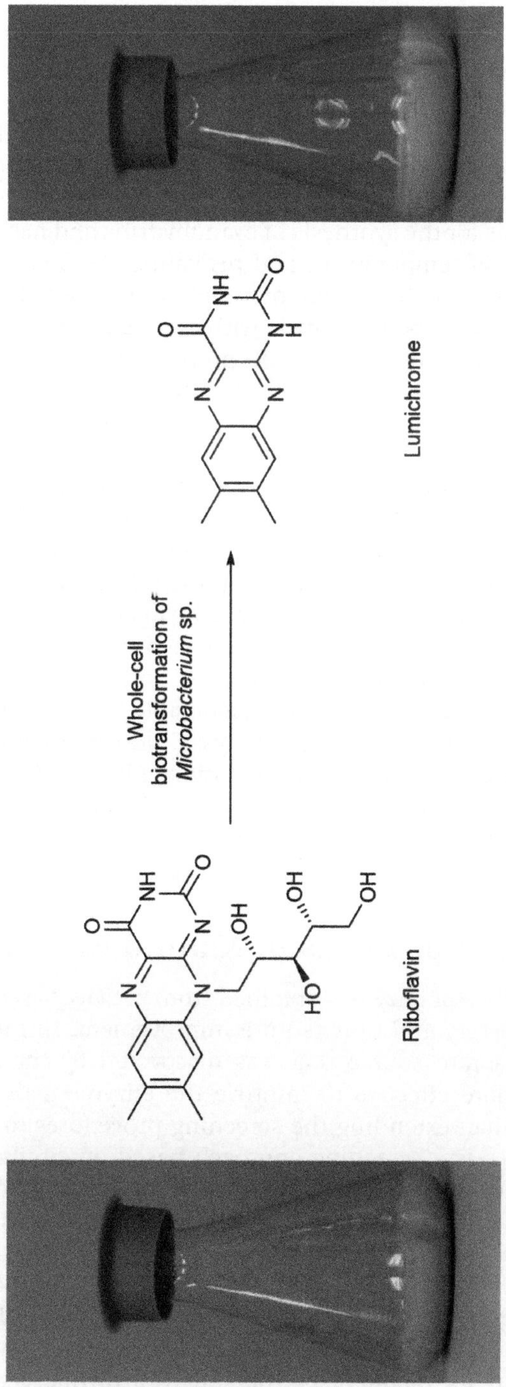

Figure 2.1 Enzymatic transcrystallization of a riboflavin suspension to lumichrome by *Microbacterium* sp.

measurement, utilizing plants from botanic gardens as a reservoir of genetic and enzymatic sources, similar to microbial stock cultures.[16] Furthermore, we discovered a new HNL from a hydrogen cyanide-emitting, invasive animal, *i.e.* the millipede *Chamberlinius hualienensis*. This millipede can be captured in kilogram-scale quantities because it has no natural enemies in his habitat and can therefore propagate in masses on the southern islands of Japan. We isolated and applied its HNL for the synthesis of (*R*)-mandelonitrile and other cyanohydrins.[17] The enzyme shows activities on various aromatic aldehydes for the synthesis of cyanohydrins and has a high stability over a wide range of temperatures and pH values. Because of its strikingly high specific activity (7500 units per mg), it is possible to synthesize (*R*)-mandelonitrile from benzaldehyde with an enantiomeric excess of 99%, without using any organic solvents.[17] We proposed that these animals can be valuable resources for exploring not only HNLs but also diverse, efficient, and stable biocatalysts in industrial biotechnology. We also identified a new enzyme, mandelonitrile oxidase (MOX), which catalyzes the oxidation of (*R*)-mandelonitrile to benzoyl cyanide, releasing hydrogen peroxide.[18] These enzymes are considered to play a major role in formation of volatile defense chemicals of millipedes against natural enemies (Scheme 2.4).

The natural activity of HNLs is based on the release of hydrogen cyanide. Consequently, the source of natural cyanohydrins can be traced back to the corresponding L-amino acids and the intersecting pathways offer another array of industrially valuable enzymes, *e.g.* nitrilases (Nit) and nitrile hydratases (NHase). Such an enzyme, responsible for the formation of alkyl and arylalkyl nitriles in nature, *i.e.* the lyase aldoxime dehydratase (Oxd), was discovered several years ago in acclimation cultures of *Bacillus* sp. OxB-1 and *Rhodococcus* sp. YH3-3.[19,20] Its remarkable development from first isolation to synthetic application is described in more details in Section 2.5.

2.3 Genetic Approaches to Natural Enzyme Sources

The initial properties of enzymes obtained from activity-based screening are not always satisfactory and thus require improvement. In the case that the enzyme is from a rare source that was discovered by the activity-guided approach, it is more effective to improve the enzyme activity by directed evolution rather than extending the screening procedures to other sources. Also of importance is a screening approach based on similarity of the primary protein structures. There is much more variety of microbial sources than one researcher's sampling of environmental specimens. Furthermore, it is often necessary to use a genetic approach, such as screening from cDNA expression from the transcriptomes, to investigate the metabolism of plants which exhibit special characteristics. If a unique organism has been identified by the primary sequence screening, it may contain a vast amount of target enzymes in its metabolism that are worthwhile screening. These aspects are described below, organized according to the target enzyme class.

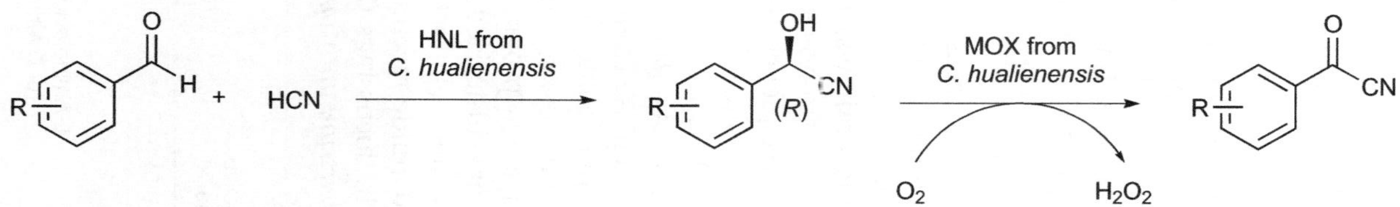

Scheme 2.4 Novel enzymes from the cyanogenic pathway of the invasive millipede *C. hualienensis*.

2.3.1 Isomerases

In silico identification of enzymes having desired functions is an attractive topic, since numerous enzymes with endless possibilities are deposited in freely accessible genetic databases. We previously discovered a new enzyme activity, *i.e.* amino acid amide-racemizing activity, in an α-amino-ε-caprolactam (ACL) racemase (EC 5.1.1.15), which had been characterized from *Achromobacter obae*.[21] By combining D-stereospecific hydrolases with such an amino acid amide racemase, the D,L-amino acid amide can be completely hydrolysed to form D-amino acid in a dynamic kinetic resolution. It is also possible to synthesize L-amino acids by using L-stereospecific hydrolases. This was demonstrated for the synthesis of D-alanine from L-alanine amide by a combination of racemization and kinetic resolution.[22,23] We also established a similar dynamic kinetic resolution for (*R*)-phenylalanine, (*R*)-tyrosine and non-natural analogues (2-fluoro, 3-fluoro- and 4-fluorophenylalanine) (Scheme 2.5).[24]

The following example describes a successful computational screening to obtain a new distribution of the amino acid amide racemase from sequence information alone. We wanted to study the metabolic and physiological role of the ACL enzyme, but the *A. obae* strain was not available and it was only possible to synthesize the gene of α-amino-ε-caprolactam (ACL) racemase using the published *A. obae* gene sequence. Instead, in order to identify other putative ACL racemases that could aid our investigation, we searched for the sequences encoding possible ACL racemase activity from the NCBI protein database using the gene of *A. obae* ACL enzyme as a template. Submitting the sequence of ACL racemase from *A. obae* to this database gave 413 putative type I pyridoxal 5′-phosphate (PLP)-dependent enzymes. By identifying Lys241 and other residues as key amino acids, 13 candidates for ACL racemase were selected using a new program, INTMSAlign.[25,26] Putative ACL racemase genes were cloned or synthesized and then expressed in *E. coli* BL21, followed by purification and characterization of the enzymes. Remarkably, ACL and amino acid amide racemizing activities were detected among ten of the candidates. The locus tags Oant_4493, Smed_5339, and CSE45_2055 derived from *Ochrobactrum anthropi* ATCC49188, *Sinorhizobium*

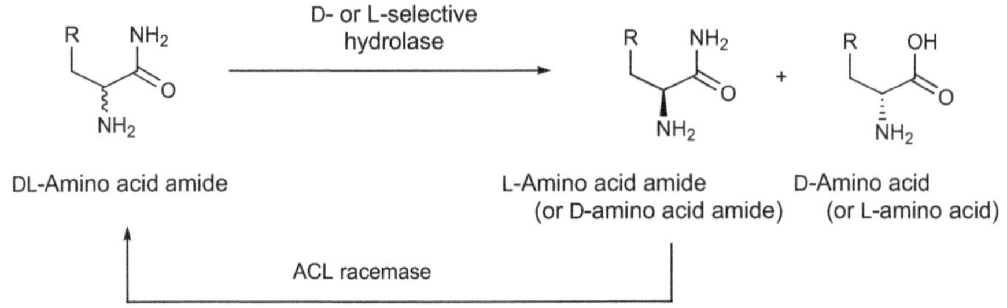

Scheme 2.5 Dynamic kinetic resolution of DL-amino acid amides with ACL racemase and stereoselective hydrolases.

medicae WSM 419, and *Citreicella* sp. SE45, respectively, showed higher racemization activity against D- and L-ACLs than the already known ACL racemase from *A. obae*. These newly discovered ACL racemases would prove to be useful for applications in dynamic kinetic resolution for D- or L-amino acid production. ACL racemase from *Citreicella* sp. SE45 was characterized in detail.[27]

2.3.2 Oxidoreductases

The stability of NAD^+-dependent L-tryptophan dehydrogenase (TrpDH, EC1.4.1.19) from *Nostoc punctiforme* ATCC29133, useful for the assay of L-tryptophan, was improved by growth-dependent genetic alterations. We previously showed the merits of screening enzymes evolved for biosynthesis for high substrate specificities useful in the micro-determination of biological samples. We additionally found a new tryptophan dehydrogenase (TrpDH) in the scytonemin biosynthetic pathway[28] of *N. punctiforme* ATCC29133.[29] The enzyme catalyzes the reversible oxidative deamination and the reductive amination between L-tryptophan and indole-3-pyruvic acid (Scheme 2.6).

TrpDH exhibited high specificity toward L-tryptophan, but poor stability. We obtained the mutant enzyme TrpDH L59F/D168G/A234D/I296N with enhanced thermal stability by re-isolation of *E. coli* transformants after prolonged cultivation with indole-3-pyruvic acid as the sole carbon source.[29] The transformants were harboring various mutant genes, which were generated by error-prone PCR using complementation in an L-tryptophan auxotroph of *E. coli*. The specific activity and stability of this isolated mutant enzyme were higher than those of the wild type enzyme (Figure 2.2). We also revealed that in these four mutation points, Asp168 and Ile296 residues contributed to increase the enzyme stability, and Leu59 and Ala234 residues increased its specific activity. Growth of the strain harboring the gene of the above four point-mutated enzyme was accelerated and, due to enhanced performance, it was easy to isolate the mutant as the major transformant inside the culture. We further demonstrated that TrpDH L59F/D168G/A234D/I296N was available for determination of L-tryptophan in human plasma. This is a good example of how "enrichment culture"

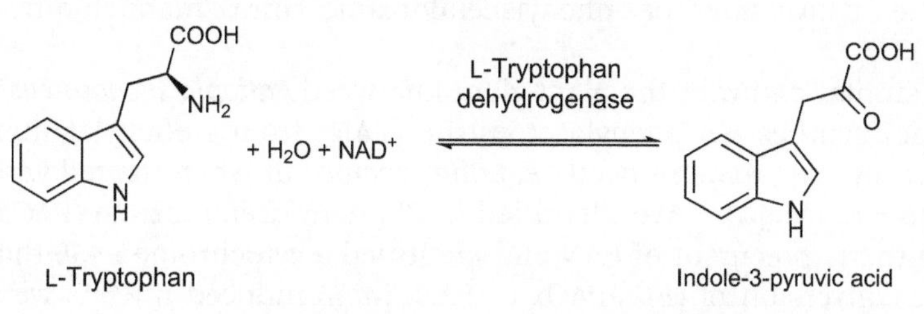

Scheme 2.6 Oxidative deamination of L-tryptophan catalyzed by L-tryptophan dehydrogenase from *Nostoc punctiforme* in the biosynthesis of scytonemin.

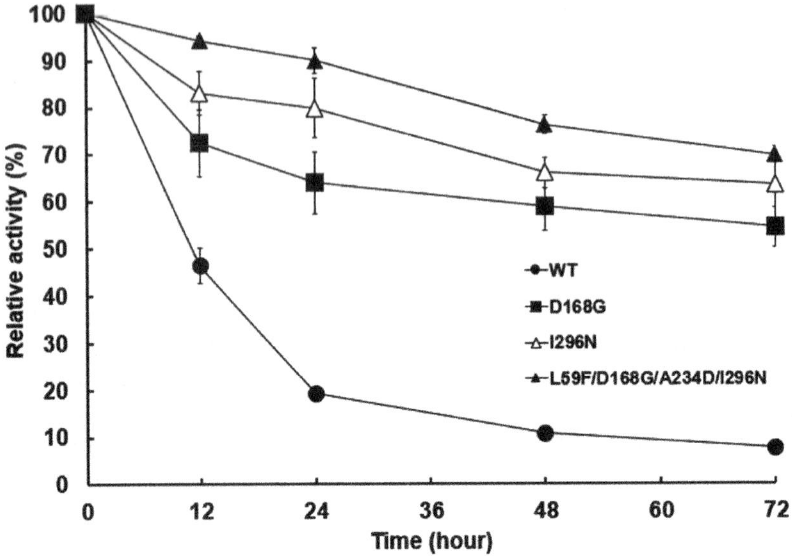

Figure 2.2 Improvement of the stability of L-tryptophan dehydrogenase from *Nostoc punctiforme* by growth-dependent genetic selections.

techniques can be used with a specifically designed medium and auxotrophic mutants of *E. coli*.

Another example of application of a genetic approach is the use of the transcriptome of an organism exhibiting a desired activity or forming a desired product, in this case mandelonitrile. Once the HNL producers were identified, further basic studies on their cyanogenesis were carried out, *e.g.* with the Japanese apricot *Prunus mume*, which has been known to produce the cyanogenic glycosides prunasin and amygdalin.[30] Cytochrome P450s catalyzing the conversion of L-phenylalanine into phenylacetaldoxime and further into mandelonitrile were identified based on their activity from *E. coli* clones of cDNA from transcriptomic mRNA. Full-length cDNAs encoding CYPs were cloned from publicly available *P. mume* RNA-sequencing data. Enzyme activity of these cytochrome P450s expressed in *Saccharomyces cerevisiae* was evaluated by GC-MS. CYP79D16 was found to catalyze the conversion of L-phenylalanine into phenylacetaldoxime, while CYP71AN24 catalyzed the conversion of phenylacetaldoxime into mandelonitrile (Scheme 2.7).[31]

In a quite similar pathway, the plant giant knotweed *Fallopia sachalinensis* (Polygonaceae) synthesizes phenylacetonitrile (PAN) from L-phenylalanine when infested by the Japanese beetle *Popillia japonica* or when treated with methyl jasmonate (MeJA).[32] We identified (*E*/*Z*)-phenylacetaldoxime (PAOx) as the biosynthetic precursor of PAN and identified a cytochrome P450 that catalyzes the conversion of (*E*/*Z*)-PAOx to PAN. MeJA-induced mRNAs were analyzed by differential expression analysis using a next-generation sequencer. Of the 74 329 contigs obtained from RNA-seq and *de novo* assembly,

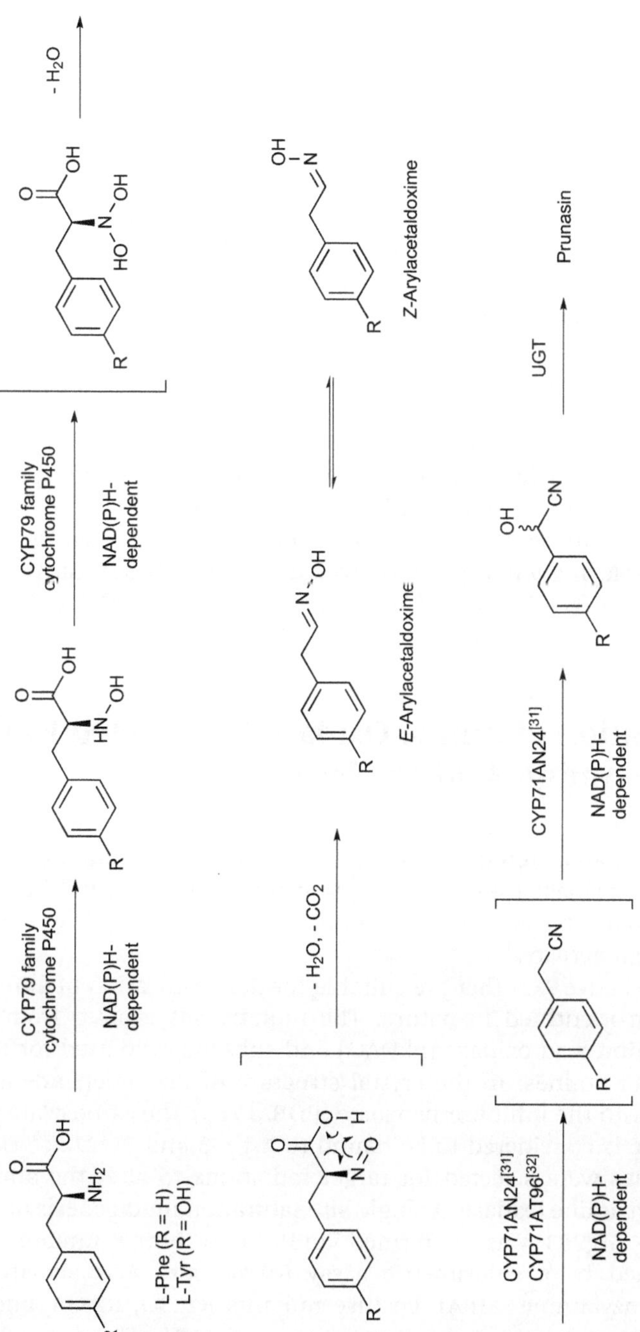

Scheme 2.7 Biosynthesis of organic nitriles and cyanoglucosides in plants catalyzed by cytochrome P450s in *Prunus mume* Sieb. et Zucc. (R = H) and *Fallopia sachalinensis* (R = OH).

252 contigs were induced by MeJA treatment. Full-length cDNAs encoding MeJA-induced cytochrome P450s CYP71AT96, CYP82AN1, CYP82D125 and CYP715A35 were cloned using 5′- and 3′-RACE and were expressed using a baculovirus expression system.[33] Among these cytochrome P450s, CYP71AT96 catalyzed the conversion of (*E*/*Z*)-PAOx to PAN in the presence of NhuaADPH and a cytochrome P450 reductase (Scheme 2.7). It is interesting that the overall reaction of the enzymatic system is the same as for bacterial aldoxime dehydratases. It also acts on (*E*/*Z*)-4-hydroxyphenylacetaldoxime and (*E*/*Z*)-indole-3-acetaldoxime to form the corresponding nitriles. The broad substrate specificity of CYP71AT96 is similar to that of aldoxime-metabolizing cytochrome P450s. Quantitative RT-PCR analysis showed that CYP71AT96 expression was highly induced because of treatment with MeJA as well as feeding by the Japanese beetle. These results indicate that CYP71AT96 contributes to the herbivore-induced PAN biosynthesis in *F. sachalinensis*.[32]

Within the same natural "aldoxime-nitrile" pathway, however now in the case of bacteria, using the *oxd* gene of *Rhodococcus erythropolis* N-771 (OxdRE) as a probe for southern hybridization and PCR, another aldoxime dehydratase from *Pseudomonas* sp. K-9 (OxdK) was identified (76.9% identity), displaying a new primary structure with highest sequence identity to the one from *Pseudomonas chlororaphis* A23 (OxdA, 90.3%) and the lowest towards OxdB from *Bacillus* sp. OxB-1 (32.7%).[34]

2.4 Creation of Amine Oxidase by Directed Evolution of D-Amino Acid Oxidase

To examplify the third strategy of screening for new enzymes, we describe the creation of an *R*-stereoselective amine oxidase (AOx) from porcine kidney D-amino acid oxidase (pkDAO) using structural information. The mutation was carried out very successfully and yielded an engineered *R*-stereoselective AOx with high activity.[35]

R-stereoselective AOx that are suitable for deracemization of amines have not yet been identified in nature. The mutant was evolved from porcine kidney D-amino acid oxidase (pkDAO) and subsequently used for the deracemization of amines. In the crystal structure of flavin-dependent pkDAO complexed with the inhibitor benzoate (PDB: 1VE9), the carboxylate group of the benzoate is considered to be bound to Arg283 and Tyr228.[36] Therefore, these residues were selected for target mutations to alter the amino acid oxidase to an amine oxidase. A single-site saturation mutagenesis of residues Tyr228 and Arg283 was performed and the resulting mutant libraries were screened by a colorimetric assay to measure AOx activity toward α-methyl benzylamine (MBA). Positive mutants R283G, R283A, and R283C catalyzed the oxidation of the *R*-enantiomer of MBA. These mutants were used as parents for the second round of saturation mutagenesis of Tyr228. Y228L/R283G mutant was shown to have markedly improved MBA oxidative

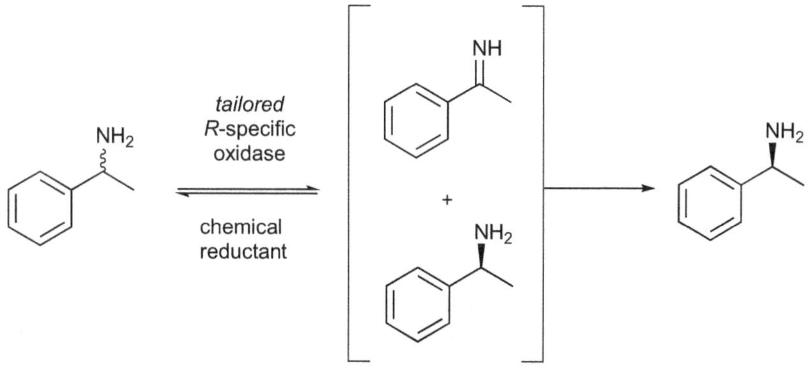

Scheme 2.8 Deracemization of α-methylbenzylamine to form (*S*)-α-methylbenzylamine with *R*-amine oxidase, *i.e.* a mutant Y228L/R283G derived from porcine D-amino acid oxidase, in the presence of a chemical reductant.

activity, while single Y228L mutant and other variants were inactive on MBA and poorly active on D-phenylalanine. The engineered pkDAO obtained by directed evolution displayed a markedly changed substrate specificity towards (*R*)-amines and totally lost its activity toward D-amino acids. The enzyme exhibited a high preference towards α-methyl benzylamine and was used to synthesize the corresponding (*S*)-amine through deracemization (Scheme 2.8).

2.5 From Discovery to Application – Aldoxime Dehydratases

Following initial observations of biological activities for transformation of aldoximes into nitriles by several groups,[37,38] Asano and Kato succeeded in isolating the first aldoxime dehydratases (Oxd) from bacteria grown *via* acclimation and enrichment cultures over a period of several months.[19,20] The proteins from *Bacillus* sp. OxB-1 (OxdB)[19,39] and *Rhodococcus erythropolis* N-771 (OxdRE)[40] were the cornerstone for a new era in nitrile metabolism based on the "aldoxime-nitrile pathway". In the following years, several other aldoxime dehydratases were isolated and characterized, *i.e.* from *Pseudomonas chlororaphis* B23 (OxdA),[41] *Pseudomonas* sp. K-9 (OxdK),[42] *Rhodococcus globerulus* A-4 (OxdRG)[43] and *Fusarium graminearum* (OxdFG),[44] the properties and applications of which are described in the following section.

2.5.1 Isolation, Characterization and Comparison of Aldoxime Dehydratases

Microorganisms possessing aldoxime dehydratase activity were isolated from soil by acclimation or enrichment cultures, exchanging the culture medium in regular intervals with fresh medium containing the synthetic substrate. This slow accumulation of an artificial nitrogen source promoted an evolutionary response and higher expression of a protein capable of degrading this new nutrient towards ammonia. By using the substrate *Z*-phenylacetaldoxime, a novel lyase was isolated from *Bacillus* sp. OxB-1 (OxdB).[19,45] In parallel, by supplementing *E*-pyridine-3-carbaldoxime according to the same method, aldoxime dehydratase activity in *Rhodococcus* sp. YH3-3 was identified.[20] Other microorganisms possessing aldoxime dehydratase activity were found by screening several hundred strains of bacteria, actinomycetes, yeasts and fungi *via* substrate-based assays (Table 2.1).[46] Crystal structure analysis of OxdRE[47] and OxdA[48] as well as sequence similarity between isolated Oxds from bacteria revealed that the active site residues are highly conserved, underpinning its evolutionary relevance in the "aldoxime-nitrile pathway". Expression of Oxds was induced primarily by the presence of aldoximes and also a wide range of structural analogues with different polarity, *e.g.* OxdRE was comparably induced by *E*-pyridine-3-carbaldoxime, its N^1-oxide, N^1-methyl pyridinium iodide as well as the product pyridine-3-carbonitrile and several other nitriles and aldoximes. This rationalized also the findings that Oxds were genetically clustered up- or downstream next to nitrilases (Nit), nitrile hydratases (NHase), amidases (Ami) and acyl-CoA synthetases (AcsA), for which nitriles were already known as inducers.[42,43,49–51]

2.5.2 Iron Heme Redox Catalysis and Mechanistic Studies

Biological aldoxime dehydration is not limited to Oxds, as it was determined that several plants contain cytochrome P450 enzymes capable of forming

Table 2.1 Comparison of Oxds from *Bacillus* sp. OxB-1 (OxdB), *R. erythropolis* N-771 (OxdRE), *R. globerulus* A-4 (OxdRG), *P. chlororaphis* B23 (OxdA), *Pseudomonas* sp. K-9 (OxdK), and *F. graminearum* MAFF305135 (OxdFG).[44]

Properties	OxdB[a]	OxdRE[a]	OxdRG[a]	OxdA	OxdK[a]	OxdFG[a]
Molecular weight per kDa						
Native	42 000	80 000	80 000	76 400	85 000	34 100
Sequence	40 972	44 794	44 817	40 127	44 511	44 070
Number of subunits	1	2	2	2	2	1
Soret peak (nm) (ferric form)	407	409	409	408	408	420
(ferrous form)	432	428	428	428	428	431
Specific activity ($U\,mg^{-1}$)[b]	851	562	633	197[c]	2.25	24.8
Optimum pH[d]	7.0	8.0	8.0	5.5[c]	7.0	5.5
Optimum Temp. (°C)[e]	30	30	30	45[c]	20	25
Stability pH[d]	6.5–8.0	6.0–9.5	6.0–9.5	6.0–8.0	5.5–6.5	4.5–8.0
Stability Temp. (°C)	<45	<40	<40	<40	<30	<20

[a]With *N*-terminal His_6-tag.
[b]*Z*-phenylacetaldoxime was used as substrate.
[c]*n*-Butyraldoxime was used as substrate.
[d]0.1 M buffers were used.
[e]0.1 M KPB (pH 7.0) between 20 °C and 80 °C was used.

nitriles from L-amino acids *via* aldoxime intermediates as part of their chemical defense against herbivores.[52,53] Yamaguchi *et al.* identified two P450 enzymes which catalyze the dehydration of phenylacetaldoxime in the biosynthesis of phenylacetonitrile, *i.e.* CYP71AN24 from the japanese apricot *Prunus mume* Sieb. et Zucc.[31] and CYP71AT96 from the giant knotweed *Fallopia sachalinensis*.[32] Although the substrates and products of these enzymes are identical with those of Oxds, P450 enzymes require additional NAD(P)H or FAD cofactors for their activity, probably to reduce the iron heme into an active catalytic species, while Oxds can dehydrate aldoximes with a redox process that is intrinsically balanced by the substrate, leading to an overall "formal" two-electron shift from carbon to nitrogen. Initial reports about Oxds stated the necessity of additional flavin mononucleotide (FMN) or reducing agents, *e.g.* dithionite, to achieve high activity.[54] However, no evidence was found that FMN acts as a cofactor for the enzyme. UV/Vis and Resonance-Raman spectroscopy investigations revealed that Oxds are expressed mainly in their ferric heme form due to rapid oxidation, while their active form is in the ferrous state.[55–57] Since the overall reaction under anaerobic conditions did not lead to ferric heme, the reduction is supposedly part of the reductive elimination of the nitrile from an intermediary ferryl (Fe^{IV}) complex, which was previously proposed for the aldoxime dehydration catalyzed by iron porphyrin model complexes.[58] A generally accepted mechanism involving the stepwise two-electron reduction of the aldoxime nitrogen by the ferrous heme,[59,60] activation of the aldoxime oxygen by a protonated distal histidine to eliminate water,[61] and reductive elimination of the ferryl complex is depicted in Scheme 2.9. It was further determined that exogenous ligands like 2-methyl imidazole could, depending on their electron-donor strength, reconstitute or even increase Oxd activity in holoenzymes lacking the proximal histidine ligand of the heme.[62]

2.5.3 Substrate Spectrum of Aldoxime Dehydratases

In nature, aldoxime dehydratases, including Oxd and P450 enzymes, metabolize aldoximes that originated from the degradation of α-amino acids, *e.g.* L-phenylalanine,[63] L-tyrosine, L-tryptophan, or certain elongated homologs.[64] However, their wide tertiary structure around the ferrous heme enables Oxds to accommodate and convert a broad range of synthetic aldoximes with aliphatic, aromatic, heterocyclic and annulated substituents (Scheme 2.10).[20,43,45]

2.5.4 Application of Oxds for Nitrile Synthesis

Due to self-sufficient redox catalysis without need of additional cofactors, broad substrate compatibility and overall stoichiometry from aldehyde to nitrile with two equivalents of water as the only by-product, aldoxime dehydratases are likely to become important biocatalysts in the cyanide-free production of nitriles. This was not only demonstrated in academia with the

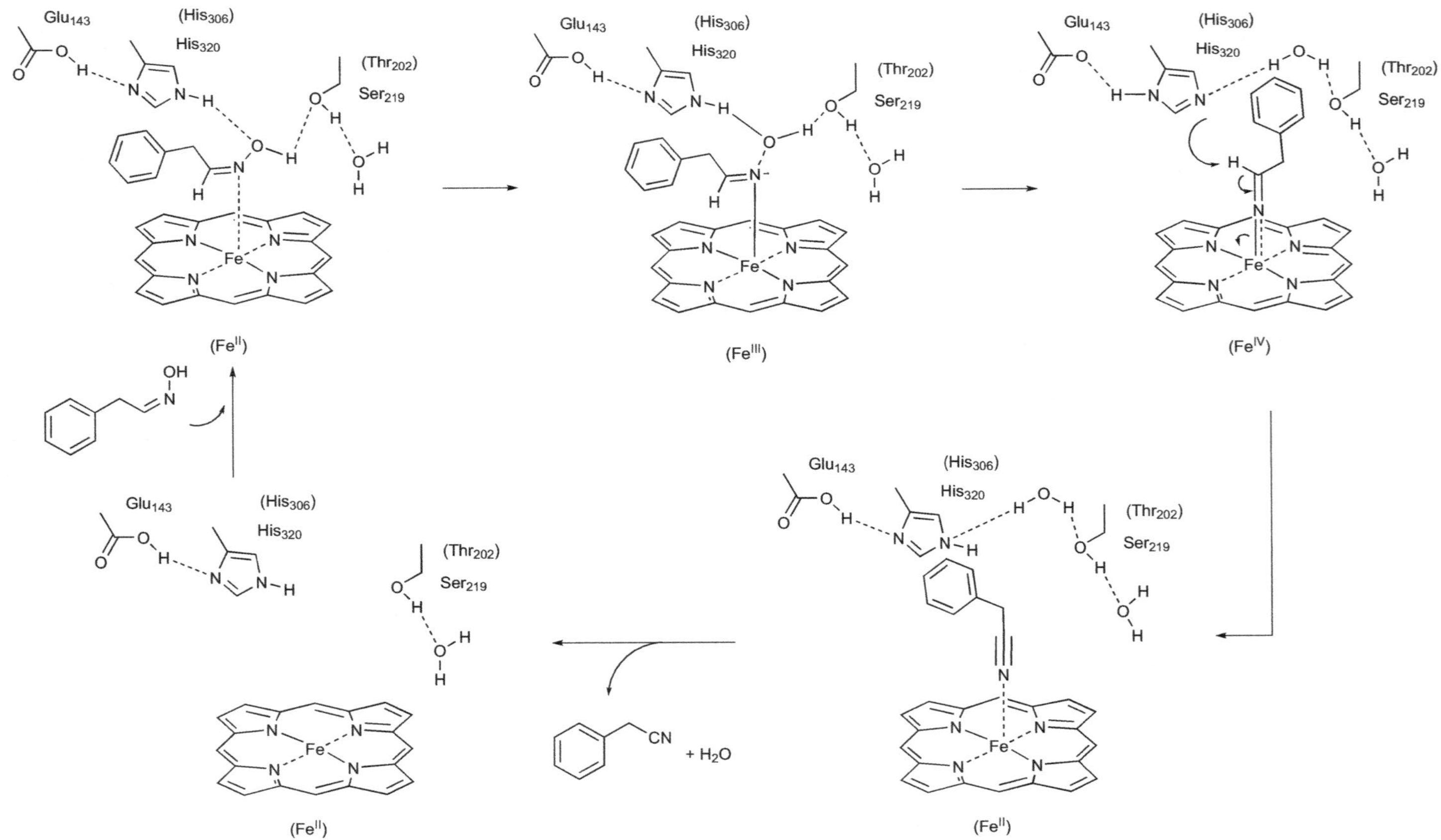

Scheme 2.9 Proposed mechanism of *Z*-phenylacetaldoxime dehydration catalyzed by Oxds based on spectroscopic and mutational analysis of the active site by several groups.

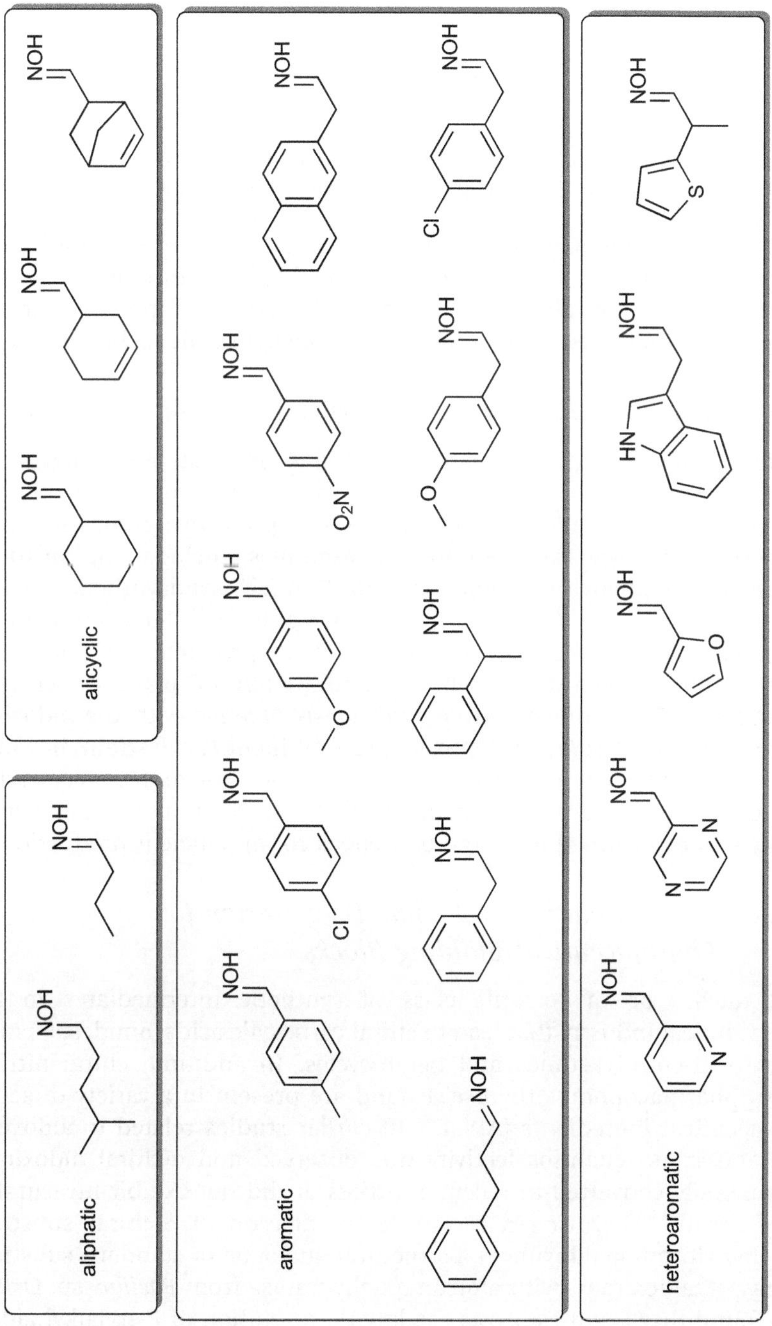

Scheme 2.10 Selected examples of synthetic aldoximes converted to their corresponding nitriles by aldoxime dehydratases (Oxds).

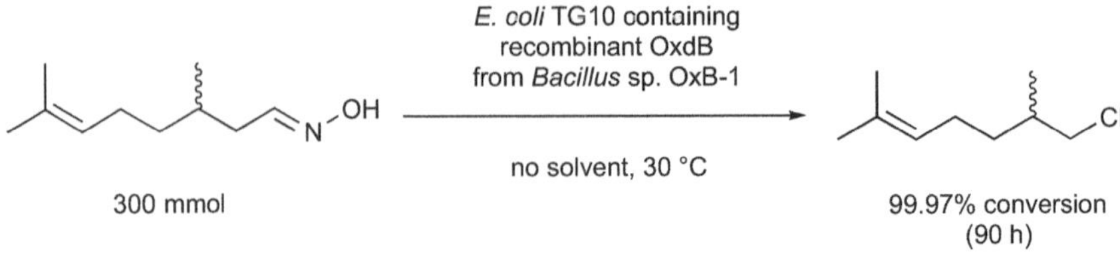

Scheme 2.11 Industrial production of fragrance compound citronellyl nitrile using OxdB from *Bacillus* sp. OxB-1 as whole-cell catalyst.

gram-scale synthesis of several bulk nitriles such as phenylacetonitrile,[46] 3-phenylpropionitrile,[65] pyridine-3-carbonitrile,[20] 2-furonitrile[66] and *n*-butyronitrile,[47] but also in a recent BASF patented process for the production of fragrance compounds like citronellyl nitrile (Scheme 2.11).[67]

2.5.4.1 *Artificial Biosynthesis of Nitriles from α-Amino Acids*

Metabolic engineering has gained increasing interest since high-value products can be produced from cheap and abundant resources like D-glucose or amino acids, without the need of protective group-chemistry or intermediate isolation. Prominent examples include the artificial biosynthesis of phenylpropanoic acids in *E. coli*,[68] semisynthetic opioids in recombinant yeasts[69] and vanillin from D-glucose.[70] Such an artificial biosynthetic pathway was also established for the production of phenylacetonitrile from L-phenylalanine by a combination of plant cytochrome CYP79A2 and CYP reductase from *Arabidopsis thaliana* with the aldoxime dehydratase from *Bacillus* sp. OxB-1 in a recombinant *E. coli* strain, in which whole cells produced phenylacetonitrile in about 4.9 mM concentration under optimized conditions (Scheme 2.12).[71] The P450-catalyzed oxidation of L-Phe was determined to be the bottleneck for an efficient production.

2.5.4.2 *Enantioselective Aldoxime Dehydration for Pharmaceutical Building Blocks*

Chiral nitriles are a versatile class of synthetic intermediates in the pharmaceutical industry that lead to chiral carboxylic acids, amides, β-chiral amines or α-chiral ketones and heterocycles. In addition, chiral nitriles are also pharmacophores themselves and are present in a variety of active pharmaceutical ingredients (APIs).[72] In earlier studies related to aldoxime dehydratases, no enantioselectivity was observed and α-chiral aldoximes were generally converted to racemic nitriles or did not exhibit measurable activity at all.[43,45] During a recent reinvestigation of α-chiral substrate *E*/*Z*-2-phenylpropionaldoxime – a structural analogue of standard substrate *Z*-phenylacetaldoxime – with aldoxime dehydratase from *Bacillus* sp. OxB-1, it was found that OxdB catalyzes the kinetic resolution of *E*-arylalkyl aldoximes, and that previous observations were due to parallel thermodynamic *E*/*Z*-isomerization coupled with nearly no enantioselectivity for the

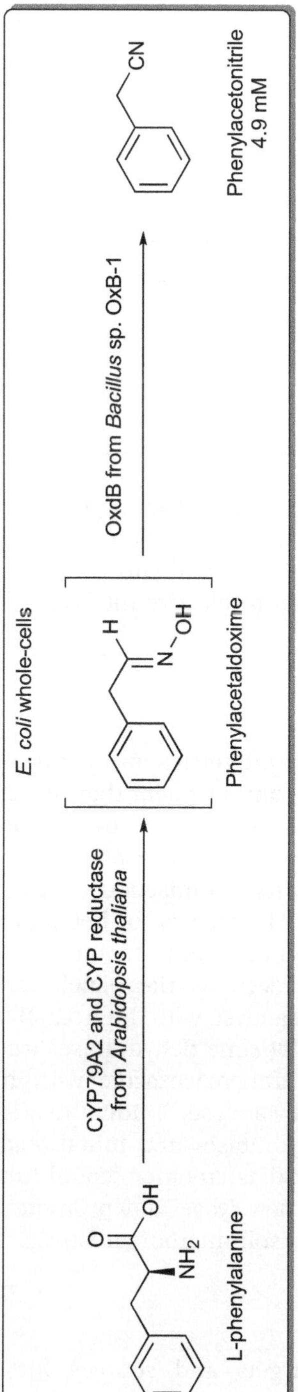

Scheme 2.12 Artificial metabolic pathway for the production of phenylacetonitrile from L-phenylalanine using whole-cell catalysts.

Table 2.2 Thermodynamically controlled kinetic resolution of *rac-E/Z*-2-phenylpropionaldoxime catalyzed by aldoxime dehydratase from *Bacillus* sp. OxB-1 (OxdB).

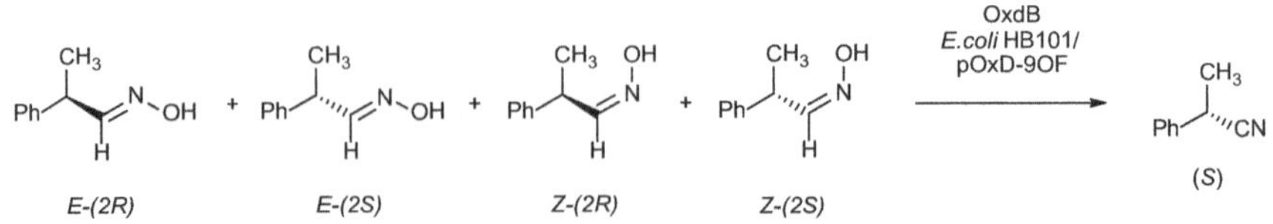

Entry	*E*/*Z* ratio	*T*/°C	Conversion (%)	$ee_{product}$ (%)
1	4 : 1	30	99	0
2	4 : 1	8	60	65 (*S*)
3	9 : 1	8	53	88 (*S*)
4	99 : 1	8	50	98 (*S*)
5	1 : 11.5	8	15	67 (*R*)
6	1 : 11.5	8	96	4 (*S*)

Z-diastereomers.[73] By lowering the reaction temperature and separating *E*- and *Z*-diastereomers *via* column chromatography, (*S*)-2-phenylpropionitrile was produced in 50% conversion and with an enantioselectivity of 98% ee, demonstrating the first enantioselective nitrile synthesis without the use of toxic cyanide (Table 2.2).

2.5.5 Perspectives

Taking all of the presented properties and applications of aldoxime dehydratases into account, it becomes evident that this enzyme class, discovered from screening of soil microorganisms over a prolonged period, might play a major role in future biocatalytic processes. This was very recently highlighted by exploiting its promiscuous activity against non-natural, structurally related, 1,2-benzisoxazoles for heteroaromatic N–O bond cleavage, a reaction better known in organic chemistry as the Kemp elimination. This reaction is currently seen as the benchmark for artificial enzyme design, since no natural catalyst with high catalytic efficiency had been found yet.[74–76] Wild-type aldoxime dehydratases were found to catalyze the Kemp elimination of 5-nitrobenzisoxazole with high catalytic efficiency similar to that of evolved catalysts.[77] Identification of the promiscuous potential of aldoxime dehydratases and mutational analysis for increased enantioselectivity against *Z*-diastereomers could further promote the application of this fascinating new lyase and promote the discovery of useful enzymes by screening from soil microorganisms.

2.6 Conclusion

We showed here that targets and sources for enzyme screening are expanding, not only among microorganisms, but also towards plants and animals, with the development of gene cloning techniques. The target enzymes that we seek vary according to the purpose of our studies, such as enzymatic transformation, enzymatic determination of L-amino acids for health monitoring, and most importantly basic studies to clarify metabolic routes in organisms. Without studies on metabolic pathways catalyzed by enzymes in non-model organisms, there would be no progress in the exploitation of new enzymes as tools for further applications. Used host organisms, in addition to *Escherichia coli,* also include yeasts such as *Pichia pastoris*, and baculovirus. We discovered that the primary structure of the hydroxynitrile lyase from millipede *C. hualinensis* is totally unrelated to any other known protein, thus the genetic database which we are currently relying on is still incomplete. Focusing on nitrile forming enzymes upstream of nitrile hydratase, we discovered aldoxime dehydratases converting aldoximes into nitriles in bacteria, and studied their substrate specificity, reaction mechanism and stereoselectivity with synthetic substrates, as well as X-ray structure. We also discovered a plant cytochrome P450 that catalyzes the same reaction as bacterial aldoxime dehydratases. Thus, exploitation of natural enzymes is still evolving based on our interests, leading to the discovery of novel protein structures and new applications.

Acknowledgements

This work was supported by the Exploratory Research for Advanced Technology (ERATO) Asano Active Enzyme Molecule Project of the Japan Science and Technology Agency (JST).

References

1. Y. Asano, *J. Biotechnol.*, 2002, **94**, 65.
2. Y. Asano, Tools for Enzyme Discovery, in *Manual of Industrial Microbiology and Biotechnology*, ed. R. H. Baltz, J. E. Davies, and A. Demain, American Society for Microbiology, 3rd Edn, 2010, pp. 441–452.
3. Y. Asano, Hydrolysis of nitriles to amides, in *Science of Synthesis: Biocatalysis in Organic Synthesis*, ed. K. Faber, W.-D. Fessner, and N. Turner, Georg Thieme Verlag KG, 2015, vol. 1, pp. 255–276.
4. K. Fuhshuku, S. Watanabe, T. Nishii, A. Ishii and Y. Asano, *J. Mol. Catal. B: Enzym.*, 2014, **102**, 115.
5. K. Fuhshuku, S. Watanabe, T. Nishii, A. Ishii and Y. Asano, *Biosci., Biotechnol., Biochem.*, 2015, **79**, 1587.
6. N. M. Shaw, A. Naughton, K. Robins, A. Tinschert, E. Schmid, M. Hischier, V. Venetz, J. Werlen, T. Zimmermann, W. Brieden, P. Riedmatten, J. Roduit, B. Zimmermann and R. Neumüller, *Org. Process Res. Dev.*, 2002, **6**, 497.
7. S. Schriek, C. Rückert, D. Staiger, E. K. Pistorius and K.-P. Michel, *BMC Genomics*, 2007, **8**, 437.
8. C. L. Fan and V. W. Rodwell, *J. Bacteriol.*, 1975, **124**, 1302.

9. D. Matsui, A. Terai and Y. Asano, *Enzyme Microb. Technol.*, 2016, **82**, 151.
10. H. Onaka, S. Taniguchi, Y. Igarashi and T. Furumai, *J. Antibiot.*, 2002, **55**, 1063.
11. C. J. Balibar and C. T. Walsh, *Biochemistry*, 2006, **45**, 15444.
12. M. Kameya, H. Onaka and Y. Asano, *Anal. Biochem.*, 2013, **438**, 124.
13. K. Kino, T. Kobayashi, E. Arima, R. Komori, T. Kobayashi and H. Miyazawa, *Bioorg. Med. Chem. Lett.*, 2009, **19**, 2070.
14. W. J. Foster, *J. Bacteriol.*, 1944, **47**, 27.
15. K. Yamamoto and Y. Asano, *Appl. Environ. Microbiol.*, 2015, **81**, 7360.
16. Y. Asano, K. Tamura, N. Doi, T. Ueatrongchit, A. H-Kittikun and T. Ohmiya, *Biosci., Biotechnol., Biochem.*, 2005, **69**, 2349.
17. M. Dadashipour, Y. Ishida, K. Yamamoto and Y. Asano, *Proc. Natl. Acad. Sci. U. S. A.*, 2015, **112**, 10605.
18. Y. Ishida, Y. Kuwahara, M. Dadashipour, A. Ina, T. Yamaguchi, M. Morita, Y. Ichiki and Y. Asano, *Sci. Rep.*, 2016, **6**, 26998.
19. Y. Asano and Y. Kato, *FEMS Microbiol. Lett.*, 1998, **158**, 185.
20. Y. Kato, R. Ooi and Y. Asano, *Arch. Microbiol.*, 1998, **170**, 85.
21. S. A. Ahmed, N. Esaki, H. Tanaka and K. Soda, *Biochemistry*, 1986, **25**, 385.
22. Y. Asano and S. Yamaguchi, *J. Mol. Catal. B: Enzym.*, 2005, **36**, 22.
23. Y. Asano and S. Yamaguchi, *J. Am. Chem. Soc.*, 2005, **127**, 7696.
24. K. Yasukawa and Y. Asano, *Adv. Synth. Catal.*, 2012, **354**, 3327.
25. W. Payoungkiattikun, S. Okazaki, S. Nakano, A. Ina, A. H-Kittikun and Y. Asano, *Appl. Biochem. Biotechnol.*, 2015, **176**, 1303.
26. S. Nakano and Y. Asano, *Sci. Rep.*, 2015, **5**, 8193.
27. W. Payoungkiattikun, S. Okazaki, A. Ina, A. H-Kittikun and Y. Asano, *J. Ind. Microbiol. Biotechnol.*, 2016, **150**, 370.
28. E. P. Balskus and C. T. Walsh, *J. Am. Chem. Soc.*, 2008, **130**, 15260.
29. D. Matsui, S. Okazaki, M. Matsuda and Y. Asano, *J. Biotechnol.*, 2015, 196–197.
30. H. Terada and Y. Sakabe, *J. Hyg. Chem.*, 1988, **34**, 36.
31. T. Yamaguchi, K. Yamamoto and Y. Asano, *Plant Mol. Biol.*, 2014, **86**, 215.
32. T. Yamaguchi, K. Noge and Y. Asano, *Plant Mol. Biol.*, 2016, **91**, 229.
33. H. R. Lo and Y. C. Chao, *Biotechnol. Prog.*, 2004, **20**, 354.
34. Y. Kato, S. Yoshida and Y. Asano, *FEMS Microbiol. Lett.*, 2005, **246**, 243.
35. K. Yasukawa, S. Nakano and Y. Asano, *Angew. Chem., Int. Ed.*, 2014, **53**, 4428.
36. C. Setoyama, R. Miura, Y. Nishina, K. Shiga, H. Mizutani, I. Miyahara and K. Hirotsu, *J. Biochem.*, 1996, **119**, 1114.
37. S. Mahadevan, *Arch. Biochem. Biophys.*, 1963, **100**, 557.
38. B. V. Milborrow, *Biochem. J.*, 1963, **87**, 255.
39. T. Yamaguchi and Y. Asano, *Genome Announce.*, 2015, **3**, e00025-15.
40. Y. Kato, S. Yoshida, S.-X. Xie and Y. Asano, *J. Biosci. Bioeng.*, 2004, **97**, 250.

41. K.-I. Oinuma, Y. Hashimoto, K. Konishi, M. Goda, T. Noguchi, H. Higashibata and M. Kobayashi, *J. Biol. Chem.*, 2003, **278**, 29600.
42. Y. Kato and Y. Asano, *Appl. Microbiol. Biotechnol.*, 2006, **70**, 92.
43. S.-X. Xie, Y. Kato, H. Komeda, S. Yoshida and Y. Asano, *Biochemistry*, 2003, **42**, 12056.
44. Y. Kato and Y. Asano, *Biosci., Biotechnol., Biochem.*, 2005, **69**, 2254.
45. Y. Kato, K. Nakamura, H. Sakiyama, S. G. Mayhew and Y. Asano, *Biochemistry*, 2000, **39**, 800.
46. Y. Kato, R. Ooi and Y. Asano, *Appl. Environ. Microbiol.*, 2000, **66**, 2290.
47. H. Sawai, H. Sugimoto, Y. Kato, Y. Asano, Y. Shiro and S. Aono, *J. Biol. Chem.*, 2009, **284**, 32089.
48. J. Nomura, H. Hashimoto, T. Ohta, Y. Hashimoto, K. Wada, Y. Naruta, K.-I. Oinuma and M. Kobayashi, *Proc. Natl. Acad. Sci. U. S. A.*, 2013, **110**, 2810.
49. Y. Kato, T. Tsuda and Y. Asano, *Eur. J. Biochem.*, 1999, **263**, 662.
50. Y. Hashimoto, H. Hosaka, K.-I. Oinuma, M. Goda, H. Higashibata and M. Kobayashi, *J. Biol. Chem.*, 2005, **280**, 8660.
51. T. Sakashita, Y. Hashimoto, K.-I. Oinuma and M. Kobayashi, *J. Bacteriol.*, 2008, **190**, 4210.
52. L. Du, J. Lykkesfeldt, C. E. Olsen and B. A. Halkier, *Proc. Natl. Acad. Sci. U. S. A.*, 1995, **92**, 12505.
53. R. Schuhegger, M. Nafisi, M. Mansourova, B. L. Petersen, C. E. Olsen, A. Svatos, B. A. Halkier and E. Glawischnig, *Plant Physiol.*, 2006, **141**, 1248.
54. Y. Kato and Y. Asano, *Protein Expression Purif.*, 2003, **28**, 131.
55. K. Kobayashi, S. Yoshioka, Y. Kato, Y. Asano and S. Aono, *J. Biol. Chem.*, 2005, **280**, 5486.
56. K. Kobayashi, B. Pal, S. Yoshioka, Y. Kato, Y. Asano, T. Kitagawa and S. Aono, *J. Inorg. Biochem.*, 2006, **100**, 1069.
57. K. Konishi, T. Ohta, K.-I. Oinuma, Y. Hashimoto, T. Kitagawa and M. Kobayashi, *Proc. Natl. Acad. Sci. U. S. A.*, 2006, **103**, 564.
58. J. Hart-Davis, P. Battioni, J.-L. Boucher and D. Mansuy, *J. Am. Chem. Soc.*, 1998, **120**, 12524.
59. X.-L. Pan, F.-C. Cui, W. Liu and J.-Y. Liu, *J. Phys. Chem. B*, 2012, **116**, 5689.
60. R.-Z. Liao and W. Thiel, *J. Phys. Chem. B*, 2012, **116**, 9396.
61. K. Konishi, K. Ishida, K.-I. Oinuma, T. Ohta, Y. Hashimoto, H. Higashibata, T. Kitagawa and M. Kobayashi, *J. Biol. Chem.*, 2004, **279**, 47619.
62. K. Kobayashi, M. Kubo, S. Yoshioka, T. Kitagawa, Y. Kato, Y. Asano and S. Aono, *ChemBioChem*, 2006, 7, 2004.
63. K. Luck, J. Jirschitzka, S. Irmisch, M. Huber, J. Gershenzon and T. G. Köllner, *BMC Plant Biol.*, 2016, **16**, 479.
64. S. Irmisch, A. Clavijo McCormick, G. A. Boeckler, A. Schmidt, M. Reichelt, B. Schneider, K. Block, J.-P. Schnitzler, J. Gershenzon, S. B. Unsicker and T. G. Kollner, *Plant Cell*, 2013, **25**, 4737.
65. S.-X. Xie, Y. Kato and Y. Asano, *Biosci., Biotechnol., Biochem.*, 2001, **65**, 2666.

66. Y. Kato, R. Ooi and Y. Asano, *J. Mol. Catal. B: Enzym.*, 1999, **6**, 249.
67. A. Piatesi, W. Siegel, K. Baldenius, *U. S. Pat.* 9 080 191, 2015.
68. S.-Y. Kang, O. Choi, J. Lee, B. Hwang, T.-B. Uhm and Y.-S. Hong, *Microb. Cell Fact.*, 2012, **11**, 153.
69. K. Thodey, S. Galanie and C. D. Smolke, *Nat. Chem. Biol.*, 2014, **10**, 837.
70. J. Ni, F. Tao, H. Du and P. Xu, *Sci. Rep.*, 2015, **5**, 13670.
71. Y. Miki and Y. Asano, *Appl. Environ. Microbiol.*, 2014, **80**, 6828.
72. F. F. Fleming, L. Yao, P. C. Ravikumar, L. Funk and B. C. Shook, *J. Med. Chem.*, 2010, **53**, 7902.
73. R. Metzner, S. Okazaki, Y. Asano and H. Gröger, *ChemCatChem*, 2014, **6**, 3105.
74. D. Röthlisberger, O. Khersonsky, A. M. Wollacott, L. Jiang, J. DeChancie, J. Betker, J. L. Gallaher, E. A. Althoff, A. Zanghellini, O. Dym, S. Albeck, K. N. Houk, D. S. Tawfik and D. Baker, *Nature*, 2008, **453**, 190.
75. O. Khersonsky, G. Kiss, D. Rothlisberger, O. Dym, S. Albeck, K. N. Houk, D. Baker and D. S. Tawfik, *Proc. Natl. Acad. Sci. U. S. A.*, 2012, **109**, 10358.
76. R. Blomberg, H. Kries, D. M. Pinkas, P. R. E. Mittl, M. G. Grütter, H. K. Privett, S. L. Mayo and D. Hilvert, *Nature*, 2013, **503**, 418.
77. Y. Miao, R. Metzner and Y. Asano, *ChemBioChem*, 2017, **18**, 451.

CHAPTER 3

Artificial Metalloenzymes

JEAN-PIERRE MAHY,*[a] WADIH GHATTAS,[a] THIBAULT DI MÉO[a,b] AND RÉMY RICOUX[a]

[a] Institut de Chimie Moléculaire et des Matériaux, UMR 8182, Laboratoire de Chimie Organique et Bioinorganique, Université Paris-Saclay, Université Paris-Sud, CNRS, 91405, Orsay Cedex, France; [b] Institut de Biologie Intégrative de la Cellule, UMR 9198, Laboratoire de Modélisation et Ingénierie des Protéines, Université Paris-Saclay, Université Paris-Sud, CNRS, 91405, Orsay Cedex, France

*Email: jean-pierre.mahy@u-psud.fr

3.1 Introduction

During the last century, the evolution of chemistry has enabled us to obtain a plethora of molecules and materials. Present in all scientific fields, chemistry has thus profoundly modified our daily life. Today, however, it is necessary to respond to new problems. Firstly, the future is emerging with limited oil resources, which indicates that we must think of chemistry in a more sober way. Secondly, the ecological factor must be taken into account, which implies that the production of chemicals must not only resort to catalytic procedures that lower wastes, but also that use harmless solvents under mild reaction conditions. From these observations was born the concept of green chemistry, introduced in the 1990s by P. Anastas and J. C. Warner,[1] which advocates the use of less energy-consuming and more selective reactions. To achieve these goals, there is thus a real need for finding cheap catalysts that would be able to work in non-polluting solvents, water being the ideal one, under smooth conditions of pressure, temperature and pH. From this point of view, enzymes are undoubtedly the

Catalysis Series No. 32
Modern Biocatalysis: Advances Towards Synthetic Biological Systems
Edited by Gavin Williams and Mélanie Hall

Published by the Royal Society of Chemistry, www.rsc.org

first-choice candidates as they are able to catalyze selective chemical reactions, in aqueous medium and under mild conditions with a limited formation of side products. These biocatalysts are then more suitable than chemical catalysts, which are often used in organic solvents. However, the application of biocatalysis for the synthesis of fine and heavy chemicals still suffers from several drawbacks, including low efficiency with their non-natural substrates (in terms of activity and/or selectivity) and the limited number of naturally occurring enzymes to cover key synthetic reactions needed in chemical industries such as hydroformylations, Diels–Alder cyclizations, and palladium-catalyzed carbon–carbon bond forming reactions, such as Suzuki–Miyaura coupling and allylic alkylation.

Considering the case of reactions catalyzed by metalloenzymes, a large selection of metal ions, ligands and substrates are involved to achieve a large set of reactivities.[2] In those biocatalysts, the first, second and outer coordination spheres of the metal ion play a crucial role in tuning reactivity. The electronic and steric control of the metal site as well as a full set of molecular interactions (H bonding, electrostatic, π-stacking...) position the substrate into the active site, facilitate the reaction by controlling its stereo-, chemo- and regioselectivity and help to release the products. Over the last 15 years, chemists have proven to be able to develop very sophisticated ligands that were used to synthesize metal complexes, which possess catalytic properties similar to those of metalloenzymes. The importance of such metal complexes as catalysts to be used in organic synthesis has been particularly highlighted by the three Nobel Prizes that have been awarded to Knowles, Noyori and Sharpless for metal-catalyzed asymmetric hydrogenations and oxidations in 2001,[3–5] Chauvin, Grubbs and Schrock for the work on olefin metathesis, based on the use of ruthenium, molybdenum and tungsten complexes as catalysts in 2005,[6–8] and Heck, Negishi and Suzuki for palladium-catalyzed cross-coupling in 2010.[9] The reactivity of such coordination metal complexes is, like in enzymes, mainly based on tuning the redox properties, substrate accessibility and/or Lewis acidity of their metal center. These can be achieved by modulating the donor/acceptor character of the ligands, and by controlling their geometry and arrangement around the metal center. The use of metal complexes as catalysts however has also several drawbacks. Firstly, the multi-step synthesis of the targeted ligands potentially represents significant economical and ecological concerns and might hamper the accessibility of metal complexes. Secondly, the conditions under which the catalysis is performed are far from being eco-friendly, involving high pressure and/or temperature, often in organic solvents.

Industry has already considered strategies that take advantage of the complementarity of biocatalysts and chemical catalysts on many aspects: range of substrates and reactions, operating conditions and stereoselectivity. Many industrial processes thus rely on either organometallic catalysis or biocatalysis[10] to produce enantiopure amino acids, amino alcohols, amines, alcohols and epoxides.[11] The optimal procedure is selected by examining several factors such as the quantity of product needed, the cost of the catalyst, its efficiency, its stability and its stereoselectivity. It appeared to be of major interest to conceive catalysts that would combine the complementary advantages of biocatalysts *i.e.* function with high selectivities, under mild conditions and in aqueous medium, with those of synthetic metal complexes *i.e.* robustness and wide range of reactions. As a result, many teams have recently been working on the conception of hybrid biocatalysts that are obtained by the association of a protein with a synthetic catalyst. Of the several strategies used to expand the scope of biocatalytic reactions, the construction of artificial metalloenzymes seems to be one of the most promising.[12]

This chapter describes major examples of artificial metalloenzymes obtained by following the three strategies: (i) direct insertion of an inorganic metal ion into a protein, (ii) supramolecular anchoring of a metal cofactor in a protein *i.e.* "Host–Guest" and "Trojan horse" strategies, and (iii) covalent binding of a metal cofactor to a protein. This chapter is restricted to examples of artificial metalloenzymes that have been obtained using natural proteins and ends up with a paragraph that portrays the catalysis of cascade reactions by artificial metalloenzymes.

3.2 Direct Insertion of Inorganic Metal Ions into Proteins

3.2.1 Carbonic Anhydrase

The metal binding sites of metalloproteins have naturally evolved to become well suited for hosting metal ions. Consequently, apo-metalloproteins were the prime target for the insertion of non-native metal ions. Many other proteins however contain histidine, lysine, aspartic acid, glutamic acid, tyrosine, cysteine and methionine residues comprising metal-binding N, O, and S functional groups in well-defined three-dimensional chiral structures that could enable the coordination of metal cations.

This inspired Kaiser *et al.* in 1976 when they replaced the natural Zn(II) ion of carboxypeptidase A (CPA) by Cu(II), which resulted in the complete loss of peptidase and esterase activities and in the gain of the ability to oxidize ascorbic acid into dehydroascorbic acid following Michaelis–Menten kinetics.[13] Similarly, Soumillion and Kazlauskas *et al.* replaced the natural Zn(II) ion of human carbonic anhydrase (hCAII) by either Mn(II), to catalyze enantioselective alkene epoxidation (up to 66.5% ee and 12.5% conversion)[14–16] or Rh(I), to catalyze both chemo-selective alkene hydrogenation (up to 80.5% conversion) and regio-selective alkene hydroformylation (up to 8 : 1 linear : branched and 74% conversion) (Figure 3.1).[17,18] The X-ray crystal structures of hCAII substituted with Co(II), Cu(II), Ni(II), and Mn(II) revealed that the primary coordination sphere of each of these is uniquely perturbed relative to the native Zn(II), although the overall metal binding motif remains intact, *i.e.* three histidine residues.[19]

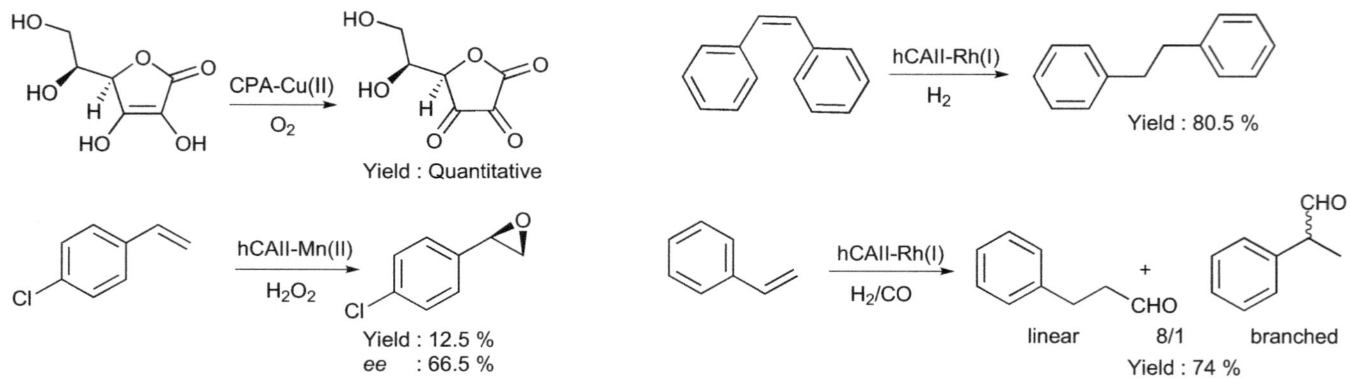

Figure 3.1 Reactions catalyzed by artificial metalloenzymes prepared by combining carboxypeptidase A and copper(II) or human carbonic anhydrase II (hCAII) and various metal cations.[13–18]

3.2.2 Ferritin

The iron storage protein ferritin was also reconstituted with a range of metal ions. Watanabe *et al.* reported that apo-ferritin captured allylpalladium(II) chloride dimer at 48 different sites in the ferritin interior.[20] Site-directed mutagenesis was used to alter Pd coordination environments and stoichiometry, as confirmed by X-ray crystallography.[21] The artificial enzymes were used to catalyze Suzuki–Miyaura coupling reaction between 4-iodoaniline and phenylboronic acid and conversions correlated with palladium loading.[22]

3.2.3 Serum Albumins

Serum albumins are transport proteins with a broad cavity, which allows accommodation of a great variety of hydrophobic compounds including fatty acids, bilirubin and steroids.[23] This makes them very interesting systems for the design of new metalloenzymes by insertion of metal ions. The first team reported to have used them for this purpose is Okano *et al.*, who in 1983 inserted osmium tetroxide in bovine serum albumin (BSA).[24] The formed artificial metalloenzyme required *tert*-butyl hydroperoxide (*t*-BuOOH) to catalyze the stereoselective *cis*-bis-hydroxylation of up to 40 equivalents of α-methylstyrene with 68% ee in favor of (*S*)-diol. Bertucci *et al.* have also exploited the ability of the serum albumins to complex metal ions. They have built an artificial metalloenzyme by association of human serum albumin (HSA) with an excess of rhodium(I) salt. The formed complex catalyzed the hydroformylation of styrene and olefins under high pressure of carbon monoxide and hydrogen (1 : 1) with good yields. Moreover, this reaction showed a high regioselectivity in favor of the branched aldehyde.[25] Similar complexes obtained with chicken egg albumin and papain were also shown to be able to catalyze styrene hydroformylation, with similar conversion rate and regioselectivity (Figure 3.2).[26]

3.2.4 Phytase and Other Proteins

Vanadium chloroperoxidases are non-heme metalloenzymes that are more resistant to oxidative degradation than their heme analogues.[27] However,

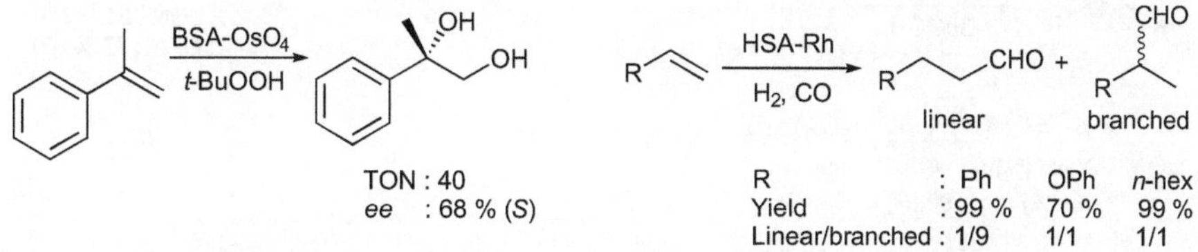

Figure 3.2 *cis*-Hydroxylation and hydroformylation reactions catalyzed by human (HSA) or bovine serum albumin (BSA)-based artificial metalloenzymes.[24,25]

due to their relatively small active site, they can only accommodate small substrates, which decreases their potential for use in asymmetric synthesis. Their active site shows very high similarities with that of phytases, which are metal-free.[28] Inspired by this observation, Sheldon *et al.* had the idea of building a new protein by inserting vanadate in phytase. The new artificial metalloprotein proved to have a catalytic activity similar to that of natural vanadium chloroperoxidase.[29] It quantitatively catalyzed the sulfoxidation of thioanisole by H_2O_2, with up to 66% ee. In later experiments, they tried to modulate the system by varying the host protein (acid phophatase, phospholipase, sulfatase, apo-ferritin, BSA) and the metal moiety (Mo, Re, W, Se, Os).[30] For example, vanadate-loaded streptavidin catalyzed enantioselective thioether sulfoxidation by *t*-BuOOH with up to 93% ee in favor of the (*R*)-product and 96% conversion,[31] and osmium tetroxide-loaded streptavidin catalyzed enantioselective olefin *cis*-dihydroxylation with up to 97% ee in favor of the (*R*)-product at ~20 TON in the presence of potassium ferricyanide (Figure 3.3).[32]

Reetz *et al.* engineered a His/His/Asp copper-binding site in the thermostable synthase subunit tHisF of the glutaminase synthase enzyme protein complex.[33] After binding copper, the artificial enzyme catalyzed quantitative Diels-Alder cycloaddition of cyclopentadiene and azachalcones with up to 46% ee and 13 : 1 endo/exo ratio.

Ward *et al.* employed the "Search for Three dimensional Atom Motifs in Protein Structure" (STAMPS) algorithm[34] to identify non-metalated two-histidine one-carboxylate metal binding motifs as well as three-histidine metal binding motifs within the structurally characterized proteins of the protein data bank (PDB).[35] Thirteen enzymes were found, six of which could be cloned and overexpressed in *E. coli* then purified relying on a Strep-tag II. The proteins were tested for their peroxidase activity in the presence of various transition metal salts including $VOSO_4$, $MnCl_2$, $FeCl_2$, $CoSO_4$, $NiCl_2$ and $CuSO_4$. Only 6-phospho-gluconolactonase was active in presence of

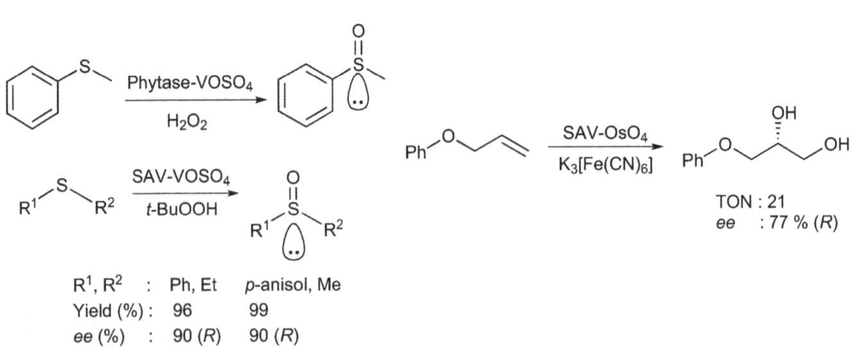

Figure 3.3 Left, thioether oxidation catalyzed by the artificial metalloenzymes vanadate-phytase[29] and vanadate-streptavidin.[31] Right, olefin *cis*-dihydroxylation catalyzed by the artificial metalloenzymes osmium tetroxide-streptavidin.[32]

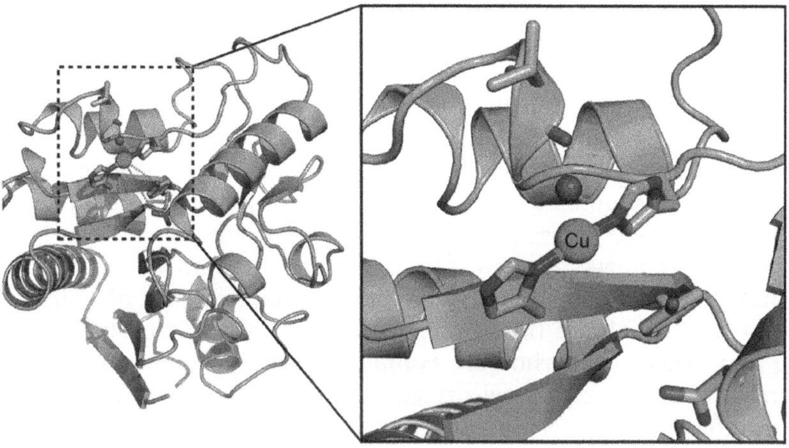

Figure 3.4 Crystal structure of 6-phospho-gluconolactonase binding copper(II). Under physiological conditions, the protein is copper free but in the presence of copper, this isomerase becomes also a peroxidase.[36] Adapted from ref. 36 with permission from The Royal Society of Chemistry.

$CuSO_4$ reaching ~25 TON. Further studies including the crystal structure and various mutations yielded ~250 TON and demonstrated that the active site was formed by a copper(II) coordinating two histidine residues in a T-shaped [2 + 1] geometry and displaying a weak contact with a water molecule (Figure 3.4).[36]

3.3 Supramolecular Anchoring of Metal Cofactors

Two strategies, namely the "Trojan-Horse" strategy and the "Host–Guest" strategy, have been used for the non-covalent anchoring of metal-containing cofactors in the cavities of proteins. The former strategy employs covalent attachment of a metal complex moiety to an inhibitor or substrate that possesses a high affinity for the protein and that is then used as a "Trojan-Horse" to drive the metal cofactor inside the active site, while the latter one relies on the inherent affinity of a protein for a cofactor.

3.3.1 The "Trojan-Horse" Strategy

3.3.1.1 Avidin, Streptavidin, Neutravidin, and Burkavidin

In the field of artificial metalloenzymes built by supramolecular anchoring of cofactors, avidin, neutravidin and streptavidin have been extensively studied. These proteins consist of four identical sub-units, each able to bind biotin with a dissociation constant $K_D \approx 10^{-15}$ M, which is, to date, one of the lowest measured for a protein–ligand complex. This very strong affinity allows almost irreversible – although non-covalent – insertion of biotin in the cavity of the protein, which several groups have taken advantage of to obtain artificial metalloenzymes.

The first published works in this area were those of Wilson and Whitesides,[37] who described in 1978 that, when associated with avidin, a rhodium complex covalently bound to biotin (Figure 3.5) could quantitatively and enantioselectivly catalyze the hydrogenation of α-acetamidoacrylic acid. When this complex was used with avidin pre-incubated with biotin, the measured efficiency drastically dropped, and the enantiomeric excess disappeared. This showed that the observed enantioselectivity was induced by the fixation of the Rh complex in the biotin-binding pocket of avidin.

Ward *et al.* employed this strategy to develop hybrid biocatalysts associating new biotinylated rhodium complexes with avidin, streptavidin and neutravidin. They used several streptavidin mutants and modulated the size and nature of the spacer separating biotin from the rhodium complex, to obtain artificial metalloenzymes that could catalyze the quantitative hydrogenation of α-acetamidoacrylic acid with up to 96% ee in favor of the (*R*)-product (Figure 3.6).[38] Interestingly, the introduction of a chiral amino acid spacer – phenylalanine or proline – between the biotin anchor and the flexible aminodiphosphine moiety, combined with saturation mutagenesis at position S212 of streptavidin, led to selectivity never achieved before with such hybrid biocatalyst in favor of the (*S*)-product (91% ee)[39] together with higher reaction rates and an improved tolerance toward organic solvents (up to 45% DMSO or EtOAc with maximal erosion of 12%). Additionally, the biohybrid catalyst was immobilized on Biotin-Sepharose beads while maintaining selectivity. Furthermore, burkavidin, secreted by the human pathogen *Burkholderia pseudomallei*, was isolated, purified, characterized, and used – following the same strategy – as a scaffold to prepare artificial metalloenzymes for catalyzing hydrogenation reactions. The burkavidin-based metalloenzymes led to only 21% conversion rate of α-acetamidoacrylic acid and 65% ee in favor of the (*S*)-isomer of the reduced species. This modest enantiomeric enrichment suggests however that the catalysis occurred in the biotin-binding pocket and that this artificial enzyme has good potential for further optimization.[40]

Over the past few years, the same group has extended this strategy to other metal-catalyzed reactions. In particular, with ruthenium-based artificial enzyme catalysts, the reduction of several aliphatic ketones by hydrogen gas was attained with both quantitative conversion rates and enantiomeric excesses of up to 97% ee in favor of the (*R*)-product (Figure 3.6).[11] The design of the catalyst was further optimized based on information derived from the X-ray crystal structure of an artificial enzyme formed by streptavidin combined to a biotinylated piano-stool ruthenium complex. The structure indicated that positions Lysine 121 and Leucine 124 might interact with arenes of incoming substrates. As a result, adapted mutants were prepared and used to obtain artificial metalloenzymes that were found to be able to reduce both alkyl and aryl ketones with good conversion rates and selectivity for the desired enantiomer.[41]

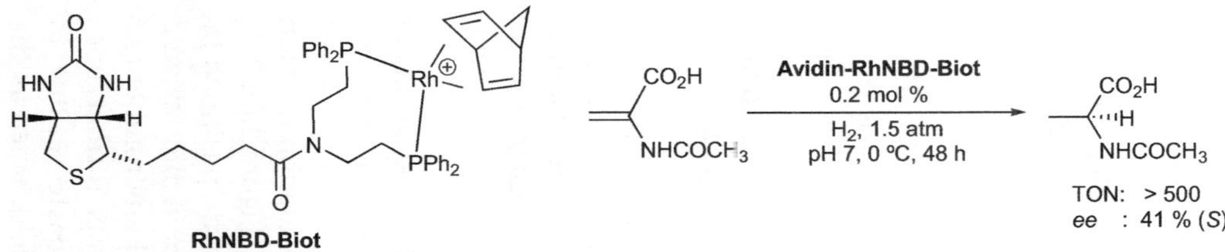

Figure 3.5 Left, RhNBD-Biot complex (NBD = 2,5-norbornadiene and Biot = biotin) used by Wilson and Whitesides. Right, hydrogenation of α-acetamidoacrylic acid catalyzed by the Avidin-RhNBD-Biot complex system.[35] See text for higher ee obtained for either (*R*)- or (*S*)-products after chemo/genetic optimization.

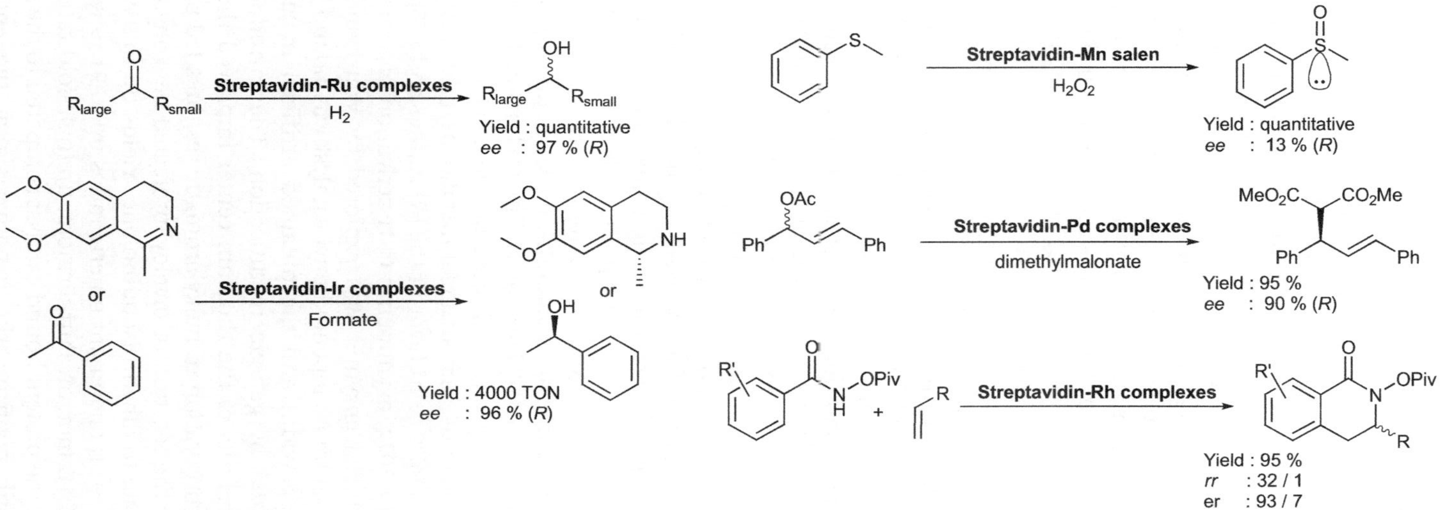

Figure 3.6 Example of various reactions that could be catalyzed with artificial enzymes based on the streptavidin–biotin system. See text for details including various substrates and chemo/genetic optimizations allowing the inversion of selectivity and better yields.

More recently, they focused their attention on the asymmetric transfer hydrogenation (ATH) of imines and ketones catalyzed by artificial metalloenzymes obtained by the combination of streptavidin with either d^5 or d^6 biotinylated piano-stool ruthenium, iridium and rhodium complexes.[42] In contrast to the reduction of ketones by hydrogen, iridium-based ATHases proved to be superior to their Ru analogues. (*R*)-Selectivity up to 96% ee was obtained with a small amino acid like glycine or alanine at position 112 (Figure 3.6). On the contrary, cationic residues like lysine or arginine at position S112 resulted in up to 78% ee (*S*)-selectivity. Consequently, both reduction product enantiomers could be obtained by a single point mutation on S112. On the basis of structural data, it was proposed that both the ketone and imine reduction proceeded through the same mechanism and Lys121 was identified as playing a major role in the protonation step. Finally, ATHases remained active after incorporation into biocompatible vesicles. This nanoparticle system remained active under highly dilute conditions and was stable under aerobic physiological conditions for several months.[43] In 2016, Ward *et al.* immobilized an artificial imine reductase consisting of biotinylated iridium cofactor incorporated in streptavidin in silica nanoparticles. The system displayed remarkable catalytic activity under physiological conditions, with TON > 46 000 in aqueous media and TON > 4000 in the presence of cellular debris, which demonstrated interesting perspectives for applications in green chemistry.[44]

In 2009, the group incorporated achiral biotinylated manganese-salen complexes into streptavidin and five other mutants to obtain artificial metalloenzymes for catalysis of enantioselective sulfoxidation. The resulting enzymes showed moderate conversions (up to 56%) and low enantioselectivities (up to 13%) for the sulfoxidation of thioanisole using hydrogen peroxide as oxidant in water (Figure 3.6).[45]

An artificial palladium enzyme for asymmetric allylic alkylation of 1,3-diphenylallylacetate by dimethylmalonate was obtained by using the same technology and optimized by chemical and genetic modifications. The most efficient of such artificial metalloenzymes catalyzed this annulation reaction with 90% ee in favor of the (*R*) isomer and 95% conversion (Figure 3.6).[46]

In 2012, asymmetric C–H activation of aromatic and vinylic positions by artificial enzymes was explored by Ward and Rovis *et al.* A biotinylated rhodium cyclopentadienyl catalyst was bound in the chiral environment of streptavidin and the obtained artificial metalloenzyme was further optimized by genetic modification. The introduction of a basic residue in proximity to the rhodium center improved the reaction yield, whereas the incorporation of bulky aromatic residues led to improved enantioselectivity for the reaction of pivaloyl-protected benzohydroxamic acid and methyl acrylate to dihydroisoquinolone. Yields up to 95%, regioisomeric ratios of up to 19 : 1, and enantiomeric ratios up to 91 : 9 were observed with the S112Y-K121E mutant. Furthermore, up to 92-fold rate accelerations were observed for the protein-bound catalyst compared to the free catalyst (Figure 3.6). This result confirms the hypothesis that the engineered carboxylate residue within the active site is key for generating a highly active and selective artificial metalloenzyme.[47]

3.3.1.2 Other Proteins

Ligands of other proteins have been bound to metal complexes to obtain artificial metalloenzymes following the "Trojan-Horse" strategy. In 2013, Ward *et al.* took advantage of the high affinity of *para*-substituted aryl-sulfonamides to the catalytic Zn ion of human carbonic anhydrase to form an ATHase. Thus, they tethered an aryl-sulfonamide to a bidentate N–N ligand and used it to anchor an iridium cyclopentadienyl catalyst within carbonic anhydrase. This artificial metalloenzyme efficiently reduced 6,7-dimethoxy-1-methyl-3,4-dihydroisoquinoline under transfer hydrogenation conditions with up to 68% ee.[48]

Additionally, following the same strategy, Mahy *et al.* have recently exploited the neocarzinostatin (NCS) variant NCS-3.24 that possesses affinity for testosterone, to prepare novel artificial enzymes. A water-soluble anionic iron-porphyrin–testosterone conjugate was synthesized and subsequently associated to the NCS-3.24 variant.[49] The protein appeared to have a better affinity for the testosterone conjugate ($K_D = 1.6$ μM) than for testosterone itself and the obtained Fe(III)-porphyrin–testosterone-NCS-3.24 artificial metalloenzyme was able to catalyze the chemo- and enantio-selective (ee = 13%) sulfoxidation of thioanisole by H_2O_2. Molecular modeling studies revealed a synergy between the binding of the steroid moiety and that of the porphyrin macrocycle into the protein binding-site, thus explaining both the observed better affinity for the conjugate and the selectivity of the catalysis. Starting from these preliminary observations, the NCS-3.24 variant was associated with a testosterone moiety bound to a much smaller zinc-*bis*-pyridyl-testosterone complex. Molecular modeling studies showed that this complex was bound deeper in the protein and the resulting artificial metalloenzyme turned out to be among the best artificial catalysts ($k_{cat}/K_M = 3.6$ M^{-1} s^{-1} at pH = 7) for the hydrolysis of the RNA model *p*-nitrophenyl-2-hydroxypropyl-phosphate (HPNP).[50] Finally, this team synthesized a copper(II)-cofactor coupled to a testosterone anchor and associated with a neocarzinostatin NCS-3.24 variant to generate a new artificial metalloenzyme that was able to efficiently catalyze the Diels–Alder cycloaddition of cyclopentadiene and azachalcones. The artificial enzymes led to an increase in the formation of the *exo*-products, without the induction of any enantiomeric excess.[51]

Another important approach for the generation of metalloenzymes relies on the presence of a large hydrophobic cavity in proteins, into which hydrophobic guests, such as fatty acid aliphatic chain residues, can be incorporated with submicromolar dissociation constants. In this aspect, Salmain *et al.* chose bovine β-lactoglobulin to prepare new Ru- and Rh-metalloenzymes for the transfer hydrogenation (reduction) of 2,2,2-trifluoroacetophenone. Rh(III) bio-hybrids provided up to quantitative conversions and 32% ee in favor of the (*R*)-enantiomer.[52]

3.3.2 The "Host–Guest" Strategy

Non-covalent supramolecular anchoring of a cofactor in the cavity of a protein, also named the "Host–Guest" strategy, constitutes the method that has by far been the most widely used to obtain artificial metalloenzymes. The very high affinity of antibodies for their antigen makes them host-proteins of choice to build artificial metalloenzymes by supramolecular anchoring of a metal cofactor in their binding site. Artificial metalloenzymes formed by using antibodies generated specifically against metal cofactors have been extensively reviewed recently[53–55] and therefore are not taken into consideration in this chapter.

3.3.2.1 *Serum Albumins*

Human serum albumin (HSA) was reported to bind heme within a narrow cavity with a strong affinity ($K_a = 1.1 \times 10^8\ M^{-1}$)[23,56–59] suggesting that it could accommodate compounds with structures close to that of heme. This property was widely used for the generation of new efficient O_2 binding artificial hemoproteins but also for preparing artificial metalloenzymes.[60,61]

Gross *et al.* inserted bis-sulfonated Ga-corrole into HSA[62] and iron and manganese complexes of the same corrole were also inserted into human, bovine (BSA), porcine (PSA), rabbit (RSA) and sheep (SSA) serum albumins to build new artificial metalloenzymes. These were found to be able to catalyze the sulfoxidation of thioanisole derivatives by hydrogen peroxide with conversions up to 98% and up to 74% ee.[63] Generally, better activities, selectivities as well as stabilities were obtained with albumin-Mn-corrole complexes with respect to their iron counterparts. More recently, the BSA-Mn-corrole artificial metalloenzyme was revealed to be able to catalyze light induced enantioselective oxidation of thioanisole into sulfoxide with 20% ee using water as the oxygen atom source in the presence of a ruthenium complex as photosensitizer (Figure 3.7).[64]

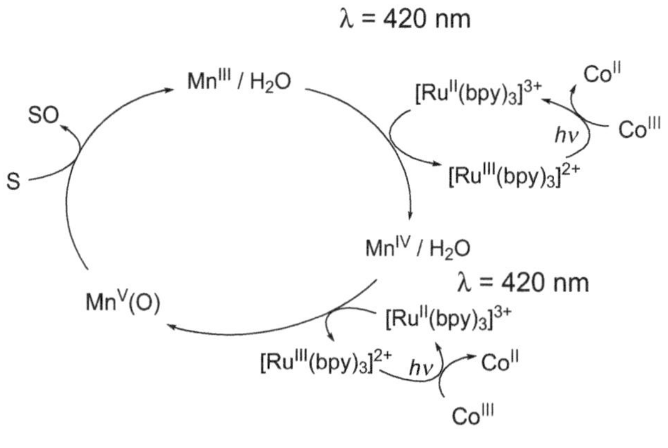

Figure 3.7 Reaction scheme for the light-driven oxidation of an organic substrate by water in the presence of manganese–corrole–BSA conjugate, ruthenium chromophore and cobalt-based sacrificial electron acceptor.[64]

In former studies, Reetz *et al.* showed that BSA could accommodate copper phthalocyanine as a cofactor, building an artificial metalloenzyme that was found able to catalyze the Diels–Alder cycloaddition of cyclopentadiene and azachalcones, with over 70% conversion and up to 98% ee.[65] Similar results were obtained with other serum albumins such as HSA, PSA and SSA, whereas RSA and chicken serum albumin (CSA) gave lower ee's (7% and 1% resp.). The non-covalent insertion of metal complexes of salen and salophen ligands into serum albumins has been widely investigated for many purposes. However, the non-covalent association of metal–salen or –salophen complexes with serum albumins produced only three reports on artificial metalloenzymes to date. The first one was about an artificial superoxide dismutase that was prepared by the non-covalent conjugation of BSA with the manganese(III) complex of *N*,*N*′-bis(salicylidene)-3,4-diaminobenzoic acid (salophen). The BSA-Mn-salophen hybrids displayed up to about 7-fold higher superoxide dismutase (SOD) activity than the complexes alone.[66] The second one was about a series of artificial metalloenzymes based on the incorporation of Mn-salen into HSA. While the oxidation of thioanisole by NaOCl catalyzed by the Mn-salen complexes led to about 60% conversion with the almost exclusive formation of methylphenylsulfone, the same reaction catalyzed by HSA-Mn-salen artificial metalloenzymes led to 90–100% conversion with the almost exclusive formation of methylphenylsulfoxide.[67] Finally, in the third report, an artificial metalloenzyme based on the incorporation of a cobalt(II) Schiff base complex {CoL, $H_2L = 2,2'$-[(1,2-ethanediyl)bis(nitrilopropylidyne)]bisphenol} in BSA was synthesized and characterized. The catalytic activity of this artificial metalloenzyme for the enantioselective oxidation of a variety of sulfides with H_2O_2 was studied as a function of pH, temperature and the concentration of catalyst and oxidant. Under optimal conditions, the BSA–CoL hybrid biocatalyst appeared to be efficient for the enantioselective oxidation of a series of sulfides into the corresponding sulfoxides and reached excellent conversion (up to 100%), chemoselectivity (up to 100%) and good enantiomeric purity (up to 87% ee).[68]

3.3.2.2 Myoglobin

Several teams used the oxygen-binding hemoprotein, myoglobin, to build artificial metalloenzymes. Its oxygen-binding site consists of a 10 Å diameter cavity containing an iron-heme prosthetic group that is non-covalently linked to the protein by hydrophobic interactions, electrostatic interactions *via* its two carboxylate moieties, and binding of the iron by the imidazole of the histidine 93.[69] Due to this non-covalent anchoring, the heme can be removed from the protein to yield apo-myoglobin, with a free cavity able to accommodate another metallic cofactor (Figure 3.8).[70]

Watanabe *et al.* used this strategy to insert synthetic chromium salophens into apo-myoblogin and mutants. The first catalytic assays showed that such systems catalyzed the stereoselective sulfoxidation of thioanisole, with rather low turnover frequencies ($TOF \leq 0.13\ min^{-1}$) and enantiomeric excesses ($ee \leq 30\%$ in favor of the (*S*)-product) with the H64D/A71G double mutant.[66]

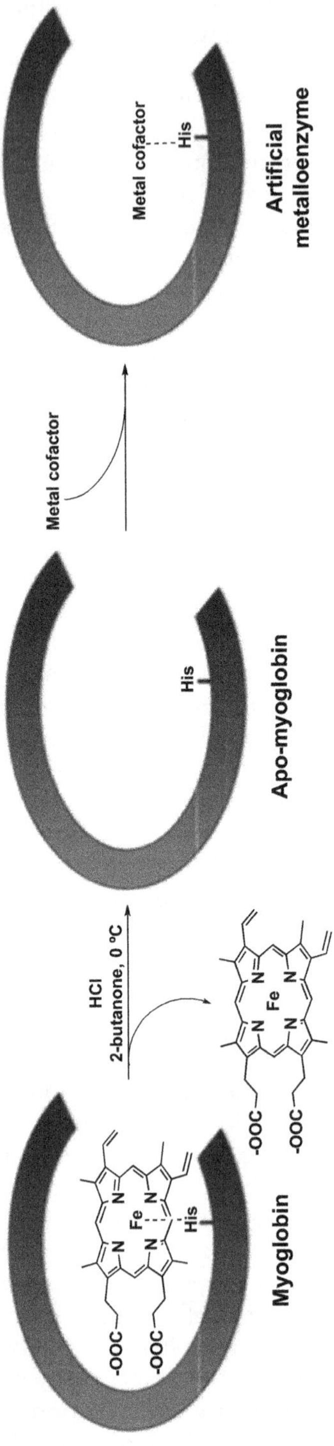

Figure 3.8 Preparation of an artificial metalloenzyme from myoglobin by the ''Host–Guest'' strategy.

The structures of two biohybrids formed by association of manganese and chromium salophens with the A71G mutant of the apo-myoglobin revealed that the accessibility to the active site was lowered by the bulky phenylenediamine moiety of the salophen cofactors.[71] Consequently, they chose to use metal–salen cofactors instead, which had a rather low effect on the efficiency of the chromium cofactors (TOF = 0.21 min^{-1}) but induced a noticeable increase on the efficiency of the manganese cofactor (TOF = 2.7 min^{-1}). In addition, the selectivity of the reaction was modulated by mutations such as the H64D/A71G double mutant, which led to 30% ee in favor of the (*S*)-product, and the A71G single mutant, which, like the wild type protein, led to 27% ee in favor of the (*R*)-product.[72]

Hayashi and Hiseada *et al.* replaced the native heme of the myoglobin by other porphyrins comprising anionic carboxylate moieties. The obtained artificial metalloenzymes had peroxidase activity and catalyzed oxidation reactions with rates up to 11-fold higher than native myoglobin in the case of catechol as substrate.[70,73] The team subsequently reported that myoglobin reconstituted with a manganese–porphycene cofactor led to a new biocatalyst that was able to catalyze the regioselective hydroxylation of ethylbenzene into 1-phenylethanol by H_2O_2. Fe-tetrapyrrolic compounds inserted into apo-myoglobin were used to catalyze the oxidation of thioanisole by H_2O_2 and led to up to 38% ee in favor of the (*S*)-sulfoxide. It is noteworthy that the enantiomeric excess can be increased by covalent anchoring of metal complexes into apo-myoglobin and mutants[74–76] or by changing the metal such as, for example, replacing iron by manganese.[63,76]

3.3.2.3 *Xylanase*

An early report from Nakamura and Tsushida *et al.* showed that the non-covalent incorporation of Fe(II)-α_4-tetra-*o*-pivalamidophenylporphyrin into this protein led to heat-resistant hemoproteins that are able to bind and release O_2 in aqueous media.[77] Therefore, xylanase A from *Streptomyces lividans* (Xln10A), a thermostable β-1,4-endoxylanase glycoside hydrolase that hydrolyzes β-1,4 bonds in the main chain of xylan,[78] was chosen to build up new artificial hemoproteins.

Iron(III)-*meso*-tetra(4-carboxyphenyl)porphyrin (Fe(TpCPP)) and iron(III)-*meso*-tetra(4-sulfonatophenyl)porphyrin (Fe(TpSPP)) were inserted into Xln10A to lead to new artificial hemoproteins Fe(TpCPP)-Xln10A and Fe(TpSPP)-Xln10A that were found to have peroxidase activity.[79] Thus, the oxidation of thioanisole by H_2O_2 was performed in the presence of either Fe(TpCPP) or Fe(TpSPP) alone and in the presence of the Xln10A. The use of Fe(III)(TpCPP) and Fe(III)(TpSPP) as catalysts led respectively to 45% and 33% yields, and respective turnovers frequencies of 0.56 and 0.41 min^{-1}, but no enantiomeric excess was observed. The insertion of both Fe-porphyrins into Xln10A led to a decrease in the yields to about 24% and in the turnover frequencies to about 0.30 min^{-1}, but enantiomeric excesses of up to 36% and 24% were observed in favor of the (*S*)-sulfoxide with Fe(TpCPP)-Xln10A and Fe(TpSPP)-Xln10A, respectively (Figure 3.9).

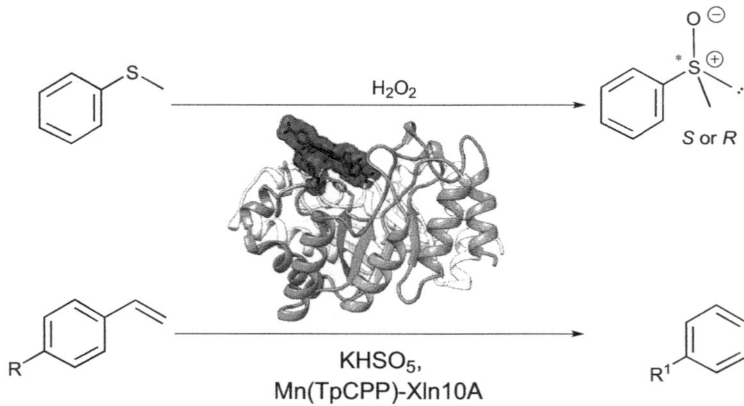

Catalyst	Yield (%)	Turnovers (Min^{-1})	*ee* (%)
Fe(TpSPP)-Xln10A	24	0.30	24 (*S*)
Fe(TpCPP)-Xln10A	24	0.30	36 (*S*)
Fe(TpCPP)-Xln10A + 100 eq. ImH	**85**	**1.09**	**40 (*S*)**

R	Total yield (%)	Epoxide yield (%)	*ee* (%)
CH_3	39.5	23	25 (*S*)
OCH_3	**49.5**	**16**	**80 (*R*)**

Figure 3.9 Selective oxidation of either thioanisole by H_2O_2[80,81] or styrene derivatives by $KHSO_5$[82] catalyzed by M(TpCPP)-Xln10A complexes. ImH: imidazole.

It is noteworthy that better yield and turnover frequency (85% and 1.09 min^{-1}) as well as a better enantiomeric excess in favor of the (*S*)-sulfoxide (ee = 40%) (Figure 3.9) could be obtained in the presence of 100 equivalents of imidazole as co-catalyst, which coordinates Fe-porphyrins in the axial position.[80,81]

Further work was carried out to apply these strategies to the selective oxidation of other substrates such as alkenes. Mn(III)(TpCPP) was non-covalently inserted into Xln10A and the oxidation of various substituted styrene derivatives by various oxidants such as hydroperoxides (H_2O_2 and *t*-BuOOH), sodium hypochlorite (NaOCl), sodium periodate ($NaIO_4$) and oxone® ($KHSO_5$) were assayed in the presence of the Mn(III)(TpCPP)-Xln10A biohybrid. Positive results were only observed with $KHSO_5$ as oxidant, together with the highest catalytic activity reported so far for the oxidation of styrene catalyzed by artificial metalloproteins. However, rather low chemo- and enantio-selectivities (max. ee of 25% in favor of the (*S*)-product) were observed for the epoxidation of styrene and poorly activated styrene derivatives. On the contrary, a high opposite enantioselectivity (80% in favor of the (*R*)-isomer) was observed for the epoxidation by $KHSO_5$ of *para*-methoxystyrene (Figure 3.9), which constitutes the best enantioselectivity ever reported for epoxidation reactions catalyzed by artificial metalloenzymes.[82]

3.3.2.4 NikA

NikA is a periplasmic nickel-binding protein involved in the transport of nickel in *E. coli* and other related Gram-negative bacteria. X-Ray diffraction studies have shown that NikA *in vivo* required a specific metallophore that may be butane-1,2,4-tricarboxylate (BTC)[83] and that it was also able to bind FeEDTA(H_2O).[84] These initial results led Ménage *et al.* to insert into NikA an EDTA-like inorganic metal complex catalyst of the N_2Py_2 type *i.e.* Fe[*N*-benzyl-*N′*-(2-hydroxybenzyl)-*N*,*N′*-ethylenediaminediacetic acid], which mimics the metal environment of iron oxygenases. By combining model chemistry and protein X-ray crystallography, they were able to study the intramolecular dihydroxylation by the artificial metalloenzyme of one of the *N*-benzyl substituents of the ligand. Indeed, the bound complex was able to activate dioxygen in the presence of a reductant (dithiothreitol, DTT), leading to the formation of catechol as the sole product. The X-ray diffraction structure determination of four of the catalytic cycle intermediates and the end product showed that the hydroxylation reaction implicated an iron peroxo, which was also observed in natural iron monooxygenases (Figure 3.10).[85]

Taking advantage of NikAs ability to bind iron complexes of the N_2Py_2 type, they further prepared nine artificial metalloenzymes. The stable NikA-Fe-N_2Py_2 hybrids were characterized, including by X-ray crystallography, which showed that the ligands interact with NikA through salt bridges with arginine residues and π-stacking with a tryptophan residue. These interactions were further characterized using quantum mechanics calculations. Overall, this study aimed at the complete characterization of the factors that

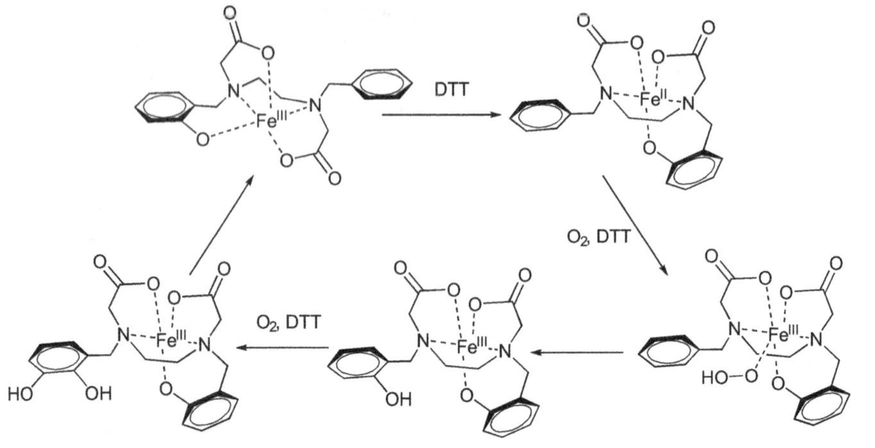

Figure 3.10 Fe[*N*-benzyl-*N*′-(2-hydroxybenzyl)-*N*,*N*′-ethylenediaminediacetic acid] complex and mechanism of the intramolecular oxidation of the ligand by O_2, catalyzed by the NikA-Fe-ligand followed by structural characterization of the four intermediates.[85]

condition the interaction of an artificial ligand and a protein and their implications for catalysis. It constituted a new approach that should be useful not only in the synthesis of artificial monooxygenases, but more generally in the field of artificial metalloenzymes.[86] Finally, molecular docking calculations were conducted to screen sulfides, of the C_6H_5–S–CH_2–X type, for catalytic oxidation by the series of iron complex NikA hybrids. A set of 374 potential sulfide substrates was identified, among which six potential substrates had a common R^1–S–CH_2–CONH–R^2 motif. The catalytic properties of each hybrid were determined using the six substrates to validate the docking simulations. This study allowed the design of one reference substrate for the NikA–Fe–N_2Py_2 artificial enzymes. The best catalytic results were thus obtained for the oxidation of the 4-CH_3–Ph–S–CH_2–CONH–Ph sulfide by NaOCl that led, in the presence of one NikA–Fe–N_2Py_2 catalyst, to the chemoselective formation of sulfoxide in 78% conversion with a TON of 199 but only 5% ee. This study constituted a nice proof-of-concept for the design of a substrate family, which allowed Ménage *et al.* to define a new kind of artificial oxygenase for the catalysis of sulfoxidation reactions.[87]

3.3.2.5 Beta-lacto Globulin

β-Lacto globulin (β-LG) is the most abundant protein in the whey of cow's milk. Its structure revealed 8-stranded antiparallel β-barrel folds forming a calyx lined with hydrophobic residues. It is now well established that a range of hydrophobic ligands such as fatty acids bind to β-LG with their aliphatic chains occupying the β-LG's hydrophobic cavity and the carboxyl function directed towards the calyx entrance.[88] Consequently, half-sandwich Ru(II)/Rh(III) complexes were bound to either lauric acid or palmitic acid and embedded within β-LG. The resulting artificial metalloproteins were able to catalyze the asymmetric transfer hydrogenation (reduction) of trifluoroacetophenone (TFAP) by formate in water with 14–72% conversions and fair ee's (16–26%) in favor of the (*R*)-alcohol (Figure 3.11). The crystal structure of bovine β-LG capturing a rhodium complex bound to lauric acid[89] confirmed that the fatty chain was buried in the β-barrel folds of the protein and that the metal complex established interactions with several residues of the protein. This work was further extended to a series of Ru(II) and Rh(III) complexes bound to saturated and unsaturated fatty acids including a dipyridylamine (dpa) or a bispyridylmethane (bpm) scaffold as a metal chelating moiety.[90] Incorporation of these complexes within bovine β-LG was studied and noticeable differences were observed between the saturated and unsaturated fatty acid derivatives. The ability of the protein hybrids to catalyze the transfer hydrogenation of TFAP was demonstrated and up to quantitative conversions to the corresponding alcohol were reached with rhodium(III)-based hybrids, and up to 32% ee in favor of the (*R*)-product were obtained with the rhodium(III) cofactors derived from palmitic and stearic acids incorporated into the A isoform of β-LG.[52]

3.4 Covalent Linking of Metallic Cofactors to a Protein

To expand the range of catalysts that can be incorporated into proteins, researchers have developed the covalent bonding of transition metal cofactors to proteins. This approach often relies on reactive protein residues such as serine, lysine and cysteine. In 1987, Sigman *et al.* used iminothiolane to modify the tryptophan repressor protein of *E. coli*, which has four lysine residues. A subsequent sulfur alkylation with 5-iodoacetamide-1,10-phenanthroline followed by insertion of Cu(II) yielded an artificial enzyme that catalyzed the specific cleavage of a DNA fragment containing the aroH transcription unit naturally recognized by the Trp repressor protein.[91] Cleavage of DNA fragments is however not relevant for chemical synthesis but, earlier, in 1984, Kaiser *et al.* conferred an oxidizing activity to the protease papain by covalently grafting a flavin on the unique cysteinyl residue, which also is serendipitously located in its active site. No metals were involved, but the bioconjugation generated a hybrid catalyst and paved the way for preparing artificial enzymes for stereoselective catalysis by binding metal complexes in papain as well as in other proteins.[92,93]

Relying on Michael addition, Reetz *et al.* covalently bound to papain a manganese salen complex, as well as a bis-pyridiniumyl rhodium complex. Both artificial enzymes appeared to be able to catalyze reactions of epoxidation and hydrogenation, respectively, but both with weak enantiomeric excesses of about 10%.[94,95] Panella *et al.* covalently modified papain by coupling its cysteinyl residue to phosphorylated ligands that were subsequently used to complex 1,5-cyclooctadienerhodium(I). The hybrid

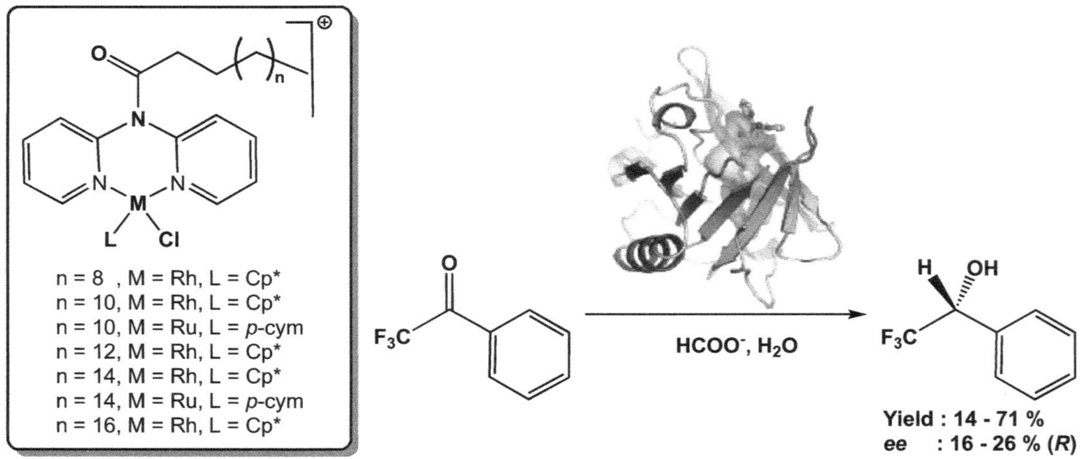

Figure 3.11 Various half-sandwich Ru(II) and Rh(III) fatty acid derivatives inserted in β-lactoglobulin and used as catalysts for the asymmetric transfer hydrogenation of trifluoroacetophenone (TFAP) by formate in water.[89,90]

biocatalyst catalyzed the quantitative hydrogenation of methyl-2-acetamidoacrylate with only 10% ee.[26] Salmain *et al.* covalently attached achiral organometallic d^6 transition metal piano stool cofactors into papain for the catalysis of enantioselective ketone transfer hydrogenations.[96–98] Results revealed low enantioselectivities (up to 15%) in contrast to Eppinger *et al.* where ee reached 64% in similar reactions. Eppinger *et al.* however employed cofactors that bound to papain both covalently *via* a reactive epoxide and noncovalently *via* recognition elements *i.e.* amino acids.[99]

The catalytic antibody 38C2 has a hydrophobic cavity, which contains a lysinyl residue. Janda *et al.* took advantage of this residue to covalently bind a bis-imidazole ligand selectively in the cavity of the protein.[100] After complexation with Cu(II), the built metalloenzyme catalyzed the hydrolysis of 4-nitrophenyl picolinate with a conversion rate improved by a factor 3.5×10^4, compared to the complex formed between the isolated ligand and copper.

Adipocyte lipid-binding protein (ALBP) is a small protein of 131 residues that has a single cysteinyl residue inside a cavity of 600 Å^3. Distefano *et al.* exploited ALBP by selectively and covalently binding to its cysteine a 1,10-phenanthroline derivative, which they subsequently relied on to bind copper(II) ions. The obtained artificial enzyme was characterized and its crystal structure revealed that little structural perturbation resulted from cofactor bioconjugation. It was used for the catalysis of the hydrolysis of non-activated amino acids esters (Figure 3.12) that resulted in a maximum of only 7.6 turnovers but an enantioselectivity reaching 86% in favor of the (*S*)-product *i.e.* L-tyrosine when a mixture of DL-tyrosine methyl ester was the substrate.[101,102]

Watanabe *et al.* also showed that apo-ferritin reacts with 2,5-norbornadiene-rhodium(I) chloride dimer. The crystal structure of the biohybrid revealed three unique rhodium binding sites per subunit, one of which involved covalent attachment of the 2,5-norbornadiene of the rhodium complex to a cysteinyl residue. The artificial enzyme catalyzed phenylacetylene polymerization, and observation of differences in the properties of produced polymers led to the conclusion that catalysis occurred within the interior of the protein.[103]

Lu *et al.* revealed that a dual-point attachment of manganese salen complex to a cysteine double mutant of apo-myoglobin (apo-Mb Y103C/L72C) led to improved selectivity (51% ee) in thioanisole sulfoxidation relative to the analogous single-point mutant (Y103C, 12% ee).[75,76]

Kamer *et al.* modified the unique cysteinyl residue in photoactive yellow protein (PYP) using phosphine ligands substituted with a 1,1′-carbonyldiimidazole activated carboxylic acid moiety. Phosphine-substituted PYP was metalated by an allylpalladium(II) chloride dimer, and the resulting artificial enzyme catalyzed non-selectively the allylic amination of racemic 1,3-diphenylprop-2-enyl acetate with benzylamine (Figure 3.13).[104,105]

Roefles *et al.* showed that using bidentate ligands, such as phenanthroline and bipyridine derivatives, the dimeric transcription repressor LmrR could be alkylated at cysteinyl residues fitted at the hydrophobic dimer

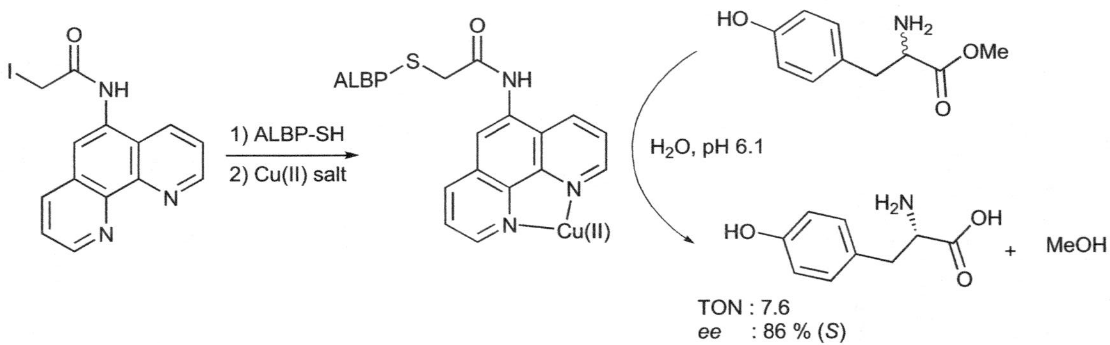

Figure 3.12 Preparation of an artificial metallohydrolase by the covalent coupling of 1,10-phenanthroline ligand to the thiol moiety of the cysteinyl residue followed by the complexation of copper(II) and application in the catalysis of the hydrolysis of esters of amino acids.[101,102]

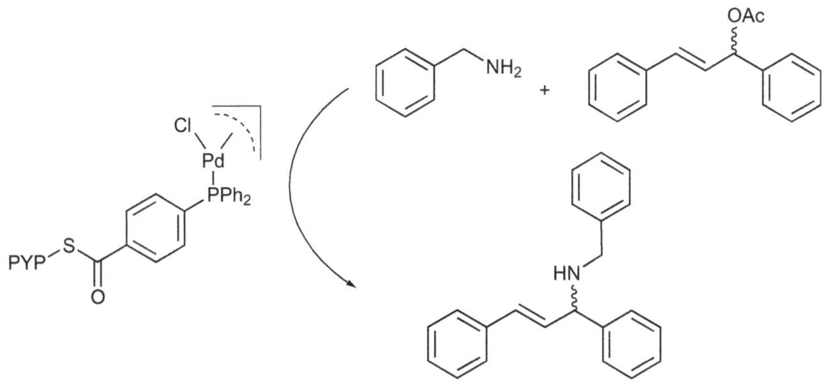

Figure 3.13 Catalysis of allylic aminations by an artificial enzyme resulting from the covalent binding of a palladium complex in photoactive yellow protein (PYP) *via* its cysteinyl residue.[104,105]

interface.[106] This approach allowed metalation of LmrR with Cu(II), and the resulting artificial metalloenzyme catalyzed the Diels–Alder cycloaddition of cyclopentadiene and azachalcones with up to 97% ee. Interestingly, the phenanthroline-containing artificial enzyme provided one enantiomer while the bipyridinyl analogue provided the other enantiomer. Additionally, the same artificial enzymes catalyzed different reactions *i.e.* the hydration of azachalcones with up to 84% ee and Friedel–Crafts alkylation reaction of 5-methoxy-1*H*-indole with 1-(1-methyl-1*H*-imidazol-2-yl)but-2-en-1-one with up to 66% ee (Figure 3.14). This demonstrated that depending on the substrates, different reactions could be catalyzed by the same class of artificial enzymes.[107]

Finally, Mahy *et al.* covalently and selectively grafted a non-heme Fe(II) polyazadentate complex into bovine β-lactoglobulin on the accessible C121. The biohybrid activated hydrogen peroxide to oxidize thioanisole into phenylmethylsulfoxide as the sole product with an ee of 20%. The mechanism of the reaction was studied and a high spin ($S = 5/2$) $Fe^{III}(\eta^2\text{-}O_2)$ intermediate was characterized and proposed to be responsible for the catalytic sulfoxidation (Figure 3.15).[108]

3.5 Cascade Reactions

In living organisms, enzymes have evolved to function in harmony with other biomacromolecules including other enzymes, and in the presence of small molecules other than their substrates. Combinations of enzymes have been successfully used in catalyzing cascades of reactions,[109,110] but the combination of transition metal catalysts with enzymes has often led to mutual inhibition,[111–114] and pushed scientists to overcome this challenge by relying on compartmentalization strategies.[115–118] It is however conceivable that artificial metalloenzymes could operate in harmony with other

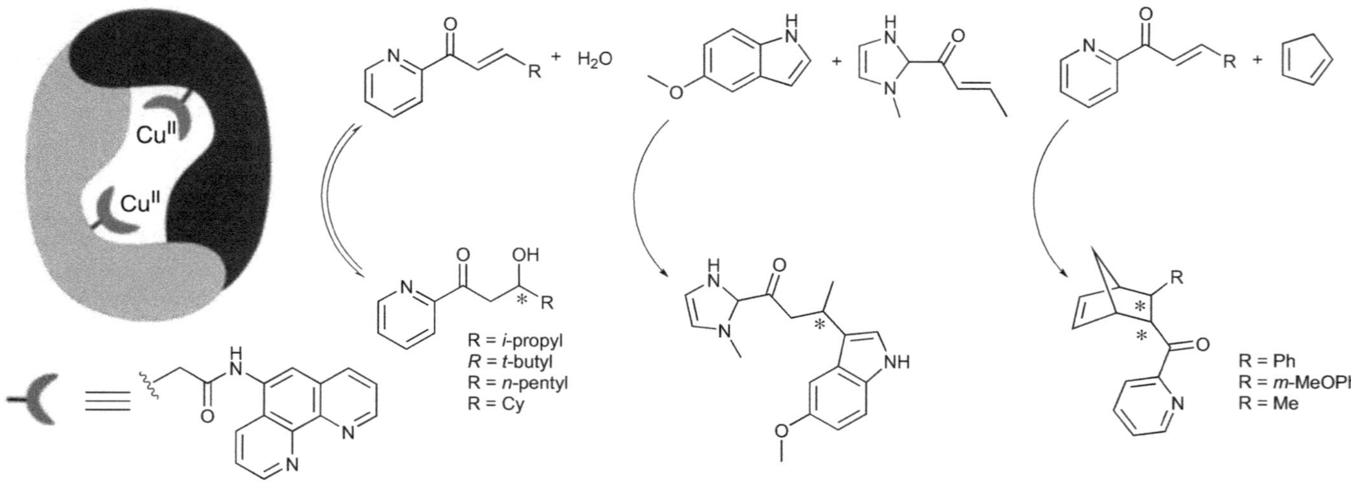

Figure 3.14 The same artificial enzyme able to catalyze hydrations, Friedel–Crafts alkylations and Diels–Alder cycloadditions depending on the chosen substrates.[106,107]

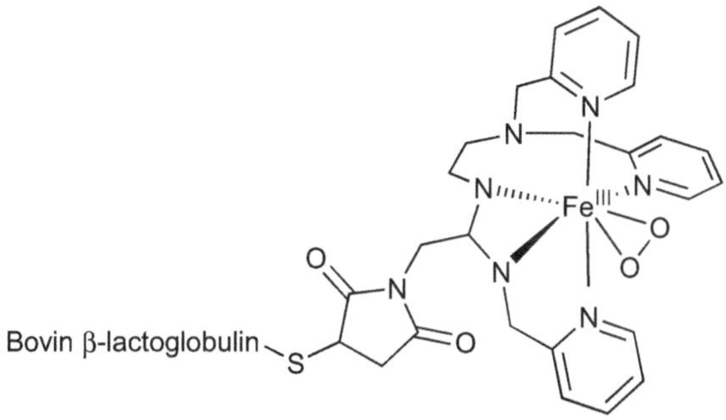

Figure 3.15 $Fe^{III}(\eta^2\text{-}O_2)$ intermediate formed upon addition of H_2O_2 on an artificial enzyme formed by covalently coupling an iron complex in β-lactoglobulin *via* its accessible cysteinyl residue.[108]

enzymes to achieve a synthetic goal. After all, the protein body of an artificial metalloenzyme offers a protective environment for the metal catalyst. The employment of artificial enzymes for the catalysis of cascade reactions remains exploratory and to the best of our knowledge only a few of such examples were reported in literature thus far.

The first example of combination of artificial and natural enzymes for catalysis of cascade of reactions resulted from the collaboration of the laboratories of F. Hollmann, N. J. Turner and T. R. Ward.[119] By following the "Trojan–Horse" strategy, a d^6-piano stool iridium complex was bound to biotin and incorporated within streptavidin affording an ATHase that catalyzed imine reduction by formate. The ATHase was combined with monoaminoxidase-N-9 (MAO-N-9), which is a natural amine oxidase that uses dioxygen as oxidant, displays near perfect (*S*)-selectivity toward a variety of amines, and produces the corresponding imines and hydrogen peroxide. The later was found to be detrimental to the ATHase activity, therefore either catalase or peroxidase was also added to the enzyme combination. The ATHase reduced imines into racemic amine and only the (*S*)-isomer was reoxidized by MAO-N-9 into imine leading to the accumulation of the (*R*)-amine isomer. The ATHase/MAO-N-9/catalase combination was thus very efficient in the dynamic kinetic resolution of various chiral amines starting from either racemic amines or imines as substrates (Scheme 3.1, left). It is noteworthy that mutual inactivation of metal catalysts and enzymes was observed when the free iridium complex was used, demonstrating the power of encapsulation in an artificial enzyme.

In a similar approach, Ward *et al.* prepared an ATHase by incorporating 4,7-dihydroxy-1,10-phenanthroline d^6-piano stool iridium complex bound to biotin in streptavidin. The formed ATHase was able to reduce imines using NADPH. In contrast, glucose dehydrogenase (GDH) uses glucose to reduce $NADP^+$ thus recycling NADPH. A combination of the ATHase/MAO/catalase and GDH showed efficient dynamic kinetic resolution of various chiral amines and using simply glucose and dioxygen (Scheme 3.1, center).[120]

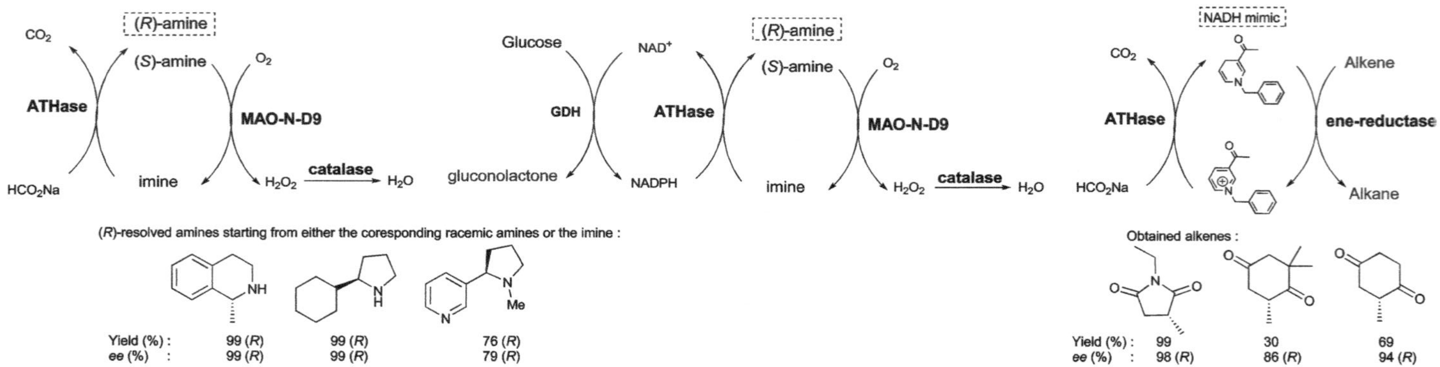

Scheme 3.1 Left, artificial metalloenzyme ATHase and natural MAO catalyzing a cascade of two reactions leading to production of (*R*)-amines, starting from either racemic amines or imines and using oxygen and formate. Catalase was used to protect the enzymes from hydrogen peroxide.[119] Center, artificial metalloenzyme ATHase working in harmony with two natural enzymes to produce (*R*)-amines using only oxygen and glucose.[120] Right, artificial metalloenzyme ATHase using formate to recycle NADH, which was used in cascade by ene-reductase to reduce various alkenes.[121]

Finally, ATHase was also used in the catalysis of cascade reactions for the regeneration of NADH mimics (mNADHs). mNADHs have been shown to accelerate and orthogonally activate ene-reductase-catalyzed reaction but existing regeneration methods of NADPH fail for mNADHs. ATHase was combined with ene-reductase-catalyzed asymmetric reduction of α,β-unsaturated compounds. TTNs reached 2000 with respect to ATHase, which regenerated mNADH by using formate and therefore only catalytic amounts of mNADH were needed for ene-reductase to catalyze the reduction (Scheme 3.1, right).[121]

3.6 Conclusion

The present chapter summarizes the impressive progress achieved in the past fifteen years in the area of artificial metalloenzymes. The preparation of these new biocatalysts was discussed from the chemist's point of view, by presenting the incorporation of metal cofactors and/or metal ions within a variety of natural protein scaffolds using either covalent or supramolecular anchoring strategies. Prominent examples were provided, highlighting the wide diversity of applications in synthesis that can arise from these hybrid biocatalysts.

At this point, only a few of the systems elaborated can compete with naturally occurring enzymes in terms of kinetics as most of the processes presented above are only complete after a few hours to a few days.[122] Up to now, their interest mainly resides in two fields. Firstly, it is clearly established that the catalytic properties of a metal complex can be highly enhanced by the stereochemical induction brought by biomolecules. Compared with standard coordination catalysts, artificial metalloenzymes thus offer a very interesting alternative as they can induce stereoselectivity without requiring the production of sophisticated ligands, which is generally expensive, uses organic solvents and generates waste. In contrast, the use of biosynthesized proteins further increases the sustainability factor brought by catalysis. Artificial metalloenzymes thus offer a unique opportunity for the preparation of chiral compounds, which are of up-most importance for pharmaceutical and food and feed industries and moreover their value is projected to rise. Secondly, it is clear that artificial metalloenzymes also offer a very interesting alternative to chemical catalysts for key synthetic reactions needed in chemical industries such as Diels–Alder cyclizations, hydroformylation, and Suzuki–Miyaura coupling reactions and for which efficient naturally occurring enzymes have not been discovered yet. The case of the Diels–Alder reaction is particularly striking, insofar as a handful of artificial metalloenzymes are capable of catalyzing this reaction[51,65,107,122–126] with yields of up to 98% and adjustable diastereo- and enantio-selectivities, as a function of protein and metal complex.

Despite ample efficient examples as described above, artificial metalloenzymes are not employed at the industrial scale thus far. They compete with various other methods to obtain chiral compounds but could offer the breakthrough that meets the newly mandated highly demanding environmental requirements. The development of artificial metalloenzymes, however, still appears to remain one of the most promising strategies used to expand the scope of biocatalytic reactions in the near future. Indeed, the fact that it stands at the crossroads between two disciplines, chemistry and biology, undoubtedly constitutes an advantage. It can benefit from the technological developments from the two fields. These include the constant improvement of our understanding of protein–substrate interactions as well as the progress in the generation of synthetic proteins[127,128] and should lead to an optimized design of the protein partner. Relying on chemistry, at least two ways to improve the existing biocatalysts can be envisioned. The first way is the simple change of ligands or metals, which usually constitutes the first optimization step but affords moderate efficiencies. The second way is the improvement in the anchoring methods,[107,129,130] which, with the help of computational modeling and directed evolution,[131] usually allows the optimization of most of the existing systems. On the biology side, the improvement of the existing systems requires an optimization of the protein partner by genetic optimization techniques, based on molecular modeling studies and on the resolution of the crystallographic structures that are crucial for the determination of relevant mutations. In this respect, a recent review by M. Dürrenberger and T. R. Ward[132] reported powerful strategies involving the redesign of proteins by applying sophisticated computational methods, or by exploiting catalytic promiscuity to unravel abiotic reactivities from natural enzymes. It also discussed directed evolution protocols to optimize a nascent catalytic activity and lead to catalytic efficiencies approaching that of natural enzymes, which is remarkable when considering the youth of the field of artificial metalloenzymes.

Finally, in the search for the "ideal host biomolecule", it appears that several candidates are still to be explored including auto-assembling proteins, such as, for example, amyloid type fibers that have been recently shown to be able to act as Zn^{2+}-dependent esterases,[133] and foldamers that are already being used for organocatalysis[134–136] and can coordinate metal ions.[137] In the latter case, although no example of new artificial metalloenzyme based on foldamer scaffolds was described so far, it is likely to occur in the very near future. The case of DNA as a hosting biomolecule is notable as it provided efficient artificial metalloenzymes for the catalysis of Diels–Alder cycloadditions,[138,139] oxa-Michael additions,[140] carbene transfer reactions[141] and Friedel–Crafts conjugate additions.[142]

Finally, the generation of catalysts inside cells constitutes a real promise for the future, which opens the way to possible applications in therapy, as recently an artificial metalloenzyme was shown to be active inside living cells.[143]

References

1. P. T. Anastas and J. C. Warner, *Green Chemistry: Theory and Practice*, Oxford University Press, Oxford, 2000, p. 135.
2. M. Hoarau, C. Hureau, E. Gras and P. Faller, *Coord. Chem. Rev.*, 2016, **308**, 445.
3. W. S. Knowles, *Angew. Chem., Int. Ed. Engl.*, 2002, **41**, 1999.
4. K. B. Sharpless, *Angew. Chem., Int. Ed.*, 2002, **41**, 2024.
5. R. Noyori, *Angew. Chem., Int. Ed.*, 2002, **41**, 2008.
6. Y. Chauvin, *Angew. Chem., Int. Ed.*, 2006, **45**, 3740.
7. R. H. Grubbs, *Angew. Chem., Int. Ed.*, 2006, **45**, 3760.
8. R. R. Schrock, *Angew. Chem., Int. Ed.*, 2006, **45**, 3748.
9. Research News, *Angew. Chem., Int. Ed.*, 2010, **49**, 8300.
10. G. Lelais and D. W. C. MacMillan, *Aldrichimica Acta*, 2006, **32**, 79.
11. C. Letondor, A. Pordea, N. Humbert, A. Ivanova, S. Mazurek, M. Novic and T. R. Ward, *J. Am. Chem. Soc.*, 2006, **128**, 8320.
12. M. Breuer, K. Ditrich, T. Habicher, B. Hauer, M. Keßeler, R. Stürmer and T. Zelinski, *Angew. Chem., Int. Ed.*, 2004, **43**, 788.
13. K. Yamamura and E. T. Kaiser, *J. Chem. Soc., Chem. Commun.*, 1976, 830.
14. K. Okrasa and R. J. Kazlauskas, *Chem. - Eur. J.*, 2006, **12**, 1587.
15. A. Fernández-Gacio, A. Codina, J. Fastrez, O. Riant and P. Soumillion, *ChemBioChem*, 2006, 7, 1013.
16. Q. Jing, K. Okrasa and R. J. Kazlauskas, in *Bio-inspired Catalysts*, ed. T. R. Ward, Springer Berlin Heidelberg, Berlin, Heidelberg, 2009, vol. 25, p. 45.
17. Q. Jing, K. Okrasa and R. J. Kazlauskas, *Chem. - Eur. J.*, 2009, **15**, 1370.
18. Q. Jing and R. J. Kazlauskas, *ChemCatChem*, 2010, **2**, 953.
19. K. Håkansson, A. Wehnert and A. Liljas, *Acta Crystallogr., Sect. D: Biol. Crystallogr.*, 1994, **50**, 93.
20. S. Abe, J. Niemeyer, M. Abe, Y. Takezawa, T. Ueno, T. Hikage, G. Erker and Y. Watanabe, *J. Am. Chem. Soc.*, 2008, **130**, 10512.
21. L. Toussaint, L. Bertrand, L. Hue, R. R. Crichton and J.-P. Declercq, *J. Mol. Biol.*, 2007, **365**, 440.
22. Z. Wang, Y. Takezawa, H. Aoyagi, S. Abe, T. Hikage, Y. Watanabe, S. Kitagawa and T. Ueno, *Chem. Commun.*, 2011, 170.
23. N. H. Fairley, *Nature*, 1938, **142**, 1156.
24. T. Kokubo, T. Sugimoto, T. Uchida, S. Tanimoto and M. Okano, *J. Chem. Soc., Chem. Commun.*, 1983, 769.
25. C. Bertucci, C. Botteghi, D. Giunta, M. Marchetti and S. Paganelli, *Adv. Synth. Catal.*, 2002, **344**, 556.
26. L. Panella, J. Broos, J. Jin, M. W. Fraaije, D. B. Janssen, M. Jeronimus-Stratingh, B. L. Feringa, A. J. Minnaard and J. G. de Vries, *Chem. Commun.*, 2005, 5656.
27. F. van de Velde, I. W. C. Arends and R. A. Sheldon, *J. Inorg. Biochem.*, 2000, **80**, 81.

28. W. Hemrika, R. Renirie, H. L. Dekker, P. Barnett and R. Wever, *Proc. Natl. Acad. Sci. U. S. A.*, 1997, **94**, 2145.
29. F. van de Velde, L. Könemann, F. van Rantwijk and R. A. Sheldon, *Chem. Commun.*, 1998, 1891.
30. F. van de Velde, I. W. C. E. Arends and R. A. Sheldon, *Top. Catal.*, 2000, **13**, 259.
31. A. Pordea, M. Creus, J. Panek, C. Duboc, D. Mathis, M. Novic and T. R. Ward, *J. Am. Chem. Soc.*, 2008, **130**, 8085.
32. V. Köhler, J. Mao, T. Heinisch, A. Pordea, A. Sardo, Y. M. Wilson, L. Knörr, M. Creus, J.-C. Prost, T. Schirmer and T. R. Ward, *Angew. Chem., Int. Ed.*, 2011, **50**, 10863.
33. J. Podtetenieff, A. Taglieber, E. Bill, E. J. Reijerse and M. T. Reetz, *Angew. Chem., Int. Ed.*, 2010, **49**, 5151.
34. G. Debret, A. Martel and P. Cuniasse, *Nucleic Acids Res.*, 2009, **37**, W459.
35. B. Amrein, M. Schmid, G. Collet, P. Cuniasse, F. Gilardoni, F. P. Seebeck and T. R. Ward, *Metallomics*, 2012, **4**, 379.
36. N. Fujieda, J. Schätti, E. Stuttfeld, K. Ohkubo, T. Maier, S. Fukuzumi and T. R. Ward, *Chem. Sci.*, 2015, **6**, 4060.
37. M. E. Wilson and G. M. Whitesides, *J. Am. Chem. Soc.*, 1978, **100**, 306.
38. J. Collot, J. Gradinaru, N. Humbert, M. Skander, A. Zocchi and T. R. Ward, *J. Am. Chem. Soc.*, 2003, **125**, 9030.
39. M. Skander, C. Malan, A. Ivanova and T. R. Ward, *Chem. Commun.*, 2005, 4815.
40. A. Sardo, T. Wohlschlager, C. Lo, H. Zoller, T. R. Ward and M. Creus, *Protein Expression Purif.*, 2011, 77, 131.
41. M. Creus, A. Pordea, T. Rossel, A. Sardo, C. Letondor, A. Ivanova, I. LeTrong, R. E. Stenkamp and T. R. Ward, *Angew. Chem., Int. Ed.*, 2008, **47**, 1400.
42. M. Dürrenberger, T. Heinisch, Y. M. Wilson, T. Rossel, E. Nogueira, L. Knörr, A. Mutschler, K. Kersten, M. J. Zimbron, J. Pierron, T. Schirmer and T. R. Ward, *Angew. Chem., Int. Ed.*, 2011, **50**, 3026.
43. T. Heinisch, K. Langowska, P. Tanner, J.-L. Reymond, W. Meier, C. Palivan and T. R. Ward, *ChemCatChem*, 2013, **5**, 720.
44. M. Hestericová, M. R. Correro, M. Lenz, P. F.-X. Corvini, P. Shahgaldian and T. R. Ward, *Chem. Commun.*, 2016, 9462.
45. A. Pordea, D. Mathis and T. R. Ward, *J. Organomet. Chem.*, 2009, **694**, 930.
46. J. Pierron, C. Malan, M. Creus, J. Gradinaru, I. Hafner, A. Ivanova, A. Sardo and T. R. Ward, *Angew. Chem., Int. Ed.*, 2008, **47**, 701.
47. T. K. Hyster, L. Knorr, T. R. Ward and T. Rovis, *Science*, 2012, **338**, 500.
48. F. W. Monnard, E. S. Nogueira, T. Heinisch, T. Schirmer and T. R. Ward, *Chem. Sci.*, 2013, **4**, 3269.
49. E. Sansiaume-Dagousset, A. Urvoas, K. Chelly, W. Ghattas, J.-D. Maréchal, J.-P. Mahy and R. Ricoux, *Dalton Trans.*, 2014, **43**, 8344.
50. A. Urvoas, W. Ghattas, J.-D. Maréchal, F. Avenier, F. Bellande, W. Mao, R. Ricoux and J.-P. Mahy, *Bioorg. Med. Chem.*, 2014, **22**, 5678.

51. W. Ghattas, L. Cotchico-Alonso, J.-D. Maréchal, A. Urvoas, M. Rousseau, J.-P. Mahy and R. Ricoux, *ChemBioChem*, 2016, **17**, 433.
52. A. Chevalley, M. V. Cherrier, J. C. Fontecilla-Camps, M. Ghasemi and M. Salmain, *Dalton Trans.*, 2014, **43**, 5482.
53. J.-P. Mahy, J.-D. Maréchal and R. Ricoux, *Chem. Commun.*, 2015, 2476.
54. J.-P. Mahy, J.-D. Maréchal and R. Ricoux, *J. Porphyrins Phthalocyanines*, 2014, **18**, 1063.
55. J.-P. Mahy and R. Ricoux, in *Handbook of Porphyrin Science: With Applications to Chemistry, Physics, Materials Science, Engineering, Biology and Medicine*, World Scientific Publishing Company, 2016, vol. 38, p. 101.
56. P. A. Adams and M. C. Berman, *Biochem. J.*, 1980, **191**, 95.
57. U. Muller-Eberhard and W. T. Morgand, *Ann. N. Y. Acad. Sci.*, 1975, **244**, 624.
58. M. Wardell, Z. Wang, J. X. Ho, J. Robert, F. Ruker, J. Ruble and D. C. Carter, *Biochem. Biophys. Res. Commun.*, 2002, **291**, 813.
59. P. A. Zunszain, J. Ghuman, T. Komatsu, E. Tsuchida and S. Curry, *BMC Struct. Biol.*, 2003, **3**, 6.
60. A. Nakagawa, T. Komatsu, M. Iizuka and E. Tsuchida, *Bioconjugate Chem.*, 2008, **19**, 581.
61. T. Komatsu, A. Nakagawa and X. Qu, *Drug Metab. Pharmacokinet.*, 2009, **24**, 287.
62. A. Mahammed, H. B. Gray, J. J. Weaver, K. Sorasaenee and Z. Gross, *Bioconjugate Chem.*, 2004, **15**, 738.
63. A. Mahammed and Z. Gross, *J. Am. Chem. Soc.*, 2005, **127**, 2883.
64. C. Herrero, A. Quaranta, R. Ricoux, A. Trehoux, A. Mahammed, Z. Gross, F. Banse and J.-P. Mahy, *Dalton Trans.*, 2016, **45**, 706.
65. M. T. Reetz and N. Jiao, *Angew. Chem., Int. Ed.*, 2006, **45**, 2416.
66. V. Oliveri and G. Vecchio, *Eur. J. Med. Chem.*, 2011, **46**, 961.
67. P. Rousselot-Pailley, C. Bochot, C. Marchi-Delapierre, A. Jorge-Robin, L. Martin, J. C. Fontecilla-Camps, C. Cavazza and S. Ménage, *ChemBioChem*, 2009, **10**, 545.
68. J. Tang, F. Huang, Y. Wei, H. Bian, W. Zhang and H. Liang, *Dalton Trans.*, 2016, **45**, 8061.
69. C. L. Hunter, E. Lloyd, L. D. Eltis, S. P. Rafferty, H. Lee, M. Smith and A. G. Mauk, *Biochemistry*, 1997, **36**, 1010.
70. T. Hayashi and Y. Hisaeda, *Acc. Chem. Res.*, 2002, **35**, 35.
71. T. Ueno, T. Koshiyama, M. Ohashi, K. Kondo, M. Kono, A. Suzuki, T. Yamane and Y. Watanabe, *J. Am. Chem. Soc.*, 2005, **127**, 6556.
72. T. Ueno, T. Koshiyama, S. Abe, N. Yokoi, M. Ohashi, H. Nakajima and Y. Watanabe, *J. Organomet. Chem.*, 2007, **692**, 142.
73. T. Hayashi, D. Murata, M. Makino, H. Sugimoto, T. Matsuo, H. Sato, Y. Shiro and Y. Hisaeda, *Inorg. Chem.*, 2006, **45**, 10530.
74. M. Ohashi, T. Koshiyama, T. Ueno, M. Yanase, H. Fujii and Y. Watanabe, *Angew. Chem., Int. Ed.*, 2003, **42**, 1005.

75. J. R. Carey, S. K. Ma, T. D. Pfister, D. K. Garner, H. K. Kim, J. A. Abramite, Z. Wang, Z. Guo and Y. Lu, *J. Am. Chem. Soc.*, 2004, **126**, 10812.
76. J.-L. Zhang, D. K. Garner, L. Liang, Q. Chen and Y. Lu, *Chem. Commun.*, 2008, 1665.
77. T. Komatsu, S. Ishihara, E. Tsuchida, H. Nishide, C. Morokuma and S. Nakamura, *Biomacromolecules*, 2005, **6**, 1489.
78. V. Ducros, S. J. Charnock, U. Derewenda, Z. S. Derewenda, Z. Dauter, C. Dupont, F. Shareck, R. Morosoli, D. Kluepfel and G. J. Davies, *J. Biol. Chem.*, 2000, **275**, 23020.
79. R. Ricoux, R. Dubuc, C. Dupont, J.-D. Marechal, A. Martin, M. Sellier and J.-P. Mahy, *Bioconjugate Chem.*, 2008, **19**, 899.
80. J.-P. Mahy, Q. Raffy, M. Allard and R. Ricoux, *Biochimie*, 2009, **91**, 1321.
81. R. Ricoux, M. Allard, R. Dubuc, C. Dupont, J.-D. Maréchal and J.-P. Mahy, *Org. Biomol. Chem.*, 2009, 7, 3208.
82. M. Allard, C. Dupont, V. Muñoz Robles, N. Doucet, A. Lledós, J.-D. Maréchal, A. Urvoas, J.-P. Mahy and R. Ricoux, *ChemBioChem*, 2012, **13**, 240.
83. M. V. Cherrier, C. Cavazza, C. Bochot, D. Lemaire and J. C. Fontecilla-Camps, *Biochemistry*, 2008, **47**, 9937.
84. M. V. Cherrier, L. Martin, C. Cavazza, L. Jacquamet, D. Lemaire, J. Gaillard and J. C. Fontecilla-Camps, *J. Am. Chem. Soc.*, 2005, **127**, 10075.
85. C. Cavazza, C. Bochot, P. Rousselot-Pailley, P. Carpentier, M. V. Cherrier, L. Martin, C. Marchi-Delapierre, J. C. Fontecilla-Camps and S. Ménage, *Nat. Chem.*, 2010, **2**, 1069.
86. M. V. Cherrier, E. Girgenti, P. Amara, M. Iannello, C. Marchi-Delapierre, J. C. Fontecilla-Camps, S. Ménage and C. Cavazza, *JBIC, J. Biol. Inorg. Chem.*, 2012, **17**, 817.
87. C. Esmieu, M. V. Cherrier, P. Amara, E. Girgenti, C. Marchi-Delapierre, F. Oddon, M. Iannello, A. Jorge-Robin, C. Cavazza and S. Ménage, *Angew. Chem., Int. Ed.*, 2013, **52**, 3922.
88. A. Chevalley and M. Salmain, *Chem. Commun.*, 2012, 11984.
89. M. V. Cherrier, S. Engilberge, P. Amara, A. Chevalley, M. Salmain and J. C. Fontecilla-Camps, *Eur. J. Inorg. Chem.*, 2013, **2013**, 3596.
90. A. Chevalley, M. V. Cherrier, J. C. Fontecilla-Camps, M. Ghasemi and M. Salmain, *Dalton Trans.*, 2014, **43**, 5482.
91. C. Chen and D. Sigman, *Science*, 1987, **237**, 1197.
92. E. Kaiser and D. Lawrence, *Science*, 1984, **226**, 505.
93. H. L. Levine, Y. Nakagawa and E. T. Kaiser, *Biochem. Biophys. Res. Commun.*, 1977, **76**, 64.
94. M. T. Reetz, *Tetrahedron*, 2002, **58**, 6595.
95. M. T. Reetz, M. Rentzsch, A. Pletsch, M. Maywald, P. Maiwald, J. J.-P. Peyralans, A. Maichele, Y. Fu, N. Jiao, F. Hollmann, R. Mondière and A. Taglieber, *Tetrahedron*, 2007, **63**, 6404.

96. B. Talbi, P. Haquette, A. Martel, F. de Montigny, C. Fosse, S. Cordier, T. Roisnel, G. Jaouen and M. Salmain, *Dalton Trans.*, 2010, **39**, 5605.
97. P. Haquette, B. Talbi, L. Barilleau, N. Madern, C. Fosse and M. Salmain, *Org. Biomol. Chem.*, 2011, **9**, 5720.
98. N. Madern, B. Talbi and M. Salmain, *Appl. Organomet. Chem.*, 2013, **27**, 6.
99. T. Reiner, D. Jantke, A. N. Marziale, A. Raba and J. Eppinger, *ChemistryOpen*, 2013, **2**, 50.
100. K. M. Nicholas, P. Wentworth, C. W. Harwig, A. D. Wentworth, A. Shafton and K. D. Janda, *Proc. Natl. Acad. Sci.*, 2002, **99**, 2648.
101. R. R. Davies and M. D. Distefano, *J. Am. Chem. Soc.*, 1997, **119**, 11643.
102. J. J. Ory, A. Mazhary, H. Kuang, R. R. Davies, M. D. Distefano and L. J. Banaszak, *Protein Eng., Des. Sel.*, 1998, **11**, 253.
103. S. Abe, K. Hirata, T. Ueno, K. Morino, N. Shimizu, M. Yamamoto, M. Takata, E. Yashima and Y. Watanabe, *J. Am. Chem. Soc.*, 2009, **131**, 6958.
104. W. Laan, B. K. Muñoz, R. den Heeten and P. C. J. Kamer, *ChemBioChem*, 2010, **11**, 1236.
105. P. J. Deuss, G. Popa, C. H. Botting, W. Laan and P. C. J. Kamer, *Angew. Chem.*, 2010, **122**, 5443.
106. J. Bos, F. Fusetti, A. J. M. Driessen and G. Roelfes, *Angew. Chem., Int. Ed.*, 2012, **51**, 7472.
107. I. Drienovská, A. Rioz-Martínez, A. Draksharapu and G. Roelfes, *Chem. Sci.*, 2015, **6**, 770.
108. C. Buron, K. Sénéchal-David, R. Ricoux, J.-P. Le Caër, V. Guérineau, P. Méjanelle, R. Guillot, C. Herrero, J.-P. Mahy and F. Banse, *Chem. – Eur. J.*, 2015, **21**, 12188.
109. E. Ricca, B. Brucher and J. H. Schrittwieser, *Adv. Synth. Catal.*, 2011, **353**, 2239.
110. E. O'Reilly and N. J. Turner, *Perspect. Sci.*, 2015, **4**, 55.
111. S. Betanzos-Lara, Z. Liu, A. Habtemariam, A. M. Pizarro, B. Qamar and P. J. Sadler, *Angew. Chem., Int. Ed.*, 2012, **51**, 3897.
112. E. Wingstrand, A. Laurell, L. Fransson, K. Hult and C. Moberg, *Chem. – Eur. J.*, 2009, **15**, 12107.
113. C. Simons, U. Hanefeld, I. W. C. E. Arends, T. Maschmeyer and R. A. Sheldon, *Top. Catal.*, 2006, **40**, 35.
114. B. Wieczorek, A. Träff, P. Krumlinde, H. P. Dijkstra, M. R. Egmond, G. van Koten, J.-E. Bäckvall and R. J. M. K. Gebbink, *Tetrahedron Lett.*, 2011, **52**, 1601.
115. B. Worsdorfer, K. J. Woycechowsky and D. Hilvert, *Science*, 2011, **331**, 589.
116. K. Engström, E. V. Johnston, O. Verho, K. P. J. Gustafson, M. Shakeri, C.-W. Tai and J.-E. Bäckvall, *Angew. Chem., Int. Ed.*, 2013, **52**, 14006.

117. J. Latham, J.-M. Henry, H. H. Sharif, B. R. K. Menon, S. A. Shepherd, M. F. Greaney and J. Micklefield, *Nat. Commun.*, 2016, **7**, 11873.
118. H. Sato, W. Hummel and H. Gröger, *Angew. Chem., Int. Ed.*, 2015, **54**, 4488.
119. V. Köhler, Y. M. Wilson, M. Dürrenberger, D. Ghislieri, E. Churakova, T. Quinto, L. Knörr, D. Häussinger, F. Hollmann, N. J. Turner and T. R. Ward, *Nat. Chem.*, 2012, **5**, 93.
120. Y. Okamoto, V. Köhler and T. R. Ward, *J. Am. Chem. Soc.*, 2016, **138**, 5781.
121. Y. Okamoto, V. Köhler, C. E. Paul, F. Hollmann and T. R. Ward, *ACS Catal.*, 2016, **6**, 3553.
122. P. Dydio, H. M. Key, A. Nazarenko, J. Y.-E. Rha, V. Seyedkazemi, D. S. Clark and J. F. Hartwig, *Science*, 2016, **354**, 102.
123. E. M. Stocking and R. M. Williams, *Angew. Chem., Int. Ed.*, 2003, **42**, 3078.
124. P. J. Deuss, G. Popa, A. M. Z. Slawin, W. Laan and P. C. J. Kamer, *ChemCatChem*, 2013, **5**, 1184.
125. J. Bos, W. R. Browne, A. J. M. Driessen and G. Roelfes, *J. Am. Chem. Soc.*, 2015, **137**, 9796.
126. M. T. Reetz, *Chem. Rec.*, 2012, **12**, 391.
127. A. Urvoas, M. Valerio-Lepiniec and P. Minard, *Trends Biotechnol.*, 2012, **30**, 512.
128. A. Urvoas, A. Guellouz, M. Valerio-Lepiniec, M. Graille, D. Durand, D. C. Desravines, H. van Tilbeurgh, M. Desmadril and P. Minard, *J. Mol. Biol.*, 2010, **404**, 307.
129. J. M. Palomo, *Org. Biomol. Chem.*, 2012, **10**, 9309.
130. H. Yang, P. Srivastava, C. Zhang and J. C. Lewis, *ChemBioChem*, 2014, **15**, 223.
131. I. D. Petrik, J. Liu and Y. Lu, *Curr. Opin. Chem. Biol.*, 2014, **19**, 67.
132. M. Dürrenberger and T. R. Ward, *Curr. Opin. Chem. Biol.*, 2014, **19**, 99.
133. C. M. Rufo, Y. S. Moroz, O. V. Moroz, J. Stöhr, T. A. Smith, X. Hu, W. F. DeGrado and I. V. Korendovych, *Nat. Chem.*, 2014, **6**, 303.
134. M. M. Müller, M. A. Windsor, W. C. Pomerantz, S. H. Gellman and D. Hilvert, *Angew. Chem., Int. Ed.*, 2009, **48**, 922.
135. C. Mayer, M. M. Müller, S. H. Gellman and D. Hilvert, *Angew. Chem., Int. Ed.*, 2014, **53**, 6978.
136. A. Schmoldt, H. F. Benthe and G. Haberland, *Biochem. Pharmacol.*, 1975, **24**, 1639.
137. G. Maayan, M. D. Ward and K. Kirshenbaum, *Chem. Commun.*, 2009, 56.
138. J. Oelerich and G. Roelfes, in *Progress in Inorganic Chemistry*, ed. K. D. Kenneth, John Wiley & Sons, Inc., Hoboken, NJ, 2012, vol. 57, p. 353.
139. C. Wang, G. Jia, J. Zhou, Y. Li, Y. Liu, S. Lu and C. Li, *Angew. Chem., Int. Ed.*, 2012, **51**, 9352.

140. J. S. Willemsen, R. P. Megens, G. Roelfes, J. C. M. van Hest and F. P. J. T. Rutjes, *Eur. J. Org. Chem.*, 2014, **2014**, 2892.
141. A. Rioz-Martínez, J. Oelerich, N. Ségaud and G. Roelfes, *Angew. Chem., Int. Ed.*, 2016, **55**, 14136.
142. A. García-Fernández, R. P. Megens, L. Villarino and G. Roelfes, *J. Am. Chem. Soc.*, 2016, **138**, 16308.
143. M. Jeschek, R. Reuter, T. Heinisch, C. Trindler, J. Klehr, S. Panke and T. R. Ward, *Nature*, 2016, **537**, 661.

CHAPTER 4

Computational Enzyme Design: Successes, Challenges, and Future Directions

ANTONY D. ST-JACQUES,[a,b] OLIVIER GAGNON[a,b] AND ROBERTO A. CHICA*[a,b]

[a] Department of Chemistry and Biomolecular Sciences, University of Ottawa, 10 Marie-Curie, Ottawa, Ontario K1N 6N5, Canada; [b] Centre for Catalysis Research and Innovation, University of Ottawa, 30 Marie-Curie, Ottawa, Ontario K1N 6N5, Canada
*Email: rchica@uottawa.ca

4.1 Introduction

The ability of enzymes to catalyze chemical reactions with high rate acceleration ($k_{cat}/k_{uncat} = 10^7$–10^{26})[1,2] and unmatched selectivity makes them highly attractive for use as biocatalysts in a large number of industrial syntheses. However, the finite repertoire of naturally-occurring enzyme activities and the fact that few enzymes inherently catalyze desired reactions under industrially-relevant conditions restrict their applicability in many industrial processes. To make enzymes practical for industrial syntheses, it is often necessary to tailor their catalytic activity to the intended application using protein engineering. Protein engineering techniques such as rational design and directed evolution have yielded many improved enzymatic activities (see Chapter 7),[3–5] yet the creation of new enzymes that possess desired catalytic properties remains very challenging due to our incomplete

Catalysis Series No. 32
Modern Biocatalysis: Advances Towards Synthetic Biological Systems
Edited by Gavin Williams and Mélanie Hall

Published by the Royal Society of Chemistry, www.rsc.org

understanding of the structural determinants of enzyme catalysis and the necessity for high-throughput screening of large mutant libraries. Thus, the development of a broadly-applicable computational method for the design of enzymes that can catalyze any desired chemical reaction with high efficiency under defined conditions would bolster the use of industrial biocatalysis while opening the door to biocatalytic processes that are currently inaccessible using natural enzymes.

Computational enzyme design, which is the construction of *de novo* enzymes using structure-based computational protein design (CPD) methodologies,[6,7] is a promising technology for the rational design of new enzymatic activities on demand. Computational enzyme design is a specialized application of the CPD algorithms that have been developed over the past 20 years[8,9] to facilitate the identification of protein sequences that can adopt a specific three-dimensional structure and possess a desired property such as increased stability,[10] improved binding affinity,[11] or altered specificity.[12] Most CPD algorithms consist of three steps (Figure 4.1A): (i) a side-chain placement step where discrete side-chain rotamers[13] are threaded onto specified positions on a predefined protein backbone template, (ii) an energy calculation where interaction energies between pairs of rotamers and between each rotamer and the backbone template are computed using a potential energy function,[8,14–17] and (iii) a sequence optimization step where combinations of rotamers are optimized using a search algorithm[18–23] that explores both rotamer and sequence space in order to identify optimal sequences. At the conclusion of this process, a list of sequences along with their corresponding rotamer-optimized structures are generated (Figure 4.1B), and each sequence is ranked according to an energy value that reflects its predicted stability on the target protein template structure.

In computational enzyme design, CPD algorithms are used to create a catalytic site in a protein scaffold that does not natively bind the target substrate and catalyze the desired reaction. This is a very challenging design problem because the CPD procedure must simultaneously create a binding pocket specific to the desired substrate(s), introduce the required catalytic machinery for the chemical reaction of interest, and preserve the overall structure and stability of the template protein fold. From a calculation perspective, computational enzyme design differs from standard CPD methodologies by the inclusion of an additional step where a user-defined active site comprising one or more catalytic side chains and a high-energy reaction intermediate, such as a transition state, is placed onto a pre-existing protein scaffold according to a set of predefined geometrical constraints postulated to be necessary for the designed chemical reaction. Active site construction begins by identifying positions in the protein scaffold where all necessary catalytic contacts can be made to a transition state derived from quantum mechanics (QM) calculations. This can be done using either a ligand placement[24] or matching[25] algorithm (Figure 4.2). Once the position of catalytic residues has been determined, amino acids at neighbouring positions in the active site are designed to optimize substrate recognition

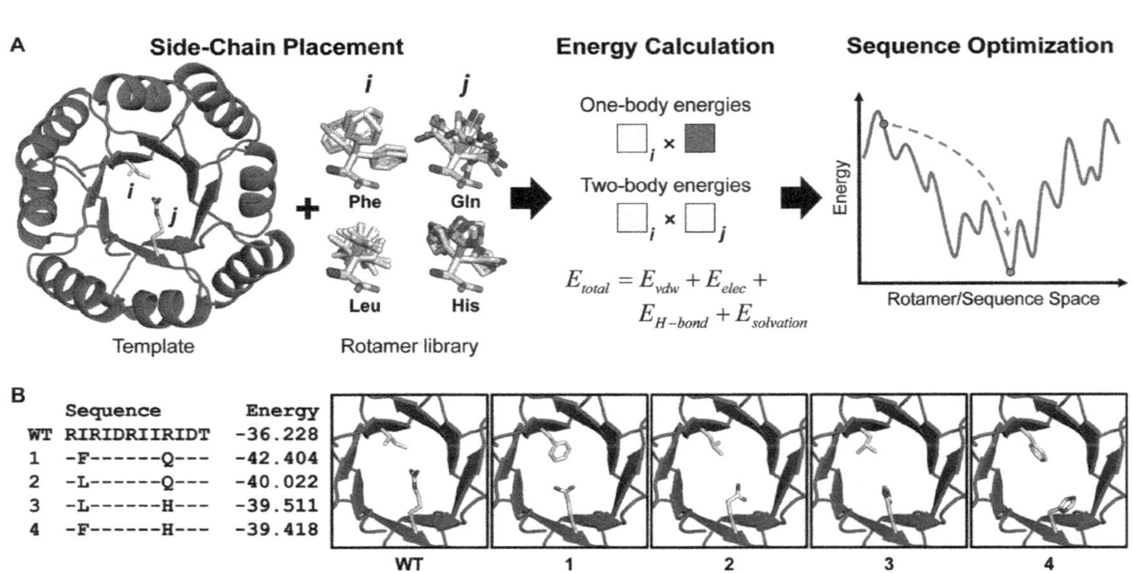

Figure 4.1 Computational protein design. (A) The general methodology of computational protein design (CPD) consists of three steps: (1) a side-chain placement step where amino acids at user-defined positions (*i* and *j*) in the protein template are replaced by a library of rotamers for each allowed amino-acid type (Phe and Leu, or Gln and His at positions *i* and *j*, respectively), (2) an energy calculation where interaction energies between pairs of rotamers (two-body energies) and between each rotamer and the backbone template (one-body energies) are computed using a potential energy function that can include van der Waals (vdw), electrostatics (elec), hydrogen bonding, and solvation terms, and (3) a sequence optimization step where a search algorithm is used to optimize combinations of rotamers in order to identify optimal sequences. (B) The CPD procedure returns a ranked list of scored sequences along with their corresponding rotamer-optimized structures. For clarity, side chains of residues at non-designed positions are omitted. Each sequence is ranked according to an energy value that reflects its predicted stability on the target protein template structure. WT indicates the wild-type sequence.

and binding in a step called packing. Stabilization of the transition state *via* additional non-bonding interactions introduced during the packing step is crucial to ensure that the designed protein can perform catalysis, as transition state stabilization is one of the key features of enzyme catalysis. Sequences predicted to efficiently stabilize the transition state are then experimentally validated. It should be noted that CPD methods optimize sequence stability, and not catalytic activity. However, by identifying sequences predicted to efficiently stabilize the transition state of a chemical reaction of interest, catalytic activity can be inferred.

In this chapter, we review examples of the use of CPD methodologies for the *de novo* design of enzyme activities. We focus on studies where enzymes that can transform specific organic molecules in predefined chemical reactions were designed computationally (Figure 4.3A). A discussion on the current challenges of computational enzyme design and future directions to improve this promising technology are also presented.

4.2 Examples of Computational Enzyme Design

4.2.1 Phase I: Initial Application of CPD Towards Enzyme Design

The introduction of enzymatic activity into catalytically-inert proteins has been a long-standing objective of protein design.[26,27] Although early design of metal binding sites into proteins using CPD[28] yielded nascent metalloenzymes capable of performing a variety of redox reactions on molecular oxygen,[29,30] the first example of the use of CPD to design a biocatalyst capable of transforming a predefined organic substrate was reported by Bolon and Mayo in 2001.[31] In this study, ester hydrolysis activity was designed into *E. coli* thioredoxin, a protein natively devoid of esterase activity. For their design, Bolon and Mayo chose the nucleophilic hydrolysis of the activated ester *p*-nitrophenylacetate (pNPA) as a model reaction (Figure 4.4A). This reaction proceeds *via* a two-step mechanism in which a covalent acyl–enzyme intermediate is first generated by nucleophilic attack of the enzyme on the substrate, followed by its hydrolysis to regenerate the free enzyme.[32] As the efficiency of nucleophilic catalysis is expected to increase with an unstable acyl–enzyme intermediate, the *p*-nitrobenzoyl–histidine intermediate was chosen for the design as it is less stable than either the *p*-nitrobenzoyl–serine or *p*-nitrobenzoyl–cysteine intermediates. According to transition state theory, stabilization of the highest-energy transition state along the reaction coordinate should accelerate the reaction rate by decreasing the energy barrier of the rate-determining step. To simplify their computational procedure, Bolon and Mayo instead used the tetrahedral intermediate formed by nucleophilic attack of histidine on pNPA (Figure 4.5A) as a proxy for the transition state, and postulated that stabilization of this high-energy intermediate *via* non-bonding interactions would bestow esterase activity to the thioredoxin scaffold.

Using individual calculations with the ORBIT protein design software,[33,34] each amino-acid position in the thioredoxin structure was replaced by a "superrotamer" library[35,36] of the tetrahedral intermediate. To accommodate the intermediate, all other positions were either mutated to alanine or allowed to sample alternate rotamers of the wild-type amino acid. The goal of this procedure was to identify positions in the thioredoxin structure where an active site could be built. Active-site designs were ranked based on packing interactions made by the protein to the *p*-nitrobenzoyl moiety of the tetrahedral intermediate. The top two designs were experimentally tested and exhibited esterase activity significantly above background, however the rate enhancement obtained for the most active variant (Table 4.1, PDZ2) was 9–11 orders of magnitude lower than that of natural esterases.[37] The marginal catalytic activity of PDZ2 ($k_{cat}/k_{uncat} \approx 10^2$) is unsurprising given that only two alanine substitutions in addition to the catalytic histidine (Table 4.2, Figure 4.6A) were incorporated into thioredoxin to accommodate the high-energy intermediate, likely resulting in sub-optimal shape and charge complementarities compared to those found in natural enzymes. Nevertheless, this pioneering study demonstrated that CPD could be applied to the creation of *de novo* enzymes capable of catalyzing desired organic transformations.

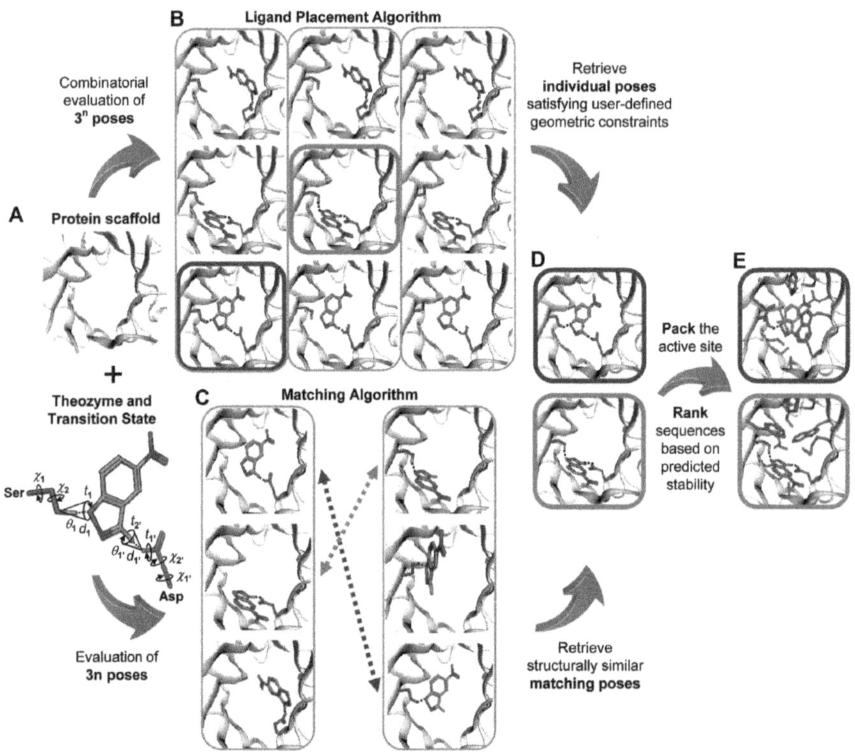

The low catalytic activity of the biocatalysts designed by Bolon and Mayo[31] highlighted limitations in the computational procedure that needed to be addressed in order to design more efficient enzymes. Specifically, a transition state proxy was used instead of a QM-derived transition state with idealized geometry, sampling of rotational, translational, and conformational degrees of freedom was limited, packing of the high-energy intermediate was not optimized, and only one protein scaffold was used as a design template. To address these limitations, ligand placement[24] and matching[25] algorithms (Figure 4.2) were developed. These algorithms enabled the incorporation of transition states and theozymes (or "theoretical enzymes")[38] (Figure 4.5), which are combinations of functional groups arranged in a unique geometry predicted by QM to stabilize a transition state, into the design procedure. In addition to the development of these algorithms, advances in computational power made tractable the consideration of an increased number of translational, rotational, and conformational degrees of freedom during transition state binding pose generation, and of a large number of protein scaffolds to be used as templates for design.

Figure 4.2 Ligand placement and matching algorithms used in computational enzyme design. (A) A protein scaffold and a theozyme with its associated transition state (TS) are required as inputs for either algorithm. Although the theozyme and TS are QM-derived, several theozyme rotamers and TS poses are sampled during calculation to compensate for the use of a fixed protein scaffold during design. Theozyme rotamers are generated according to user-defined torsional degrees of freedom (χ) while TS poses are generated from each theozyme rotamer using translational (d), rotational (θ), and torsional (t) degrees of freedom to maintain user-defined catalytic contacts (dashed lines). The various theozyme rotamers and TS poses are placed inside the active-site pocket of the protein scaffold using either a ligand placement (B) or matching (C) algorithm. For clarity purposes, only three rotamers each of the designed catalytic residues Asp and Ser, as well as a single TS binding pose are shown. In ligand placement (B), the single TS binding pose is generated relative to each of the three catalytic Asp rotamers (thin-lined boxes). For each of these Asp and TS combinations, the three Ser rotamers are tested to identify which one(s) form(s) user-defined catalytic contacts (dashed lines). Active sites containing TS poses that form both catalytic contacts (thick-lined boxes) are optimized *via* a packing step, and are ranked according to their predicted stability. In this example, the bottom left active site ranks better than the middle one. In matching (C), each Asp and Ser rotamer of the theozyme is placed in the active site, and the TS pose is built off of each to create the user-defined catalytic contact. The coordinates of each rotamer and TS pose combination are saved in a hash map corresponding to the catalytic contact formed. Structurally similar TS poses forming different catalytic contacts are matched (dashed arrows). Following matching, the active sites are optimized by a packing step and ranked. In this toy example, both algorithms yield the same results, however the matching algorithm is more computationally efficient given that it scales linearly (3 rotamers$\times$2 catalytic residues$=$6 active sites) instead of exponentially ($3^2=9$ active sites), as is the case in ligand placement.

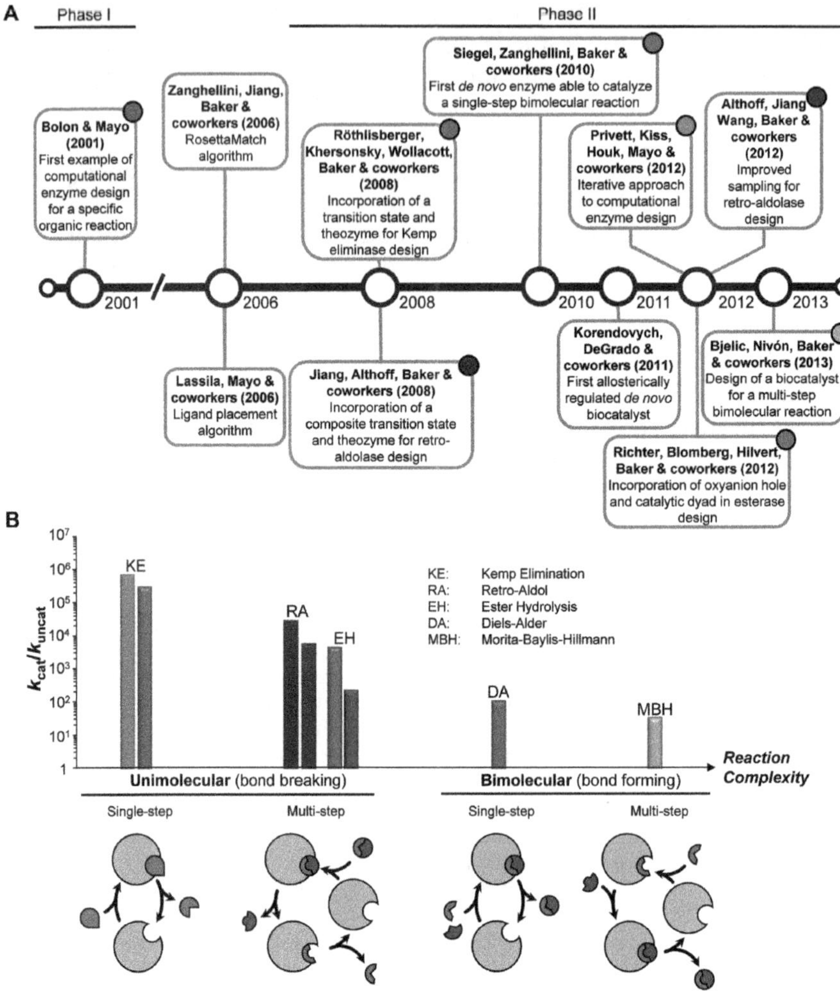

Figure 4.3 Examples of computational enzyme design. (A) Timeline highlighting computational enzyme design studies discussed herein, which are grouped into one of two phases (see text for detail). (B) Rate enhancements (k_{cat}/k_{uncat}) achieved by the various *de novo* enzymes designed in the studies listed in panel A are shown as bars colored according to the study where they are described. Enzymes are grouped according to their catalyzed reactions and reactions are classified based on their complexity, which we define as a function of the reaction order and the number of steps in the proposed mechanism. Cartoons showing the general mechanism of the enzymatic reactions are shown below the graph, with the enzyme in grey and substrates and products in green and/or magenta.

This opened the door to a very active second phase of computational enzyme design from 2008 to 2013 (Figure 4.3A) that enabled the development of *de novo* enzymes capable of catalyzing a variety of single- or multi-step bond cleaving and bond forming reactions (Figure 4.3B).

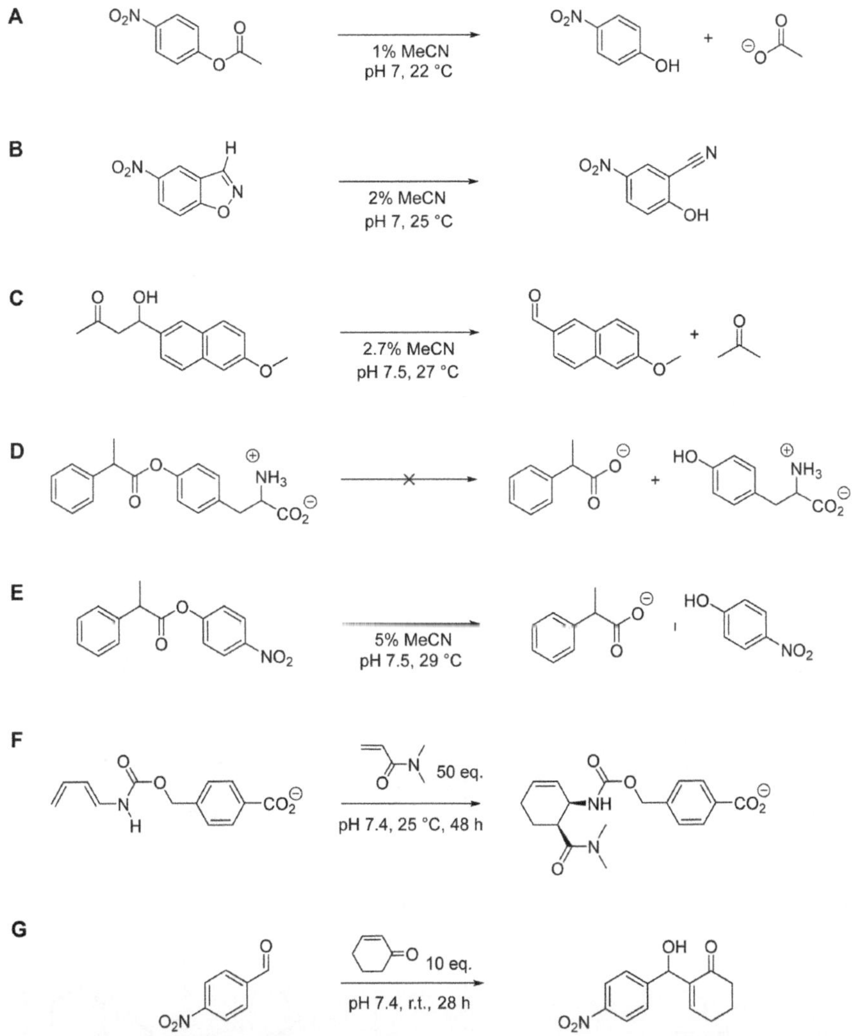

Figure 4.4 Chemical reactions catalyzed by *de novo* enzymes. (A) Hydrolysis of *p*-nitrophenylacetate catalyzed by PDZ2. (B) Kemp elimination of 5-nitrobenzisoxazole catalyzed by KE59, HG-2, or HG-3. (C) Retro-aldol decomposition of 4-hydroxy-4-(6-methoxy-2-napthyl)-2-butanone catalyzed by RA60 or RA95. (D) Hydrolysis of tyrosyl ester could not be catalyzed by ECH13. (E) Hydrolysis of *p*-nitrophenyl-2-phenylpropanoate catalyzed by ECH13. (F) Diels–Alder [4 + 2] cycloaddition of 4-carboxybenzyl-*trans*-1,3-butadiene-1-carbamate on *N*,*N*-dimethylacrylamide catalyzed by DA_20_10. (G) Morita–Baylis–Hillman reaction between 4-nitrobenzaldehyde and 2-cyclohexenone catalyzed by BH32.

4.2.2 Phase II: Incorporation of Transition States and Theozymes into Computational Enzyme Design

4.2.2.1 *Single-step Bond Cleaving Reaction – The Kemp Elimination*

In 2008, Röthlisberger, Khersonsky, Wollacott *et al.*[39] designed several enzymes able to catalyze the Kemp elimination of 5-nitrobenzisoxazole

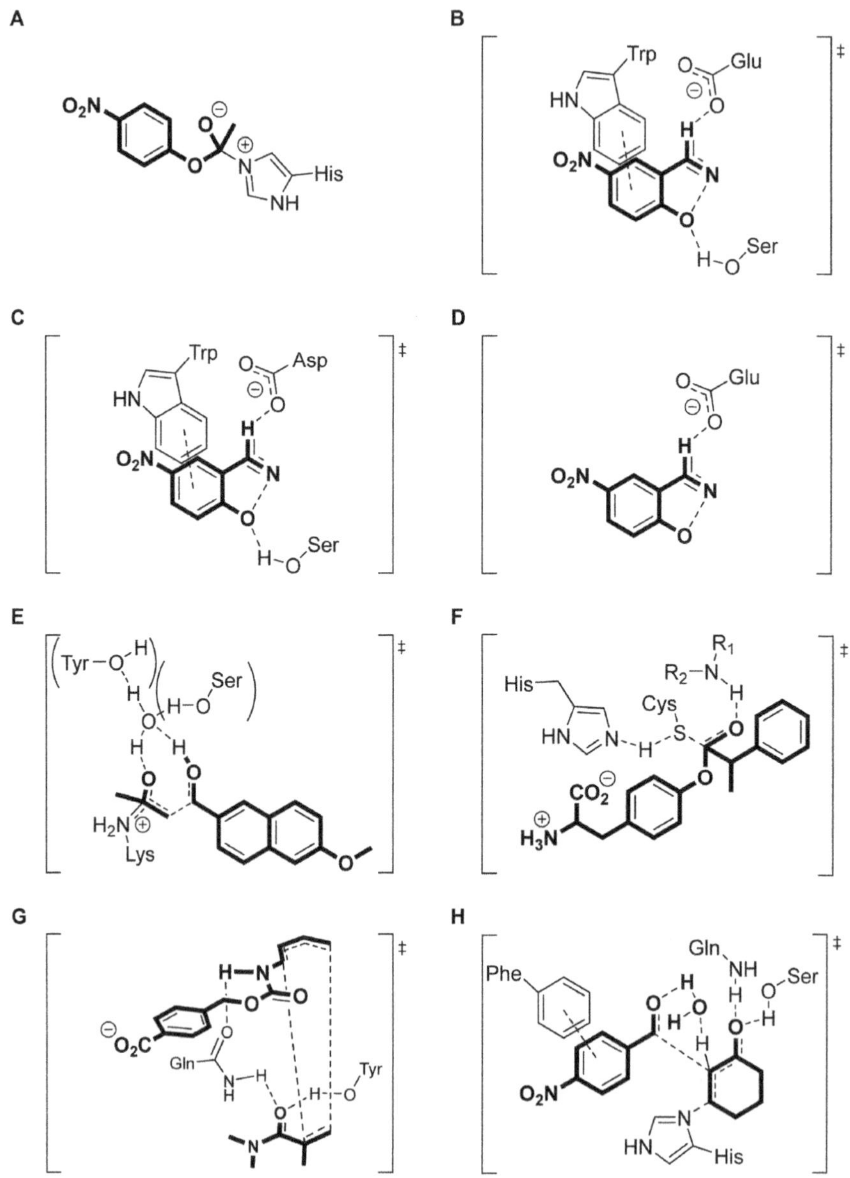

(Figure 4.4B), a well-studied model reaction for the deprotonation of carbon that proceeds *via* a single transition state. Instead of modeling a high-energy intermediate along the reaction coordinate to approximate the transition state as was done by Bolon and Mayo,[31] a QM-derived transition state bound by various theozymes containing a general base to deprotonate the acidic proton and an H-bond donor to stabilize the negative partial charge on the phenolic oxygen (Figure 4.5B) was used. To provide additional transition state stabilization, a π-stacking interaction to the substrate benzene ring was included using idealized geometries.[40] With the RosettaMatch matching algorithm,[25] the theozymes and transition state were placed on 87 different scaffolds that include protein folds containing ligand-binding pockets such as TIM barrels, β-propellers, and jelly rolls, among others. Following the matching step, active site residues were designed for improved packing using RosettaDesign[41] in order to identify additional stabilizing interactions to the transition state. A total of 59 designs on 17 unique scaffolds were predicted to be active and were experimentally tested. Eight of the designs displayed Kemp eliminase activity, of which the most active (KE59) contained 10 mutations including the designed theozyme residues Glu231, Ser180, and Trp110, (Table 4.2 and Figure 4.6B) and displayed a rate enhancement of approximately 10^5 (Table 4.1). Although the rate enhancement

Figure 4.5 High-energy intermediate, or theozymes and transition states used in various *de novo* computational enzyme designs. In all cases the high-energy intermediate or transition state (TS) is in bold. (A) Tetrahedral intermediate used in the design of PDZ2 for hydrolysis of *p*-nitrophenylacetate showing the catalytic His. (B and C) QM-calculated theozymes and TS used in the design of KE59 and HG-2 showing a Glu or Asp side chain as general base for Kemp elimination of 5-nitrobenzisoxazole, respectively. In both cases, Ser serves as an H-bond donor to stabilize the phenolic oxygen while Trp stabilizes the TS *via* π-stacking. (D) QM-calculated theozyme and TS used in the design of AlleyCat for Kemp elimination of 5-nitrobenzisoxazole. Glu is the catalytic base. (E) Superposition of multiple QM-calculated theozymes to bind the composite TS used in the design of retro-aldol enzymes RA60 and RA95. Lys is shown as the catalytic nucleophile, while H-bonds are mediated by an explicit water molecule as well as Ser and Tyr residues. In the case of RA95, the theozyme did not include Ser or Tyr. (F) QM-calculated theozyme and TS used in the design of the ester hydrolase ECH13. A His–Cys catalytic dyad is shown as well as an oxyanion hole formed by a backbone amide. (G) QM-calculated theozyme and TS used in the design of the DA_20_00 Diels–Alderase. Gln as well as Tyr are depicted as H-bond acceptors and/or donors. For clarity purposes, only the transition state component of the composite transition state used in the design procedure (see text for details) is shown. (H) Superposition of multiple QM-calculated theozymes and transition states along the reaction pathway used in the design of BH32, a Morita–Baylis–Hillman catalyst for the reaction of 4-nitrobenzaldehyde and cyclohexenone. His is shown as the catalytic nucleophile while Gln and Ser act as H-bond donors for enolate stabilization. Phe stabilizes the benzaldehyde moiety through π-stacking, and a water molecule (in bold) facilitates proton shuttling.

Table 4.1 Kinetic parameters of computationally-designed enzymes before and after optimization by directed evolution or rational design.

Enzyme[a]	k_{cat}/s^{-1}	$K_M/\mu M$	$k_{cat}/K_M/$ $M^{-1} s^{-1}$	$k_{cat}/$ k_{uncat}	Ref.
Ester hydrolysis					
PDZ2	0.00046	1.7×10^2	2.7	1.8×10^2	31
ECH13	0.018	57	3.1×10^2	4.4×10^3	52
Kemp elimination					
KE59	0.29	1.8×10^3	1.6×10^2	2.5×10^5	39
$KE59.13_{evolved}$	9.5	1.6×10^2	5.9×10^4	8.2×10^6	68
HG-2	—[b]	—[b]	1.2×10^2	—	42
$HG\text{-}3_{improved}$	0.68	1.6×10^3	4.3×10^2	5.9×10^5	42, 69
$HG\text{-}3.17_{evolved}$	7.0×10^2	3×10^3	2.3×10^5	6.0×10^8	69
AlleyCat	—[b]	—[b]	5.8	—	48
$AlleyCat7_{evolved}$	1.9	2.0×10^3	1.3×10^3	1.6×10^6	70
Retro-aldol					
RA60	0.00016	5.1×10^2	0.30	2.4×10^4	50, 51
RA95	0.000033	5.4×10^2	0.055	4.8×10^3	51
$RA95.5\text{-}8_{evolved}$	0.17	2.0×10^2	8.5×10^2	2.6×10^7	72
Diels–Alder					
DA_20_00	0.000028	$3.5\times10^3_{(diene)}$ $1.5\times10^5_{(dienophile)}$	0.06[c]	4.4[d]	53, 73
$DA\_20\_10_{improved}$	0.00058	$1.3\times10^3_{(diene)}$ $7.3\times10^4_{(dienophile)}$	6.1[c]	94[d]	53, 73
$DA\_20\_20\times10_{evolved}$	0.0013	$1.0\times10^3_{(diene)}$ $3.7\times10^4_{(dienophile)}$	34[c]	2.0×10^2 [d]	73
Morita–Baylis–Hillman					
BH32 Q128H	—[e]	—[e]	—	54	54
Average reaction					
Natural enzymes	$1\text{–}10^2$	$10^1\text{–}10^3$	$10^3\text{–}10^6$	$10^7\text{–}10^{19}$	1, 63

[a]Enzymes optimized by directed evolution or rational design are indicated by the evolved or improved subscripts, respectively.
[b]Could not be determined as saturation with the substrate was not achieved within its solubility limit.
[c]Units are $M^{-2} s^{-1}$.
[d]Units are M.
[e]Not reported in the article.

obtained was approaching that of natural enzymes, the success rate was low (14%) as many false positive predictions were made by the computational procedure.

Because of the low success rate of computational enzyme design, experiments that aim to understand the causes of failure are crucial to the continued improvement of this technology. In 2012, Privett *et al.*[42] aimed to design Kemp elimination activity (Figure 4.4B) onto the *Thermoascus aurantiacus* xylanase scaffold (Figure 4.6C) using the PHOENIX protein design software[43,44] and its ligand placement algorithm[24] with theozymes and a

Table 4.2 Mutations found in computationally-designed enzymes before and after optimization by directed evolution or rational design.

Enzyme[a]	PDB ID	Designed mutations[b]	Improved/evolved mutations	Ref.
Ester hydrolysis				
PDZ2	—	F12A, **L17H**, Y70A		31
ECH13	3U13	F20L, F42Y, W43K, V44A, **S45C**, I60R, W63A, **I100H**	—	52
Kemp elimination				
KE59	—	E51V, S81V, **K110W**, L131G, L157A, E159V, G178I, **W180S**, E210A, **L231E**		39
KE59.13$_{evolved}$	5UZJ		K9E, L14R, L16Q, F21V, N33K, I48M, S69A, Y75G, A76V, V80A, T94D, I104V, F111I, Y151L, N160H, S179T, R181H, K190N, A208V, R222Y, S233T, L247Q	68
HG-2[c]	3NYD	Q42M, **T44W**, R81G, H83G, T84M, N130G, N172M, A234S, T236L, E237M, **T265S**, W267F		42
HG-3$_{improved}$	—		S265T	42, 69
HG-3.17$_{evolved}$	4BS0		V6I, Q37K, N47E, K50Q, G82A, M84C, S89N, Q90F, T105I, A125T, T142N, T208M, S265T, F267M, W275A, R276F, T279S, D300N	69
AlleyCat	2KZ2	**F92E**		48
AlleyCat7$_{evolved}$	2M3S		M144R, H107I, L112R, I85L, A128T, M124L, A88Q	70
Retro-aldol				
RA60	—	N46W, **V48K**, N72T, Y74W, **E87S**, Y89S, T119Y, R121W, **F133Y**, Q135S, E176V, Y178V		50, 51

Table 4.2 (*Continued*)

Enzyme[a]	PDB ID	Designed mutations[b]	Improved/evolved mutations	Ref.
RA95	4A29	E51V, K53E, L83T, K110S, E159L, N180M, L184F, L187G, **E210K**, S211L, G233S		51
RA95.5-8$_{evolved}$	—		R23H, V51Y, E53T, F72Y, T83K, T95M, S110N, K135N, G178V, M180F, R182M, D183N, G212D	72
Diels–Alder				
DA_20_00	3I1C[d]	E21A, E36Y, N120A, **D121Y**, Y144F, R146I, M148L, F173A, N175A, **T195Q**, E225K, D229A, N272A		53, 73
DA_20_10	—		A21T, A74I, Q149R, A173C, S271A, A272N	53, 73
DA_20_20×10$_{evolved}$	—		R50H, V96I, T197R, D232V, H274L, E288D, L309S	73
Morita–Baylis–Hillman				
BH32[e]	3U26	F9S, V10L, L14N, E19A, T22S, I64L, E68L, **H91S**, H95S, **Y128Q**, L129A, **H132F**		54
BH32 Q128H$_{improved}$	—		Q128H	54

[a]Enzymes optimized by directed evolution or rational design are indicated by the evolved or improved subscripts, respectively.
[b]Designed catalytic residues are in bold.
[c]The catalytic base is the wild-type residue D127.
[d]The structure contains an additional mutation (A74I).
[e]The catalytic nucleophile is the wild-type residue H23.

transition state (Figure 4.5C) similar to those used by Röthlisberger, Khersonsky, Wollacott *et al.*[39] The top-scoring design, HG-1, was experimentally tested and displayed no Kemp elimination activity on 5-nitrobenzisoxazole, even though circular dichroism spectroscopy indicated that HG-1 was folded similarly to the xylanase scaffold. To understand the

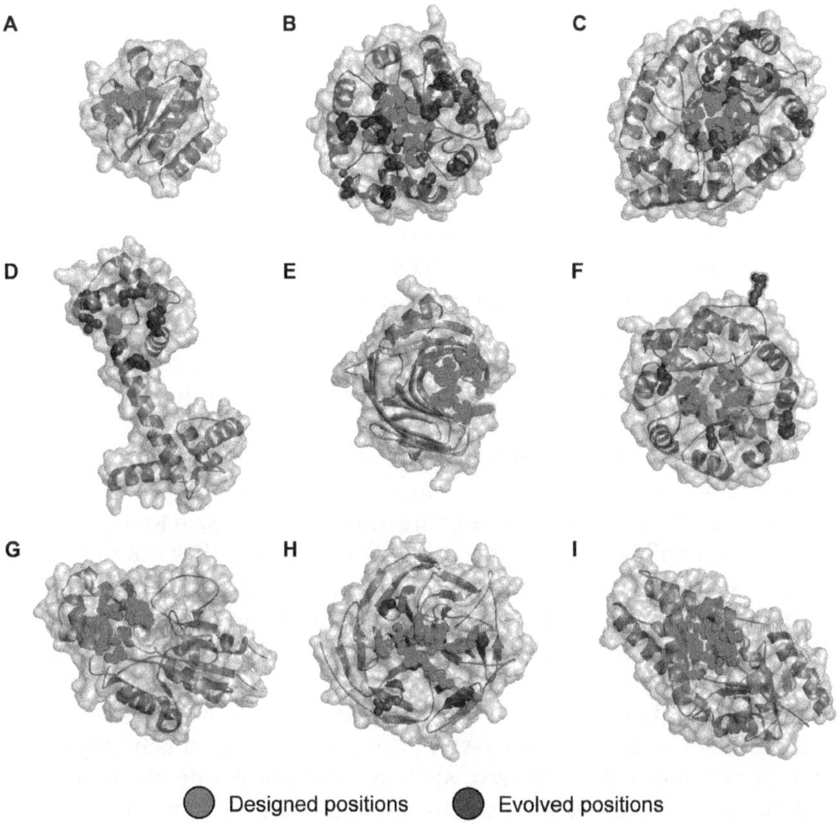

Figure 4.6 Structural models of *de novo* enzymes. In all cases, residue positions in the protein structure that were designed computationally or mutated during directed evolution are colored green or magenta, respectively. (A) The *E. coli* thioredoxin scaffold (PDB ID: 2TRX[103]) used in the design of the PDZ2 ester hydrolase. (B) The *S. solfataricus* indole-3-glycerolphosphate synthase scaffold (PDB ID: 1A53[104]) used in the design and evolution of the KE59 Kemp eliminase. (C) The *T. auranticus* xylanase scaffold (PDB ID: 1GOR[105]) used in the design and evolution of the HG-3 Kemp eliminase. (D) The *P. tetraurelia* calmodulin scaffold (PDB ID: 1EXR[106]) used for the design and evolution of the switchable Kemp eliminase AlleyCat. (E) The *N. flexuosa* β-1,4-xylanase scaffold (PDB ID: 1M4W[107]) used in the design of the RA60 retro-aldolase. (F) The *S. solfataricus* indole-3-glycerolphosphate synthase scaffold (PDB ID: 1LBL[104]) used in the design and evolution of the RA95 retro-aldolase. (G) The *H. sapiens* mitochondrial deoxyribonucleotidase scaffold (PDB ID: 1Q91[108]) used in the design of the ECH13 ester hydrolase. (H) The *L. vulgaris* diisopropylfluorophosphatase scaffold (PDB ID: 1E1A[109]) used in the design and evolution of the DA_20_10 Diels–Alderase. (I) The *P. horikoshii* haloacid dehalogenase scaffold (PDB ID: 1X42[110]) used in the design of BH32, a Morita–Baylis–Hillman catalyst.

cause of the catalytic inactivity, HG-1 was crystallized. Molecular dynamics (MD) simulations on the HG-1 crystal structure unveiled a highly mobile and solvent-exposed active site that could not bind the substrate in a catalytically productive binding pose. As the Kemp elimination is expected to be more efficient with a catalytic base present in a hydrophobic environment,[45] the authors performed a second iteration of computational enzyme design, but this time with an active site buried 7 Å deeper into the xylanase scaffold. This procedure yielded HG-2, an active variant (Table 4.1) containing 12 mutations (Table 4.2). MD simulations performed on the HG-2 crystal structure revealed both productive and unproductive substrate binding poses, suggesting a reason for the inability to achieve enzyme saturation within the solubility limit of the substrate that prevented K_M to be measured. In a third iteration, Privett *et al.* introduced into HG-2 the S265T mutation predicted by MD to stabilize the substrate in the productive binding pose. The resulting HG-3 variant (Table 4.2) could now be saturated with the substrate ($K_M = 1.6$ mM), suggesting that the S265T mutation increased productive binding. HG-3 is the most active *de novo* Kemp eliminase designed rationally to date (Table 4.1), demonstrating that the iterative approach to computational enzyme design developed by Privett *et al.* is useful for obtaining active designs and increasing the success rate.

The examples described above demonstrate that computational enzyme design can be used to create biocatalysts capable of catalyzing a reaction for which no naturally occurring enzyme has evolved. However, the 5-nitrobenzisoxazole substrate used in this model reaction is highly reactive in the presence of a base catalyst, such as amino-acid side chains found in the naturally-occurring proteins ketosteroid isomerase[45] and bovine serum albumin,[46,47] which have both been shown to catalyze the Kemp elimination. This observation led Korendovych *et al.*[48] to create a switchable Kemp eliminase out of the calcium-binding protein calmodulin by introducing a single catalytic base at the bottom of the protein's C-terminal hydrophobic cavity (Figure 4.6D), which was previously shown to bind aromatic side chains of peptides.[49] The creation of an active site in calmodulin was done using a three-step procedure consisting of (i) a CPD step to identify residue positions in the cavity that can accommodate either a glutamate or aspartate mutation without significantly destabilizing the calmodulin scaffold, (ii) a docking step to identify productive binding poses of the 5-nitrobenzisoxazole substrate on the single mutants remaining from the first step, and (iii) a computational enzyme design step where a "superrotamer" library[35,36] of a glutamate carboxylate fused to the transition state (Figure 4.5D) was optimized at each position. The top-scoring mutant, F92E, was experimentally characterized and displayed Kemp eliminase activity in the presence of 10 mM Ca^{2+}, an activity that decreased by over 25-fold in the absence of calcium. This single mutant was named AlleyCat (ALLostEricallY Controlled cATalyst) as it was the first *de novo* enzyme that is allosterically regulated. Interestingly, AlleyCat displays a catalytic efficiency (Table 4.1, k_{cat}/K_M) comparable to that of several of the lower-activity Kemp eliminases designed with the more elaborate computational approach of Röthlisberger, Khersonsky, Wollacott *et al.*,[39] suggesting that significant rate enhancements for this model reaction can be achieved with few non-specific interactions with the substrate. In the next section, examples of the use of computational enzyme design to create biocatalysts that can catalyze more challenging multi-step bond breaking reactions are presented.

4.2.2.2 Multi-step Bond Cleavage Reactions – Retro-aldol and Ester Hydrolysis Reactions

In parallel to Röthlisberger, Khersonsky, Wollacott *et al.*,[39] Jiang, Althoff *et al.*[50] used a similar procedure to design several enzymes able to catalyze the retro-aldol decomposition of 4-hydroxy-4-(6-methoxy-2-napthyl)-2-butanone into 6-methoxy-2-naphthaldehyde and acetone (Figure 4.4C). As this reaction involves multiple steps, a composite transition state that is representative of key intermediates along the reaction coordinate was constructed by combining QM models of each of these intermediates. Using the composite transition state, an ensemble of theozymes that also include a nucleophilic lysine, a base (either a polar amino acid or a water molecule), and a residue forming a π-stacking interaction (Figure 4.5E), were generated. The theozymes and composite transition state were placed on a library of 71 different scaffolds using RosettaMatch.[25] Following packing by RosettaDesign,[41] 72 designs on 10 of these scaffolds were predicted to stabilize the composite transition state, and 32 of these (44% success rate) displayed retro-aldol activity. The most active variant displayed a rate enhancement of 2.4×10^4 (Table 4.1, RA60), resulting from 12 mutations (Table 4.2, Figure 4.6E), including the predicted nucleophilic lysine (K48), base (water H-bonded to S87) and π-stacking residue (Y133). To demonstrate that computational enzyme design using RosettaMatch and RosettaDesign is robust and reproducible, Althoff, Wang, Jiang *et al.*[51] designed a new set of retro-aldolases using a similar procedure as described above, but with improved sampling of the composite transition state, catalytic residues, and binding-interaction residues. This procedure yielded 42 designs on 13 scaffolds, of which 33 (75% success) had a 10-fold greater retro-aldol activity than background. Although the rate enhancements of the most active variant RA95 (Tables 4.1 and 4.2, and Figure 4.6F) were lower than those obtained previously,[50] the increased conformational sampling led to an improved success rate, demonstrating that the computational procedure is indeed robust and reproducible.

Following these studies, Richter, Blomberg *et al.*[52] aimed to design a biocatalyst capable of hydrolyzing a tyrosyl ester (Figure 4.4D). The authors attempted to introduce a QM-derived transition state of the acylation step stabilized by a 5-residue theozyme consisting of a Cys, His, and Asn/Gln catalytic triad as well as an oxyanion hole formed by two backbone NH groups into a library of 214 protein scaffolds, but it could not be placed on any scaffold using RosettaMatch. To increase the probability of matching, various simplified 3-residue theozymes containing instead a Cys–His dyad and a single backbone NH group, water molecule, or H-bonding amino acid to stabilize the oxyanion formed during the acylation step were prepared (Figure 4.5F). Following matching and packing, 55 designs were chosen for experimental characterization, but none hydrolyzed the intended tyrosyl ester substrate. However, 4 designs hydrolyzed the more reactive, activated ester *p*-nitrophenyl-2-phenylpropanoate (Figure 4.4E). The most active design (Table 4.1, ECH13) contained 8 mutations, including the designed Cys45–His100 catalytic dyad (Table 4.2, Figure 4.6G). Although ECH13 could not turn over the target tyrosyl ester substrate, the designed oxyanion hole and catalytic dyad contributed to catalysis of *p*-nitrophenyl-2-phenylpropanoate hydrolysis, as demonstrated by mutagenesis experiments. ECH13 displayed a rate enhancement that is one order of magnitude greater than the one obtained by Bolon and Mayo[31] for a similar multi-step reaction but still 8–10 orders of magnitude lower than those of natural esterases.[37] Overall, the studies described above demonstrate that computational enzyme design can successfully be used to design novel biocatalysts for bond cleavage reactions, but that more efficient *de novo* enzymes are obtained for single-step than for multi-step processes (Figure 4.3B). An even more challenging objective is the design of enzymes catalyzing bimolecular bond-forming reactions, which are discussed in the next section.

4.2.2.3 Multi-step Bond Formation Reactions – Diels–Alder and Morita–Baylis–Hillman Reactions

Chemical reactions that form carbon–carbon bonds are important in organic chemistry as they enable the synthesis of complex molecules from simple building blocks. Thus, the *de novo* design of biocatalysts that can catalyze carbon–carbon bond-forming reactions with high stereoselectivity is very desirable. However, these reactions are challenging to design because the relative orientation of both substrates dictates the efficiency of catalysis and stereoselectivity, making the accurate placement of substrates relative to each other crucial to the success of the design procedure. In 2010, Siegel, Zanghellini *et al.*[53] reported the design of a *de novo* Diels–Alderase capable of catalyzing the [4 + 2] cycloaddition of 4-carboxybenzyl-*trans*-1,3-butadiene-1-carbamate on *N*,*N*-dimethylacrylamide (Figure 4.4F). This reaction is pericyclic, which simplifies the design procedure as a single transition state can be modeled. In this case, the *endo-re* transition state (Figure 4.5G) leading to the 3*R*,4*S* *endo* product was chosen to create an enzyme capable of selectively catalyzing the synthesis of this stereoisomer, which makes up less than half of the total product mixture obtained without catalysis. A composite transition state prepared by overlaying the QM-derived *endo-re* transition state, the two substrates bound in a transition-state-like orientation and conformation, and the 3*R*,4*S* *endo* product, was generated. Several theozymes comprising hydrogen bond acceptors and donors (Figure 4.5G) were generated to stabilize the composite transition state so as to reduce the

HOMO–LUMO gap and thus lower the activation barrier. Following matching of these theozymes and composite transition state on a large library of 207 protein scaffolds, 84 designs were selected for experimental validation.

Two designs displayed activity (2% success rate) as reported by a liquid chromatography–tandem mass spectrometry (LC–MS) assay. The most active variant (Table 4.1, DA_20_00) contained 13 mutations including the D121Y and T195Q substitutions that were predicted to hydrogen bond with the diene and dienophile (Table 4.2, Figure 4.6H). To understand the causes of the low catalytic efficiency of DA_20_00, an MD simulation was performed on the design model, which revealed that the catalytic tyrosine can adopt an alternative conformation that is not conducive to catalysis. Based on the MD results, an additional 6 mutations predicted to buttress the catalytic residues and transition state were introduced to yield the DA_20_10 mutant (Table 4.2). DA_20_10 displayed a catalytic efficiency (k_{cat}/K_M) that is approximately 100-fold greater than that of the computationally-designed DA_20_00 (Table 4.1) but that is still orders of magnitude lower than that of natural enzymes. However, DA_20_10 is highly stereoselective as the 3*R*,4*S* *endo* product now comprises >97% of the total product mixture, as revealed by chiral LC–MS. These results highlight the promise of computational enzyme design, which is to not only create *de novo* enzymes that can catalyze a desired chemical transformation with high efficiency but that also control the stereochemical outcome of a reaction.

Following the successful design of the first *de novo* enzyme able to catalyze a bimolecular reaction, Bjelic, Nivón *et al.*[54] tackled the challenging design of an enzyme that catalyzes the more complex, multi-substrate and multi-step Morita–Baylis–Hillman (MBH) reaction between 4-nitrobenzaldehyde and 2-cyclohexenone (Figure 4.4G). This carbon–carbon bond-forming reaction is of high industrial interest because it generates densely functionalized products that can be used in subsequent synthetic steps. However, it is not catalyzed by any known enzyme or catalytic antibody,[54] and existing small molecule catalysts tend to be slow, often requiring one or more days to complete the reaction in high yields.[55] As this reaction proceeds through multiple steps, Bjelic, Nivón *et al.*[54] used a composite transition state composed of overlaid QM- and MD-derived transition states and reaction intermediates. Various theozymes were prepared, which consisted of a nucleophilic cysteine activated by a lysine or a histidine held in place by an aspartate or glutamate, pairs of backbone amide and/or H-bond donor amino acids to stabilize the enolate formed from 2-cyclohexenone, aromatic amino acids to stabilize 4-nitrobenzaldehyde *via* π-stacking, and a water molecule to facilitate the final deprotonation (Figure 4.5H). Following matching of the theozymes and composite transition state on a library of 244 different scaffolds and packing by RosettaDesign,[41] 48 designs were predicted to catalyze the MBH reaction. LC–MS analysis revealed that only two of these (4% success rate) displayed any MBH activity. The most active variant, BH32, was generated from a haloacid dehalogenase scaffold (Figure 4.6I) and contains 12 mutations (Table 4.2). Following visual inspection of the BH32 crystal structure and *in silico* mutagenesis with Rosetta, several point mutations predicted to enhance activity were identified. This procedure led to the more active BH32 Q128H variant, which however displayed a rate enhancement of only 54-fold (Table 4.1). The low rate enhancement was attributed to a partially formed active site based on results obtained by microsecond MD simulations and fluorescence assays to detect formation of reaction intermediates. The low success rate and catalytic activities obtained for the MBH reaction emphasize the challenges associated with the computational design of *de novo* enzymes for complex multi-step, bond-forming reactions using current protocols (Figure 4.3B).

4.3 Challenges in Computational Enzyme Design

The examples described above demonstrate that computational enzyme design can be used to create *de novo* enzymes capable of catalyzing a variety of organic transformations. However, the success rate of the computational procedure remains low as many of the failed designs simply do not fold,[52,53] while others are folded but cannot bind the substrate.[42] Failed designs can be attributed to several approximations made in the computational methodologies to ensure tractability. For example, the protein backbone is held fixed during design, and translational, rotational, and conformational sampling of the transition state and amino-acid side chains is discretized, preventing the computational procedure from achieving the sub-Ångstrom level of precision that is required for highly efficient catalysis.[56] Additionally, potential energy functions used in computational enzyme design contain inaccuracies in their treatment of non-bonded interactions.[33,57–59] Particularly problematic is the accurate description of long-range electrostatics and atom polarization,[60] which are crucial to enzymatic catalysis.[61] These inaccuracies have thus far prevented kinetic parameters of enzymes (*i.e.*, k_{cat} and K_M) predicted by CPD to be strongly correlated with experimental values.[62]

Even when computational enzyme design yields active biocatalysts, catalytic efficiencies (k_{cat}/K_M) obtained are orders of magnitude lower than those of natural enzymes.[1] These modest catalytic efficiencies are mostly due to k_{cat} values that are 1–6 orders of magnitude lower than those of natural enzymes,[1,63] and not to Michaelis constants (K_M), which fall within the range of those found in natural biocatalysts (Table 4.1). This result demonstrates the ability of computational enzyme design to create active sites capable of productively binding a substrate but that this does not ensure highly efficient catalysis. Indeed, enzymatic reaction rates are determined by the rate-limiting step of the complete catalytic process, which may not necessarily be the chemical transformation that is modeled with a transition state and theozyme during design. For example, it has been shown that for many enzymes, product release[64] or conformational exchange between various sub-states during the catalytic cycle[65] are rate-limiting. It is thus possible that such processes slow down the rate of reactions catalyzed by *de novo* enzymes.

Because of the difficulties in designing *de novo* enzymes displaying catalytic efficiencies comparable to those of natural enzymes, many of these biocatalysts were subjected to directed evolution[66] in order to improve their activity. Kemp eliminases,[39,67–71] retro-aldolases,[51,72] and Diels–Alderases[73] have all been enhanced following multiple rounds of directed evolution, which yielded improved variants containing many additional mutations (Table 4.2) that increase catalytic efficiencies to the level found in natural enzymes (Table 4.1). Interestingly, significant improvements to k_{cat} were obtained in all cases while K_M remained mostly unaffected, suggesting that the additional mutations found during the evolution process contribute to rate enhancement without significantly affecting productive binding of the substrate. In all but two cases,[71,72] the mutations introduced by the evolution process did not significantly remodel the active site or overall structure of the enzyme and instead created a widened active site entrance[68] or new oxyanion hole,[69] improved shape complementarity of the active site to the substrate,[71] or eliminated unproductive binding poses.[69,70] Unsurprisingly, most of the beneficial mutations obtained by directed evolution are far from the active site (Figure 4.6), making their prediction by computational enzyme design very challenging. Typically, designed residue positions in the protein are restricted to those in the active site. Yet, distant mutations have been shown to have a large impact on the catalytic properties of many enzymes,[74,75] making their prediction by computational enzyme design essential if we are to create highly active enzymes.

4.4 Future Directions

In order to improve the success rate of computational enzyme design, continued development of CPD force fields and sampling methodologies will be vital. Recently developed electrostatic potential energy terms[76] as well as polarizable force fields[77,78] could be incorporated into future computational enzyme design methodologies to better model electron distribution during the catalytic step. Use of flexible backbone design algorithms,[79–82] as well as continuous rotamer[83,84] and backbone[85] flexibility may lead to improved sampling of the protein scaffold during design. As well, methods for multistate design (MSD) with large structural ensembles[86,87] may contribute to increase the success rate of computational enzyme design by helping to address the fixed backbone approximation.[88] In addition to improvements to the computational methodology, it will be crucial for researchers to optimize residue positions distant to the active site during design. However, it remains very difficult to know *a priori* which positions to consider. Potential computational approaches to identify designable distal sites include a self-consistent mean field entropy-based method,[89] which identifies residue positions in the protein that may be tolerant to mutation, and the Monte Carlo side chain entropy method,[90] which could be used to calculate side-chain fluctuations and their coupling across residues in order to identify distal sites that can affect catalysis.

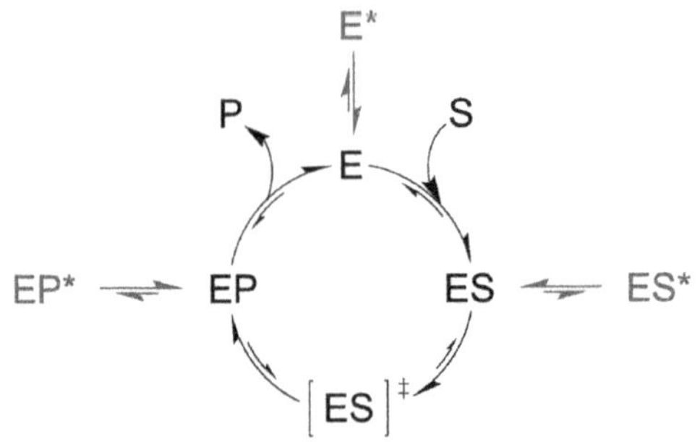

Figure 4.7 States adopted by an enzyme as it undergoes its catalytic cycle. As an enzyme (E) undergoes its catalytic cycle, it must bind the substrate (S) to form the Michaelis complex (ES), leading to catalysis by transition state stabilization ($[ES]^{\ddagger}$). Following the reaction, the enzyme-product complex is formed (EP), and product (P) is released to regenerate the free enzyme. Potential competing states in grey include unproductive conformations of the enzyme (E*) and unproductive binding poses of the substrate or product to the enzyme (ES* and EP*, respectively).

Ultimately, the design of *de novo* enzymes with catalytic efficiencies matching those of natural enzymes will likely require a holistic view of enzymatic catalysis where every possible state that the enzyme adopts as it undergoes its catalytic cycle (Figure 4.7) is included in the calculation. This could be achieved by the implementation of "multistate" approaches[91–94] in computational enzyme design, which to the best of our knowledge have not yet been developed. In a multistate computational enzyme design process, a minimum of four states should be considered for the creation of biocatalysts catalyzing simple, single-step mechanisms: the free enzyme, the Michaelis complex, the enzyme-bound transition state intermediate, and the product-bound state. Consideration of these states can ensure that mutations designed to create the active site and stabilize the transition state do not inadvertently result in over-stabilization of the enzyme–product intermediate, leading to decreased rate due to a slow product release step, or over-packing of the transition state, which may reduce catalytic efficiency by creating a high-energy barrier preventing substrate entry and product exit into and from the active site (Figure 4.8). Thus, an ideal multistate computational enzyme design calculation should aim to (i) facilitate substrate entry and binding in the active site, (ii) destabilize the substrate ground state by active site preorganization, (iii) stabilize the transition state, and (iv) destabilize the enzyme–product complex to facilitate product release (Figure 4.9). Achieving this delicate balance between the relative energies of these states could theoretically lead to more catalytically efficient *de novo* enzymes in the future.

In addition to the four chemical states described above, alternate conformations of the unbound enzyme need also be considered, as enzymes are intrinsically dynamic molecules that can sample numerous conformations.[65,95] The ability of enzymes to exchange between multiple

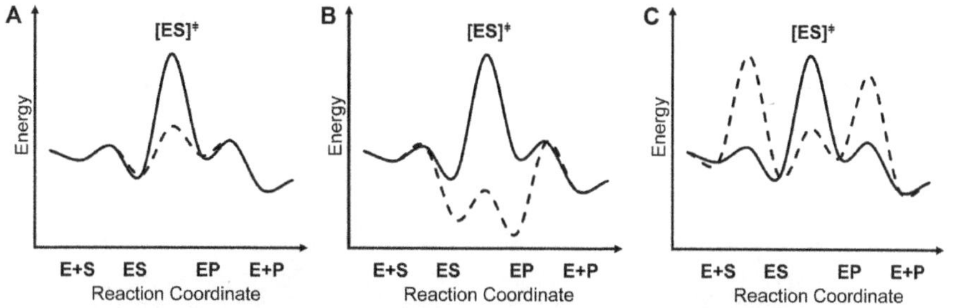

Figure 4.8 Hypothetical consequences of transition state stabilization by computational enzyme design. In all cases, reaction coordinates before and after transition state stabilization are shown as full or dashed lines, respectively. (A) Transition state stabilization by computational enzyme design aims to lower the energy barrier of the rate-limiting step and thus increase the rate of the reaction. However, transition state stabilization may also result in the unintended over-stabilization of the enzyme–substrate and enzyme–product complexes (B), which would result in a rate-limiting product-releasing step that would decrease catalytic efficiency. Alternatively, the computational procedure may result in transition state over-packing (C), which could create a high barrier for substrate entry in the active site or product release and thereby decrease catalytic efficiency.

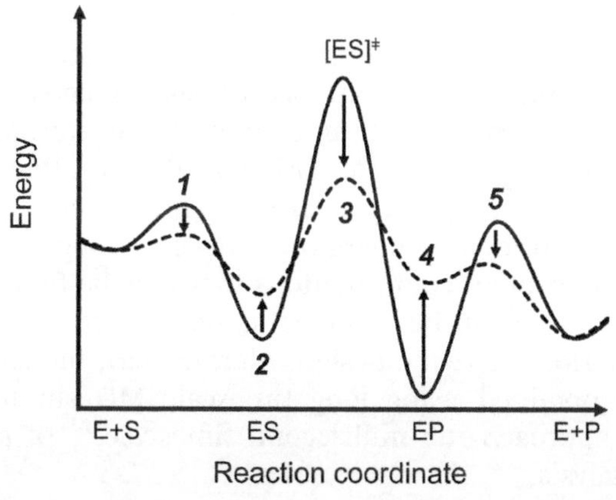

Figure 4.9 Multistate computational enzyme design. The multistate computational design procedure should aim to (1) facilitate substrate entry and binding in the active site, (2) destabilize the substrate ground state by active site preorganization, (3) stabilize the transition state, and (4) destabilize the enzyme–product complex to (5) facilitate product release. Full and dashed lines show theoretical reaction coordinates of low and high efficiency enzymes, respectively.

conformations is vital to catalysis,[96] as different conformational sub-states are needed to bind the substrate(s), stabilize the transition state, and release the product(s). However, dynamics can also be detrimental to catalysis, as motions that cause sampling of unproductive conformations can limit the

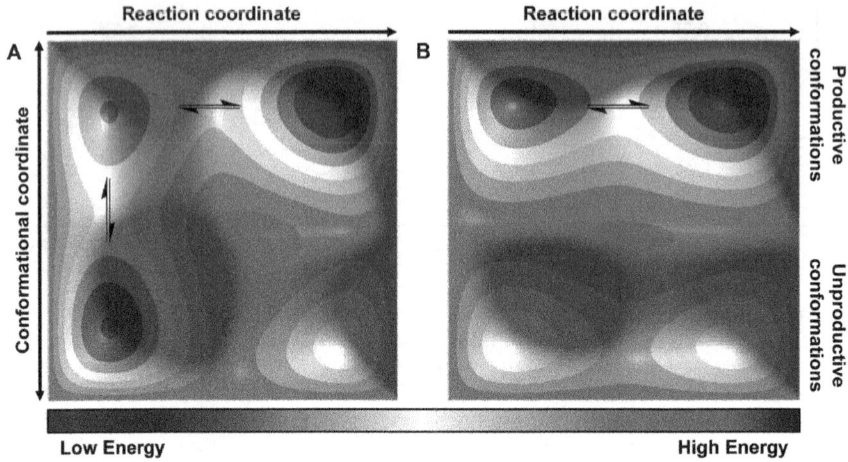

Figure 4.10 Potential energy surfaces of hypothetical enzymes. An enzyme that exchanges between productive and unproductive conformations (A) is expected to be less catalytically efficient than an enzyme that cannot exchange to catalytically unproductive conformations (B).

rate of enzyme turnover.[97,98] For example, an enzyme that exchanges between catalytically productive and unproductive conformations would be less active than an enzyme where unproductive conformations are inaccessible (Figure 4.10). Thus, the multistate computational enzyme design procedure should aim to eliminate sequences predicted to stabilize unproductive conformations, which can be done by negative MSD.[86] However, the structural features of the conformational sub-states that the enzyme can adopt throughout the catalytic cycle are difficult to define. Fortunately, recent developments in room temperature X-ray crystallography have revealed hidden alternative protein conformations that may be functionally relevant to catalysis, and these can be modeled as discrete structures[95,99,100] to be used in computational enzyme design. Alternatively, models of such structures could be obtained using long timescale MD simulations,[97] which are starting to approach the millisecond timescale[101] of protein motions pertinent to catalysis.[102]

4.5 Conclusions

Over the past two decades, the development and application of computational enzyme design methodologies has led to the successful creation of several *de novo* enzymes capable of catalyzing a variety of bond-breaking and bond-forming reactions. Although none of these designed biocatalysts are as active as natural enzymes, they have served as excellent starting points for improvement by directed evolution, leading to the creation of highly efficient biocatalysts able to catalyze chemical reactions for which no known natural enzyme has been identified. However, continued development of computational enzyme design methodologies will be essential to realize the on-demand design of enzymes that can catalyze any desired chemical reaction with high efficiency and selectivity under defined conditions, and without the need to rely on directed evolution. We expect that fulfilment of this long-term objective will require the design of complete enzymatic catalytic cycles for reactions of interest. This objective may be tackled following the development and validation of multistate approaches to computational enzyme design, which could open the door to the next phase of *de novo* biocatalyst design.

Acknowledgements

We gratefully acknowledge support from the Natural Sciences and Engineering Research Council of Canada (NSERC) through Discovery Grant RGPIN-2016-04831 (R.A.C.). O.G. is the recipient of an Ontario Graduate Scholarship and a Master's scholarship from the *Fonds de recherche du Québec – Nature et technologies (FRQNT).*

References

1. R. Wolfenden and M. J. Snider, *Acc. Chem. Res.*, 2001, **34**, 938.
2. D. R. Edwards, D. C. Lohman and R. Wolfenden, *J. Am. Chem. Soc.*, 2012, **134**, 525.
3. R. D. Gupta, M. Goldsmith, Y. Ashani, Y. Simo, G. Mullokandov, H. Bar, M. Ben-David, H. Leader, R. Margalit, I. Silman, J. L. Sussman and D. S. Tawfik, *Nat. Chem. Biol.*, 2011, 7, 120.
4. U. T. Bornscheuer and M. Pohl, *Curr. Opin. Chem. Biol.*, 2001, **5**, 137.
5. J. J. Onuffer and J. F. Kirsch, *Protein Sci.*, 1995, **4**, 1750.
6. A. Zanghellini, *Curr. Opin. Biotechnol.*, 2014, **29**, 132.
7. G. Kiss, N. Celebi-Olcum, R. Moretti, D. Baker and K. N. Houk, *Angew. Chem., Int. Ed. Engl.*, 2013, **52**, 5700.
8. B. I. Dahiyat and S. L. Mayo, *Protein Sci.*, 1996, **5**, 895.
9. K. T. Simons, C. Kooperberg, E. Huang and D. Baker, *J. Mol. Biol.*, 1997, **268**, 209.
10. S. M. Malakauskas and S. L. Mayo, *Nat. Struct. Biol.*, 1998, **5**, 470.
11. M. D. Altman, E. A. Nalivaika, M. Prabu-Jeyabalan, C. A. Schiffer and B. Tidor, *Proteins*, 2008, **70**, 678.
12. J. Ashworth, J. J. Havranek, C. M. Duarte, D. Sussman, R. J. Monnat, Jr., B. L. Stoddard and D. Baker, *Nature*, 2006, **441**, 656.
13. R. L. Dunbrack, Jr., *Curr. Opin. Struct. Biol.*, 2002, **12**, 431.
14. A. G. Street and S. L. Mayo, *Structure*, 1999, 7, R105.
15. F. E. Boas and P. B. Harbury, *J. Mol. Biol.*, 2008, **380**, 415.
16. M. Schmidt Am Busch, A. Lopes, D. Mignon and T. Simonson, *J. Comput. Chem.*, 2008, **29**, 1092.
17. C. A. Rohl and D. Baker, *J. Am. Chem. Soc.*, 2002, **124**, 2723.

18. J. Desmet, M. De Maeyer, B. Hazes and I. Lasters, *Nature*, 1992, **356**, 539.
19. P. Koehl and M. Delarue, *J. Mol. Biol.*, 1994, **239**, 249.
20. P. Koehl and M. Delarue, *Nat. Struct. Biol.*, 1995, **2**, 163.
21. J. Desmet, J. Spriet and I. Lasters, *Proteins*, 2002, **48**, 31.
22. B. D. Allen and S. L. Mayo, *J. Comput. Chem.*, 2006, **27**, 1071.
23. B. Kuhlman and D. Baker, *Proc. Natl. Acad. Sci. U. S. A.*, 2000, **97**, 10383.
24. J. K. Lassila, H. K. Privett, B. D. Allen and S. L. Mayo, *Proc. Natl. Acad. Sci. U. S. A.*, 2006, **103**, 16710.
25. A. Zanghellini, L. Jiang, A. M. Wollacott, G. Cheng, J. Meiler, E. A. Althoff, D. Rothlisberger and D. Baker, *Protein Sci.*, 2006, **15**, 2785.
26. T. Sasaki and E. T. Kaiser, *J. Am. Chem. Soc.*, 1989, **111**, 380.
27. K. Johnsson, R. K. Allemann, H. Widmer and S. A. Benner, *Nature*, 1993, **365**, 530.
28. H. W. Hellinga and F. M. Richards, *Proc. Natl. Acad. Sci. U. S. A.*, 1994, **91**, 5803.
29. A. L. Pinto, H. W. Hellinga and J. P. Caradonna, *Proc. Natl. Acad. Sci. U. S. A.*, 1997, **94**, 5562.
30. D. E. Benson, M. S. Wisz and H. W. Hellinga, *Proc. Natl. Acad. Sci. U. S. A.*, 2000, **97**, 6292.
31. D. N. Bolon and S. L. Mayo, *Proc. Natl. Acad. Sci. U. S. A.*, 2001, **98**, 14274.
32. H. Gutfreund and J. M. Sturtevant, *Biochem. J.*, 1956, **63**, 656.
33. S. L. Mayo, B. D. Olafson and W. A. Goddard, *J. Phys. Chem.*, 1990, **94**, 8897.
34. B. I. Dahiyat and S. L. Mayo, *Science*, 1997, **278**, 82.
35. L. Regan and N. D. Clarke, *Biochemistry*, 1990, **29**, 10878.
36. N. D. Clarke and S. M. Yuan, *Proteins*, 1995, **23**, 256.
37. R. Wolfenden, C. A. Lewis, Jr. and Y. Yuan, *J. Am. Chem. Soc.*, 2011, **133**, 5683.
38. D. J. Tantillo, J. Chen and K. N. Houk, *Curr. Opin. Chem. Biol.*, 1998, **2**, 743.
39. D. Röthlisberger, O. Khersonsky, A. M. Wollacott, L. Jiang, J. DeChancie, J. Betker, J. L. Gallaher, E. A. Althoff, A. Zanghellini, O. Dym, S. Albeck, K. N. Houk, D. S. Tawfik and D. Baker, *Nature*, 2008, **453**, 190.
40. K. M. Misura, A. V. Morozov and D. Baker, *J. Mol. Biol.*, 2004, **342**, 651.
41. B. Kuhlman, G. Dantas, G. C. Ireton, G. Varani, B. L. Stoddard and D. Baker, *Science*, 2003, **302**, 1364.
42. H. K. Privett, G. Kiss, T. M. Lee, R. Blomberg, R. A. Chica, L. M. Thomas, D. Hilvert, K. N. Houk and S. L. Mayo, *Proc. Natl. Acad. Sci. U. S. A.*, 2012, **109**, 3790.
43. B. D. Allen and S. L. Mayo, *J. Comput. Chem.*, 2010, **31**, 904.
44. R. A. Chica, M. M. Moore, B. D. Allen and S. L. Mayo, *Proc. Natl. Acad. Sci. U. S. A.*, 2010, **107**, 20257.
45. V. Lamba, E. Sanchez, L. R. Fanning, K. Howe, M. A. Alvarez, D. Herschlag and M. Forconi, *Biochemistry*, 2017, **56**, 582.

46. F. Hollfelder, A. J. Kirby and D. S. Tawfik, *Nature*, 1996, **383**, 60.
47. K. Kikuchi, S. N. Thorn and D. Hilvert, *J. Am. Chem. Soc.*, 1996, **118**, 8184.
48. I. V. Korendovych, D. W. Kulp, Y. Wu, H. Cheng, H. Roder and W. F. DeGrado, *Proc. Natl. Acad. Sci. U. S. A.*, 2011, **108**, 6823.
49. J. L. Urbauer, J. H. Short, L. K. Dow and A. J. Wand, *Biochemistry*, 1995, **34**, 8099.
50. L. Jiang, E. A. Althoff, F. R. Clemente, L. Doyle, D. Rothlisberger, A. Zanghellini, J. L. Gallaher, J. L. Betker, F. Tanaka, C. F. Barbas, 3rd, D. Hilvert, K. N. Houk, B. L. Stoddard and D. Baker, *Science*, 2008, **319**, 1387.
51. E. A. Althoff, L. Wang, L. Jiang, L. Giger, J. K. Lassila, Z. Wang, M. Smith, S. Hari, P. Kast, D. Herschlag, D. Hilvert and D. Baker, *Protein Sci.*, 2012, **21**, 717.
52. F. Richter, R. Blomberg, S. D. Khare, G. Kiss, A. P. Kuzin, A. J. Smith, J. Gallaher, Z. Pianowski, R. C. Helgeson, A. Grjasnow, R. Xiao, J. Seetharaman, M. Su, S. Vorobiev, S. Lew, F. Forouhar, G. J. Kornhaber, J. F. Hunt, G. T. Montelione, L. Tong, K. N. Houk, D. Hilvert and D. Baker, *J. Am. Chem. Soc.*, 2012, **134**, 16197.
53. J. B. Siegel, A. Zanghellini, H. M. Lovick, G. Kiss, A. R. Lambert, J. L. St Clair, J. L. Gallaher, D. Hilvert, M. H. Gelb, B. L. Stoddard, K. N. Houk, F. E. Michael and D. Baker, *Science*, 2010, **329**, 309.
54. S. Bjelic, L. G. Nivón, N. Celebi-Olcum, G. Kiss, C. F. Rosewall, H. M. Lovick, E. L. Ingalls, J. L. Gallaher, J. Seetharaman, S. Lew, G. T. Montelione, J. F. Hunt, F. E. Michael, K. N. Houk and D. Baker, *ACS Chem. Biol.*, 2013, **8**, 749.
55. D. Basavaiah, A. J. Rao and T. Satyanarayana, *Chem. Rev.*, 2003, **103**, 811.
56. P. A. Sigala, D. A. Kraut, J. M. M. Caaveiro, B. Pybus, E. A. Ruben, D. Ringe, G. A. Petsko and D. Herschlag, *J. Am. Chem. Soc.*, 2008, **130**, 13696.
57. O. Alvizo and S. L. Mayo, *Proc. Natl. Acad. Sci. U. S. A.*, 2008, **105**, 12242.
58. Y. Song, M. Tyka, A. Leaver-Fay, J. Thompson and D. Baker, *Proteins*, 2011, **79**, 1898.
59. A. Leaver-Fay, M. J. O'Meara, M. Tyka, R. Jacak, Y. Song, E. H. Kellogg, J. Thompson, I. W. Davis, R. A. Pache, S. Lyskov, J. J. Gray, T. Kortemme, J. S. Richardson, J. J. Havranek, J. Snoeyink, D. Baker and B. Kuhlman, *Methods Enzymol.*, 2013, **523**, 109.
60. K. Vanommeslaeghe and A. D. MacKerell, Jr., *Biochim. Biophys. Acta*, 2015, **1850**, 861.
61. A. Warshel, P. K. Sharma, M. Kato, Y. Xiang, H. Liu and M. H. Olsson, *Chem. Rev.*, 2006, **106**, 3210.
62. D. A. Carlin, R. W. Caster, X. Wang, S. A. Betzenderfer, C. X. Chen, V. M. Duong, C. V. Ryklansky, A. Alpekin, N. Beaumont, H. Kapoor, N. Kim, H. Mohabbot, B. Pang, R. Teel, L. Whithaus, I. Tagkopoulos and J. B. Siegel, *PLoS One*, 2016, **11**, e0147596.

63. A. Bar-Even, E. Noor, Y. Savir, W. Liebermeister, D. Davidi, D. S. Tawfik and R. Milo, *Biochemistry*, 2011, **50**, 4402.
64. W. W. Cleland, *Acc. Chem. Res.*, 1975, **8**, 145.
65. S. J. Kerns, R. V. Agafonov, Y. J. Cho, F. Pontiggia, R. Otten, D. V. Pachov, S. Kutter, L. A. Phung, P. N. Murphy, V. Thai, T. Alber, M. F. Hagan and D. Kern, *Nat. Struct. Mol. Biol.*, 2015, **22**, 124.
66. F. H. Arnold and A. A. Volkov, *Curr. Opin. Chem. Biol.*, 1999, **3**, 54.
67. O. Khersonsky, D. Rothlisberger, O. Dym, S. Albeck, C. J. Jackson, D. Baker and D. S. Tawfik, *J. Mol. Biol.*, 2010, **396**, 1025.
68. O. Khersonsky, G. Kiss, D. Rothlisberger, O. Dym, S. Albeck, K. N. Houk, D. Baker and D. S. Tawfik, *Proc. Natl. Acad. Sci. U. S. A.*, 2012, **109**, 10358.
69. R. Blomberg, H. Kries, D. M. Pinkas, P. R. Mittl, M. G. Grutter, H. K. Privett, S. L. Mayo and D. Hilvert, *Nature*, 2013, **503**, 418.
70. O. V. Moroz, Y. S. Moroz, Y. Wu, A. B. Olsen, H. Cheng, K. L. Mack, J. M. McLaughlin, E. A. Raymond, K. Zhezherya, H. Roder and I. V. Korendovych, *Angew. Chem., Int. Ed. Engl.*, 2013, **52**, 6246.
71. O. Khersonsky, D. Rothlisberger, A. M. Wollacott, P. Murphy, O. Dym, S. Albeck, G. Kiss, K. N. Houk, D. Baker and D. S. Tawfik, *J. Mol. Biol.*, 2011, **407**, 391.
72. L. Giger, S. Caner, R. Obexer, P. Kast, D. Baker, N. Ban and D. Hilvert, *Nat. Chem. Biol.*, 2013, **9**, 494.
73. N. Preiswerk, T. Beck, J. D. Schulz, P. Milovnik, C. Mayer, J. B. Siegel, D. Baker and D. Hilvert, *Proc. Natl. Acad. Sci. U. S. A.*, 2014, **111**, 8013.
74. D. Gagne and N. Doucet, *FEBS J.*, 2013, **280**, 5596.
75. D. Gagne, R. L. French, C. Narayanan, M. Simonovic, P. K. Agarwal and N. Doucet, *Structure*, 2015, **23**, 2256.
76. M. J. O'Meara, A. Leaver-Fay, M. D. Tyka, A. Stein, K. Houlihan, F. DiMaio, P. Bradley, T. Kortemme, D. Baker, J. Snoeyink and B. Kuhlman, *J. Chem. Theory Comput.*, 2015, **11**, 609.
77. S. D. LuCore, J. M. Litman, K. T. Powers, S. Gao, A. M. Lynn, W. T. Tollefson, T. D. Fenn, M. T. Washington and M. J. Schnieders, *Biophys. J.*, 2015, **109**, 816.
78. J. W. Ponder, C. Wu, P. Ren, V. S. Pande, J. D. Chodera, M. J. Schnieders, I. Haque, D. L. Mobley, D. S. Lambrecht, R. A. DiStasio, Jr., M. Head-Gordon, G. N. Clark, M. E. Johnson and T. Head-Gordon, *J. Phys. Chem. B*, 2010, **114**, 2549.
79. J. J. Havranek and D. Baker, *Protein Sci.*, 2009, **18**, 1293.
80. P. S. Huang, Y. E. Ban, F. Richter, I. Andre, R. Vernon, W. R. Schief and D. Baker, *PLoS One*, 2011, **6**, e24109.
81. G. D. Friedland, A. J. Linares, C. A. Smith and T. Kortemme, *J. Mol. Biol.*, 2008, **380**, 757.
82. M. Babor, D. J. Mandell and T. Kortemme, *Protein Sci.*, 2011, **20**, 1082.
83. P. Gainza, K. E. Roberts and B. R. Donald, *PLoS Comput. Biol.*, 2012, **8**, e1002335.
84. K. E. Roberts and B. R. Donald, *Proteins*, 2015, **83**, 1151.

85. M. A. Hallen, D. A. Keedy and B. R. Donald, *Proteins*, 2013, **81**, 18.
86. J. A. Davey, A. M. Damry, C. K. Euler, N. K. Goto and R. A. Chica, *Structure*, 2015, **23**, 2011.
87. B. D. Allen, A. Nisthal and S. L. Mayo, *Proc. Natl. Acad. Sci. U. S. A.*, 2010, **107**, 19838.
88. J. A. Davey and R. A. Chica, *Proteins*, 2014, **82**, 771.
89. C. A. Voigt, S. L. Mayo, F. H. Arnold and Z. G. Wang, *Proc. Natl. Acad. Sci. U. S. A.*, 2001, **98**, 3778.
90. A. Bhowmick and T. Head-Gordon, *Structure*, 2015, **23**, 44.
91. J. A. Davey and R. A. Chica, *Protein Sci.*, 2012, **21**, 1241.
92. A. Leaver-Fay, R. Jacak, P. B. Stranges and B. Kuhlman, *PLoS One*, 2011, **6**, e20937.
93. A. M. Sevy, T. M. Jacobs, J. E. Crowe, Jr. and J. Meiler, *PLoS Comput. Biol.*, 2015, **11**, e1004300.
94. S. C. Howell, K. K. Inampudi, D. P. Bean and C. J. Wilson, *Structure*, 2014, **22**, 218.
95. D. A. Keedy, L. R. Kenner, M. Warkentin, R. A. Woldeyes, J. B. Hopkins, M. C. Thompson, A. S. Brewster, A. H. Van Benschoten, E. L. Baxter, M. Uervirojnangkoorn, S. E. McPhillips, J. Song, R. Alonso-Mori, J. M. Holton, W. I. Weis, A. T. Brunger, S. M. Soltis, H. Lemke, A. Gonzalez, N. K. Sauter, A. E. Cohen, H. van den Bedem, R. E. Thorne and J. S. Fraser, *Elife*, 2015, **4**, e07574.
96. D. D. Boehr, D. McElheny, H. J. Dyson and P. E. Wright, *Science*, 2006, **313**, 1638.
97. G. Jimenez-Oses, S. Osuna, X. Gao, M. R. Sawaya, L. Gilson, S. J. Collier, G. W. Huisman, T. O. Yeates, Y. Tang and K. N. Houk, *Nat. Chem. Biol.*, 2014, **10**, 431.
98. E. Campbell, M. Kaltenbach, G. J. Correy, P. D. Carr, B. T. Porebski, E. K. Livingstone, L. Afriat-Jurnou, A. M. Buckle, M. Weik, F. Hollfelder, N. Tokuriki and C. J. Jackson, *Nat. Chem. Biol.*, 2016, **12**, 944.
99. H. van den Bedem, A. Dhanik, J. C. Latombe and A. M. Deacon, *Acta Crystallogr.*, 2009, **D65**, 1107.
100. P. T. Lang, H. L. Ng, J. S. Fraser, J. E. Corn, N. Echols, M. Sales, J. M. Holton and T. Alber, *Protein Sci.*, 2010, **19**, 1420.
101. K. Lindorff-Larsen, P. Maragakis, S. Piana and D. E. Shaw, *J. Phys. Chem. B*, 2016, **120**, 8313.
102. G. Bhabha, J. Lee, D. C. Ekiert, J. Gam, I. A. Wilson, H. J. Dyson, S. J. Benkovic and P. E. Wright, *Science*, 2011, **332**, 234.
103. S. K. Katti, D. M. LeMaster and H. Eklund, *J. Mol. Biol.*, 1990, **212**, 167.
104. M. Hennig, B. D. Darimont, J. N. Jansonius and K. Kirschner, *J. Mol. Biol.*, 2002, **319**, 757.
105. L. Lo Leggio, S. Kalogiannis, K. Eckert, S. C. Teixeira, M. K. Bhat, C. Andrei, R. W. Pickersgill and S. Larsen, *FEBS Lett.*, 2001, **509**, 303.
106. M. A. Wilson and A. T. Brunger, *J. Mol. Biol.*, 2000, **301**, 1237.

107. N. Hakulinen, O. Turunen, J. Janis, M. Leisola and J. Rouvinen, *Eur. J. Biochem.*, 2003, **270**, 1399.
108. A. Rinaldo-Matthis, C. Rampazzo, J. Balzarini, P. Reichard, V. Bianchi and P. Nordlund, *Mol. Pharmacol.*, 2004, **65**, 860.
109. E. I. Scharff, J. Koepke, G. Fritzsch, C. Lucke and H. Ruterjans, *Structure*, 2001, **9**, 493.
110. R. Arai, M. Kukimoto-Niino, C. Kuroishi, Y. Bessho, M. Shirouzu and S. Yokoyama, *Protein Sci.*, 2006, **15**, 373.

Section II: Understanding and Engineering Enzymes

CHAPTER 5

Computational Techniques for Efficient Biocatalysis

IONNIS V. PAVLIDIS,[a] NATALIE M. HENDRIKSE[b,c] AND PER-OLOF SYRÉN*[b,c]

[a] University of Crete, Department of Chemistry, Enzyme Technology Laboratory, Voutes University Campus, Heraklion, 70013, Greece; [b] KTH Royal Institute of Technology, School of Chemical Science and Engineering, Teknikringen 30, Stockholm, 100 44, Sweden; [c] KTH Royal Institute of Technology, Science for Life Laboratory, Tomtebodavägen 23, Solna, 171 65, Sweden

*Email: per-olof.syren@biotech.kth.se

5.1 Introduction to Computational Biocatalysis

Mild and sustainable generation of chemicals and polymeric materials[1] by the biocatalytic upcycling[2] of renewable synthons requires an expansion of the current available biochemical reaction space.[3] Affording biocatalysts harboring novel chemistries – whether inside or outside the cell[4] – is dependent on the discovery of novel enzymes[5] and metabolic pathways,[6] together with the *de novo* design of enzymes[7] and directed evolution.[3]

Evolution can be considered to be an endless protein engineering experiment for which the impact of introduced mutations on organism fitness is constantly evaluated.[8] The outcome of this process, for instance accelerated promiscuous activities for the development of new metabolic functions,[9] is submerged in the vast amount of protein sequences that we observe today. During the last decade, the astonishing development of high-throughput biotechnology methods has allowed approximately 135 million sequences to

Catalysis Series No. 32
Modern Biocatalysis: Advances Towards Synthetic Biological Systems
Edited by Gavin Williams and Mélanie Hall

Published by the Royal Society of Chemistry, www.rsc.org

be collected in GenBank that now contains close to 200 million sequences.[10] On the protein level, the UniProt database[11] encompasses close to 70 million entries, out of which roughly 39 000 distinct protein sequences have a crystal structure deposited in the Protein Data Bank (PDB).[12] Computational methods can take advantage of the rich and available sequence information for the discovery (see Chapter 1) and engineering of biocatalysts with

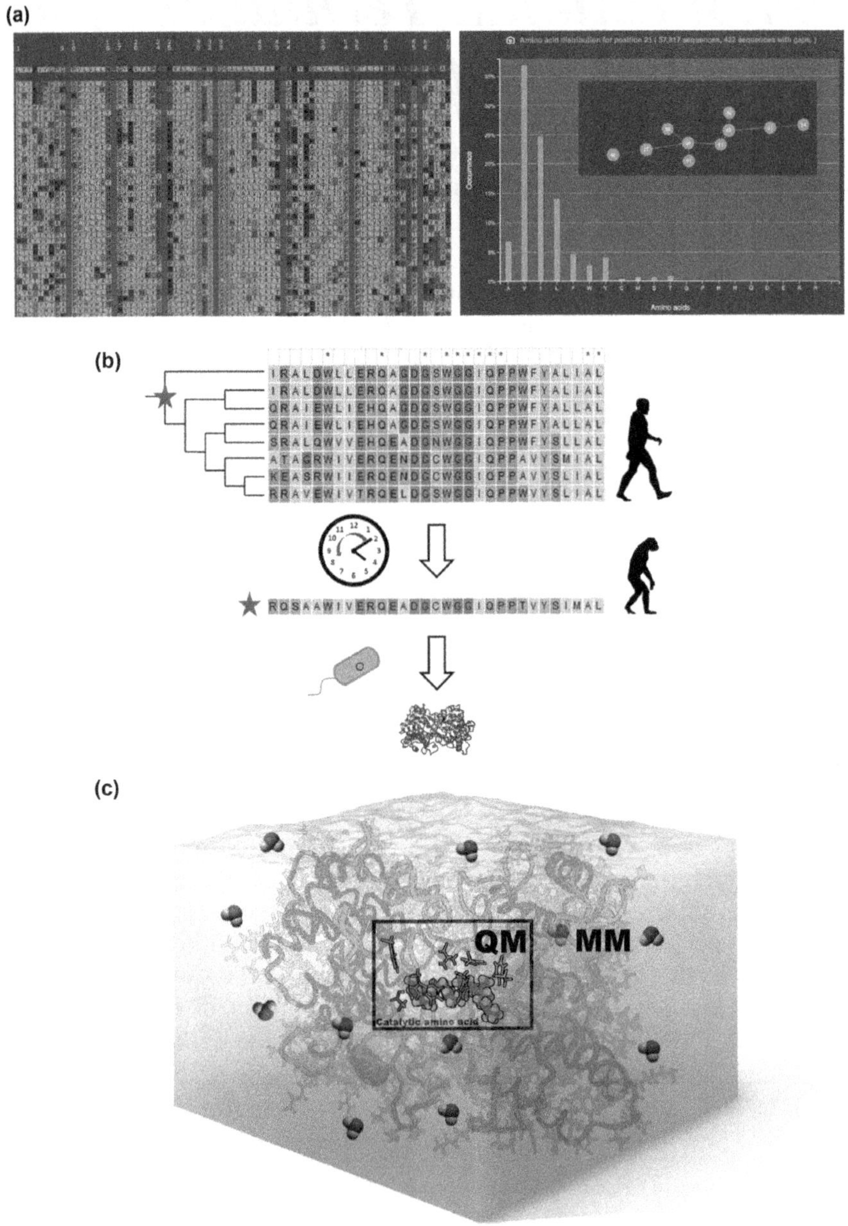

increased catalytic versatility (Figure 5.1a). Indeed, massive functional analysis of sequence space[13] can reveal how biocatalysis is governed by specific amino acid patterns in proteins without the need for a crystal structure. Bioinformatic sequence-inferred methods can further be exploited to create enzymes solely based on sequence data (Figure 5.1b). The availability of a protein crystal structure, or that of a close homolog, allows for *in silico* computer modelling of the biocatalytic transformation (Figure 5.1c) and the associated reaction mechanism. This constitutes a cornerstone for the identification of "hot-spot" residues that can be targeted by (semi-) rational design and directed evolution to obtain improved biocatalysts and new reactivities that were not yet fully developed in nature.[3]

In this chapter, we describe the basis of present and emerging sequence- and structure-based computational methods, with a focus on bioinformatics, quantum mechanics and molecular dynamics simulations. Moreover, their successful implementation in biocatalysis will be highlighted. We will emphasize how sequence-based bioinformatic tools in concert with *in silico* structure-based molecular modelling provides a framework towards a better understanding of enzymes and their reaction mechanisms. The emerging topic of *de novo* enzyme design is discussed in other chapters of the book (see in particular Chapter 4).

Figure 5.1 Current and emerging sequence- and structure-based protein engineering methods. (a) Analysis of the natural variability of structurally and/or functionally related residues can guide protein engineering efforts (left). For instance, the least represented amino acids are potentially discarded *via* natural evolutionary mechanisms and could thus be avoided when performing multiple-site mutagenesis to minimize the size of libraries. Data from sequence alignment analysis can also serve as a foundation to discover correlated mutations and their interaction networks (right). (b) In ancestral gene resurrection, a phylogenetic tree is constructed based on aligned sequences of present day enzymes. Different statistical methods, such as Maximum Likelihood (ML), Parsimony or Neighbor Joining, can then be used to infer the sequences at branch points in the tree, which represent putative ancestral proteins. The reconstructed ancestors may be expressed, purified and experimentally characterized to study the functional evolution of the protein of interest. (c) Molecular modelling founded on a protein crystal structure, or that of a close homolog, can be based on different levels of theory. Computationally demanding quantum mechanical (QM) calculations allows for a detailed study of transition state structures, intermediates and their energies which are usually studied in a limited part of the enzyme structure (represented by the QM region, box). The QM region typically comprises the substrate (shown in balls), amino acids directly participating in biocatalysis (labeled) and surrounding residues (highlighted). In QM/MM hybrid methods, Newtonian-based molecular mechanics (MM) algorithms are used to account for the influence of the remaining part of the protein (*i.e.* the MM-region, shaded) on catalysis. Optionally, the biocatalytic simulation can be represented by integrating Newton's second law of motion, which is the basis of molecular dynamics (MD) simulations that usually are performed in a periodic box (shown) with explicit solvent (water represented as balls).

Using the methodologies discussed herein, an accelerated biosynthesis of chemicals,[14,15] pharmaceuticals,[16] biofuels[17] and monomeric building blocks[18] is envisioned.

5.1.1 Bioinformatic Tools

Present-day sequencing technologies enable reading of genomes and metagenomic libraries from a dazzling array of organisms at a rate that shifts the bottleneck of the analysis to the annotation and functional characterization of the proteins encoded by the genes identified.[19,20] For instance, just one metagenome sequencing project from the human gut identified 3.3 million non-redundant microbial genes.[21] It is easily understandable that such huge numbers of proteins cannot be expressed and properly characterized individually, thus tools for accurate annotation and identification of interesting biocatalysts for desired reactions are required.

Recently, two very interesting benchmark reviews centered on the speed and accuracy of metagenome analysis tools using realistic data were published.[22,23] Interestingly, the comparison of the analyzed tools showed that the most accurate methods are not the most frequently used, and – probably against popular belief – the most accurate tools are not necessarily the most time-consuming.[22] For instance CLARK[24] and Kraken[25] are among the fastest and reliable methodologies. Still, it needs to be highlighted that significant variations regarding their efficiency may occur,[23] and thus for each set of sequences the most appropriate tool needs to be identified. However, this is often not practically achievable and groups that are active in the field rely on public databases, such as NCBI. The NCBI's non-redundant protein database captures only a single annotation[26] and hence contains accumulated sequences that are often misannotated. Annotations (and even sequences) from third-party databases like KEGG,[27] SEED[28] and IMG[29] are disregarded. A first way to tackle the accelerating problem of potential discrepancies is to be able to include as many annotations as possible. In order to take advantage of all available information, researchers have developed the non-redundant protein database M5nr, which combines sequences and annotations from several sources.[26] The database consists of two parts: the data, which are the unique sequences from the databases the user decides to include, and the metadata, which are all the annotation information found for a given sequence. The two sets are coupled to a unique identifier, which is a 32-character hexadecimal code, based on an MD5 checksum algorithm.[30] As only one set is included from each unique sequence, while at the same time sequences not deposited in NCBI are available for analysis, duplicate sequence entries are eliminated. Since not all databases use the same annotation tool, different predictions may be given for the same sequence. The identification of such discrepancies in the metadata of a sequence is a big advantage of M5nr and an important clue for detecting misannotation.

To identify enzymes with novel properties from the sequence space, a simple sequence similarity screening is not sufficient. When improved catalytic performances are targeted, such as higher turnover number and substrate scope, the screening should be focused on conserved areas of the enzymes directly related to the desired catalytic activity. In this case, the screening can start from the catalytic residues by checking if they display a specific motif which would facilitate the identification of enzymes of a specific class. For instance, some hydrolases have a typical α/β fold,[31] and – in most cases – a nucleophile-base-acid (Ser–His–Asp/Glu) catalytic triad responsible for a diversified set of enzymatic activity (*e.g.* esterase, epoxide hydrolase).[32] The catalytic serine is located at the end of the β5 strand, followed by the acidic residue at the loop after the β7 strand and the histidine after the β8 strand. Analysis of the α/β fold hydrolase database[33] revealed two conserved regions that can guide the identification of desired hydrolytic biocatalysts: the catalytic elbow with the nucleophile confined as well as the oxyanion hole.[32] The catalytic serine can be readily identified from the sequence motif GXSXG of the catalytic elbow. In addition, Kourist *et al.* identified five conserved patterns which could be assigned to the substrate specificity of the hydrolases. For instance, the GHSXGG group contains epoxide hydrolases and dehalogenases, while GNSMGG enzymes are hydroxynitrile lyases.[32] However, there are also hydrolases without this motif, including lipases and esterases that do not have the nucleophile elbow, but a flexible active site GDSL that changes its conformation during catalysis.[34] This subfamily includes thioesterases, proteases, arylesterases, and lysophospholipases. It is obvious that the identification of the catalytic nucleophile is not sufficient for accurate prediction of the substrate scope. For this reason, motifs in the binding pocket of the substrate are crucial. The oxyanion hole plays a critical role in the coordination of the substrate during catalysis and the GGGX motif has been correlated experimentally with the acceptance of sterically demanding tertiary substrates.[35] Taken together, these motifs, along with some other minor features (such as the topology of secondary structure elements, the presence of a lid *etc.*) can guide the identification of novel hydrolases with high accuracy.

The identification of potent biocatalysts from sequence databases is facilitated in the case of enzymes that require cofactors. This is because the coordination of cofactors usually requires conserved motifs in the structure of the protein such as the Rossmann fold; a structural motif correlated with several enzyme classes that require cofactors.[36] This structural element is responsible for the coordination of nicotinamide adenine dinucleotide (NADH), nicotinamide adenine dinucleotide phosphate (NADPH) as well as flavin adenine dinucleotide (FAD).[37] The most thoroughly studied sequence motif of the Rossmann fold is XZXZGXGXXGXXXZXXZ$(X)_8$ZXZE(D), where X is any residue and Z is any hydrophobic residue.[37] This sequence is significant for the coordination of the ribose and the phosphate groups. An interesting feature here is the small variations in sequence space that can guide different activities. For instance, *S*-adenosyl methionine (SAM)-dependent enzymes also adopt a Rossmann fold for the coordination of the cofactor.[38] However, as SAM does not contain phosphate groups, the glycine rich part of the pattern is not conserved; a characteristic that differentiates this enzyme class from others with a typical Rossmann fold. Other cofactors also require specific motifs; for instance, the [Fe–S] cluster in enzymes such as oxidoreductases requires four cysteine residues, and the motif responsible for the coordination of the cluster is CXXCXXXCX$_n$C.[39,40] However, the number of residues between the two last cysteines is not constant; an NADH-quinone oxidoreductase had 27 residues in between,[40] whereas the enoate reductase from *Eubacterium ramulus* and NADH flavinoxidase from *Eubacterium* sp. showed a significant difference ($n = 22$ and $n = 11$, respectively).[39]

Based on the aforementioned examples, it is essential to have some information concerning the targeted enzyme; the more information about the catalytic mechanism, the better. However, it is not always the case that information regarding conserved areas for the targeted enzyme is available in the literature. In that case, several tools have been developed to aid protein engineering by guiding the enzyme evolution process and identifying the desired biocatalysts. In this chapter we will present two emerging tools, the 3DM platform and peptide pattern recognition. However, novel methods develop continuously and tailor-made solutions for specific enzymes classes may already exist, so we motivate the reader to search the most suitable tool for their research.

3DM[33] is a platform which combines several tools that help the user both in the discovery of new enzymes, as well as in the rational design of improved biocatalysts. The distinguishing feature of this tool is that the database is based on a structure-based alignment of sequences, which allows alignment of sequences with significant difference in length with high accuracy.[41] The database uses sequences of proteins with available structures in the PDB[42] as "seeds" in order to produce super families from the structure-based alignment. The platform identifies core areas which are conserved in most sequences, and variable areas which are not conserved and can differ a lot – probably due to insertion, deletion and duplication events. This analysis comprises a significant tool for protein engineering. Useful information that can be extracted includes the natural variability in a specific position in the structure (see Figure 5.1a, left). Under the hypothesis that that natural evolution is not biased, the amino acids that are highly represented are the ones that lead to soluble and active proteins. This hypothesis seems to be right, as it has been shown that saturation (NNK) libraries introducing the most often occurring amino acids into proteins had a higher ratio of active variants in contrast to libraries that contained the "non-allowed" amino acids.[43] This knowledge accessible through available platforms results in a significant reduction of the size of libraries where multiple positions are targeted, as is discussed later in Section 5.2. However, here the user needs to be careful: natural variability helps us to identify active variants, but this does not mean that the variants will have the desired properties. If the reaction that is targeted is commonly occurring in nature, the chances to identify an interesting variant are high. In the case that the desired activity is more "exotic", then the mutation that leads to this activity may be underrepresented, or even not present in currently known sequences in nature. For this reason, it is interesting to note that the 3DM platform performs an indexing of literature and provides information on mutations that have already been performed at a specific position – very useful information, as it includes data not only from the wild-type sequences, but also from variants that were already characterized. Another interesting function that is available in the platform is the possibility to analyze correlated mutation networks.[44] The platform can calculate the co-evolution of two positions (the frequency of how often specific pairs occur) and provide a network of the correlated mutations (Figure 5.1a, right). This information does not only provide a useful hint on how to mutate several positions simultaneously in order to retain activity, but at the same time suggests which amino acids should be selected at each position.

When seeking to expand the substrate scope of an enzyme, we usually target the first shell of amino acids which interacts directly with the bound substrate. Thus, we tend to neglect contributions from the secondary shell which are not in direct contact with the substrate. RINalyzer is a software that enables the depiction of the residue interaction network (RIN) of a protein in a two-dimensional map, after analysis of a structure file.[45,46] When performing this analysis, refined experimental crystallographic data are preferred, but homology models can also be used if the model is of sufficient quality. The combination of the previously mentioned 3DM correlated mutation network analysis with RIN can provide significant insight into why these interactions are significant and guide our efforts in (semi-) rational design.

Another tool that can accelerate the identification of novel biocatalysts from sequence space is the peptide pattern recognition (PPR) methodology.[47] This tool is based on the identification of short motifs, normally located around the active site, that are responsible for catalytic activity. An algorithm was developed to sort sequences according to the patterns that can be recognized with respect to a specific function. This method has the benefit that even proteins with low sequence similarity (as low as 6% sequence identity) can be properly annotated.[47] The output of the algorithm is a grouping of the sequences under investigation. Of course, to predict the activity of each group, at least one of the enzymes of the group should have been biochemically characterized. This methodology is relatively new, and we believe that its broader implementation could provide new patterns for several enzyme classes that can guide the identification of novel biocatalysts.

One benefit of the aforementioned methods is that they are typically performed online and do thus not have a specific hardware requirement. The M5nr requires a sufficiently fast internet connection to be able to download all the sequences and metadata from the databases. The assembly of the database is not time consuming, neither demanding. Thus, any average server system can support this in-house database, although a high-end server would be preferred. The data are available *via* any commonly available suite for DNA and protein analysis, such as NTi Vector or GENEious. Such software can also be used to examine the sequences for desired motifs. Once the sequences with the putative activity are selected, the user can search the metadata for the annotation of these sequences. The metadata are stored in an SQL format and can be easily accessed *via* a unique identifier. In general, some programming skills are desirable for the setup of an M5nr analysis, but it does not require the skills of an expert in the field. Optionally, the user can employ the NCBI database for the same purposes, but we strongly suggest that other databases should be searched with the same criteria, to get as much feedback as possible.

The 3DM platform is completely online, thus no hardware specifications are needed. Here, two software packages that can be useful are Utopia Documents and YASARA. Utopia Documents connects with the 3DM database and can read data in PDF files concerning specific mutations. Comparison of amino acid alterations with other sources in literature or the natural variability is possible. The other software that can be useful is the molecular-modelling and -simulation program YASARA. YASARA also interacts with the 3DM database and can visualize all the information obtained from 3DM such as the core/conserved areas, as well as the correlated mutation networks.

The RINalyzer is not a self-sufficient software, but instead functions as a plug-in. The details for the installation and implementation are described in the literature.[45,46] The user needs to install the Cytoscape software for network analysis, where the RINalyzer is added as a plug-in. To visualize the network within a 3D structure, the software Chimera is required, with the additional plug-in StructureViz for Cytoscape. To depict the RIN in the topology of the protein, the Pro-origami web service can also be used. The authors established this tool as freeware to provide access to the general public.

Concerning PPR, the setup is quite simple: the PPR installer (.exe file) is available online.[48] All the user needs to do is to provide all sequences under investigation in a strict FASTA format. To accelerate the preparation of the dataset there is another executable file that helps with the preparation of the FASTA file. The user can decide the specific details for the pattern recognition from the parameters available (*e.g.*, the size of the pattern). The output of a PPR analysis constitutes files containing groups of protein sequences together with lists of recognition peptides that were identified within the determined groups. The proteins are also scored based on the occurrence of identified patterns in their sequence which forms the basis for further clustering. Although a general method cannot be provided for each unique case, in Figure 5.2 we provide a general flow chart of how enzymes with certain functions could be discovered from sequence space.

5.1.2 Ancestral Gene Resurrection

Sequence-based protein engineering can also be performed by investigating the sequence–function relationship of proteins from an evolutionary point

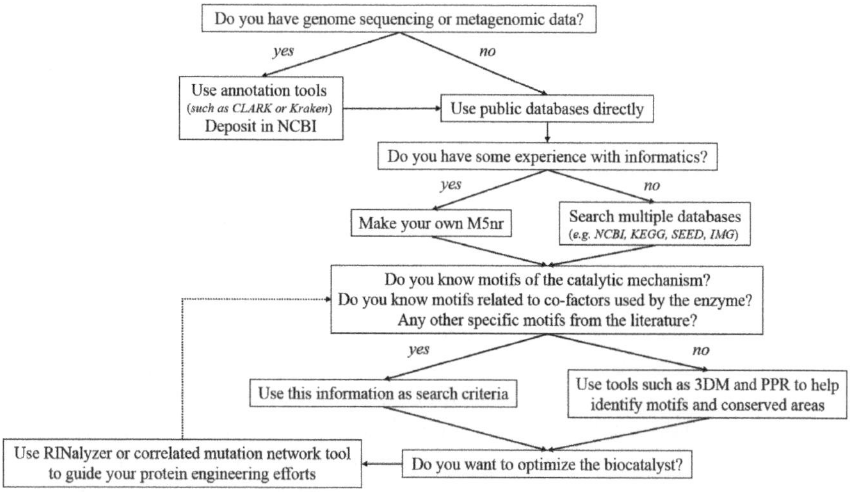

Figure 5.2 Flow chart of the identification of novel and improved biocatalysts based on the level of information available at each stage. Together with feedback from experimental data, a better understanding of the interactions in the catalytic active site and/or the binding pocket can provide valuable insight that can guide a second round of screening.

of view. Ancestral gene resurrection in particular is a developing field which, in recent years, has proven to be a valuable tool for understanding enzyme function and evolution. Several examples exist for which previously unknown active enzymes, representing putative ancestors, were produced based on the sequences of modern day enzymes.[49,50] This has led to an enhanced understanding of reaction mechanisms and the discovery of promiscuous enzymes with extended catalytic function.[51–53] In brief, reconstruction of ancestral enzymes is based on phylogenetic trees which are based on an alignment of amino acid or nucleotide sequences that are assumed to be homologous. Assuming that evolution proceeds *via* an accumulation of point mutations, ancestral sequences can be inferred from available present day sequences, using statistical methods such as Maximum Likelihood (ML), Parsimony or Neighbor Joining.[54] Internal branch points of the tree represent potential ancestral sequences, which would correspond to ancient enzymes that presumably prevailed in the past.

Besides accounting for the topology of the tree, certain statistical methods address factors such as the likelihood of different amino acid substitutions and assumptions about evolutionary rate variation among different sites in a protein (*e.g.* surface residues *vs.* active site residues).[55] This is the case for Maximum Likelihood (ML), which is a method commonly used today. In contrast, while Parsimony reconstructs the ancestral state based on minimizing the number of changes required, the Maximum Likelihood method evaluates the probability that the set of observed modern day sequences would evolve given a certain ancestral state, the topology of the tree and chosen evolutionary models. Bayesian Inference is a recent version of Maximum Likelihood for which the most likely tree, given the data, is evaluated.[54] The most widely used method for estimating the reliability of an ML-constructed tree is the bootstrap technique. This is essentially a repeated sampling that measures reliability as the probability that a given branch or node is reproduced in a large number of computed trees (typically between 100 and 1000, referred to as bootstrap steps).[54] It is important to emphasize that the quality of a phylogenetic tree and the associated inferred ancestral sequences can be no better than the quality of the underlying alignment. A certain level of homology is required to obtain reliable alignments: it has previously been found that a decrease in accuracy can be observed below 20% sequence identity, where less than 50% of residues are correctly aligned by most programs.[16]

A main advantage of the described method is its accessibility: it can be done by anyone with at least basic knowledge of bioinformatics, on a commonly used PC using well-established freeware such as MEGA or MrBayes.[56,57] In our laboratory, we typically perform ancestral sequence reconstruction using MEGA on a 12-core Intel Xeon in-house mini cluster. Challenges and important considerations of the methodology include which and how many sequences to use, how to deal with gaps in the alignment and which algorithms and evolutionary models to use. Many of these decisions will be made based on experience as well as knowledge about the particular protein under study. Moreover, the estimated probability of an ancestral sequence being a correct representation of evolution is equal to the product of the probabilities of all individual observed mutated residues being "true", which can result in a small probability for longer sequences. Therefore, rather than the true ancestor, an inferred ancestral sequence can be thought of as a representation that is likely to be similar to the historical ancestor.[53] This has implications for experimental biocatalysis: multiple possible ancestors should be considered and reconstructed at a certain node in a phylogenetic tree. These representative sequences should then be expressed, purified and experimentally characterized. If all the resurrected biocatalysts behave in a similar way, it is more likely that they are in fact a close approximation of the "true" ancestor that once existed.[52]

5.1.3 Structure-based Computational Tools

Today, many biocatalysts have a high-resolution crystal structure deposited in the protein data bank. Optionally, starting from the roughly 1400 currently known protein folds,[12] a representative model could be built by homology modelling.[58] The accuracy of such structural prediction will greatly benefit from the tremendous advancement in protein design.[59] The most appropriate structure-based simulation methodology is entirely dependent on the actual research question and the related timescales. Quantum mechanical (QM) calculations provide a framework for detailed elucidation of potential reaction pathways, their associated transition state structures, intermediates and energetics on timescales of a bond vibration (*i.e.* femtoseconds).

In contrast, force field-dependent molecular dynamics (MD) methodologies can currently access timescales up to milliseconds (ms) through rapid assessment of the energy of the system and its components. Still, for practical reasons, MD simulations are typically run on the 100 nanosecond (ns) timescale. Although this only corresponds to a fraction of the time of an enzyme-catalyzed event, it is sufficient to account for some dynamical effects.

5.1.4 QM Computational Tools

First-principles calculations[58] constitute a cornerstone for *de novo* design of enzymes (see Chapter 4)[60] and elucidation of reaction mechanisms[61] and are in the simplest case based on solving the time-independent Schrödinger equation:

$$\hat{H}\Psi = E\Psi \tag{5.1}$$

Here the Hamiltonian energy operator (**H**), composed of kinetic and potential energy terms, corresponds to the total energy (*i.e.* the eigenvalue E) of the system given by:

$$E = \frac{\int \psi^* \hat{H} \psi d\tau}{\int \psi^* \psi d\tau} \tag{5.2}$$

As no exact solutions to the Schrödinger equation exist for molecules with multiple electrons and nuclei (*i.e.* many-body systems), quantum mechanical computational tools use sophisticated algorithms to find approximate bonding and non-bonding molecular orbitals (represented by the wave functions Ψ, eqn (5.1)) by minimizing the energy of the system (eqn (5.2)). Two commonly used algorithms to locate a local energy minimum include Steepest Descent minimization and Newton–Rapson. For Steepest Descent, the gradient of the energy is analyzed to take a "step" towards the energy minimum whereas Newton–Rapson and related methods depend on evaluating the computationally more expensive second derivative of the energy function. Advanced quantum mechanical calculations, such as perturbation theory (*e.g.* MP2), do not assume electron motion in a mean field created by the other electrons. Density functional theory (DFT) constitutes one post Hartree–Fock method which has become an indispensable tool for chemists and biologists alike to simulate complex biochemical systems, such as metalloenzymes, with enhanced computational efficiency and accuracy.

The identification of electronic configurations and positions of nuclei of lowest energy from QM calculations is greatly facilitated by key assumptions including:

(i) The Born–Oppenheimer (BO) approximation[58] which, in simplified terms, states that electrons move much faster than the heavy nuclei and their motions can thus be described separately.

(ii) The Linear Combination of Atomic Orbitals (LCAO) approximation, which states that the molecular orbitals (corresponding to the wave function **Ψ**) can be represented by a linear combination of a chosen set of atomic orbitals representing the so-called basis set of the calculation.

QM models of biocatalytic systems are usually constructed based on dissecting a small part of an enzyme crystal structure representing its active site (Figure 5.1c, box). The substrate in these models, which have been referred to as "theozymes"[60,62] or cluster models,[63] can either be built and placed in the truncated and "capped" model from scratch based on mechanistic intuition, or re-modelled from a co-crystallized ligand. For the actual energy minimization of a biocatalyst model, the user typically provides a starting geometry representing an initial guess of the configuration of the nuclei. Based on a careful selection of a pre-defined set of mathematical functions (*i.e.* the basis set) normally representing the well-known atomic orbitals (*e.g.* **s**,**p**), the modelling program performs an iterative solving process for which the coefficients of the basic functions on each atom are optimized to find the local energy minima of the molecule. To avoid unrealistic motion of the protein backbone or its side chains during calculations on the truncated model, certain selected positions are typically frozen.[63] After geometry optimization of the starting complex, QM calculations allow for detailed study of bond-forming and -breaking processes in the model system. Geometries and energies of transition states and intermediates associated with different reaction mechanisms can thus be obtained and verified based on a so called vibrational frequency calculation. Moreover, the most reasonable biocatalytic cycle – corresponding to the one with lowest activation energy – can be identified. The effect of dielectric constants can be addressed by including continuum solvent models or, optionally, by the computationally more expensive addition of explicit solvent molecules. Importantly, a suggested biocatalytic reaction mechanism can be scrutinized on the atomic level by analyzing the so called intrinsic reaction coordinate (IRC), which should yield the expected intermediates in both the forward and backward direction starting from the transition state structure.

Although the availability of computational power has seen a tremendous growth during the last decade, QM calculations do not scale linearly with system size. Thus, quantum mechanics for biocatalysis is practically limited to a couple of hundred atoms when using high-level of theory that typically display errors in the estimated activation energy of up to 5 kcal mol^{-1},[64] but that in some cases can approach chemical accuracy.[65] Several program suites for performing quantum mechanical calculations exist, including Jaguar, Gaussian and GAMESS to mention a few. In our laboratory we typically use US-GAMESS,[66] a freeware for academic groups built upon a highly parallelizable code. For systems consisting of roughly 100 atoms, we use an in-house cluster consisting of 12 Intel Xeon processors. For larger and more complex systems, including metalloenzymes, which require special treatment of unpaired electrons, we take advantage of the PDC Center for High Performance Computing at the KTH Royal Institute of Technology. In our experience, QM calculations are quite challenging to learn (>2 years of experience) due to a complex underlying theoretical framework. However, user friendly software greatly facilitates these calculations also for non-experts. For instance, US-GAMESS comes in a Microsoft Windows version based on a 64-bit code that should make these tools more accessible to the biocatalytic community.

5.1.5 Molecular Mechanics and Molecular Dynamics Computational Tools

Molecular mechanics (MM) refers to Newtonian mechanics applied to molecules by approximating atoms as hard spheres and chemical bonds as springs. This allows for rapid determination of the energy of the protein and its ligands (Figure 5.1c, eqn (5.3)):

$$E(\vec{r}) = \nu_{\text{stretch}} + \nu_{\text{bend}} + \nu_{\text{dihedral}} + \nu_{\text{electrostatics}} + \nu_{\text{Lennard-Jones}} + \cdots. \quad (5.3)$$

Here the total potential energy function E, referred to as the force field, is evaluated based on deviations of bond lengths (*i.e.* the stretching potential energy ν_{stretch}) and angles (*e.g.* bending) from their equilibrium value. Force fields are typically empirically derived to fit observed experimental data and/or high-level quantum mechanical calculations. The geometrical positions of the atoms in the molecule (represented by the positional vector **r**) further dictate Coulomb and van der Waals energies. Computer modelling suites typically use a pre-defined cut-off distance for pairwise electrostatic interactions between atoms, as calculations would otherwise be prohibitively slow. Instead, sophisticated algorithms such as the point mesh Ewald method[67] account for long range electrostatics. The point charges residing on the atoms, as well as other key force field parameters, can either be obtained automatically from semi-empirical quantum mechanical calculations implemented in the modelling software,[68] or provided by the user from high-level quantum mechanical calculations. Today, force field parameters are typically provided in the modelling software for many different types of covalent bonds, ligands and even for transition metals.

The initial geometry of the molecular mechanics model is obtained from a crystal structure or, optionally, from homology modelling. After manually adding hydrogens, whose spatial positions can usually not be retrieved from crystallographic electron density, energy minimization is performed – usually starting with hydrogens while keeping the protein backbone and its sidechains fixed. After relaxation of solvent water molecules, the ligand and finally the protein are minimized. It should be emphasized that X-ray crystallography can introduce structural artefacts arising from packing interactions in the crystal. To escape from "trapped" local energy minimum conformations, simulated annealing could be used for which a heating phase is followed by slow cooling.

Analysis of molecular mechanics over time is referred to as molecular dynamics (MD),[69,70] which is based on integrating Newton's second law of motion (eqn (5.4)):

$$\vec{F} = m\vec{a} \tag{5.4}$$

Here, the force acting on each atom at time point t can be obtained as the negative gradient of the associated potential energy. Assuming a constant force during the short time step Δt (typically 1–2 fs) allows for calculating the new spatial positions of the atoms at the time point $t+\Delta t$ through numeral integration of eqn (5.4). MD-simulations are usually performed in a periodic water box (represented in Figure 5.1c) neutralized by NaCl (or analogous counter ions) under the canonical NPT ensemble (constant number of particles, pressure and temperature). In some cases, simulations on the ms scale[71,72] have been pursued, which is made possible as the computational cost for MD roughly follows a linear function of system size. At these time scales, study of protein folding and large conformational changes in biocatalysts are feasible. Analysis of the relative abundance of key conformational states, for instance the extent of productive substrate binding, from the resulting MD-trajectory allows for evaluation of their relative energies according to a Boltzmann distribution. The main weakness associated with MD-simulations is the fact that the formation and breaking of covalent bonds, a paradigm of biocatalysis, cannot be directly studied. However, careful selection of adequate high-energy intermediates, for instance the tetrahedral intermediate formed during serine hydrolase catalysis,[73] allows for performing MD simulations of transition state mimics of a biocatalytic transformation.

Today, user friendly molecular modelling suites for molecular dynamics exist, for example YASARA[74] which can be used even by non-experts. In our laboratory, we typically run MD-simulations on 24 Intel Xeon cores using a Linux-based version of YASARA. The length of the trajectories we generate to study reaction mechanisms and water dynamics inside proteins typically amounts to 100 ns. In our experience, it is instrumental to run MD-simulations in duplicate on these timescales when analyzing key statistical properties, such as the probabilities to form a certain conformational state. This is readily achieved by a small perturbation of the simulation temperature (*e.g.* by 0.001 K) to obtain different initial random velocities at the onset of the simulation.

5.1.6 QM/MM

The combination of the power of quantum mechanics with molecular mechanics is referred to as hybrid QM/MM methods.[75,76] Briefly, for QM/MM a small part of the enzyme is treated using quantum mechanics, whereas the remaining part is considered by molecular mechanics. Strikingly, this allows for the study of bond-forming and breaking processes or other phenomena such as fluorophore excitation within proteins, their associated transition state structures, reaction pathways and energies under the influence of the protein environment. In the most accurate implementation of the QM/MM formalism, electrostatic embedding assures continuous interaction between point charges in the MM and QM regions during the minimization processes.

QM/MM calculations for biocatalysis are practically performed through a code linking a QM interface with an MM program. QM/MM simulations require significant computational expertise and experience. Major challenges include how the QM region should be chosen and capped by so called link atoms, and how to account for the interaction between the dissected QM-region (box, Figure 5.1c) and the MM-region, which can influence the obtained geometries and their energies.[76] Furthermore, residues distant to the QM regions are typically frozen for practical reasons, which could potentially mask protein dynamical effects.

5.2 Implementation of Computational Tools in Biocatalysis

5.2.1 Identification of Novel Biocatalysts from Sequence Space

Hypothesis-driven screening of the sequence space displays a huge potential to identify biocatalysts with desired properties. The first discovery of (*R*)-selective amine transaminases[77] by Höhne *et al.* constitutes an important case. Based on a hypothesis on the evolution of (*R*)-selective transaminases, they identified the key positions correlated to specific activities and controlling the enantioselectivity. By excluding certain amino acids at these positions, the authors managed to identify a class of PLP-dependent enzymes which, when experimentally characterized, had the desired activity and the preference towards the (*R*)-enantiomer.

In a more recent example, we could show that not only the activity, but also the substrate scope, can be predicted. Amine transaminases are interesting biocatalysts for the reductive amination of ketones, with significant impact on the pharmaceutical industry, *e.g.* in the production of sitagliptin.[78] However, their active site consists of a small and a big binding pocket, which is the reason for their high selectivity. The challenge here is that the small binding pocket is typically only able to accommodate a methyl group, whereas the industrially relevant applications would require larger substituents. In order to be able to identify a toolbox of transaminases that act on bulky substrates, we initially performed rational design based on *in silico* analysis of the most energetically demanding transition state of the catalytic mechanism.[79] This way, we identified five previously unacknowledged positions that enabled the acceptance of bulky substrates in a scaffold that exhibited some basal activity. Interestingly, most of these mutations represented a change of the native amino acid for one that existed in the most well-studied wild-type aminotransferases (from *Vibrio fluvialis* and *Chromobacterium violaceum*), which, however, did not have any activity towards the bulky substrates under investigation. Thus, we formed the hypothesis that it is the combination of the scaffold and the residues in the right position that enables the desired activity. With this basic understanding of the acceptance and the hypothesis we had formed, we searched the M5nr database available in our institution and were able to identify that the combination (*i.e.* scaffold and specific mutations) did not exist in nature (or at least within the known sequences). To prove our theory, we selected suitable template sequences with identities as low as 70% and with the key motif incorporated. Interestingly, all soluble enzymes expressed from the corresponding ordered synthetic genes exhibited the predicted activity. Thus, based on basic understanding, a motif that facilitates the easy identification of (*S*)-selective amine transaminases active on bulky substrates was identified.

For another enzyme class, the PPR strategy was successfully used for the classification and screening of carbohydrate-active enzymes.[47] The method was used multiple times, with different peptide lengths as input, in order to identify the ideal peptide pattern length. Short peptides down to two amino acids did not provide accurate predictions, as they can, besides for evolutionary reasons, be found in the sequence simply by chance. Yet the peptides used for functional annotation cannot be too long, as not all amino acids could be expected to be conserved. Busk *et al.* found an optimum peptide length of 4 residues that provided more than 90% accuracy in the prediction of the specificity of the enzyme.[47] Based on this work, Huang *et al.* mined the genome of *Mucor circinelloides* for genes encoding for β-glucosidases and glycoside hydrolases using the PPR methodology.[80] In this work, they identified twelve novel genes that were assigned to different families. Two of the biocatalysts that were assigned to the GH3 family were experimentally validated and found to have the desired activity. These examples highlight the potential to use bioinformatic computational tools for the identification of novel enzymes from sequence space.

5.2.1.1 Guidance of Protein Engineering for Improved Biocatalysts

The 3DM platform was used successfully in several cases to guide rational design efforts to acquire efficient biocatalysts.[81–83] For instance, the development of transaminases that are active towards branched substrates was guided from the *in silico* analysis by 3DM. Docking analysis revealed seven positions that were of interest to be mutated.[84] Simultaneous screening of saturation mutagenesis libraries for seven positions requires a tremendous effort (about 100 billion colonies should have been screened for 95% coverage for an NNK library). However, considering the hypothesis that only hydrophobic residues can contribute to the acceptance of the substrates under investigation as well as the natural variability at the given positions, the library size was reduced to only 50 000 variants. More importantly, Genz *et al.* did not even need to screen these 50 000 variants, since the high hit rate triggered by the proper design of the library rapidly provided several enzymes acting on branched-chain substrates. In another study, Steffen-Munsberg *et al.* performed a thorough analysis of the whole PLP-dependent enzyme superfamily based on the 3DM platform, from which they could determine a sequence function matrix to correlate the sequence space with reaction and substrate specificities.[85]

The RINalyzer tool has proven helpful in our hands in a recent promiscuity study. Although amidases and esterases share the same catalytic triad, there is a clear difference between the two activities. In our efforts centered on rational design of amidase activity in an esterase scaffold, we took advantage of interacting amino acid networks for the identification of a mutation that improved the promiscuous amidase activity. This mutation was located in the second shell of amino acids, and it did not directly interact with the amide substrate. The analysis of the wild type and the improved variant *via* RINalyzer provided the interaction network that was responsible for the improved activity and provided insight used to continue our rational design efforts. In the end, we established a three residue interaction π–π network, which stabilized one of the residues in the catalytically active conformation for the amidase reaction.[86]

5.2.2 Ancestral Enzyme Reconstruction for the Study and Engineering of Binding and Catalysis

Phylogenetic analysis enables the tracking of evolutionary paths backwards in time, which can provide valuable insight into how proteins acquire specification and new function at the resolution of specific amino acid alterations. Moreover, phylogenetic trees based on the protein of interest can be linked to the tree of life[87] to infer estimates of the corresponding timeline. Of particular relevance for biocatalysis is the methodology of ancestral gene resurrection, which allows researchers to infer the sequences of potential historical ancestors. Besides having a great potential to guide protein engineering for enhanced catalytic versatilities, the generation of ancestral proteins in the laboratory could provide access to interesting biocatalytic templates that could be subjected to further evolution in order to access evolutionary paths that were not yet explored in nature.

The high potential of using ancestral gene resurrection to shed light on the evolution of binding specificity is nicely illustrated in a recent work by Wilson *et al.*[51] To understand the mechanism behind the blockbuster drug Gleevec's 3000-fold higher affinity for the tyrosine kinase Abl over the structurally similar Src and other closely related enzymes of this class, four ancestral enzymes were resurrected and characterized. This was achieved by a phylogenetic tree of the Tec, Abl and Src kinase families using a marginal likelihood function in Bayesian Inference. Interestingly, the degree of inhibition by Gleevec was found to increase along the evolutionary path connecting Src and Abl. Furthermore, the inhibition displayed by the last common ancestor was intermediate to that of Abl and Src. A detailed analysis of the mutations found in the different ancestral enzymes unraveled a set of residues that were involved in a unique hydrogen bond network in Src and its close ancestors; one that was not present in Abl. This hydrogen bonding network prevented the so called P-loop in Src from forming a kink and closing over the active site, a pre-requisite of tight inhibition due to holding Gleevec in place. Wilson *et al.* concluded that the motion of this loop was part of a rate limiting "induced-fit" step of the Gleevec binding mechanism. They hypothesized that such an induced-fit mechanism may have evolved by exerting an evolutionary pressure for differential regulation among members of the structurally highly similar "kinome".

Convergent evolution refers to the phenomenon of a trait evolving independently and simultaneously in different species (Figure 5.3). In a recent study of convergent evolution of increased hemoglobin–oxygen (Hb–O_2) affinity, Natarajan *et al.* compared 28 pairs of high- and low-altitude taxa of birds in terms of their hemoglobin sequences and Hb–O_2 affinity.[88] By utilizing ancestral gene resurrection, it was found that the increase in Hb–O_2 affinity observed in high-altitude taxa was nearly entirely due to an increase in oxygen affinity caused by parallel substitutions in residues that were previously not considered as key residues. These results suggest that several possible mutations can lead to an evolutionary increase in Hb–O_2 affinity, which is in accordance with the notion that several genotypes can lead to the same biochemical phenotype. In addition, Natarajan *et al.* reconstructed four ancestral hemoglobins using the maximum likelihood method and investigated the effect of the observed mutations in different species against the inferred ancestral genetic backgrounds. It was found that certain mutations that clearly enhanced Hb–O_2 affinity in a certain lineage did not do so in another. This supports the hypothesis that the phenotypic effect of a mutation is dependent on the genetic background leading to lineage specific substitutions when adapting to selection pressure during evolution.

Ancestral gene reconstruction may also be used to explore more general concepts in evolution, such as epistasis, which is the study of how complex biophysical interactions between amino acids determine protein function.[89] This approach was used in a study by Bridgham *et al.* who used ancestral reconstruction, mutational studies and crystallography to investigate the evolution of cortisol binding specificity of glucocorticoid receptors.[90] They found that cortisol-specific receptors evolved from more promiscuous ancestors that could be activated by multiple steroids through five subsequent restrictive mutations which enabled the specificity shift.

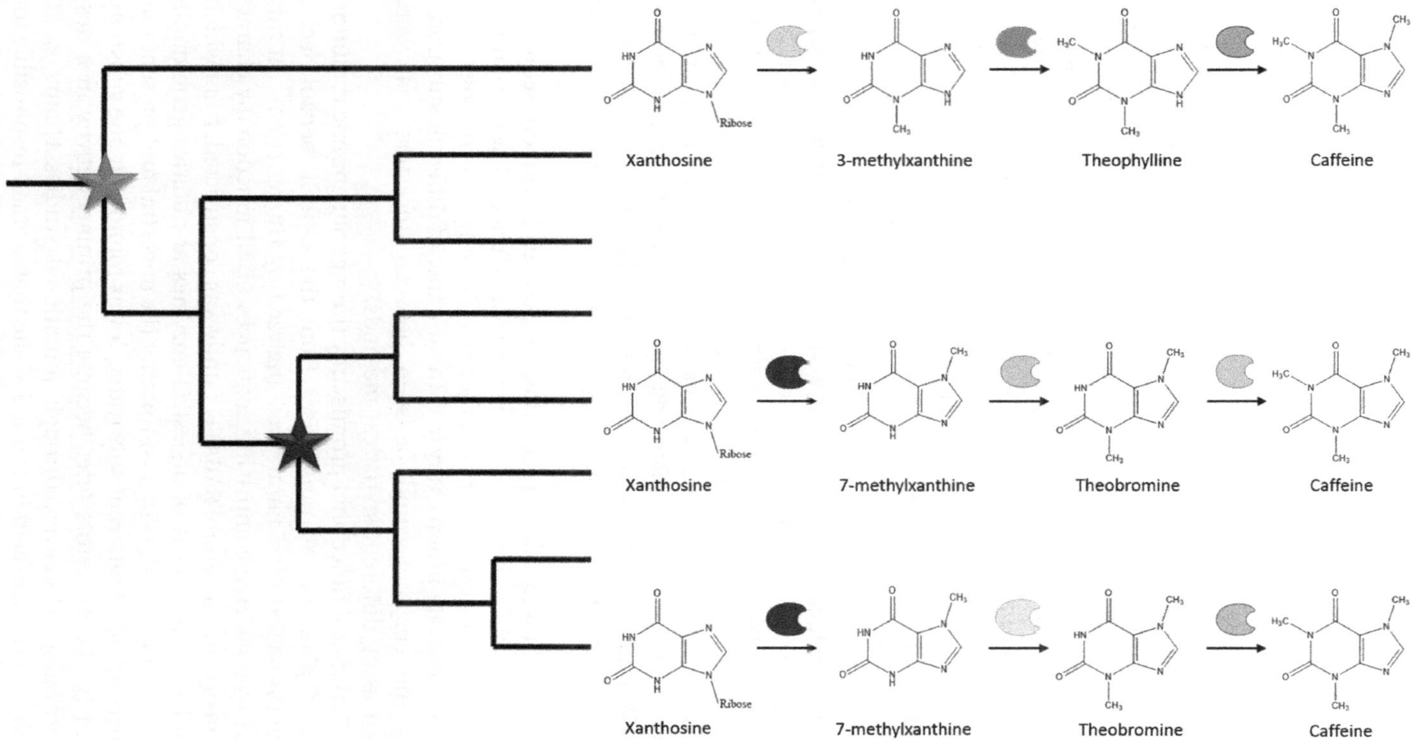

Figure 5.3 Convergent evolution illustrated by caffeine biosynthesis in plants. Convergent traits evolve independently and in parallel in different lineages and may be based on identical biochemical pathways catalyzed by different enzymes (see bottom part of figure) or different pathways (see top part of figure). This results in great genetic and metabolic diversity. The stars represent divergence points in the phylogenetic tree, *i.e.* the last common ancestors. Inspired by ref. 92.

Catalytic promiscuity refers to the ability of an enzyme to catalyze multiple biochemical reactions through the stabilization of the corresponding transition state structures in its active site. Promiscuity is key to the development of new chemistries, which is of paramount importance for extending the synthetic applicability of enzymes in biocatalytic processes.[14] Efforts on identifying enzymes harboring interesting promiscuous activities could be centered on screening naturally occurring biocatalysts (see Section 5.2.1). However, as promiscuous activity could be considered an accidental feature in plastic active sites, analyzing branch points in an evolutionary tree, in which enzymes with different functions diverge, could be a more viable option. It has been shown that this is indeed the case for several reconstructed ancestral enzymes on the evolutionary path along which esterases and hydroxynitrile lyases diverge.[52] In their study, Devamani *et al.* inferred the most probable sequence of the most recent common ancestor of esterases and hydroxynitrile lyases as well as three additional divergence points in the phylogenetic tree that was constructed using Neighbor Joining, Parsimony and Maximum Likelihood methods, or combinations thereof. Present-day and ancestral enzymes differed by roughly 50 amino acids, corresponding to a changed sequence identity of about 20%. Only a few of the mutations were located in or near the active site. The four reconstructed ancestral enzymes were expressed, purified and characterized for promiscuous activity using six esters and cyanohydrins with varying architectures. All but one ancestral enzyme were found to catalyze both ester hydrolysis and cyanohydrin cleavage. Thus, ancestral esterases had mainly esterase-like activities but also displayed hydroxynitrile lyase activity and *vice versa*. Strikingly, enzymes that are present today showed little or no promiscuous activity when challenged with the same substrate set. The substrate preferences were found to be reflected in several identified key residues, which allowed some prediction of activity based on sequence alone. In addition to active site residues, enzyme function was found to depend on amino acids outside or even far from the active site. Interestingly, a difference in substrate promiscuity (*e.g.* the number of different substrates accepted by the enzyme) could be seen between ancestors that were reconstructed using different statistical methods.

The role of ancestral functional promiscuity throughout enzyme evolution for a group of plant methyltransferases from the salicylic/benzoic acid/theobromine (SABATH) gene family was analyzed by Huang *et al.*[91] In this study, a phylogenetic tree comprising sequences of all functionally characterized enzymes from the entire SABATH family was constructed, from which the most probable sequences of ancestral enzymes at putative gene duplication points were inferred. When comparing the ancestral and present day enzyme activity using 18 different substrates, it was found that one substrate not preferred by the ancestors later became the primary activity in a descendant biocatalyst. Moreover, through mutational studies Huang *et al.* could pinpoint one amino acid in the active site that seemed responsible for a large part of the enzyme substrate specificity. The authors hypothesized that an ancestral ability to convert multiple substrates can be refined along the evolutionary path through several mutations, resulting in the shift of enzyme functionality. This approach to understand and fine-tune enzyme substrate preference could be of great use when engineering enzymes for specific biocatalytic pathways.

In a different study, Huang *et al.* investigated the evolution of biocatalytic pathways, focusing on the convergent evolution of caffeine biosynthesis in plants.[92] Convergent traits may or may not be based on identical or similar biochemical pathways (Figure 5.3), which is associated with great metabolic and genetic diversity. For instance, caffeine biosynthesis can be performed *via* 12 potential and different pathways, only one of which has been characterized for plants. The identified pathway makes use of either caffeine synthase methyltransferases (*e.g.* in *Camellia*, tea) or xanthine methyltransferase-type (XMT) enzymes (*e.g.* in *Coffea*, coffee). Huang *et al.* found two previously uncharacterized biochemical pathways for caffeine biosynthesis in chocolate, citrus and guarana plants, which proceed *via* different intermediates than in coffee and tea. To investigate how XMT-like enzymes in *Citrus* and *Coffea* could have evolved to catalyze two completely different pathways, Huang *et al.* used ancestral gene resurrection of XMT enzymes and analyzed their catalytic activities with respect to different substrates. They observed that some ancestral activities were completely lost in modern descendants, which were much more specialized with respect to their function. They hypothesized that convergent evolution of caffeine biosynthesis was possible because the ancient XMT enzymes were maintained for alternative biochemical functions and that their exaptation towards caffeine biosynthesis only became relevant after the initial reactions in the caffeine pathway evolved.

Besides being a very useful tool to study and tune substrate specificity, ancestral gene resurrection could potentially also be used to obtain thermostable enzymes – a paradigm for industrial biocatalysis. It has previously been shown that ancestral enzymes are more likely to be thermostable, probably due to the thermophilicity of ancestral life.[93] For instance, Watanabe *et al.* have shown that it is possible to enhance thermal stability of 3-isopropylmalate dehydrogenases (IPMDH) by introducing mutations found in a reconstructed ancestral sequence.[94] They constructed, expressed, and characterized 12 variants that each contained one amino acid substitution found in the ancestral sequence, and six variants showed higher thermal stability than wild-type IPMDH. The likelihood of thermophilic ancestral Precambrian life is also illustrated by the work of Risso *et al.*, who resurrected 2–3 billion year old β-lactamases and characterized their activity and thermal stability.[95] β-Lactamases infer resistance toward β-lactam antibiotics through hydrolysis of the acylated enzyme–lactam intermediate made possible by nucleophilic attack of a water molecule, which is prohibited in the structurally related penicillin binding proteins. A phylogenetic tree was constructed by Bayesian inference using 75 sequences of class-A β-lactamases covering different bacterial phyla. The reconstructed ancestors differed from extant enzymes by roughly 100 mutations, corresponding to roughly 30–40% of change at the sequence level. Crystallization of two ancestral lactamases allowed for determination of their 3D structure and despite the large number of changed residues, both ancestors shared the present-day lactamase fold. Risso *et al.* studied the thermal stability of the Precambrian β-lactamases and found that their denaturation temperatures were about 35 °C higher than those of present β-lactamases. In addition, they characterized the ancestral capability to infer resistance to different antibiotics, which revealed promiscuous activities at similar catalytic rates compared to modern β-lactamases. Further exploration of this technique could prove it to be an efficient method for designing thermostable enzymes for biocatalytic processes.

5.3 Structural-based *In Silico* Modelling for Efficient Biocatalysis

Upon the availability of a representative protein structure, different levels of theory can be exploited for an in-depth study of biocatalysis and the associated reaction mechanisms. In a recent study by Liao *et al.*, QM calculations on representative dissected active site models of the tungsten-dependent acetylene hydratase enzyme in a continuum solvent were compared with hybrid QM/MM calculations.[96] Interestingly, it was found that the two complementary approaches predicted different rate-limiting biochemical steps, which was partially attributed to the fact that QM-calculations on the dissected and "frozen" active site models failed to represent adequate dynamical aspects of the protein structure.

5.3.1 QM Is a Powerful Tool for the Elucidation of Reaction Mechanisms and Associated Energies in "Frozen" Models of the Active Site

The high-potential of using QM towards an enhanced understanding of enzyme catalysis and to guide protein engineering efforts is exemplified by the low promiscuous amidase activity displayed by some esterases. By using high-level QM calculations on a small model representing a dissected serine hydrolase catalytic machinery, one that captured the Ser–His–Asp catalytic triad and the oxyanion hole, the rate limiting step for amide bond hydrolysis was identified to be nitrogen inversion.[97] The inversion of the single lone pair sitting on the reacting nitrogen atom of the substrate is essential to yield a productive second tetrahedral intermediate capable of abstracting a proton from the catalytic histidine, which onsets the rupture of the scissile C–N bond. Interestingly, all analyzed proteases and amidases[98] have evolved to harbor a key hydrogen bond involving the reacting amide NH-group of the substrate and an enzyme – or substrate-assisted hydrogen bond acceptor – with a spatial arrangement that would correspond to a close-to-perfect stabilization of the quantum mechanically calculated transition state structure.

Protein engineering efforts centered on facilitating nitrogen inversion in esterases, for which an enzyme-assisted hydrogen bond acceptor is lacking, led to a seven-fold increase in absolute activity and a 50-fold shift in the amidase over esterase reaction specificity.[99]

The concertedness of complex biosynthetic reactions is of great biocatalytic interest as potential intermediates could be intercepted to instead funnel reaction pathways towards new chemistries and products (Figure 5.4a). QM calculations were used to shed light on the protonation-initiated electrophilic enzymatic polycyclization cascade of the C30 squalene to pentacyclic scaffolds catalyzed by triterpene cyclases.[100] By obtaining a geometry of the pre-folded polyisoprene substrate from a crystal structure representing the enzyme–substrate complex, Smentek *et al.* were able to show that the first three ring closures through carbon–carbon bond formation were synchronous *in vacuo*.[100] Interestingly, the generation of arabidiol by triterpene cyclases through water addition to the tricyclic malabaricane cation generated from cyclization of (3*S*)-2,3-oxidosqualene has been observed experimentally (Figure 5.4a).[101]

Quantum mechanical calculations are indispensable within the emerging fields of metabolic engineering.[102] Here, QM can provide meticulous information on viable mechanistic pathways in complex biosynthetic machineries. In a recent work, Zi *et al.* pursued a synthetic biology approach combined with QM calculations to elucidate the biocatalytic generation of the diterpenoid (+)-lycosantalonol starting from the relatively simple – but unusual – *cisoid* linear isoprene (*Z*,*Z*,*Z*)-nerylneryl diphosphate.[103] High-level QM analysis was used to identify the high-energy carbocationic intermediates generated along the reaction trajectory, which provided detailed insight into the energetics of the formation of the highly-strained and peculiar tricyclene core confined within the final product.[103] Furthermore, the calculations suggested that the diphosphate group that departs from the isoprene substrate during the generation of the initial carbocation played a stabilizing role during catalysis *in vacuo*.[103] In another study, Potter *et al.* used molecular modelling and rational design to exchange the hypothesized histidine catalytic base that presumably quenches the final carbocation in diterpene cyclases involved in the biosynthesis of labdane-type of natural products.[104] Strikingly, substitution for larger amino acids resulted in the formation of new products that were not accessible by the wild-type enzyme. QM calculations shed light on the reaction mechanism of the associated complex rearrangements towards the formation of the novel bi-cyclized scaffold displayed by the variants.[104]

5.3.2 Molecular Dynamics Simulation Allows for the Study of Dynamical Aspects of Biocatalysis

In a recent study by Zhang *et al.*, docking was used to elucidate the molecular reasons for the observed switch in enantioselectivity from (*S*) to (*R*) in the

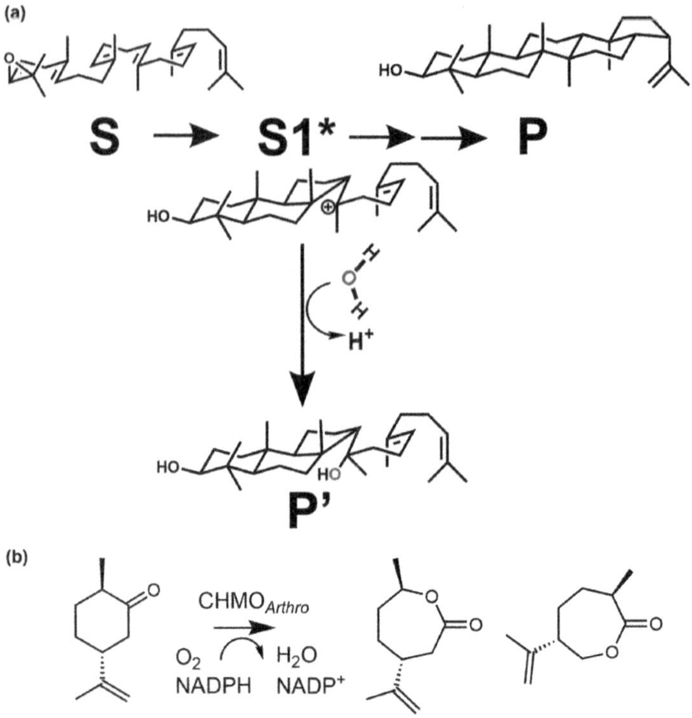

	Relative distribution	
wild type	0	100
F299A/F330A /F485A	99	1

Figure 5.4 Engineered biocatalysts can display different reactivities with respect to high-energy intermediates generated during the catalytic cycle. (a) A triterpene cyclase-catalyzed conversion of (3*S*)-2,3-oxidosqualene (**S** in figure) is schematically shown, for which concerted cyclization[100,113] yields the high-energy tricyclic malabaricane-like cation (**S1***) that is susceptible to react further with available nucleophiles: either an isoprene, leading to **P** (top) or a water molecule leading to **P′** (bottom). Biocatalysts from different organisms[101] and engineered variants[119] display different active site geometries and hydration,[120] which affects the preferred pathway and thus the product specificity (*i.e.* **P** *vs.* **P′**). P refers to the hopane skeleton and P′ to the malabaricane skeleton. Adapted from ref. 101 and 119. (b) The conversion of (+)-*trans*-dihydrocarvone by wild type and a designed variant of cyclohexanone monooxygenase from *Arthrobacter* sp. BP2 (abbreviated as $\text{CHMO}_{\textit{Arthro}}$) is shown. *In silico* enzyme design in concert with experimental biocatalysis has been utilized[106] to remodel the spatial arrangement of the migrating group in relation to the peroxy bond in the generated Criegee intermediate, which afforded a switch in the lactone product distribution. The thermodynamically favored most-substituted product is shown to the left and the "abnormal" lactone is shown to the right. Adapted from ref. 106.

sulfoxidation of thioethers displayed by a quadruple mutant of phenylacetone monooxygenase (PAMO).[105] PAMO and other flavin-dependent Baeyer–Villiger monooxygenases (BVMO) depend on a key (hydro)peroxy–flavin intermediate for the insertion of oxygen next to a plethora of functional groups. Pursuing a flexible docking approach, Zhang *et al.* were able to explain the stereo-preference displayed by wild-type PAMO and engineered variants thereof obtained in a directed evolution program.[105] Based on the observed docking poses *in silico*, the authors concluded that the distance between the peroxy–flavin intermediate and the closest sulfur lone pair of the pro-chiral thioether substrate correlated with the experimentally-determined enantioselectivity. Thus, their MM approach revealed the impact of the introduced mutations on the shape of the active site pocket and on the space available for the substrate, which was found to impact the degree of over-oxidation through sulfone formation. In another study, docking was used to study the binding poses of the two possible lactone products formed by the oxidation of the monoterpenoid (+)-*trans*-dihydrocarvone, catalyzed by cyclohexanone monooxygenase from *Arthrobacter* sp.[106] For this purpose, Balke *et al.* utilized homology modelling capitalizing on a known crystal structure of the closely related (84% sequence similarity) BVMO from *Rhodococcus* sp. HI-31. MM was used to identify amino acids in and around the active site that could potentially affect product specificity by assisting in directing the migration of the bond situated antiperiplanar to the oxygen–oxygen bond of the peroxide in the high-energy Criegee intermediate. Based on *in silico* modelling, rational protein engineering afforded a triple mutant with an impressive complete switch in regioselectivity from the "abnormal" least-substituted lactone to the more thermodynamically favored product in 39% yield and 99% selectivity (Figure 5.4b).[106]

The potential green recycling of polyesters through mild hydrolysis catalyzed by the cutinase from *Thermobifida fusca* (TfCut2) was analyzed in a recent work by Roth *et al.*[107] through experimental biochemistry and molecular dynamics simulations on the 50 ns timescale. Focusing on protein dynamics, regions of high mobility in the outskirts of the active site were identified *in silico*, which were suggested to constitute "hot-spot" residues. These residues could be targeted by protein engineering in order to impact accommodation of the polymeric substrate and the efficiency of catalysis.[107]

The evolution of the function of biocatalysts is subject to the aqueous environment within the cell. Thus, the interaction between water molecules, protein surfaces and transition state structures and their associated dynamics could be envisaged to impact biocatalysis regardless of protein fold or enzyme class. Importantly, MD-simulation allows in principle for a detailed study of networks of "spectator" water molecules that do not actively participate in catalysis, as well as the dynamical coupling between fluctuations in protein structure and water dynamics.[108] Based on this strategy, we demonstrated a key role for spectator water molecules in driving the unusual entropy-assisted biosynthesis of polycyclic molecular architectures catalyzed by an extremophilic triterpene cyclase.[109] Specifically, a number of previously unknown water tunnels in the structure were identified *in silico.* It was suggested that water passage between the bulk and the protein interior through the tunnels was linked to the experimentally observed high-entropy of activation (16 kcal mol^{-1} at 328 K). The 20 ns *in silico* simulations served as a foundation to engineer variants with obstructed water tunnels by rational protein design.[109] Experimental biocatalysis demonstrated that reconfigured water networks through the closing of tunnels had a large impact of the thermodynamics of catalysis as well as on product specificity.[109] This was further corroborated in a recent study by Hammer *et al.* for which the synthesis of substrate probes in combination with site-directed mutagenesis was used to demonstrate the impact of altered water networks on product distribution in the triterpene-catalyzed cyclization of unnatural substrates.[110] Inspired by the potential impact of water on all aspects of biocatalysis, enzyme engineering was used to rewire a water cluster in an esterase to afford enhanced transition state stabilization representing promiscuous amidase activity.[111] Specifically, one point mutation afforded an unshielded protein backbone that allowed facilitated access of water to the protein surface which enabled hydrogen bonding to the transition state in the variant. Experimental characterization in both light and heavy water confirmed the importance of water in the 34-fold enhanced catalytic efficiency and 200-fold change in reaction specificity displayed by the variant.[111] These findings highlight the potential to use MD computational tools to develop novel protein engineering technologies.

5.3.3 A Combination of QM and MM Allows for an Enhanced Understanding of Enzymes

The simplest implementation of QM and MM is perhaps through independent calculations and subsequent combination of the retrieved results. This strategy has been fruitful in studying the docking of high-energy intermediates and transition state structures to biocatalysts. In their recent work, Schrepfer *et al.* used MD to build the closed and active form of the diterpene cyclase taxadiene synthase from the crystal structure that resides in a catalytically non-productive open form.[112] Based on quantum mechanical calculations, productive conformations of the substrate as well as of high-energy carbocationic intermediates and transition states – which would be difficult to obtain through MM only – were docked into the generated model.[112] Using this approach, Schrepfer *et al.* could successfully use protein engineering for the interception of key intermediates by proton transfer, which led to altered product distribution. Furthermore, this study illuminated a key role for the departing diphosphate in acting as the final base of the cyclization reaction.[112] By bridging QM and MM, Chen *et al.*[113] investigated the biocatalytic reaction mechanism catalyzed by human oxidosqualene cyclase that generates lanosterol, a key intermediate in steroid metabolism. Using QM/MM hybrid simulations, a calculated barrier close to

22 kcal mol^{-1} was found with protonation of the epoxide being the rate-limiting step.[113] Interestingly, driven by the protonation event, the formation of the first three rings towards the polycyclic protosteryl cation was suggested to occur in a concerted fashion without any significant barrier. The incorporation of dynamical effects on the picosecond time scale by the QM/MM calculations allowed for a detailed study of the life-time of key intermediates of uttermost importance for product distribution. In another study, Faus *et al.*[114] used high-level QM/MM to probe the geometry of the peculiar $[4Fe–4S]^{2+}$ cluster confined within the LytB/IspH protein. This metallo-center is instrumental for the final step of the methylerythritol phosphate pathway in pathogenic bacteria, for which LytB/IspH catalyzes the formation of the universal terpenoid precursors isopentenyl diphosphate and dimethylallyl diphosphate. Fascinatingly, the QM/MM calculations on an $[4Fe–4S]^{2+}$ cluster ligated by three water molecules successfully reproduced experimental spectroscopic measurements.[114] Lonsdale *et al.* used QM/MM to study the concertedness of the reduction of prochiral α,β-unsaturated ketones trough *trans*-addition of hydrogen catalyzed by ene-reductases, a paradigm in the enantiomeric generation of valuable building blocks.[115] Using the ene-reductase YqjM from *Bacillus subtilis* as a model system, it was found that hydride transfer from the reduced flavin to the β-carbon of the 2-cyclohexen-1-one model substrate contributed significantly to the overall barrier. Importantly, hydride transfer occurred prior to the *trans*-addition of a proton from the opposite side of the double bond by the catalytic tyrosine.[115] Simulations on a "flipped" substrate conformation with respect to the spatial location of the double bond of the substrate in relation to the co-factor, demonstrated a higher activation energy.[115] This information is useful for the design of variants capable of the stereospecific reduction of substituted substrates for which the binding pose of lowest energy is perhaps not accessible due to steric limitations.

Sugar-based biochemistries are of high-relevance in the biocatalytic generation of building blocks for the manufacturing of platform chemicals, biofuels and medicines. Recently, Alonso-Gil *et al.* used QM/MM calculations to shed light on the reaction mechanism displayed by *exo*-1,6-α-mannosidases.[116] Specifically, the authors were able to decipher the distortedness of the reacting sugar substrate in the Michaelis complex; crucial information that was obscured in the crystal structure available at that time containing the inhibitor 1,6-α-thiomannobiose. The high-level calculations and the suggested mechanism were confirmed by experimentally using crystallography.[116]

Oxidation biochemistries constitute one instrumental biotransformation to tune bioactivities of natural products and fine chemical synthons. In a study by Rokob, QM/MM calculations were used to study potential biocatalytic reaction mechanisms for arene oxidation displayed by non-heme diiron enzymes.[117] The calculations were performed using epoxidation of aromatic rings as a model system which gave insight into the geometry and electronic features of reactive iron–oxygen species generated in the biocatalyst active site.[117] Another prime example is the regioselective hydroxylation mediated by P450 tailoring enzymes. In a recent work by Dubey *et al.*,[118] MD on the sub μs timescale was used to study the gating mechanism responsible for the conformationally selective juxtaposition of fatty acid substrates in proximity of the catalytic heme iron. QM/MM calculations on relevant snapshots from the MD-simulations shed mechanistic light on the experimentally observed regioselectivity and enantioselectivity of hydroxylation.[118]

5.4 Conclusion

In silico computer simulation can shed light on the molecular mechanisms underpinning the evolution of enzyme regioselectivity, substrate specificity and activity. Available and emerging *in silico* tools will form a new paradigm in enzyme and protein design for efficient biocatalysis.

Disclosure

The authors declare no financial interest and named commercial program suites and technologies are given as examples to facilitate method implementation for the reader.

Acknowledgements

N. M. Hendrikse greatly acknowledges funding from the Swedish Foundation for Strategic Research (SSF) through grant#ID15-0036. P.-O. Syrén greatly acknowledges the Swedish Research Council (VR) for financial support through a young investigator grant #621-2013-5138 and the PDC Center for High Performance Computing at the KTH Royal Institute of Technology for the MD-simulations.

References

1. O. Hauenstein, S. Agarwal and A. Greiner, *Nat. Commun.*, 2016, 7, 11862.
2. M. Lek, K. J. Karczewski, E. V. Minikel, K. E. Samocha, E. Banks, T. Fennell, A. H. O'Donnell-Luria, J. S. Ware, A. J. Hill, B. B. Cummings, T. Tukiainen, D. P. Birnbaum, J. A. Kosmicki, L. E. Duncan, K. Estrada, F. Zhao, J. Zou, E. Pierce-Hoffman, J. Berghout, D. N. Cooper, N. Deflaux, M. De Pristo, R. Do, J. Flannick, M. Fromer, L. Gauthier, J. Goldstein, N. Gupta, D. Howrigan, A. Kiezun, M. I. Kurki, A. L. Moonshine, P. Natarajan, L. Orozco, G. M. Peloso, R. Poplin, M. A. Rivas, V. Ruano-Rubio, S. A. Rose, D. M. Ruderfer, K. Shakir, P. D. Stenson, C. Stevens, B. P. Thomas, G. Tiao, M. T. Tusie-Luna, B. Weisburd, H.-H. Won, D. Yu, D. M. Altshuler, D. Ardissino, M. Boehnke, J. Danesh, S. Donnelly, R. Elosua, J. C. Florez, S. B. Gabriel,

G. Getz, S. J. Glatt, C. M. Hultman, S. Kathiresan, M. Laakso, S. McCarroll, M. I. McCarthy, D. McGovern, R. McPherson, B. M. Neale, A. Palotie, S. M. Purcell, D. Saleheen, J. M. Scharf, P. Sklar, P. F. Sullivan, J. Tuomilehto, M. T. Tsuang, H. C. Watkins, J. G. Wilson, M. J. Daly and D. G. MacArthur, *Nature*, 2016, **536**, 285–291.

3. H. Renata, Z. J. Wang and F. H. Arnold, *Angew. Chem., Int. Ed.*, 2015, **54**, 3351–3367.
4. S. M. Pearsall, C. N. Rowley and A. Berry, *ChemCatChem*, 2015, **7**, 3078–3093.
5. Q. Li, X. Qin, J. Liu, C. Gui, B. Wang, J. Li and J. Ju, *J. Am. Chem. Soc.*, 2016, **138**, 408–415.
6. Y.-L. Yang, H. Zhou, G. Du, K.-N. Feng, T. Feng, X.-L. Fu, J.-K. Liu and Y. Zeng, *Angew. Chem., Int. Ed.*, 2016, **55**, 5463–5466.
7. P.-S. Huang, S. E. Boyken and D. Baker, *Nature*, 2016, **537**, 320–327.
8. R. E. Lenski, M. J. Wiser, N. Ribeck, Z. D. Blount, J. R. Nahum, J. J. Morris, L. Zaman, C. B. Turner, B. D. Wade, R. Maddamsetti, A. R. Burmeister, E. J. Baird, J. Bundy, N. A. Grant, K. J. Card, M. Rowles, K. Weatherspoon, S. E. Papoulis, R. Sullivan, C. Clark, J. S. Mulka and N. Hajela, *Proc. R. Soc. London, Ser. B*, 2015, **282**, 20152292.
9. O. Khersonsky and D. S. Tawfik, *Annu. Rev. Biochem.*, 2010, **79**, 471–505.
10. GenBank statistics, https://www.ncbi.nlm.nih.gov/genbank/statistics/. Accessed 1 November 2016.
11. A. Weckbecker and W. Hummel, *Biocatal. Biotransform.*, 2006, **24**, 380–389.
12. PDB statistics, http://www.rcsb.org/pdb/static.do?p=general_information/pdb_statistics/index.html. Accessed 1 November 2016.
13. S. Govindarajan, B. Mannervik, J. A. Silverman, K. Wright, D. Regitsky, U. Hegazy, T. J. Purcell, M. Welch, J. Minshull and C. Gustafsson, *ACS Synth. Biol.*, 2015, **4**, 221–227.
14. U. T. Bornscheuer, G. W. Huisman, R. J. Kazlauskas, S. Lutz, J. C. Moore and K. Robins, *Nature*, 2012, **485**, 185–194.
15. B. M. Nestl, S. C. Hammer, B. A. Nebel and B. Hauer, *Angew. Chem., Int. Ed.*, 2014, **53**, 3070–3095.
16. I. M. Slaymaker, L. Gao, B. Zetsche, D. A. Scott, W. X. Yan and F. Zhang, *Science*, 2016, **351**, 84–88.
17. P. Xu, K. Qiao, W. S. Ahn and G. Stephanopoulos, *Proc. Natl. Acad. Sci. U. S. A.*, 2016, **113**, 10848–10853.
18. S. Schmidt, C. Scherkus, J. Muschiol, U. Menyes, T. Winkler, W. Hummel, H. Groeger, A. Liese, H.-G. Herz and U. T. Bornscheuer, *Angew. Chem., Int. Ed.*, 2015, **54**, 2784–2787.
19. J. M. Heather and B. Chain, *Genomics*, 2016, **107**, 1–8.
20. S. Goodwin, J. D. McPherson and W. R. McCombie, *Nat. Rev. Genet.*, 2016, **17**, 333–351.
21. J. Qin, R. Li, J. Raes, M. Arumugam, K. S. Burgdorf, C. Manichanh, T. Nielsen, N. Pons, F. Levenez, T. Yamada, D. R. Mende, J. Li, J. Xu, S. Li, D. Li, J. Cao, B. Wang, H. Liang, H. Zheng, Y. Xie, J. Tap,

P. Lepage, M. Bertalan, J.-M. Batto, T. Hansen, D. Le Paslier, A. Linneberg, H. B. Nielsen, E. Pelletier, P. Renault, T. Sicheritz-Ponten, K. Turner, H. Zhu, C. Yu, S. Li, M. Jian, Y. Zhou, Y. Li, X. Zhang, S. Li, N. Qin, H. Yang, J. Wang, S. Brunak, J. Dore, F. Guarner, K. Kristiansen, O. Pedersen, J. Parkhill, J. Weissenbach, P. Bork, S. D. Ehrlich and J. Wang, *Nature*, 2010, **464**, 59–65.
22. S. Lindgreen, K. L. Adair and P. P. Gardner, *Sci. Rep.*, 2016, **6**, 19233.
23. R. J. Randle-Boggis, T. Helgason, M. Sapp and P. D. Ashton, *FEMS Microbiol. Ecol.*, 2016, **92**, fiw095.
24. R. Ounit, S. Wanamaker, T. J. Close and S. Lonardi, *BMC Genomics*, 2015, **16**, 236.
25. D. E. Wood and S. L. Salzberg, *Genome Biol.*, 2014, **15**, R46.
26. A. Wilke, T. Harrison, J. Wilkening, D. Field, E. M. Glass, N. Kyrpides, K. Mavrommatis and F. Meyer, *BMC Bioinf.*, 2012, **13**, 141.
27. M. Kanehisa, M. Araki, S. Goto, M. Hattori, M. Hirakawa, M. Itoh, T. Katayama, S. Kawashima, S. Okuda, T. Tokimatsu and Y. Yamanishi, *Nucleic Acids Res.*, 2008, **36**, D480–D484.
28. R. Overbeek, T. Begley, R. M. Butler, J. V. Choudhuri, H.-Y. Chuang, M. Cohoon, V. de Crécy-Lagard, N. Diaz, T. Disz, R. Edwards, M. Fonstein, E. D. Frank, S. Gerdes, E. M. Glass, A. Goesmann, A. Hanson, D. Iwata-Reuyl, R. Jensen, N. Jamshidi, L. Krause, M. Kubal, N. Larsen, B. Linke, A. C. McHardy, F. Meyer, H. Neuweger, G. Olsen, R. Olson, A. Osterman, V. Portnoy, G. D. Pusch, D. A. Rodionov, C. Rückert, J. Steiner, R. Stevens, I. Thiele, O. Vassieva, Y. Ye, O. Zagnitko and V. Vonstein, *Nucleic Acids Res.*, 2005, **33**, 5691–5702.
29. F. Meyer, D. Paarmann, M. D'Souza, R. Olson, E. M. Glass, M. Kubal, T. Paczian, A. Rodriguez, R. Stevens, A. Wilke, J. Wilkening and R. A. Edwards, *BMC Bioinf.*, 2008, **9**, 386.
30. R. Rivest, *RFC*, 1992, **1321**, 1–21.
31. D. L. Ollis, E. Cheah, M. Cygler, B. Dijkstra, F. Frolow, S. M. Franken, M. Harel, S. J. Remington, I. Silman, J. Schrag, J. L. Sussman, K. H. G. Verschueren and A. Goldman, *Protein Eng., Des. Sel.*, 1992, **5**, 197–211.
32. R. Kourist, H. Jochens, S. Bartsch, R. Kuipers, S. K. Padhi, M. Gall, D. Böttcher, H.-J. Joosten and U. T. Bornscheuer, *ChemBioChem*, 2010, **11**, 1635–1643.
33. 3DM Platform, https://3dm.bio-prodict.nl. Accessed 1 November 2016.
34. C. C. Akoh, G.-C. Lee, Y.-C. Liaw, T.-H. Huang and J.-F. Shaw, *Prog. Lipid Res.*, 2004, **43**, 534–552.
35. E. Henke, J. Pleiss and U. T. Bornscheuer, *Angew. Chem., Int. Ed.*, 2002, **41**, 3211–3213.
36. M. G. Rossmann, D. Moras and K. W. Olsen, *Nature*, 1974, **250**, 194–199.
37. O. Dym and D. Eisenberg, *Protein Sci.*, 2001, **10**, 1712–1728.
38. P. Laurino, A. Toth-Petroczy, R. Meana-Paneda, W. Lin, D. G. Truhlar and D. S. Tawfik, *PLoS Biol.*, 2016, **14**, e1002396.

39. M. Gall, M. Thomsen, C. Peters, I. V. Pavlidis, P. Jonczyk, P. P. Grünert, S. Beutel, T. Scheper, E. Gross, M. Backes, T. Geißler, J. P. Ley, J.-M. Hilmer, G. Krammer, G. J. Palm, W. Hinrichs and U. T. Bornscheuer, *Angew. Chem., Int. Ed.*, 2014, **53**, 1439–1442.
40. E. Nakamaru-Ogiso, T. Yano, T. Ohnishi and T. Yagi, *J. Biol. Chem.*, 2002, **277**, 1680–1688.
41. R. K. Kuipers, H.-J. Joosten, W. J. H. van Berkel, N. G. H. Leferink, E. Rooijen, E. Ittmann, F. van Zimmeren, H. Jochens, U. Bornscheuer, G. Vriend, V. A. P. Martins, dos Santos and P. J. Schaap, *Proteins: Struct., Funct., Bioinf.*, 2010, **78**, 2101–2113.
42. PDB Homepage, http://www.rcsb.org/pdb/home/home.do. Accessed 1 November 2016.
43. H. Jochens and U. T. Bornscheuer, *ChemBioChem*, 2010, **11**, 1861–1866.
44. R. K. P. Kuipers, H.-J. Joosten, E. Verwiel, S. Paans, J. Akerboom, J. van der Oost, N. G. H. Leferink, W. J. H. van Berkel, G. Vriend and P. J. Schaap, *Proteins: Struct., Funct., Bioinf.*, 2009, **76**, 608–616.
45. N. T. Doncheva, K. Klein, J. H. Morris, M. Wybrow, F. S. Domingues and M. Albrecht, *BMC Proc.*, 2014, **8**, S2.
46. N. T. Doncheva, Y. Assenov, F. S. Domingues and M. Albrecht, *Nat. Protoc.*, 2012, **7**, 670–685.
47. P. K. Busk and L. Lange, *Appl. Environ. Microbiol.*, 2013, **79**, 3380–3391.
48. Peptide Pattern Recognition http://vbn.aau.dk/en/publications/peptide-pattern-recognition(1400c5df-fa69-4701-8d67-ec5c38cc963b).html. Accessed 1 November 2016.
49. Y. Gumulya and E. M. J. Gillam, *Biochem. J.*, 2017, **474**, 1–19.
50. R. Merkl and R. Sterner, *Biol. Chem.*, 2016, **397**, 1–21.
51. C. Wilson, R. V. Agafonov, M. Hoemberger, S. Kutter, A. Zorba, J. Halpin, V. Buosi, R. Otten, D. Waterman, D. L. Theobald and D. Kern, *Science*, 2015, **347**, 882–886.
52. T. Devamani, A. M. Rauwerdink, M. Lunzer, B. J. Jones, J. L. Mooney, M. A. Tan, Z. J. Zhang, J. H. Xu, A. M. Dean and R. J. Kazlauskas, *J. Am. Chem. Soc.*, 2016, **138**, 1046–1056.
53. J. W. Thornton, *Nat. Rev. Genet.*, 2004, **5**, 366–375.
54. B. G. Hall, *Phylogenetic Trees Made Easy: A How-to Manual*, Sinauer Associates, Sunderland, MA, 3rd edn, 2008.
55. J. Echave, S. J. Spielman and C. O. Wilke, *Nat. Rev. Genet.*, 2016, **17**, 109–121.
56. B. G. Hall, *Mol. Biol. Evol.*, 2013, **30**, 1229–1235.
57. J. P. Huelsenbeck and F. Ronquist, *Bioinformatics*, 2001, **17**, 754–755.
58. A. R. Leach, *Molecular Modelling: Principles and Applications*, Pearson Education Limited, 2nd edn, 2001.
59. S. E. Boyken, Z. Chen, B. Groves, R. A. Langan, G. Oberdorfer, A. Ford, J. M. Gilmore, C. Xu, F. DiMaio, J. H. Pereira, B. Sankaran, G. Seelig, P. H. Zwart and D. Baker, *Science*, 2016, **352**, 680–687.
60. F. Richter, R. Blomberg, S. D. Khare, G. Kiss, A. P. Kuzin, A. J. T. Smith, J. Gallaher, Z. Pianowski, R. C. Helgeson, A. Grjasnow, R. Xiao,

J. Seetharaman, M. Su, S. Vorobiev, S. Lew, F. Forouhar, G. J. Kornhaber, J. F. Hunt, G. T. Montelione, L. Tong, K. N. Houk, D. Hilvert and D. Baker, *J. Am. Chem. Soc.*, 2012, **134**, 16197–16206.
61. G.-J. Cheng, X. Zhang, L. W. Chung, L. Xu and Y.-D. Wu, *J. Am. Chem. Soc.*, 2015, **137**, 1706–1725.
62. D. J. Tantillo, *Nat. Prod. Rep.*, 2011, **28**, 1035–1053.
63. P. E. M. Siegbahn and F. Himo, *J. Biol. Inorg. Chem.*, 2009, **14**, 643–651.
64. C. R. A. Catlow, *Angew. Chem., Int. Ed.*, 2016, **55**, 9132–9133.
65. G. M. Piccini, M. Alessio and J. Sauer, *Angew. Chem., Int. Ed.*, 2016, **55**, 5235–5237.
66. M. W. Schmidt, K. K. Baldridge, J. A. Boatz, S. T. Elbert, M. S. Gordon, J. H. Jensen, S. Koseki, N. Matsunaga and K. A. Nguyen, *et al.*, *J. Comput. Chem.*, 1993, **14**, 1347–1363.
67. U. Essmann, L. Perera, M. L. Berkowitz, T. Darden, H. Lee and L. G. Pedersen, *J. Chem. Phys.*, 1995, **103**, 8577–8593.
68. A. Jakalian, D. B. Jack and C. I. Bayly, *J. Comput. Chem.*, 2002, **23**, 1623–1641.
69. J. A. McCammon and M. Karplus, *Proc. Natl. Acad. Sci. U. S. A.*, 1979, **76**, 3585–3589.
70. M. Levitt, *J. Mol. Biol.*, 1976, **104**, 59–107.
71. D. E. Shaw, P. Maragakis, K. Lindorff-Larsen, S. Piana, R. O. Dror, M. P. Eastwood, J. A. Bank, J. M. Jumper, J. K. Salmon, Y. Shan and W. Wriggers, *Science*, 2010, **330**, 341–346.
72. H. Nguyen, J. Maier, H. Huang, V. Perrone and C. Simmerling, *J. Am. Chem. Soc.*, 2014, **136**, 13959–13962.
73. R. J. Kazlauskas, *Curr. Opin. Chem. Biol.*, 2000, **4**, 81–88.
74. E. Krieger, T. Darden, S. B. Nabuurs, A. Finkelstein and G. Vriend, *Proteins: Struct., Funct., Bioinf.*, 2004, **57**, 678–683.
75. A. Warshel and M. Levitt, *J. Mol. Biol.*, 1976, **103**, 227–249.
76. H. M. Senn and W. Thiel, *Angew. Chem., Int. Ed.*, 2009, **48**, 1198–1229.
77. M. Höhne, S. Schätzle, H. Jochens, K. Robins and U. T. Bornscheuer, *Nat. Chem. Biol.*, 2010, **6**, 807–813.
78. C. K. Savile, J. M. Janey, E. C. Mundorff, J. C. Moore, S. Tam, W. R. Jarvis, J. C. Colbeck, A. Krebber, F. J. Fleitz, J. Brands, P. N. Devine, G. W. Huisman and G. J. Hughes, *Science*, 2010, **329**, 305.
79. I. V. Pavlidis, M. S. Weiß, M. Genz, P. Spurr, S. P. Hanlon, B. Wirz, H. Iding and U. T. Bornscheuer, *Nat. Chem.*, 2016, **8**, 1076–1082.
80. Y. Huang, P. K. Busk, M. N. Grell, H. Zhao and L. Lange, *Enzyme Microb. Technol.*, 2014, **67**, 47–52.
81. A. Nobili, Y. Tao, I. V. Pavlidis, T. van den Bergh, H.-J. Joosten, T. Tan and U. T. Bornscheuer, *ChemBioChem*, 2015, **16**, 805–810.
82. A. Nobili, M. G. Gall, I. V. Pavlidis, M. L. Thompson, M. Schmidt and U. T. Bornscheuer, *FEBS J.*, 2013, **280**, 3084–3093.
83. H. Yu and H. Huang, *Biotechnol. Adv.*, 2014, **32**, 308–315.
84. M. Genz, O. Melse, S. Schmidt, C. Vickers, M. Dörr, T. van den Bergh, H.-J. Joosten and U. T. Bornscheuer, *ChemCatChem*, 2016, **8**, 3199–3202.

85. F. Steffen-Munsberg, C. Vickers, H. Kohls, H. Land, H. Mallin, A. Nobili, L. Skalden, T. van den Bergh, H.-J. Joosten, P. Berglund, M. Höhne and U. T. Bornscheuer, *Biotechnol. Adv.*, 2015, **33**, 566–604.
86. S. Hackenschmidt, E. J. Moldenhauer, G. A. Behrens, M. Gand, I. V. Pavlidis and U. T. Bornscheuer, *ChemCatChem*, 2014, **6**, 1015–1020.
87. T. A. Williams, P. G. Foster, C. J. Cox and T. M. Embley, *Nature*, 2013, **504**, 231–236.
88. C. Natarajan, F. G. Hoffmann, R. E. Weber, A. Fago, C. C. Witt and J. F. Storz, *Science*, 2016, **354**, 336–339.
89. M. J. Harms and J. W. Thornton, *Nat. Rev. Genet.*, 2013, **14**, 559–571.
90. J. T. Bridgham, E. A. Ortlund and J. W. Thornton, *Nature*, 2009, **461**, 515–519.
91. R. Huang, F. Hippauf, D. Rohrbeck, M. Haustein, K. Wenke, J. Feike, N. Sorrelle, B. Piechulla and T. J. Barkman, *Proc. Natl. Acad. Sci. U. S. A.*, 2012, **109**(2966–2971), S2966/2961–S2966/2969.
92. R. Huang, A. J. O'Donnell, J. J. Barboline and T. J. Barkman, *Proc. Natl. Acad. Sci. U. S. A.*, 2016, **113**, 10613–10618.
93. S. Akanuma, Y. Nakajima, S. Yokobori, M. Kimura, N. Nemoto, T. Mase, K. Miyazono, M. Tanokura and A. Yamagishi, *Proc. Natl. Acad. Sci. U. S. A.*, 2013, **110**, 11067–11072.
94. K. Watanabe, T. Ohkuri, S. Yokobori and A. Yamagishi, *J. Mol. Biol.*, 2006, **355**, 664–674.
95. V. A. Risso, J. A. Gavira, D. F. Mejia-Carmona, E. A. Gaucher and J. M. Sanchez-Ruiz, *J. Am. Chem. Soc.*, 2013, **135**, 2899–2902.
96. R.-Z. Liao and W. Thiel, *J. Chem. Theory Comput.*, 2012, **8**, 3793–3803.
97. P.-O. Syren and K. Hult, *ChemCatChem*, 2011, **3**, 853–860.
98. P.-O. Syren, *FEBS J.*, 2013, **280**, 3069–3083.
99. P.-O. Syren, P. Hendil-Forssell, L. Aumailley, W. Besenmatter, F. Gounine, A. Svendsen, M. Martinelle and K. Hult, *ChemBioChem*, 2012, **13**, 645–648.
100. L. Smentek and B. A. Hess, *J. Am. Chem. Soc.*, 2011, **132**, 17111–17117.
101. V. Domingo, J. F. Arteaga, J. F. Quilez, del Moral and A. F. Barrero, *Nat. Prod. Rep.*, 2009, **26**, 115–134.
102. J. B. Siegel, A. L. Smith, S. Poust, A. J. Wargacki, A. Bar-Even, C. Louw, B. W. Shen, C. B. Eiben, H. M. Tran, E. Noor, J. L. Gallaher, J. Bale, Y. Yoshikuni, M. H. Gelb, J. D. Keasling, B. L. Stoddard, M. E. Lidstrom and D. Baker, *Proc. Natl. Acad. Sci. U. S. A.*, 2015, **112**, 3704–3709.
103. J. Zi, Y. Matsuba, Y. J. Hong, A. J. Jackson, D. J. Tantillo, E. Pichersky and R. J. Peters, *J. Am. Chem. Soc.*, 2014, **136**, 16951–16953.
104. K. C. Potter, J. Zi, Y. J. Hong, S. Schulte, B. Malchow, D. J. Tantillo and R. J. Peters, *Angew. Chem., Int. Ed.*, 2016, **55**, 634–638.
105. Z.-G. Zhang, R. Lonsdale, J. Sanchis and M. T. Reetz, *J. Am. Chem. Soc.*, 2014, **136**, 17262–17272.
106. K. Balke, S. Schmidt, M. Genz and U. T. Bornscheuer, *ACS Chem. Biol.*, 2016, **11**, 38–43.

107. C. Roth, R. Wei, T. Oeser, J. Then, C. Foellner, W. Zimmermann and N. Straeter, *Appl. Microbiol. Biotechnol.*, 2014, **98**, 7815–7823.
108. M. J. Fink and P.-O. Syren, *Curr. Opin. Chem. Biol.*, 2017, **37**, 107–114.
109. P.-O. Syren, S. C. Hammer, B. Claasen and B. Hauer, *Angew. Chem., Int. Ed.*, 2014, **53**, 4845–4849.
110. S. C. Hammer, P.-O. Syren and B. Hauer, *ChemistrySelect*, 2016, **1**, 3589–3593.
111. P. Hendil-Forssell, M. Martinelle and P.-O. Syren, *Chem. Commun.*, 2015, **51**, 17221–17224.
112. P. Schrepfer, A. Buettner, C. Goerner, M. Hertel, J. van Rijn, F. Wallrapp, W. Eisenreich, V. Sieber, R. Kourist and T. Brueck, *Proc. Natl. Acad. Sci. U. S. A.*, 2016, **113**, E958–E967.
113. N. Chen, S. Wang, L. Smentek, B. A. Hess, Jr. and R. Wu, *Angew. Chem., Int. Ed.*, 2015, **54**, 8693–8696.
114. I. Faus, A. Reinhard, S. Rackwitz, J. A. Wolny, K. Schlage, H.-C. Wille, A. Chumakov, S. Krasutsky, P. Chaignon, C. D. Poulter, M. Seemann and V. Schuenemann, *Angew. Chem., Int. Ed.*, 2015, **54**, 12584–12587.
115. R. Lonsdale and M. T. Reetz, *J. Am. Chem. Soc.*, 2015, **137**, 14733–14742.
116. S. Alonso-Gil, A. Males, P. Z. Fernandes, S. J. Williams, G. J. Davies and C. Rovira, *J. Am. Chem. Soc.*, 2017, **139**, 1085–1088.
117. T. A. Rokob, *J. Am. Chem. Soc.*, 2016, **138**, 14623–14638.
118. K. D. Dubey, B. Wang and S. Shaik, *J. Am. Chem. Soc.*, 2016, **138**, 837–845.
119. T. Hoshino, K. Shimizu and T. Sato, *Angew. Chem., Int. Ed.*, 2004, **43**, 6700–6703.
120. M. Jia and R. J. Peters, *Front. Plant Sci.*, 2016, 7, 1765.

CHAPTER 6

Modulating Enzyme Activity via *Incorporation of Non-canonical Amino Acids*

JULES KIRK, THOMAS HUBER AND COLIN J. JACKSON*

Research School of Chemistry, Australian National University, Canberra, ACT 2601, Australia
*Email: colin.jackson@anu.edu.au

6.1 Introduction

Naturally occurring proteins are linear polypeptides that are generally comprised of the twenty canonical amino acids. There are two notable examples of additional genetically-encoded amino acids, selenocysteine[1] and pyrrolysine,[2] as well as a large array of post-translational modifications of amino acids, such as phosphorylation,[3] sulfonation[4] and glycosylation.[5,6] Proteins can adopt a variety of structures and spontaneously fold in a multitude of shapes and conformations that allow them to carry out their diverse physiological roles. These roles range from being purely structural proteins, to receptors, catalysts (enzymes), and regulatory proteins, such as transcription factors.

Enzymes, which are the focus of this chapter, have been studied for over a century owing to their remarkable ability to catalyse chemical reactions. They increase the rates of chemical reactions essential to sustain life and allow them to take place on biologically relevant timescales. Over many years, our understanding of enzymatic catalysis has become increasingly sophisticated, with concepts such as transition state theory, catalytic

Catalysis Series No. 32
Modern Biocatalysis: Advances Towards Synthetic Biological Systems
Edited by Gavin Williams and Mélanie Hall

Published by the Royal Society of Chemistry, www.rsc.org

preorganization, conformational change and even quantum tunnelling, all contributing to our now relatively mature understanding of these extraordinary molecular machines.[7,8] That being said, the exact biophysical and biochemical basis of enzymatic catalysis is not without controversy. Rational engineering and design, which is the ultimate test of our understanding of protein structure and function, still has a failure rate that is too high for us to truly say that we understand their function. The use of protein mutagenesis is one of the best available approaches to expand our understanding, allowing us to probe the function of proteins by altering their structure.

Owing to their remarkable catalytic properties, enzymes rapidly became the focus of applied research and are frequently used in industrial and medical applications. For instance, enzymes have been used in the production of cheese and as digestive aids since the 19th century. Their use greatly increased through the development of recombinant gene technology, which allowed the facile production of large amounts of specific enzymes through fermentation. Large-scale production led to their widespread use in industry, for example as detergents such as subtilisin,[9] or in medicine, for example tissue plasminogen activator, which is produced recombinantly to break down blood clots.[10] The development of enzymes in biocatalysis has been greatly facilitated through the advent of engineering techniques, from rational design,[11] through to random approaches like directed evolution,[12] and onto computational protein design (see Chapters 4, 5 and 7).[13,14] This has resulted in an explosion in the use of designer enzymes, particularly in the industrial production of valuable chemical precursors.[15]

In parallel to intensive efforts to genetically engineer enzymes, in recent years there has also been increasing focus on new methodologies to chemically modify proteins. The lure of chemical modification has always been the ability to introduce functionalities that go beyond what is available among the canonical amino acids. Traditionally, the most commonly applied form of chemical modification utilizes thiol groups in a protein reacting with maleimides to form covalent bonds with other molecules.[16] This has been used to label proteins with dyes,[17] metal chelators that can act as paramagnetic lanthanide tags,[18] and metal chelators that allow the formation of entirely novel metalloenzymes.[19] A second frequently used avenue to chemically modify proteins takes advantage of reactive amine groups to allow covalent attachment of bulk chemicals. For instance, protein modification with polyethylene glycol (PEGylation) is widely used in the pharmaceutical industry,[20] because it has been shown to significantly increase the circulatory lifetime of biotherapeutics. Finally, the isoelectric point (pI) of proteins can be drastically modified by chemical modification of amino acids,[21] which can facilitate the electrostatic attachment of polymers to create hybrid proteins.[22]

Despite the advances in our understanding of enzyme function and the technical progression of our abilities to genetically engineer improved enzymes and modify them with new chemical functionalities, these areas of research have their limitations. In terms of advancing our understanding of enzymes, the ubiquitous nature of the canonical amino acids makes it difficult to use particular amino acids as mechanistic probes, or "handles" in biophysical characterization; in genetic engineering we are fundamentally limited by the functionalities and reactivity inherent in the canonical amino acids; in chemical modification, we are limited by the reactivities present in the canonical amino acids and since thiol and amine groups are present in most/all proteins, it is impossible to make specific chemical modifications in complex mixtures.

Non-proteogenic or "unnatural" amino acid (UAA) mutagenesis was first developed as a method to allow global, or residue-specific replacement of amino acids either through natural metabolic incorporation of structural analogues,[23] or chemical misacylation of tRNA,[24] to yield proteins with UAAs.[25] This approach was improved by taking advantage of suppressor read-through, allowing site-specific incorporation of amino acids using bioorthogonal machinery derived from distantly related organisms.[26–28] UAA mutagenesis is a potential solution to many of the challenges one faces when engineering enzymes, because it allows the introduction of new bioorthogonal reactive groups or biophysically unique probes at specific positions. In the following sections, we provide a brief outline to the residue-specific and site-specific approaches, before highlighting recent prominent examples of the use of UAAs for the study and engineering of enzymes. Finally, we provide some perspectives for the field in coming years, focusing in particular on the challenges to be addressed and the opportunities that are available.

6.2 Residue-specific *vs.* Site-specific UAA Incorporation

6.2.1 Residue-specific UAA Incorporation

The purpose of residue-specific UAA incorporation is to replace a given endogenous amino acid with a specific UAA analogue. Pauling[29] proposed that compounds isosteric to a given amino acid may be bound and acylated to a cognate tRNA by the respective aaRS.[30,31] This phenomenon is observed where structurally similar aliphatic amino acids isoleucine, methionine and valine often compete with the cognate leucine amino acid for activation and acylation by leucyl-tRNA synthetase (LeuRS).[32] Further, mistranslation can be an adaptive response for organism survival: in the hyperthermophilic archaea *Aeropyrum pernix*, MetRS misacylates $tRNA^{Leu}$ with methionine at low temperatures because substituting leucine with methionine in the proteome increases the flexibility of proteins, which is advantageous at low temperatures.[33] Thus, by replacing a given amino acid with an isosteric UAA analogue, the UAA can often be successfully incorporated by the endogenous aminoacyl-tRNA synthetase (aaRS):tRNA pair for the particular amino acid.[34] Alternatively, chemical misacylation of tRNA can allow incorporation of UAAs that are unable to be charged to tRNA using native tRNA synthetases (aaRSs).[24] Because most amino acids are present at multiple positions throughout a protein's primary sequence (and throughout the proteome), replacement of an amino acid in the cell or in cell-free translation media with an isosteric analogue results in global substitution of that amino acid throughout the protein.[35]

Selective Pressure Incorporation (SPI) is the most commonly used methodology for *in vivo* residue specific UAA incorporation.[36] The procedure is founded upon work in the late fifties, which successfully replaced methionine with the isosteric analogue selenomethionine throughout the *E. coli* proteome.[37] For selenomethionine incorporation, auxotrophic cells that have been genetically engineered to lack the biosynthetic pathways required for the synthesis of methionine are most commonly used, making protein translation dependent on the availability of methionine (or an isosteric analogue) in the growth medium. SPI has been further refined by recombinant DNA technology, which can amplify the expression of a target gene under the control of an artificial promoter.[38] This minimizes any detrimental effects to cell growth and is more economical because the isosteric amino acid analogue is supplied only when cells have reached the appropriate growth stage and overexpression of the recombinant protein has been initiated.[39] The combination of these metabolic engineering advances has allowed for almost any gene to be expressed in conditions where a canonical amino acid has been replaced with its isosteric analogue to allow residue-specific incorporation. It is important to recognize that the SPI methodology depends on an endogenous aaRS misacylating tRNA with a UAA. If this process is inefficient, low UAA incorporation is the result. This can be addressed by simple overexpression of specific aaRSs or through directed evolution or rational design of aaRS for improved UAA affinity.[40]

The primary advantage of residue specific incorporation is the ability to incorporate multiple amino acid substitutions throughout the protein sequence.[41] This multiplicity of substitutions can result in cumulative, or synergistic effects of the UAAs, yielding much larger improvements to certain properties than can be achieved with single amino acid substitutions. For example, this has been exploited to increase the thermal stability of coiled-coil proteins *via* introducing hydrophobic side chains at the helix interface.[42] However, multiple mutations in a protein can also be problematic because a single deleterious substitution may cause the enzyme to become structurally unstable or lose catalytic activity, and thereby masking the effects of other neutral or beneficial substitutions. In general, the probability of the protein misfolding and exhibiting reduced activity increases with the difference between the physical and chemical properties between the canonical amino acids and the UAA replacement.[35]

6.2.2 Site-specific UAA Incorporation

Owing to the lack of control over which amino acids are substituted with UAAs using residue-specific incorporation, site-specific UAA incorporation is a valuable alternative methodology. Site-specific UAA incorporation is comparatively more involved than residue-specific incorporation in terms of the underlying methodology. It requires a "bio-orthogonal pair" of an UAA-specific aminoacyl-tRNA synthetase (uaaRS) and a cognate tRNA that recognises a nonsense or frameshift codon at which the UAA is inserted (Figure 6.1).[43] For example, a bio-orthogonal suppressor tRNA that is complementary to an amber stop codon will allow read-through of the TAG stop codon,[27,44] with UAA incorporation occurring due to the bioorthogonal tRNA having been aminoacylated with an UAA by a cognate bio-orthogonal aaRS.[45,46] tRNA–aaRS pairs from different domains of life are often bio-orthogonal and provide the platform to expand the genetic code of an organism with additional UAAs. The most commonly used approaches for the genetic encoding of UAAs in bacterial cells utilize mutants of the tyrosyl-RS and $tRNA^{Tyr}$ pair from the archaeon *Methanococcus jannaschii*, which do not cross-react in bacterial cells due to a unique recognition of the tRNA C1-G72 base-pair by *aa*RS.[47]

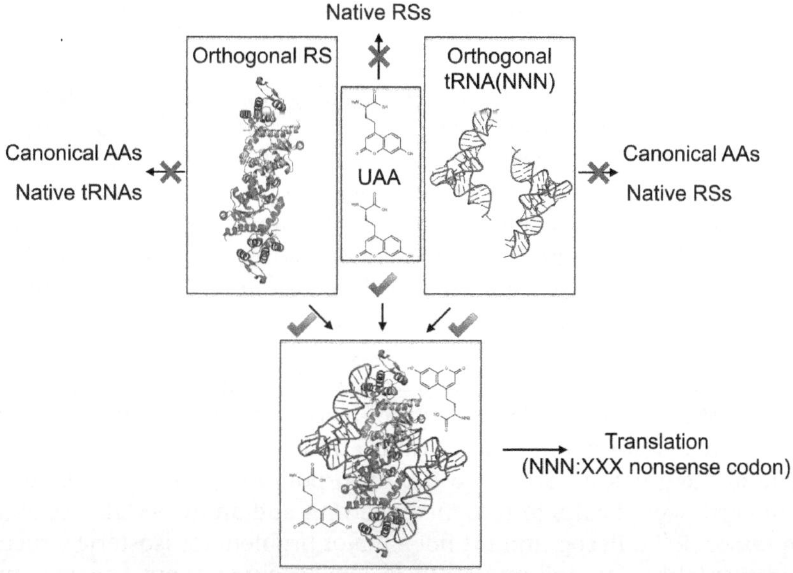

Figure 6.1 Orthogonal translational machinery for *in vivo* UAA incorporation. A tRNA synthetase from a distantly related organism is introduced to the cell and is bio-orthogonal because it does not interact with endogenous tRNA molecules, and is engineered to interact with UAAs in place of canonical/endogenous amino acids. An orthogonal tRNA from a distant related organism is introduced that is recognised by the orthogonal tRNA synthetase, but not recognised by endogenous tRNA synthetases. The tRNA is modified to possess a (typically) nonsense anticodon that will recognise a complementary nonsense codon. This orthogonal tRNA synthetase:tRNA pair then charges the orthogonal tRNA with the UAA; the nonsense codon (usually stop codon, such as UAG) results in suppression of translation termination, resulting in incorporation of the UAA.

6.2.3 *In Vitro* and *In Vivo* UAA Incorporation

Most of the work described above focuses on UAA incorporation *via* natural systems, such as mammalian cells,[48,49] yeast,[50] and prokaryotes.[27] *In vitro* site specific incorporation by cell-free protein synthesis is an alternative and increasingly popular avenue for mutagenesis.[51] The method is generally considered to be less time consuming and more economical, while simultaneously providing more control over components in the reaction mixture.[52] To this end, cell-free protein synthesis can be even applied in cases when the UAA is toxic to cells,[53] and is well suited when chemically misacylated tRNAs are used.[24] While exact details of the protocol may vary to satisfy individual goals, the essential steps are obtaining and adding all the components necessary for RNA transcription and protein translation to a reaction mixture containing an energy source and amino acids.[53]

6.3 Engineering tRNA Synthetases for UAA Incorporation

No discussion of the role of UAAs in enzymology would be complete without first covering how the enzymes that are responsible for incorporating UAAs have themselves been modified to carry out this role. Amino acids are 'charged' to their cognate tRNAs *via* a two-step process catalysed by aaRSs. The amino acid is first adenylated by an ATP molecule in the active site of the aaRS before the amino acid moiety of the resulting amino acid–AMP complex is transferred to the 2′ or 3′ hydroxyl group of the ribose moiety on the 3′ end of the terminal adenine of the tRNA molecule.[54] The various aaRSs and tRNAs form cognate pairs that each correspond to one of the twenty canonical amino acids. Specific and selective binding of amino acids in the aaRSs combined with an elaborate proof-reading process facilitates accurate tRNA aminoacylation and ensures translational fidelity.

One of the greatest barriers to UAA-incorporation is the requirement for the bio-orthogonal aaRS to recognise the UAA and aminoacylate the amber-suppressor tRNA. Recognition is not a major problem for isosteric structural analogues (although this can result in low specificity), but becomes problematic when the structure of the UAA diverges significantly from the natural amino acid. In these instances, it is necessary to use protein engineering to alter the substrate specificity of the aaRS. Directed evolution has been one of the most successful approaches, and begins with the generation of a large library of mutants that have been generated from a bio-orthogonal wild-type protein, for instance through the use of error-prone PCR.[55,56] Alternatively, computational design is initiated with docking and energy minimization programs, such as Rosetta.[57] Here, substrates can be superimposed on a known aaRS structure, as obtained through X-ray crystallography. Atomic interactions between residues around the active site and the substrate are then scored. Favourable interactions are hypothesized to increase binding affinity to the UAA without negatively impacting on the protein, and mutations with low energy scores are then chosen for experimental testing.[58,59] These methods are thought to be complementary when designing aaRSs that are specific for UAA, because rational design can efficiently explore larger variations in sequence whereas directed evolution is very apt to improve on basal function.[60]

In the variant screening procedure developed by Schultz *et al.*,[61] mutant sequences go through a double sieve selection process to determine their function and translation fidelity *in vivo*. Initially, the antibiotic resistance of the bacteria assesses positive selection; resistance is achieved if the UAA-charged tRNA, aminoacylated by the mutant aaRS, effectively suppresses a nonsense or frameshift codon in the chloramphenicol acetyl transferase gene. Remaining sequences subsequently undergo negative selection. In a growth medium that contains only the 20 proteogenic amino acids, the uaaRS/$tRNA^{UAA}$ pair will be co-expressed with a toxic barnase gene containing nonsense or frameshift codons at permissive sites. If the engineered aaRS misacylates an endogenous amino acid to the $tRNA^{UAA}$, and the amino acid successfully fills the permissive sites in the toxic gene, cell death ensues.[46]

6.4 Enzyme Engineering with UAAs

6.4.1 UAAs for Increased Protein Thermostability

Since the first years of protein engineering, researchers have been seeking ways to increase the thermostability of enzymes, particularly for industrial applications and developing more robust and longer-lived enzymes for use in biocatalysis.[62] As with traditional protein mutagenesis and chemical modification, UAA incorporation has been used to increase protein stability (Figure 6.2).

The fluorous effect has been widely exploited in medicinal chemistry and industry to yield molecules with altered physicochemical properties, with one of the most prominent examples being Teflon.[63] Similarly, one of the most common applications of UAAs in enzyme engineering is the incorporation of fluorinated analogues.[64] Incorporation of fluorinated amino acids is particularly suited to residue-specific approaches because of the minimal structural changes that fluorination causes, which makes the isosteric amino acid analogues compatible with existing translation machinery, while the novel physicochemical properties of fluorinated UAAs can result in marked changes to the target enzymes. Several proteins containing one, two, or three different simultaneously fluorinated amino acids have been produced. For example, Budisa *et al.* have produced a lipase from *Thermoanaerobacter thermohydrosulfuricus* with (4*S*)-fluoroproline, 4-fluorophenylalanine, and 6-fluorotryptophan analogues in place of proline, phenylalanine and tryptophan, respectively.[65] They successfully incorporated monofluorinated amino acids at 24 different positions in a single

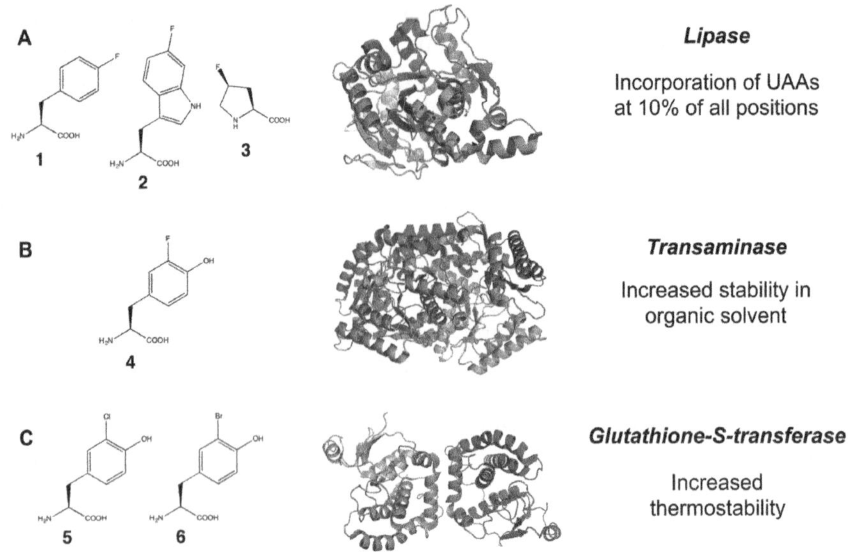

Figure 6.2 The use of UAA in engineering enzymes with enhanced stability. (A) Phenylalanine, tryptophan and proline were replaced by 4-fluorophenylalanine (**1**), 6-fluorotryptophan (**2**), (4*S*)-fluoroproline (**3**), respectively, in a lipase from *T. thermohydrosulfuricus* (represented by the structurally similar lipase from *Geobacillus* sp. SBS-4S, 3AUK),[115] resulting is incorporation of fluorinated UAAs at almost 10% of the sequence positions in the enzyme with no loss of stability. (B) Tyrosine was replaced by the UAA 3-fluorotyrosine (**4**) in a residue-specific manner in the *Vibrio fluvialis* ω-transaminase (4E3Q),[116] resulting in significantly enhanced stability in organic solvents. (C) 3-Chloro- (**5**) and 3-bromo-L-tyrosine (**6**) UAAs were used to replace native tyrosine residues in glutathione-*S*-transferase from *Schistosoma japonicum* (4WR4),[68] leading to significant increases in the thermostability of the enzyme.

expression, corresponding to approximately 10% of the amino acid sequence. Notably, the enzyme retained native structure and significant activity. Yun *et al.* have shown that, by non-specifically incorporating 3-fluorotyrosine into ω-transaminase in place of tyrosine, the fluorinated UAA was able to enhance the thermostability of the enzyme.[66] They also demonstrated that the effects of fluorotyrosine led to increased stability in organic solvent, for instance, increasing the half-life of the enzyme in 20% (v/v) DMSO from less than 10 hours to approximately 50 hours. However, in contrast to these positive reports of the effects of fluorinated UAAs on stability, Holzberger and Marx have shown that multifluorination of DNA polymerase resulted in very little change to the catalytic activity, but did cause loss of thermostability.[67]

Recently, Sakamoto *et al.* have shown that the thermostability of enzymes can also be enhanced by incorporating bulky halogenated UAAs, such as 3-chloro- and 3-bromo-L-tyrosines, at certain sites in glutathione *S*-transferase.[68] The level of stabilization that was achieved was significant (up to 5.6 kcal mol^{-1}). In this study, the author explored the biophysical basis for the stabilizing effect through the use of X-ray crystallography, which showed that the additional steric bulk from the halogen moieties provided a stabilizing effect by filling internal cavities and forming additional van der Waals interactions. Filling internal cavities in proteins is a well-established mechanism of enzyme stabilization.[69] This mechanistic understanding of the stabilizing effect of these UAAs allowed the authors to extend this work to include the facile stabilization of an industrially relevant azoreductase in the same study.

Altogether, these examples reinforce two observations: (i) at certain positions, mutations can be highly destabilizing, and (ii) UAAs can also provide stabilizing effects in excess of those that can be conferred by canonical amino acids. This suggests that, as our knowledge of the fluoro-effect develops alongside the development of our understanding of the epistatic, context dependent, effects of mutations,[70,71] we are likely to see fluorinated amino acids incorporated more widely, particularly in industrial biocatalysts, owing to their ability to increase stability and resilience to harsh industrial environments.

6.4.2 UAAs for Increased Catalytic Efficiency

When it comes to enzyme engineering, one of the foremost motivations is to increase the activity of these catalysts. As such, there are thousands of reports of better combinations of canonical amino acids within and around the active site of an enzyme, which led to increased activity. A number of approaches have been utilized to achieve this, from rational design,[72] through directed evolution approaches,[73] and most recently, computational design (see Chapters 4 and 7).[13,14] However, these approaches are often limited by the intrinsic reactivity of canonical amino acids; for instance, serine, one of the most common nucleophiles in enzymes, is a relatively poor nucleophile at physiological pH unless it is part of a catalytic triad including histidine and aspartic or glutamic acid. When additional reactivity is required, enzymes have evolved the ability to incorporate non-covalently bound organic or inorganic cofactors, which complicate the development of designer enzymes due to the need to retain, synthesise or recycle these cofactors. UAA mutagenesis is a potential route to producing cofactor-free enzymes with unique chemical properties that can extend beyond those found in nature (Figure 6.3).

Using a residue-specific approach, Antranikian and colleagues have investigated UAA incorporation in a lipase from *Thermoanaerobacter thermohydrosulfiricus.*[74] They showed that global replacement of methionine residues throughout the enzyme with the UAA norleucine resulted in an ~10-fold increase in the enzymatic activity without thermal activation. This was suggested to be due to the significantly greater hydrophobicity of norleucine than that of methionine, which could enhance the hydrophobicity of the lid domain and interactions with hydrophobic lipids.

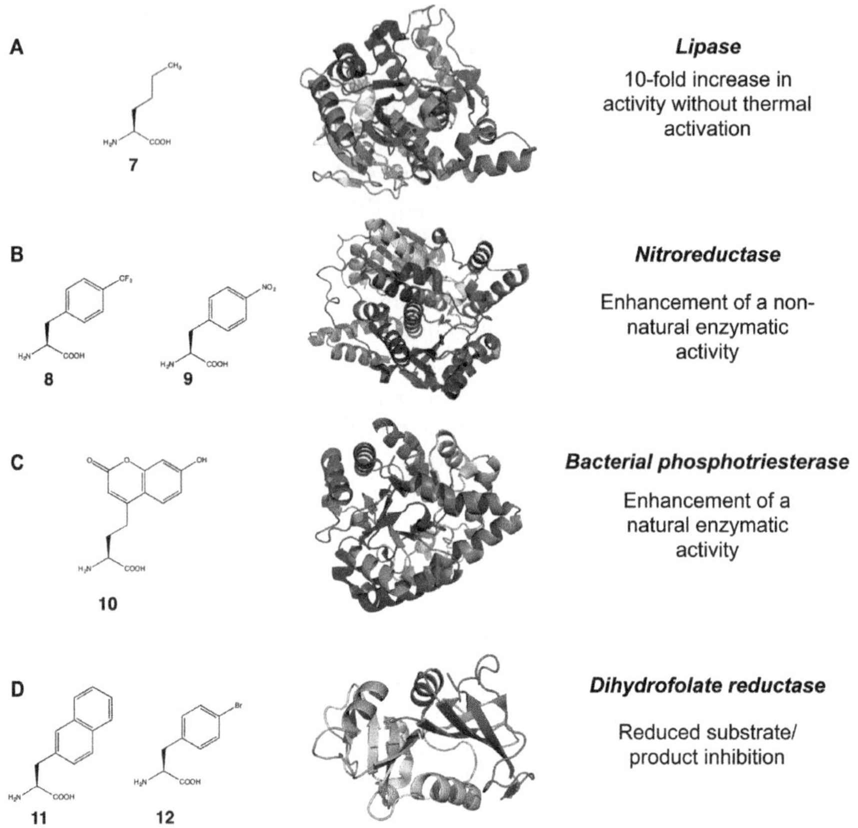

Figure 6.3 The use of UAA in engineering enzymes with improved catalytic activity. (A) Residue-specific replacement of methionine residues with the UAA norleucine (7) in a lipase from *T. thermohydrosulfuricus* (represented by the structurally similar lipase from *Geobacillus* sp. SBS-4S, 3AUK),[115] resulted in a 10-fold increase in catalytic activity without enzyme thermal activation. (B) In investigating whether UAA mutagenesis could lead to better catalytic activity than mutagenesis with naturally occurring amino acids, it was found that replacement of a phenylalanine amino acid by either *p*-trifluoromethylphenylalanine (**8**) or *p*-nitrophenylalanine (**9**) in a nitroreductase from *E. coli* (1IDT)[117] was significantly superior than mutation to Tyr, Lys or Asn. (C) When a tyrosine residue within the substrate binding site of a bacterial phosphotriesterase (2R1N)[118] was mutated to a hydroxycoumarinyl amino acid (**10**), the activity was increased to a larger extent that could be achieved by screening thousands of different randomly generated natural mutants of the enzyme. (D) A phenylalanine at position 31 was substituted by 2-napthylphenylalanine (**11**) and *para*-bromo-phenylalanine (**12**) in murine dihydrofolate reductase (1U70),[119] resulting in significantly lower substrate and product inhibition.

Site specific mutagenesis has also been used in enzyme engineering: Mehl and coworkers have shown that the use of UAAs can be advantageous in enzyme engineering and allows rapid improvement of enzyme activity.[75]

In that work, they studied a nitroreductase from *Escherichia coli* that can activate a prodrug used in cancer therapy, CB1954,[76] showing that improvement of the enzyme activity was not only possible (by >30-fold) using UAAs, but that UAAs also out-performed native amino acids at this position. Substitution of phenylalanine at position 124 with *p*-trifluoromethylphenylalanine or *p*-nitrophenylalanine was significantly superior than mutation to Tyr, Lys or Asn. The work also demonstrated that UAAs can lead to the development of enzymes that have greater catalytic efficiency than those that can be generated using the naturally occurring amino acids. Similarly, Fasan and colleagues have investigated the effect of UAAs on the regio-selectivity of cytochromes P450 using a site-specific approach.[77] The addition of *p*-aminophenylalanine, in particular, resulted in a 5-fold increase in the oxidation rate of (+)-nootkatone. Importantly, the functional changes that were observed could not be replicated through the use of any of the 20 canonical amino acids, reinforcing the idea that the UAAs provided new and otherwise inaccessible structure and function to these modified enzymes.

In contrast to the previous studies, which investigated enzymatic catalysis of synthetic substrates or substrates for which the enzyme had not evolved under strong selective pressure to turn over, recent work examined a naturally evolved and highly efficient ($k_{cat}/K_M \approx 10^7$ M^{-1} s^{-1}) phosphotriesterase[78,79] with the ability to detoxify and protect against neurotoxic organophosphates.[80,81] This enzyme was engineered using UAAs to test whether highly efficient natural enzymes can be improved further through UAAs, *i.e.* whether the catalytic potential of the enzyme can be advanced beyond what is possible through natural evolution. Several efforts to increase the activity of the phosphotriesterase through rational design and directed evolution with canonical amino acids, effectively sampling hundreds of thousands of different mutations, yielded only a modest improvement in activity (4-fold) with the substrate paraoxon.[82] In contrast, a single tyrosine > hydroxycoumarinyl amino acid replacement at the substrate binding site yielded an order of magnitude greater activity, establishing that the catalytic limits observed in nature can indeed be exceeded through the use of UAA mutagenesis.[83]

Another area of interest in the context of increasing enzyme efficiency and usefulness involves reducing the level of substrate or product inhibition that enzymes undergo, as substrate/product inhibition has the potential to lessen substrate loading and/or reduce total substrate turnover in biocatalytic applications.[84] The unique attributes of UAAs also have potential in this area. Zheng and Kwon have used site-specific UAA incorporation into murine dihydrofolate reductase (mDHFR) as a model system to control the level of inhibition by methotrexate.[85] Structural analysis highlighted the role that phenylalanine 31 played in substrate and inhibitor binding. This residue was then mutated to *p*-bromophenylalanine and 2-napthylphenylalanine. These unnatural variants of mDHFR displayed interesting kinetic characteristics, with the *p*-bromophenylalanine and 2-napthylphenylalanine variants exhibiting ∼2-fold and ∼4-fold higher dissociation constants (K_d) for the inhibitor methotrexate. Surprisingly, both variants also exhibited higher affinity as measured by the Michaelis constant (K_M) for the substrate dihydrofolate (6.5 μM for the wild-type *vs.* 3.4 and 4.8 μM for the respective variants). Although this was accompanied by a trade-off in turnover number (k_{cat}) in the case of the 2-napthylphenylalanine variant, the *p*-bromophenylalanine variant actually displayed enhanced substrate turnover leading to a 4-fold increase in the ratio of inhibitor affinity to substrate affinity (K_d/K_M).

6.4.3 UAAs to Alter Specificity and Selectivity

The desire to increase the catalytic efficiency of enzymes is partly academic—to understand how we can improve enzymes in a general sense—but also derives from a need to make more efficient catalysts for medical and industrial applications. However, any efforts to increase catalytic efficiency assume that the enzyme already has the necessary substrate specificity. In many medical and industrial applications, the target substrate is synthetic and no enzymes have naturally evolved to recognise it. Thus, there is significant interest in engineering enzymes for altered substrate specificity (Figure 6.4). A particular subfield within this topic is the intense research into modulating enzymatic stereo/regioselectivity, which is of particular relevance to the chemical and pharmaceutical industry, where production of the correct stereoisomer can be of particular importance.[86]

Using a residue-specific approach, Fasan and colleagues have investigated the effect of UAAs on the regioselectivity of cytochromes P450.[77] In this work, a number of different aromatic UAAs were introduced at the substrate-binding site and these variants were screened against (*S*)-ibuprofen and (+)-nootkatone. The addition of the UAAs in place of Ala82 dramatically increased the rate of oxidation of (+)-nootkatone to form an allylic alcohol (from 4% oxidation to 62% oxidation in the case of a *p*-acetylphenylalanine) and replacement of Ala78 with *p*-acetylphenylalanine catalysed the oxidation of (+)-nootkatone to (9*R*)-hydroxynootkatone with absolute stereoselectivity, when this reaction did not proceed at detectable rates with the native enzyme. Similarly, replacement of Leu75 with *p*-aminophenylalanine resulted in a significant increase in the oxidative conversion of (*S*)-ibuprofen to produce a tertiary alcohol (from 38% to 86%). Additional residue-specific work to alter enzyme specificity includes the global incorporation of *m*-fluorophenylalanine, replacing phenylalanine, in a lipase from *Thermoanaerobacter thermohydrosulfiricus*, which also resulted in increased substrate range,[74] with the lipase displaying greater activity with a range of triglyceride chains, both greater and shorter than the wild-type enzyme, which is relatively specific for C6–C8 substrates.

A notable example of the use of site-specific UAA mutagenesis to enhance the enantioselectivity of diketoreductase has been recently reported by Chen and colleagues.[87] Diketoreductase is an important biocatalyst that is capable of stereoselectively reducing β,δ-diketo esters to dihydroxy products.[88]

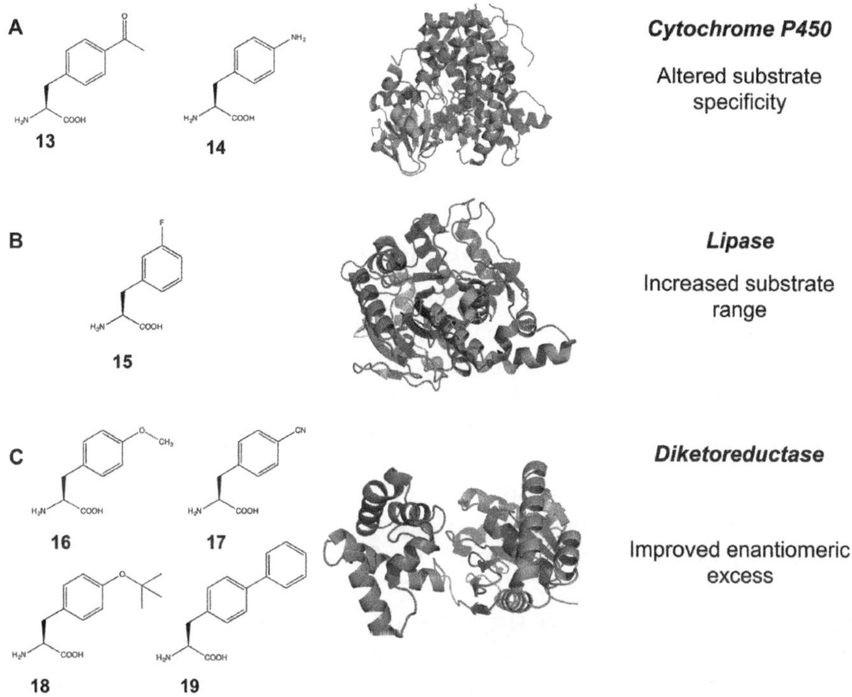

Figure 6.4 The use of UAA in engineering enzymes with altered substrate specificity. (A) The substrate specificity of a cytochrome P450 (2IJ2)[120] was altered by introducing the UAAs *para*-acetyl-phenylalanine (**13**) and *para*-amino-phenylalanine (**14**) into the active site of the enzyme. (B) The lipase from *T. thermohydrosulfuricus* (represented by the structurally similar lipase from *Geobacillus* sp. SBS-4S, 3AUK)[115] was engineered by the residue-specific substitution of phenylalanine with *m*-fluorophenylalanine (**15**), leading to an increase in the substrate range. (C) By introducing a series of bulky UAAs (4-methoxy-L-phenylalanine (**16**), 4-cyano-L-phenylalanine (**17**), *O*-*tert*-butyl-L-tyrosine (**18**) and 4-phenyl-L-phenylalanine (**19**)) in place of a native tryptophan at position 222 in a diketoreductase (4E12),[89] the enantiomeric excess of the enzyme could be altered.

By using structure-guided rational design, they pinpointed Trp149 and Trp222 as important residues for substrate binding.[89] Trp222 was replaced by several canonical amino acids (valine, leucine, methionine, phenylalanine, tyrosine) or UAAs (4-cyano-L-phenylalanine, 4-methoxy-L-phenylalanine, 4-phenyl-L-phenylalanine, *O*-*tert*-butyl-L-tyrosine). The kinetic behaviour of the variants revealed that the enantiomeric excess of the wild-type enzyme for the *R*-isomer of 2-chloro-1-phenylethanone could be increased from 9.1% for the wild-type enzyme to 10.5%, 29.7% and 33.7% when tryptophan was replaced, by 4-methoxy-L-phenylalanine, 4-phenyl-L-phenylalanine and *O*-*tert*-butyl-L-tyrosine, respectively. This is an excellent example of an UAA providing properties not otherwise accessible by the 20 canonical amino acids. In this case, the greater enantiomeric excess was a result of greater steric bulk of the residue at this position as it controls access of the respective isomers into the active site. Thus, tryptophan, which is the largest canonical amino acid by volume (147.1 Å^3), could not be improved upon by any canonical amino acid, but could be more favourably replaced by the larger UAAs 4-methoxy-L-phenylalanine (166.7 Å^3), 4-phenyl-L-phenylalanine (181.9 Å^3) and *O-tert*-butyl-L-tyrosine (185.3 Å^3).

6.4.4 UAAs to Probe Enzyme Function and Mechanism

One of the most valuable applications of UAA mutagenesis and incorporation, given the practical challenges that still need to be overcome for broad utilization in industrial biocatalysis,[90] is for the UAAs to be used as mechanistic probes to allow us to better understand enzyme structure, function and catalytic mechanisms. One of the earliest examples of the use of UAAs to probe enzyme structure and function was from Dupureur *et al.*, who used residue-specific incorporation of *o*-, *m*- or *p*-fluorophenylalanine into the *Pvu*II restriction endonuclease.[91] Notably, the incorporation of *m*-fluorophenylalanine did not affect conformational stability but caused a doubling of specific activity. Analysis of the effects that substitutions that were remote from the active site had on the catalytic activity of the enzyme allowed the authors to conclude that relatively subtle changes in enzyme conformation due to remote mutations can have significant impact on endonuclease activity and suggests novel ways to influence catalytic behaviour (Figure 6.5).

In a recent and particularly elegant study, Boxer and colleagues have used UAA mutagenesis to dissect the proton delocalization in the hydrogen bond network of the model enzyme ketosteroid isomerase.[92] This work was carried out to examine the role of extended hydrogen bond networks in enzymes due to their common occurrence. Extending work that suggested quantum delocalization of protons across a triad of Tyr16, Tyr32, and Tyr57 residues with very close bonding distances and an unusually low pK_a value in the ketosteroid isomerase active site was an important factor in the understanding of the catalytic mechanism of the enzyme.[93,94] Using site-specific UAA mutagenesis, the tyrosine residues were replaced, separately with 3-chlorotyrosine analogues, to investigate the delocalization of protons in the triad (since the reduced pK_a value of the 3-chlorotyrosine disrupts the proton affinity balance, but not the H-bonding). Biophysical analysis with X-ray crystallography, UV-Vis spectroscopy and ^{13}C-NMR revealed that the proton delocalization within the triad was affected by the replacements, highlighting the complexity of enzymes and the role of quantum effects such as proton delocalization in enzyme catalysis.

Recently, Otting and colleagues have used site-specific UAA incorporation to introduce *O-tert*-butyltyrosine into the DnaB helicase from *Stearothermophilus.*[95] DnaB unwinds double stranded DNA during DNA replication and functions as part of the larger replisome complex.[96] *O-tert*-Butyltyrosine was particularly useful as a NMR probe, because the nine

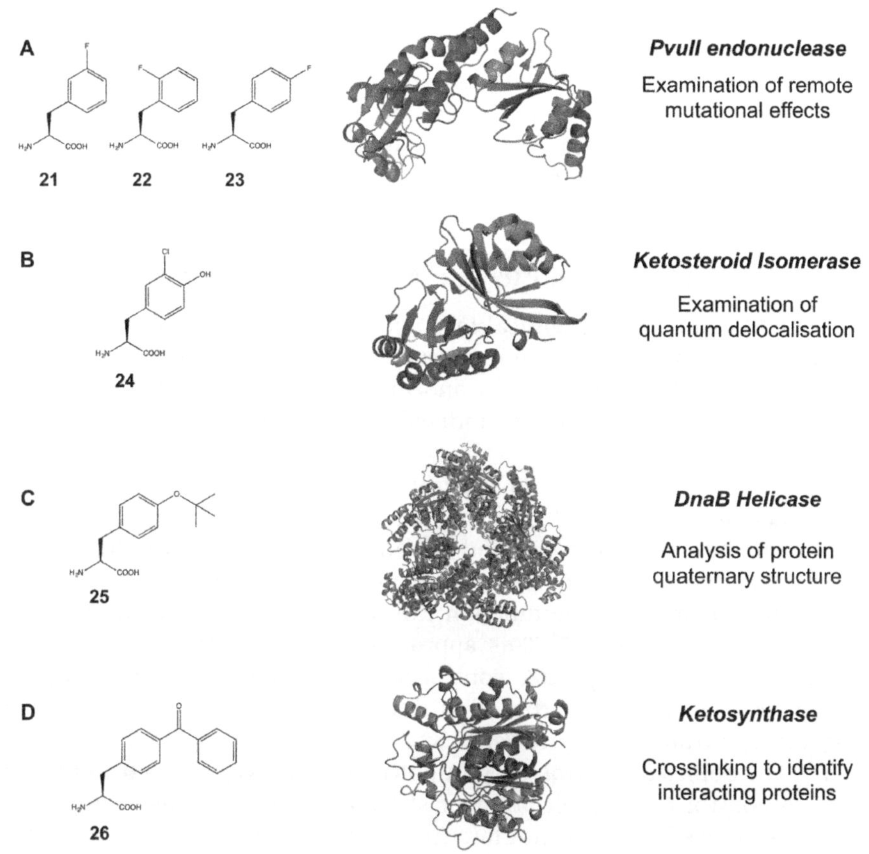

Figure 6.5 The use of UAAs in engineered enzymes to study structure and function. (A) The residue-specific incorporation of *o*-, *m*- or *p*-fluorophenylalanine (**21, 22, 23**) in place of phenylalanine in the *Pvu*II restriction endonuclease (1PVU)[121] allowed for analysis of subtle, remote, structural effects on enzyme activity. (B) In studying the contribution of quantum delocalization to the catalytic activity of a ketosteroid isomerase (5D82),[92] the replacement of certain tyrosine residues with 3-chlorotyrosine (**24**) was useful because the UAA has a significantly lower pK_a than the native amino acid, allowing selective disruption of the proton delocalization. (C) The use of the UAA *O*-*tert*-butyltyrosine (**25**) in the study of the DnaB helicase from *E. coli* (represented here by DnaB from *A. aeolicus*)[122] allowed the quaternary structure of the enzyme to studied by NMR spectroscopy. (D) Using *p*-benzoyl-L-phenylalanine (**26**), an UAA capable of undergoing photo-induced crosslinking, the interaction between acyl carrier proteins and ketosynthase (represented by the ketosynthase-chain length factor heterodimer; 1KAS)[123] could be studied in unprecedented detail.

protons of a solvent-exposed *tert*-butyl group produce a narrow and intense single signal which stands out from the background of other methyl resonances in the protein. This unique chemical property means that the

O-tert-butyl group can be easily observed in one-dimensional proton NMR spectra, eliminating the need for expensive isotope labeling and time consuming multi-dimensional NMR experiments. Remarkably, even though NMR is most often used for analysis of relatively small proteins (<50 kDa), the 320 kDa hexamer of DnaB could be shown to adopt 3-fold, rather than 6-fold, symmetry in solution.

6.4.5 UAAS to Control Enzyme Activity and Interactions

Just as UAAs can act as chemical tags and handles that allow us to gain a deeper understanding of a protein's structure and mechanism, they can also be used to control enzyme activity and interactions. This is of tremendous value in allowing us to better understand enzyme function in complex biological environments.

Photo-labile UAAs have found application in the study of enzyme function due to their ability to act as figurative "cages" for chemical reactivity: in their caged form they have no activity and it is only after the UAA is irradiated with UV light that it undergoes chemical decomposition to yield the active amino acid. Consequently, enzyme function can be tightly controlled through the use of photo-chemistry.[97,98] This approach has recently been used in enzymology, allowing the precise control of T7 RNA polymerase with a photocaged lysine derivative that was incorporated through the use of engineered, orthogonal pyrrolysine synthetase/tRNA pair in mammalian cells.[99] The ability to control T7 RNA polymerase activity allowed Hemphill *et al.* to control the transcription of short RNA hairpins that could subsequently be used for RNA interference of other genes, creating a powerful optogenetic system for light-controlled manipulation of biology. This work has been extended further with the development of small molecule caging approaches. In work by Deiters *et al.*, phosphine-mediated Staudinger reduction was used to uncage an unnatural azidobenzyloxycarbonyl amino acid to activate protein function.[100] This UAA was again genetically encoded using an orthogonal pyrrolysyl tRNA synthetase/tRNA pair in mammalian cells. In this application, the authors successfully demonstrated control over the activity of luciferase to induce fluorescence, as well CRISPR/Cas9 gene editing.

UAAs have also found a niche in the study of enzyme:protein interactions, primarily through their applicability to photo-cross-linking. For example, Williams *et al.* have used the photo-cross-linking UAA *p*-benzoyl-L-phenylalanine to probe the interactions between acyl carrier proteins and ketosynthase subunit in fatty acid synthase.[101] By using photo-cross-linking, a variety of acyl carrier proteins were tested, highlighting the role of residues within the conserved helix II region of the acyl carrier proteins. This work is notable because it demonstrated that UAAs are of particular value in moderate throughput screening experiments owing to their specificity and bio-orthogonality.

6.4.6 UAAs for Chemical Modification of Enzymes

Protein chemical modification is of increasing interest to the enzymology community. For example, chemical modification of non-catalytic scaffolds can allow the attachment of reactive groups to generate a catalyst,[102] specific labelling of an enzyme with a fluorescent tag to monitor function,[103] and PEGylation to protect enzymes and increase their circulatory lifetime and stability.[104] One of the biggest challenges in the area of enzyme chemical modification is the need for specificity—it is often necessary to chemically modify a single protein in a complex mixture, or a single position in a protein. The development of new labelling methodologies through UAA mutagenesis appears to have become one of the largest sub-fields in the applications of UAAs to enzyme engineering. The value of UAAs in this approach is their bio-orthogonality: whereas almost all proteins include amine groups, and most include thiol groups, the addition of an UAA results in the incorporation of a unique amino acid, with reactivity not otherwise present among the proteome, thereby allowing specific labelling (Figure 6.6).[105]

One of the most commonly used chemical modifications of enzymes is PEGylation, which has been shown to increase the stability and circulatory lifetime of medically useful enzymes.[104] In one of the earliest examples of UAA incorporation, Schultz and colleagues have shown that *p*-azidophenylalanine, which can be incorporated in a site-specific manner in proteins, can be used in a [3 + 2] cycloaddition reaction with an alkyne derivatised PEG reagent for PEGylation of superoxide dismutase,[106] building on previous work demonstrating that an azide–alkyne pair can be used in copper-mediated Huisgen [3 + 2] cycloaddition.[107]

Similar work has also exploited copper-catalysed alkyne–azide cycloaddition click reactions to allow labelling of T4 lysozyme with fluorescent probes. By labelling one cysteine amino acid with a fluorescent dye using maleimide chemistry, and incorporating UAAs such as propargyllysine, which contains an alkyne functionality that reacts with azide-containing dyes,[105] two dyes could be attached that can then undergo Förster Resonance Energy Transfer, or FRET (previous work utilized *p*-acetylphenylalanine, which contains a ketone functionality that reacts with hydroxylamine-containing dyes)[108] to allow the enzyme to be studied through single molecule spectroscopy. This provided new insight into the folding landscape of T4 lysozyme.

Enzyme function in many biological situations is dependent on post-translational modification, such as phosphorylation or acetylation of important amino acids. Studying the effects of post-translational modification is extremely challenging owing to the heterogeneity of biological systems, where there is often a mixture of differently modified enzymes. To define the effects of the modification on enzyme activity, or to study the biological effects of the mutation, it is necessary to generate a homogeneous population of the modified protein. Thus, UAA incorporation is a powerful technique that allows specific amino acids within a protein to be replaced with an UAA that mimics or contains the post-translational modification, generating a

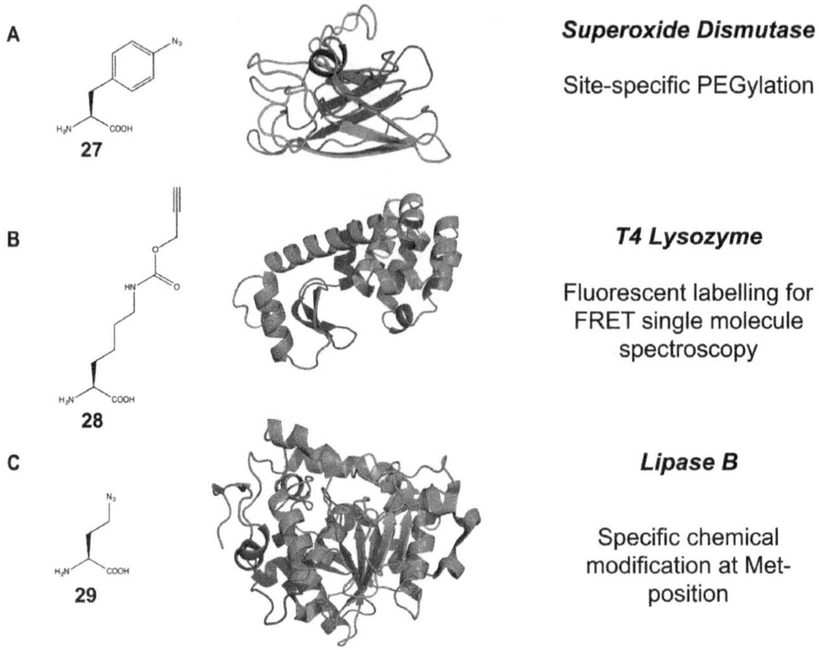

Figure 6.6 The use of UAA in engineering enzymes to facilitate chemical modification. (A) The site-specific introduction of *p*-azidophenylalanine (**27**) into superoxide dismutase (1SOS)[124] allowed site-specific PEGylation of the enzyme. (B) Single-molecule FRET spectroscopy of T4 lysozyme (2LZM)[125] was made possible by the site-specific attachment of fluorescent dyes at positions where the UAA propargyllysine (**28**) was incorporated. (C) Chemical modification at specific positions of lipase B from *Candida antarctica* (4ZV7)[126] was made possible through the residue-specific replacement of methionine by azidohomoalanine (**29**), exploiting the fact that only one methionine residue was in a solvent exposed position.

homogeneous sample. Successes in this area include the incorporation of *N*(ε)-acetyllysine in the enzyme manganese superoxide dismutase by Chin *et al.* using an orthogonal pyrrolysine synthetase/tRNA pair from *Methanosarcina barkeri* in *E. coli.*[109] This allowed them to measure and define the effect of lysine acetylation on manganese superoxide dismutase, revealing that the enzyme kinetics were essentially the same as the unacetylated form and suggesting acetylation must affect another aspect of its function. More recently, Chin *et al.* have engineered an orthogonal pathway for phosphoserine and demonstrated it was successfully incorporated into ubiquitin and the kinase Nek7.[110] The technological advance that allowed the challenging incorporation of phosphoserine involved exploiting the mechanism used by some methanogenic bacteria for cysteine incorporation, which involves aminoacylation of phosphoserine onto $tRNA^{Cys}{}_{GCA}$ by the tRNA synthetase SepRS (conversion of phosphoserine to cysteine happens subsequently).[111]

This orthogonal archael SepRS/tRNA pair was introduced into *E. coli* after converting the GCA anticodon to CUA, which significantly reduced the efficiency of amino acylation. Thus, extensive evolution and manipulation of the regions around the anticodon in the tRNA was performed to facilitate recognition and decoding of the tRNA by the bacterial ribosome, and SepRS was also subsequently evolved for improved recognition of the new anticodon.

Finally, van Hest and colleagues have used UAA mutagenesis to site-specifically modify the commonly used industrial lipase B from *Candida antarctica*.[112] By using residue-specific replacement of methionine by azidohomoalanine, they were able to carry out specific functionalisation of these UAAs. Notably, four methionines in this enzyme are buried in the hydrophobic core of the enzyme and are therefore not solvent accessible. The one remaining and solvent accessible methionine was therefore targeted for replacement by UAA mutagenesis. Again using a copper-catalysed cycloaddition reaction, a monofunctionalised, and active, enzyme could be obtained.

6.4.7 Engineering Enzymes for UAA Biosynthesis

One of the greatest hurdles to the large-scale use of UAAs is the need to chemically synthesise the majority of the UAAs that are incorporated, which vastly increases the cost of the experiment and requires that these UAAs be effectively fed to the organism or cell-free system into which they will be incorporated.[90] In contrast, most naturally occurring amino acids are inexpensively biosynthesised. To address this, Turner and colleagues have performed engineering of the bacterial ammonia lyase EncP, which allowed them to biosynthetically produce in high enantiomeric excess, a range of phenylalanine derivatives.[113] Similar work has been carried out to allow the whole-cell biosynthesis of L-homoalanine.[114] Although the primary focus of that work was to reduce the cost of L-homoalanine production and use as a chiral precursor for the pharmaceuticals levetiracetam, brivaracetam, and ethambutol, it demonstrated that engineering bacteria to produce additional UAAs is feasible. In that work, they altered the substrate specificity of glutamate dehydrogenase to work more efficiently with 2-ketobutyrate, rather than the natural substrate, 2-ketoglutarate. Through coexpression with a threonine dehydratase in a modified threonine-hyperproducing *Escherichia coli* strain, substantial amounts of L-homoalanine could be produced.

6.5 Future Perspectives

In writing this chapter, we were surprised at the relatively small number of examples of UAAs being used to improve our understanding of enzymes, as well as their catalytic activity and physical properties. In contrast to the phenomenal growth of enzyme engineering in general, the application of UAAs in this field appears to be lagging, despite the enormous potential that they have. The reasons for this are probably two-fold: (i) the technique of UAA mutagenesis, especially site-specific UAA mutagenesis, is still significantly more demanding than traditional mutagenesis, which has led to its use being generally confined to research groups that have developed expertise in UAA mutagenesis; (ii) the approach is unfortunately limited by economic factors, especially the need to chemically synthesise UAAs and the intrinsically low yields of the enzymes that are produced through site-specific mutagenesis. This has mostly limited their application to academic research or proof-of-principle style studies that have, importantly, demonstrated the potential of UAA incorporation, or yielded new insight into enzyme function, but cannot be translated into industrial scale without difficulty, owing to the yield and cost.

There are therefore two areas that need particular attention in the coming years to further advance the use of UAAs in enzymology. First, the continuing efforts to simplify and make the technique widely available should see it become adopted by a wider range of research groups and industry, as a standard laboratory approach, rather than a specialist technique. Second, the economic limitations of the technique must continue to be addressed, perhaps through the development of biosynthetic pathways for UAAs to allow them to be produced as cheaply and simply, *in vivo*, as canonical amino acids. However, this requires the development of separate biosynthetic pathways for every amino acid, although this might be less demanding than it initially appears, since the majority of UAAs are derivatives of tyrosine and lysine.

6.6 Conclusion

In conclusion, UAA incorporation and site-specific mutagenesis are ideally suited to use in modern enzymology, from providing unique biochemical and biophysical handles and probes to advance our understanding of enzymes, through to more applied outcomes, such as exploiting the novel chemistry that can be incorporated into UAAs to push back the limits of catalysis and improving the physical properties of enzymes. The use of UAAs could potentially lead to the development of true designer enzymes with vastly superior catalytic power and biophysical robustness or versatility than we see in the natural repertoire of enzymes. However, to get to this point, more work is required to increase the usability of this technique and grow the size of the community of UAA–enzyme engineers, as well as to address some of the economic barriers that currently limit the industrial applications of this technique.

Acknowledgements

C. J. J. and T. H. acknowledge support from the Australian Research Council (DP170100181, FT140101059, DP150100383).

References

1. B. J. Lee, P. J. Worland, J. N. Davis, T. C. Stadtman and D. L. Hatfield, *J. Biol. Chem.*, 1989, **264**, 9724.
2. B. Hao, W. Gong, T. K. Ferguson, C. M. James, J. A. Krzycki and M. K. Chan, *Science*, 2002, **296**, 1462.
3. W. Eckhart, M. A. Hutchinson and T. Hunter, *Cell*, 1979, **18**, 925.
4. F. R. Bettelheim, *J. Am. Chem. Soc.*, 1954, **76**, 2838.
5. R. Harwood, M. E. Grant and D. S. Jackson, *Biochem. J.*, 1975, **152**, 291.
6. B. Imperiali and S. E. O'Connor, *Curr. Opin. Chem. Biol.*, 1999, **3**, 643.
7. A. Warshel, *Annu. Rev. Biophys. Biomol. Struct.*, 2003, **32**, 425.
8. E. Campbell, M. Kaltenbach, G. J. Correy, P. D. Carr, B. T. Porebski, E. K. Livingstone, L. Afriat-Jurnou, A. M. Buckle, M. Weik, F. Hollfelder, N. Tokuriki and C. J. Jackson, *Nat. Chem. Biol.*, 2016, **12**, 944.
9. C. von der Osten, S. Branner, S. Hastrup, L. Hedegaard, M. D. Rasmussen, H. Bisgard-Frantzen, S. Carlsen and J. M. Mikkelsen, *J. Biotechnol.*, 1993, **28**, 55.
10. F. Van, de Werf, S. R. Bergmann, K. A. Fox, H. de Geest, C. F. Hoyng, B. E. Sobel and D. Collen, *Circulation*, 1984, **69**, 605.
11. G. Dalbadie-McFarland, L. W. Cohen, A. D. Riggs, C. Morin, K. Itakura and J. H. Richards, *Proc. Natl. Acad. Sci. U. S. A.*, 1982, **79**, 6409.
12. W. P. Stemmer, *Nature*, 1994, **370**, 389.
13. D. Rothlisberger, O. Khersonsky, A. M. Wollacott, L. Jiang, J. DeChancie, J. Betker, J. L. Gallaher, E. A. Althoff, A. Zanghellini, O. Dym, S. Albeck, K. N. Houk, D. S. Tawfik and D. Baker, *Nature*, 2008, **453**, 190.
14. O. Khersonsky, D. Röthlisberger, O. Dym, S. Albeck, C. J. Jackson, D. Baker and D. S. Tawfik, *J. Mol. Biol.*, 2010, **396**, 1025.
15. N. J. Turner and M. D. Truppo, *Curr. Opin. Chem. Biol.*, 2013, **17**, 212.
16. A. Karlin, *J. Gen. Physiol.*, 1969, **54**, 245.
17. H. Bozler, K. D. Jany and G. Pfleiderer, *Biochim. Biophys. Acta*, 1983, **749**, 238.
18. X.-C. Su, T. Huber, N. E. Dixon and G. Otting, *ChemBiochem*, 2006, **7**, 1599.
19. C. Mayer, D. G. Gillingham, T. R. Ward and D. Hilvert, *Chem. Commun.*, 2011, **47**, 12068.
20. N. V. Katre, *J. Immunol.*, 1990, **144**, 209.
21. A. Koyama, H. Inage, M. Kobayashi, H. Nakamura, M. Narita and S. Tojo, *Immunology*, 1986, **58**, 535.
22. F.-X. Gallat, A. P. S. Brogan, Y. Fichou, N. McGrath, M. Moulin, M. Härtlein, J. Combet, J. Wuttke, S. Mann, G. Zaccai, C. J. Jackson, A. W. Perriman and M. Weik, *J. Am. Chem. Soc.*, 2012, **134**, 13168.
23. A. Ochoa-Solano and C. Gitler, *J. Nutr.*, 1968, **94**, 243.
24. S. M. Hecht, B. L. Alford, Y. Kuroda and S. Kitano, *J. Biol. Chem.*, 1978, **253**, 4517.
25. W. A. Hendrickson, J. R. Horton and D. M. LeMaster, *EMBO J.*, 1990, **9**, 1665.

26. S. K. Blight, R. C. Larue, A. Mahapatra, D. G. Longstaff, E. Chang, G. Zhao, P. T. Kang, K. B. Green-Church, M. K. Chan and J. A. Krzycki, *Nature*, 2004, **431**, 333.
27. C. J. Noren, S. J. Anthony-Cahill, M. C. Griffith and P. G. Schultz, *Science*, 1989, **244**, 182.
28. J. D. Bain, E. S. Diala, C. G. Glabe, T. A. Dix and A. R. Chamberlin, *J. Am. Chem. Soc.*, 1989, **111**, 8013.
29. L. Pauling, *Arbeiten aus dem Gebiet der Naturstoffchemie, Festschrift Arthur Stroll.*, Birkhauser Verlag, Basel, Switzerland, 1957, pp. 597–602.
30. K. E. Splan, K. Musier-Forsyth, M. T. Boniecki and S. A. Martinis, *Methods*, 2008, **44**, 119.
31. N. Cvetesic and I. Gruic-sovulj, *Methods*, 2016, 1.
32. A. P. Mascarenhas and S. A. Martinis, *FEBS Lett.*, 2009, **583**, 3443.
33. M. H. Schwartz and T. Pan, *Nucleic Acids Res.*, 2016, **44**, 294.
34. A. J. Link, M. K. S. Vink, N. J. Agard, J. A. Prescher, C. R. Bertozzi and D. A. Tirrell, *Proc. Natl. Acad. Sci. U. S. A.*, 2006, **103**, 10180.
35. S. Zheng and I. Kwon, *Biotechnol. J.*, 2012, 7, 47.
36. C. Minks, S. Alefelder, L. Moroder, R. Huber and N. Budisa, *Tetrahedron*, 2000, **56**, 9431.
37. G. N. Cohen and D. B. Cowie, *C. R. Hebd. Seances Acad. Sci.*, 1957, **244**, 680.
38. A. J. Link and D. A. Tirrell, *Methods*, 2005, **36**, 291.
39. Y. Ravikumar, S. P. Nadarajan, T. Hyeon Yoo, C. S. Lee and H. Yun, *Biotechnol. J.*, 2015, **10**, 1862.
40. K. L. Kiick, J. C. M. van Hest and D. A. Tirrell, *Angew Chem. Int. Ed.*, 2000, **39**, 2148.
41. Y. Ravikumar, S. P. Nadarajan, T. Hyeon Yoo, C. S. Lee and H. Yun, *Trends Biotechnol.*, 2015, **33**, 462.
42. Y. Tang, G. Ghirlanda, W. A. Petka, T. Nakajima, W. F. DeGrado and D. A. Tirrell, *Angew Chem. Int. Ed.*, 2001, **40**, 1494.
43. L. Wang and P. G. Schultz, *Annu. Rev. Biophys. Biomol. Struct.*, 2005, **44**, 34.
44. J. D. Bain, E. S. Diala, C. G. Glabe, D. A. Wacker, M. H. Lyttle, T. A. Dix and A. R. Chamberlin, *Biochemistry*, 1991, **30**, 5411.
45. R. Furter, *Protein Sci.*, 1998, 7, 419.
46. L. Wang, A. rock, B. Herberich and P. G. Schultz, Science, 2001, **292**, 498.
47. T. Kobayashi, O. Nureki, R. Ishitani, A. Yaremchuk, M. Tukalo, S. Cusack, K. Sakamoto and S. Yokoyama, *Nat. Struct. Biol.*, 2003, **10**, 425.
48. R. J. Ernst, T. P. Krogager, E. S. Maywood, R. Zanchi, V. Beranek, T. S. Elliott, N. P. Barry, M. H. Hastings and J. W. Chin, *Nat. Chem. Biol.*, 2016, **12**, 776.
49. W. Liu, A. Brock, S. Chen, S. Chen and P. G. Schultz, *Nat. Methods*, 2007, **4**, 239.
50. S. Chen, P. G. Schultz and A. Brock, *J. Mol. Biol.*, 2007, **371**, 112.
51. I. N. Arthur, J. E. Hennessy, D. Padmakshan, D. J. Stigers, S. Lesturgez, S. A. Fraser, M. Liutkus, G. Otting, J. G. Oakeshott and C. J. Easton, *Chem. - Eur. J.*, 2013, **19**, 6824.

52. F. Katzen, G. Chang and W. Kudlicki, *Trends Biotechnol.*, 2005, **23**, 150.
53. K. Ozawa, M. J. Headlam, P. M. Schaeffer, B. R. Henderson, N. E. Dixon and G. Otting, *Eur. J. Biochem.*, 2004, **271**, 4084.
54. R. W. Alexander, in *Encyclopedia of Biological Chemistry*, 2013, pp. 441–444.
55. M. Abou-Nader and M. J. Benedik, *Bioeng. Bugs*, 2010, **1**, 337.
56. R. C. Cadwell and G. F. Joyce, *PCR Methods Appl.*, 1992, **2**, 28.
57. J. Meiler and D. Baker, *Proteins: Struct., Funct., Genet.*, 2006, **65**, 538.
58. W. Ren, T. M. Truong and H. Ai, *Sci. Rep.*, 2015, **5**, 12632.
59. R. Sun, H. Zheng, Z. Fang and W. Yao, *Biochem. Biophys. Res. Commun.*, 2010, **391**, 709.
60. A. Currin, N. Swainston, P. J. Day and D. B. Kell, *Chem. Soc. Rev.*, 2015, **44**, 1172.
61. J. Xie and P. G. Schultz, *Nat. Rev. Mol. Cell Biol.*, 2006, **7**, 775.
62. R. D. Socha and N. Tokuriki, *FEBS J.*, 2013, **280**, 5582.
63. E. Neil and G. Marsh, *Chem. Biol.*, 2000, **7**, R153.
64. C. Odar, M. Winkler and B. Wiltschi, *Biotechnol. J.*, 2015, **10**, 427.
65. L. Merkel, M. Schauer, G. Antranikian and N. Budisa, *ChemBioChem*, 2010, **11**, 1505.
66. K. Deepankumar, M. Shon, S. P. Nadarajan, G. Shin, S. Mathew, N. Ayyadurai, B. G. Kim, S. H. Choi, S. H. Lee and H. Yun, *Adv. Synth. Catal.*, 2014, **356**, 993.
67. B. Holzberger and A. Marx, *J. Am. Chem. Soc.*, 2010, **132**, 15708.
68. K. Ohtake, A. Yamaguchi, T. Mukai, H. Kashimura, N. Hirano, M. Haruki, S. Kohashi, K. Yamagishi, K. Murayama, Y. Tomabechi, T. Itagaki, R. Akasaka, M. Kawazoe, C. Takemoto, M. Shirouzu, S. Yokoyama and K. Sakamoto, *Sci. Rep.*, 2015, **5**, 9762.
69. B. Borgo and J. J. Havranek, *Proc. Natl. Acad. Sci. U. S. A.*, 2012, **109**, 1494.
70. E. Sugrue, P. D. Carr, C. Scott and C. J. Jackson, *Biochemistry*, 2016, **55**, 6304.
71. C. M. Miton and N. Tokuriki, *Protein Sci.*, 2016, **25**, 1260.
72. C. Mitchinson and J. A. Wells, *Biochemistry*, 1989, **28**, 4807.
73. C. J. Jackson, C. W. Coppin, P. D. Carr, A. Aleksandrov, M. Wilding, E. Sugrue, J. Ubels, M. Paks, J. Newman, T. S. Peat, R. J. Russell, M. Field, M. Weik, J. G. Oakeshott and C. Scott, *Appl. Environ. Microbiol.*, 2014, **80**, 4003.
74. M. G. Hoesl, C. G. Acevedo-Rocha, S. Nehring, M. Royter, C. Wolschner, B. Wiltschi, N. Budisa and G. Antranikian, *ChemCatChem*, 2011, **3**, 213.
75. J. C. Jackson, S. P. Duffy, K. R. Hess and R. A. Mehl, *J. Am. Chem. Soc.*, 2006, **128**, 11124.
76. W. A. Denny, *Curr. Pharm. Des.*, 2002, **8**, 1349.
77. J. N. Kolev, J. M. Zaengle, R. Ravikumar and R. Fasan, *ChemBioChem*, 2014, **15**, 1001.
78. C. J. Jackson, J. L. Foo, N. Tokuriki, L. Afriat, P. D. Carr, H. K. Kim, G. Schenk, D. S. Tawfik and D. L. Ollis, *Proc. Natl. Acad. Sci. U. S. A.*, 2009, **106**, 21631.

79. C. J. Jackson, K. Weir, A. Herlt, J. Khurana, T. D. Sutherland, I. Horne, C. Easton, R. J. Russell, C. Scott and J. G. Oakeshott, *Appl. Environ. Microbiol.*, 2009, **75**, 5153.
80. C. J. Jackson, C. Scott, A. Carville, K. Mansfield, D. L. Ollis and S. B. Bird, *Biochem. Pharmacol.*, 2010, **80**, 1075.
81. C. J. Jackson, A. Carville, J. Ward, K. Mansfield, D. L. Ollis, T. Khurana and S. B. Bird, *Toxicology*, 2014, **317**, 1.
82. C. J. Jackson, J.-L. Foo, N. Tokuriki, L. Afriat, P. D. Carr, H.-K. Kim, G. Schenk, D. S. Tawfik and D. L. Ollis, *Proc. Natl. Acad. Sci. U. S. A.*, 2009, **106**, 21631.
83. I. N. Ugwumba, K. Ozawa, Z.-Q. Xu, F. Ely, J.-L. Foo, A. J. Herlt, C. Coppin, S. Brown, M. C. Taylor, D. L. Ollis, L. N. Mander, G. Schenk, N. E. Dixon, G. Otting, J. G. Oakeshott and C. J. Jackson, *J. Am. Chem. Soc.*, 2011, **133**, 326.
84. P.-Y. Kim, D. J. Pollard and J. M. Woodley, *Biotechnol. Prog.*, 2007, **23**, 74.
85. S. Zheng and I. Kwon, *Biotechnol. Bioeng.*, 2013, **110**, 2361.
86. R. Wohlgemuth, *Curr. Opin. Microbiol.*, 2010, **13**, 283.
87. H. Ma, X. Yang, Z. Lu, N. Liu and Y. Chen, *PLoS One*, 2014, **9**, e103792.
88. Y. Chen, C. Chen and X. Wu, *Chem. Soc. Rev.*, 2012, **41**, 1742.
89. M. Lu, Y. Huang, M. A. White, X. Wu, N. Liu, X. Cheng and Y. Chen, *Chem. Commun.*, 2012, **48**, 11352.
90. W. H. Zhang, G. Otting and C. J. Jackson, *Curr. Opin. Struct. Biol.*, 2013, **23**, 581.
91. M. A. J. Dominguez, K. C. Thornton, M. G. Melendez and C. M. Dupureur, *Proteins*, 2001, **45**, 55.
92. Y. Wu, S. D. Fried and S. G. Boxer, *Biochemistry*, 2015, **54**, 7110.
93. A. T. Fafarman, P. A. Sigala, J. P. Schwans, T. D. Fenn, D. Herschlag and S. G. Boxer, *Proc. Natl. Acad. Sci. U. S. A.*, 2012, **109**, E299.
94. L. Wang, S. D. Fried, S. G. Boxer and T. E. Markland, *Proc. Natl. Acad. Sci. U. S. A.*, 2014, **111**, 18454.
95. W.-N. Chen, K. V. Kuppan, M. D. Lee, K. Jaudzems, T. Huber and G. Otting, *J. Am. Chem. Soc.*, 2015, **137**, 4581.
96. J. S. Lewis, S. Jergic and N. E. Dixon, *Enzymes*, 2016, **39**, 31.
97. H.-M. Lee, D. R. Larson and D. S. Lawrence, *ACS Chem. Biol.*, 2009, **4**, 409.
98. A. Deiters, D. Groff, Y. Ryu, J. Xie and P. G. Schultz, *Angew. Chem., Int. Ed. Engl.*, 2006, **45**, 2728.
99. J. Hemphill, C. Chou, J. W. Chin and A. Deiters, *J. Am. Chem. Soc.*, 2013, **135**, 13433.
100. J. Luo, Q. Liu, K. Morihiro and A. Deiters, *Nat. Chem.*, 2016, **8**, 1027.
101. Z. Ye and G. J. Williams, *Biochemistry*, 2014, **53**, 7494.
102. T. Matsuo and S. Hirota, *Bioorg. Med. Chem.*, 2014, **22**, 5638.
103. M. J. Jezewska, S. Rajendran, D. Bujalowska and W. Bujalowski, *J. Biol. Chem.*, 1998, **273**, 10515.
104. G. DeSantis and J. B. Jones, *Curr. Opin. Biotechnol.*, 1999, **10**, 324.

105. S. Tyagi and E. A. Lemke, *Curr. Opin. Struct. Biol.*, 2015, **32**, 66.
106. A. Deiters, T. A. Cropp, D. Summerer, M. Mukherji and P. G. Schultz, *Site-specific PEGylation of Proteins Containing Unnatural Amino Acids*, 2004, vol. 14.
107. A. Deiters, T. A. Cropp, M. Mukherji, J. W. Chin, J. C. Anderson and P. G. Schultz, *J. Am. Chem. Soc.*, 2003, **125**, 11782.
108. E. M. Brustad, E. A. Lemke, P. G. Schultz and A. A. Deniz, *J. Am. Chem. Soc.*, 2008, **130**, 17664.
109. H. Neumann, S. Y. Peak-Chew and J. W. Chin, *Nat. Chem. Biol.*, 2008, **4**, 232.
110. D. T. Rogerson, A. Sachdeva, K. Wang, T. Haq, A. Kazlauskaite, S. M. Hancock, N. Huguenin-Dezot, M. M. K. Muqit, A. M. Fry, R. Bayliss and J. W. Chin, *Nat. Chem. Biol.*, 2015, **11**, 496.
111. A. Sauerwald, W. Zhu, T. A. Major, H. Roy, S. Palioura, D. Jahn, W. B. Whitman, J. R. Yates, M. Ibba and D. Söll, *Science*, 2005, **307**, 1969.
112. S. Schoffelen, M. H. L. Lambermon, M. B. van Eldijk and J. C. M. van Hest, *Bioconjugate Chem.*, 2008, **19**, 1127.
113. N. J. Weise, F. Parmeggiani, S. T. Ahmed and N. J. Turner, *J. Am. Chem. Soc.*, 2015, **137**, 12977.
114. K. Zhang, H. Li, K. M. Cho and J. C. Liao, *Proc. Natl. Acad. Sci. U. S. A.*, 2010, **107**, 6234.
115. S. Angkawidjaja, C. Tayyab, M. Rashid, N. Kanaya, Protein Data Bank, 2012.
116. K. S. Midelfort, R. Kumar, S. Han, M. J. Karmilowicz, K. McConnell, D. K. Gehlhaar, A. Mistry, J. S. Chang, M. Anderson, A. Villalobos, J. Minshull, S. Govindarajan and J. W. Wong, *Protein Eng., Des. Sel.*, 2013, **26**, 25.
117. E. Johansson, G. N. Parkinson, W. A. Denny and S. Neidle, *J. Med. Chem.*, 2003, **46**, 4009.
118. C. J. Jackson, J. L. Foo, H. K. Kim, P. D. Carr, J. W. Liu, G. Salem and D. L. Ollis, *J. Mol. Biol.*, 2008, **375**, 1189.
119. V. Cody, J. R. Luft and W. Pangborn, *Acta Crystallogr., Sect. D: Biol. Crystallogr.*, 2005, **61**, 147.
120. H. M. Girvan, H. E. Seward, H. S. Toogood, M. R. Cheesman, D. Leys and A. W. Munro, *J. Biol. Chem.*, 2007, **282**, 564.
121. A. Athanasiadis, M. Vlassi, D. Kotsifaki, P. A. Tucker, K. S. Wilson and M. Kokkinidis, *Nat. Struct. Biol.*, 1994, **1**, 469.
122. M. S. Strycharska, E. Arias-Palomo, A. Y. Lyubimov, J. P. Erzberger, V. L. O'Shea, C. J. Bustamante and J. M. Berger, *Mol. Cell*, 2013, **52**, 844.
123. W. Huang, J. Jia, P. Edwards, K. Dehesh, G. Schneider and Y. Lindqvist, *EMBO J.*, 1998, **17**, 1183.
124. H. E. Parge, R. A. Hallewell and J. A. Tainer, *Proc. Natl. Acad. Sci. U. S. A.*, 1992, **89**, 6109.
125. L. H. Weaver and B. W. Matthews, *J. Mol. Biol.*, 1987, **193**, 189.
126. P. Strzelczyk, G. D. Bujacz, P. Kielbasinski and J. Blaszczyk, *Acta Biochim. Pol.*, 2016, **63**, 1065.

CHAPTER 7

Enhancing Enzymatic Performance via *Restricted Sequence Space Approaches*

ANDREAS S. BOMMARIUS*[a,b] AND BETTINA R. BOMMARIUS[a]

[a] School of Chemical and Biomolecular Engineering, Georgia Institute of Technology, Atlanta, GA 30332, USA; [b] School of Chemistry and Biochemistry, Georgia Institute of Technology, Atlanta, GA 30332, USA
*Email: andreas.bommarius@chbe.gatech.edu

7.1 Background and Scope

The application of enzymes, notably for processes to enantiomerically pure amino acids[1] and as additives to laundry detergents for fighting blood, grease, and grass stains,[2] became significantly more popular in the 1970s and 1980s. However, only the development of recombinant DNA technology after 1979 enabled heterologous expression of enzymes in suitable hosts at much higher expression levels than shown by the native environment, rendering pure enzymes available in large quantities (see Chapter 16). Nevertheless, the use of enzymes in industry has been slow to develop, owing to their often perceived and sometimes true limitations, such as lack of availability, high cost, instability under process conditions, narrow substrate specificity, and lack of competitiveness with established chemical processes.[3] These perceptions are changing now; especially, the lack of competitiveness is now overcome as biocatalysts are increasingly applied in synthesis routes to fine chemicals and pharmaceuticals.[3,4] One big accelerator was the development

Catalysis Series No. 32
Modern Biocatalysis: Advances Towards Synthetic Biological Systems
Edited by Gavin Williams and Mélanie Hall

Published by the Royal Society of Chemistry, www.rsc.org

of new mutational strategies to improve desirable and promising enzymes at hand, such as to incorporate the improved versions into industrial processes. *In vitro*, *in vivo* and/or *in silico* methods for the evolutionary improvements of proteins provide for the generation of active enzymes with new or improved properties. When generating such improved proteins, the wild-type enzyme undergoes mutagenesis of its gene and/or recombination of its gene with related ones to produce a library of variants that display changes in amino acid sequence relative to the wild-type. Successful screening of the variants for a desired property, such as an improvement in enzyme activity or stability, change in stereoselectivity or activity against new substrates, followed by nucleic acid sequencing allows for the identification of amino acid residues associated with the desired property.[5–7]

The principal problem of protein engineering remains the exceptionally large (but not infinite!) number of possible variants for a wild-type enzyme sequence; that number is provided by eqn (7.1).

$$\# = 19^M\ [N!/\{(N-M)!M!\}], \qquad (7.1)$$

with N = number of amino acid residues in the target protein, M = number of mutated residues, and # = number of resulting amino acid sequences.

There are two principle *strategies* for protein engineering in light of the large number of possible enzyme variants: (i) *increasing the library size* – this can be achieved either *via* improved screening modes, such as high-throughput well plate fluorescence assay, fluorescence-activated cell sorting (FACS sorting), or microemulsion droplet screening, or *via* selection, such as through auxotrophy, growth-based selection, or antibiotic resistance markers; (ii) *restricting the sequence space* – this goal can be achieved through picking a restricted set of amino acid residues to mutate (such as a set lining a substrate binding pocket) and/or a restricted set of amino acid residues to code for (*i.e.* not all 19 proteinogenic amino acids besides the wild-type).

Only after the choice of strategy between increasing library size or effective sequence space can the *approach* be picked, such as whether to employ rational design, combinatorial design ('directed evolution'), or data-driven protein engineering. For each approach, there are a range of *methods* available: (i) *rational design*: site-specific mutagenesis, site-saturation mutagenesis (SSM); (ii) *combinatorial design*: random mutagenesis, DNA recombination, or circular permutation; (iii) *data-driven design*: mutagenic plasmid amplification, (structure-guided) consensus approach, iterative saturation mutagenesis (ISM), or Combinatorial Active Site Saturation Testing (CASTing). Lastly, a multitude of experimental *protocols* have been developed that allow reduction to practice of the methods.

The complete mutagenesis of a gene to produce all possible single-amino acid sequence variants has been demonstrated in Diversa's Gene Site Saturation Mutagenesis (GSSM®)[8,9] but remains a challenge due to the large sequence space at issue for a typical polypeptide (see eqn (7.1)). The large number of library members renders it difficult both to prepare such complete libraries and to screen them. This 'numbers problem' in directed evolution[10] has been addressed by taking into account knowledge regarding protein structure, sequence similarity of homologous proteins, information regarding protein function, or relatedness of amino acids. This information is used both to limit the number of residues that are mutated and what amino acids the mutant residues encode.

To afford increases in activity, selectivity, or specificity of an enzyme, amino acid residues involved in the catalytic cycle should best be left untouched. Instead, saturation mutagenesis of residues lining the binding pocket of the substrate represents the gold standard for targeted protein evolution. However, picking the optimum number of residues in such a binding pocket to mutate as well as choosing the right number of amino acid residues (out of the remaining 19 choices of proteinogenic amino acids besides the wild type) to code for during mutagenesis with the goal of optimizing labor-intensive screening efforts remains a critical issue.

This chapter focuses especially on the *restriction of sequence space* to enable efficient variant generation with minimal effort and to allow maximum coverage with a given library size. Variant generation and testing of such variants should be viewed as embedded in a process that starts from picking a smart protein template, ideally one that already incorporates mechanistic analogues or even steps of the desired reaction, and evolves *via* stabilization of the template to imparting the substrate specificity and selectivity, often viewed as the key step, to fine-tuning the trait in question, often with the help of computational tools.[11]

7.2 Variant Generation *via* Restriction of Sequence Space

7.2.1 Restricted Sequence Space *via* Restricted Codon Libraries

A variety of logical filters have been applied to reduce the number of mutations that are made to explore a given sequence space. Rationales for selecting a subset of the 20 proteinogenic amino acids include: (i) groupings of amino acids based on structural similarity or charge (polar, hydrophobic, anionic/cationic), (ii) sequence alignment, and (iii) docking with the help of crystal structures or at least homology structures.

7.2.1.1 Algorithms

In addition to reducing the number of amino acid residues picked for site mutations in a variant library, its codons can also be selected to reduce the number of variants and thereby also increase the frequency by which a certain amino acid is mutated (*restricted codon libraries*). One conventional example of this approach is the use of "NNK" or "NNS" degenerate codon sets, where N = A, C, G, or T (*i.e.* all four codons), K = G or T and S = C or G.[12]

Degenerate codons that encode selected amino acids at particular sites can also be automatically determined, thus simplifying library construction. Two prominent examples of such library design programs are 'Libdesign' by Mena and Daugherty using a Monte Carlo protein design algorithm[12] and 'Caster' developed by the Reetz group.[13] Especially the NNK degenerate codon set or the NDT set (D = A, G or T; T = thymine) have been used successfully in several examples for site-saturation mutagenesis to reduce library complexity.[14] NNK encodes all 20 amino acids (plus one stop codon), using 32 codons; NDT encodes 12 amino acids using 12 codons (no stop codons). Kille *et al.* also demonstrated that low codon coverage oligonucleotides can be combined to achieve a genetic code redundancy with a codon to amino acid ratio of 22 : 20, which they called the 22ctrick.[14]

7.2.1.2 Grouping of Amino Acids

Similarities between amino acids can be exploited and amino acids grouped together to simplify sequences by reducing the amino acid alphabet.[15] Akanuma *et al.* constructed a fully functional orotate phosphoribosyltransferase in a growth-dependent selection scheme that did not contain seven amino acid residues in its sequence (Cys, His, Ile, Met, Asn, Gln, and Trp).[16] Nine of the remaining 13 proteinogenic amino acid residues occupied 88% of the 213 residues making up the protein (Ala, Asp, Gly, Leu, Pro, Arg, Thr, Val, and Tyr). Overall, combining 22 rounds of combinatorial mutagenesis of small sets of residues with 6 rounds of random mutagenesis resulted in 73 amino acid exchanges. Hilvert's group constructed a fully functional chorismate mutase with just 9 amino acids, including four polar (Asp, Glu, Asn, Lys) and four apolar residues (Phe, Ile, Leu, Met) as well as Arg.[17] The resulting enzyme was nearly as active as the native protein but not as tightly packed, and thus not as stable.

7.2.1.3 Analysis of Sequence Alignment

Sequence alignments of homologous phenylacetone monooxygenases (PAMOs) have also been used to produce selected amino acid alphabets in a four-amino acid loop next to the substrate binding site, specifically the use of 2, 6, 9, and 7 amino acid alphabets in each position 441 to 444.[18] The resulting 1700 transformants yielded more enantioselective but equally stable PAMO variants compared to the wild-type.

7.2.1.4 Inspection of Crystal Structures

Knowledge regarding the structure of a target polypeptide can be used to integrate structure-guided design with some degree of evolutionary randomization, such as by iterative saturation mutagenesis (ISM).[10,14,19]

The Bäckvall group intended to redesign the substrate binding pocket of *Candida antarctica* lipase A (CALA) for further enhancement of activity and enantioselectivity in the hydrolysis of ibuprofen esters after already having applied CASTing (see Section 7.2.2.1). Aided by insights from inspection of CALA crystal structures, the authors picked nine residues located around the binding pocket. Complete saturation coverage at all nine sites would require $20^9 = 5\times10^{11}$ colonies, resulting in impossibly huge libraries to screen. With the help of crystal structures and modelling, a single amino acid was picked at each of the nine sites to replace with, resulting in a library size of $2^9 = 512$ colonies to screen.[11] The most improved variant, with six-fold activity enhancement and increase of the enantiomeric ratio *E* from 3.4 to 100, contained 5 changes in the 9 targeted residues, a finding hard to reproduce with other techniques, such as CASTing or ISM (see Sections 7.2.2.1 and 7.2.2.2), which screen less broadly but more deeply.

The Bommarius group developed an amine dehydrogenase (AmDH) from a corresponding amino acid dehydrogenase (AADH), which enables the stereoselective reductive amination of ketones to chiral (*R*)-amines (Figure 7.1). Based on structural and mechanistic knowledge obtained on phenylalanine dehydrogenase (PheDH) from *Rhodococcus rhodocrous*,[20–23] the group found initial AmDH activity from LeuDH K68M in the first round of protein engineering. In ten subsequent rounds of either single-site saturation mutagenesis with NNK libraries or CASTing across two to four residues, employing nine different combinations of nucleotides, a quadruple variant gave the best results for reductive amination ('L-AmDH': K68S/E114V/N261I/V291C).[24] Starting with PheDH from *B. badius*,[22] application of a two-site K77DDK-N276DDK library with 969 colonies resulted in 36 active F-AmDH variants, the best of which (K77S/N276L) gave a k_{cat} value at 225 mM $NH_{3/4}{}^{(+)}$ of 2.8 s^{-1} (Figure 7.1).[25]

7.2.1.5 Triple-code Saturation Mutagenesis (TCSM)

A recently applied technique, structure-guided triple-code saturation mutagenesis (TCSM), allows the pick of a rather wide number of residues to mutate but limits the choice of amino acid residues to code for to very few, between two and four; those amino acids are chosen rationally with the help of crystal structures and consensus arguments and depend on the steric and electronic features of the enzyme as well as the trait to be optimized. While for an example of a 10-site library, NNK degeneracy coding of all 20 proteinogenic amino acids for 95% library coverage would lead to required screening of more than $3\cdot(20^{10}) = 3\times10^{13}$ transformants, application of TCSM with two amino acids to code for only requires screening of $3\cdot(2^{10}) = 3000$ transformants.

The Reetz group applied TCSM to the optimization and inversion of stereoselectivity of a limonene epoxide hydrolase (LEH).[26] Meso-type epoxides were successfully hydrolytically desymmetrized with formation of either (*R*,*R*)- or (*S*,*S*)-configurated diols.

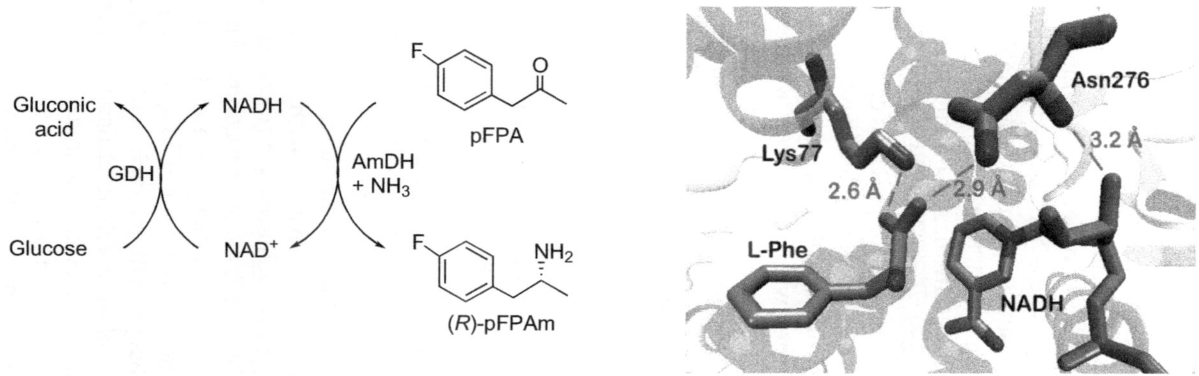

Figure 7.1 Left: Amine dehydrogenase (AmDH) catalysed reaction with glucose dehydrogenase (GDH) cofactor regeneration (pFPA: *p*-F-phenylacetone, (*R*)-pFPAm: (*R*)-*p*-F-phenylisopropylamine).[25] Right: Interaction of proximate residues Lys77 and Asn276 observed in the crystal structure of PheDH from *Rhodococcus* sp. M4 (PDB: 1C1D).
Reprinted with permission from *Adv. Synth. Catal.* 2013, **355**, 1780.[25] © 2013 WILEY-VCH Verlag GmbH & Co. KGaA, Weinheim.

7.2.2 Restricted Sequence Space *via* Selection of Mutation Sites: Focused Libraries

Focused libraries were first developed around the year 2000 to deal with the vast sequence space of protein mutations and to increase efficacy in laboratory methods of enzyme evolution. Random mutagenesis at defined positions is used to enhance catalytic activity and stereoselectivity with the aid of structural information. There are two accepted and most commonly used approaches, CASTing and iterative saturation mutagenesis (ISM), both developed by the Reetz group. These methods aim to improve fitness of the target enzyme, either for substrate specificity (CASTing, ISM)[13,18,27,28] or regio- and stereoselectivity[29] and thermostability.[19] There are several successful examples in the literature based on the use of such methods, which display major advantages, especially in comparison to previously established methods of directed evolution, such as error-prone PCR. Overall, the structure-guided approach to protein engineering becomes a valuable tool, and with the increasing number of structures deposited in the protein database (PDB) – being deposited without characterization – data mining of this database can be a valuable source for additional enzymatic functions.[30]

7.2.2.1 Combinatorial Active-site Saturation Testing (CASTing)

An alternative for expanding substrate space in enzymes was introduced with the method of CASTing (Combinatorial Active-site Saturation Testing).[28] The method's key is to simultaneously and randomly mutate two, three, or possibly even four critical amino acids close to the binding pocket that have been identified *via* structural information.[13] Mutating two or three amino acids simultaneously accounts for potential synergistic beneficial conformational or electrostatic effects. The amino acids involved are usually closely neighbored and only span the immediate binding pocket; repetition across various parts of the protein enables covering more relevant sequence space than with conventional random mutagenesis (Figure 7.2, left). The key is to generate relatively small libraries even though several amino acids are mutated at once. This goal can be achieved *via* restricted codons. An early example involved redesign of *Pseudomonas aeruginosa* lipase (PAL), where the authors changed the substrate specificity towards fatty acids with bulkier side chains, which are not accepted by the wild type enzyme.[13] Clouthier *et al.* also designed a new Baeyer–Villiger cyclopentanone monooxygenase (CPMO) using this method, and were able to enhance the enantioselectivity of the mutant towards substituted cyclohexanones using a NDT library.[31] CASTing requires knowledge of the substrate binding and, even better, the mechanism of catalysis of the enzyme and the associated amino acids in the active site. In the case of CPMO, previous knowledge of sites gained through rational design and modelling helped developed the subsequent mutations. This is a limitation of CASTing, which relies primarily on solved crystal structures of the target enzymes and helps keeping library sizes small. In

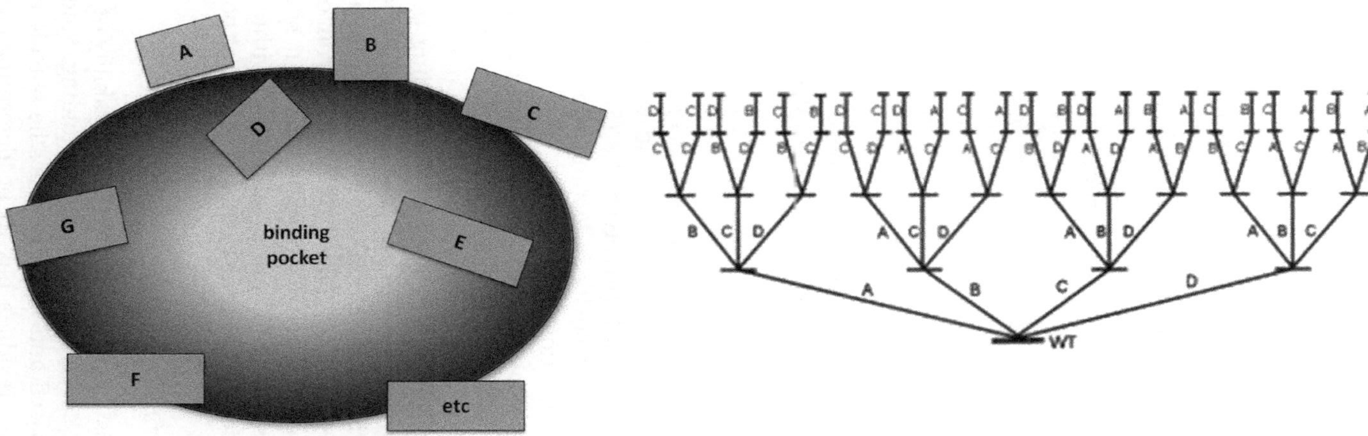

Figure 7.2 Left: General scheme for Combinatorial Active-Site Saturation Testing (CASTing). The sites A, B, C, *etc.* align the binding pocket and can be composed of one or more amino acid positions. The side chains of the amino acids are in general inside the binding pocket. Reprinted with permission from *Angew Chem. Int. Ed.* 2006, **45**, 1236.[28] Copyright 2006 Wiley-VCH. Right: Iterative Saturation Mutagenesis (ISM) employing four sites A, B, C and D, each site in a given upward pathway being visited only once. Reprinted with permission from *Angew. Chem. Int. Ed.* 2011, **50**, 138.[23] © 2011 WILEY-VCH Verlag GmbH & Co. KGaA, Weinheim.

addition, CASTing mostly focuses on amino acids in close proximity to the binding pocket and to one another, which limits the range of mutational space.

7.2.2.2 Iterative Saturation Mutagenesis (ISM)

Iterative Saturation Mutagenesis (ISM) is an extension of CASTing where iterative cycles of CAST are carried out to generate daughter libraries from the parent libraries. It relies on the 3D structure of the protein and it constitutes a symbiosis of rational design derived from the structure and random mutagenesis based on the chosen primer set. The crucial part of ISM revolves around the fact that the best hits from each round serve as the new templates for the next round and several examples of successful improvement of a target feature have been demonstrated (Figure 7.2, right).[24,25,32] Abrahamson *et al.* demonstrated the first successful generation of an amine dehydrogenase from an amino acid dehydrogenase, which is not only an extension of substrate specificity of a known reaction, but a total switch of activity. The switch from leucine dehydrogenase from *Bacillus megaterium* to an L-amine dehydrogenase was achieved in eleven rounds of site-saturation mutagenesis using rational design derived from crystal structure combined with ISM.[24]

Enantioselectivity switch can also be achieved using such a method, although examples are fewer. Zhang *et al.* achieved a complete switch in enantioselectivity from (*S*) to (*R*) with phenyl acetone monooxygenase (PAMO), the final variant showing four amino acid exchanges.[33] Interestingly, all four single mutations still preserved the (*S*)-selectivity, only the quadruple mutant showed (*R*)-selectivity, indicating a strong synergistic and cooperative mechanism of enantioselectivity.

Building from these approaches, several codons can be targeted which are spread out over the distance of the gene, and expanded outside of the binding pocket. Essentially, a combination of two methods, 'iterative CASTing',[28] is used on several amino acids of the binding pocket simultaneously, and variants are created in iterative cycles and screened for improvement. Since selection is based on improved variants in each cycle, cooperative interactions between non-related amino acids from one round to the next one cannot be investigated. Importantly, without site-saturation mutagenesis, none of these methods are possible, and without structural insight, an intelligent choice of mutational sites is often not possible.

7.2.2.3 B-factor Iterative Test

The B-factor iterative test is used to improve thermostability of enzymes by increasing their rigidity at surface sites for tighter folding. Since this technique is not catalytically-based and target residues lie outside of the active site, the selection of target residues has to be determined *via* crystallographic B factor data. Similar to the previously described methods, a 3D structure of the protein is needed to apply this method. The B factor reflects the degree to which the measured electron density of selected atoms in a protein structure spreads out. It is influenced by thermal fluctuations and freedom of mobility for a particular atom. The higher the B-factor, the more mobile the atom, so proteins with high B-factors for surface amino acids tend to be less stable and reduction of the B-factor at these sites increases rigidity and therefore stability. Identified sites for improved B-factor are submitted to iterative saturation mutagenesis (ISM, see Section 7.2.2.2) until improvement has been identified. Positive variants are then chosen for the next round at the next appropriate site and the cycle is repeated in an iterative manner.[19,34] One application is the enhancement of thermostability of lipase A from *Bacillus subtilis*, a mesophilic enzyme. After picking mutation sites from the crystal structure, site-saturation mutagenesis was performed on these sites and the best hits were chosen for the next round of mutagenesis. In the case of lipase A, five cycles involving 8000 clones yielded two hyperthermophilic variants. Successful evolution of a target enzyme importantly depends on a good screen and an easy substrate to test. For lipase A, a non-natural substrate (*p*-nitrophenylcaprylate) was chosen over the natural triglycerides to simplify the high-throughput screening methodology. Thermostability assay involves a straightforward screen by simply heating up well plates to find surviving variants. For most other changes that involve substrate change or activity change, a more robust screen is a must.

7.2.2.4 Targeted Sequential Saturation Mutagenesis (SSM)

This method is similar to a combination of CASTing and ISM and involves individual site-saturation mutagenesis (SSM) of each active site residue. The positive variant is selected as the parent for the next round of improvement on the next site. This can be carried out until no further improvement is achieved or additional mutations prove deleterious. The disadvantage of this method is that combinatorial effects are likely neglected since every single site is explored individually and only the additive effect of sequential mutation is intentionally explored. A rare application of protein engineering towards engineering the substrate diversity of a human nuclear receptor was developed in the Zhao lab: a stepwise individual SSM within the ligand–receptor contacts and interaction was used to generate a ligand–receptor pair with specificity towards 4,4′-dihydroxybenzil.[35] Not only is this rare because a human nuclear receptor was chosen for directed evolution, but also because phenotypic evaluation had to be performed with a yeast-two-hybrid system, which differs from the usual assays and growth in *E. coli* normally applied in directed evolution. This feature points towards another current limitation of applications for protein engineering: in case the target enzyme can only be expressed in higher organisms for functional activity, high-throughput assays and many of the methods described here are more difficult to achieve.

7.3 Variant Generation *via* Molecular Biology Protocols

7.3.1 Gene Splicing by Overlap Extension

Gene splicing by overlap extension (gene SOEing)[38] is a sequence-independent method of recombining DNA sequences without the need for restriction sites. Initial PCRs generate overlapping gene segments; those segments serve as templates for subsequent PCR to create a full-length gene. Internal primers utilize these complementary overhanging ends on intermediate segments to introduce nucleotide substitutions for site-directed mutagenesis, insertions or deletions (Figure 7.3).[36,37]

Two regions (A, C in Figure 7.3) flanking the region to be deleted (B) are fused by SOEing PCR. The fusion is mediated by an overlap of the two strands of PCR products that were created with the use of primers (outer forward primer, inner reverse primer, inner forward primer, outer reverse primer). The 5′-region of inner reverse primer used in the amplification of region A is complementary to the segment of region C. The 5′-region of inner forward primer used in the amplification of region C is complementary to the segment of region A. A recombinant product is formed when this overlap is extended in a subsequent reaction, and this recombined product is amplified with outer primers.

A multiplexed gene SOEing methodology employs selected variants that have a set or combinations of sets of mutations and subjects them to SOEing, which is a technique to combine fragments from genes that have

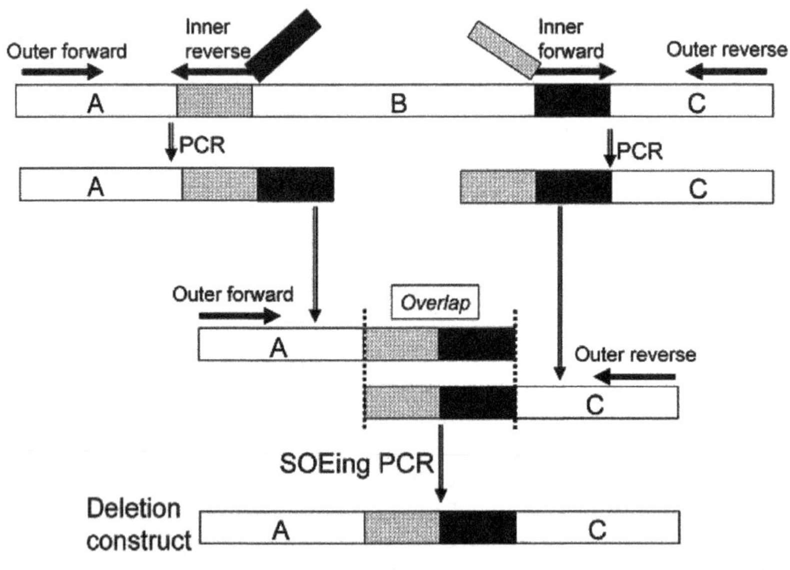

Figure 7.3 Schematic process of gene splicing by overlap extension (gene SOEing). Adapted by permission from Macmillan Publishers Ltd: *Pediatr. Res.* (ref. 38), copyright 2007.

been spliced, and then introduces mutations at the splicing site. The process has been automated by Codexis[39] to robotically manage between 100 and several 100s of variants in parallel (automated parallel SOEing, APS). The resulting beneficial mutations can then be further optimized. Nowadays, combining and testing several protein engineering methods on a large scale to cover as much sequence space as possible is very advantageous, as throughput scales with the level of automation.

With combined automated parallel SOEing (combined APS or CAPS), variants are pooled before screening. Given the need of oversampling, the screening efforts with CAPS are increased, but synthesis efforts decreased. CAPS allows quick surveying of up to hundreds of positions, which is very important in an industrial environment with tight timelines and milestones.

7.3.2 *In Vivo* Overlap Extension

In vivo overlap extension (IVOE)[40] involves the mutational exchange of sequence space between targeted species using a combinatorial approach of site-directed mutagenesis and recombination methods (Figure 7.4). This is particularly useful in yeasts where the extensive homologous recombination machinery can be exploited. One can expect a broad variety in libraries generated through such recombination–DNA shuffling combinations. Alcalde *et al.* demonstrated *in vivo* overlap extension to enhance the turnover rate of fungal laccase 3-fold by combinatorial site-saturation mutagenesis of two sites, S510 and L513. That rate enhancement was attributed to the opening up of the C-terminal plug, which led to higher accessibility toward the copper cluster.[33] Libraries were constructed in *Saccharomyces cerevisiae*. IVOE is particularly useful in yeast by avoiding intermediate PCR reaction and *in vitro* ligation with all the possible side reactions.

7.3.3 Omnichange

Since focused libraries mainly target and improve localized properties like activity or selectivity, codon changes are limited to a few sites and are rarely changed all simultaneously due to the increase in library size. Moreover, the methods described here depend on sequence and even more importantly, on protein structure and thereby limit the accessible pool of proteins to evolve. Dennig *et al.* approached multi-site mutagenesis from a different angle and reported a simultaneous five codon change independent of sequence and called their sequence-independent method Omnichange.[41] They chose spread-out positions within a phytase from *Yersinia mollaretii* and introduced NNK codons *via* PCR to improve thermal stability. All 5 codons were fully saturated and had only minimal wild-type reoccurrence at some position; good diversity was demonstrated in randomly picked 48 clones. While this method does not particularly rely on sequence knowledge *per se*, the choice of codons to improve was dependent on previous SSM experiments using traditional directed evolution methods.

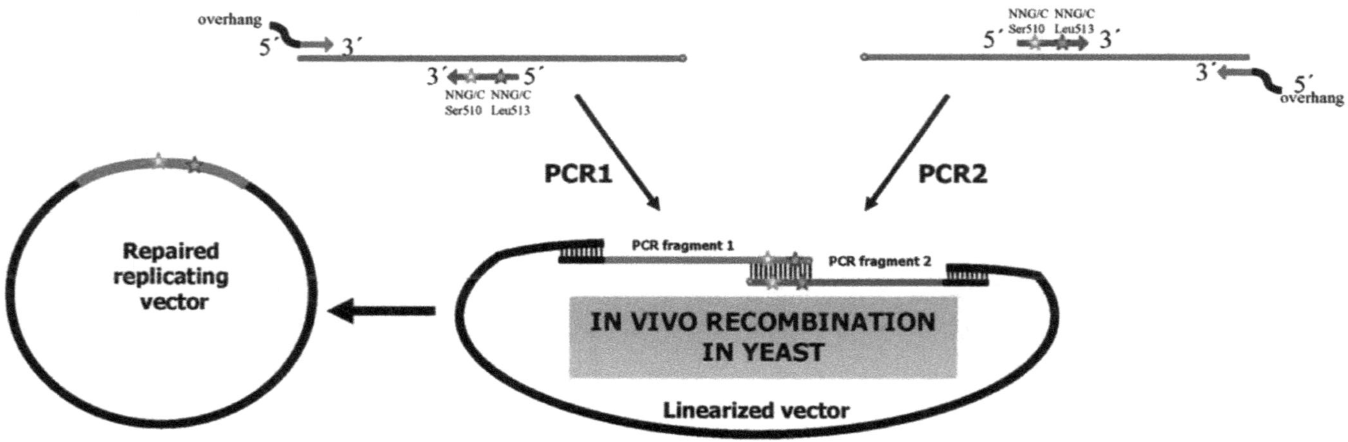

Figure 7.4 *In vivo* overlap extension (IVOE) strategy for combinatorial saturation mutagenesis. Reproduced from *Comb. Chem. High Throughput Screen.* 2006, **9**, 719,[40] Eureka Science Ltd. Reprinted by permission of Eureka Science Ltd. Copyright (2006) Eureka Science Ltd.

7.3.4 Circular Permutation

Circular permutation is a technique used to alter the position of the N- and C-termini in a protein while not changing the amino acid sequence.[42] The original N- and C-termini in a gene are fused, then the now circular gene is cut at a position different from the original one, establishing new N- and C-termini. This technique is applicable most straightforwardly in proteins having their original termini in close proximity. Not only can novel proteins be generated (important for intellectual property reasons) but often circular permutants feature improved properties, such as increased activity[43] or increased resistance against proteolysis.[44] Thus, several proteins have been extensively or even exhaustively circularly permutated; examples include Green Fluorescent Proteins (GFPs)[45] or *Candida antarctica* lipase B (CALB).[43]

7.4 Some Computational Design Tools

Computational design is becoming a necessary component in data-driven enzyme design[46,47] and in enzyme design from scratch to develop novel enzyme function. In the latter case, the discrepancy between purely computationally designed and naturally evolved enzymes is still large: mostly, wild-type activities and other properties of natural enzymes exceed the man-made counterparts. However, new computational strategies are aiming to change this (see Chapters 4 and 5).[48] A good current overview on computational tools for protein evolution is given by Damborsky and Brezovsky.[49]

7.4.1 Statistical Coupling Analysis (SCA)

Amino acids separated by distance on a polypeptide chain can still influence each other and affect activity, specificity and stereoselectivity of the enzyme. Interactions can alter the geometry or electrostatic properties as well as flexibility of amino acids in the active site. Statistical Coupling Analysis (SCA) estimates the co-evolution and therefore co-dependency of amino acids in the multiple sequence alignment of protein families. This approach is purely sequence-based and no structural information is needed. Amino acids are divided up into protein sectors that represent amino acids that are physically contiguous. This is most useful for potential allosteric regulation and SCA can reveal networks of small subset of residues that link distal sites and their cooperative interactions.[50]

7.4.2 ProSAR

ProSAR (Protein Sequence Activity Relationships) enriches experimental protein engineering methods with statistical analysis of the relationship between the sequence space and the activity of a protein. Thereby, the mutation space is enhanced with beneficial mutations within variants of lesser fitness,[51] but ProSAR does not require three-dimensional structural information. Examples of the successful application of ProSAR to the development of industrially viable biocatalysts by Codexis are ketoreductase for montelucast[52] and transaminase for sitagliptin[53] (see Section 7.5).

The ProSAR platform is based on existing first-round mutations, whether resulting in activity or not, and aims to provide a rational basis for designing further mutant libraries without resorting to structural knowledge. The method uses statistical models based on experimental data sets provided by the user.[54] For the method to be applicable, it has to contain variants with several mutations in a library, but no insertion/deletions. The sequences from this dataset together with the experimentally obtained activity measurements form the training data for the ProSAR tool, followed by the learning dataset, which contains a machine learning algorithm to build predictive models (Figure 7.5). Independent contribution of residues and epistatic coupling between residue-pairs derived from sequence information are combined. The independent contributions by single residues are captured *via* linear regression-based models, while epistatic interactions are captured by a genetic algorithm. The knowledge gained from the learning dataset then can lead to a predictive dataset, where new mutations are suggested to enhance the desired criteria in the next round of directed evolution. The obtained results consist in the prediction of the new sequences of mutants, the identification of the best mutants from the library, an estimation of the prediction error, and the relative influence of each residue in order to identify the beneficial, neutral or detrimental mutations. In this regard, ProSAR selects more than just hits, but rather selects hits and then recombines them into rapidly improving variants.

7.4.3 Structure-guided Recombination (SCHEMA)

Structure-guided recombination (SCHEMA) dissects proteins into domains with the goal of minimal structural disruption (Figure 7.6). Recombination of such domains has been demonstrated to yield functional proteins in far higher proportion than with recombination of random dissection.[55] SCHEMA is particularly advantageous for the stabilization of proteins.

Cellulases are crucial biocatalysts for the production of biofuels; owing to a complex mechanism, these enzymes are hard to improve *via* traditional protein engineering.[56,57] However, thermostable cellulases feature higher levels of stability and inhibition of microbial growth; in addition, higher thermostability often correlates with enhanced solvent, salt, and extreme pH tolerance. The Arnold group developed a highly successful structure-guided recombination (SCHEMA) approach, based on analyzing thermostability of chimeras from sequence blocks recombined at spots with the least expected structural disruption. Recombinations of three fungal cellobiohydrolases 2 (CBH2) yielded additive stability that followed a linear mathematical model.[58] Manipulation of cysteines turned out to be most promising: mutation C313S was successful in the case of CBH2.[59] S329G enhanced the half-life of *Clostridium thermocellum* Cel8A 8.5-fold[60] and *Talaromyces*

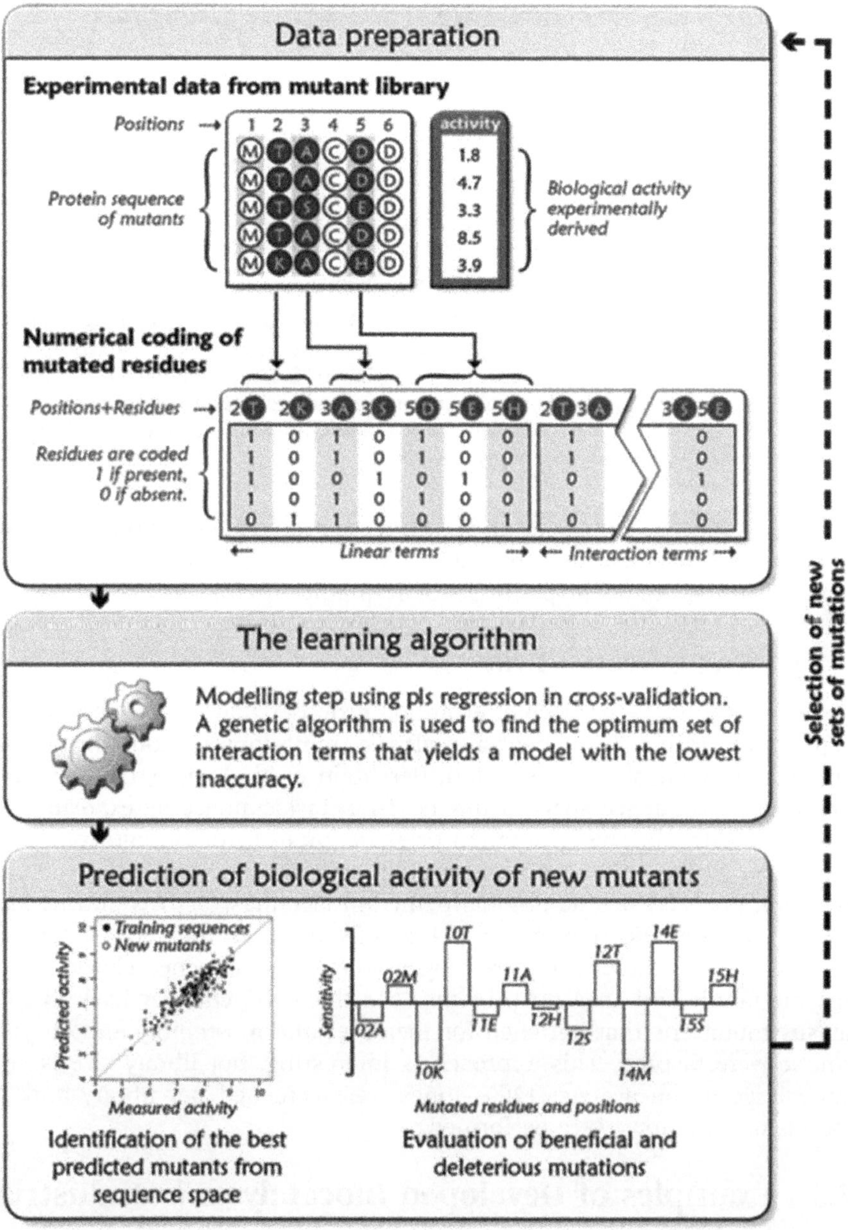

Figure 7.5 Three main steps of ProSAR: (i) data preparation, (ii) learning process and (iii) prediction.
Reproduced from M. Berland, *et al.*, A web-based tool for rational screening of mutants libraries using ProSAR, *Protein Eng. Des. Sel.*, **2014**, 27, 375,[54] by permission of Oxford University Press.

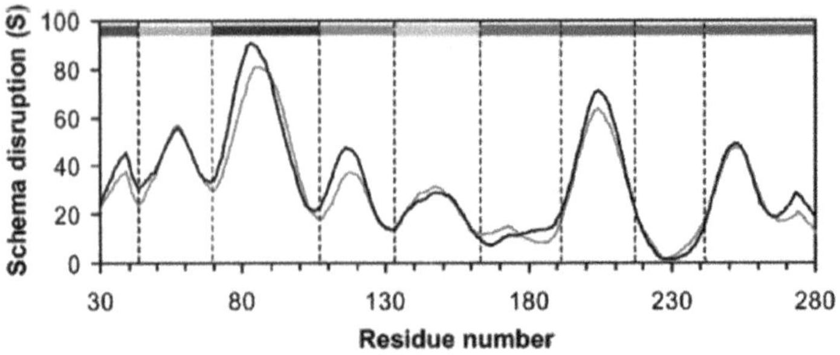

Figure 7.6 SCHEMA disruption profile along the residues of β-lactamases TEM-1 (gray) and PSE-4 (black). Dissection of the protein at disruption profile minima retains a maximum number of interactions within a domain; preferred dissection would occur around residues 165 and 227. Reprinted by permission from Macmillan Publishers Ltd: *Nat. Struct. Mol. Biol.* (ref. 55), copyright 2002.

emersonii CBH1 variants with up to 37 altered amino acid positions resulted in hydrolysis of solid cellulose at 70 °C, 5 °C higher than the working temperature of the wild-type for six of 16 predicted chimeras.[61]

7.4.4 Strong Neutral Drift

The neutral theory of molecular evolution proposes that genetic variation in populations results from neutral drift, *i.e.* mutations without directional pressure, instead of from selection. Bershtein *et al.* demonstrated the enrichment of sequence space using TEM-1 beta lactamase by exposing the enzyme to extremely high mutation pressure, up to 1 million-fold compared to nature, while maintaining the protein original function and structure. Selection pressure was its original penicillin lactamase activity and neutral drift implicates the purging of deleterious mutants and enrichment of beneficial ones, thereby approaching a consensus sequence closer to the enzyme family and ancestor proteins.[62] In this case, changes back to consensus mutations that enriched for increased thermostability and enzyme kinetic were favored. This approach is interesting, but library size is still quite large as on average 1200 clones were screened per library and 18 rounds of mutations were performed.

7.5 Examples of Developed Biocatalysts in Industrial Setting

7.5.1 Sitagliptin (Combination of ProSAR and CAPS)

Sitagliptin (Januvia®) is a first-in-class dipeptidyl peptidase 4 inhibitor used to treat type II diabetes. It is an oral drug and contains a chiral amine, a moiety present in 33% of the largest 200 pharmaceuticals sold in the USA, which highlights the importance of chiral amines and the necessity to develop biocatalytic routes towards those building blocks. The traditional chemical route to sitagliptin[63] requires two recrystallization steps to achieve the required enantiomeric excess value, which adds complexity and cost. The benefits of a recently introduced biocatalytic route[53] include elimination of high pressure hydrogenation and its expensive equipment, avoidance of enamine production, which is a severe road block, elimination of heavy metal use, and the avoidance of a second recrystallization step with its ensuing yield loss, all resulting in higher initial enantioselectivity. The enzyme of choice was a transaminase; the equilibrium limitations were overcome by elimination *via* pervaporation of the volatile by-product. Protein engineering was necessary because none of the transaminases tested showed activity towards the pro-sitagliptin ketone. Merck and Codexis applied substrate-walking starting from the methyl ketone **1** and directed evolution to achieve a transaminase that would accept ketositagliptin (**2**) (Figure 7.7). The CAPS method was aided by ProSAR and 576 mutations from 7 homologs, resulting in 26 mutations, yielded 1.25–5-fold improvements at a much higher rate than random mutagenesis. Evolution rounds 3–7 focused on increasing enzyme activity and process stability, aided by ProSAR (sorting of variants) and proposing the hits for the next round. The successive rounds were screened under more stringent process conditions, such as substrate loading, solvent concentration, pH and temperature. The final catalyst contained 25 mutations. This process is now applied on large scale and features product concentrations of up to 400 $g\,L^{-1}$.[53]

7.5.2 Montelukast (Singulair®) Using ProSAR

A catalyst developed by Codexis was used in the synthesis of the key intermediate of montelukast (the (*R*)-isomer), the active ingredient in Singulair®, Merck's anti-asthma drug. A ketoreductase (KRED) was engineered *via* directed evolution to produce the desired (*S*)-alcohol used in the synthesis of montelukast (Figure 7.8).

The new KRED is able to function in the presence of 70% organic solvent at 45 °C to accept the water-insoluble substrate (100 $g\,L^{-1}$), while being highly regio- and stereoselective. The process is now currently run at >200 kg scale where the product is obtained in >95% yield in >99.9% ee and >98.5% chemical purity, and offers a greener alternative to the process initially developed by Merck. In addition to featuring lower cost and generating far less waste, the enzyme-catalyzed step produced the desired alcohol in greater yield and higher stereochemical purity, employing KRED-catalyzed oxidation of isopropanol to acetone for regeneration of the nicotinamide cofactor. By screening several NAD(P)-dependent ketoreductases from available panels, some were found to be active on the substrate and displayed great stereoselectivity (yielding the (*S*)-product with ee >99.9%) but had low activity, were unstable at high organic solvent concentration and needed to be improved. Directed evolution techniques allowed a 3000-fold improvement

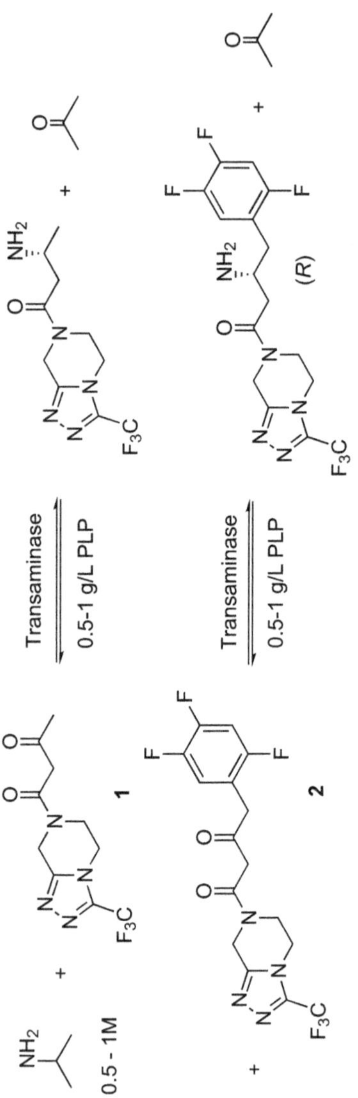

Figure 7.7 Substrate walking from methylketone (**1**) to ketositagliptin (**2**). From *Science*, 2010, **329**, 305.[53] Reprinted with permission from AAAS.

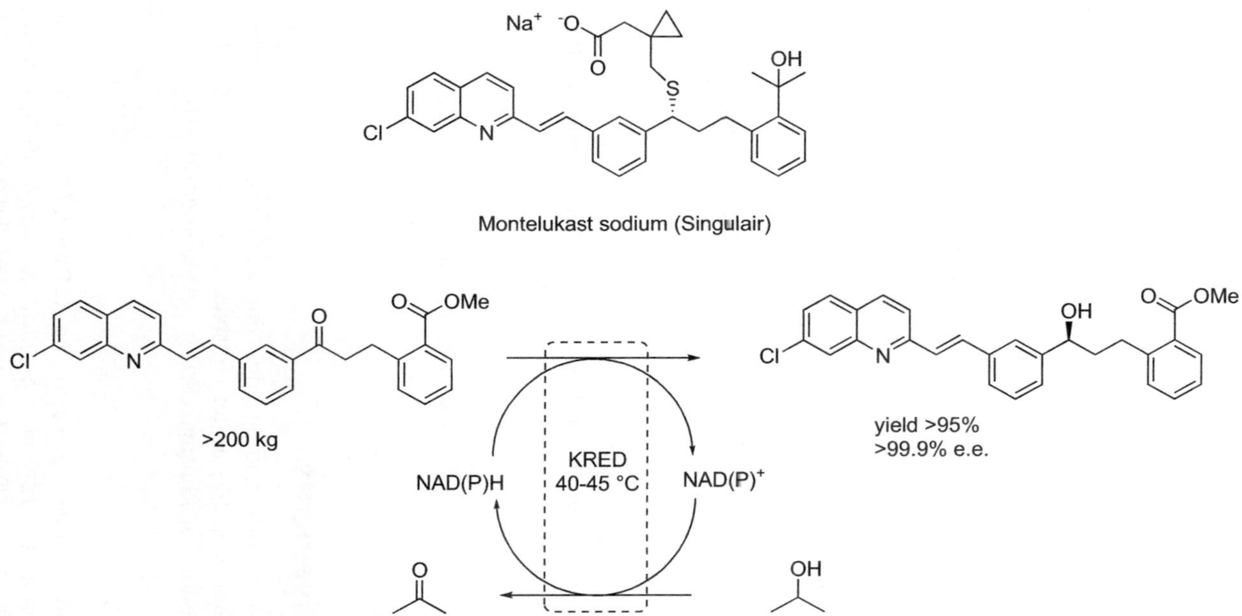

Figure 7.8 Biocatalytic route to montelukast, developed by Codexis. Reprinted with permission from *Chem. Rev.* 2011, **111**, 4088.[64] Copyright 2011 American Chemical Society.

in activity to produce the final biocatalyst with 19 mutations obtained from the parent enzyme (and guided with ProSAR).[52,64]

7.6 Conclusions

Restricting the sequence space to be investigated for improvement of fitness of a wild-type or base-case protein seems to be an effective and efficient way to achieve such improvement with a much smaller library than utilized in the case of unrestricted such space. Both available strategies – picking a restricted set of residues to mutate and restricting the number of amino acids to code for at a particular residue – require structural and/or mechanistic knowledge or at least intuition.

So far, the last two examples discussed are the only commercially available large processes known to the authors that implement newer protein engineering tools described in this chapter. There are a lot of different enzymes developed for and used in pharmaceutical processes that still rely on catalyst improvement *via* sequence alignments, gene shuffling, and site-saturation mutagenesis, or simply metabolic engineering or expression optimization. Nevertheless, several examples show that using smart small libraries can also lead to fast success. Almost always good structural and (for activity improvement) mechanistic knowledge are key to overcoming limitations of throughput. Just relying on sequence alone and utilizing small libraries will most likely not lead to improved biocatalysts. This is particularly problematic for enzyme functions that are new and for newly discovered enzyme groups for which no structure yet exists, but novel enzyme function needs to be improved for successful biotransformation.

While there is not enough evidence yet, some studies suggest that if a constant overall library size is given, severe restriction of the number of amino acids to code for at a particular residue while picking a larger set of residues to mutate is a better strategy than the inverse tactic: screen less deeply but rather more widely.

Acknowledgements

The authors gratefully acknowledge financial support from the National Science Foundation (NSF) under contracts IIP-1540017 and IIP-0969003 to the I/UCRC Center for Pharmaceutical Development (CPD).

References

1. A. S. Bommarius, M. Schwarm and K. Drauz, *Chimia*, 2001, **55**, 50.
2. H. S. Olsen and P. Falholt, *J. Surfactants Deterg.*, 1998, **1**, 555.
3. M. T. Reetz, *J. Am. Chem. Soc.*, 2013, **135**, 12480.
4. D. J. Pollard and J. M. Woodley, *Trends Biotechnol.*, 2007, **25**, 66.
5. G.-D. Roiban, R. Agudo, A. Ilie, R. Lonsdale and M. T. Reetz, *Chem. Commun.*, 2014, **50**, 14310.

6. S. Bershtein and D. S. Tawfik, *Curr. Opin. Chem. Biol.*, 2008, **12**, 151.
7. E. G. Hibbert and P. A. Dalby, *Microb. Cell Fact.*, 2005, **4**, 29.
8. K. A. Gray, T. H. Richardson, K. Kretz, J. M. Short, F. Bartnek, R. Knowles, L. Kan, P. E. Swanson and D. E. Robertson, *Adv. Synth. Catal.*, 2001, **343**, 607.
9. G. DeSantis, K. Wong, B. Farwell, K. Chatman, Z. L. Zhu, G. Tomlinson, H. J. Huang, X. Q. Tan, L. Bibbs, P. Chen, K. Kretz and M. J. Burk, *J. Am. Chem. Soc.*, 2003, **125**, 11476.
10. M. T. Reetz, D. Kahakeaw and R. Lohmer, *ChemBioChem*, 2008, **9**, 1797.
11. A. G. Sandstrom, Y. Wikmark, K. Engstrom, J. Nyhlen and J. E. Backvall, *Proc. Natl. Acad. Sci. U. S. A.*, 2012, **109**, 78.
12. M. A. Mena and P. S. Daugherty, *Protein Eng., Des. Sel.*, 2005, **18**, 559.
13. M. T. Reetz, J. D. Carballeira, J. Peyralans, H. Hobenreich, A. Maichele and A. Vogel, *Chemistry*, 2006, **12**, 6031.
14. S. Kille, C. G. Acevedo-Rocha, L. P. Parra, Z. G. Zhang, D. J. Opperman, M. T. Reetz and J. P. Acevedo, *ACS Synth. Biol.*, 2013, **2**, 83.
15. T. Li, K. Fan, J. Wang and W. Wang, *Protein Eng.*, 2003, **16**, 323.
16. S. Akanuma, T. Kigawa and S. Yokoyama, *Proc. Natl. Acad. Sci. U. S. A.*, 2002, **99**, 13549.
17. K. U. Walter, K. Vamvaca and D. Hilvert, *J. Biol. Chem.*, 2005, **280**, 37742.
18. M. T. Reetz and S. Wu, *Chem. Commun.*, 2008, 5499.
19. M. T. Reetz and J. D. Carballeira, *Nat. Protoc.*, 2007, **2**, 891.
20. J. L. Vanhooke, J. B. Thoden, N. M. Brunhuber, J. S. Blanchard and H. M. Holden, *Biochemistry*, 1999, **38**, 2326.
21. N. M. Brunhuber, J. B. Thoden, J. S. Blanchard and J. L. Vanhooke, *Biochemistry*, 2000, **39**, 9174.
22. Y. Asano, A. Nakazawa and K. Endo, *J. Biol. Chem.*, 1987, **262**, 10346.
23. M. T. Reetz, *Angew. Chem., Int. Ed.*, 2011, **50**, 138.
24. M. J. Abrahamson, E. Vazquez-Figueroa, N. B. Woodall, J. C. Moore and A. S. Bommarius, *Angew. Chem., Int. Ed.*, 2012, **51**, 3969.
25. M. W. Abrahamson, J. W. Wong and A. S. Bommarius, *Adv. Synth. Catal.*, 2013, **355**, 1780.
26. Z. Sun, R. Lonsdale, G. Li and M. T. Reetz, *ChemBioChem*, 2016, **17**, 1865.
27. Y. Gumulya, J. Sanchis and M. T. Reetz, *ChemBioChem*, 2012, **13**, 1060.
28. M. T. Reetz, L. W. Wang and M. Bocola, *Angew. Chem., Int. Ed.*, 2006, **45**, 1236.
29. G. D. Roiban, R. Agudo and M. T. Reetz, *Angew. Chem., Int. Ed.*, 2014, **53**, 8659.
30. T. Davids, M. Schmidt, D. Bottcher and U. T. Bornscheuer, *Curr. Opin. Chem. Biol.*, 2013, **17**, 215.
31. C. M. Clouthier, M. M. Kayser and M. T. Reetz, *J. Org. Chem.*, 2006, **71**, 8431.
32. H. Kohls, F. Steffen-Munsberg and M. Hohne, *Curr. Opin. Chem. Biol.*, 2014, **19**, 180.

33. Z. G. Zhang, R. Lonsdale, J. Sanchis and M. T. Reetz, *J. Am. Chem. Soc.*, 2014, **136**, 17262.
34. J. D. Carballeira, P. Krumlinde, M. Bocola, A. Vogel, M. T. Reetz and J. E. Backvall, *Chem. Commun.*, 2007, 1913.
35. K. Chockalingam, Z. Chen, J. A. Katzenellenbogen and H. Zhao, *Proc. Natl. Acad. Sci. U. S. A.*, 2005, **102**, 5691.
36. K. L. Heckman and L. R. Pease, *Nat. Protoc.*, 2007, **2**, 924.
37. M. Anbar and E. A. Bayer, *Methods Enzymol.*, 2012, **510**, 261.
38. K. Izumi, M. Aramaki, T. Kimura, Y. Naito, T. Udaka, M. Uchikawa, H. Kondoh, H. Suzuki, G. Cho, Y. Okada, T. Takahashi, J. A. Golden and K. Kosaki, *Pediatr. Res.*, 2007, **61**, 660.
39. J. Colbeck, B. Mijts, L. J. Giver and R. J. Fox, Combined automated parallel synthesis of polynucleotide variants, *Pat. Appl.* WO 2010144103 A1; Dec 16, 2010.
40. M. Alcalde, M. Zumarraga, J. Polaina, A. Ballesteros and F. J. Plou, *Comb. Chem. High Throughput Screening*, 2006, **9**, 719.
41. A. Dennig, A. V. Shivange, J. Marienhagen and U. Schwaneberg, *PLoS One*, 2011, **6**, e26222.
42. Y. Yu and S. Lutz, *Trends Biotechnol.*, 2011, **29**, 18.
43. T. A. Whitehead, L. M. Bergeron and D. S. Clark, *Protein Eng., Des. Sel.*, 2009, **22**, 607.
44. Z. Qian and S. Lutz, *J. Am. Chem. Soc.*, 2005, **127**, 13466.
45. G. S. Baird, D. A. Zacharias and R. Y. Tsien, *Proc. Natl. Acad. Sci. U. S. A.*, 1999, **96**, 11241.
46. J. F. Chaparro-Riggers, K. M. Polizzi and A. S. Bommarius, *Biotechnol. J.*, 2007, **2**, 180.
47. U. T. Bornscheuer, G. W. Huisman, R. J. Kazlauskas, S. Lutz, J. C. Moore and K. Robins, *Nature*, 2012, **485**, 185.
48. R. J. Fox and G. W. Huisman, *Trends Biotechnol.*, 2008, **26**, 132.
49. J. Damborsky and J. Brezovsky, *Curr. Opin. Chem. Biol.*, 2014, **19**, 8.
50. G. M. Suel, S. W. Lockless, M. A. Wall and R. Ranganathan, *Nat. Struct. Biol.*, 2003, **10**, 59.
51. R. J. Fox, S. C. Davis, E. C. Mundorff, L. M. Newman, V. Gavrilovic, S. K. Ma, L. M. Chung, C. Ching, S. Tam, S. Muley, J. Grate, J. Gruber, J. C. Whitman, R. A. Sheldon and G. W. Huisman, *Nat. Biotechnol.*, 2007, **25**, 338.
52. J. Liang, J. Lalonde, B. Borup, V. Mitchell, E. Mundorff, N. Trinh, D. A. Kochrekar, R. Nair Cherat and G. G. Pai, *Org. Process Res. Dev.*, 2010, **14**, 193.
53. C. K. Savile, J. M. Janey, E. C. Mundorff, J. C. Moore, S. Tam, W. R. Jarvis, J. C. Colbeck, A. Krebber, F. J. Fleitz, J. Brands, P. N. Devine, G. W. Huisman and G. J. Hughes, *Science*, 2010, **329**, 305.
54. M. Berland, B. Offmann, I. Andre, M. Remaud-Simeon and P. Charton, *Protein Eng., Des. Sel.*, 2014, **27**, 375.
55. C. A. Voigt, C. Martinez, Z.-G. Wang, S. L. Mayo and F. H. Arnold, *Nat. Struct. Mol. Biol.*, 2002, **9**, 553.

56. M. Anbar and E. A. Bayer, *Methods Enzymol.*, 2012, **510**, 261.
57. A. S. Bommarius, M. Sohn, Y. Kang, J. H. Lee and M. J. Realff, *Curr. Opin. Biotechnol.*, 2014, **29**, 139.
58. P. Heinzelman, C. D. Snow, I. Wu, C. Nguyen, A. Villalobos, S. Govindarajan, J. Minshull and F. H. Arnold, *Proc. Natl. Acad. Sci. U. S. A.*, 2009, **106**, 5610.
59. P. Heinzelman, C. D. Snow, M. A. Smith, X. L. Yu, A. Kannan, K. Boulware, A. Villalobos, S. Govindarajan, J. Minshull and F. H. Arnold, *J. Biol. Chem.*, 2009, **284**, 26229.
60. M. Anbar, R. Lamed and E. A. Bayer, *ChemCatChem*, 2010, **2**, 997.
61. P. Heinzelman, R. Komor, A. Kanaan, P. Romero, X. L. Yu, S. Mohler, C. Snow and F. Arnold, *Protein Eng., Des. Sel.*, 2010, **23**, 871.
62. S. Bershtein, K. Goldin and D. S. Tawfik, *J. Mol. Biol.*, 2008, **379**, 1029.
63. K. B. Hansen, Y. Hsiao, F. Xu, N. Rivera, A. Clausen, M. Kubryk, S. Krska, T. Rosner, B. Simmons, J. Balsells, N. Ikemoto, Y. Sun, F. Spindler, C. Malan, E. J. J. Grabowski and J. D. Armstrong, *J. Am. Chem. Soc.*, 2009, **131**, 8798.
64. M. Hall and A. S. Bommarius, *Chem. Rev.*, 2011, **111**, 4088.

Section III: Enzymes from Secondary Metabolism

CHAPTER 8

Customizing Transcription-factor Biosensors for Modern Biotechnology

CHRISTIAN KASEY[a,†] AND GAVIN J. WILLIAMS*[a,b]

[a] Department of Chemistry, NC State University, Raleigh, NC 27695-8204, USA; [b] Comparative Medicine Institute, NC State University, Raleigh, NC, USA

*Email: gjwillia@ncsu.edu

8.1 Introduction

The rational design of proteins and biosynthetic pathways can prove a successful approach for engineering alternative phenotypes that support the production of pharmaceuticals, commodity chemicals, chemical feedstocks, biofuels, and food additives.[1–5] Usually, structural information, mechanistic insight, and data from mutagenesis are critical to successfully alter enzyme activity for a specific application.[6–8] Without sufficient information to support this *a priori* approach, engineering a target enzyme can quickly become intractable. The *a priori* design and optimization of *in vivo* biosynthetic pathways can be extremely challenging due to the necessary multiple complex interactions between various components of the target biosynthetic pathway and the supporting metabolic processes. To address these challenges, modern approaches for the design and optimization of biocatalysts

[†]Present address: Zymergen, Emeryville, CA 94608, USA

Catalysis Series No. 32
Modern Biocatalysis: Advances Towards Synthetic Biological Systems
Edited by Gavin Williams and Mélanie Hall

Published by the Royal Society of Chemistry, www.rsc.org

(see Chapter 7) and "microbial factories" are now leveraging high-throughput screening and selection tools for the rapid identification of improved enzymes and pathways from large libraries of variants.

There is a plethora of technologies available for construction of variant libraries. For example, libraries of protein or pathway variants can be generated using semi-rational (*e.g.* multi-site saturation mutagenesis[9,10]), random (*e.g.* DNA shuffling[11,12] and error-prone polymerase chain reaction [epPCR][13–15]), or automated and continuous mutagenic methods such as Multiplexed Automated Genome Engineering (MAGE)[16] and Phage-Assisted Continuous Evolution (PACE).[17] The library size that these methods generate span 10^3–10^9 variants, making traditional low-throughput methods such as nuclear magnetic resonance (NMR) and chromatography (GC–MS, LC–MS, HPLC–UV–Vis, HPLC–ELSD) only suitable for searching through small portions of such libraries. Many approaches have been developed to overcome this key limitation. A variety of high-throughput screens have been utilized that generate a colorimetric or fluorescent signal to indicate the turnover of a substrate and/or consumption of a reaction pathway cofactor. These reactions often rely on a coupled enzyme and/or substrate to produce the reporter output.[18] While effective, these reactions are reliant on suitable coupling reagents for analysing the enzyme or pathway of interest and are often limited in throughput. The throughput of fluorescence-based screens has been improved by compartmentalising the screening components in droplets. Droplet-based screening methods leverage the ability of flow cytometers to measure the level of reporter output in each droplet. Screening droplets by flow cytometry or in more complex micro-fabricated devices can achieve throughputs of millions of droplets per day.[19,20] Droplet screening does depend on the stability and compatibility of the reaction of interest with the droplet formulation being utilized, and is often limited to screening single enzyme-catalyzed reactions, often utilizing *in vitro* transcription/translation kits. Yeast cell surface display has been adapted for the evolution of bond-forming enzymes by displaying the enzyme variants on the cell surface and detecting the enzyme-catalyzed product with single-cell precision *via* bio-orthogonal chemical ligation.[21,22] While powerful and potentially very high-throughput, this technique is perhaps not well suited for measuring catalysis as it does not measure multiple enzymatic turnovers, the enzyme must be active when displayed on the cell surface, and the enzyme substrate needs to be tagged with a suitably reactive handle or the product should be differentiated from the substrate in some other way. Genetic selections have been used to good effect but are usually limited to the directed evolution of enzymes or pathways that can be easily coupled to the survival of the host strain.[23–28] While the strategies described above have proven effective for various directed evolution and high-throughput engineering campaigns, they lack the ability to be broadly applied for reporting the activity of any given target enzyme- or pathway-catalyzed transformation.

A potentially broadly applicable and highly tuneable approach to screening the activity of enzyme and pathway variants is based on DNA-binding

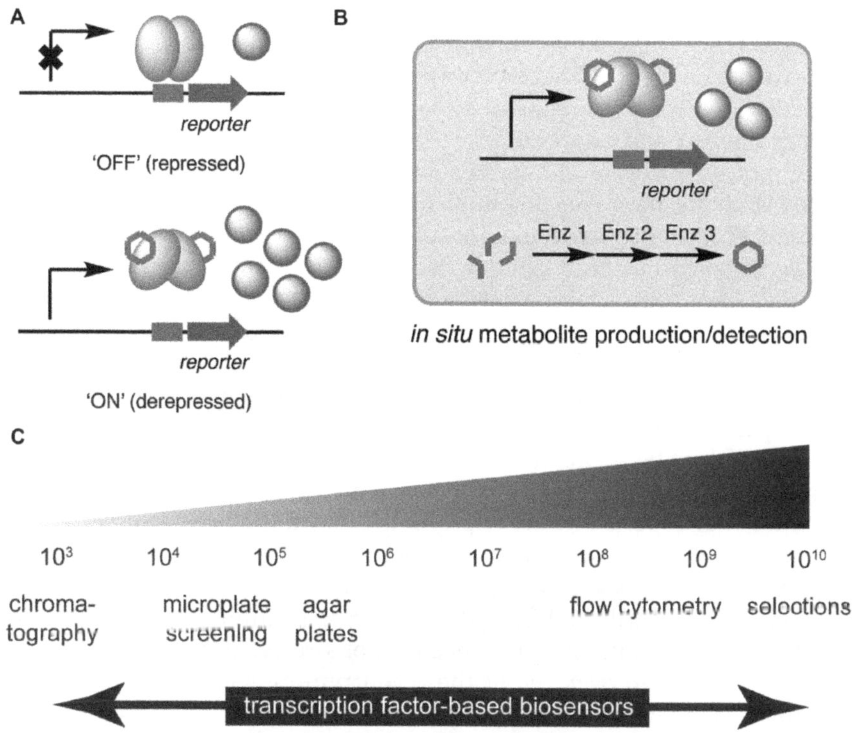

Figure 8.1 Biosensor-guided high-throughput metabolic engineering and synthetic biology. (A) A biosensor in a repressed (top) and de-repressed (bottom) state based on the presence of an inducer molecule. (B) A key application of engineered transcription factors is to report production of enzymatically generated small molecules. (C) Methods for screening and selecting enzyme/pathway variant libraries.

allosteric transcription factors (aTFs) that control gene expression in response to ligand binding. aTFs are one-component regulatory systems that mediate the majority of signal transduction in prokaryotes (Figure 8.1A).[29] One-component aTFs rely on multiple interactions (ligand–protein, protein–DNA, and in the case of multimeric transcriptional regulators protein–protein) and allosteric signal transfer for correct function in their natural or engineered context. aTFs usually adopt one of three overall mechanisms to control either the activation or repression of transcription, ultimately by impacting the interaction between RNA polymerase (RNAP) and the promoter sequence. Most aTFs that have been utilized as biosensors are repressor proteins that inhibit transcription by preventing binding of the RNAP to the promoter or by preventing RNAP from transcribing by sequestering the ribosome binding site (RBS). Less commonly, aTFs have been employed that are activator proteins[30] or act as both activators and repressors.[31] In spite of their mechanistic complexity and requisite multitude of protein interactions, the function and properties of aTFs have been the target of engineering to create tailored biosensors. Indeed, the wide variety of ligands that induce allosteric transcriptional factors and the relative ease with which aTFs can be engineered (particularly by directed evolution) and coupled to a variety of reporter systems have contributed to the rapidly growing interest and adoption of aTFs as biosensors that can guide high-throughput engineering of enzymes and pathways. Several other biological sensory systems have the potential to be reconfigured as biosensors for detection of specific molecules or biochemical events such as protein–protein interactions and cellular localization. Biosensor systems other than aTFs have been superbly reviewed elsewhere.[32–34]

While natural transcription factors exist that respond to a wide variety of natural and non-natural ligands, they are not always well suited to the needs of protein engineers, metabolic engineers, or synthetic biologists, especially with respect to their sensitivity or specificity for a desired compound. This chapter will highlight the customization of transcription factors to create designer biosensors for detection of a variety of small molecules. Structural and functional aspects of transcription factors will be briefly surveyed, and the common reporter systems and operational characteristics that can be used to describe aTFs will be introduced. Several examples of transcription factor engineering will highlight the successes in tuning biosensors to respond to desired molecules at concentrations relevant to the targeted application. Finally, an overview of these contributions and a future outlook will provide a summary of the field and the prospects for using biosensors in synthetic biology and metabolic engineering.

8.2 Introduction to Transcription Factor Engineering

Synthetic biologists, metabolic engineers, protein engineers, and chemical biologists are beginning to create a toolbox of engineered aTFs for a myriad of applications that involve engineering biological systems. The next section describes the availability of a wide range of aTFs that serve as components for biosensing natural molecules and as templates for engineering biosensors that detect non-natural compounds.

8.2.1 Diversity of Transcription Factor Structure, Function, and Ligand Repertoire

Transcription factors are responsible for the temporal regulation of gene expression in organisms driven by environmental or internal cues. In particular, aTFs are a large and diverse subset of transcription factors and are ubiquitous in prokaryotes and eukaryotes. In prokaryotes, aTFs comprise transcriptional repressors and activators and are readily identified by a signature helix-turn-helix DNA-binding motif present in nearly 95% of described DNA-binding transcription factors protein structures.[35] In addition to a DNA-binding domain (DBD) motif, aTFs also include a ligand-binding domain (LBD) responsible for the binding of an effector molecule which allosterically impacts transcription. Despite the structural similarities between aTFs, the amino acid sequences of LBDs are not well conserved, and this reflects the extremely diverse set of effector structures including antibiotics, terpenoids, cofactors, cell-signalling lactones, amino acids, metal ions, flavonoids, sugars, and fatty acids. There are about 20 major classes of aTF based on amino acid conservation in the DBD, each typically named after the first member that was characterized.[36] Among the best characterized prokaryotic regulatory families are the AraC activators,[37] TetR repressors,[35] LacI repressors,[38] LuxR activators,[39] MarR activator/repressors,[40] and the *Streptomyces* antibiotic regulatory proteins (SARP) activators.[41,42] This important but by no means exhaustive list of examples of well-studied transcriptional regulator families has been the focus of numerous structural and mechanistic studies to understand ligand binding, DNA binding, and the conformational changes that occur due to these binding events.

Several studies have investigated the structural aspects of ligand binding to aTFs. The LuxR-type protein CviR was characterized with a variety of homoserine lactones (HSLs) that induced conformational changes captured by X-ray crystallography.[43] AraC has been thoroughly investigated, including individual components such as the DNA binding domain[44] and the dimerization motif,[45] and as a complete aTF.[46] aTFs undergo distinct conformational changes upon ligand binding, as evidenced by studies of the apo and holo forms of several aTFs.[47–49] DNA-bound crystal structures of aTFs have also shed light on the significant structural differences from the DNA-free form.[44,50] Taken together, these investigations have provided extensive information to researchers interested in repurposing aTFs for customized responses to target molecules.

8.2.2 Reporter Systems, Range of Throughput, and Scope of Biosensor Hosts

The potential value of reprogramming aTFs with customized components to generate biosensors that enable high-throughput selection or screening is immense (Figure 8.1B).[32,51] Thousands to millions of cells that harbour variant enzymes or pathways can be analysed by engineered aTFs that respond to a target small molecule. This in turn allows the enrichment of a cellular population displaying the desired output or phenotype or the direct identification of the best performing strains. An often employed aTF design to leverage their power to search large libraries of variants is to configure a survivability mechanism under the control of the transcriptional regulator.[52] Reduced to practice, this is often achieved by placing an antibiotic resistance gene downstream of the cognate operator DNA of a specific aTF repressor protein. Accumulation of the target biomolecule *in vivo* leads to expression of the antibiotic resistance gene *via* the aTF and thereby confers survivability to the host cell when grown in the presence of the antibiotic marker. Other popular formats include inserting a colorimetric[53] or fluorescent[54] reporter in place of or in conjunction with a survivability marker to facilitate visual inspection or measuring fluorescent output, respectively (Figure 8.1C). Fluorescent reporters can be especially versatile given the typically low detection limit and a variety of assay formats and instruments available, including microplate readers, trans-illuminators, and flow cytometers (Figure 8.1C). Although colorimetric reporters usually suffer from high detection limits, this is a signal output that can be easily visualized. Notably, directed evolution can be used to reduce the detection limit of whole-cell biosensors that use colorimetric reporters.[53] The design and implementation of aTFs as biosensors is often carried out in microorganisms that facilitate DNA manipulation such as *Escherichia coli* (*E. coli*) and *Streptomyces coelicolor* (*S. coelicolor*). Notably though, aTFs from highly diverse microorganisms have been characterized, engineered, and reprogrammed to function in a variety of hosts.[55–57] In this regard, aTFs have proven to be remarkably modular and versatile. For example, recently Keasling and coworkers leveraged an *Acinetobacter* aTF from the LysR-type transcriptional regulator (LTTR) superfamily to successfully detect production of *cis,cis*-muconic acid in yeast.[30] They also demonstrated the successful transplantation of other LTTR family members from *Herbaspirillum seropedicae* (FdeR), *Sinorhizobium meliloti* (PcaQ), *E. coli* (ArgP) and *Klebsiella pneumonia* (MdcR) into yeast and demonstrated their utility as biosensors for naringenin, protocatechuic acid, L-arginine and malonic acid, respectively. This work significantly expands the scope of small molecules that can be detected in eukaryotes and subsequently sets the stage for high-throughput approaches to engineering biocatalysts and pathways in such hosts. In another example, the ChnR/Pb transcription factor–promoter pair from *Acinetobacter* was transplanted into *E. coli* to enable the detection of lactams.[58] Following further improvements of the biosensor, a variety of lactams, including butyro-, valero-, and caprolactam could be detected in a dose-dependent manner using mCherry as the reporter. The *E. coli* protein Ada is an aTF that detects methylating compounds by sensing aberrant methyl adducts on the phosphodiester backbone of DNA. The biosensor was first characterized and engineered in *E. coli*, and then fused to the yeast Gal4 transcriptional activation domain to afford a functional biosensor for methyl phosphotriester adducts in *Saccharomyces cerevisiae*.[59] The native function of the engineered Ada system was retained in yeast. In another example, Joensson and coworkers demonstrated that fine-tuning the expression of the transcription factor PadR from *Bacillus subtilis* by random ribosome binding site (RBS) mutagenesis was necessary to transplant it as a functional biosensor into *E. coli* and *Corynebacterium glutamicum*.[60]

Biosensors typically measure the concentrations of effector compounds that are intracellular. This feature can make designing robust sensors that function in multiple hosts technically demanding, as highlighted above. In many cases, the extracellular concentration of a small molecule is more important than that inside the cells. This is particularly important for microbial cell factories that excrete antibiotics or other biologically active natural products into the culture media. To address these issues, Joensson and co-workers ingeniously encapsulated yeast *p*-coumaric acid-producing cells with *E. coli*-biosensing cells in picolitre droplets. Subsequently, a microfluidic device was used to rapidly sort droplets containing yeast cells that produced high amounts of extracellular target compound using the *E. coli* fluorescent biosensor signal. This not only provides a versatile screening platform that can be coupled to a broad range of producing strains, but also enables ultra high-throughput screening *via* flow cytometry.

8.2.3 Overview of Biosensor-driven Applications

Transcriptional biosensors based on aTFs are employed in three main areas that are most relevant to synthetic biology and biocatalysis: (i) high-throughput identification of microbial strains with desired phenotypes, (ii) high-throughput identification of enzyme variants with desired properties, and (iii) pathway control and precursor supply. The application of high-throughput transcriptional regulator-mediated selection to identify strain variants that support improved production of naringenin and glucaric acid in engineered *E. coli* strains exemplifies the first category of applications. In this study, Church and co-workers[61] used MAGE to carry out targeted genome-wide mutagenesis to vary the expression of pathway genes involved in the biosynthesis of naringenin and glucaric acid. After up to four rounds of evolution and biosensor-guided screening, the production of naringenin and glucaric acid was increased by 36- and 22-fold, respectively. Key to this achievement was the use of the TolC membrane protein in the design and implementation of a toggle selection that preserved library complexity while minimizing evolutionary escapes during each round of selection. In an earlier example, Keasling and co-workers improved short-chain alcohol production in *E. coli* by 35% through optimization of enzyme expression levels by deploying a survival-based selection. Clones that produced low amounts of alcohol were eliminated by linking host cell growth in the presence of tetracycline to 1-butanol production.[52] Related to the second category of applications—screening the activity of biocatalysts—Cirino and co-workers used an engineered AraC transcriptional protein to report triacetic acid lactone production catalysed by 2-pyrone synthase in *E. coli*.[62] Subsequently, after two rounds of mutagenesis and biosensor-guided screening, an enzyme variant that supported 20-fold increased TAL production was identified. Notably, the endogenous AraC had to be knocked out for the success of this approach. Another engineered AraC variant that detects ectoine was used to identify variants of L-2,4-diaminobutyric acid aminotransferase that supported improved ectoine production in *E. coli*.[63] To date, few engineered aTFs have been used to screen libraries of biocatalysts, perhaps in large part because the engineered biosensor must be able to distinguish the enzyme substrate, product, and related intermediates (see Section 8.5). Finally, engineered aTFs have been frequently applied as regulatory proteins to control complex metabolic pathways and to support the *in vivo* production of end-products by detecting and regulating key metabolic precursor compounds. The expression of heterologous biosynthetic pathways often leads to the production of intermediates that are toxic to the host cell. Accumulation of such intermediates could be avoided by using biosensors to dynamically control the pathway. For example, the Keasling group have used transcriptome analysis to identify promoters that are regulated by stress-inducing metabolites such as farnesyl pyrophosphate (FPP). These promoters were then used to regulate FPP production in the isoprenoid biosynthetic pathways in *E. coli*, leading to a 2-fold improvement in production of amorphadiene.[64] A malonyl-CoA biosensor has been developed using the malonyl-CoA sensitive repressor protein, FapR from *B. subtilis*. This has been used to control and optimize various biosynthetic pathways that depend on malonyl-CoA as a precursor, including the overproduction of 3-hydroxypropanoic acid in yeast[65] and fatty acids in *E. coli*.[31,66]

8.2.4 Quantitative Description of the Activity and Properties of Transcriptional Factors

A dose–response curve is usually obtained so that a variety of quantitative parameters can be extracted to describe the activity and properties of a given biosensor system. In this way, a transfer function is usually used to describe the mathematical relationship between the biosensor input (ligand concentration) and output (reporter response).[67] The dynamic range is the difference in biosensor output between the maximum and background states. Typically, a large dynamic range is desired so that the presence of the target ligand is distinguished more easily than its absence. The linear range of detection is the range of input ligand concentrations that can be linearly correlated with the reporter signal. The $K_{1/2}$ (sometimes called the "response threshold") is the concentration of ligand required for half maximal reporter signal. The desired linear range of detection and corresponding $K_{1/2}$ depends on the expected range of small molecule concentration in the producing strain. Finally, closely related to the linear range of detection, the sensitivity of a given biosensor system reflects the linear slope of the corresponding response curve at the $K_{1/2}$. This is proportional to the cooperativity (how the subunits of a transcription factor–effector pair cooperate during regulation) and can be measured by extraction of the Hill coefficient. A specific set of such biosensor characteristics are often required for a given biosensor application. For example, during initial rounds of screening libraries of enzyme or pathway variants, the linear range of detection (and sensitivity) might be required to be very low in order to identify those rare variants that produce just a small increase in product titer (*e.g.* variants that support 3% conversion from those that support 0–2% conversion). During later stages of screening, the linear range of detection might need to be several-fold higher in order to distinguish the productivity of variants when their average activities is very high (*e.g.* distinguish a variant with 90% conversion from ones that support 80% conversion). In other instances, a toggle switch might be required whereby the reporter is expressed at high levels only when the compound of interest reaches a certain minimum concentration. This might be particularly useful when the throughput of a genetic selection is required but only a small number of variants with well-defined phenotypes are required.

8.3 Allostery in Transcriptional Regulators

Allostery is the process by which proteins transmit the effect of ligand binding at one site to another, enabling the regulation of activity over some distance. Allostery is a phenomenon that has been described in many proteins[68,69] including receptors,[70] thrombin,[71] protein kinases,[72] heat-shock proteins,[73] and transport proteins.[74] For transcriptional regulators, such signal transduction is crucial for the correct operation of the regulator. Given that binding of ligand causes a conformational change in activator and repressor proteins, allosteric signal transduction must take place between the ligand-binding domain and DNA-binding domain of the regulator. Accordingly, allosteric regulation in transcription factors has been extensively studied,[61,75,76] and a recent perspective by Church and co-workers discussed in depth the topic of engineering allosteric signal transduction using the well-studied LacI system as a focal point.[51] Clearly, understanding and engineering the allosteric behaviour of transcription factors is often a crucial first step to manipulate the activity of these proteins for new functions, such as biosensing.

Allostery in transcription factors constitutes a complex signal transduction cascade that depends on multiple amino acid residues that undergo concerted rearrangements for successful transfer of the input signal (ligand binding) to output action (*e.g.* transcription). In some transcription factors, such as the LacI transcription factor,[77–79] allostery has been fully decoupled from ligand binding by the mutation of just a single amino acid. Once decoupled in this way, the original output action can be restored or even inverted by amino acid mutagenesis to provide an alternative signalling pathway.[77]

A particularly interesting example that targeted allostery involved engineering the LacI repressor protein by Wilson and co-workers.[77] Wild-type LacI represses transcription in the absence of the effector compound, isopropyl β-D-1-thiogalactopyranoside (IPTG). In this study, a complete reversal of the natural signal transduction pathway was achieved by engineering LacI variants that instead fully repress transcription in the presence of IPTG and allow transcription in its absence ("anti-lac"). In total, eleven LacI variants were constructed that displayed this anti-lac behaviour.[80,81] The first step in the inversion of transcription factor activity was to engineer LacI into a 'super-repressor' phenotype by destroying the native allostery while retaining the ability to bind IPTG. This was achieved by introducing the mutations D275F, K84A, V95A, V95F after leveraging previously published mutagenesis data.[79] One round of epPCR and screening on solid media was then utilized to furnish compensatory mutations that either restored the wild-type response or introduced the anti-lac phenotype. Other repressor proteins, such as TetR, have been the subject of similar campaigns of intensive mutagenesis and screening to explore how to modulate allosteric behaviour and to reverse transcription factor function.[82–84]

8.4 Engineering the Sensitivity and Dynamic Range of Transcription Factors

Often, a transcription factor capable of detecting the target small molecule or a closely related one is known, but limitations in sensitivity or dynamic range limit the usefulness of the transcription factor as a biosensor for a given application. To address these shortcomings and to make use of transcription factors platforms as made-to-order biosensors, the dynamic range and/or sensitivity of a transcriptional regulator may be modified to provide a biosensor with the desired characteristics. Sensitivity may need to be increased or decreased to suit a particular application but in a way that does not interfere with the transfer function behaviour of the transcription factor. The dynamic range usually needs to be increased and several strategies are available to improve it. Not surprisingly, the dynamic range and sensitivity of an aTF are intricately linked, and manipulating one usually affects the other.

The most direct approach to manipulate sensitivity and dynamic range is to alter the expression level of the aTF itself. For example, Keasling and co-workers modified the dynamic range of the aTF BenM by placing it under control of different yeast promoters and by mutagenesis of the BenM gene.[30] Ultimately, the most improved variant was a biosensor with nearly four times the fluorescence output intensity of the wild-type biosensor with *cis,cis*-muconic acid. Similarly, the 2010 Peking International Genetically Engineered Machine competition (iGEM) team described how varying promoter strengths that control expression of the MerR transcription factor greatly affected its response profile to mercury.[85] Among the findings were that the level of expression of MerR directly determines the sensitivity threshold to mercury. Also, changes in the promoter controlling MerR affected the slope of the dose–response to mercury and impacted the maximum reporter protein output. In another example, Ching and co-workers developed an organophosphorus biosensor based on the aTF DmpR and a colorimetric reporter.[53] Because organophosphorus compounds have been used as nerve agents and agricultural compounds, there is much interest in being able to rapidly determine their presence in environmental samples. However, because these compounds are present in environmental samples in trace quantities, the goal of their work was to engineer the aTF biosensor to detect the target analytes at lower concentrations than the prototype design.

By introducing random mutations throughout the entire DmpR gene *via* epPCR, the authors identified several variants that were able to detect parathion in whole-cell biosensor strains at concentrations 100-fold lower than the best previously-reported prototype.[53] The authors suggested that the improvement in the lower detection limit was a result of elevated basal reporter expression. Other approaches have been explored to alter the expression level of the aTF. The RBS strength of the aTF component of a biosensor was successfully modified by Joensson and co-workers to generate biosensors with increased sensitivity to *p*-coumaric acid.[60] Random mutagenesis of the upstream RBS of the PadR gene resulted in biosensors with up to a 130-fold induction in the presence of the effector. Interestingly, the dynamic ranges of a small subset of operational RBS mutants correlated with the predicted translation initiation rates calculated by the RBS calculator.[86] The expression level of the aTF can also be impacted by manipulation of the copy number of the plasmid that carries the aTF. Plasmid copy number is determined primarily by the strength of antibiotic selection (for plasmids harbouring a selectable resistance gene) and the origin of replication (*ori*). The identity of the *ori* can determine the plasmid copy number spanning several orders of magnitude. Church and co-workers investigated the effects of high (pUC origin, 100–500 copies per cell) and low copy-number plasmids (SC101 origin, 2–5 copies) on the characteristics of the AcuR, CdaR, MphR, and TtgR aTFs that respond to acrylate, glucarate, erythromycin, and naringenin, respectively.[87] While copy number did not have a particularly dramatic effect (maximum of 2-fold difference) on the Hill coefficient of the systems tested, a high copy-number plasmid was consistently associated with an increased fold-induction and higher $K_{1/2}$.

An additional consideration that may impact transcription factor sensitivity is background signal reduction. For example, an arsenite-responsive biosensor based on the ArsR repressor protein initially displayed high reporter signal in the absence of the inducer.[88] This problem was addressed by adding a second copy of the ArsR operator binding site at a position downstream of the ArsR gene and close to the reporter gene. This modification significantly reduced the background signal without interfering with the dose–response behaviour of the biosensor.

One aspect of reporter output that is easily overlooked is the stability of the reporter itself. The stability of the reporter protein in terms of signal strength and duration is an important factor in biosensor development. Tarayre and co-workers discuss the mechanisms of degradation a GFP reporter may encounter and the intrinsic noise that accompanies different strength promoters.[89]

One method to control sensitivity that is completely decoupled from both the expression of the transcription factor and its interaction with the operator is to modify the strength of the reporter protein output. Zhuang and co-workers provide an example of this that relied on modifying the strength of the RBS of the reporter gene that was part of the 2-chlorotoluene biosensor, BTEX.[90] The luminescence output of the wild-type BTEX and a variant dubbed BTEX-SD (containing the strong RBS from the T7 phage *g10* gene) was determined with 2-chlorotoluene, revealing that the BTEX-SD variant was 50% more sensitive than its wild-type counterpart. With the addition of an enhancer sequence upstream of the RBS, the signal-to-noise ratio was improved even further and the total luminescence output was improved.

Similarly, the reporter promoter can be decoupled from the transcription factor operator. Instead, the reporter gene may be placed under the control of a highly active promoter, which is in turn activated by a cognate activating protein. By placing the activating protein under the control of the transcription factor promoter, an otherwise weak signal may be amplified.[91]

Shedding some light onto the success and limitations of these experimental approaches, a recent combination of modelling and strain characterization sought to provide a quantitative understanding of how tuneable parameters shape biosensor function.[92] Their analysis found that mutations in the operator sequence simultaneously affect the basal output, dynamic range, and $K_{1/2}$ of the dose–response curve, rendering a quantitative guideline crucial. It was also found that biosensor dynamic range and $K_{1/2}$ can be controlled orthogonally with the promoter dynamic range and the ligand binding affinity. Promoter libraries with combinatorial designs for operator site and promoter strength cover a large fraction of the design space for the dynamic range, while aTF engineering can help to control the $K_{1/2}$ of leaky transcription factors.

8.5 Engineering the Ligand Specificity of Transcription Factors

The ability to tailor a biosensor for a molecule of interest greatly expands the potential to utilize high-throughput screening to engineer biocatalysts and producing organisms. The specificity of an aTF for a compound or set of compounds may be modified, much in the same way that the substrate specificity of enzymes has been modified. Along these lines, many of the same techniques that have been applied to engineering enzyme substrate specificity may be utilized for aTFs, including structure-guided mutagenesis,[93] random mutagenesis,[94] and *in silico* screening.[95] Most commonly, the ligand effector specificity of aTFs has been altered to include ligands poorly detected by the naturally occurring aTF. Less commonly, aTFs have been engineered to respond to small molecule ligands that are not recognized at all by the template transcription factor or by any known aTF. Other aspects of engineering aTF ligand specificity include altering the specificity between closely related ligand structures such that a biosensor is tailored to recognize the target ligand in the presence of potential competing small molecules. Indeed, in the field of biosensor engineering, examples are emerging of tuning specificity of aTFs for ligands of interest, and taking the challenging step of tailoring the selectivity of those biosensors to discriminate between closely related compounds.

8.5.1 Expanding Specificity Towards Non-native Ligands

Many high-value small molecules that are the products of engineered biocatalysts or metabolic pathways are not detected by known naturally occurring or engineered aTFs. Although genome mining has proven useful for the identification of several prokaryotic aTFs capable of integration into complex biological circuits,[96] and several databases are available that compile transcription factors–effector pairs,[97–100] it is unclear whether mining alone can identify aTFs with the requisite ligand specificity or selectivity, particularly when non-natural compounds are concerned. Other solutions to this limitation include enzymatically converting the target compound of interest to a derivative that can be detected by a known transcriptional factor.[101] In the case where a transcription factor for a given small molecule target is not known, Keasling and co-workers explored a clever strategy to identify suitable alternative candidates. Termed "Analogue Generation toward Catabolizable Chemicals" (AGTC), a list of chemical analogues of the target compound are computationally generated.[58] This shortlist is then cross-referenced with the BRENDA database,[102] with the goal of identifying analogues that play known or putative roles in microbial metabolism-related enzymatic reactions. For the target compound valerolactam, this process returned cyclohexanone as a catabolizable analogue. The cyclohexanone catabolic gene cluster includes a putative transcription factor/promoter pair that is induced by cyclohexanone. After some optimization, the biosensor was used to sense the desired target compounds (see Section 8.2.2). A potentially more general and efficient strategy is to customize existing aTFs for the detection of target molecules by engineering their ligand binding capabilities.

When a crystal structure with bound ligand is available, this provides an opportunity to test our ability to rationally or semi-rationally design new effector specificities *via* site-directed and site-saturation mutagenesis. The effector specificity of the quorum sensing aTF TraR was successfully engineered by a structure-guided approach.[103] TraR is natively induced by the pheromone *N*-3-oxooctanoyl-L-homoserine lactone (OOHSL). Structural studies of the TraR:OOHSL:DNA complex revealed four residues that made hydrogen bonds to the ligand. These were mutated to specific residues by site-directed mutagenesis to generate a series of mutants with altered inducer specificity. Most notably, several mutants displayed enhanced preference for shorter chain lactones. A structure-guided strategy was used to great effect to expand the ligand specificity of AraC to include effectors with structures distantly related to its native activator, L-arabinose (L-ara). For example, while wild-type AraC does not display detectable induction by triacetic acid lactone (TAL), it was engineered to detect TAL and was subsequently integrated into a biosensor that was used to improve the production of TAL in *E. coli*.[62] Simultaneous saturation mutagenesis at five key residues in the ligand binding cavity of AraC (P8, T24, H80, Y82, and H93) afforded a large combinatorial library of AraC variants (3.2×10^6) which was then screened in the presence of TAL by flow cytometry. The resulting AraC variant could detect TAL with an apparent $K_{1/2}$ of ~4 mM. The engineered sensor was poorly induced by L-ara and mevalonate, presumably because they were present in the host cells during negative sorting. The biosensor was sufficiently sensitive to detect TAL that was biosynthesized by a polyketide synthase, 2-pyrone synthase, in *E. coli*. Using a similar approach, the same group had previously engineered AraC to be induced by mevalonate, a ligand that does not activate the wild-type biosensor.[104] The engineered variant included mutations at four of the five sites targeted by multi-site saturation mutagenesis. Although the $K_{1/2}$ with mevalonate was quite high (~150 mM), the authors noted that the linear range of detection (10–100 mM mevalonate) was ideally suited to the intended application of screening engineered isoprenoid metabolic pathways. Together with other examples of engineering the ligand repertoire of AraC (see Section 8.5.2), these examples highlight the remarkable capacity to expand the ligand specificity of AraC towards effectors very different in structure to the native inducer.

Notably, in the absence of an experimental protein structure, structural bioinformatics were used to broaden the effector specificity of PobR to include 3,4-dihydroxybenzoate (34DHB) in addition to the natural effector, 4-hydroxybenzoate (4HB). More specifically, a comprehensive program of comparative modelling of dimer structures, *ab initio* loop reconstruction, and ligand docking was used to select likely ligand-binding residues for mutagenesis.[105] After using flow cytometry to screen a library of ~10^7 variants, one particular mutant that was induced by 34DHB was characterized, even though this ligand does not induce the wild-type transcription factor. Interestingly, no sensitivity was observed for a non-targeted but similar ligand, 2-hydroxybenzoate.

The ligand specificity of aTFs has been expanded *via* elaborate mutagenesis strategies. For example, to facilitate the bioremediation and detection of explosives, de Lorenzo and colleagues have developed a series of biosensors for the detection of trinitrotoluene (TNT) and dinitrotoluene (DNT) in soil samples. While methods such as LC–MS and GC–MS have been developed to detect these compounds, solutions based on a microbial biosensor have also been explored.[106] The transcription factor XylR from *Pseudomonas putida* detects toluene as well as other aromatic compounds including 1,3-dimethylbenzene (*m*-xylene) and is responsible for regulating degradation pathways of these compounds.[107,108] A two-pronged approach was employed to engineering the detection capabilities of XylR.[109] First, domain-specific epPCR was used to introduce point mutations in the effector-binding domain of XylR. Second, XylR was subjected to DNA shuffling with the phenol-responsive homolog DmpR. Both library strategies yielded hits that displayed increased sensitivity to DNT, as well as other distinct phenotypes. In the case of the clones from the epPCR library, they possessed the additional trait of decreased activation with the natural effector, toluene, compared to the wild-type XylR. Background reporter output was also higher (measured by β-galactosidase activity). In contrast, the clones from the DNA shuffling library had lower basal activity and increased activation with toluene compared to the epPCR clones. Further engineering of XylR to display effector selectivity is described in Section 8.5.2. The ligand specificity of the quorum-sensing aTF LuxR from *Vibrio fischeri* was also engineered by a combination of epPCR and DNA shuffling, resulting in variants with expanded effector specificity and a 100-fold improved sensitivity towards one non-native ligand.[110] Although random mutations were introduced throughout the entire LuxR gene, most of the residues in the evolved aTFs were located in the N-terminal LBD, and could not have been predicted from previous extensive loss-of-function mutational studies. Chimeragenesis has also been explored to expand the effector specificity of aTFs. For example, a combinatorial library composed of shuffled N-terminal A domains of the homologous regulators DmpR, XylR, and TbuT was reassembled within the XylR structure. After screening, several protein variants were obtained that displayed expanded inducer range that included new effectors that failed to induce any of the wild-type transcription factors.[111] Notably, chimeric transcription factors with hybrid DBD and LBDs that have mismatched interfaces have been found to retain ligand specificity of the LBD.[112] The concept of chimeric transcription factors was taken to the extreme by Keasling and co-workers who created a fusion between an enzyme that binds the target ligand, a linker that affects signal transduction, and a DBD that responds to ligand binding by regulating binding of RNA polymerase. Random mutagenesis was used to fine-tune the biosensor for use in dynamic feedback control.[113]

Recently, Williams and co-workers used directed evolution to engineer the effector specificity of a macrolide sensing aTF, MphR.[114] Macrolides are an important group of natural products that display broad and potent biological activities and are biosynthesized by giant mega-enzyme assembly lines (see Chapter 10). New synthetic routes to macrolides and unnatural derivatives are urgently required for drug discovery, drug manufacture, and probe development.[115] The complexity of macrolide biosynthetic pathways and our limited ability to rationally engineer them has hampered efforts to reprogram the biosynthesis of macrolides in various microbial hosts, notable successes notwithstanding (see Chapter 10). High-throughput synthetic biology and directed evolution approaches could overcome this problem, but the lack of methods that can identify mutant enzymes, pathways, and strains that produce the desired macrolide target are not generally available. Williams *et al.* demonstrated that the promiscuous macrolide-sensing transcription factor MphR[49] is a powerful platform for engineering variants with tailored properties. The application of random mutagenesis, structure-guided saturation mutagenesis, and positive/negative screening led to the identification of MphR variants with improved sensitivity towards the antibiotic erythromycin, tailored inducer specificity, and significantly improved sensitivity to macrolides that were very poor effectors of the wild-type MphR biosensor (Figure 8.2). Such designer macrolide biosensors will enable high-throughput synthetic biology and directed evolution to be applied to macrolide biosynthesis to solve long-standing problems in the area of natural

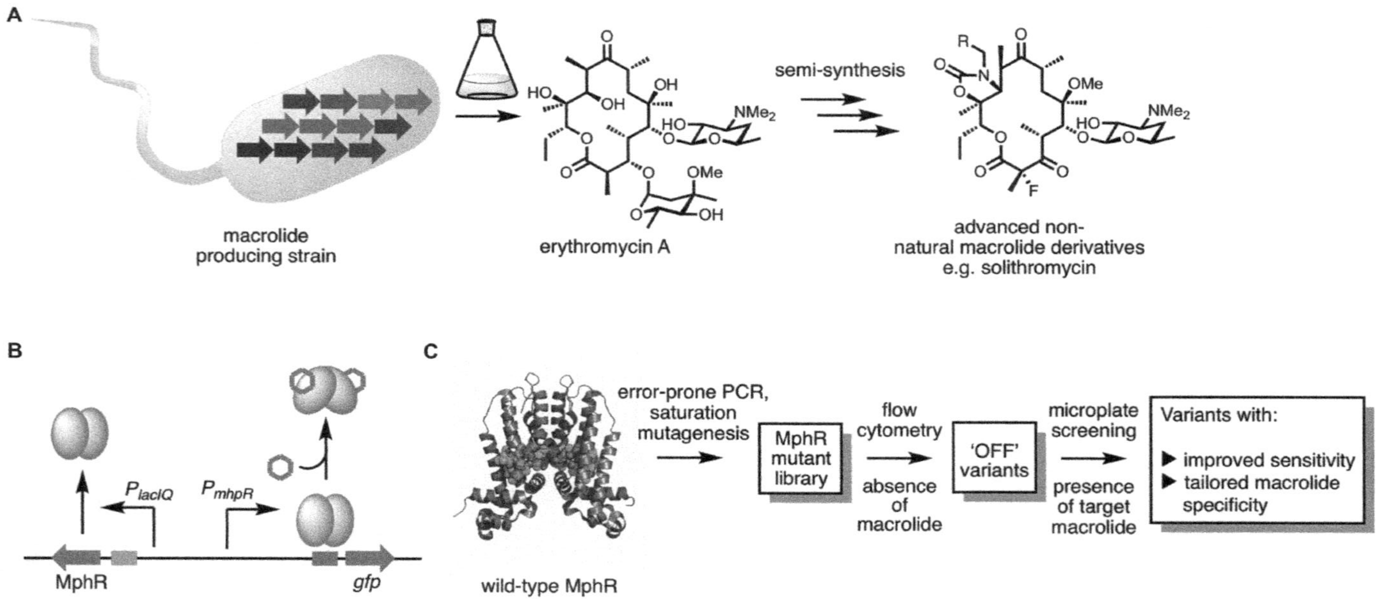

Figure 8.2 Engineered MphR biosensors for natural product synthetic biology. (A) Macrolides are produced by complex biosynthetic pathways and are often subjected to semi-synthesis to afford advanced unnatural derivatives for drug discovery. (B) MphR is a transcription factor that detects macrolides. (C) Scheme illustrating the process employed to tailor the sensitivity and specificity of MphR.

product biosynthetic engineering. This study also suggests that the identification of a suitable transcription factor with inherent plasticity or capacity to adapt to new effectors is critical for directed evolution of new ligand specificities. Indeed, the native role of MphR is to switch on host self-resistance to macrolide antibiotics, and this might be key to its remarkable capacity for adaptability.

The success of directed evolution methods to expand the ligand specificity of transcription factors notwithstanding, the capacity to evolve the ligand specificity of transcription factors towards ligands that are distantly related to the native effector is still being explored. To complement such evolutionary approaches, structural data and computational algorithms are increasingly being employed. Indeed, tremendous successes in protein design have advanced the construction of protein ligand binding[116] with successes in biocatalysis,[117] including allosterically-regulated biocatalysis,[118] therapeutic activity,[119] and more recently in the recognition of new effectors for allosteric transcription factors.[120] Such efforts are fuelled by the increasing availability of transcription factor crystal structures, particularly those that include bound effectors.

The Rosetta suite of tools from the Baker group has proven an invaluable resource in several transcription factor engineering efforts.[121,122] The Rosetta software computationally models static and dynamic macromolecules allowing informed decisions regarding protein structure and structural and functional changes caused by protein alteration (see Chapter 4). Raman and co-workers used Rosetta design to engineer the LacI repressor protein to be induced by several compounds that do not serve as inducers of the wild-type transcription factor (Figure 8.3A).[61] In this study, three different strategies were explored to sensitize LacI to gentiobiose, fucose, lactitol, and sucralose. In the first, a computational approach utilized ligand conformer sampling using OpenEye Omega software which generated hundreds of conformational possibilities for three ligands most dissimilar to the native effector: fucose, lactitol, and sucralose. The conformers were docked into the ligand-binding pocket of the LacI structure. Then, several rounds of *in silico* ligand position perturbations coupled with combinatorial mutagenesis and backbone minimizations were conducted for optimal binding pocket interactions. By targeting three regions of LacI that spanned most of the native ligand-binding site (Figure 8.3B), several solutions for fucose, lactitol, and sucralose-responsive LacI variants were generated in this manner, and key residue changes were identified. Most of the best hits contained multiple mutations. For sucralose, for example, the best variant contained four mutations, which the authors suggest is significantly complex enough to not be discovered from an epPCR library (Figure 8.3C, D). Interestingly, when three portions of the protein were targeted simultaneously, variant designs emerged that were nearly all unable to repress transcription. This suggests that allostery may be negatively affected by such a heavy mutational load. Accordingly, each region of the ligand-binding site was targeted individually. In the second approach, a library of 6000 single-site saturation mutagenesis

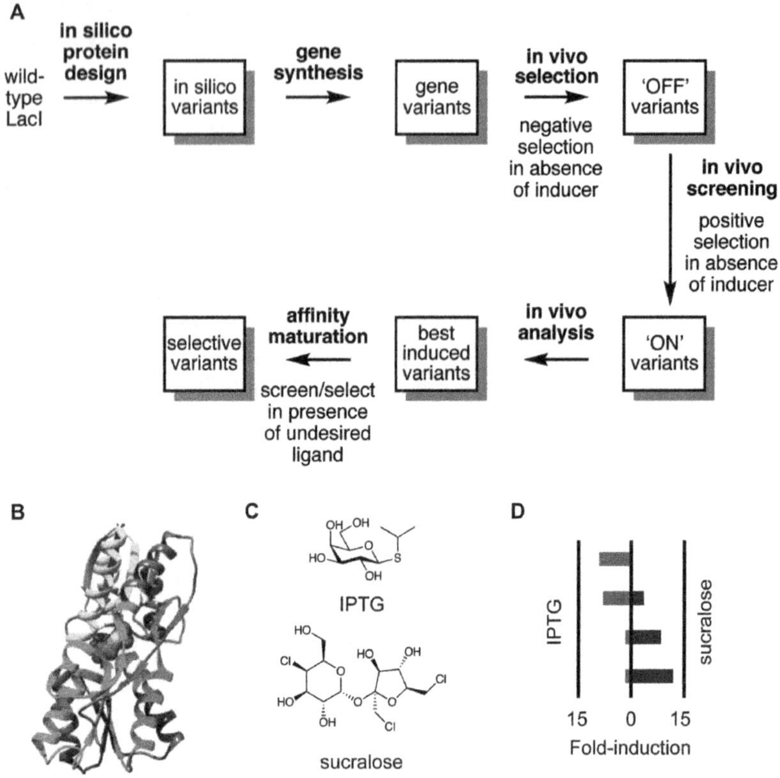

Figure 8.3 *In silico* guided engineering of altered ligand specificity in an allosteric transcription factor. (A) The design and engineering steps from *in silico* protein redesign to physical testing of selected variant(s). (B) Structure of LacI highlighting various segments targeted by Rosetta-design. (C) Structures of IPTG and sucralose. (D) Development of a sucralose specific sensor through computational design and screening (round 1) and two rounds of activity maturation (rounds 3 and 4) with inverted ligand selectivity.

LacI variants was constructed and screened for novel effector specificities. Several clones were found to be responsive to gentiobiose. Surprisingly, several of the most responsive clones had additional accidental mutations in the dimerization interface of the protein monomer. In the third approach, an epPCR library yielded variants that responded to fucose or lactitol but not to sucralose, but displayed ~2-fold higher $K_{1/2}$'s and lower induction factors than the best computationally-derived hits. The authors proposed that since epPCR failed to yield variants that were as good as the computationally derived mutants, in general epPCR might be insufficient to expand specificity of aTFs towards ligands that are distant in structure to the native effector. However, it was not clear whether the authors exhaustively screened the epPCR libraries or what the experimental mutation rate was. In addition, the plethora of examples already highlighted here is not consistent with this argument. Interestingly, none of the initial hits from any of the three approaches displayed a complete inversion of ligand selectivity or optimal induction. To address this, a process dubbed "activity maturation" was employed. This involved subsequent rounds of mutagenesis to decrease binding of the undesired ligand and to combine mutations beneficial for the target ligand. To increase induction of sucralose-responsive variants, their genes were shuffled and reassembled to achieve a 22-fold increase in induction by sucralose. The final sucralose-responsive variant demonstrated a nearly complete inversion in ligand induction *versus* IPTG (Figure 8.3).

In another computational *tour de force*, a vanillin-responsive biosensor was constructed using the QacR repressor protein as a starting point.[120] Wild-type QacR is activated by a variety of cationic small molecules including alkaloids (*e.g.* berberine and avicin), dyes (*e.g.* ethidium, proflavin, rhodamine 6G), quaternary ammonium compounds (*e.g.* benzalkonium and acetylpyridinium), and diamidines (*e.g.* DAPI and amicarbalide).[92] Murray and co-workers utilized computational protein design and cell-free transcription–translation (dubbed cell-free "TX–TL") to first generate *in silico* libraries that were then rapidly screened by TX–TL for induction.[120] Vanillin is a valuable fragrance and flavour chemistry additive produced by lignin hydrolysis or from chemical synthesis from guaiacol and glyoxylic acid. In terms of both size and structure, vanillin differs significantly from the ligands known to activate QacR, and therefore represents a challenging test of the computational approach. QacR was selected as a scaffold in part because of the availability of several crystal structures in various ligand bound states that aided the *in silico* ligand docking and modelling studies. Utilizing the TX–TL system, several QacR sensor variants were developed and ported to an *in vivo* system configured so that QacR expression was under control of tetracycline. Interestingly, QacR and vanillin were shown to be somewhat toxic to *E. coli*—features that could not be predicted by the cell-free TX–TL screen. After extensive normalization of the fluorescence response, one variant displayed a fluorescence response five-times higher than that of the wild-type repressor at 1 mM vanillin. The number of amino acid mutations in this variant (eleven) *versus* the wild-type QacR makes it very unlikely to have been discovered from an error-prone or saturation mutagenesis library, although it is not clear whether every one of these mutations was required or if it was optimal. Clearly, computational redesign offers a powerful approach to reduce the number of variants that need to be experimentally screened to manipulate ligand specificity.

8.5.2 Engineering the Ligand Selectivity of Transcription Factors

Notably, the examples in the previous section highlighted our ability to expand the *specificity* of transcription factors towards non-native and non-natural effectors. However, in many cases it may be necessary to engineer the *selectivity* of a transcription factor to be able to discriminate between various potential effectors. This could be particularly important when a transcription factor is used as a biosensor *in vivo* where competing ligands may be present at varying concentrations. In another scenario, a transcription factor that is capable of discriminating between the substrate and product of an enzyme might prove useful for evolving the enzyme activity in the absence of a more traditional screening strategy. Consistent with the rapidly increasing use of transcription factor-based biosensors for metabolic engineering and synthetic biology, more examples of highly selective engineered aTFs are now emerging, and advances in this area are expected to continue at a rapid pace.

An early example of mutagenesis to alter the ligand selectivity of a transcription factor involved TetR—a particularly well-known repressor protein.[123] Using directed evolution, the specificity of TetR for its natural inducer (tetracycline) and four closely related tetracycline-like compounds was dramatically altered using a β-galactosidase screen to identify TetR variants that allowed transcription (Figure 8.4A). Several TetR variants that exhibited unique induction profiles were identified and characterized in detail. Notably, the triple mutant H64K/E114Q/S135L, a product of four rounds of directed evolution, was fully induced by 4-de(dimethylamino)-6-demethyl-6-deoxy-tetracycline (cmt3) but not by tetracycline (Figure 8.4B), representing a complete inversion of specificity for the effectors, as compared to the wild-type protein.[124] The H64K mutation was found in over half of the reported TetR variants. Some of the key residues were subsequently used in templates for structure-guided randomization that led to other new

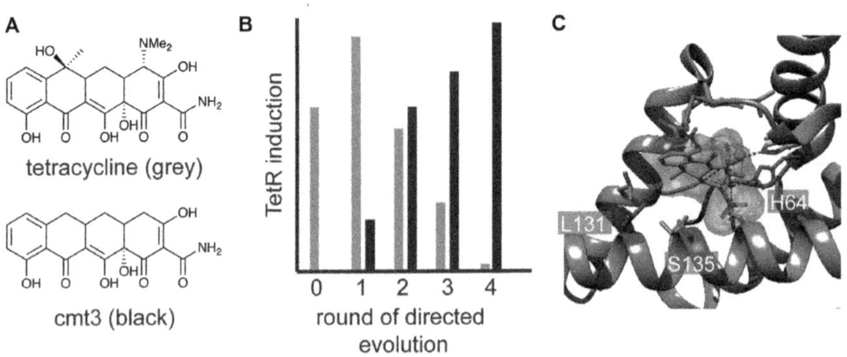

Figure 8.4 Directed evolution of the effector selectivity of TetR. (A) The structures of tetracycline and cmt3. (B) The directed evolution of TetR for selective activation by cmt3 (dark) instead of its natural inducer, tetracycline (grey). (C) The structure of TetR (PDB code 2VKE) and residues within a 5 Å shell of tetracycline. Residues H64, L131, and S135 are shown as sticks and tetracycline is shown as sticks with a semi-transparent surface. Hydrogen bonding between tetracycline and nearby residues is depicted by dashed lines.

effector specificities.[125] Interestingly, a series of crystal structures showed that the ligand binding position, as well as the positions of the residues that lined the binding site, are very well conserved, regardless of the ligand structure.[126] Selective recognition was achieved by fine-tuned hydrogen-bonding constraints, in addition to a combination of the hydrophobic effect and removal of an unfavourable electrostatic interaction (Figure 8.4C). These examples demonstrate the utility of applying protein evolution techniques to design a customized biosensor capable of detecting distinct tetracycline-class compounds.

The arabinose regulatory transcriptional unit, AraC, has been the subject of extensive manipulation. One example of AraC engineering shifted the selectivity of the AraC regulatory protein to D-arabinose (D-ara) from its natural effector L-ara.[127] The ability to subtly tweak the capability of AraC to discriminate between enantiomers and invert selectivity was achieved through dual FACS selections from saturation mutagenesis AraC libraries that targeted key residues in the ligand binding pocket. This strategy revealed several improved variants from two libraries. Several of the variants with altered ligand-binding properties contained residue changes from non-charged sidechains to charged (P8R, F15W, T24D, T24E, Y82Q), charged to non-charged sidechains (H80T, H80I, H80A), and changes in side-chain size and aromaticity (T24P, Y82T, Y82S). In particular, one variant displayed a complete switch in effector specificity while maintaining tight repression in the absence of ligand. Moreover, the evolved AraC biosensor was not activated by other sugars, as might be expected given the screening conditions.

Building on a comprehensive program of XylR engineering, de Lorenzo and co-workers sought to not only expand the promiscuity of XylR to include TNT and DNT (see Section 8.5.1), but to tailor its selectivity away from the natural effector (*m*-xylene) and towards the compound of interest, DNT.[128] This approach used a previously discovered XylR variant that had shown the best response to DNT as the starting point. Interestingly, this variant possesses two amino acid mutations, F48I and L222R, and mutagenesis at both of these positions supported at least some role in broadening the effector promiscuity of XylR. However, L222R, which lies in a domain that connects the ligand-binding domain and DNA-binding domain, contributed the most to effector promiscuity. The double mutant was used as a template for random mutagenesis. After screening a large library of variants, a new variant was identified that displayed an effector selectivity that was shifted towards DNT but not yet fully inverted towards it. This shift was mainly due to loss of activation with the undesired effector, *m*-xylene. The authors commented that while arriving at a promiscuous transcription factor might be quite facile, creating a new specificity (*e.g.* an inversion of selectivity) is more difficult due to the requirement of finding a sequential and well-orchestrated series of amino acid changes.

As described earlier (see Section 8.5.1), activity maturation has been used to improve specificity and induction in a series of computationally designed

LacI variants.[61] This was required because the computational approach employed did not include steps to eliminate variants with broad effector specificity (this would be difficult). For example, in the case of the gentobiose responsive variant (Q291H), the fold-induction with gentobiose was identical to that of IPTG. To render the transcription factor more selective towards gentobiose, 44 variants with reduced IPTG induction were combined with Q291H. After screening, chimeras were identified that were no longer induced by IPTG, and actually showed greater induction by gentobiose.[61]

8.6 Conclusions and Future Outlook

As high-throughput approaches to protein, pathway, and microbial engineering continue to mature and expand to address previously intractable targets, new methods to quickly and precisely identify variants with the desired productivity must keep pace. Frequently this has required detection systems capable of analysing thousands to millions of variants in the span of days to hours. While naturally occurring transcription factors that display useful effector specificities can be used as biosensors, there are numerous examples—some highlighted here—of situations where a biosensor for identification or quantification of the target compound is not currently known. This challenge is being addressed head-on by engineering biosensors capable of detecting a target analyte using existing transcription factor scaffolds. The feasibility of this strategy has been demonstrated several times by evolving or designing novel biosensors for sensing target ligands from transcription factor families including LacI, AraC, and TetR-type proteins.

Such evolutionary methods incorporate screening or selections of mutant transcription factor libraries, often (but not always[120]) housed in a host cell for reporting activation by the target ligand, and are perhaps best suited when the effector specificity of a target transcription factor requires expansion to other ligands that are similar in structure to the native effector of the transcription factor template. Cell-free screening might prove useful for quickly testing the activity of small panels of candidate engineered aTFs or refactored/re-configured biosensors, but effective *in vitro* biosensors may not always translate to effective *in vivo* biosensors. If the target ligand is distantly related in structure to the native effector, a computational design step may precede this process[61,105] to greatly reduce the design space that has to be physically screened. However, the ability of computational methods to drive the identification of transcription factors with new effector selectivities is largely unexplored and might challenge the power of this approach. At the same time, evolutionary approaches to engineering selectivity of transcription factors are now slowly emerging. Nevertheless, as illustrated by the examples here and by growing investments in this area, engineering transcription factor-based biosensors for specific uses is a very attractive strategy given its potential generality and simplicity.

As the field of synthetic biology expands and redefines classical approaches to environmental chemical sensing, chemical manufacture,[129,130] and drug discovery,[5,131] the need for biosensors that detect chemicals of interest—especially non-natural chemicals—at specific concentrations becomes apparent due to the relative ease with which custom made aTFs can be integrated into high-throughput screens/selections. The potential for "plug and play" customization of transcriptional unit biosensors may be developed where one or more sensitivity and selectivity elements are combined to develop sensors for precise response and output characteristics. This chapter has identified several key aspects of sensitivity, specificity, and selectivity engineering and how they can be combined.

The "design-build-test" cycle of synthetic biology is becoming an integral approach to engineering biology for biotechnology.[132–134] The paradigm that swappable and modular "parts" (promoters, RBS, transcriptions factors, enzymes, circuits) in biology can be integrated into a new or existing biotechnological process is perhaps best exemplified by the construction of biosensors that can sense chemicals of interest and be used to guide production in engineered biological systems. As advances continue, customized biosensors have the potential to become indispensable tools in modern biotechnology.

Acknowledgements

We gratefully acknowledge support from the National Institutes of Health (GM104258, G.J.W.) and the National Science Foundation (NSF CAREER Award, CHE-1151299, G.J.W.).

References

1. H. J. Frasch, M. H. Medema, E. Takano and R. Breitling, *Curr. Opin. Biotechnol.*, 2013, **24**, 1144–1150.
2. B. Pscheidt and A. Glieder, *Microb. Cell Fact.*, 2008, 7, 25.
3. L. T. Cordova and H. S. Alper, *Curr. Opin. Chem. Biol.*, 2016, **35**, 37–42.
4. H. Guo, S. Su, C. Madzak, J. Zhou, H. Chen and G. Chen, *Appl. Microbiol. Biotechnol.*, 2016, **100**, 9875–9884.
5. A. R. Awan, W. M. Shaw and T. Ellis, *Adv. Drug Delivery Rev.*, 2016, **105**, 96–106.
6. R. A. Sheldon and P. C. Pereira, *Chem. Soc. Rev.*, 2017, **46**, 2678–2691.
7. S. K. Padhi, *ChemBioChem*, 2017, **18**, 152–160.
8. V. Tiwari, *Front. Chem.*, 2016, **4**, 39.
9. J. Wang, S. Zhang, H. Tan and Z. K. Zhao, *J. Microbiol. Methods*, 2007, **71**, 225–230.
10. E. E. Wrenbeck, J. R. Klesmith, J. A. Stapleton, A. Adeniran, K. E. Tyo and T. A. Whitehead, *Nat. Methods*, 2016, **13**, 928–930.
11. W. P. Stemmer, *Nature*, 1994, **370**, 389–391.
12. A. J. Ruff, A. Dennig and U. Schwaneberg, *FEBS J.*, 2013, **280**, 2961–2978.

13. E. O. McCullum, B. A. Williams, J. Zhang and J. C. Chaput, *Methods Mol. Biol.*, 2010, **634**, 103–109.
14. R. C. Cadwell and G. F. Joyce, *PCR Methods Appl.*, 1992, **2**, 28–33.
15. T. Vanhercke, C. Ampe, L. Tirry and P. Denolf, *Anal. Biochem.*, 2005, **339**, 9–14.
16. H. H. Wang, F. J. Isaacs, P. A. Carr, Z. Z. Sun, G. Xu, C. R. Forest and G. M. Church, *Nature*, 2009, **460**, 894–898.
17. K. M. Esvelt, J. C. Carlson and D. R. Liu, *Nature*, 2011, **472**, 499–503.
18. J. P. Goddard and J. L. Reymond, *Trends Biotechnol.*, 2004, **22**, 363–370.
19. V. Taly, B. T. Kelly and A. D. Griffiths, *ChemBioChem*, 2007, **8**, 263–272.
20. A. F. Oliveira, A. C. Pessoa, R. G. Bastos and L. G. de la Torre, *Biotechnol. Prog.*, 2016, **32**, 1372–1389.
21. E. T. Boder and K. D. Wittrup, *Methods Enzymol.*, 2000, **328**, 430–444.
22. I. Chen, B. M. Dorr and D. R. Liu, *Proc. Natl. Acad. Sci. U. S. A.*, 2011, **108**, 11399–11404.
23. A. Crameri, S. A. Raillard, E. Bermudez and W. P. Stemmer, *Nature*, 1998, **391**, 288–291.
24. D. R. Liu, T. J. Magliery, M. Pastrnak and P. G. Schultz, *Proc. Natl. Acad. Sci. U. S. A.*, 1997, **94**, 10092–10097.
25. S. W. Santoro, L. Wang, B. Herberich, D. S. King and P. G. Schultz, *Nat. Biotechnol.*, 2002, **20**, 1044–1048.
26. T. Gaj, A. C. Mercer, C. A. Gersbach, R. M. Gordley and C. F. Barbas, 3rd, *Proc. Natl. Acad. Sci. U. S. A.*, 2011, **108**, 498–503.
27. E. M. Young, A. Tong, H. Bui, C. Spofford and H. S. Alper, *Proc. Natl. Acad. Sci. U. S. A.*, 2014, **111**, 131–136.
28. B. Worsdorfer, K. J. Woycechowsky and D. Hilvert, *Science*, 2011, **331**, 589–592.
29. L. E. Ulrich, E. V. Koonin and I. B. Zhulin, *Trends Microbiol.*, 2005, **13**, 52–56.
30. M. L. Skjoedt, T. Snoek, K. R. Kildegaard, D. Arsovska, M. Eichenberger, T. J. Goedecke, A. S. Rajkumar, J. Zhang, M. Kristensen, B. J. Lehka, S. Siedler, I. Borodina, M. K. Jensen and J. D. Keasling, *Nat. Chem. Biol.*, 2016, **12**, 951–958.
31. P. Xu, L. Li, F. Zhang, G. Stephanopoulos and M. Koffas, *Proc. Natl. Acad. Sci. U. S. A.*, 2014, **111**, 11299–11304.
32. J. K. Rogers, N. D. Taylor and G. M. Church, *Curr. Opin. Biotechnol.*, 2016, **42**, 84–91.
33. R. Mahr and J. Frunzke, *Appl. Microbiol. Biotechnol.*, 2016, **100**, 79–90.
34. L. Eggeling, M. Bott and J. Marienhagen, *Curr. Opin. Biotechnol.*, 2015, **35C**, 30–36.
35. J. L. Ramos, M. Martinez-Bueno, A. J. Molina-Henares, W. Teran, K. Watanabe, X. Zhang, M. T. Gallegos, R. Brennan and R. Tobes, *Microbiol. Mol. Biol. Rev.*, 2005, **69**, 326–356.
36. L. Cuthbertson and J. R. Nodwell, *Microbiol. Mol. Biol. Rev.*, 2013, **77**, 440–475.
37. R. G. Martin and J. L. Rosner, *Curr. Opin. Microbiol.*, 2001, **4**, 132–137.

38. L. Swint-Kruse and K. S. Matthews, *Curr. Opin. Microbiol.*, 2009, **12**, 129–137.
39. J. Chen and J. Xie, *J. Cell. Biochem.*, 2011, **112**, 2694–2702.
40. S. P. Wilkinson and A. Grove, *Curr. Issues Mol. Biol.*, 2006, **8**, 51–62.
41. A. Wietzorrek and M. Bibb, *Mol. Microbiol.*, 1997, **25**, 1181–1184.
42. G. Liu, K. F. Chater, G. Chandra, G. Niu and H. Tan, *Microbiol. Mol. Biol. Rev.*, 2013, **77**, 112–143.
43. G. Chen, L. R. Swem, D. L. Swem, D. L. Stauff, C. T. O'Loughlin, P. D. Jeffrey, B. L. Bassler and F. M. Hughson, *Mol. Cell*, 2011, **42**, 199–209.
44. M. E. Rodgers and R. Schleif, *Proteins*, 2009, **77**, 202–208.
45. S. M. Soisson, B. MacDougall-Shackleton, R. Schleif and C. Wolberger, *J. Mol. Biol.*, 1997, **273**, 226–237.
46. S. M. Soisson, B. MacDougall-Shackleton, R. Schleif and C. Wolberger, *Science*, 1997, **276**, 421–425.
47. V. Saridakis, D. Shahinas, X. Xu and D. Christendat, *J. Mol. Biol.*, 2008, **377**, 655–667.
48. L. Ni, N. K. Tonthat, N. Chinnam and M. A. Schumacher, *Nucleic Acids Res.*, 2013, **41**, 1998–2008.
49. J. Zheng, V. Sagar, A. Smolinsky, C. Bourke, N. LaRonde-LeBlanc and T. A. Cropp, *J. Mol. Biol.*, 2009, **387**, 1250–1260.
50. A. Di Fiore, G. Fiorentino, R. M. Vitale, R. Ronca, P. Amodeo, C. Pedone, S. Bartolucci and G. De Simone, *J. Mol. Biol.*, 2009, **388**, 559–569.
51. S. Raman, N. Taylor, N. Genuth, S. Fields and G. M. Church, *Trends Genet.*, 2014, **30**, 521–528.
52. J. A. Dietrich, D. L. Shis, A. Alikhani and J. D. Keasling, *ACS Synth. Biol.*, 2013, **2**, 47–58.
53. H. Chong and C. B. Ching, *ACS Synth. Biol.*, 2016, **5**, 1290–1298.
54. C. S. Frei, Z. Wang, S. Qian, S. Deutsch, M. Sutter and P. C. Cirino, *Protein Sci.*, 2016, **25**, 804–814.
55. D. Liu, T. Evans and F. Zhang, *Metab. Eng.*, 2015, **31**, 35–43.
56. S. A. Morgan, D. C. Nadler, R. Yokoo and D. F. Savage, *Curr. Opin. Chem. Biol.*, 2016, **35**, 150–158.
57. M. Schallmey, J. Frunzke, L. Eggeling and J. Marienhagen, *Curr. Opin. Biotechnol.*, 2014, **26**, 148–154.
58. J. Zhang, J. F. Barajas, M. Burdu, T. L. Ruegg, B. Dias and J. D. Keasling, *ACS Synth. Biol.*, 2017, **6**, 439–445.
59. F. Moser, A. Horwitz, J. Chen, W. Lim and C. A. Voigt, *ACS Synth. Biol.*, 2013, **2**, 614–624.
60. S. Siedler, N. K. Khatri, A. Zsohar, I. Kjaerbolling, M. Vogt, P. Hammar, C. F. Nielsen, J. Marienhagen, M. O. A. Sommer and H. N. Joensson, *ACS Synth. Biol.*, 2017, **6**, 1860–1869.
61. N. D. Taylor, A. S. Garruss, R. Moretti, S. Chan, M. A. Arbing, D. Cascio, J. K. Rogers, F. J. Isaacs, S. Kosuri, D. Baker, S. Fields, G. M. Church and S. Raman, *Nat. Methods*, 2016, **13**, 177–183.
62. S. Y. Tang, S. Qian, O. Akinterinwa, C. S. Frei, J. A. Gredell and P. C. Cirino, *J. Am. Chem. Soc.*, 2013, **135**, 10099–10103.

63. W. Chen, S. Zhang, P. Jiang, J. Yao, Y. He, L. Chen, X. Gui, Z. Dong and S. Y. Tang, *Metab. Eng.*, 2015, **30**, 149–155.
64. R. H. Dahl, F. Zhang, J. Alonso-Gutierrez, E. Baidoo, T. S. Batth, A. M. Redding-Johanson, C. J. Petzold, A. Mukhopadhyay, T. S. Lee, P. D. Adams and J. D. Keasling, *Nat. Biotechnol.*, 2013, **31**, 1039–1046.
65. S. Li, T. Si, M. Wang and H. Zhao, *ACS Synth. Biol.*, 2015, **4**, 1308–1315.
66. D. Liu, Y. Xiao, B. S. Evans and F. Zhang, *ACS Synth. Biol.*, 2015, **4**, 132–140.
67. J. A. Dietrich, A. E. McKee and J. D. Keasling, *Annu. Rev. Biochem.*, 2010, **79**, 563–590.
68. V. J. Hilser, J. O. Wrabl and H. N. Motlagh, *Annu. Rev. Biophys.*, 2012, **41**, 585–609.
69. V. J. Hilser, *Science*, 2010, **327**, 653–654.
70. M. C. Limon, R. Rodriguez-Ortiz and J. Avalos, *Appl. Microbiol. Biotechnol.*, 2010, **87**, 21–29.
71. P. S. Gandhi, Z. Chen, F. S. Mathews and E. Di Cera, *Proc. Natl. Acad. Sci. U. S. A.*, 2008, **105**, 1832–1837.
72. J. Kim, L. G. Ahuja, F. A. Chao, Y. Xia, C. L. McClendon, A. P. Kornev, S. S. Taylor and G. Veglia, *Sci. Adv.*, 2017, **3**, e1600663.
73. A. Rehn, E. Moroni, B. K. Zierer, F. Tippel, G. Morra, C. John, K. Richter, G. Colombo and J. Buchner, *J. Mol. Biol.*, 2016, **428**, 4559–4571.
74. A. Sikora, B. Joseph, M. Matson, J. R. Staley and D. S. Cafiso, *Biophys. J.*, 2016, **111**, 1908–1918.
75. D. Beckett, *Proc. Natl. Acad. Sci. U. S. A.*, 2009, **106**, 22035–22036.
76. S. E. Reichheld, Z. Yu and A. R. Davidson, *Proc. Natl. Acad. Sci. U. S. A.*, 2009, **106**, 22263–22268.
77. D. H. Richards, S. Meyer and C. J. Wilson, *ACS Synth. Biol.*, 2017, **6**, 6–12.
78. P. Markiewicz, L. G. Kleina, C. Cruz, S. Ehret and J. H. Miller, *J. Mol. Biol.*, 1994, **240**, 421–433.
79. J. Suckow, P. Markiewicz, L. G. Kleina, J. Miller, B. Kisters-Woike and B. Muller-Hill, *J. Mol. Biol.*, 1996, **261**, 509–523.
80. S. Meyer, R. Ramot, K. Kishore Inampudi, B. Luo, C. Lin, S. Amere and C. J. Wilson, *Protein Eng., Des. Sel.*, 2013, **26**, 433–443.
81. C. J. Marx, *Cell*, 2011, **146**, 350–352.
82. E. M. Henssler, R. Bertram, S. Wisshak and W. Hillen, *FEBS J.*, 2005, **272**, 4487–4496.
83. O. Scholz, E. M. Henssler, J. Bail, P. Schubert, J. Bogdanska-Urbaniak, S. Sopp, M. Reich, S. Wisshak, M. Kostner, R. Bertram and W. Hillen, *Mol. Microbiol.*, 2004, **53**, 777–789.
84. M. Resch, H. Striegl, E. M. Henssler, M. Sevvana, C. Egerer-Sieber, E. Schiltz, W. Hillen and Y. A. Muller, *Nucleic Acids Res.*, 2008, **36**, 4390–4401.
85. P. U. iGEM Team, iGEM, 2010, http://2010.igem.org/Team:Peking/Project/Biosensor, (accessed May 2017).
86. T. Tian and H. M. Salis, *Nucleic Acids Res.*, 2015, **43**, 7137–7151.

87. J. K. Rogers, C. D. Guzman, N. D. Taylor, S. Raman, K. Anderson and G. M. Church, *Nucleic Acids Res.*, 2015, **43**, 7648–7660.
88. J. Stocker, D. Balluch, M. Gsell, J. Feliciano, S. Daunert, K. A. Malik and J. R. van der Meer, *Environ. Sci. Technol.*, 2004, **38**, 2959.
89. F. Delvigne, H. Pecheux and C. Tarayre, *Front. Bioeng. Biotechnol.*, 2015, **3**, 147.
90. Q. Yu, Y. Li, A. Ma, W. Liu, H. Wang and G. Zhuang, *Anal. Bioanal. Chem.*, 2011, **401**, 2891–2898.
91. U. o. C. IGEM Team, iGEM, 2009, http://2009.igem.org/Team:Cambridge/Project/Amplification, (accessed May 2017).
92. A. A. Mannan, D. Liu, F. Zhang and D. A. Oyarzun, *ACS Synth. Biol.*, 2017, **6**, 1851–1859.
93. P. Bubner, M. Klimacek and B. Nidetzky, *FEBS Lett.*, 2008, **582**, 233–237.
94. G. J. Williams, C. Zhang and J. S. Thorson, *Nat. Chem. Biol.*, 2007, **3**, 657–662.
95. S. M. Lippow, T. S. Moon, S. Basu, S. H. Yoon, X. Li, B. A. Chapman, K. Robison, D. Lipovsek and K. L. Prather, *Chem. Biol.*, 2010, **17**, 1306–1315.
96. B. C. Stanton, A. A. Nielsen, A. Tamsir, K. Clancy, T. Peterson and C. A. Voigt, *Nat. Chem. Biol.*, 2014, **10**, 99–105.
97. M. J. Cipriano, P. N. Novichkov, A. E. Kazakov, D. A. Rodionov, A. P. Arkin, M. S. Gelfand and I. Dubchak, *BMC Genomics*, 2013, **14**, 213.
98. P. S. Novichkov, A. E. Kazakov, D. A. Ravcheev, S. A. Leyn, G. Y. Kovaleva, R. A. Sutormin, M. D. Kazanov, W. Riehl, A. P. Arkin, I. Dubchak and D. A. Rodionov, *BMC Genomics*, 2013, **14**, 745.
99. S. Gama-Castro, H. Salgado, A. Santos-Zavaleta, D. Ledezma-Tejeida, L. Muniz-Rascado, J. S. Garcia-Sotelo, K. Alquicira-Hernandez, I. Martinez-Flores, L. Pannier, J. A. Castro-Mondragon, A. Medina-Rivera, H. Solano-Lira, C. Bonavides-Martinez, E. Perez-Rueda, S. Alquicira-Hernandez, L. Porron-Sotelo, A. Lopez-Fuentes, A. Hernandez-Koutoucheva, V. Del Moral-Chavez, F. Rinaldi and J. Collado-Vides, *Nucleic Acids Res.*, 2016, **44**, D133–D143.
100. G. Carbajosa, A. Trigo, A. Valencia and I. Cases, *Nucleic Acids Res.*, 2009, **37**, D598–D602.
101. V. Libis, B. Delepine and J. L. Faulon, *ACS Synth. Biol.*, 2016, **5**, 1076–1085.
102. S. Placzek, I. Schomburg, A. Chang, L. Jeske, M. Ulbrich, J. Tillack and D. Schomburg, *Nucleic Acids Res.*, 2017, **45**, D380–D388.
103. Y. Chai and S. C. Winans, *Mol. Microbiol.*, 2004, **51**, 765–776.
104. S. Y. Tang and P. C. Cirino, *Angew. Chem., Int. Ed. Engl.*, 2011, **50**, 1084–1086.
105. R. K. Jha, S. Chakraborti, T. L. Kern, D. T. Fox and C. E. Strauss, *Proteins*, 2015, **83**, 1327–1340.
106. B. Shemer, N. Palevsky, S. Yagur-Kroll and S. Belkin, *Front. Microbiol.*, 2015, **6**, 1175.

107. F. Behzadian, H. Barjeste, S. Hosseinkhani and A. R. Zarei, *Curr. Microbiol.*, 2011, **62**, 690–696.
108. D. Devos, J. Garmendia, V. de Lorenzo and A. Valencia, *Environ. Microbiol.*, 2002, **4**, 29–41.
109. J. Garmendia, A. de las Heras, T. C. Galvao and V. de Lorenzo, *Environ. Microbiol.*, 2008, **1**, 236–246.
110. Y. Tashiro, Y. Kimura, M. Furubayashi, A. Tanaka, K. Terakubo, K. Saito, S. Kawai-Noma and D. Umeno, *J. Gen. Appl. Microbiol.*, 2016, **62**, 240–247.
111. J. Garmendia, D. Devos, A. Valencia and V. de Lorenzo, *Mol. Microbiol.*, 2001, **42**, 47–59.
112. S. Meinhardt, M. W. Manley, Jr., N. A. Becker, J. A. Hessman, L. J. Maher, 3rd and L. Swint-Kruse, *Nucleic Acids Res.*, 2012, **40**, 11139–11154.
113. H. H. Chou and J. D. Keasling, *Nat. Commun.*, 2013, **4**, 2595.
114. C. M. Kasey, M. Zerrad, Y. Li, T. A. Cropp and G. J. Williams, *ACS Synth. Biol.*, 2017, DOI: 10.1021/acssynbio.1027b00287.
115. I. B. Seiple, Z. Zhang, P. Jakubec, A. Langlois-Mercier, P. M. Wright, D. T. Hog, K. Yabu, S. R. Allu, T. Fukuzaki, P. N. Carlsen, Y. Kitamura, X. Zhou, M. L. Condakes, F. T. Szczypinski, W. D. Green and A. G. Myers, *Nature*, 2016, **533**, 338–345.
116. C. E. Tinberg, S. D. Khare, J. Dou, L. Doyle, J. W. Nelson, A. Schena, W. Jankowski, C. G. Kalodimos, K. Johnsson, B. L. Stoddard and D. Baker, *Nature*, 2013, **501**, 212–216.
117. D. Hilvert, *Annu. Rev. Biochem.*, 2013, **82**, 447–470.
118. O. V. Makhlynets and I. V. Korendovych, *Methods Enzymol.*, 2016, **580**, 191–202.
119. A. Morin, J. Meiler and L. S. Mizoue, *Trends Biotechnol.*, 2011, **29**, 159–166.
120. E. L. de los Santos, J. T. Meyerowitz, S. L. Mayo and R. M. Murray, *ACS Synth.* Biol., 2016, **5**, 287–295.
121. R. F. Alford, A. Leaver-Fay, J. R. Jeliazkov, M. J. O'Meara, F. P. DiMaio, H. Park, M. V. Shapovalov, P. D. Renfrew, V. K. Mulligan, K. Kappel, J. W. Labonte, M. S. Pacella, R. Bonneau, P. Bradley, R. L. Dunbrack, Jr., R. Das, D. Baker, B. Kuhlman, T. Kortemme and J. J. Gray, *J. Chem. Theory Comput.*, 2017, **13**, 3031–3048.
122. R. Das and D. Baker, *Annu. Rev. Biochem.*, 2008, **77**, 363–382.
123. R. Bertram and W. Hillen, *Environ. Microbiol.*, 2008, **1**, 2–16.
124. O. Scholz, M. Kostner, M. Reich, S. Gastiger and W. Hillen, *J. Mol. Biol.*, 2003, **329**, 217–227.
125. E. M. Henssler, O. Scholz, S. Lochner, P. Gmeiner and W. Hillen, *Biochemistry*, 2004, **43**, 9512–9518.
126. M. A. Klieber, O. Scholz, S. Lochner, P. Gmeiner, W. Hillen and Y. A. Muller, *FEBS J.*, 2009, **276**, 5610–5621.
127. S. Y. Tang, H. Fazelinia and P. C. Cirino, *J. Am. Chem. Soc.*, 2008, **130**, 5267–5271.

128. A. de Las, Heras and V. de Lorenzo, *Mol. Microbiol.*, 2011, **79**, 1248–1259.
129. S. Yuzawa, J. D. Keasling and L. Katz, *J. Antibiot.*, 2017, **70**, 378–385.
130. J. D. Keasling, *Science*, 2010, **330**, 1355–1358.
131. J. R. King, S. Edgar, K. Qiao and G. Stephanopoulos, *F1000Research*, 2016, **5**(397), 1–11.
132. C. M. Agapakis, *ACS Synth. Biol.*, 2014, **3**, 121–128.
133. P. Carbonell, A. Currin, A. J. Jervis, N. J. Rattray, N. Swainston, C. Yan, E. Takano and R. Breitling, *Nat. Prod. Rep.*, 2016, **33**, 925–932.
134. S. Poust, A. Hagen, L. Katz and J. D. Keasling, *Curr. Opin. Biotechnol.*, 2014, **30C**, 32–39.

CHAPTER 9

Exploiting Biosynthetic Pathways in Fungi: Opportunities for Enhanced or Novel Production

KATE M. J. DE MATTOS-SHIPLEY AND ANDY M. BAILEY*

School of Biological Sciences, University of Bristol, 24 Tyndall Avenue, Bristol BS8 1TQ, UK
*Email: andy.bailey@bristol.ac.uk

9.1 Introduction

Fungi have a long history of exploitation, be it directly as foodstuffs such as the cultivated mushroom *Agaricus bisporus*, exploitation of *Saccharomyces cerevisiae* (baker's yeast) as a rising agent for bread dough or for production of alcohol from sugary liquids, or the use of various *Penicillum* species for modifying the textures and flavours of foods such as in the maturation of cheeses or salami. It wasn't until the mid-20th century, as there became an increased understanding of biochemistry and fermentation that fungi really came into their own. The first notable success was with citric acid biosynthesis by *Aspergillus niger*,[1,2] then more famously the production of penicillins by *Penicillium notatum* and *P. chrysogenum*.[3] The comparative ease of fungal growth, at least for some species, coupled with their propensity for efficient secretion of

Catalysis Series No. 32
Modern Biocatalysis: Advances Towards Synthetic Biological Systems
Edited by Gavin Williams and Mélanie Hall

Published by the Royal Society of Chemistry, www.rsc.org

extracellular enzymes, has led to a substantial interest in these species for various biotechnological purposes.

With the discovery of penicillin, there was substantial interest in the discovery of other antibiotics from microorganisms, and whilst *Streptomycete* bacteria proved to be a more productive group of organisms for such compounds, considerable research efforts continued looking for new potential pharmaceuticals from fungi. This led to many notable products of fungal origin, not least immuno-modulators such as cyclosporins;[4] antifungals for medical use, such as griseofulvin[5] and echinocandins,[6] or for agriculture such as the strobilurins;[7] and targeting cholesterol biosynthesis, one of the biggest selling classes of drugs of recent years: the statins.[8] The majority of these compounds were discovered from simple screens, whereby partially purified culture filtrates were used to detect useful biological activities. The rationale of many such screens was to assess as many isolates as possible in the hope of identifying new compounds, but perhaps predictably, this led to the problem of dereplication, whereby the same or similar compounds were found time and time again, eventually leading to a decline in the popularity of such screening methods. Nevertheless, several thousand fungal natural products were discovered by such routes, even if their biological activities were not always readily apparent.

Despite the incredible diversity seen in fungal natural products, the vast majority can be placed into only a few chemical classes, based on the chemical building blocks from which they are constructed and the class-specific enzymes which are responsible for catalysing their biosynthesis.[9] Thus, terpenes in fungi are all isoprenoid-derived and generally require the action of a prenyltransferase and terpene synthase, such as a cyclase. Polyketide scaffolds are synthesised by polyketide synthases (PKSs, see also Chapter 10), which in fungi are large iterative multi-domain enzymes that collectively produce an incredibly diverse and chemically complex class of compounds. Non-ribosomal peptides are also produced by large multi-domain enzymes called non-ribosomal peptide synthetases (NRPSs, see also Chapter 10), but unlike PKSs, these tend to function in a modular function in fungi with a separate module responsible for the incorporation each amino acid. Small peptide natural products can also be produced ribosomally in fungi, and are then known as RiPPs; ribosomally synthesised post-translationally modified peptides. Large hybrid enzymes known as PKS–NRPSs produce molecules that are fusions of non-ribosomal peptides with polyketides. A slightly less clearly defined group is the alkaloids, which are compounds that contain one or more basic nitrogen atoms in a heterocyclic ring. Being defined by this structural feature rather than their biosynthetic route, alkaloids are not synthesised by a specific enzyme, but can be loosely divided into four groups depending on their origin: amino acid-derived, purine-alkaloids, aminated terpenes and polyketide alkaloids. Representative compounds of the different classes are shown in Figure 9.1.

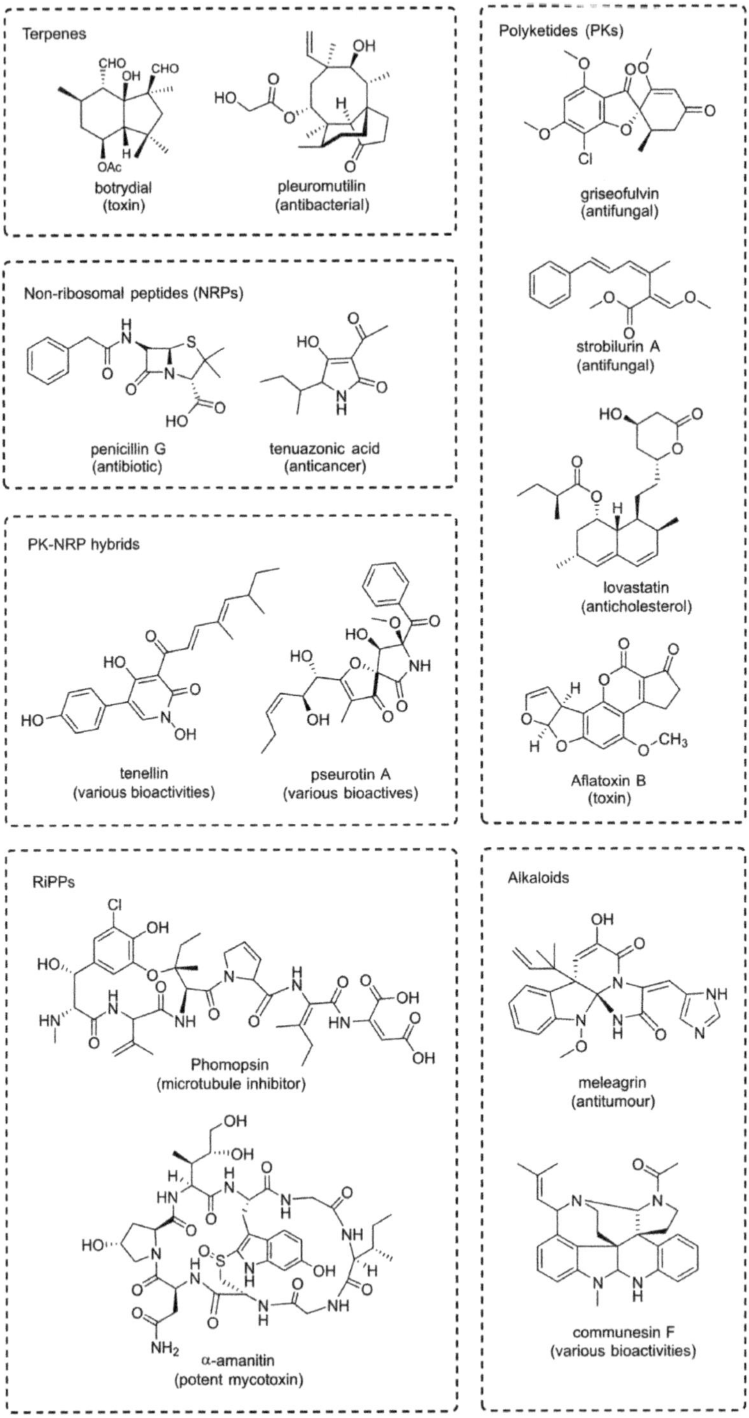

Figure 9.1 Representatives of the main classes of fungal natural products.

9.2 Impacts of the Genetic and Then Genomic Era on Pathway Discovery

The advent of molecular genetic techniques for the manipulation of fungi in the early 1980s coincided with a decline in natural product screening programs, so most focus was initially on the development of genetic tools for model species such as *Neurospora crassa* and *Aspergillus nidulans*. Whilst the former had a very restricted repertoire for secondary metabolism, *A. nidulans* somewhat fortuitously was a producer of penicillin and closely related to many of the species already used in fungal biotechnology such as *A. niger* and *A. oryzae*. A number of research groups quickly identified first the *ipns* gene for isopenicillin-*N*-synthase, then showed that the three core genes for penicillin biosynthesis were located together in the genome.[10] This highlighted that like in bacteria, fungal genomes contained clusters of genes, whereby those for a particular secondary metabolite tend to be co-located. Studies of industrially improved isolates of *P. chrysogenum* also indicated that there had been a substantial increase in gene copy number for the biosynthetic genes in such strains,[11,12] showing that titres of useful products could be enhanced by having higher gene copy number and hence transcript and enzyme levels, pointing to a direction for rational strain improvement. In addition to the analysis within a naturally producing host, the three genes for penicillin biosynthesis were transferred as a genomic cosmid clone into *N. crassa* and *A. niger*, in both cases giving rise to transformants capable of penicillin biosynthesis[13] even if at low titres, raising the idea of deliberate heterologous expression for natural products.

As the tools for manipulation of fungi became more widespread, so the numbers of fungi studied for secondary metabolism increased, initially focussing on the best-known metabolites such as the gibberellins or mycotoxins, such as aflatoxin, trichothecenes and fusarins. In each case the genes were conveniently co-located in gene clusters easing their identification, highlighting that this was not unique to non-ribosomally synthesised products such as penicillin, but also applied to the full range of typical fungal secondary metabolites, including polyketides, terpenes and alkaloids. What also became apparent was that each of the fungi concerned was capable of producing a wider range of secondary metabolites than had initially been realised. Subtle changes in growth conditions could alter a metabolite profile considerably. Using a range of differing growth conditions, one strain could produce a number of different metabolites, a process termed OSMAC: One Strain, Many Active Compounds.[14] This led to the realisation that the random high throughput screens deployed for lead generation had perhaps only scratched the surface of the true potential of fungi for secondary metabolism. Advances in genome sequencing of fungi also meant that genome sequence data was readily available for many species, indeed with the 1000 genomes project (http://1000.fungalgenomes.org/) fungi are likely to be the best sampled of the eukaryotes. Coupled with the availability of genome data, bioinformatics tools have been developed and robust search algorithms such as SMURF[15] and ANTISMASH[16] meant that secondary metabolite gene clusters became comparatively simple to identify. This confirmed that even in well studied fungi, often only 10–20% of the secondary metabolism gene clusters had a defined product, and that ascomycete fungi typically have 40–60 possible secondary metabolite gene clusters. This plethora of potential secondary metabolism showed that even the detailed OSMAC-based fermentation approaches had failed to access all of the potential compounds within an isolate and thus there was significant scope for further research. The ease of genome sequencing also gave rise to the idea that with so many genes available, it might become possible to make designer pathways by a simple mix and match approach. To do this effectively and with confidence, the genes would need well-defined activities and specificities, which is currently beyond the reach of available bioinformatic tools. However, with the ever increasing number of well-characterised secondary metabolite enzymes, the concept of designer gene clusters holds promise.

9.3 Reshaping Regulatory Networks

One of the challenges facing natural product research has been the variable yields of such metabolites, often produced in vanishingly low titres under laboratory conditions. Being secondary metabolites, these compounds are not directly involved in the normal growth or survival of the producing organism and so are not constitutively produced. They are thought to be produced in nature for a wide range of reasons, including as chemical weaponry against other organisms with which they are competing, or for communication with symbiotic partners. Thus, they are generally tightly regulated and produced only when required. The idea that fungi are capable of producing a much greater repertoire of natural products than is observed in the laboratory has been borne out with the advent of genome sequencing: genome mining has consistently identified far more natural product pathways than could have been predicted by metabolically profiling the organism,[17] as with strict regulation of expression, many of the pathways are silent under conventional growth conditions.

Various approaches have been developed to activate these cryptic gene clusters, or increase the production of naturally low yielding compounds. Systematically varying the culture conditions has had some success, for example with the marine fungi *Ascotricha* sp. ZJ-M-5[18] and the endophyte *Dothideomycete* sp. CRI7.[19] The discovery of novel compounds from *A. nidulans* highlights the success such an approach can have, even with well-known and intensively studied species.[20] A slightly more sophisticated approach, co-culturing the fungal species with other organisms that may trigger their defence or communication pathways, has also led to the activation of silent gene clusters and production of novel compounds[21–23] but finding the correct microbial partner to trigger expression is not always obvious, limiting such an approach. The development of the various tools discussed in this review has allowed slightly more directed approaches to be used.

Perhaps the most direct approach involves promoter swaps (Figure 9.2), which can be used to target specific genes in a biosynthetic pathway. Ahuja *et al.*[24] conducted promoter replacements for a variety of nrPKS enzymes in *A. nidulans*, along with genes required for the starter units or

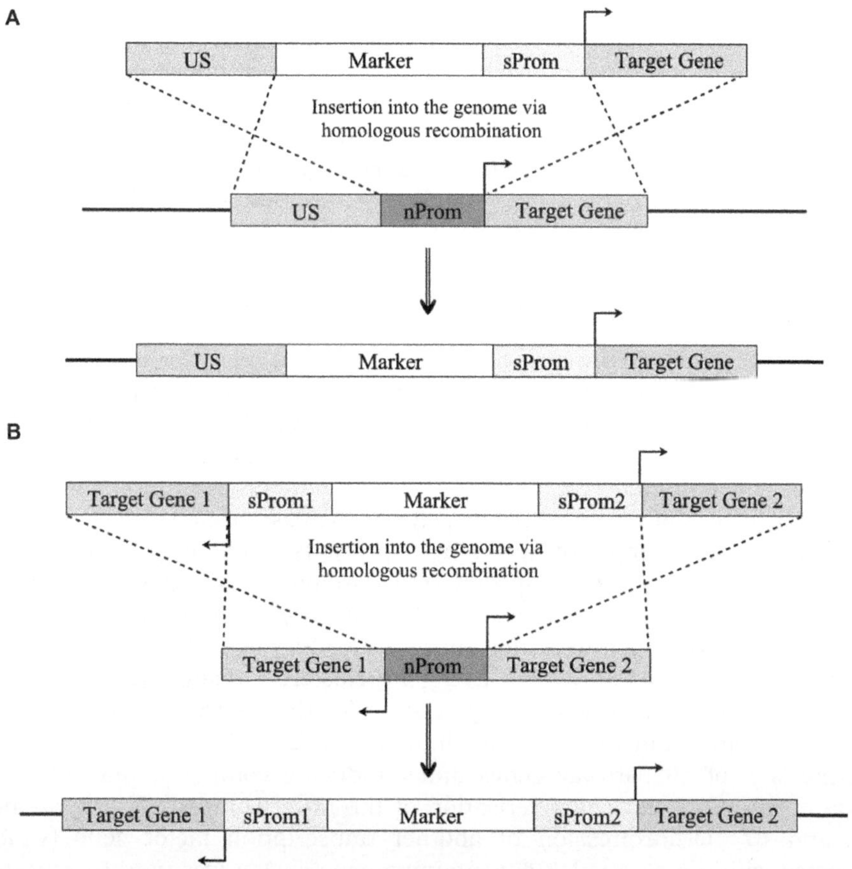

Figure 9.2 Targeted promoter replacement to alter the expression of secondary metabolite genes in the natural producer. When making a single promoter replacement (**A**), the transforming sequence consists of DNA homologous to the genomic DNA upstream of the native promoter (the *Up**S**tream* sequence: US), a suitable selectable marker, your selected promoter (sProm) and a 5′ portion of the target gene. If this is integrated into the genome by homologous recombination, the native promoter (nProm) will be replaced with the selectable marker and your chosen promoter, placing the target gene under the control of the latter. If two genes are divergently transcribed, as is often seen for key structural genes in secondary metabolism, a dual promoter replacement can be achieved in one step (**B**). This strategy was used by Ahuja *et al.*[24] with significant success.

release of the PKS and other key genes in the pathways. The native promotors were replaced with the inducible promoter alcA, allowing the researchers to culture the fungus under conditions guaranteed to deliver efficient gene expression, thus maximising the likelihood of metabolite production. This approach was very fruitful, leading to the characterisation of products from eight *A. nidulans* PKS genes, seven of which were novel. It also led to the identification of the PKS genes responsible for the known compounds cichorine and alternariol. The main limitation of such an approach is that only a few promoters can be replaced at a time, so it is limited to metabolites that require few biosynthetic genes, or to overproduction of the initial starter molecule, but perhaps without its full subsequent tailoring.

Alternatively, the various regulatory pathways can be manipulated, including pathway-specific regulation, global regulation and the chromatin environment. For a deeper understanding of these systems, we direct readers to the recent and thorough review.[25]

9.3.1 Exploiting Pathway-specific Regulators

The presence of a pathway specific transcriptional regulator is a common, but not absolute, trait of fungal SM gene clusters.[26] The first pathway specific transcription factor to be identified was AflR,[27] which binds to the promoters of sterigmatocystin and aflatoxin biosynthetic genes.[28] Deletion of this gene eliminates all expression of genes under its control,[29] whereas multiple copies of this gene in the genome increases expression.[30]

The earliest example of activating a previously cryptic gene cluster by overexpressing the pathway specific transcription factor is the aspyridones. Genome sequencing of *A. nidulans* had revealed many cryptic gene clusters, the products of which were not known, including one gene cluster which contains a PKS–NRPS. Within this gene cluster, a regulatory gene *apd*R was identified, which was then placed under the control of an inducible promoter and transformed back into *A. nidulans*. This resulted in the expression of all pathway genes under inducing conditions and allowed the purification and characterisation of the novel compounds aspyridone A and B.[31] Overexpression of another transcription factor gene (*scp*R) located within a cryptic NRPS containing gene cluster was used to activate the NRPS pathway. Unexpectedly this also led to activation of a silent asperfuranone gene cluster on a separate chromosome, providing the first evidence for secondary metabolite cross-talk between chromosomes. It is hypothesised that this cross-pathway interaction occurs *via* direct activation of the asperfuranone activator by the transcription factor scpR.[32] It is important to be aware that these systems are complex and transcriptional regulators do not always positively regulate gene expression. A recent study has reported the unexpected finding that in the beauvericin pathway, a pathway specific transcription factor negatively regulates the production of beauvericin, with mutants for this gene demonstrating a higher beauvericin yield.[33]

The ability of pathway specific transcription factors to induce very high gene expression levels has recently been exploited in a rather elegant way to construct a novel type of expression system.[34] The TerA transcription factor from the terrein pathway of *Aspergillus terreus*, which supports very high terrein yields (>1 $g L^{-1}$),[35] was transformed into *A. niger* under the control of the inducible *amy*B promoter. A plasmid containing a TerR target promoter could then be transformed into the TerR containing strain, leading to high expression levels of any gene placed under its control. The levels of expression seen, and yields of resulting metabolites, were significantly higher than when placed directly under the control of the *gpd*A promoter, a strong promoter often used in secondary metabolism research. Although this system was tested using only one TerR target promoter, as the terrein pathway contains 7 genes, this system could theoretically support the expression of multiple gene simultaneously.

9.3.2 Exploiting Genome-wide Regulators

In addition to often being regulated by pathway specific transcription factors, secondary metabolites have been shown to be impacted by various global regulatory pathways. Global regulation can be defined as the conversion of environmental cues into the required changes in gene expression across the genome – including changes in primary metabolism pathways but also many secondary metabolite biosynthetic pathways. Known global transcription factors now include AreA/AreB (nitrogen),[36] pacC (pH),[37] CreA (carbon)[38] and the CCAAT-binding complex (iron).[39] A globally acting transcription factor from *A. flavus* called *nsdC* has been identified, which regulates asexual development and secondary metabolism.[40] However, the stimulus it responds to is unknown. Likewise a negative global regulator of secondary metabolism *mcrA* has been identified in *A. nidulans*, but again the stimuli controlling this are somewhat unclear, as is its more general applicability to fungal secondary metabolite biosynthesis.[41] Various studies have now demonstrated that these broad-domain transcription factors impact a great number of secondary metabolite gene clusters, acting as both activators and repressors of SM biosynthesis. Slightly more complex to manipulate are the large transcriptional complexes which have been shown to regulate both primary and secondary metabolism in fungi, the best known being the multiprotein velvet complex, which integrates the light dependant regulation of sexual development and secondary metabolism. The OSMAC approach discussed earlier is a way of triggering changes in these global pathways by providing a range of stimuli, and has the advantage that no knowledge of the genetic basis of these pathways is required. However, as the pathways are better understood, it will be increasingly possible to permanently alter the pathways by genetic means – thus negating the need to screen large numbers of potentially expensive or problematic fermentation conditions.

The most intensively studied global regulator of fungal secondary metabolism is LaeA ('*L*oss of *a*flR *e*xpression'). LaeA was first identified in

A. nidulans, and has since been shown to regulate approximately 50% of secondary metabolite gene clusters in *A. nidulans*, *A. fumigatus* and *A. flavus*.[42–47] LaeA is a putative methyltransferase, although its exact mode of action is still elusive. Bayram *et al.*[48] demonstrated that LaeA forms a heterotrimeric protein complex with the velvet transcription factors VeA and VelB. The velvet complex is part of a regulatory system which responds to light levels in fungi, playing a crucial role in coordinating developmental and morphological changes with the required changes in metabolic output.[49] In darkness and with the correct environmental stimuli, a VeA–VelB heterodimer enters the nucleus where it can form a heterotrimeric protein complex with LaeA and thus trigger secondary metabolism production. VeA and VelB have now been shown to form complexes with a large number of other regulatory proteins in a pathway that is incredibly complex and has diverse effects on cellular biology.[50] But the importance of the velvet complex and LaeA in regulating secondary metabolism is now well established, and overexpression of LaeA has been used as a tool for increasing secondary metabolite formation in many fungi, such as *Fusarium graminearum*.[51] In *Monascus pilosus*, a fungus used to make *Monascus*-fermented red rice in Asia, LaeA overexpression led to an increase in yields of known compounds as well as the production compounds not produced in the wild-type.[52] A similar result with the production of previously unseen metabolites was observed for *Aspergillus fumisynnematus*,[53] and overexpression of LaeA in *P. citrinum* enhanced mevastatin production 5.8 fold.[54] Indirectly engineering the velvet complex was achieved by deleting a gene called LlmF.[55] LlmF has been shown to negatively regulate sterigmatocystin production by inhibiting the movement of VeA into the nucleus. The LlmF deletion removed this repression by allowing an accumulation of VeA in the nucleus and a subsequent increase in sterigmatocystin production.

Although the exact mechanism by which the velvet complex regulates secondary metabolism is unknown, there is evidence that LaeA regulation of SM gene clusters occurs *via* chromatin remodelling, by converting heterochromatin (compacted untranscribed DNA) to euchromatin (loosely compacted DNA). The structure of chromatin, which is the complex of DNA with the histone proteins, depends on various protein modifications, mainly specific methylations and acetylations on the histones. A role for LaeA in chromatin remodelling is supported by the fact that many of the LaeA-controlled gene clusters are sub-telomeric (*i.e.* near the ends of the chromosomes), in areas with a high amount of heterochromatin.[45] It has been shown that LaeA has a positional bias for local gene regulation: placing aflR outside the sterigmatocystin gene cluster removed its LaeA regulation, and moving other genes into the ST gene cluster placed them under LaeA control.[44] Indeed, such location-based co-regulation of genes has been suggested as an evolutionary reason for the maintenance of biosynthetic pathways in gene clusters. In LaeA mutants of *A. nidulans*, the sterigmatocystin locus was found to have increased methylation of H3K9 and contain high levels of heterochromatin protein-1 (HepA), which directly linked the function of LaeA to Histone H3 methylation.[56] It has been proposed that LaeA mediates the removal of heterochromatic marks by inhibiting an H3K9 methyltransferase, leading to the conversion of defined cluster regions to euchromatin, allowing transcription.[57]

As a mechanism for gene regulation, the chromatin environment can be directly altered by targeting the various enzymes involved in modifying the histones. This can be done chemically – methyltransferase and histone deacetylase inhibitors, such as 5-azacytidine (5AC) or suberoylanilide hydroxamic acid (SAHA) respectively, have been reported to increase SM production in several fungi,[58] but the inhibitors may impact fungal growth and have variable efficacy. An alternative is to achieve chromatin remodelling by genetic means. The deletion of HEP1, a chromatin protein that mediates the establishment and maintenance of heterochromatin, has been shown to upregulate multiple SM gene clusters in *A. nidulans* and strongly elevate the production of aurofusarin in *F. graminearum.*[59] Unexpectedly, the same deletion in *F. graminearum* led to a loss of deoxynivalenol *via* repression of two genes within the deoxynivalenol gene cluster, thought to be due to the chromatin upregulation of a deoxynivalenol repressor.[59] In a similar approach, disruption of the histone deacetylases has also proved a fruitful means to upregulate some secondary metabolite pathways, *e.g.* in *F. fujikuroi.*[60] The relationship between chromatin remodelling and natural product biosynthesis is an ever growing and complex field of research.[61]

9.3.3 Spatial Regulation

A final example of reshaping fungal regulatory networks comes from a study by Herr and Fischer,[62] where penicillin production was used as a model for investigating directed strain improvement. In addition to increasing yields simply by increasing expression of the biosynthetic genes, they also targeted the spatial regulation of the pathway, meaning the positioning of the required substrates and enzymes within certain subcellular locations. Penicillin biosynthesis is known to occur both in the cytoplasm and peroxisomes: the penicillin NRPS and IPN synthase function in the cytoplasm to produce isopenicillin N. A bifunctional enzyme then catalyses the final two reactions of the pathway within a peroxisome to produce penicillin. They demonstrated that if AcvA, the *A. nidulans* penicillin NRPS, is targeted to the peroxisome, an increase in yield of 320% is achieved. Interestingly, if the cyclase IpnA is also targeted to the peroxisome, a complete loss of production occurs, suggesting that either the oxidative environment of the peroxisome inactivates the enzyme or the necessary substrates or cofactors are not present in the peroxisome. Additionally, yields were improved by increasing the number of peroxisomes *via* the overexpression of *pexK.* Combining two of these three approaches further increased yields, but combining all three had no further effects, showing that a different rate-limiting factor, perhaps substrate availability, came into play. One of the challenges of manipulating regulatory systems to increase yields is first predicting the limiting factors of the pathway. Of course, all of the above methods of altering the regulation of a pathway are only feasible if the fungus is comparatively easy to culture and is amenable to genetic modification, and unfortunately this in not true for many fungi.

9.4 Heterologous Production of Compounds

Although various approaches have been developed to modify the production of secondary metabolites in their natural hosts, a lack of genetic tools often hampers this. Additionally, yields in natural producers often remain low, for example due to substrate limitations, and the species may not be suitable for large scale fermentation. Therefore, an alternative approach is to reconstruct the biosynthetic pathways in a suitable alternative host. This can facilitate the discovery and elucidation of biosynthetic pathways for academic purposes. It can also increase the commercial potential of the compounds produced, both because of the pathway knowledge and also the ease of production if expressed in a more suitable host species.

9.4.1 *S. cerevisiae* as a Host

Much of the heterologous production of natural products has been carried out using *S. cerevisiae*, the baker's yeast, as the heterologous host.[63] Being eukaryotic, yeasts such as *S. cerevisiae* may have advantages over bacterial hosts for the production of compounds from filamentous fungi and plants, due to issues over codon usage for example. The work of many decades has also developed a complex set of molecular tools for *S. cerevisiae*, including transformation protocols, the development of auxotrophic strains, expression of heterologous genes either from plasmids or after genomic integration, highly site-specific targeted integration, and even the production of the first artificial chromosome.[64] Additionally *S. cerevisiae* has no significant secondary metabolism so there is little background. This both aids detection and purification of resulting compounds and minimises the chance of undesirable shunt products through crosstalk with endogenous pathways. One factor that it is important to be aware of when using yeasts as heterologous hosts however, is the absence of certain machinery or cellular components needed for secondary metabolism. For example, megasynthases such as PKSs and NRPSs require posttranslational activation *via* the activity of a 4-phosphopantetheinyl transferase,[65] an enzyme not naturally present in *S. cerevisiae*. This absence of an endogenous PPTase can be overcome by using a strain expressing a PPTase from another organism. In 1998, for example, 6-MSA was successfully produced in yeast by expressing both the PKS from *Penicillium patulum* and a phosphopantetheinyl transferase (PPTase) from *Bacillus subtilis.*[66] More recently Gao *et al.*[67] coexpressed a 6-MSA synthase from *A. terreus* and the *A. nidulans* PPTase gene resulting in yields of 6-MSA as high as 2.2 $g L^{-1}$ in another yeast species called *Pichia pastoris* – an impressive yield for a polyketide.

Complex, multi-gene pathways have also been reconstructed in yeast. One such example is artemisinin, which is used to make the leading malaria treatments called artemisinin-based combination therapies (ACTs). Artemisinin is a sesquiterpene produced by the sweet wormwood plant *Artemisia annua*. Supply of artemisinin from *Artemisia annua* crops has suffered from significant instability, leading to efforts to develop an alternative supply of the compound by synthetic biology means. In 2006, Keasling and coworkers[68] reported the development of a yeast strain capable of producing artemisinic acid. This strain was engineered to allow an accumulation of farnesyl diphosphate, the sesquiterpene precursor, which was then converted, *via* the heterologous expression of the amorphadiene synthase and a cytochrome P450, to artemisinic acid. Methods were also developed for efficient chemical conversion of artemisinic acid to artemisinin.[69] Further engineering and optimisation of the system, including the expression of an additional P450 and a dehydrogenase, led to yields of 25 g of artemisinic acid per litre of culture.[69,70]

Work done to produce plant-derived morphine precursors such as reticuline and various related benzylisoquinoline alkaloids in *S. cerevisiae* by Smolke and coworkers[71] involved some sophisticated design aspects. The initial coexpression of three plant biosynthetic genes on separate plasmids (an *N*-methyltransferase and two *O*-methyltransferases) created a yeast strain capable of carrying out the biotransformation of fed norlaudanosoline to reticuline. Further transformation of the yeast strain with a chimeric biosynthetic pathway consisting of genes from multiple organisms including various plant genes and a human P450, led to the biosynthesis of a range of benzylisoquinoline alkaloids. But the most notable aspect of this project was the fine tuning of the pathway to control metabolic flux through the pathway. A tuneable GAL1-10 promoter was used to work out the optimum expression for each of the genes to produce the desired products. A promoter of appropriate strength was then chosen for production of the entire biosynthetic pathway. When metabolic flux is not controlled, bottlenecks in biosynthetic pathway often occur, leading to the accumulation of intermediates and the production of shunt products. This side effect can actually be advantageous if the desire is to generate novel compounds or identify pathway intermediates, but if the aim of the work is to produce high quantities of a pure final product, the streamlining of a pathway is worthy of consideration.

Yeast has also been shown to be a suitable host for expressing, and thus characterising, cryptic unannotated PKS and NRPS genes. Rugbjerg *et al.*[72] successfully reconstructed the pathway for the polyketide-derived rubrofusarin from the fungus *Fusarium graminearum*, building up the pathway in a stepwise manner in yeast, although with comparatively low titres of the mature product. In an improved approach, Watanabe and co-workers engineered a yeast strain to express a PPTase along with a malonyl-CoA synthetase, to supplement the host's supply of malonyl-CoA and meet any increased demand brought about by elevated PKS production.[73] Six expression plasmids containing cryptic PKS or NRPS genes from various fungal species were then constructed by homologous yeast recombination and transformed into the host strain, resulting in the isolation of 8 compounds, one of which was novel. In a similar way, Wang *et al.* reported enhanced titres of polyketides in yeast following co-expression with a plant-derived malonyl CoA synthase.[74]

9.4.2 Filamentous Fungi as Hosts

An alternative to using a yeast strain is to use a filamentous fungus, with *Aspergillus* species such as *A. oryzae* and *A. nidulans* being typical choices.[75] As mentioned earlier, the natural inability of yeasts to produce secondary metabolites means extra work may be required to develop a strain suitable for biosynthesis, for example by expressing a PPTase, cytochrome P450 reductases, or other genes required to produce the necessary substrates. Another element of natural product biosynthesis that makes filamentous fungi a more attractive host is the subcellular localisation of catalytic activities seen in some pathways such as the involvement of specialised vesicles in alflatoxin biosynthesis[76] and peroxisomes in penicillin production.[62] Filamentous fungi can also act as surprisingly efficient protein production factories; demonstrated by the fact that approximately 50% of fermentation-produced enzymes are produced from filamentous fungi.[75]

In the same way that 6-MSA has been produced by various yeast strains, a 6-MSA synthase called atX from *A. terreus* was expressed in *A. nidulans*, but without the requirement of PPTase coexpression due to the endogenous *A. nidulans* PPTase.[77] Various PKSs have been expressed and characterised in *Aspergillus* hosts since that time, allowing some of the chemical complexities of these large synthase enzymes to be elucidated. For example, expressing the 3-methylorcinaldehyde synthase (MOS) from *Acremonium strictum* in *A. oryzae* identified a reductive release mechanism for the first time.[78]

The first example of the heterologous production of a fungal secondary metabolite from a multi-gene pathway was the transfer of the penicillin gene cluster from *P. chrysogenum* to *N. crassa* and *A. niger.*[13] In this case a cosmid clone containing the entire gene cluster was used and it was somewhat fortuitous that all three promoters were active in their alternate hosts and that detection of penicillin is very sensitive and straightforward. The transfer of an entire intact gene cluster from one species to another is not always a viable approach, depending on the strains ability to recognise the foreign promoter sequences and correctly splice introns to form mature RNA. The first of these potential hurdles can be overcome by combining the transfer of the gene cluster with some of the techniques discussed in 'reshaping regulatory networks'. Overexpressing a pathway specific transcription factor from the gene cluster in the heterologous host has been shown to significantly increase yields.[79,80] In another example, overexpression of LaeA has been combined with heterologous expression of whole clusters in *A. oryzae* to produce monacolin K and terrequinone A gene clusters from *Monascus pilosus* and *A. nidulans*, respectively.[81]

An alternative approach is to reconstruct the pathway almost from first principles, placing each individual gene within a designed promoter/terminator cassette known to function in the host organism. This is often done using inducible promoters, allowing control over expression of the pathway to circumvent any potential toxicity issues. Building the pathway in this way rather than simply transferring a fragment of genomic DNA containing the cluster is quite work intensive but has the added advantage of allowing cDNA or synthetic DNA to be used rather than genomic DNA, avoiding potential spicing issues and allowing codon optimisation. Furthermore, expressing a well-defined and characterised set of genes can provide more insight into the minimal requirements of a pathway and would also be readily available for formation of hybrid pathways. The first example of this strategy was the reconstruction of the four-gene tenellin biosynthetic pathway *via* three sequential transformations.[82] Later this pathway was also reconstructed as a test system to demonstrate the applicability of a newly developed multi-gene expression system where the expression of all four pathway genes was achieved in a single transformation step.[83]

The medicinally important antibiotic pleuromutilin recently became the first example of a basidiomycete pathway being successfully reconstructed in an ascomycete host.[84] The gene cluster for the antibiotic pleuromutilin was identified in the basidiomycete *Clitopilus passeckerianus* and using the multigene expression vector system designed by Lazarus and co-workers,[83] the seven-gene pathway was reconstructed and expressed in *A. oryzae*. This involved refactoring all seven cDNA with suitable regulatory elements and co-expression in *A. oryzae*. This served to confirm the repertoire of genes needed for pleuromutilin biosynthesis and coincidentally to increase pleuromutilin yields.[84]

Heterologous hosts can also be used to elucidate the catalytic roles of individual genes in the pathway, and the production of multiple compounds from one biosynthetic gene cluster can be achieved by expressing different combinations of genes, as exemplified by the production of 8 different compounds by heterologously expressing genes from the *A. nidulans* aspyridone biosynthetic cluster in *A. oryzae.*[85] More recently, Oikawa's group has had considerable success using a similar approach to identify the roles of the genes involved in various secondary metabolism pathways such as paxilline, aflatrem, penitrem and shearinine.[86–91] Whilst heterologous expression in *A. oryzae* has proven to be an effective research platform, aided by the quadruple auxotrophic strain NSAR1[92] allowing incorporation of four different transformation constructs into the host, the titres obtained with this species have not yet reached those seen in some of the more notable natural product producers, and certainly do not rival the levels seen in commercial penicillin production. It has not escaped our attention that *A. oryzae* was initially chosen as it has GRAS status, as a fungus used in food preparation, and as such, abundant secondary metabolite production may well have been under counter-selection. Other heterologous platforms have been used successfully and in some cases with higher titres than seen in *A. oryzae*.

Robinson and Panaccione[93] elucidated a key step in ergot alkaloid biosynthesis by expressing candidate genes in *A. fumigatus.* This work also highlighted that native genes from the host strain can act on the heterologously produced compounds to produce novel end products. When the agroclavine producing enzyme easA from *Epichloë* sp. Lp1 was expressed in *A. fumigatus*, native *A. fumigatus* enzymes led to the production of two novel prenylated ergot alkaloids. This was particularly significant in the absence of the oxidase responsible for converting agroclavine to lysergic acid, where an accumulation of agroclavine therefore occurred.

As expression systems in filamentous fungi do not usually involve self-replicating plasmids, but rely on integration into the genome, issues can arise due to unexpected pleiotropic effects such as mutations in, or changes to, expression of neighbouring genes at the integration site. Additionally, the insertion of multiple copies of the construct can result in genomic instability and loss of production over time. To address this issue, an expression system was developed for *A. nidulans* which involved targeting exogenous constructs to a well-characterised location (named Insertion Site 1: *IS*1),[94] known to support good gene expression with no inherent impacts on strain fitness (Figure 9.3). As proof of concept, *mpaC* from *P. brevicompactum* was integrated into *IS*1 and successfully expressed, conclusively identifying this gene

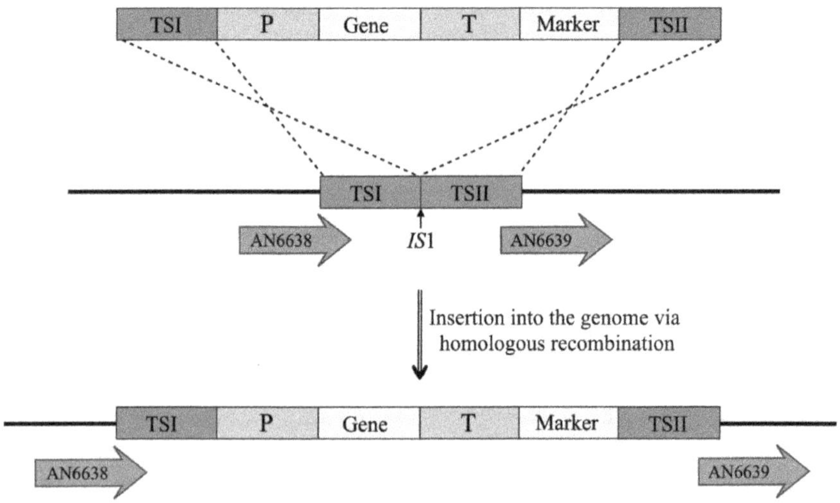

Figure 9.3 Integration of exogenous DNA into the insertion site *IS*1 of *A. nidulans*, as designed by Mortensen and co-workers.[94] *IS*1 is situated between two genes, AN6638 and AN6639, on chromosome I. This example shows the insertion of an expression cassette, with a promoter (P), terminator (T), and gene to be expressed. A selectable marker is also included in the DNA fragment, to allow for selection of transformants after successful integration. The targeting sequences (TSI and TSII) allow the fragment to be inserted into IS1 by homologous recombination. Insertion of an entire gene cluster, over two sequential insertion events, has also been achieved at this locus.[95]

as encoding the first biosynthetic enzyme in the pathway for the medicinally important immunosuppressant compound mycophenolic acid.

Nielsen *et al.*[95] went on to reconstruct the entire 25 kb *A. terreus* geodin gene cluster within *IS*1. Their strategy was to reconstruct the whole cluster over two constructs, with the only modification to the cluster being the placement of the cluster transcription factor under the control of the strong constitutive gpdA promoter. The two constructs were then sequentially integrated into *IS*1 leading to the biosynthesis of geodin. The presence of a functional geodin gene cluster in an amenable host then allowed the generation of gene deletion strains to probe the roles of individual genes in the cluster.

Chiang *et al.* designed a strategy where PKS genes were targeted to the *wA* locus of *A. nidulans.*[96] As *wA* produces the green pigment present in the conidia of *A. nidulans,*[97] transformants were readily identified due to their white colouration. In total 9 PKS genes from *A. terreus* were expressed using this system, allowing the identification of 6 products. This study highlighted two aspects of the heterologous production of polyketides which must be considered. Firstly, there must be a release mechanism present, for releasing the polyketide from the synthase. Many polyketide synthases have a final thioesterase domain responsible for releasing the polyketide chain, but this is not always the case and those that do not may require a trans-acting enzyme. A bioinformatic analysis determined that for two of the polyketides expressed, the domain was indeed missing from the PKS, but co-expression of a cluster encoding a β-lactamase-like gene resulted in production of the polyketide product. The elucidation of a biosynthetic pathway for the important maleidride compounds similarly found that the activity of a hydrolase responsible for chain release was necessary for the pathway to function.[98] The second potential pit-fall exemplified in the work by Chiang *et al.* is the provision of unusual starter units.[96] In one case, the co-expression of the target PKS with a second PKS that was predicted to provide a starter unit led to the successful production of an asperfuranone intermediate.

Various synthetic biology approaches were elegantly combined to transfer and express the biosynthetic genes for penicillin production in an *A. nidulans* strain.[99] The three necessary coding regions were expressed *via* a single large polycistronic mRNA. By using viral 2A peptide sequences between each open reading frame (ORF), cotranslational cleavage into the individual pathway enzymes was achieved. This had the benefit of using just one promoter to drive the expression of the polycistronic mRNA, rather than having to re-engineer each gene separately, although it did mean expression levels of the individual enzymes were all the same, which may not be optimal for such a pathway. The construct was built using synthetic DNA to produce two of the biosynthetic genes (*pcbC* and *penDE)* with N-terminal viral 2A sequences attached, with only the largest gene (*pcbAB*) being amplified by PCR. These components were then combined using yeast homologous recombination to place the large polycistronic gene under the control of a xylose-inducible promoter.

Some fungi are renowned for producing high titres of secondary metabolites, so may prove useful production platforms. An example of this was shown by Kakule *et al.*[100] with *Fusarium heterosporum* which is normally an abundant producer of equisitin, and using the same regulatory elements to drive heterologous production of the lovastatin precursor polyketide, or the full pathway of three genes for the polyketide pyrolocin in an equisitin deficient strain.

Similarly, very elegant use of *Penicillium chrysogenum* as a heterologous host was demonstrated by McLean *et al.* for pravastatin production.[101] The first step involved taking an industrially improved penicillin G producing strain DS17690, capable of giving very high titres of penicillin, and through several rounds of manipulation, deleting the entire penicillin gene cluster from this strain.[102] The rationale was that this host strain was optimised for both high secondary metabolite production, and also for efficient growth in conventional fermentation conditions. Into this host they first transferred the gene cluster for compactin biosynthesis, obtaining a strain giving good titres from the native promoters. Into this strain they then added the P450 sca-2 gene from *Streptomyces carbophilus*, fused with the FMN reductase of RhF from *Rhodococcus*, all expressed under the penicillin pcbC promoter and terminator regions. The P450 converted compactin to the more valuable pravastatin, yielding an impressive 550 mg L^{-1} of pravastatin.[101] Had they used penicillin-pathway promoters for compactin production, there is the possibility that yields may have been much higher in this system, nevertheless this clearly demonstrates how a hybrid metabolic pathway can be generated and efficiently expressed in a heterologous host.

Thus, the heterologous production of compounds in fungi, either a yeast or a filamentous fungi depending on specific requirements, allows characterisation of known or novel biosynthetic pathways, as well as preparing for the demands of downstream applications such as commercial production.

9.5 Developing Novel Products

9.5.1 Manipulation of the Core Megasynthases for New Core Molecules

Megasynthases from fungi, including polyketide synthases (PKSs), non-ribosomal synthases (NRPSs), and indeed PKS–NRPS hybrids, share many structural characteristics, with fairly consistent domain architecture within each class of enzyme, yet the chemical products they synthesise are astoundingly diverse. This has made them the focus of decades of intensive research to uncover the roles of individual domains in determining aspects of the final structures. Although there is still much unknown about where precise control and specificity lie, enough is now understood about the domain architecture of these enzymes to explore the possibility of creating functional chimeric enzymes.

The first domain found in fungal polyketides, located at the N-terminal of the protein, is known as the starter unit:ACP transacylase (SAT) domain. SAT domains are large N-terminal domains which are responsible for initiating polyketide biosynthesis by transferring a starter unit, for which they have very high selectivity, onto the acyl protein carrier (ACP) domain. Liu *et al.*[103] used this understanding to create a hybrid polyketide synthase by swapping the SAT domain from the asperfuranone nonreducing-PKS (AfoE) with the same domain from the sterigmatocystin PKS (StcA). The production of a chimeric AfoE–StcA PKS effectively engineered AfoE to accept the hexanoyl starter unit usually accepted by StcA, thus resulting in the production of a novel compound and again confirming the role of the SAT domain in controlling starter unit specificity (Figure 9.4). Interestingly, the novel compound had the same chain length as the native product, despite the fact that the chimeric enzyme was accepting a shorter starter unit, showing that the ketosynthase domain of the PKS is programmed to control the final chain length, rather than to carry out a fixed number of chain extensions. This theory was borne out by further work that engineered AfoE to accept a longer starter unit, which again resulted in the same final chain length as the native AfoE.[104]

Work has also focussed on the final domains of fungal polyketides. The release of fungal polyketides from their synthases can be mediated by a range of different final domain classes including reductase (R), thioesterase (TE) or

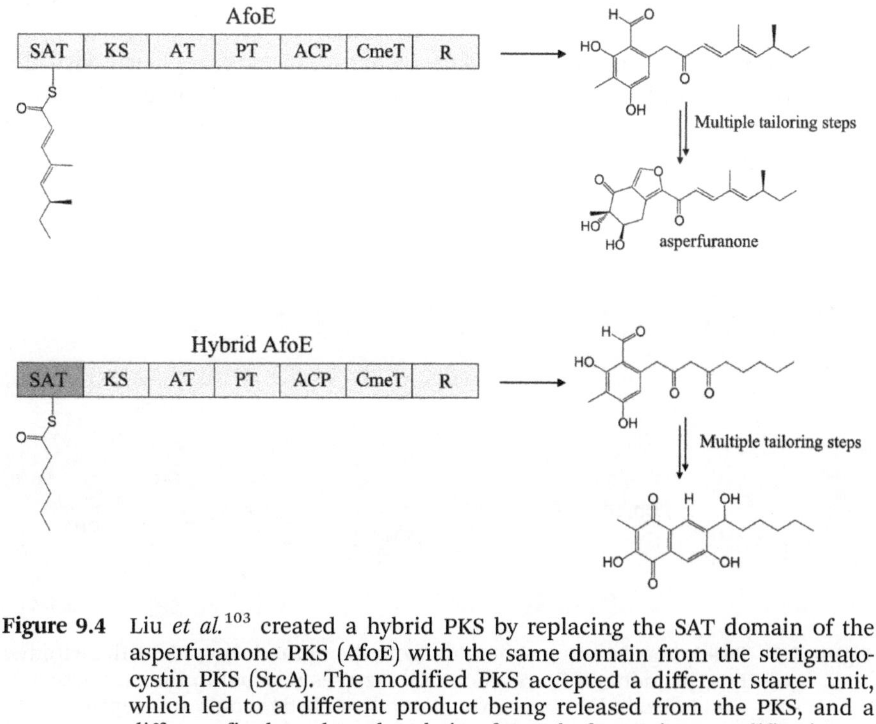

Figure 9.4 Liu *et al.*[103] created a hybrid PKS by replacing the SAT domain of the asperfuranone PKS (AfoE) with the same domain from the sterigmatocystin PKS (StcA). The modified PKS accepted a different starter unit, which led to a different product being released from the PKS, and a different final product then being formed after various modifications.

thioesterase/Claisen-cyclase (TE/CLC), or even by a transacting enzyme such as a hydrolase. Yeh *et al.*[105] swapped the final reductase (R) domain of a PKS from *A. nidulans* called DtbA with the final thioesterase (TE) domains from two different *A. nidulans* PKS genes, with both swaps resulting in the production of two metabolites which contained carboxylic acids rather than the original corresponding aldehydes (Figure 9.5). It is worth noting that the two donor PKS genes were chosen based on their phylogenetic proximity to DtbA, and when swaps were conducted with more phylogenetically distant PKS genes, no metabolites were detected. It was suggested that this was due to required domain–domain interactions within the polyketide, but it has also been shown that TE domains effectively act as decision gates, with the shape and size of the compound offered to the TE domain being the crucial determinant in whether the final product is successfully released.[106] Thus, part of the rational design strategy for producing desired "unnatural products" from chimeric PKSs is the selection of a suitable TE domain.

An elegant example of engineering chimeric PKS–NRPS enzymes came from Cox and co-workers.[107] Various rational domain swaps were conducted within the PKS portions of the megasynthetases responsible for synthesising the closely related compounds tenellin and desmethylbassianin. Analysis of the products of these hybrid synthetases in the heterologous host *A. oryzae* allowed changes in the methylation patterns and chain length to be mapped to the domain swaps, thus providing a vital insight into how these large enzymes are programmed. To demonstrate that the chimeric PKS–NRPS genes could also be used to reconstruct entire pathways, a hybrid gene, where the tenellin ketoreductase (KR) domain had been replaced with the desmethylbassianin KR, was co-expressed with two cytochrome P450s from the tenellin gene cluster and led to the production of bassianin, a compound for which there is no longer a natural source due to the loss of the original

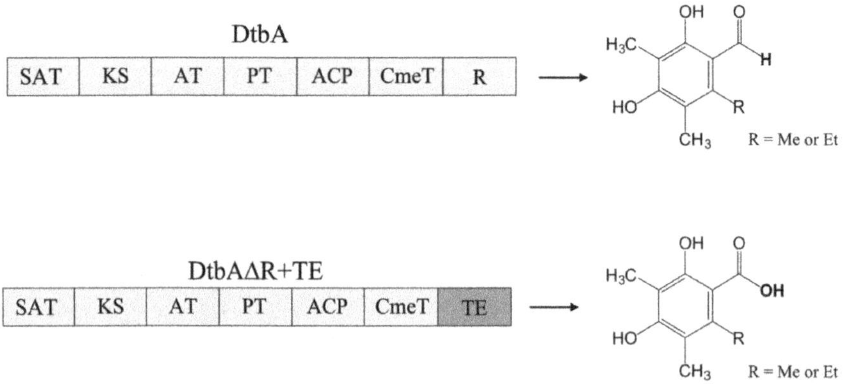

Figure 9.5 Swapping the native R domain from the nonreducing-PKS DtbA with the final TE domain from other PKS enzymes resulted in the production of novel metabolites where the natural aldehyde moiety was replaced with carboxylic acid.[105]

producing strain. In a comprehensive study into the potential of recombinant PKS–NRPS enzymes, Kakule *et al.*[108] expressed a total of 57 different chimeric enzymes, totalling 34 individual module swaps, and succeeded in producing 6 novel compounds as well as providing an initial set of ground rules for the successful engineering of such enzymes. Another advance in this field came from Nielsen *et al.* who uncovered an unexpected level of flexibility in the functionality of inter-modular PKS–NRPS linkers, with linker variants demonstrating high tolerance to changes in both content and length.[109]

Little success has been reported thus far investigating the potential of domain swapping in fungal NRPS enzymes, but one study[110] expressed recombinant NRPS-like enzymes. In these enzymes, rather than recognising and activating an amino acid, the A domains recognise and activate α-keto acids. Swaps were conducted between three such enzymes, resulting in the production of a novel metabolite, and demonstrating that the A domain of these enzymes controls the substrate selection whereas the TE domain determines the pattern of cyclisation (Figure 9.6). Thus, the production of recombinant megasynthetases, both to probe certain aspects of enzymatic control and to produce natural product analogues, is proving to be a promising field of research. Currently there is an element of trial and error,

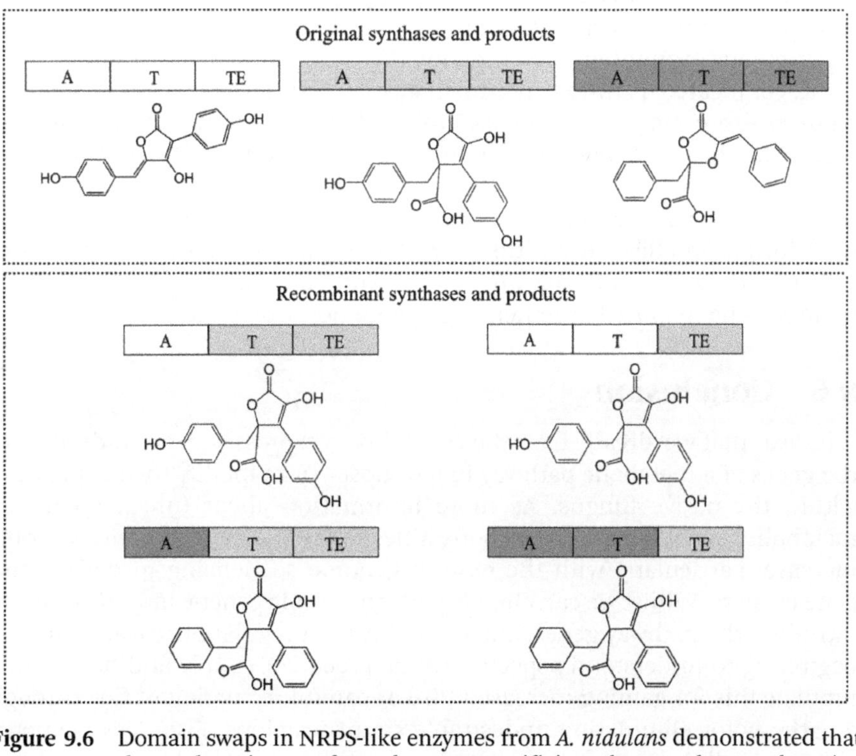

Figure 9.6 Domain swaps in NRPS-like enzymes from *A. nidulans* demonstrated that the A domains confer substrate specificity whereas the TE domains determine the cyclisation pattern.

with many combinations failing to produce functional enzymes. However, as additional structural data for these large enzymes are generated, as the complexities of domain–domain linkages and interactions are better understood, and as the secrets of enzyme programming – particularly in iterative polyketides – are uncovered, the potential of such synthetic biology will only grow.

9.5.2 Novelty *via* Feeding or Semi-synthesis

An interesting area for the future may well be that of combining synthetic biology with chemical synthesis. Biological conversions are routinely performed to generate high value steroids using whole fungal cells to deliver the required stereoconversions, whilst penicillin G or V biosynthesis (rather than the natural penicillin) has long been achieved by feeding phenylacetate or phenoxyacetate respectively to the culture media during penicillin fermentation.[111] This shows that not only can fungi be efficient producers of novel compounds, but that they can also take up a range of substrates from their growth media and perform valuable reactions using such starter molecules. Hawkins and Smolke[71] fed precursors of the morphine pathway into yeast and got the desired biotransformations when the yeast strain was expressing the appropriate enzymes, showing this compound could be readily transported into the cell and hence into the biosynthetic pathway. In other experiments examining PKS functionality, it has been common to feed SNAC-conjugated pathway intermediates or pathway derivatives to fungal cultures expressing a relevant PKS in the hope that they would be further metabolised to give mature or indeed novel polyketides.[103] Together these data show that it may well be possible to feed novel starter molecules or indeed novel pathway intermediates into fungal expression platforms, to then further modify these within the fungal cells to give new derivatives of the mature compound, allowing both synthetic chemistry and synthetic biology to be deployed together to generate novel products.

9.6 Conclusion

It is clear that we already have the capability to express some or indeed all of the genes of a metabolic pathway in new hosts, or to modify their expression within the native fungus. As more information about fungal secondary metabolism is obtained, the opportunities to exploit these systems will only increase, particularly with the ease of genome sequencing to deliver new gene clusters. Whilst we can already perform simple genetic modifications to introduce the pathway genes, and to modify their regulation, we are only just beginning to understand aspects such as precursor supply and how to manipulate this for gaining increased titre or careful regulation of flux through a pathway for eliminating undesirable shunt reactions. Even less is known about subcellular compartmentalisation of the enzymes and how to engineer this for yield, or how to optimise transport or efflux of the products, showing there is plenty of fundamental research yet to be performed on fungal secondary metabolism.

In the future we predict that there will be considerable opportunities to develop rationally designed host strains as expression platforms, along with a catalogue of promoters with known expression levels, or even tuneable promoters so that comparative expression levels of each gene in a pathway can be optimised. The possibilities of coupling this optimised expression with modified recombinant megasynthases delivering new core molecules, and with a cocktail of selected tailoring enzymes derived from a range of organisms, could deliver the holy grail of biosynthetic pathways for designer metabolites. Whatever the case, we are still in the infancy of this technology and expect there to be some exciting times ahead.

Acknowledgements

We gratefully acknowledge the support of the BBSRC and MRC in funding recent research under awards BB/K002341/1 and MR/N029909/1.

References

1. J. N. Currie, *J. Biol. Chem.*, 1917, **31**, 15.
2. M. Papagianni, *Biotechnol. Adv.*, 2007, **25**, 244.
3. G. Ozcengiz and A. L. Demain, *Biotechnol. Adv.*, 2013, **31**, 287.
4. P. R. Twentyman, *Biochem. Pharmacol.*, 1992, **43**, 109.
5. E. Finkelstein, B. Amichai and M. H. Grunwald, *Int. J. Antimicrob. Agents*, 1996, **6**, 189.
6. D. W. Denning, *J. Antimicrob. Chemother.*, 2002, **49**, 889.
7. D. W. Bartlett, J. M. Clough, J. R. Godwin, A. A. Hall, M. Hamer and B. Parr-Dobrzanski, *Pest Manage. Sci.*, 2002, **58**, 649.
8. M. Manzoni and M. Rollini, *Appl. Microbiol. Biotechnol.*, 2002, **58**, 555.
9. N. P. Keller, G. Turner and J. W. Bennett, *Nat. Rev. Microbiol.*, 2005, **3**, 937.
10. A. P. MacCabe, M. Riach, S. E. Unkles and J. R. Kinghorn, *EMBO J.*, 1990, **9**, 279.
11. D. J. Smith, J. H. Bull, J. Edwards and G. Turner, *Mol. Gen. Genet.*, 1989, **216**, 492.
12. D. Smith, M. Burnham, J. Bull, J. Hodgson, J. Ward, P. Browne, J. Brown, B. Barton, A. Earl and G. Turner, *EMBO J.*, 1990, **9**, 741.
13. D. J. Smith, M. K. Burnham, J. Edwards, A. J. Earl and G. Turner, *Nat. Biotechnol.*, 1990, **8**, 39.
14. H. B. Bode, B. Bethe, R. Höfs and A. Zeeck, *ChemBioChem*, 2002, **3**, 619.
15. N. Khaldi, F. T. Seifuddin, G. Turner, D. Haft, W. C. Nierman, K. H. Wolfe and N. D. Fedorova, *Fungal Genet. Biol.*, 2010, **47**, 736.
16. T. Weber, K. Blin, S. Duddela, D. Krug, H. U. Kim, R. Bruccoleri, S. Y. Lee, M. A. Fischbach, R. Müller and W. Wohlleben, *Nucleic Acids Res.*, 2015, **43**, W237–W243.

17. D. O. Inglis, J. Binkley, M. S. Skrzypek, M. B. Arnaud, G. C. Cerqueira, P. Shah, F. Wymore, J. R. Wortman and G. Sherlock, *BMC Microbiol.*, 2013, **13**, 91.
18. W.-J. Wang, D.-Y. Li, Y.-C. Li, H.-M. Hua, E.-L. Ma and Z.-L. Li, *J. Nat. Prod.*, 2014, 77, 1367.
19. R. T. Hewage, T. Aree, C. Mahidol, S. Ruchirawat and P. Kittakoop, *Phytochemistry*, 2014, **108**, 87.
20. K. Scherlach and C. Hertweck, *Org. Biomol. Chem.*, 2006, **4**, 3517.
21. V. Schroeckh, K. Scherlach, H.-W. Nützmann, E. Shelest, W. Schmidt-Heck, J. Schuemann, K. Martin, C. Hertweck and A. A. Brakhage, *Proc. Natl. Acad. Sci. U. S. A.*, 2009, **106**, 14558.
22. C. C. König, K. Scherlach, V. Schroeckh, F. Horn, S. Nietzsche, A. A. Brakhage and C. Hertweck, *ChemBioChem*, 2013, **14**, 938.
23. J. P. Tauber, V. Schroeckh, E. Shelest, A. A. Brakhage and D. Hoffmeister, *Environ. Microbiol.*, 2016, **18**, 5218.
24. M. Ahuja, Y.-M. Chiang, S.-L. Chang, M. B. Praseuth, R. Entwistle, J. F. Sanchez, H.-C. Lo, H.-H. Yeh, B. R. Oakley and C. C. Wang, *J. Am. Chem. Soc.*, 2012, **134**, 8212.
25. J. Macheleidt, D. J. Mattern, J. Fischer, T. Netzker, J. Weber, V. Schroeckh, V. Valiante and A. A. Brakhage, *Annu. Rev. Genet.*, 2016, **50**, 371.
26. D. Hoffmeister and N. P. Keller, *Nat. Prod. Rep.*, 2007, **24**, 393–416.
27. P.-K. Chang, J. Cary, D. Bhatnagar, T. Cleveland, J. Bennett, J. Linz, C. Woloshuk and G. Payne, *Appl. Environ. Microbiol.*, 1993, **59**, 3273.
28. M. Fernandes, N. P. Keller and T. H. Adams, *Mol. Microbiol.*, 1998, **28**, 1355.
29. J.-H. Yu, R. A. Butchko, M. Fernandes, N. P. Keller, T. J. Leonard and T. H. Adams, *Curr. Genet.*, 1996, **29**, 549.
30. P.-K. Chang, K. C. Ehrlich, J. Yu, D. Bhatnagar and T. E. Cleveland, *Appl. Environ. Microbiol.*, 1995, **61**, 2372.
31. S. Bergmann, J. Schümann, K. Scherlach, C. Lange, A. A. Brakhage and C. Hertweck, *Nat. Chem. Biol.*, 2007, **3**, 213.
32. S. Bergmann, A. N. Funk, K. Scherlach, V. Schroeckh, E. Shelest, U. Horn, C. Hertweck and A. A. Brakhage, *Appl. Environ. Microbiol.*, 2010, **76**, 8143.
33. E. M. Niehaus, L. Studt, K. W. von Bargen, W. Kummer, H. U. Humpf, G. Reuter and B. Tudzynski, *Environ. Microbiol.*, 2016, **18**, 4282.
34. M. Gressler, P. Hortschansky, E. Geib and M. Brock, *Front. Microbiol.*, 2015, **6**, 184.
35. C. Zaehle, M. Gressler, E. Shelest, E. Geib, C. Hertweck and M. Brock, *Chem. Biol.*, 2014, **21**, 719.
36. P. K. Chang, J. Yu, D. Bhatnagar and T. E. Cleveland, *Biochim. Biophys. Acta*, 2000, **1491**, 263.
37. E. A. Espeso, J. Tilburn, H. N. Arst and M. A. Peñalva, *EMBO J.*, 1993, **12**, 3947.

38. C. E. Dowzer and J. M. Kelly, *Mol. Cell. Biol.*, 1991, **11**, 5701.
39. P. Hortschansky, M. Eisendle, Q. Al-Abdallah, A. D. Schmidt, S. Bergmann, M. Thön, O. Kniemeyer, B. Abt, B. Seeber and E. R. Werner, *EMBO J.*, 2007, **26**, 3157.
40. M. K. Gilbert, B. M. Mack, Q. Wei, J. M. Bland, D. Bhatnagar and J. W. Cary, *Microbiol. Res.*, 2016, **182**, 150.
41. C. E. Oakley, M. Ahuja, W. W. Sun, R. Entwistle, T. Akashi, J. Yaegashi, C. J. Guo, G. C. Cerqueira, J. Russo Wortman and C. C. Wang, *Mol. Microbiol.*, 2017, **103**, 347.
42. J. W. Bok and N. P. Keller, *Eukaryotic Cell*, 2004, **3**, 527.
43. J. W. Bok, S. A. Balajee, K. A. Marr, D. Andes, K. F. Nielsen, J. C. Frisvad and N. P. Keller, *Eukaryotic Cell*, 2005, **4**, 1574.
44. J. W. Bok, D. Noordermeer, S. P. Kale and N. P. Keller, *Mol. Microbiol.*, 2006, **61**, 1636.
45. R. M. Perrin, N. D. Fedorova, J. W. Bok, R. A. Cramer, J. R. Wortman, H. S. Kim, W. C. Nierman and N. P. Keller, *PLoS Pathog.*, 2007, **3**, e50.
46. S. P. Kale, L. Milde, M. K. Trapp, J. C. Frisvad, N. P. Keller and J. W. Bok, *Fungal Genet. Biol.*, 2008, **45**, 1422.
47. D. Georgianna, N. D. Fedorova, J. L. Burroughs, A. L. Dolezal, J. W. Bok, S. Horowitz-Brown, C. P. Woloshuk, J. Yu, N. P. Keller and G. A. Payne, *Mol. Plant Pathol.*, 2010, **11**, 213.
48. Ö. Bayram, S. Krappmann, M. Ni, J. W. Bok, K. Helmstaedt, O. Valerius, S. Braus-Stromeyer, N.-J. Kwon, N. P. Keller and J.-H. Yu, *Science*, 2008, **320**, 1504.
49. J. W. Bok and N. P. Keller, in *Biochemistry and Molecular Biology*, Springer, edn, 2016, pp. 21–29.
50. Ö. Sarikaya-Bayram, J. M. Palmer, N. Keller, G. H. Braus and Ö. Bayram, *Front. Microbiol.*, 2015, **6**, 1.
51. H.-K. Kim, S. Lee, S.-M. Jo, S. P. McCormick, R. A. Butchko, R. H. Proctor and S.-H. Yun, *PLoS One*, 2013, **8**, e68441.
52. S. S. Lee, J. H. Lee and I. Lee, *J. Microbiol. Biotechnol.*, 2013, **23**, 959.
53. E. J. Hong, N. K. Kim, D. Lee, W. G. Kim and I. Lee, *Fungal Biol.*, 2015, **119**, 973.
54. Y. Zheng, S. Cao, Y. Huang, G. Liao and C. Hu, *Acta Microbiol. Sin.*, 2014, **54**, 1438.
55. J. M. Palmer, J. M. Theisen, R. M. Duran, W. S. Grayburn, A. M. Calvo and N. P. Keller, *PLOS Genet.*, 2013, **9**, e1003193.
56. Y. Reyes-Dominguez, J. W. Bok, H. Berger, E. K. Shwab, A. Basheer, A. Gallmetzer, C. Scazzocchio, N. Keller and J. Strauss, *Mol. Microbiol.*, 2010, **76**, 1376.
57. J. Strauss and Y. Reyes-Dominguez, *Fungal Genet. Biol.*, 2011, **48**, 62.
58. R. B. Williams, J. C. Henrikson, A. R. Hoover, A. E. Lee and R. H. Cichewicz, *Org. Biomol. Chem.*, 2008, **6**, 1895.
59. Y. Reyes-Dominguez, S. Boedi, M. Sulyok, G. Wiesenberger, N. Stoppacher, R. Krska and J. Strauss, *Fungal Genet. Biol.*, 2012, **49**, 39.

60. L. Studt, F. J. Schmidt, L. Jahn, C. M. Sieber, L. R. Connolly, E. M. Niehaus, M. Freitag, H. U. Humpf and B. Tudzynski, *Appl. Environ. Microbiol.*, 2013, **79**, 7719.
61. A. Gacek and J. Strauss, *Appl. Microbiol. Biotechnol.*, 2012, **95**, 1389.
62. A. Herr and R. Fischer, *Metab. Eng.*, 2014, **25**, 131.
63. Y. Tsunematsu, K. Ishiuchi, K. Hotta and K. Watanabe, *Nat. Prod. Rep.*, 2013, **30**, 1139.
64. M. Dunham, M. Gartenberg and G. W. Brown, *Methods in Yeast Genetics and Genomics: 2-015 Edition: A CSHL Course Manual*, CSHL Press, USA, 2015.
65. R. H. Lambalot, A. M. Gehring, R. S. Flugel, P. Zuber, M. LaCelle, M. A. Marahiel, R. Reid, C. Khosla and C. T. Walsh, *Chem. Biol.*, 1996, **3**, 923.
66. J. T. Kealey, L. Liu, D. V. Santi, M. C. Betlach and P. J. Barr, *Proc. Natl. Acad. Sci. U. S. A.*, 1998, **95**, 505.
67. L. Gao, M. Cai, W. Shen, S. Xiao, X. Zhou and Y. Zhang, *Microb. Cell Fact.*, 2013, **12**, 77.
68. D. K. Ro, E. M. Paradise, M. Ouellet, K. J. Fisher, K. L. Newman, J. M. Ndungu, K. A. Ho, R. A. Eachus, T. S. Ham, J. Kirby, M. C. Chang, S. T. Withers, Y. Shiba, R. Sarpong and J. D. Keasling, *Nature*, 2006, **440**, 940.
69. C. J. Paddon, P. Westfall, D. Pitera, K. Benjamin, K. Fisher, D. McPhee, M. Leavell, A. Tai, A. Main and D. Eng, *Nature*, 2013, **496**, 528.
70. P. J. Westfall, D. J. Pitera, J. R. Lenihan, D. Eng, F. X. Woolard, R. Regentin, T. Horning, H. Tsuruta, D. J. Melis and A. Owens, *Proc. Natl. Acad. Sci. U. S. A.*, 2012, **109**, E111.
71. K. M. Hawkins and C. D. Smolke, *Nat. Chem. Biol.*, 2008, **4**, 564.
72. P. Rugbjerg, M. Naesby, U. H. Mortensen and R. J. Frandsen, *Microb. Cell Fact.*, 2013, **12**, 31.
73. K. I. Ishiuchi, T. Nakazawa, T. Ookuma, S. Sugimoto, M. Sato, Y. Tsunematsu, N. Ishikawa, H. Noguchi, K. Hotta and H. Moriya, *ChemBioChem*, 2012, **13**, 846.
74. Y. Wang, H. Chen and O. Yu, *Appl. Microbiol. Biotechnol.*, 2014, **98**, 5435.
75. D. Lubertozzi and J. D. Keasling, *Biotechnol. Adv.*, 2009, **27**, 53.
76. L. V. Roze, A. Chanda and J. E. Linz, *Fungal Genet. Biol.*, 2011, **48**, 35.
77. I. Fujii, Y. Ono, H. Tada, K. Gomi, Y. Ebizuka and U. Sankawa, *Mol. Gen. Genet.*, 1996, **253**, 1.
78. A. M. Bailey, R. J. Cox, K. Harley, C. M. Lazarus, T. J. Simpson and E. Skellam, *Chem. Commun.*, 2007, 4053.
79. K. Sakai, H. Kinoshita, T. Shimizu and T. Nihira, *J. Biosci. Bioeng.*, 2008, **106**, 466.
80. W.-B. Yin, A. W. Reinke, M. Szilágyi, T. Emri, Y.-M. Chiang, A. E. Keating, I. Pócsi, C. C. Wang and N. P. Keller, *Microbiology*, 2013, **159**, 77.

81. K. Sakai, H. Kinoshita and T. Nihira, *Appl. Microbiol. Biotechnol.*, 2012, **93**, 2011.
82. M. N. Heneghan, A. A. Yakasai, L. M. Halo, Z. Song, A. M. Bailey, T. J. Simpson, R. J. Cox and C. M. Lazarus, *ChemBioChem*, 2010, **11**, 1508.
83. K. A. K. Pahirulzaman, K. Williams and C. M. Lazarus, *Methods Enzymol.*, 2012, **517**, 241.
84. A. M. Bailey, F. Alberti, S. Kilaru, C. M. Collins, K. de Mattos-Shipley, A. J. Hartley, P. Hayes, A. Griffin, C. M. Lazarus, R. J. Cox, C. L. Willis, K. O'Dwyer, D. W. Spence and G. D. Foster, *Sci. Rep.*, 2016, **6**, 25202.
85. Z. Wasil, K. A. Pahirulzaman, C. Butts, T. J. Simpson, C. M. Lazarus and R. J. Cox, *Chem. Sci.*, 2013, **4**, 3845.
86. K. Tagami, C. Liu, A. Minami, M. Noike, T. Isaka, S. Fueki, Y. Shichijo, H. Toshima, K. Gomi and T. Dairi, *J. Am. Chem. Soc.*, 2013, **135**, 1260.
87. C. Liu, K. Tagami, A. Minami, T. Matsumoto, J. C. Frisvad, H. Suzuki, J. Ishikawa, K. Gomi and H. Oikawa, *Angew. Chem., Int. Ed. Engl.*, 2015, **54**, 5748.
88. C. Liu, A. Minami, T. Dairi, K. Gomi, B. Scott and H. Oikawa, *Org. Lett.*, 2016, **18**, 5026.
89. Y. Ye, A. Minami, A. Mandi, C. Liu, T. Taniguchi, T. Kuzuyama, K. Monde, K. Gomi and H. Oikawa, *J. Am. Chem. Soc.*, 2015, **137**, 11846.
90. K. Tagami, A. Minami, R. Fujii, C. Liu, M. Tanaka, K. Gomi, T. Dairi and H. Oikawa, *ChemBioChem*, 2014, **15**, 2076.
91. A. Minami, C. Liu and H. Oikawa, *Heterocycles*, 2016, **92**, 397.
92. F. J. Jin, J. Maruyama, P. R. Juvvadi, M. Arioka and K. Kitamoto, *FEMS Microbiol. Lett.*, 2004, **239**, 79.
93. S. L. Robinson and D. G. Panaccione, *Appl. Environ. Microbiol.*, 2014, **80**, 6465.
94. B. G. Hansen, B. Salomonsen, M. T. Nielsen, J. B. Nielsen, N. B. Hansen, K. F. Nielsen, T. B. Regueira, J. Nielsen, K. R. Patil and U. H. Mortensen, *Appl. Environ. Microbiol.*, 2011, **77**, 3044.
95. M. T. Nielsen, J. B. Nielsen, D. C. Anyaogu, D. K. Holm, K. F. Nielsen, T. O. Larsen and U. H. Mortensen, *PLoS One*, 2013, **8**, e72871.
96. Y. M. Chiang, C. E. Oakley, M. Ahuja, R. Entwistle, A. Schultz, S. L. Chang, C. T. Sung, C. C. Wang and B. R. Oakley, *J. Am. Chem. Soc.*, 2013, **135**, 7720.
97. M. E. Mayorga and W. E. Timberlake, *Mol. Gen. Genet.*, 1992, **235**, 205.
98. K. Williams, A. J. Szwalbe, N. P. Mulholland, J. L. Vincent, A. M. Bailey, C. L. Willis, T. J. Simpson and R. J. Cox, *Angew. Chem., Int. Ed. Engl.*, 2016, **55**, 6784.
99. S. E. Unkles, V. Valiante, D. J. Mattern and A. A. Brakhage, *Chem. Biol.*, 2014, **21**, 502.
100. T. B. Kakule, R. C. Jadulco, M. Koch, J. E. Janso, L. R. Barrows and E. W. Schmidt, *ACS Synth. Biol.*, 2014, **4**, 625.

101. K. J. McLean, M. Hans, B. Meijrink, W. B. van Scheppingen, A. Vollebregt, K. L. Tee, J. M. van der Laan, D. Leys, A. W. Munro and M. A. van den Berg, *Proc. Natl. Acad. Sci. U. S. A.*, 2015, **112**, 2847.
102. D. M. Harris, Z. A. van der Krogt, P. Klaassen, L. M. Raamsdonk, S. Hage, M. A. van den Berg, R. A. Bovenberg, J. T. Pronk and J.-M. Daran, *BMC Genomics*, 2009, **10**, 75.
103. T. Liu, Y.-M. Chiang, A. D. Somoza, B. R. Oakley and C. C. Wang, *J. Am. Chem. Soc.*, 2011, **133**, 13314.
104. T. Liu, J. F. Sanchez, Y.-M. Chiang, B. R. Oakley and C. C. Wang, *Org. Lett.*, 2014, **16**, 1676.
105. H.-H. Yeh, S.-L. Chang, Y.-M. Chiang, K. S. Bruno, B. R. Oakley, T.-K. Wu and C. C. Wang, *Org. Lett.*, 2013, **15**, 756.
106. Y. Xu, T. Zhou, S. Zhang, L.-J. Xuan, J. Zhan and I. N. Molnár, *J. Am. Chem. Soc.*, 2013, **135**, 10783.
107. K. M. Fisch, W. Bakeer, A. A. Yakasai, Z. Song, J. Pedrick, Z. Wasil, A. M. Bailey, C. M. Lazarus, T. J. Simpson and R. J. Cox, *J. Am. Chem. Soc.*, 2011, **133**, 16635.
108. T. B. Kakule, Z. Lin and E. W. Schmidt, *J. Am. Chem. Soc.*, 2014, **136**, 17882.
109. M. L. Nielsen, T. Isbrandt, L. M. Petersen, U. H. Mortensen, M. R. Andersen, J. B. Hoof and T. O. Larsen, *PLoS One*, 2016, **11**, e0161199.
110. J. W. A. van Dijk, C.-J. Guo and C. C. Wang, *Org. Lett.*, 2016, **18**, 6236.
111. S. Havn Eriksen, B. Jensen, I. Schneider, S. Kaasgaard and J. Olsen, *Appl. Microbiol. Biotechnol.*, 1994, **40**, 883.

CHAPTER 10

Engineering Enzymes for Natural Product Biosynthesis and Diversification

FREDERICK TWIGG,[a] DAVID SKYRUD,[b] JEFFREY LI[a] AND WENJU ZHANG*[a]

[a] University of California, Berkeley, Department of Chemical and Biomolecular Engineering, 2151 Berkeley Way, Berkeley, CA 94720, USA;
[b] University of California, Berkeley, Department of Chemistry, 2151 Berkeley Way, Berkeley, CA 94720, USA
*Email: wjzhang@berkeley.edu

10.1 Introduction

Natural products and their derivatives present opportunities for engineering innovation and design. Because they are relevant to the medical, agrochemical, and commodity chemical fields, breakthroughs in the optimization and alteration of biosynthetic pathways may have far-reaching benefits.[1–5] Engineering biocatalysts to make new natural products or diversify existing natural product scaffolds is benefitting from low cost DNA synthesis, efficient genetic editing in model organisms, and the rapid expansion of information available through bioinformatics.[6–8] Previously, massive chemical libraries could be synthesized and screened for a particular activity, but rational design was limited by the chemical scaffolds which were simple enough to make through total synthesis.[1,4] Since many complex scaffolds already exist in the plant and microbial world, combining

Catalysis Series No. 32
Modern Biocatalysis: Advances Towards Synthetic Biological Systems
Edited by Gavin Williams and Mélanie Hall

Published by the Royal Society of Chemistry, www.rsc.org

synthetic techniques with the power of biosynthesis has expanded the freedom with which new active compounds can be made.[9]

The pharmaceutical industry has particular interest in engineering natural products because these compounds cover distinct chemical spaces compared to medicinal synthetic compounds,[10] and because natural products are among the most important pharmaceutical compounds known to mankind. Collectively, they have played important roles in treating nearly all human health conditions, including infectious, neurological, cardiovascular, metabolic, and oncological diseases.[11] It is estimated that nearly all major classes of antibiotics and >70% of medicines for cancer treatments are natural products, their derivatives, or mimics.[1] Whereas combinatorial chemistry fails to deliver leads that form the basis for development of successful new drugs, medicinally active natural products have functional group arrays and scaffold architectures that offer advanced platforms for the optimization of compound activity profiles.[12] Engineering natural products thus offers solutions to many contemporary problems such as battling multi-drug resistant pathogens or developing new anti-cancer chemotherapy with minimal side effects.[13]

The scope of natural products produced in nature includes the broad chemical categories of isoprenoids, glycosides, alkaloids, ribosomal peptides, nonribosomal peptides (NRPs), and polyketides (PKs).[14] This chapter will focus primarily on the latter two as they are biosynthesized by probably the most profound biosynthetic factories. Both NRPs and PKs are typically produced by enzymes consisting of many catalytic domains organized into modules that act on a growing product through a process similar to an assembly line. Polyketide synthases (PKSs) can further be broken down into three types. Type I PKSs are modular systems of megasynthases that act in either linear procession, as an elongated polyketide backbone is passed down from module to module in one direction where one module incorporates one building monomer, or in an iterative fashion, as one module catalyzes several elongation cycles.[15] Type II and Type III PKSs are dissociated systems of proteins that act iteratively and interact transiently upon the growing polyketide intermediate, in a carrier-protein dependent or independent manner, respectively. Nonribosomal peptide synthetases (NRPSs) generally follow the Type I linear modular arrangement. The suite of PKS and NRPS machinery is known to be comprised of a vast number of enzymatic domains and modules, with distinct monomer block selectivity (>500 for NRPSs and PKSs) and tailoring actions, rendering tremendous diversity in the molecular scaffolds of natural products. The general enzymology and logic of these systems has been well characterized and reviewed.[15–20] The abbreviations for the domains as shown in Figure 10.1 will be used throughout the chapter. Current research is primarily focused on understanding exceptions to these basic concepts and engineering these systems for making natural product derivatives or analogs with improved or altered functionalities and biological activities.

The following three sections focus on examples regarding engineering PKSs and NRPSs for natural product overproduction and diversification in microbes. While the given examples represent only a portion of natural products and their biosynthetic assembly lines, the established strategy demonstrates cutting-edge techniques used to engineer these unique enzymes and diversify natural products. The sections, in order, are divided into recent developments of *in vivo* engineering, *in vitro* reconstitution, and directed evolution of natural product biosynthetic enzymes. The first two sections distinguish the differences in working with cellular systems to create novel compounds *versus* the use of purified proteins that can be used as isolated biocatalytic parts. The final section covers how random mutagenesis can be used to understand and improve the utility of assembly line components without a complete understanding of design rules.

10.2 *In Vivo* Engineering of NRP and PK Biosynthetic Pathways

Engineering NRPSs or PKSs *in vivo* takes advantage of the complex enzymatic machinery available naturally in microbial organisms to create compounds which may be subjected to large-scale production through microbial fermentation. Working directly in the living host allows one to alter or optimize pathways without provision of expensive protein purification and supply of cofactors or substrates that are necessary for the target pathways to function. The *in vivo* engineering efforts can be conducted through precursor directed biosynthesis (PDB) based on the promiscuity of assembly lines or coupled with mutasynthesis with minimal change on the overall molecular scaffold of natural products. *In vivo* engineering can also be conducted in a more aggressive way through biosynthetic domain/module deletion, alteration, or addition, to make a new class of molecules.[21,22] While the latter was quickly found to be more difficult than hoped, domain swapping strategies and the creation of domain chimeras based upon protein homology are still commonly used. Techniques for the design of recombinant strains that can reliably produce a new natural product derivative are valuable because of the number of equivalent synthetic steps that can be skipped. Several of the complex scaffolds available through secondary metabolism in microbes are readily fermented, but would be prohibitively costly to manufacture through organic synthesis.

10.2.1 Building Monomer Alteration

In practice, the starter unit, which becomes the first building monomer incorporated into a natural product, has been easier to engineer and manipulate than extender units. Several megasynthase systems have been induced to produce novel compounds through PDB, with the provision of alternative substrates which the initial loading module can use.[23] In a cellular host, the fed precursors will compete with naturally present substrates if the production of native substrates cannot be easily abolished, but this technique still allows the installation of interesting functionalities to expand

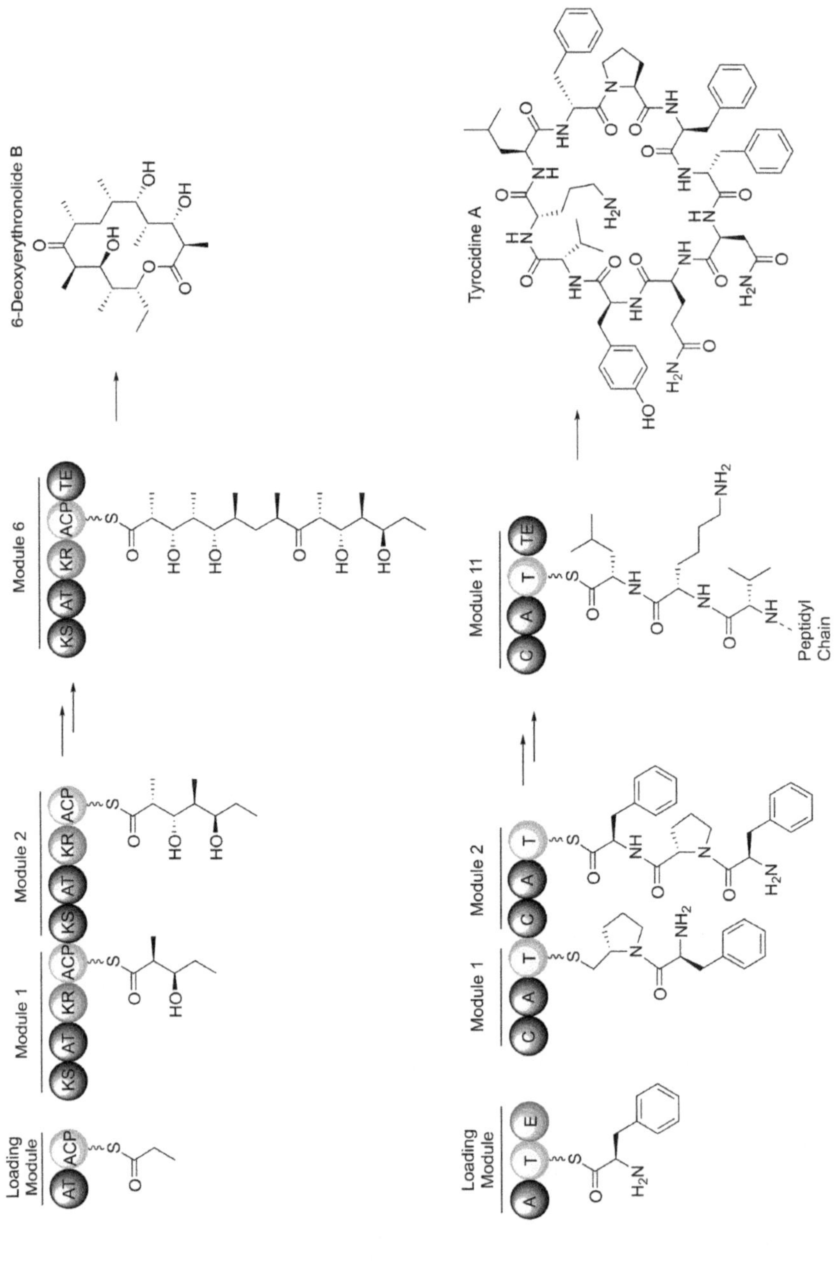
Loading Module
Module 1
Module 2
Module 6
6-Deoxyerythronolide B
Loading Module
Module 1
Module 2
Module 11
Peptidyl Chain
Tyrocidine A

structural diversity of natural products.[24] To date, nearly all different chemical groups including the naturally rare moieties such as alkyne, azido, and halogens have all been incorporated in natural product scaffolds through PDB.[21] For example, the anticancer drug epothilone, produced natively in the myxobacterium *Sorangium cellulosum*, is biosynthesized by a PKS–NRPS hybrid pathway with the encoding genes spanning over 50 kb in length. Initial attempts at reconstituting the pathway in *E. coli* only expressed the back half of the system, EpoDEF.[25] A synthetic version of the acyl intermediate that would normally be fed to EpoD on the assembly line was supplied as an acyl-*N*-acetylcysteamine thioester (SNAC) precursor to the *E. coli* culture, and the expected final product was observed. After subsequent heterologous expression of the whole PKS–NRPS hybrid pathway in *E. coli*, simpler SNAC substrates were used for obtaining the epothilone product as well as installing novel starter units to produce epothilone analogs.[26] Notably, acyl-SNACs are mimics of thio-activated biosynthetic intermediates and cell permeable, allowing the installation of variable building monomers in the thio-templated PKS and NRPS assembly lines (Figure 10.2). Borrelidin is another medically relevant macrolide where simple feeding of succinate analogs resulted in novel derivatives of borrelidin, some of which were more active than the native product.[27] More recently, alkyne-containing starter units have been incorporated into both the erythromycin pathway and the avermectin pathways through PDB approaches, which potentially

Figure 10.1 Model schematics for a type I PKS and an NRPS are shown through abbreviated arrangements of the 6-deoxyerthronolide B and tyrocidine A pathways. The most common PKS domains and their abbreviations are as follows: KS = ketosynthase, AT = acyltransferase, ACP = acyl carrier protein, DH = dehydratase, ER = enoylreductase, KR = ketoreductase, MT = methyltransferase, TE = thioesterase. For an NRPS: C = condensation, A = adenylation, T = thiolation, E = epimerization, MT = methyltransferase, TE = thioesterase. A type I PKS module must at least have a KS, AT, and ACP domain (though the KS may be absent or inactive in the first loading module of a PKS). A typical catalytic cycle begins with the translocation of the elongation product from the upstream module's ACP to the current module's KS domain. The AT domain selects the extender unit, and transfers it to the current module's ACP. The KS domain catalyzes the decarboxylative Claisen condensation between the intermediate on the KS domain and the extender unit on the ACP. The resulting elongation product, bound to the ACP, is then translocated to the downstream module's KS domain. Several auxiliary domains are commonly found in PKS modules such as KRs, DHs, and ERs which all contribute to successive reductions of the β-ketoacyl intermediate to β-hydroxyacyl, α,β-enoyl, or fully reduced forms, respectively. In NRPSs, the C, A, and T domains are analogous to a PKS's KS, AT, and ACP domains, however C domains catalyze peptide bond formation between peptidyl thioesters on the T domains, and the A domains form aminoacyl-AMP substrates through ATP-dependent activation of an amino acid carboxylate group. In both PKSs and NRPSs, the terminal TE domain frequently catalyzes intramolecular cyclization, or hydrolytically releases a free acid.

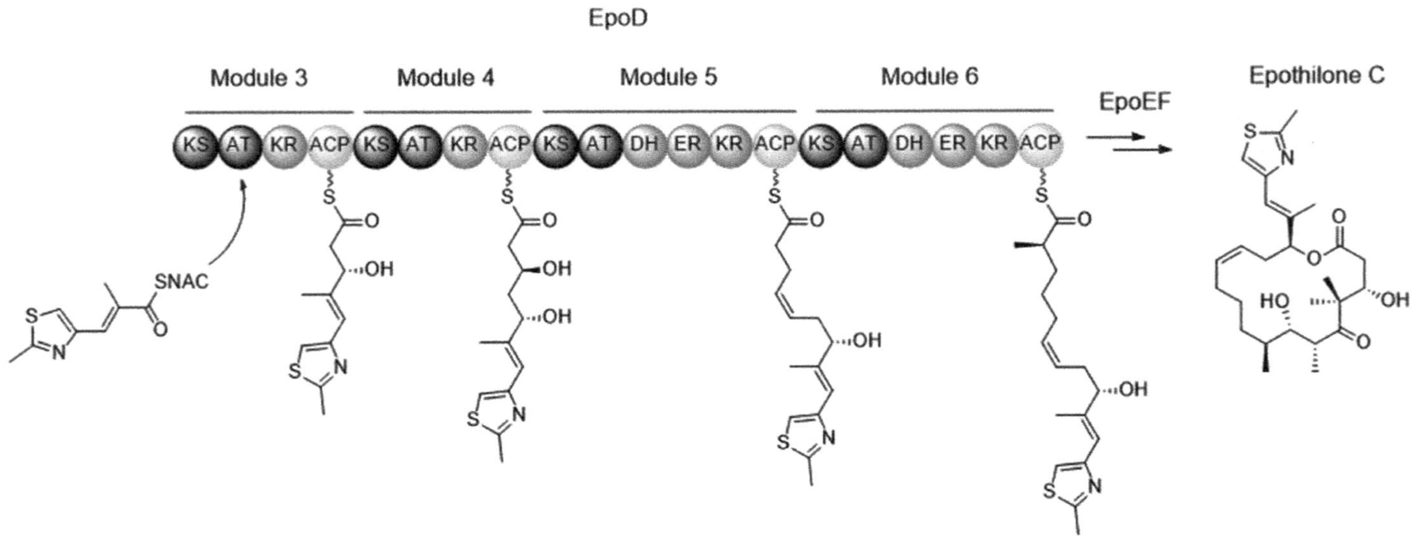

Figure 10.2 Initial reconstitution of the epothilone hybrid NRPS–PKS pathway only included the PKS portion, EpoDEF in *E. coli*. The use of acyl-SNAC substrate analogues, based upon the structure of the intermediate leaving EpoC, allows the jump starting of a truncated system. Acyl-SNAC substrates can also be accepted as non-native substrates by some engineered AT domains.

allow further chemical derivatization, visualization, and mode-of-action study of these compounds through click chemistry.[24,28–30] It is notable that in many cases, the promiscuity of the loading module allows the incorporation of alternative starter units. When such substrate tolerance is absent, engineering of the loading module to achieve the desired substrate specificity is needed. Nonetheless, PDB likely provides low-hanging fruit when cheap starter unit reagents can be supplemented into cell cultures to create more bioactive derivatives of the native natural product.

In addition to starter unit diversification, incorporation of alternative extender units could further expand access to natural product derivatives.[31] While the building monomer selection in NRPS is controlled by the A domain which often has stringent substrate specificity, recent discoveries of promiscuous AT domains or engineering of AT domains, the gate-keeping enzymes in building monomer selection in PKSs, could potentially allow the incorporation of novel extender units beyond the typical malonyl-CoA or methylmalonyl-CoA for PKSs. These engineering efforts need the companion of atypical PKS extender unit supplies, which are mainly achieved using the malonyl-CoA synthetase (such as engineered MatB) or crotonyl-CoA carboxylase/reductase (CCR) machinery (Figure 10.3).[32–35] For example, the employment of CCRs that catalyze the formation of α-substituted acylmalonyl-CoAs from α,β-unsaturated acyl-CoAs through reductive carboxylation has resulted in quite a number of novel extender units.[36] AntE and its homologs, the CCR enzymes in the antimycin-type depsipeptide biosynthetic pathway, are examples that have shown relaxed substrate specificity to produce malonyl-CoAs with various alkyl and aromatic substitutions.[37–41] Further structure-based mutagenesis of AntE allowed biosynthesis of heterocycle- and substituted arene-bearing antimycins.[38] Additional CCR enzymes, such as CinF and SalG involved in cinnabaramide and salinosporamide biosynthesis, have also been demonstrated to accept a variety of native substrates such as 2-octenoyl-CoA and chlorocrotonyl-CoA, respectively.[42] Recent systematic screening of several CCR homologues with a diverse library of potential substrates coupled with protein engineering through site-directed mutagenesis further defined the molecular basis of CCR reactivity, allowing for the prediction and manipulation of these enzymes for generating atypical extender units for PKSs.[43] Additionally, there is now evidence from stambomycin biosynthesis that atypical extender units are also accessed through direct carboxylation of acyl-CoA thioesters by acyl-CoA carboxylases, an unprecedented CCR-independent pathway.[44]

10.2.2 Chimeric Domains and Domain Swapping

One of the ultimate goals of PKS/NRPS engineering is to construct *de novo* pathways by assembling enzymes from a catalog of domains and modules capable of performing the spectrum of catalytic chemistry observed in nature's assembly lines. This has been attempted in the past three decades, but with very limited success. Early attempts to swap domains/modules in

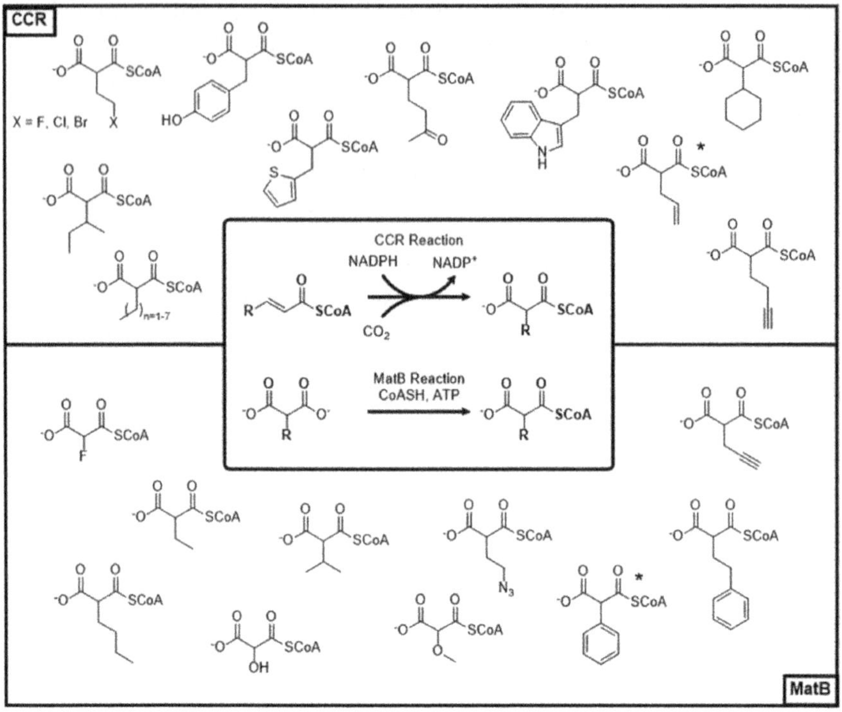

Figure 10.3 Generation of malonyl-CoA derivatives as PKS extender units *via* the MatB and CCR reactions. Wild type and mutant MatB are able to activate a broad library of malonate analogs through ATP-dependent ligation directly to CoA. The family of CCR enzymes catalyzes the reductive carboxylation of CoA linked α,β-unsaturated acyl-CoAs *via* utilization of one reducing equivalent of NADPH to afford malonyl-CoA analogs. Structures marked with * have been produced by both MatB and CCR reactions.

megasynthases were often met with no production of the expected products, or severely decreased titers. In some successful cases, it was found that C domains of NRPSs had some roles in proofreading the acceptor substrate, and the entire module (C–A–T) swapping had a higher success rate in incorporating the new amino acid into the elongation product than simply swapping at the domain level (such as A–T).[45] Daptomycin is another well-studied antibiotic of which the NRPS biosynthetic machinery has been extensively engineered and many valuable daptomycin derivatives have been made, although the majority of the mutant yields rarely exceeded 30% of the native daptomycin yield.[46–48] A 2016 report showed that protein–protein interactions, not substrate recognition, dominate the turnover of chimeric assembly line PKSs.[49] Unfortunately, the current understanding of protein–protein interactions among various catalytic domains in megasynthases is far from completion, even after several structures of megasynthases have been obtained.[50–54]

Despite the challenging nature of megasynthase engineering, recent technical development in easy manipulation of chimeric PKS/NRPS enzymes has led to some exciting results. For example, a versatile method for generating and identifying functional chimeric PKS enzymes for synthesizing custom macrolactones and macrolides was reported based on PKS evolution by homologous recombination. Pikromycin and erythromycin are both well-studied macrolides whose biosynthetic PKS machinery has been engineered to make new natural products.[55–59] Using a yeast host for its DNA repair capabilities, hybrids of the pikromycin and erythromycin PKS were made through homologous recombination.[60] After transformation of the active chimeric PKSs into *Streptomyces venezuelae*, *in vivo* production of engineered macrolactones, including a 3-dehydro-10-deoxymethynolide derivative of pikromycin was observed. While most chimeric megasynthases are still limited by poor overall titers compared to the wild type, this method of homologous recombination naturally retains highly conserved catalytic regions, which leads to a more reliable generation of functional PKS chimeras. Another successful display of full assembly line manipulation was the *in vivo* reprogramming of the aureothin type I modular PKS to biosynthesize luteoreticulin, a distinct natural product not natively produced by the aureothin producing strain.[61] All of the engineering efforts were performed using only the domains and tailoring enzymes already available in the aureothin pathway. For the design of the mutant, the first module was left intact since it produced the nitro functionality of the natural product. The second module was mostly excised, leaving only the KS and AT domains which were found to function with the ACP domain of the third module. The original third and fourth modules were recombined into a chimera which utilized the KS and ACP of the fourth module, but with the AT of the third module. It is impressive that this extensive cutting and pasting of domains to create a new distinct modular arrangement succeeded in producing luteoreticulin, though at less than one milligram per litre of culture. This work showed the possibility of morphing a given modular PKS to produce complex polyketides with different backbones.

10.2.3 Conclusion on *In Vivo* Engineering

The advantages of *in vivo* engineering include cheap costs for compound production, accessible genetic engineering tools such as CRISPR/Cas, TAR, and λ Red/ET recombineering in the common natural product-producing strains, and direct observation of metabolic changes due to genetic manipulations.[62–72] Furthermore, some pathways have proven very difficult to reconstitute *in vitro*, whereas *in vivo* experiments permit pathway manipulation without fully understanding the exact biotransformation of each step. However, *in vivo* approaches are still limited by the abundance of variables present in cell growth processes, and the incomplete understanding of NRPS and PKS logic. Characterizing pathways is best done through *in vitro* studies, where available substrates, cofactors, and particular enzymes can be tightly controlled, and substrates and products of each enzymatic step can be checked. Large modular megasynthases, however, are sometimes difficult to purify intact, such that *in vivo* approaches are still used. Additionally, where rational models fail, directed evolution provides a more brute force, but effective, solution to creating efficient mutants. Both of these approaches with examples will be discussed in the subsequent sections.

10.3 *In Vitro* Reconstitution and Engineering of NRP and PK Biosynthetic Pathways

In vitro analysis of proteins allows for enzyme-specific characterization, eliminates competitive side reactions and cellular regulation, determines necessary and minimal cofactors, and is the most direct route to incorporation of non-natural substrates. In the case of NRPS and PKS assembly line megasynthases, *in vitro* characterization provides insight into substrate specificity and promiscuity of each module or domain as well as the order of tailoring reactions and gate-keeping mechanisms.[73,74] When certain biosynthetic pathways are naturally promiscuous and produce a suite of compounds *in vivo* which are difficult to be separated during purification, *in vitro* biosynthesis can easily overcome this challenge by limiting the substrate supply. Further application also permits theoretically simple derivatization of natural products by supplementing non-natural substrates, omitting tailoring enzymes, or combining machinery from separate pathways.[75] Site-specific mutagenesis of enzymes can further lead to more relaxed substrate specificity or the elimination of gate-keeping residues for further derivatization.[76] However, there are obstacles to *in vitro* characterization and engineering of these enzymes. Most notably, the massive size of NRPSs and PKSs make expression and purification challenging. Multimodular megasynthases are often purified as dissociated modules or domains which are then recombined *in vitro* with engineered interfaces to restore pathway activity, albeit often at a lower efficiency.[77,78] Similar to *in vivo* approaches, knowledge of the mechanisms of protein–protein interactions and substrate hand off is incomplete and this limits the success of these recombined pathways.[50] When the natural products of interest are complex and require multiple enzymes to produce specific building monomers, assemble molecular scaffolds, and further modify post-assembly line, it is always difficult to realize the total biosynthesis of the fully modified compounds *in vitro*. Another obvious limitation of *in vitro* approaches lies in the difficulty in scaling-up and obtaining enough biosynthetic intermediates or products for detailed characterization. Nonetheless, more as a characterization tool, *in vitro* reconstitution and engineering have unique strengths in answering many mechanistic questions. The following section highlights the current state of the art of *in vitro* NRPS, PKS and NRPS–PKS hybrid pathway engineering and how this inspires the future of NRP and PK exploration and derivatization.

10.3.1 Nonnative Building Monomer Incorporation through *In Vitro* Biosynthesis

The most accessible engineering method in *in vitro* assembly line synthesis is PDB, similar to the given *in vivo* examples, but with more direct control over substrate availability.[21,75,76] PDB allows exploitation of the natural promiscuity of gate-keeping domains of both PKSs and NRPSs. The notable advantage of *in vitro* PDB is that the desired nonnative substrates do not need to compete with the native substrates in most cases.[75] However, this type of complete reconstitution with nonnative substrates requires a pathway's natural ability to accept a wide range of substrates, and that the downstream enzymes are tolerant enough. One successful example is the biosynthesis of analogs of pacidamycins, a group of peptidyl nucleoside antibiotics that are synthesized from a highly dissociated cluster of NRPS modules and domains.[79–83] *In vitro* total reconstitution identified a minimum set of nine NRPS enzymes that are required for generating the pacidamycin scaffold from amino acid and nucleoside monomers, and further generated several pacidamycin analogs based on the substrate promiscuity of this enzymatic assembly line and different precursor feeding. In another example, a minimum set of NRPS–PKS enzymes needed for generating the antimycin dilactone scaffold were identified by *in vitro* total biosynthesis.[37] Several antimycin analogs were further produced *in vitro* based on the substrate promiscuity of this NRPS–PKS machinery, in particular the tolerance towards various starter units and PKS extender units (Figure 10.4).[37] These examples show the benefits of natural promiscuity and how this can lead to a wide range of natural product analogs if the entire pathway is reconstituted. Other notable successes in *in vitro* derivatization involve the mutation of single or multiple residues within the core scaffolds to expand beyond the naturally activated substrates and provide a basis for rational

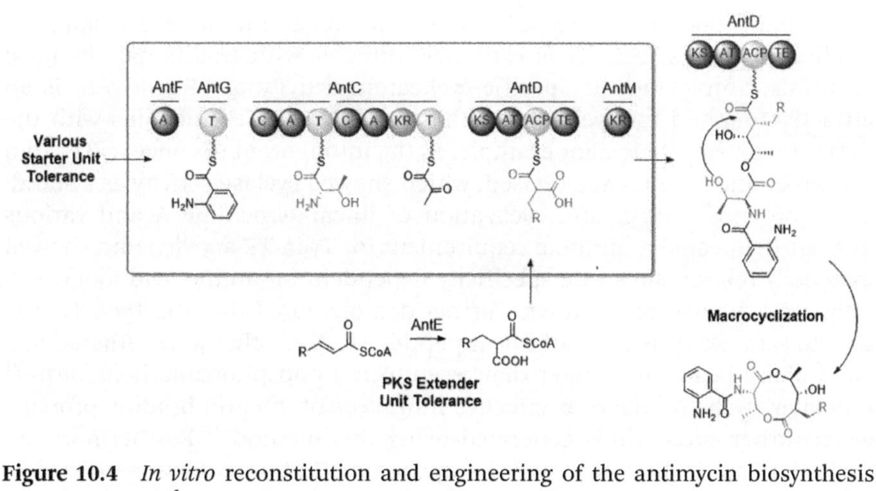

Figure 10.4 *In vitro* reconstitution and engineering of the antimycin biosynthesis pathway.

design of new natural products.[21,84–86] However, due to the difficulty of reconstituting a complete pathway, most of this site-directed mutagenesis has been conducted on single enzymes and verified *via* biochemical assays rather than the enzymatic total synthesis of natural product analogs.

Furthermore, instead of directly providing the building monomers, an emerging method called engineering precursor supply (EPS) seeks to engineer the pathways that provide the building monomers for NRPS and PKS scaffolds from the readily available precursors.[76] A successful target of EPS has been the family of CCR enzymes discussed in the previous section. *In vitro* work with these enzymes is more accessible due to their smaller protein size and the availability of diverse α,β-unsaturated acyl-CoA substrates. Another use of EPS for generating atypical building monomers for PKSs is based on the engineered malonyl-CoA synthetases (such as MatB) with broad substrate specificity. These enzymes directly activate malonate derivatives using ATP and ligate them to CoA.[87] Protein engineering has led to the construction of promiscuous malonyl-CoA synthetase variants that can be used to synthesize a broad range of acylmalonyl-CoA extender units, several of which contain handles for chemoselective ligation and are not found in natural biosynthetic systems.[33,88] Harnessing the power of these simple and promiscuous enzymes can be the easiest way to supply the necessary building monomers to create a vast library of PKs or PK–NRP hybrids.

10.3.2 *In Vitro* Biosynthesis Using Promiscuous Off-loading Enzymes for Macrocyclization

Many of the pharmacologically active PKs and NRPs are constrained in a cyclic structure arising from a C-terminal TE that catalyzes macrocyclization upon release from the final module in the assembly line. This macrocyclization is often specific and effective due to the TE's ability to promote the attack by a specific nucleophilic moiety to the thioester when there are often many potential nucleophiles on the linear PK or NRP chain.[89,90] Replicating this selectivity is extremely difficult with traditional chemical synthesis. Employing the specific cyclization activity of TE's *in vitro* is an attractive method for cyclizing synthesized linear PK/NRP chains with the desired regiospecificity. For example, as the initial proof of concept, TE from the tyrocidine A NRPS was excised, which showed cyclase activity as a standalone protein.[91] Successful cyclization of linear tyrocidine A and various analogs delineated a minimal requirement for TycA TE activity, and showed extremely relaxed substrate specificity dependent on amino acid identity at only two key positions.[91] It was further demonstrated that the TycA TE was able to catalyze cyclization of linear peptide scaffolds that were tethered to a solid phase bead *via* a linker that resembled a phosphopantetheine arm,[91] and later, cyclic products as effective inhibitors of integrin binding proteins were further successfully generated using this method.[92] Furthermore, an excised TE in combination with its upstream NRPS module was utilized to convert a chemically synthesized SNAC precursor into a cyclized anti-cancer agent of the cryptophycin family.[93] This chemoenzymatic approach was used to probe the activity of the final NRPS module in the pathway and create the cyclic product in its active form. Employing macrocyclization as an engineering tool is particularly appealing due to the small size and relative promiscuity of the TE domain. Coupled with solid phase peptide synthesis, this method may easily generate targeted cyclic peptides on a decent scale.[94]

10.3.3 Domain Swapping *In Vitro*

Similar to *in vivo* study, domain/module deletion, swapping, and addition can be performed *in vitro*, with the advantage of easy assessment of the engineering outcome with minimal interference from other cellular components. For example, in the case of DEBS reconstitution, the clever engineering of docking domains allowed for the *in vitro* reconstitution of DEBS as well as the synthesis of notable derivatives of 6-deoxyerythronolide B for the first time.[78] In particular, DEBS2 and DEBS3 (each has two modules) were purified in their intact form, while DEBS1 (loading and the initial two modules) was truncated into three modules. Each module had the docking domains of the 1–2 and 2–3 interaction grafted onto it to ensure substrate channeling. Although this reconstituted system with engineered docking domains was successful in providing a baseline for future work, the applicability of domain swapping is still not generalizable, as discussed above. Mechanism-based crosslinking is an emerging tool from the field of chemical biology for quantifying the interaction between an ACP and a KS using an ACP linked tether that covalently links the thiol of the KS at a rate that correlates to the affinity for a given ACP and KS domain.[95–97] Experimental approaches such as this one will aid in future determination of protein–protein interactions for applications in efficient modular assembly line engineering.

10.3.4 Conclusion on *In Vitro* Reconstitution and Engineering

In vitro engineering has been successful in delineating and derivatizing natural products despite the inherent difficulty arising from megasynthase pathways and poorly understood protein dynamics. Notable successes have supplied non-natural precursors to naturally promiscuous PKS and NRPS modules as well as mutated many gate-keeping domains to accept a broader range of substrates. Engineering *via* precursor supply has increased the number of PKS substrates we can generate, and malonyl-CoA and amino acid analogs have been used to insert separate functionalities into natural product backbones. In an effort to avoid the hindrances of *in vitro* reconstitution, TEs have been excised and used to probe libraries of peptide chains and late stage intermediates in order to successfully synthesize important macrocycles. As knowledge of these pathways, in particular protein–protein interactions, increases with more complex pathways reconstituted *in vitro*, the success rate of megasynthase engineering is expected to increase, leading to more efficient production of natural product analogs either *in vitro* or *in vivo*.

10.4 Directed Evolution of NRPSs and PKSs

The knowledge gap in NRPS and PKS assembly lines remains one of the major challenges in rational engineering of such biosynthetic machinery. However, directed evolution may be a viable method for fine-tuning a selected NRPS or PKS. Directed evolution methods target genetic sequences for mutagenesis and then use selection or screening to assess mutants in terms of desired assembly line or product properties. Over time, both library generation and screening methods have grown more sophisticated thanks to insights from phylogenetics, structural models, and improved biochemical understanding of NRPS and PKS domain functions. The latter have seen greater representation among NRPS studies; most notably, *in silico* work with NRPS structures has introduced the concept of a specificity-coding region in the "gate-keeper" A domains.[98,99] This has prompted a slew of studies generating focused libraries in NRPS A domains to alter their substrate specificity. Presently, the rules for PKS substrate prediction are not as well understood, and at least one study has suggested that the substrate recognition function may not be conveniently limited to the AT domains.[100] This may be changing with the availability of several structures for modular[101,102] and *trans*-ATs,[103] and several AT substrate-prediction studies for PKS extender units such as malonyl-CoA, methylmalonyl-CoA, ethylmalonyl-CoA, and methoxymalonyl-CoA.[104,105] The list will grow as more ATs are sequenced and functionally characterized.[106] Aside from adenylating domains, KR domains have been targeted by directed evolution as well. The stereochemistry determined by KR domains has been well studied using mutant libraries. Systematic screening has identified residues key to determining or even altering the expected chirality.[107,108] Increasing understanding of assembly line enzyme fundamentals has driven directed evolution efforts, which have been leveraged to improve PKS and NRPS in a range of metrics such as productivity,[109–112] substrate specificity,[113–116] and stability.[117] These developments have also moved the field towards rationally designed mutant libraries, which are less bulky than truly random mutant libraries and tend to have a higher proportion of active mutants.

10.4.1 Mutagenesis Strategies for Constructing NRPS and PKS Mutant Libraries

A variety of mutagenesis techniques have been applied to PKS and NRPS systems. In general, these approaches fall into two categories: random mutagenesis casts a wide net by targeting a whole sequence with point mutations, while rational mutagenesis can limit these to a handful of codons and avoid silent mutations if structural or bioinformatic analyses are available to guide experimental design. A classic approach for random mutagenesis is to amplify a target domain using error-prone polymerase chain reaction (PCR).[118,119] The resulting cassettes can be ligated or recombined into vector backbones to complete a mutant library that can be transformed into a host for expression. This method can easily randomize a region at a semi-tunable range of point mutation frequencies before assembly into a vector by other methods.[65] It is often used if the function being screened for is not correlated with particular amino acid residues such as those found in a binding pocket. There are several reported examples of the use of error-prone PCR to engineer NRPS A domains[111,120,121] and T-domains,[122] and a type III PKS.[117] One example of *in vivo* random mutagenesis was reported in the engineering of an NRPS/PKS hybrid-bearing vector into an *E. coli* mutator strain to optimize production titer.[112] This allowed introduction of mutations over a much longer DNA sequence than possible with PCR-based methods. After screening, the successful hits were found to contain mutations in the vector backbone rather than on the biosynthetic gene, demonstrating the benefit of targeting more positions for mutagenesis.

Rational mutagenesis can target selected residues known to be related to the desired protein function. This way, sequence diversity at that site can be fully explored in greater depth than by random mutagenesis. This is especially true for codon–codon mutations that need multiple adjacent nucleotide substitutions and therefore are uncommon results of error-prone PCR. In a typical saturation mutagenesis experiment, individual residues are explored separately and then successful mutations are combined additively for subsequent iterations of screening. Saturation libraries can be generated by PCR-based methods using synthesized DNA oligomers containing the desired sequences. The number of required oligomers can also be decreased by the use of degenerate nucleotides.[123,124] Saturation mutagenesis has been a powerful tool for diversifying enzyme binding substrates or activity. For example, the *in vivo* site-directed mutagenesis method has been used to engineer an NRPS in *E. coli*.[116] This work used a homologous recombination based method, which used a toxic gene cassette to first mark the target sequence, and then repair it with linear DNA containing the designed mutations. The caveat was unwanted recombination that can occur in the presence of the repeating homologous sequences frequently found in modular megasynthases. This limited the ratio of correct mutants to an average of 15% of the pool, limiting efficiency and providing a challenge for recovering or screening more in-depth saturation mutagenesis libraries. Synthetic shuffling is another method of creating libraries using the known diversity found in other sequenced homologs of the gene of interest.[125] Bioinformatic analyses can provide insights into which residues might tolerate mutation to other residues, generating a library with a higher active fraction than error-prone PCR or saturation mutagenesis. One reported use of synthetic shuffling resulted in a PKS library of 2×10^{11} mutants for high throughput screening, with active fraction between 0.1% and 20%.[109]

10.4.2 Screening of NRPS and PKS Mutant Libraries

A great variety of screening methods have been reported to recover optimized assembly line enzymes from mutant libraries. These can be broadly categorized into individual enzyme activity screening which optimizes one biotransformation step, or final natural product-based screening relying on the physicochemical properties or biological activity of natural products. Individual enzyme activity screening offers a broader toolkit, albeit with challenges in translating the benefits at the individual enzyme level to the full assembly line synthase. Cleverly designed natural product-based screening methods can access the overall efficiency of the engineered assembly line, but often rely on unique compound properties that can be difficult to come across.

10.4.2.1 Screening by Individual Enzyme Activity

There are many established methods in the realm of protein evolution that can screen the activity of individual enzymes.[126] Many of them have been successfully applied to engineering individual catalytic domains involved in natural product biosynthesis. For example, a MatB saturation mutagenesis library was screened for activity for making unusual PKS extender units.[33] A particular malonyl-CoA synthetase mutant was isolated by screening cell lysates in plate reader format in the presence of methylmalonate, ethylmalonate, allylmalonate, and free coenzyme A (CoASH). Ellman's reagent was added to provide a quantifiable screen for free CoASH, with enzyme activity reducing the sample readout at 412 nm. A total of 300 mutants was generated using structure-guided saturation mutagenesis and screened, with one demonstrating a 240-fold shift in specificity.

As a gate-keeping enzyme, the A domain of NRPS has been extensively subjected to engineering efforts in attempts to modify its substrate specificity. For example, a microtiter-plate-based pyrophosphate exchange assay was used to screen for A domain mutants with altered substrate specificity, and a 10^5-fold change in substrate specificity was achieved in two cases.[113,115] In addition, substrate affinity screening has been used to engineer A domains as well. One group used yeast cell-surface display to screen DhbE A domain mutants that could recognize substrates other than the native 2,3-dihydroxybenzoic acid in bacillibactin biosynthesis.[114] A chemical active-site probe that contains a substrate moiety was synthesized that provided a handle for attachment to either fluorophores or solid supports, allowing fast functional mutant identification. Several rounds of yeast cell sorting yielded clones that could bind nonnative substrates, 3-hydroxybenzoic acid and 2-aminobenzoic acid, with 11- and 6-fold increase in catalytic efficiency, and accompanying 3- and 33-fold decrease in native substrate specificity. Together, these accounted for a switch in substrate specificity of up to 200-fold. Despite the success of these efforts, challenges remain in relying on the yeast display machinery (which limits the technique to 10–60 kDa proteins), and translating the results from substrate recognition to actual incorporation. Notably, the directed evolution regimen optimized DhbE to bind new substrates, but none of the mutants could catalyze the transfer of substrate to an aryl carrier protein. However, the authors were able to restore transferase activity with a structure-guided reversion of one of the active site mutations. In a recent complementary study for mutating A domains, cell lysates containing protein mutants were screened by their binding strength to a similar active site probe.[127] The probe enabled a quantitative enzyme-linked immunosorbent assay by fixing A domain mutants to the bottom of a microtiter plate for subsequent detection by fusion antibodies. This design led to successful engineering of several A domains (GrsA, TycA, TycB1, AusA2, and EntE) to switch substrate specificity from 2,3-dihydroxybenzoic acid to salicylic acid. The resulting mutants exhibited up to a 26-fold switch in substrate specificity.

10.4.2.2 Screening by Physicochemical Properties of Natural Products

The natural products content of a mutant library of organisms or extracts may be detected and quantified based on their physicochemical properties, such as distinct colorimetric/fluorescent signals or mass signals, using the corresponding analytical equipment. Compared to colorimetric/fluorescent signals, which are only limited to certain natural products, mass spectrometry (MS)-based analytical chemistry is a superior method for detecting, quantifying and even partially characterizing nearly all PKs/NRPs with high accuracy and sensitivity, albeit with a high demand in instrumentation. These screening methods have been successfully implemented for directed evolution of both NRPS and PKS systems.

One example of colorimetric screening in an NRPS system described the engineering of single-module BspA, which produces a pigment indigoidine in *Streptomyces lavendulae*.[122] The visibility of the pigment in *E. coli* colonies allowed visual screening of functioning NRPS mutants, allowing for rapid engineering of incompatible heterologous T domains. These T domains were diversified by error-prone PCR and clones were screened for the presence of blue indigoidine in plated *E. coli* colonies. The study also succinctly demonstrated the utility of BspA in a 4′-phosphopantetheinyl transferase (PPTase) deficient background to screen metagenomic libraries for efficient PPTases, the enzyme required to catalyze the post-translational attachment of a 4′-phosphopantetheine group to a conserved serine residue in carrier protein domains of PKSs and NRPSs. Despite a few successful examples, the natural product-based colorimetric/fluorescent screening method is not widely used since most products lack such signals. A recent method in *de novo* biosynthesizing alkyne-tagged natural products[128] coupled with fluorogenic transformation offers a promising way for generalizing the detectability of polyketide and peptide products by fluorescence screening. *De novo* biosynthesis of these tagged natural products without feeding of alkyne precursors is necessary to eliminate the background of fluorescence signals, and utilization of an azido fluorogenic probe in the subsequent click reaction instead of an azido fluorescence probe is also critical for signal generation. As proof-of-principle, a platform for quantifying extracellular alkyne-tagged metabolites was built which was then used to evolve the activity of JamB, a membrane-bound bifunctional desaturase/acetylenase responsible for introducing a terminal alkyne moiety based on the titer of extracellular alkyne-tagged metabolites.[129] This led to the isolation of one JamB variant with ~20-fold increased activity in *E. coli*.

Mass spectrometry (MS) is the most sensitive and accurate screening method used for detecting natural products. This method yields both quantitative and structural data and has the lowest limit of detection among all methods discussed here. One example of MS screening was carried out in *Pantoea agglomerans*,[121] which natively produces andrimid, an antibiotic produced from the Adm genes encoding a multimodular NRPS. The native AdmK A domain was mutated at three residues in the active site through saturation mutagenesis, and subsequent MS screening of over 1.4×10^4 clones yielded producers of three new andrimid derivatives which were active inhibitors of *Staphylococcus aureus* growth. More excitingly, recent advances in MS technology allow a powerful and generalizable screening of microbial cultures, with minimal processing. Extracts can be pre-purified with solid phase extraction before fractionation by liquid chromatography and analysis by MS, with processing time as low as five to seven seconds per sample.[130,131] Improvements to *in situ* MS have been made in rapid progression towards on-the-plate screening.[132] Another key technology has been desorption electrospray ionization (DESI), which allows MS to be carried out under ambient conditions by spraying a surface with a stream of ionized solvent, and then transferring charged analyte molecules away to a mass spectrometer. Capillary tubes allow a nano-DESI method for direct sampling from smaller objects such as microbial colonies on a plate, and liquid-microjunction surface sampling probes have improved that design to increase flow rates while easily unclogging intake capillaries.[133] Finally, moving the surface while it is beneath the MS probe during sampling can allow spatial compound imaging, which can be intuitively presented in chemical heat maps.[134–136] Although only select labs currently have access to these new methods, the foreseen wide adoption of them will greatly ease natural product analysis, which will further power the engineering of natural product biosynthetic machinery (Figure 10.5).

10.4.2.3 Screening by Biological Activity of Natural Products

Polyketides and peptides frequently have biological activity, allowing unique screening opportunities for PKS/NRPS mutants with higher overall bioactivity in terms of higher titer of natural products, or stronger biological

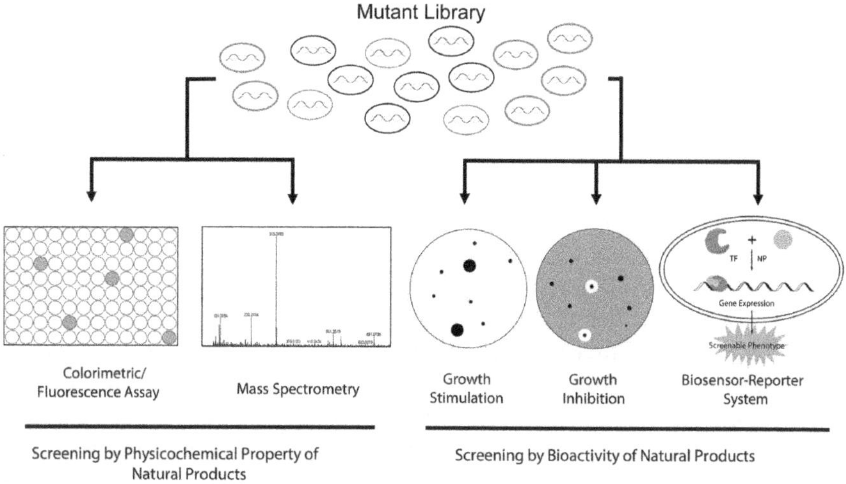

Figure 10.5 Overview of directed evolution screening methods. TF: Transcription Factor. NP: Natural Product.

activity of natural product analogs, or both. Traditionally, the growth stimulating or inhibitory activities of natural products have been explored. For example, the nutrient-acquisition function of siderophores has been used to select the optimized NRPSs involved in siderophore biosynthesis in iron-deficient media.[120,137] The antibiotic activities of andrimid towards *E. coli imp*, erythromycin towards *Bacillus subtilis*, and lovastatin towards *Neurospora crassa* were all utilized, typically in the formation of an inhibition zone (halo) on agar plates during co-culture, to screen for the best producers with evolved biosynthetic components.[111,112,120]

In addition, biosensor–reporter systems responding to the level of natural products have also been used for screening. These biological sensors typically consist of transcription factors or riboswitches that can monitor intracellular natural product signals and respond by regulating the expression of a reporter gene.[138,139] Several natural transcription factor-based sensing systems have been successfully utilized for screening bacterial libraries for mutants with improved productivity.[140–145] For NRP and PK products with no known biosensors, directed evolution can be further used to engineer a sensing transcription factor for the metabolite of interest. In one instance, an AraC regulator was engineered to bind a simple polyketide, triacetic acid lactone (TAL), rather than the wild-type arabinose.[110] The resulting AraC variant enabled the directed evolution of a type III PKS from *Gerbera hybrida* that catalyzes the biosynthesis of TAL through coupling to a β-galactosidase reporting gene to allow visual screening of the PKS mutant libraries when plated on media supplemented with 5-bromo-4-chloro-3-indolyl β-D-galactopyranoside. Two rounds of mutagenesis and screening yielded a 20-fold increase in TAL production by the improved PKS. This work demonstrates the potential of AraC and other biosensors to be customized for sensing natural products of interest.

10.4.3 Conclusions on Directed Evolution of NRPSs and PKSs

To date, directed evolution of NRPS and PKS has been used to probe the chemistry of NRPS/PKS assembly lines such as domain interactions, to improve assembly efficiency (especially chimeric assembly lines for natural product overproduction), and to change building monomer selection to produce natural product analogs. A variety of tools are available *in vitro* as well as *in vivo* to generate the diversity needed to rapidly evolve enzymes, and clever screening methods have also been developed over the years. Thanks to the growing structural and sequence data of NRPS/PKS and knowledge of their enzymology, libraries containing more targeted mutations and fewer inactive mutants can be constructed, increasing screening hit rates. Even when random mutagenesis targeting a whole sequence is performed, several successful examples of enzyme evolution have shown that modest library sizes (10^3 to 10^4) and just a few rounds of screening can yield significantly improved PKS/NRPS mutants, illustrating the high promise of this approach. Since you may only get what you screen for in directed evolution, caution is needed in selecting the appropriate screening method. In this regard, a general natural product-based, *in situ*, sensitive, quantitative, reliable, and high-throughput screening method is preferred, although one method rarely has all of these characteristics. Looking forward, natural product tagging or MS-based technique may provide a possible solution, and their applications in engineering NRPS and PKS are yet to be seen.

References

1. D. J. Newman and G. M. Cragg, *J. Nat. Prod.*, 2012, **75**, 311.
2. S. Saxena, *Allelopathy J.*, 2014, **33**, 1.
3. F. E. Dayan, D. K. Owens and S. O. Duke, *Pest Manag. Sci.*, 2012, **68**, 519.
4. L. Katz and R. H. Baltz, *J. Ind. Microbiol. Biotechnol.*, 2016, **43**, 155.
5. U. T. Bornscheuer, G. W. Huisman, R. J. Kazlauskas, S. Lutz, J. C. Moore and K. Robins, *Nature*, 2012, **485**, 185.
6. C. Khosla, *Nat. Chem. Biol.*, 2015, **11**, 438.
7. J. Chen, Q. Wu, U. W. Hawas and H. Wang, *Appl. Microbiol. Biotechnol.*, 2016, **100**, 2953.
8. P. Bernhardt and S. E. O'Connor, *Curr. Opin. Chem. Biol.*, 2009, **13**, 35.
9. A. Kirschning and F. Hahn, *Angew. Chem. – Int. Ed.*, 2012, **51**, 4012.
10. J. Rosen, J. Gottfries, S. Muresan, A. Backlund and T. I. Oprea, *J. Med. Chem.*, 2009, **52**, 1953.
11. M. S. Butler, *Nat. Prod. Rep.*, 2008, **25**, 475.
12. K. J. Weissman, *Nat. Prod. Rep.*, 2016, **33**, 203.

13. D. D. Baker, M. Chu, U. Oza and V. Rajgarhia, *Nat. Prod. Rep.*, 2007, **24**, 1225.
14. J. R. King, S. Edgar, K. Qiao and G. Stephanopoulos, *F1000Research*, 2016, **5**, 397.
15. M. A. Fischbach and C. T. Walsh, *Chem. Rev.*, 2006, **106**, 3468.
16. E. S. Sattely, M. A. Fischbach and C. T. Walsh, *Nat. Prod. Rep.*, 2008, **25**, 757.
17. Y. Katsuyama and Y. Ohnishi, *Methods Enzymol.*, 2012, **515**, 359.
18. G. H. Hur, C. R. Vickery and M. D. Burkart, *Nat. Prod. Rep.*, 2012, **29**, 1074.
19. C. Khosla, D. Herschlag, D. E. Cane and C. T. Walsh, *Biochemistry*, 2014, **53**, 2875.
20. B. Lowry, X. Li, T. Robbins, D. E. Cane and C. Khosla, *ACS Cent. Sci.*, 2016, **2**, 14.
21. C. C. Ladner and G. J. Williams, *J. Ind. Microbiol. Biotechnol.*, 2016, **43**, 371.
22. D. Sherman, *Nat. Biotechnol.*, 2005, **23**, 1083.
23. B. Moore and C. Hertweck, *Nat. Prod. Rep.*, 2002, **19**, 70.
24. C. J. B. Harvey, J. D. Puglisi, V. S. Pande, D. E. Cane and C. Khosla, *J. Am. Chem. Soc.*, 2012, **134**, 12259.
25. C. Boddy, K. Hotta, M. Tse, R. Watts and C. Khosla, *J. Am. Chem. Soc.*, 2004, **126**, 7436.
26. S. Mutka, J. Carney, Y. Liu and J. Kennedy, *Biochemistry*, 2006, **45**, 1321.
27. S. J. Moss, I. Carletti, C. Olano, R. M. Sheridan, M. Ward, V. Math, M. Nur-E-Alam, A. F. Brana, M. Q. Zhang, P. F. Leadlay, C. Mendez, J. A. Salas and B. Wilkinson, *Chem. Commun.*, 2006, **22**, 2341.
28. C. Dutton, S. Gibson, A. Goudie, K. Holdom, M. Pacey, J. Ruddock, J. Bulock and M. Richards, *J. Antibiot.*, 1991, **44**, 357.
29. J. Zhang, X. Nan, H. Yu, P. Cheng, Y. Zhang, Y. Liu, S. Zhang, G. Hu, H. Liu and A. Chen, *Eur. J. Med. Chem.*, 2016, **121**, 422.
30. Y. Zhuo, T. Zhang, Q. Wang, P. Cruz-Morales, B. Zhang, M. Liu, F. Barona-Gomez and L. Zhang, *Biotechnol. J.*, 2014, **9**, 316.
31. I. Koryakina, J. B. McArthur, M. M. Draelos and G. J. Williams, *Org. Biomol. Chem.*, 2013, **11**, 4449.
32. H. A. Crosby, K. C. Rank, I. Rayment and J. C. Escalante-Semerena, *Appl. Environ. Microbiol.*, 2012, **78**, 6619.
33. I. Koryakina and G. J. Williams, *ChemBioChem*, 2011, **12**, 2289.
34. M. C. Wilson and B. S. Moore, *Nat. Prod. Rep.*, 2012, **29**, 72.
35. Y. A. Chan, A. M. Podevels, B. M. Kevany and M. G. Thomas, *Nat. Prod. Rep.*, 2009, **26**, 90.
36. T. J. Erb, I. A. Berg, V. Brecht, M. Muller, G. Fuchs and B. E. Alber, *Proc. Natl. Acad. Sci. U. S. A.*, 2007, **104**, 10631.
37. M. Sandy, Z. Rui, J. Gallagher and W. Zhang, *ACS Chem. Biol.*, 2012, **7**, 1956.
38. L. Zhang, T. Mori, Q. Zheng, T. Awakawa, Y. Yan, W. Liu and I. Abe, *Angew. Chem., Int. Ed.*, 2015, **54**, 13462.

39. C. Chang, R. Huang, Y. Yan, H. Ma, Z. Dai, B. Zhang, Z. Deng, W. Liu and X. Qu, *J. Am. Chem. Soc.*, 2015, **137**, 4183.
40. J. Kundert and T. A. M. Gulder, *Angew. Chem., Int. Ed.*, 2016, **55**, 858.
41. T. J. Erb, V. Brecht, G. Fuchs, M. Müller and B. E. Alber, *Proc. Natl. Acad. Sci. U. S. A.*, 2009, **106**, 8871.
42. L. Ray and B. S. Moore, *Nat. Prod. Rep.*, 2016, **33**, 150.
43. D. M. Peter, L. S. von Borzyskowski, P. Kiefer, P. Christen, J. A. Vorholt and T. J. Erb, *Angew. Chem., Int. Ed.*, 2015, **54**, 13457.
44. L. Ray, T. R. Valentic, T. Miyazawa, D. M. Withall, L. Song, J. C. Milligan, H. Osada, S. Takahashi, S. C. Tsai and G. L. Challis, *Nat. Commun.*, 2016, 7, 13609.
45. M. J. Calcott and D. F. Ackerley, *Biotechnol. Lett.*, 2014, **36**, 2407.
46. R. H. Baltz, *ACS Synth. Biol.*, 2014, **3**, 748.
47. K. T. Nguyen, D. Ritz, J. Q. Gu, D. Alexander, M. Chu, V. Miao, P. Brian and R. H. Baltz, *Proc. Natl. Acad. Sci. U. S. A.*, 2006, **103**, 17462.
48. S. Doekel, M. C. Gal, J. Gu, M. Chu, R. H. Baltz and P. Brian, *Microbiology*, 2008, **154**, 2872.
49. M. Klaus, M. P. Ostrowski, J. Austerjost, T. Robbins, B. Lowry, D. E. Cane and C. Khosla, *J. Biol. Chem.*, 2016, **291**, 16404.
50. K. J. Weissman and R. Mueller, *ChemBioChem*, 2008, **9**, 826.
51. K. J. Weissman, *Nat. Prod. Rep.*, 2015, **32**, 436.
52. E. J. Drake, B. R. Miller, C. Shi, J. T. Tarrasch, J. A. Sundlov, C. L. Allen, G. Skiniotis, C. C. Aldrich and A. M. Gulick, *Nature*, 2016, **529**, 235.
53. J. R. Whicher, S. Dutta, D. A. Hansen, W. A. Hale, J. A. Chemler, A. M. Dosey, A. R. H. Narayan, K. Hakansson, D. H. Sherman, J. L. Smith and G. Skiniotis, *Nature*, 2014, **510**, 560.
54. S. Dutta, J. R. Whicher, D. A. Hansen, W. A. Hale, J. A. Chemler, G. R. Congdon, A. R. H. Narayan, K. Hakansson, D. H. Sherman, J. L. Smith and G. Skiniotis, *Nature*, 2014, **510**, 512.
55. S. Gupta, V. Lakshmanan, B. S. Kim, R. Fecik and K. A. Reynolds, *ChemBioChem*, 2008, **9**, 1609.
56. J. D. Kittendorf and D. H. Sherman, *Bioorg. Med. Chem.*, 2009, **17**, 2137.
57. H. Oh, J. Yun, K. Nah, H. Kang and D. H. Sherman, *Eur. J. Org. Chem.*, 2007, **2007**, 3369.
58. S. Kapur, B. Lowry, S. Yuzawa, S. Kenthirapalan, A. Y. Chen, D. E. Cane and C. Khosla, *Proc. Natl. Acad. Sci. U. S. A.*, 2012, **109**, 4110.
59. S. L. Ward, R. P. Desai, Z. Hu, H. Gramajo and L. Katz, *J. Ind. Microbiol. Biotechnol.*, 2007, **34**, 9.
60. J. A. Chemler, A. Tripathi, D. A. Hansen, M. O'Neil-Johnson, R. B. Williams, C. Starks, S. R. Park and D. H. Sherman, *J. Am. Chem. Soc.*, 2015, **137**, 10603.
61. Y. Sugimoto, L. Ding, K. Ishida and C. Hertweck, *Angew. Chem., Int. Ed.*, 2014, **53**, 1560.
62. X. Bian, F. Huang, F. A. Stewart, L. Xia, Y. Zhang and R. Mueller, *ChemBioChem*, 2012, **13**, 1946.

63. R. E. Cobb, Y. Wang and H. Zhao, *ACS Synth. Biol.*, 2015, **4**, 723.
64. Z. Feng, J. H. Kim and S. F. Brady, *J. Am. Chem. Soc.*, 2010, **132**, 11902.
65. E. Kim, B. S. Moore and Y. J. Yoon, *Nat. Chem. Biol.*, 2015, **11**, 649.
66. Y. Li, Z. Li, K. Yamanaka, Y. Xu, W. Zhang, H. Vlamakis, R. Kolter, B. S. Moore and P. Qian, *Sci. Rep.*, 2015, **5**, 9383.
67. Y. Luo, B. Li, D. Liu, L. Zhang, Y. Chen, B. Jia, B. Zeng, H. Zhao and Y. Yuan, *Chem. Soc. Rev.*, 2015, **44**, 5265.
68. D. Montiel, H. Kang, F. Chang, Z. Charlop-Powers and S. F. Brady, *Proc. Natl. Acad. Sci. U. S. A.*, 2015, **112**, 8953.
69. C. Osswald, G. Zipf, G. Schmidt, J. Maier, H. S. Bernauer, R. Mueller and S. C. Wenzel, *ACS Synth. Biol.*, 2014, **3**, 759.
70. A. C. Ross, L. E. S. Gulland, P. C. Dorrestein and B. S. Moore, *ACS Synth. Biol.*, 2015, **4**, 414.
71. Z. Shao, G. Rao, C. Li, Z. Abil, Y. Luo and H. Zhao, *ACS Synth. Biol.*, 2013, **2**, 662.
72. K. Yamanaka, K. A. Reynolds, R. D. Kersten, K. S. Ryan, D. J. Gonzalez, V. Nizet, P. C. Dorrestein and B. S. Moore, *Proc. Natl. Acad. Sci. U. S. A.*, 2014, **111**, 1957.
73. S. J. Kwon, M. Mora-Pale, M. Lee and J. S. Dordick, *Curr. Opin. Chem. Biol.*, 2012, **16**, 186.
74. W. Zhang and J. Liu, *F1000Research*, 2016, **5**, 208.
75. N. Tibrewal and Y. Tang, *Annu. Rev. Chem. Biomol. Eng.*, 2014, **5**, 347.
76. M. Winn, J. K. Fyans, Y. Zhuo and J. Micklefield, *Nat. Prod. Rep.*, 2016, **33**, 317.
77. J. L. Meier and M. D. Burkart, *Chem. Soc. Rev.*, 2009, **38**, 2012.
78. B. Lowry, T. Robbins, C. Weng, R. V. O'Brien, D. E. Cane and C. Khosla, *J. Am. Chem. Soc.*, 2013, **135**, 16809.
79. W. Zhang, I. Ntai, M. L. Bolla, S. J. Malcolmson, D. Kahne, N. L. Kelleher and C. T. Walsh, *J. Am. Chem. Soc.*, 2011, **133**, 5240.
80. W. Zhang, B. Ostash and C. T. Walsh, *Proc. Natl. Acad. Sci. U. S. A.*, 2010, **107**, 16828.
81. E. J. Rackham, S. Grüschow, A. E. Ragab, S. Dickens and R. J. Goss, *ChemBioChem*, 2010, **11**, 1700.
82. W. Zhang, I. Ntai, N. L. Kelleher and C. T. Walsh, *Proc. Natl. Acad. Sci. U. S. A.*, 2011, **108**, 12249.
83. S. Gruschow, E. J. Rackham, B. Elkins, P. L. A. Newilll, L. M. Hill and R. J. M. Goss, *ChemBioChem*, 2009, **10**, 355.
84. U. Sundermann, K. Bravo-Rodriguez, S. Klopries, S. Kushnir, H. Gomez, E. Sanchez-Garcia and F. Schulz, *ACS Chem. Biol.*, 2012, **8**, 443.
85. F. Wang, Y. Wang, J. Ji, Z. Zhou, J. Yu, H. Zhu, Z. Su, L. Zhang and J. Zheng, *ACS Chem. Biol.*, 2015, **10**, 1017.
86. I. Koryakina, C. Kasey, J. B. McArthur, A. N. Lowell, J. A. Chemler, S. Li, D. A. Hansen, D. H. Sherman and G. J. Williams, *ACS Chem. Biol.*, 2017, **12**, 114.
87. A. J. Hughes and A. Keatinge-Clay, *Chem. Biol.*, 2011, **18**, 165.

88. I. Koryakina, J. McArthur, S. Randall, M. M. Draelos, E. M. Musiol, D. C. Muddiman, T. Weber and G. J. Williams, *ACS Chem. Biol.*, 2012, **8**, 200.
89. M. E. Horsman, T. P. A. Hari and C. N. Boddy, *Nat. Prod. Rep.*, 2016, **33**, 183.
90. F. Kopp and M. A. Marahiel, *Nat. Prod. Rep.*, 2007, **24**, 735.
91. J. W. Trauger, R. M. Kohli, H. D. Mootz, M. A. Marahiel and C. T. Walsh, *Nature*, 2000, **407**, 215.
92. R. M. Kohli, J. Takagi and C. T. Walsh, *Proc. Natl. Acad. Sci. U. S. A.*, 2002, **99**, 1247.
93. Y. Ding, C. M. Rath, K. L. Bolduc, K. Håkansson and D. H. Sherman, *J. Am. Chem. Soc.*, 2011, **133**, 14492.
94. R. M. Kohli, C. T. Walsh and M. D. Burkart, *Nature*, 2002, **418**, 658.
95. A. S. Worthington, D. F. Porter and M. D. Burkart, *Org. Biomol. Chem.*, 2010, **8**, 1769.
96. J. Bruegger, B. Haushalter, A. Vagstad, G. Shakya, N. Mih, C. A. Townsend, M. D. Burkart and S. Tsai, *Chem. Biol.*, 2013, **20**, 1135.
97. C. Nguyen, R. W. Haushalter, D. J. Lee, P. R. L. Markwick, J. Bruegger, G. Caldara-Festin, K. Finzel, D. R. Jackson, F. Ishikawa, B. O'Dowd, J. A. McCammon, S. J. Opella, S. Tsai and M. D. Burkart, *Nature*, 2014, **505**, 427.
98. G. L. Challis, J. Ravel and C. A. Townsend, *Chem. Biol.*, 2000, 7, 211.
99. T. Stachelhaus, H. D. Mootz and M. A. Marahiel, *Chem. Biol.*, 1999, **6**, 493.
100. S. Kushnir, U. Sundermann, S. Yahiaoui, A. Brockmeyer, P. Janning and F. Schulz, *Angew. Chem., Int. Ed.*, 2012, **51**, 10664.
101. Y. Tang, C. Y. Kim, I. I. Mathews, D. E. Cane and C. Khosla, *Proc. Natl. Acad. Sci. U. S. A.*, 2006, **103**, 11124.
102. Y. Tang, A. Y. Chen, C. Kim, D. E. Cane and C. Khosla, *Chem. Biol.*, 2007, **14**, 931.
103. F. T. Wong, X. Jin, I. I. Mathews, D. E. Cane and C. Khosla, *Biochemistry*, 2011, **50**, 6539.
104. G. Yadav, R. S. Gokhale and D. Mohanty, *J. Mol. Biol.*, 2003, **328**, 335.
105. Y. Minowa, M. Araki and M. Kanehisa, *J. Mol. Biol.*, 2007, **368**, 1500.
106. B. J. Dunn and C. Khosla, *J. R. Soc., Interface*, 2013, **10**, 20130297.
107. A. P. Siskos, A. Baerga-Ortiz, S. Bali, V. Stein, H. Mamdani, D. Spiteller, B. Popovic, J. B. Spencer, J. Staunton and K. J. Weissman, *Chem. Biol.*, 2005, **12**, 1145.
108. H. M. O'Hare, A. Baerga-Ortiz, B. Popovic, J. B. Spencer and P. F. Leadlay, *Chem. Biol.*, 2006, **13**, 287.
109. W. Zha, S. B. Rubin-Pitel and H. Zhao, *Mol. BioSyst.*, 2008, **4**, 246.
110. S. Tang, S. Qian, O. Akinterinwa, C. S. Frei, J. A. Gredell and P. C. Cirino, *J. Am. Chem. Soc.*, 2013, **135**, 10099.
111. X. Gao, X. Xie, I. Pashkov, M. R. Sawaya, J. Laidman, W. Zhang, R. Cacho, T. O. Yeates and Y. Tang, *Chem. Biol.*, 2009, **16**, 1064.

112. H. Y. Lee, C. J. Harvey, D. E. Cane and C. Khosla, *J. Antibiot.*, 2011, **64**, 59.
113. B. Villiers and F. Hollfelder, *Chem. Biol.*, 2011, **18**, 1290.
114. K. Zhang, K. Nelson, K. Bhuripanyo, K. Grimes, B. Zhao, C. Aldrich and J. Yin, *Chem. Biol.*, 2013, **20**, 92.
115. H. Kries, R. Wachtel, A. Pabst, B. Wanner, D. Niquille and D. Hilvert, *Angew. Chem., Int. Ed.*, 2014, **53**, 10105.
116. X. Bian, A. Plaza, F. Yan, Y. Zhang and R. Müller, *Biotechnol. Bioeng.*, 2015, **112**, 1343.
117. G. Rao, J. Lee and H. Zhao, *Appl. Microbiol. Biotechnol.*, 2013, **97**, 5861.
118. E. O. McCullum, B. A. Williams, J. Zhang and J. C. Chaput, *In Vitro Mutagenesis Protocols*, 3rd edn, 2010, p. 103.
119. M. Zaccolo, D. M. Williams, D. M. Brown and E. Gherardi, *J. Mol. Biol.*, 1996, **255**, 589.
120. M. A. Fischbach, J. R. Lai, E. D. Roche, C. T. Walsh and D. R. Liu, *Proc. Natl. Acad. Sci. U. S. A.*, 2007, **104**, 11951.
121. B. Evans, Y. Chen, W. Metcalf, H. Zhao and N. Kelleher, *Chem. Biol.*, 2011, **18**, 601.
122. J. G. Owen, K. J. Robins, N. S. Parachin and D. F. Ackerley, *Environ. Microbiol.*, 2012, **14**, 1198.
123. M. A. Mena and P. S. Daugherty, *Protein Eng., Des. Sel.*, 2005, **18**, 559.
124. K. Miyazaki and F. H. Arnold, *J. Mol. Evol.*, 1999, **49**, 716.
125. J. E. Ness, S. Kim, A. Gottman, R. Pak, A. Krebber, T. V. Borchert, S. Govindarajan, E. C. Mundorff and J. Minshull, *Nat. Biotechnol.*, 2002, **20**, 1251.
126. M. S. Packer and D. R. Liu, *Nat. Rev. Genet.*, 2015, **16**, 379.
127. F. Ishikawa, K. Miyamoto, S. Konno, S. Kasai and H. Kakeya, *ACS Chem. Biol.*, 2015, **10**, 2816.
128. X. Zhu, J. Liu and W. Zhang, *Nat. Chem. Biol.*, 2015, **11**, 115.
129. X. Zhu, P. Shieh, M. Su, C. R. Bertozzi and W. Zhang, *Chem. Commun.*, 2016, **52**, 11239.
130. A. L. Harvey, R. Edrada-Ebel and R. J. Quinn, *Nat. Rev. Drug Discovery*, 2015, **14**, 111.
131. G. C. Adam, J. Meng, J. M. Rizzo, A. Amoss, J. W. Lusen, A. Patel, D. Riley, R. Hunt, P. Zuck, E. N. Johnson, V. N. Uebele and J. D. Hermes, *J. Biomol. Screen.*, 2015, **20**, 212.
132. C. Hsu and P. C. Dorrestein, *Curr. Opin. Biotechnol.*, 2015, **31**, 24.
133. Z. Takats, J. Wiseman, B. Gologan and R. Cooks, *Science*, 2004, **306**, 471.
134. J. Watrous, P. Roach, T. Alexandrov, B. S. Heath, J. Y. Yang, R. D. Kersten, M. van der Voort, K. Pogliano, H. Gross, J. M. Raaijmakers, B. S. Moore, J. Laskin, N. Bandeira and P. C. Dorrestein, *Proc. Natl. Acad. Sci. U. S. A.*, 2012, **109**, E1743.
135. C. Hsu, M. S. ElNaggar, Y. Peng, J. Fang, L. M. Sanchez, S. J. Mascuch, K. A. Møller, E. K. Alazzeh, J. Pikula and R. A. Quinn, *Anal. Chem.*, 2013, **85**, 7014.

136. M. E. Monge, G. A. Harris, P. Dwivedi and F. M. Fernández, *Chem. Rev.*, 2013, **113**, 2269.
137. Z. Zhou, J. R. Lai and C. T. Walsh, *Proc. Natl. Acad. Sci. U. S. A.*, 2007, **104**, 11621.
138. J. Yang, S. W. Seo, S. Jang, S. Shin, C. H. Lim, T. Roh and G. Y. Jung, *Nat. Commun.*, 2013, **4**, 1413.
139. J. Zhang, M. K. Jensen and J. D. Keasling, *Curr. Opin. Chem. Biol.*, 2015, **28**, 1.
140. S. Siedler, S. G. Stahlhut, S. Malla, J. Maury and A. R. Neves, *Metab. Eng.*, 2014, **21**, 2.
141. S. Raman, J. K. Rogers, N. D. Taylor and G. M. Church, *Proc. Natl. Acad. Sci. U. S. A.*, 2014, **111**, 17803.
142. S. Choi, E. Rha, S. J. Lee, H. Kim, K. Kwon, Y. Jeong, Y. H. Rhee, J. J. Song, H. Kim and S. Lee, *ACS Synth. Biol.*, 2013, **3**, 163.
143. S. Tang and P. C. Cirino, *Angew. Chem., Int. Ed.*, 2011, **50**, 1084.
144. S. Binder, G. Schendzielorz, N. Stäbler, K. Krumbach, K. Hoffmann, M. Bott and L. Eggeling, *Genome Biol.*, 2012, **13**, 1.
145. J. A. Dietrich, D. L. Shis, A. Alikhani and J. D. Keasling, *ACS Synth. Biol.*, 2012, **2**, 47.

CHAPTER 11

Impact of Synthetic Biology on Secondary Metabolite Biosynthesis

FREDARLA S. MILLER AND MICHAEL F. FREEMAN*

Department of Biochemistry, Molecular Biology, and Biophysics & BioTechnology Institute, University of Minnesota-Twin Cities, St. Paul, MN 55108, USA
*Email: mffreema@umn.edu

11.1 Introduction

Secondary metabolites, often falling under the broad heading of 'natural products', are specialized small molecules produced by plants, animals, fungi, and microbes that are not required for normal growth but display a wide variety of biological activities that confer selective advantages under certain environmental conditions. Humans have exploited secondary metabolites like pigments and dyes since pre-historic times. While active plant extract components from traditional medicines led to the isolation and early commercialization of natural products such as acetylsalicylic acid and morphine, the discovery of penicillin by Sir Alexander Fleming in the late 1920s ignited and revolutionized the study and manipulation of microbial secondary metabolites.[1] Currently, an amazing diversity of natural products containing unique chemical structures and properties has been discovered; a diversity which includes scaffolds of peptides, polyketides, terpenes, alkaloids and saccharides (Figure 11.1).

Catalysis Series No. 32
Modern Biocatalysis: Advances Towards Synthetic Biological Systems
Edited by Gavin Williams and Mélanie Hall

Published by the Royal Society of Chemistry, www.rsc.org

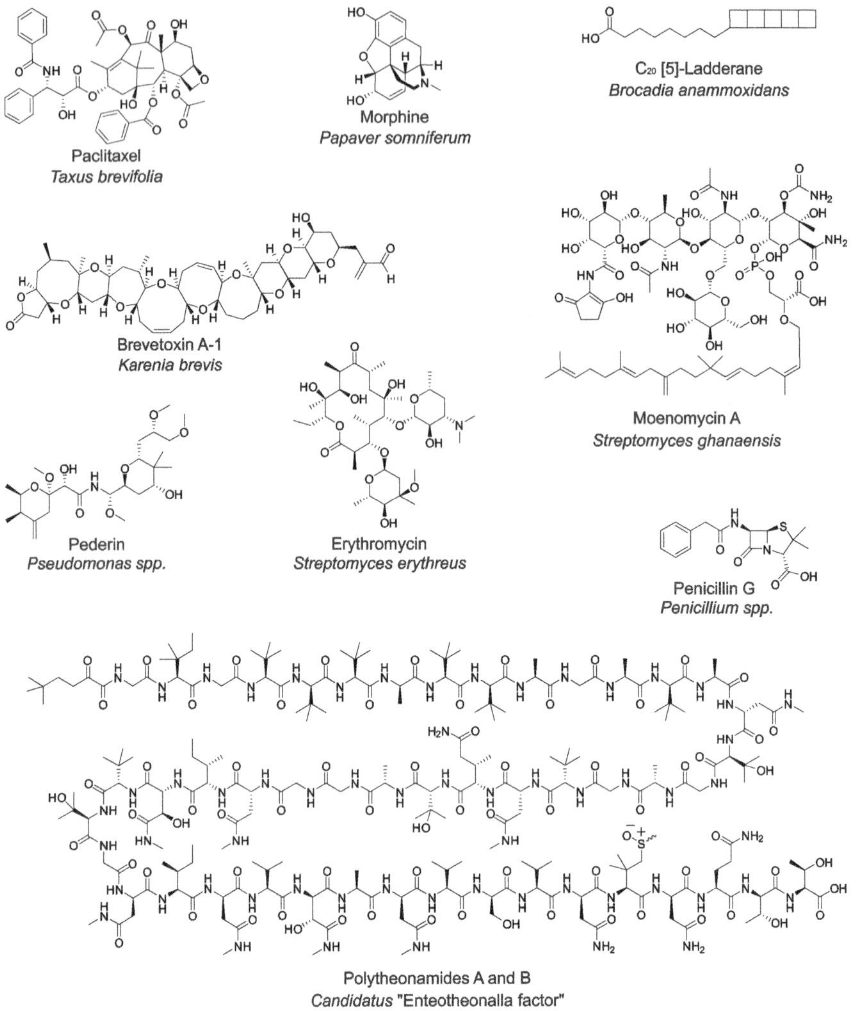

Figure 11.1 Representatives for major classes of natural products. Paclitaxel[2] is a terpene isolated from the yew tree and is used in anti-cancer therapies. Morphine[3] is an alkaloid isolated from the opium poppy and is used as an analgesic. Ladderanes[4] are fatty acids isolated from annamox bacteria and are produced to create dense membranes to prevent diffusion of toxic metabolic intermediates. Brevetoxin[5] is a polyether toxin synthesized by the dinoflagellate responsible for red tide. Meonomycin[6] is an oligosaccharide antibiotic isolated from *Streptomyces ghanaensis*. The antibiotic erythromycin[7] is a *cis*-AT polyketide derived from *Streptomyces erythreus*. Pederin,[8] a *trans*-AT polyketide, is a toxin isolated from a bacterial endosymbiont of the *Paederus* beetle. Penicillin G[9] is a non-ribosomal peptide antibiotic originally isolated from fungal *Penicillium* spp. Polytheonamides[10] are cytotoxic ribosomally synthesized post-translationally modified peptides produced by an uncultivated bacterial endosymbiont from a marine sponge.

Early work in natural products discovery focused largely on bioassay-guided fractionation, isolation, and structure elucidation to afford the desired pure metabolites. As an early example, fermentative methods were initially used and improved to enable testing the efficacy of penicillin from the fungus *Penicillium notatum*.[11] Total synthesis and chemical derivatization enriched the diversity that could be achieved on a given natural product scaffold and led to the commercialization of synthetic penicillins.[1] For the next several decades, increasingly sophisticated experimental approaches were developed to study natural product biosynthesis that included random chemical mutagenesis, cross-mutant complementation, radioisotope-labelling of metabolic precursors, and assay-guided protein purification from native producers. With the advent of polymerase chain reaction (1985) and commercially available restriction endonucleases (1975), targeted genetic manipulation and heterologous expression of biosynthetic genes and pathways reinvigorated the merger of molecular biology and natural product biosynthesis.[12]

New technologies developed in the post-genomic era of the past two decades have provided incredible advances in DNA sequencing, synthesis, and manipulation. Collectively, these technologies steadily and crucially influenced the field of natural products, creating an enormous impact on the pharmaceutical and agricultural industries. Microbes are now known to harbor numbers of natural product pathways that greatly exceed their known metabolic output, despite having been intensely studied for over 50 years.[13,14] Furthermore, deep sequencing and metagenomics have unveiled and made accessible the 'silent majority', a diverse uncultivated microbial world that dwarfs currently cultivated strains, and contains prolific natural product producers.[15,16] Given the cost effectiveness of microbial biomolecule production and the seemingly inexhaustible need for new antibiotics and biomaterials, the fusion of natural products research and the emerging field of synthetic biology was an inevitability.

At its simplest definition, synthetic biology is the study, creation, and re-design of biological systems using engineered biological parts. While mankind has been enhancing desirable traits in plants and animals through selective breeding for millennia, synthetic biological principles rely on the rational import and (re)combination of genetic parts from multiple sources to produce a desirable trait or outcome. Synthetic biology and natural products research most often intersect in the fields of molecular and systems biology, genetic, metabolic, and protein engineering, and biotechnology. These fields are often blurred together under the canopy of synthetic biology, seeking to improve natural product titers, discovery, and the creation of non-natural (custom-engineered) metabolites.[17] This chapter attempts to compile select examples that define the synthetic biological parts, technologies, and approaches currently implemented in natural product biosynthetic research. Future perspectives and limitations will also be discussed.

11.2 Host and Heterologous Strain Improvement

The choice of a suitable host is crucial to the employment of synthetic biology principles for the production and manipulation of natural products. The ability to implement rational DNA design into an organism quickly, and to subsequently assess and refine the outcome of this design in an iterative fashion, is fundamental in synthetic biology. Thus, while selective rounds of random mutagenesis of microbial strains were standard in the pharmaceutical industry and utilized with great success for many years, native producers often have non-ideal characteristics. Namely, poor parameters spanning low-production titers, insufficient phenotypic analysis, undeveloped genetic systems, and failed cultivation of 'microbial dark matter' has led to inherent limits of native host secondary metabolite biosynthesis.

Heterologous expression is an attractive and often an essential option to study natural product biosynthesis. These alternative hosts can be regarded as production chassis, for which genetic and environmental conditions are easily manipulated to suit the needs of a given secondary metabolic pathway. Heterologous expression therefore requires a detailed understanding of the metabolite's biosynthetic pathway. The genetic composition of the pathway should be bioinformatically predicted or known. Thankfully, microbes and fungi were shown early on[18] to encode biosynthetic pathways that are often clustered into a single genetic locus; while generally true, a few notable exceptions exist.[19–21] Thus, transfer of a single contiguous stretch of DNA should confer metabolite production and any necessary resistance mechanisms to the heterologous host. Conversely, the opposite trend holds true in animals and plants—only a minority of plant natural product pathways have been shown to have clustered biosynthetic genes.[22] Advances in metabolomics and transcriptomics (as transcriptional coordination of biosynthetic pathways is often observed) have overcome many of the hurdles associated with genetically dispersed plant pathways.[23] Comparative to microbial systems, relatively slow growth rates, limitations in molecular tools, compartmentalization of secondary metabolism, and the subcellular constraints of certain metabolites[24] have hindered synthetic biological approaches in plants.

11.2.1 Genome Reduction

Primary and secondary metabolisms are inherently linked, as natural product pathways often siphon or, at times, require primary metabolic precursors and cofactors like acetate, chorismate, dimethylallyl- and isopentyl-pyrophosphate (DMAPP and IPP), coenzyme A, adenosine triphosphate, heme, thiamin, and flavins. Thus, restrictions to an organism's maximum natural product output include precursor availability, cell growth, and cell density limitations. In addition, the simultaneous production of multiple secondary metabolites can limit the amount of the desired metabolite, as seen in robust natural product producers like *Streptomyces* bacteria. To alleviate potential bottlenecks, Ikeda and coworkers[25] created a suite of mutants using Cre-*loxP* and homologous recombination in the industrially relevant strain *Streptomyces avermitilis*. This eliminated many of the encoded natural product pathways including avermectin, filipin, geosmin, neopentalenolactone, and carotenoids. These genome-minimized mutants not only heterologously produced streptomycin and cephamycin C in higher titers than the original *S. avermitilis*, but they also outperformed the original producing strains.[25] A similar strategy was used to remove several secondary metabolite clusters in *Streptomyces coelicolor*.[26] Using a recycling selectable marker strategy, Wang, Oakley, and coworkers[27] deleted several natural product clusters in the filamentous fungus *Aspergillus nidulans* and developed a streamlined promoter exchange strategy to successfully produce peptide–polyketide hybrid metabolites and asperfuranone originating from *Aspergillus terreus*.

A multitude of *Escherichia coli* reduced genomes also exist as effective platforms for synthetic biology approaches to natural product biosynthesis.[28–30] Of note, use of a multiplex automated genome engineering (MAGE) technique reduced the codon degeneracy in *E. coli* and eliminated all instances of amber codons in the genome.[31] If and when combined with established tRNA recoding techniques[32] for the incorporation of non-natural amino acids into proteins, this strain could expand the chemical diversity of heterologously expressed ribosomally synthesized and post-translationally modified peptide (RiPP) natural products.

11.2.2 Regulatory Network Engineering

As stated above, bacteria and fungi often have biosynthetic pathways clustered in their genome.[18] However, unlike eukaryotes, prokaryotes further organize biosynthetic gene clusters into polycistronic (multi-gene) operons.[33] Many genes, sometimes with overlapping coding regions, are co-transcribed on a single RNA strand that can have further levels of regulation due to secondary structure effects and obstructed ribosomal binding sites. As exemplified in *Streptomyces* bacteria, the genetic architecture of multi-operon gene clusters can be quite complex, where natural product pathways are environmentally and temporally regulated during the organisms' multistage lifecycle.[34,35]

Over the last several decades, *Streptomyces* and other Actinobacteria have been shown to encode pathways for hundreds of elucidated and putative natural products. Many of these studies have investigated natural product biosynthesis in the context of this organism. As a prolific secondary metabolite-producing microorganism, *S. coelicolor*[13] temporally controls its vast and varied natural product biosynthetic pathways partially through the use of global regulators involved in morphological development. Chater and coworkers[36] discovered a mutation in *bldA* encoding the tRNA specific for the rare leucine UUA codon. This mutation was shown to have a direct effect on the biosynthesis and transport of the antibiotic actinorhodin, notably demonstrating pathway-specific regulation accomplished by global translational control.[37] Biosynthetic pathways are often further regulated by pathway-specific regulatory proteins. As an example, the biosynthetic gene cluster responsible for the production of hormaomycins, structurally unique depsipeptides that act as bacterial hormones, contains three encoded pathway-specific regulators in *S. griseoflavus* W-384.[38] Manipulation of these genes resulted in increased titers of six known congeners in addition to a 140-fold improvement of hormaomycin A production. Disregulation of the cluster also resulted in the production of a novel hormaomycin analog.[39]

In addition to the above translational and transcriptional regulatory enzymes employed on a global or localized scale, biosynthetic pathways and associated operons can be modulated further by direct mutation of promoters, ribosome binding sites (RBSs), and DNA coding regions. Engineered and synthetically designed promoters and RBSs are now commonly being used for engineering pathways to fine-tune expression levels of individual genes. Salis and coworkers[40] demonstrated that the strength of an RBS is directly affected by its entire genetic context, including the mRNA sequence upstream of the binding site and the coding sequence of the gene itself. They were able to impart controllable protein expression in *E. coli* that encompasses a 100 000-fold range and developed an RBS calculator for engineering custom ribosome binding sites for specified gene sequences.[41]

Identification and characterization of the promoter from the erythromycin resistance gene *ermE* in the mid-1980s[42] began an effective strategy of promoter replacement for increasing or inducing natural product biosynthesis in Actinobacteria. For instance, the *tipA* promoter from *Streptomyces lividans* was shown to be activated in response to thiostrepton and was one of the first widely used inducible promoters for *Streptomyces* heterologous expression.[43] In another study, testing and implementation of *Streptomyces* endogenous, exogenous, and synthetic promoters lead to increased secretion of the α-amylase inhibitor tendamistat from 0.5 $mg L^{-1}$ to 500 $mg L^{-1}$ in *S. lividans*.[44] More recently, a *Streptomyces* promoter library was created and analyzed through conferred neomycin resistance, demonstrating synthetic promoter strengths varying up to 12-fold through select changes to the −35 and −10 boxes of the promoter region.[45]

Beyond *Streptomyces*, mutation of endogenous promoters in *Bacillus subtilis* to more closely resemble consensus promoter sequences of the strain resulted in increased transcription and subsequently elevated protein levels of recombinant nattokinase.[46] In the interest of developing prokaryotes for nuanced expressions and development of synthetic switches, Voigt and coworkers[47] used genome mining to develop a set of 20 orthogonal sigma factor/promoter pairs in *E. coli* with cognate anti-sigma factors that function as repressors. Collins and coworkers[48] developed a rationally-designed promoter library using a combination of random nucleotide bases to alter promoter strength and fixed motifs to ensure basic promoter function. The authors demonstrated promoters with predictable strengths are capable of functioning as biological timers by controlling the temporal flocculation of yeast, an important characteristic for the industrial production of beer, bioethanol, and other such commodities.[49]

Non-genetic methods for activating silent gene clusters (those that are not transcriptionally active in their native hosts) have also been proven successful.[50] High-throughput screenings using reporter genes with media supplements revealed that doping bacterial cultures with molecules such as antibiotics can elicit the activation of silent gene clusters and result in novel metabolite production.[51] In line with these findings, an intriguing study was conducted to activate natural product synthesis in the soil-dwelling anaerobe, *Clostridium celluloyticum.* Traditional culturing methods and media supplementation were initially unsuccessful in eliciting secondary metabolite production in *C. celluloyticum.* However, doping the culture with extracts derived from the surrounding soil of the isolate led to the discovery of the first natural product from an obligate anaerobe, closthioamide.[52] Despite the above successes to activate pathways of interest, there are far more unreported failures than successes in eliciting natural product biosynthesis as metabolites are only natively produced in very specific, and often unknown, circumstances.[53]

11.3 Genetic Refactoring

In recent years, the field of synthetic biology has been advancing towards the complete removal and rewiring of gene cluster regulatory elements to rationally deconstruct their operons into independently expressed genes.[54] This method, termed genetic refactoring, is an approach where individual synthetic components are compiled and reassembled to predictably control the expression of individual genes in a given pathway. This strategy has emerged as a consequence of the difficulties in expressing heterologous biosynthetic pathways that can exceed 150 kb in length.[55] More importantly, a large number of silent gene clusters, or those identified in uncultivated microorganisms, are often not expressed in their native or heterologous hosts in the context of their native regulatory constraints. Common regulatory mechanisms involve the temporal induction of promoters of varying strengths, diverse sequences for ribosome binding sites, and transcription terminators with different read-through frequencies. Together with tailored codon usage for genes, differences in DNA or mRNA secondary structures, and pathway encoded regulatory systems, the native host can finely control the expression of all the requisite components in a pathway. Thus, these layers of regulation, in the context of the global metabolism of the host, are complicated and often unpredictable. The successful refactoring of a gene cluster gives tunable control over the production of individual proteins encoded within the pathway to ultimately optimize the production of the desired metabolite(s).

To address these complexities, genetic refactoring first relies on identifying the boundaries of all open reading frames within a selected pathway. Once established, the codons for each gene are randomized to disrupt potential secondary structures or other unknown regulatory mechanisms (Figure 11.2A). Indeed, codon usage can be rationally manipulated to increase or decrease protein expression since the degeneracy of the genetic code allows for a subset of codons that can slow translation rates.[56] In an impressive study conducted by Hunt and coworkers,[57] more than 6000 experiments demonstrated the effect of mRNA sequence on protein expression. This data was subsequently used to test the effects of synonymous codon usage in engineered genes, and through *in vitro* biochemical analyses,

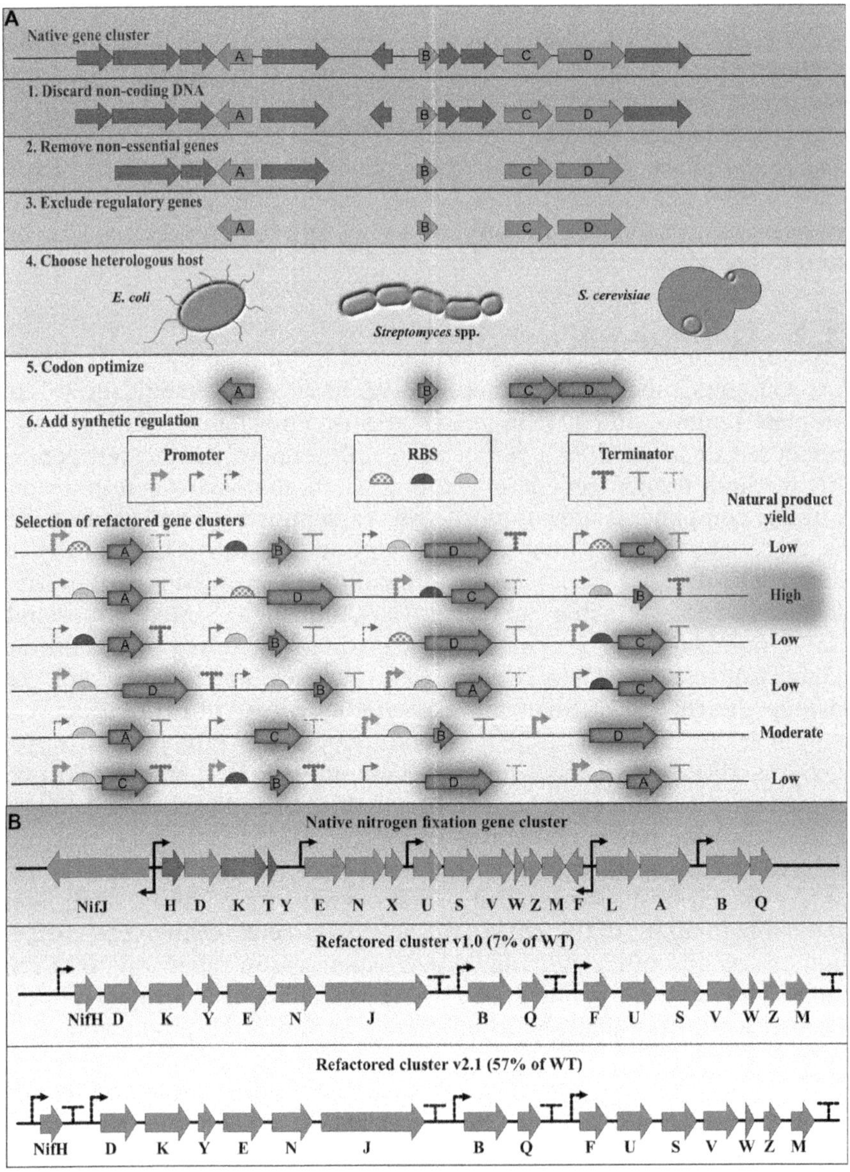

confirmed the positive effect of so-called codon optimization. An additional notable finding established the importance of the first nine codons in a given sequence, a notion mentioned in several previous studies. This region of the mRNA is physically occluded by the ribosome initiation complex;[40,58] adenosine and guanine have the most effect on increasing or decreasing protein expression, respectively.[59,60] There are, however, some caveats in codon randomization and optimization that must be taken into account. A single silent mutation has been shown to alter the substrate specificity of a protein. This is likely due to the change in translation rate caused by the altered codon usage that caused the protein to fold differently.[61] In this vein, the original operon structure may play a direct role in subunit association and complex formation by employing temporal and co-translational control over genes within operons.[62] In addition, operon composition and order has been shown to correlate with protein complex assembly, and is most highly associated with proteins expressed at low levels.[63]

Additional non-natural regulatory elements can also be added to a refactored system to help limit influence from neighboring synthetic operons. Synthetic upstream untranslated regions with overlapping genetic motifs allowed expressional control of green fluorescent protein in *E. coli* over a 1000-fold range with 93% reliability.[64] Ribozyme insulators have also been designed and added between promoter and coding sequences to further buffer expression output of target genes in *E. coli*[65] and *Streptomyces*.[66] Post-translational regulation can even be engineered into a system through degradation signals. Orthogonal and tunable C-terminal *ssrA*-based tags have been used for controlling protein degradation,[67] while an inducible protease can be used to cleave and expose the N-terminus of a protein to exploit the N-rule in bacteria,[68] thereby fastidiously degrading the protein of interest.[69]

Figure 11.2 Genetic refactoring of biosynthetic gene clusters. (A) Schematic demonstrating a typical refactoring workflow. Beginning with a natural gene cluster (ORFs shown as arrows), (1) the non-coding DNA is identified and removed, (2) non-essential and (3) regulatory genes are removed, (4) an appropriate heterologous host is chosen, (5) codons are randomized and optimized based on the codon frequencies of the expression host, and (6) synthetic regulation is added that is specific to the chosen heterologous host (regulation elements of different strengths indicated by disparate shadings and line representations). Variations of the refactored cluster are heterologously expressed and analyzed for natural product output (6), which will vary based on the imposed and predicted synthetic gene regulation.[54] (B) Successful refactoring of the nitrogen fixation gene cluster from *Klebsiella oxytoca*. The native gene cluster (top), was initially refactored to obtain 7% of wild-type activity (middle).[74] The cluster was further refactored to obtain 57% of wild-type (WT) activity (bottom).[75] Solid promoters and terminators indicate those found within the native cluster, dashed lines indicate synthetically imposed regulatory elements.

Some of the earliest attempts to refactor natural product gene clusters have been performed in *Streptomyces* and in yeast. The antiviral polyketide A-74528 was heterologously expressed and refactored in *Streptomyces coelicolor*.[70] Non-essential genes were removed from the cluster and the remaining genes were expressed on multiple plasmids in *S. coelicolor* to induce detectable production of A-74528. Successful refactoring of the spectinabilin gene cluster in *Streptomyces lividans* required the identification of a suite of *Streptomyces* promoters before they could be incorporated into an effective refactoring strategy.[71] Moore and coworkers[72] utilized a successful 'plug and play' method with recombination strategies in yeast to express the 67 kb non-ribosomal peptide taromycin from *Saccharomonospora* sp. CNQ-490 in *S. coelicolor*. And as an example in *S. cerevisiae*, β-amyrin production was increased approximately 200-fold by optimizing transcriptional flux through the addition of UPC2 binding sites upstream of genes in the engineered terpenoid pathway.[73]

An impressive refactoring strategy was performed by Voigt and coworkers[74] that targeted the nitrogen fixation pathway in the bacterium *Klebsiella oxytoca* (Figure 11.2B). In this study, a refactored cluster was expressed in a *K. oxytoca* nitrogen fixation pathway knockout strain, where the authors obtained 7% nitrogenase activity of wild-type, a notable achievement.[74] The nitrogen fixation pathway is known to rely on precise expression levels and as such, the group subsequently sought to optimize the refactored pathway by creating several hundred variants of the gene cluster. This approach garnered 57% reconstitution of the wild-type activity, provided data for investigating the activity of synthesized parts in different genetic contexts, and allowed for the transfer of the pathway into *E. coli*.[75] This work kept its focus on the nitrogen fixation pathway but its success has ramifications for the synthetic biology field at large. They demonstrated a high-throughput scheme for assessing the function of genetic parts in combination, a concept that may also be applied to optimizing the heterologous production of natural products. It should be noted that complementary synthetic biological approaches to refactoring are also being explored for eukaryotic systems. Brakhage and coworkers[76] recently constructed large polycistronic synthons for expression in *Aspergillus nidulans*, where viral 2A peptide sequences inserted between protein coding sequences of a synthetic penicillin cluster were co-translationally cleaved to ensure equimolar production of the biosynthetic genes to produce penicillin K.

11.4 DNA Manipulation Technologies

The impact of synthetic biology on natural product discovery and optimization is intimately entwined with new DNA synthesis and manipulation strategies. Although reviewed elsewhere,[77] within the past few decades a revolution in DNA technologies has thoroughly remodelled the traditional pipeline of natural product discovery, enabling high-throughput assembly methods beyond the landmark discovery and uses of restriction enzymes.[78]

This in part began with the development of the BioBrick™ system. Through incorporation of a defined set of restriction endonuclease cleavage sites into a set of vector backbones, this system allows for easily obtained gene substitutions.[79] Although a step forward, this method becomes less effective for reconstructing even modest-sized pathways, and always leaves a short DNA scar between components. Golden Gate cloning took the 'swappable parts' notion further by eliminating the scar site and enabling automation of genetic pathway construct building.[80] Another powerful scarless method, Gibson assembly, relies on the simultaneous combination of polymerase, exonuclease, and ligase activities to assemble multiple DNA fragments with short overlapping sequences into a vector of interest in a one-pot, fast-acting process.[81] Programmable DNA cleavage systems also revolutionized *in vivo* and *in vitro* DNA methodologies. This area of research began with the engineering of zinc-finger endonucleases and the discovery/engineering of transcription activator-like effector nucleases (TALENS),[82] but for most applications these were quickly replaced by the explosive technologies surrounding the CRISPR/Cas9 systems.[83,84]

While the cost of DNA synthesis decreased almost 100-fold in less than two decades[85] and may soon be the primary route for constructing large biosynthetic pathways, DNA capture strategies are still needed for the isolation and alteration of large gene clusters. Presently, many advances in DNA capture of natural product gene clusters were developed and utilized for metagenomic DNA acquisition of environmental samples; much of the DNA extracted from these samples originates from microbes that currently evade modern culturing methods.[86] Handelsman and coworkers[87] used bacterial artificial chromosomes (BACs) to isolate large DNA fragments up to 100 kb directly from an environmental soil sample. Functional metagenomics utilizing phage-assisted DNA packaging in *E. coli* on cosmids has also been successful in identifying novel biologically active molecules; in one study on uncultivated soil communities, 65 clones of a 700 000-member clone library were shown to possess antibacterial activity.[88] In the mid-2000s, Piel and coworkers[89] introduced a novel method to screen metagenomic cosmid/fosmid libraries derived from marine sponges. They used a semi-liquid culturing method in lieu of traditional solid media techniques; this process involved a pool-dilution strategy of spatially unbiased cell growth that resulted in the rapid isolation of a targeted clone from a 400 000-member library. By combining this process with a degenerate primer strategy to amplify conserved polyketide ketosynthase domains of predictable substrate specificity, they were able to isolate the complete biosynthetic gene cluster for the antitumor metabolite psymberin from a *Psammocinia* sponge metagenome.[90]

Direct DNA capture circumvents the need to create and screen vector libraries that may not contain the entire genetic locus of interest. Methods involving oriT-mediated DNA capture,[91] ϕBTI integrase,[92] and different variants of RED/ET recombineering[93–95] were successfully used to clone large biosynthetic gene clusters directly into a variety of hosts such as

Photorhabdus, *Bacillus*, *Streptomyces*, and *E. coli*. The Cre/loxP recombinase system was used in combination with BACs to clone pathways up to 78 kb from *Photorhabdus luminescence* and *Agrobacterium tumefaciens*.[96] Developed by Zhao and coworkers,[97] DNA assembler takes advantage of the high frequency of homologous recombination in *Saccharomyces cerevisiae*, and was initially shown to assemble gene clusters up to 19 kb in length with 70–100% efficiency. A similar technique based on yeast homologous recombination developed in the late 1990s, termed transformation assisted recombination (TAR), is capable of isolating large genomic DNA fragments up to 250 kb, albeit at a low efficiency.[98] TAR cloning has gained traction in the natural products community, where it is being combined with refactoring techniques to isolate and interchange regulatory elements in as little as one step. This process was employed for the brominated lipopeptide alterochromide[99] and silent lipopetide taromycin A.[72] A suite of orthogonally selective constitutive promoter cassettes have been used in concert with TAR cloning for *Streptomyces* heterologous expression; for example, a complete exchange of promoters in the cluster responsible for the indolotryptoline lazarimides resulted in the activation of the otherwise silent cluster.[100] Excitingly, TAR cloning has been combined with CRISPR/Cas9 multiplexing in the development of mCRISTAR, which simultaneously targets, excises, and recombines regulatory elements into refactored gene clusters in yeast to facilitate the activation and optimization of silent pathways.[101] Gibson assembly combined with in-gel cell lysis and Cas9 DNA digestion isolates gene clusters up to 100 kb in a process aptly termed CATCH.[102] Creative and powerful DNA manipulation techniques will continue to be designed in the next decade as the field of synthetic biology matures in concert with natural product discovery.

11.5 Natural Product Discovery in the Post-genomic Era

The golden age of natural products (1940s–1970s) fuelled advancements in synthetic chemistry due to the discovery of structurally unique molecular scaffolds, biochemistry with new enzymology, and the medical/biotechnology fields for producing new and valuable therapeutics and biomaterials.[103] These areas of discovery and progress have been the pillars of the natural products field. The typical discovery workflow was often initiated with an activity screen (*e.g.* cytotoxicity or bactericidal activity), followed by structure elucidation, and finally (bio)synthetic studies to both uncover the natural mechanisms of synthesis, as well as methods streamlining or altering the production and composition of the final natural product. Due to increasing metabolite rediscovery rates from these assays, pharmaceutical industry divestment began towards the end of the twentieth century. However, the post-genomic era has unveiled an exponentially growing and tantalizing pool of cryptic and silent gene clusters likely encoding yet-to-be-discovered secondary metabolites of potential chemical, enzymological, or therapeutic novelty.[103] Putative gene clusters are now bioinformatically identifiable through a multitude of available online tools such as anti-SMASH,[104] which helps identify pathways for previously defined natural product families, and ClusterFinder,[105] which outlines genetic loci for potentially undiscovered natural product families.[106] These genetic revelations have simultaneously reinvigorated the natural products field and uprooted (and in many ways reversed) the well-established workflow for metabolite discovery. New high-throughput methodologies and Big Data pipelines entwine synthetic biology and systems biology approaches, and are currently needed to find, prioritize, and link genetic loci to these as-yet unidentified chemically or therapeutically novel compounds.

In the last two decades, the discovery of new natural products has been incrementally hampered by time-consuming and increasing costs associated with dereplication—removal of known compounds from natural product discovery pipelines.[107] Advances in mass spectrometry (MS) now enable non-targeted metabolomics strategies that link genotypes to chemotypes in order to characterize novel peptide[108] and glycosylated[109] natural products. Molecular networking, a process where MS/MS spectra are related to each other through similarities in fragmentation patterns, is currently being used to tackle the biosynthesis and structure of elusive metabolites like colibactin.[110] Dorrestein and coworkers, having shown the potential of molecular networking and untargeted metabolomics,[111] have now created an interactive social molecular networking platform where MS data is compiled, cross-referenced, and networked to promote natural product discovery and foster new collaborations.[112]

High-throughput cytological profiling has been employed to determine mechanisms of actions for natural products in bacteria through using high content, image-based screening.[113] When combined with non-targeted metabolomics, Linington and coworkers[114] were able to tease out natural product identities and their respective modes of action from complex mixtures by visually integrating results from MS molecular networks and cell morphological differences. This multidisciplinary approach is called Compound Activity Mapping (CAM). These technologies and others[115] allow researchers to dereplicate complex biological samples in a medium- or high-throughput fashion such that focus may be directed towards the discovery and characterization of novel metabolites.

A more recent approach to drug target identification is reverse chemical genetics. This method contrasts well-established forward chemical–genetics methods that rely on selecting mutants treated with increasing doses of a drug to identify its target.[115] Reverse chemical genetics relies on comparing cellular hypersensitivity to a given drug to a library of genetic mutants of the treated cell type. These chemical–genetic interactions can help identify natural product pathway targets and has been observed in a proof-of-principle study that screened a number of well-defined metabolites against an ~5000 single-gene knockout library constructed in *S. cerevisiae*.[116] To evaluate synthetic genetic interactions, Boone and coworkers[117] recently expanded this approach to include an impressive ~5.4 million clone double-knockout yeast library. Within this library, double gene knockouts were shown to correlate to the cellular fitness of synonymous chemical–genetic single deletions; this work also enabled the creation of detailed genetic network maps by compiling approximately one million positive and negative genetic interactions.[118] In a complementary approach, Magarvey and coworkers[119] took advantage of the link between natural product resistance genes and cellular targets to create a retrobiosynthetic algorithm (PRISM), enabling them to determine a unique and previously unknown cardiolipin-interacting mechanism of action for the nonribosomal peptide telomycin.

11.6 Precursor Supplementation

The desire or need to heterologously express a gene cluster to produce high titers of a given metabolite often breaches the natural capacity of the chosen host organism to effectively siphon out carbon and the necessary metabolic precursors and cofactors for the metabolite of interest. The choice of heterologous host is typically first selected for other attributes such as ease of genetic manipulation, growth rate in monoculture, and biomass yield per liter of medium. Thus, much effort is placed on eliminating precursor/cofactor bottlenecks or expanding the breadth of metabolic pools for natural product biosynthetic pathway expression, either indirectly through metabolic engineering or through direct precursor supplementation. In addition, modification of templated building blocks for polyketides and peptides can significantly change the properties of the engineered metabolites.[120]

Landmark studies of the discovery and initial heterologous expressions of erythromycin biosynthetic genes from *Saccharopolyspora erythraea* paved the way for production of increased titers and engineered congeners of erythromycin and other polyketides.[121–123] As holo-polyketide synthases (PKSs) require post-translational modifications with 4-phosphopantethiene, Khosla and coworkers[124] inserted a chromosomal copy of the 4-phosphopantetheinyl transferase *sfp* into the genome of *E. coli*. Additional expression of genes responsible for (*S*)-methylmalonyl coenzyme A (CoA) production along with the three requisite polyketide synthases and a fused, non-native non-ribosomal peptide synthetase (NRPS) module resulted in high-titer production of 6-deoxyerthronolide B and the corresponding NRPS–PKS hybrid scaffold. Metabolic engineering efforts to improve erythromycin A production in *E. coli* have continued in recent years to explore multiple routes to propionyl-CoA and (*S*)-methylmalonyl-CoA production that match or exceed the previously engineered pathway.[125] Antisense RNA technology to mediate ribosome binding site occlusion has also been used to modulate and lower the expression of certain fatty acid synthesis genes in *E. coli* to improve malonyl-CoA production by 4.5 fold.[126]

Importantly, the discovery of the *matABC* locus in *Rhizobium trifolii* provided the first plastic enzymatic routes to alternative malonyl building blocks for polyketide biosynthesis.[127,128] Since then, a variety of alternative malonyl units have been enzymatically incorporated into polyketides both *in vivo* or *in vitro* that include ethylmalonyl-, hydroxymalonyl-, methoxymalonyl-,[129] fluoromalonyl-,[130] and even alkynyl-[131,132] and azido-containing-CoAs.[132] Of note, the discovery of crotonyl-CoA carboxylase/reductases more than doubled the possibilities of non-standard C-2 polyketide units to include propylmalonyl-,[133] chloroethylmalonyl-,[134] and many others.[135]

For nonribosomal peptide natural product pathways, amino acid precursors originating from primary metabolism are either directly incorporated into the natural product or are further modified by cluster-encoded enzymes later in the biosynthesis, creating a diverse set of non-proteinogenic amino acid-containing peptides.[136] Precursor-directed biosynthesis (alternative amino acids added to culture medium to compete for incorporation into the peptides) and mutasynthesis (deletion of encoded amino acid biosynthetic machinery followed by supplementation of alternative substrates) have been used successfully for decades to increase the diversity of NRPS-derived metabolites.[137] A common theme to create structural diversity in NRPS peptide natural products is to engineer non-native modifying enzymes into a pathway that either act on amino acid pools prior to NRPS-catalyzed ligation or on post-synthesized metabolites. A prime example of this synthetic biology approach has been conducted with the pacidamycin producer *Streptomyces coeruleorubidus*. A non-native tryptophan 7-halogenase gene constitutively expressed in *S. coeruleorubidus* resulted in the production of 1.0 mg L^{-1} chloropacidamycin. When the amino acid recognition code of NRPS adenylation domains was discovered at the end of the twentieth century, it opened up the possibility to determine and thus change the specificity of NRPS modules and engineer non-ribosomal peptides.[138,139] Piel and coworkers,[140] noticing an unusually high sequence identity encoded in adenylation domains within the hormaomycin biosynthetic gene cluster, were able to swap short sequences (~400 base pairs) to interchange domain substrate specificities for structurally divergent amino acids. Using a complementary approach, Hilvert and coworkers[141] structurally identified a similarly sized flavodoxin-like subdomain in the first adenylation domain of the gramicidin pathway. This subdomain was successfully replaced with a set of domains activating disparate amino acids.

Vanillin, a shikimate pathway-derived flavor compound originally isolated from the *Vanilla planifolia* seed pod, is predominantly synthesized for human consumption due to the cost of production and extraction from the vanilla orchid. As a cheaper alternative for 'natural' sources, efforts were made to produce vanillin heterologously in a variety of microbial hosts.[142] The *Amycolatopsis* sp. ATCC 39116 is used in industry as a heterologous host for vanillin production due to its toxicity tolerance to vanillin and the phenolic precursor ferulic acid. Identification and deletion of the vanillin catabolic enzyme vanillin dehydrogenase in this strain, along with chromosomal insertions of the constitutive promoter *permE** proceeding two biosynthetic genes, afforded vanillin in yields of 19.3 g L^{-1} with a molar yield of 94.9%.[143] Other work has focused on the production of vanillin in

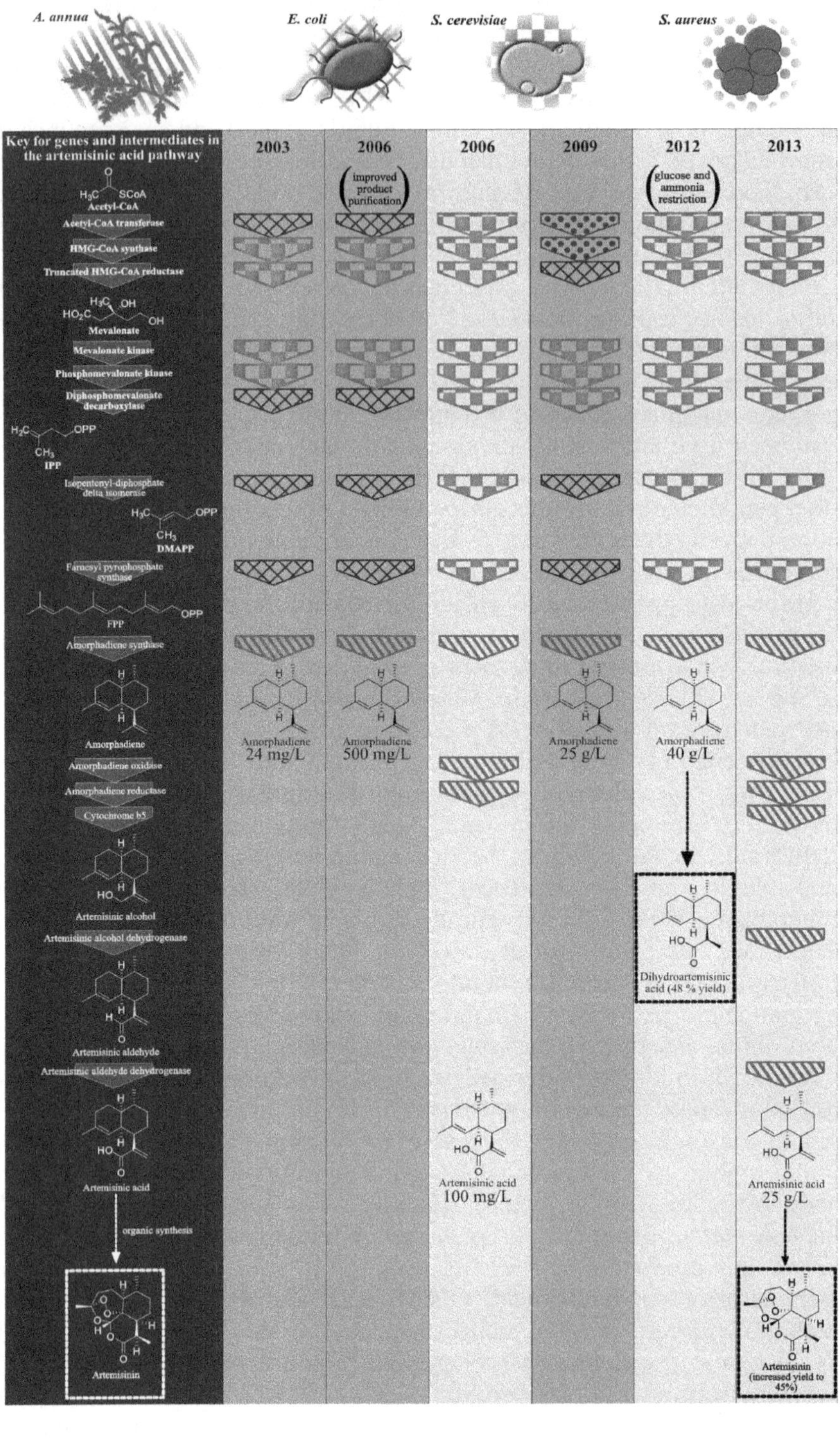
A. annua
E. coli
S. cerevisiae
S. aureus
Key for genes and intermediates in the artemisinic acid pathway
2003
2006
(improved product purification)
2006
2009
2012
(glucose and ammonia restriction)
2013
Acetyl-CoA
Acetyl-CoA transferase
HMG-CoA synthase
Truncated HMG-CoA reductase
Mevalonate
Mevalonate kinase
Phosphomevalonate kinase
Diphosphomevalonate decarboxylase
IPP
Isopentenyl-diphosphate delta isomerase
DMAPP
Farnesyl pyrophosphate synthase
FPP
Amorphadiene synthase
Amorphadiene
Amorphadiene 24 mg/L
Amorphadiene 500 mg/L
Amorphadiene 25 g/L
Amorphadiene 40 g/L
Amorphadiene oxidase
Amorphadiene reductase
Cytochrome b5
Artemisinic alcohol
Artemisinic alcohol dehydrogenase
Dihydroartemisinic acid (48 % yield)
Artemisinic aldehyde
Artemisinic aldehyde dehydrogenase
Artemisinic acid
Artemisinic acid 100 mg/L
Artemisinic acid 25 g/L
organic synthesis
Artemisinin
Artemisinin (increased yield to 45%)

engineered yeasts[144] or *E. coli*[145] because they are incapable of degrading the metabolite. Overall vanillin production from simple carbon sources like glucose, glycerol, or tyrosine resulted in far lower yields compared to those required from heterologous expression of the vanillin pathway in *Amycolatopsis* sp. ATCC 39116. This was due, in part, to toxicity of vanillin in these microorganisms.

Some of the most successful fusions of synthetic biology in natural product precursor supplementation have occurred in terpenoid biosynthesis. Artemisinin is a potent antimalarial terpenoid natural product isolated from the plant *Artemisia annua*.[146] Its increasing use to treat malaria and its low extraction yields from the native host[147] made it a prime candidate for synthetic biological approaches to heterologously express in microbes (Figure 11.3). All terpenoids are known to be biosynthesized from the precursors DMAPP and IPP through either the mevalonic (MVA) or methylerythritol-phosphate (MEP) pathways.[148] Initial work focused on overexpressing the native MEP pathway in *E. coli*;[149] however, expression of the *Saccharomyces cerevisiae* MVP pathway in *E. coli*[150] resulted in >0.5 g L^{-1} titers of the artemisinin precursor amorphadiene after fermentation optimization and headspace capture.[151] Identifying and alleviating accumulation of the toxic precursor 3-hydroxy-3-methyl-glutaryl-CoA through flux balance analysis[152] as well as marked improvement in fed-batch fermentation drastically increased amorphadiene production to >27 g L^{-1} in *E. coli*.[153] Effort was redirected to production in yeast, where manipulation of the native MVA pathway, and addition of an *A. annua* cytochrome P450, lead to the production of the more downstream artemisinin precursor artemisinic acid, albeit at modest levels of 100 mg L^{-1}.[154] However, careful selection of the yeast heterologous host with overexpression of all enzymes in the MVA pathway[155] and supplementation of an *A. annua* cytochrome and an additional P450 ultimately resulted in the impressive yield of 25 g L^{-1} artemisinic acid.[156]

Optimization of precursor supplementation in heterologous expressions has been similarly employed for the production of paclitaxel (Taxol), a potent anticancer drug isolated from the bark of the Pacific yew tree, in both *E. coli*[157] and yeast.[158,159] However, in partitioning the native MEP pathway 'module' from the module of downstream genes required for the production of the Taxol intermediate taxadiene, Stephanopoulos and coworkers[160] were able to use a condensed multivariate engineering approach to reveal

Figure 11.3 Progress in artemisinin semisynthesis from 2003 to 2013. The key for the engineered pathway to biosynthesize artemisinic acid (left column) uses enzymes from various organisms which are indicated by the arrows in each column: *Artemisia annua* (the original artemisinin producer, stripes), *E. coli* (crosshatch), *S. cerevisiae* (boxes), and *S. aureus* (circles). Dashed lines indicate multiple synthetic steps. Each column represents different progress in a heterologous host, either *E. coli* (dark grey background) or *S. cerevisiae* (light grey background), in 2003,[150] 2006,[151,154] 2009,[153] 2012,[155] and 2013.[156]

nonlinear pathway optimization that resulted in 1 g L^{-1} in *E. coli* feed-batch fermentations. To create further oxygenated taxane intermediates, the same group interestingly decided to partition the pathway between *E. coli* and yeast.[161] Using a mutualistic co-culturing approach, *E. coli* was engineered to metabolize xylose and excrete acetate, a large accumulation of which would prove toxic to *E. coli*. As yeast is unable to consume xylose, *S. cerevisiae* subsequently metabolized the acetate without producing ethanol—a compound also toxic to *E. coli*. This stable co-culture was further optimized for *S. cerevisiae* growth, maximizing production of the necessary and limiting P450, associated reductase, and an acetyltransferase, and thus producing a monoacetylated, oxygenated taxane. Realizing that P450 expression in these *E. coli* strains negatively downregulated production of previously engineered pathway modules, lowering the expression ratio of reductase to P450 alone achieved >0.5 g L^{-1} of oxygenated taxanes.

11.7 Compartmentalization

As touched upon in the previous section, eukaryotic-derived P450s often present significant challenges for microbial expression. This is in part due to their need of an appropriate reducing system that is not often genetically clustered with the P450s.[162] However, these enzymes often encode N-terminal hydrophobic leader sequences that localize the proteins to the endoplasmic reticulum and thus are not typically active in prokaryotic systems without (or even with) modification.[163] To circumvent expression problems associated with P450s, Sato and coworkers[164] devised a partitioned and alternative route to the production of benzylquinoline alkaloids. They first created a synthetic pathway for reticuline in *E. coli* using microbial and plant-derived genes. By then adding *S. cerevisiae* expressing an *N*-methyltransferase and a P450 to the *E. coli* culture, the authors were able to bypass *E. coli* expression of two P450s and succeeded in producing the alkaloid magnoflorine. Redistribution of entire natural product pathways to different organelles *in planta* has also shown some promise. The reducing power of tobacco chloroplasts were used for light-induced electron transfer to P450s for the production of the alkaloid dhurrin;[165] synonymous work was done in cyanobacterial systems to activate *Bacillus* P450s.[166]

Beyond P450s, the cellular environment/metabolic capacity of *E. coli* was shown to be insufficient to activate cobalamin-dependent radical *S*-adenosylmethionine (SAM) methyltransferases encoded in the biosynthetic pathway for polytheonamides, potent cytotoxins produced by an uncultivated bacterial symbiont in the Japanese sponge, *Theonella swinhoei*.[15,16] Presumably due to lack of sufficient cobalamin availability in *E. coli*, *Rhizobium leguminosarum* bv. *viciae* 3841 was implemented as an alternative expression host, allowing the previously illusive 17 C-methylations observed in these RiPP natural products to be observed.[167] As the trend for putative secondary metabolic pathway discovery continues to increase for uncultivated microbes, careful attention must be paid to appropriately pair the heterologous host to the pathway of interest. Online tools are now beginning to emerge that account for the metabolic framework of a heterologous host and the way in which that pertains to the design of biosynthetic pathways.[168]

Bacteria, like eukaryotes, have the ability to segregate biochemical pathways in proteinaceous organelles, or 'nano/microcompartments'.[169] Recently, the microcompartments encoding a cyanobacterial carboxysome[170] and the *Citrobacter freundii* metabolosome for propanediol utilization[171] were successfully reconstituted in *E. coli*. Through addition of an N-terminal 18 amino acid targeting sequence, pyruvate decarboxylase and alcohol dehydrogenase were successfully directed to heterologously expressed metabolosomes in *E. coli* for engineered ethanol production.[172] For a *Salmonella enterica* LT2-derived nanocompartment, Schmidt-Dannert and coworkers[173] defined two motifs required for the binding of a targeting helix in the shell protein EutS to target two different reporter proteins to the nanocompartment, each with a different signal sequence. Custom engineered microcompartments with the capacity to encapsulate natural product pathways hold great promise for both increasing local concentrations of pathway intermediates (thereby increasing overall catalytic efficiency), and preventing toxicity effects of any intermediates in the heterologous host. Artificial scaffolds have also been used to increase local concentrations of natural product biosynthetic enzymes relative to one another; non-covalent associating affibodies tethered to biosynthetic enzymes increased sesquiterpene production by 135% in *S. cerevisiae*.[174]

Cell-free and *in vitro* systems eliminate toxicity issues from metabolite production and provide several advantages and disadvantages over biological systems.[175] Additional advantages range from the ability to precisely compile native and non-native pathway constituents, eliminating any impeding genetic regulation, and circumventing the need for sterility. Thus, a biosynthetic route can be quickly teased out with *in vitro* methods through addition or deletion of proteins in bottom-up and top-down approaches. Potential disadvantages include phenomena such as enzymatic feedback inhibition, cofactor regeneration, oxygen sensitivity, and the need for a universal buffering system for all enzymes involved in the pathway. As a testament to the advantages of *in vitro* natural product pathway reconstitution, the structurally complex type II polyketides enterocin and wailupemycin were successfully reconstituted *in vitro* through heterologous expressions of 11 proteins purified from either *E. coli* or *S. coelicolor* YU105.[176] SAM inhibition of the terminal P450 reaction required the mixture to be acidified and extracted prior to addition of the P450 and regenerating enzymes for optimal activity. The use of this *in vitro* reconstitution method also allowed the authors to tease out the Favorskii-like rearrangement activity catalyzed by the flavoenzyme EncM. In a follow-up study, Moore and coworkers[177] were able to efficiently test alternative benzoate analogs to create 24 unnatural polyketide variants. Hybrid approaches, where biosynthetic pathways have been partitioned with both *in vitro* and *in vivo* components, have been successful to synthesize a suite of complex glycosylated polyketides.[178] RiPP natural products termed cyanobactins were also recently reconstituted *in vitro*; this work unveiled a previously unknown plasticity of PatG to cyclize peptides 5–22 amino acids in length.[179]

Cell-free translation systems offer an additional level of flexibility for the production of natural and engineered RiPPs.[180] There are a variety of methods available to create *in vitro* translation systems.[181,182] The development of the promiscuous aminoacyl-tRNA synthetase ribozyme coined 'flexizyme' was the pinnacle for the production of cell-free custom peptides.[183] This technology has been combined with cell-free translation into the FIT system (flexible *in vitro* translation system) or additionally with mRNA display to incorporate peptide selection strategies in the RaPID system (random non-standard peptide integrated discovery).[184] Non-proteinogenic amino acid incorporation and post-translational processing including macrocyclization are accessible with these technologies. An interesting advance in using cell-free translation applications involves the production and regulation of genetic cascades in minimal synthetic cells: complex mixtures of metabolites and cell-free components encapsulated in liposomes. Boyden and coworkers[185] show that externally regulated gene circuits (through using a membrane-bound pore) could be used to promote catalytic turnover in one-, two-, or three-part circuits for luciferase production. Additionally, genetic circuits with unique microenvironments encapsulated in different liposomes were engineered to perform controlled fusion using SNARE complexes. This proof of concept demonstrated that capacity for genetic liposome modules to be mixed and matched, creating ever more complex gene circuits and pathways. Time will tell whether this technology can or will be utilized for complex natural product pathway assembly, where limitations concerning targeted, membrane-bound proteins could be overcome in a 'cell-free' environment.

11.8 Combinatorial Biosynthesis

Either in a cell-free, *in vitro* setting or through *in vivo* expressions, the fusion of enzymes from different metabolic pathways to invoke chemical and structural diversity in natural products has long been termed combinatorial biosynthesis.[186] The term 'synthetic biology' has recently gained popularity but the concepts are closely interwoven. For polyketides, some of the first *in vivo* combinatorial approaches were performed on iterative PKSs, such as those that produce actinorhodin, through genetic hybrids made in *Streptomyces coelicolor*.[120] More than a decade later, Townsend and coworkers published a series of seminal papers[187–191] successfully deconstructing and identifying individual domains of fungal iterative PKSs that facilitated the study and production of non-native iterative PKS metabolites. More recently, combinatorial biosynthesis was employed in yeast with these iterative PKS systems in an *in vivo* 'plug and play' fashion to create a library of benzenediol lactones.[192] KOSAN Biosciences was the early pioneer in type II modular PKS engineering, where the erythromycin biosynthetic pathway was subject to most study. Domain deletions in the erythromycin system along with replacements of acyltransferase domains from PKSs involved in rapamycin biosynthesis enabled the creation of an *in vivo* combinatorial library of over 50 erythromycin scaffold analogs in *S. lividans* and *S. coelicolor*.[193] Later efforts pushed forward the idea of designer polyketides with detailed studies of mixing and matching modules to create bi- and triketide libraries using a synthetic biology-like genetic approach, resulting in the identification of some important rules for module connectivity.[194] Interestingly, Piel and coworkers[195] revealed a natural combinatorial evolution strategy for the less-studied *trans*-AT PKSs[20] that provided compelling evidence for differences in these systems compared to the canonical *cis*-AT PKSs such as erythromycin. Despite major advances in our understanding of templated natural products like polyketides, the process of making complex, programmable custom natural products has yet to be truly realized.

NRPS-derived peptide natural products are another class of templated metabolites in which the evolutionary ramifications of their synthesis has long been studied.[196] Similar to PKSs, the non-structured linkers between domains and modules of NRPSs are important for catalytic fidelity and function.[197] However, like polyketides, non-ribosomal peptides often contain more than one NRPS megasynthase to create a given natural product. In support of this, interacting and complementary communication (COM) domains were first identified and combinatorially shuffled using the NRPSs involved in tyrocidine biosynthesis.[198,199] Engineering of the synonymous docking domains in PKS biosynthesis helped create a suite of 154 combinations of erythromycin-like PKS hybrids, approximately half of which were active.[200] Domain boundaries bridging PKS modules and NRPS modules of hybrid natural products have also been combinatorially designed and assessed for fungal systems.[201,202] Much of the early work on NRPS combinatorial biosynthesis was performed by Baltz and coworkers at Cubist Pharmaceuticals on their blockbuster drug Cubicin (daptomycin) produced by *Streptomyces roseoporus*.[203–205] In some exciting recent work, Bode and coworkers[206] discovered, characterized, and exploited the biosynthetic machineries producing rhabdopeptides[207] and xenortides,[208] which exist as natural chemical libraries produced by various *Photorhabdus* and *Xenorhabdus* strains symbiotically associated with nematodes. These pathways encode iterative monomodular NRPSs with highly flexible terminal condensation domains that accept a variety of amine substrates and plastic methyltransferase domains able to produce a suite of differentially methylated metabolites. By combining these NRPS modules *in vitro*, further combinatorial flexibility was observed and achieved through adjusting NRPS stoichiometries along with additional variability instigated by module skipping and crosstalk between non-native modules.

Another very effective route to the diversification of non-ribosomal peptides is through post-peptide synthesis tailoring reactions exemplified by the enormous complexity achieved though glycodiversification.[209] Multiple routes to activated sugars have been described and used in combination with promiscuous glycosyltransferases to create a 'sweet' suite of (non)native natural products that can have markedly different properties such as solubility, bioavailability, and target specificity.[210] *In vitro* enzymatic synthesis of complex carbohydrates used in chemoenzymatic approaches successfully added to natural product diversity through combinatorial glycosylation methods.[211] Solid phase synthesis of peptides followed by *in vitro* cyclization and release by NRPS-derived thioesterases was also a successful chemoenzymatic approach to increase the structural space of peptide natural products.[212] In an effort to circumvent the need for heterologous expression in natural product discovery, Brady and coworkers[213] combined *in silico* prediction of NRPS pathways and solid-phase synthesis in an attempt to directly discover new bioactive peptide scaffolds. This method may hold great promise in the future as more structure and target predictive software become robust and available, especially if and when combined with any ecologically relevant inferences of occurrence or activity.

Ribosomally synthesized and post-translationally modified peptides (RiPPs) are another class of secondary metabolites that naturally display diverse profiles as seen with conotoxins[214] and cyanobactins.[215] Due to the genetic coding of their peptide scaffolds, the potential to create large combinatorial peptide libraries is immense for this natural product class. In a prime example, van der Donk and coworkers[216] show that 29 unique precursors for lantipeptides named prochorosins were found spread throughout the genome of the cyanobacterium *Prochlorococcus* MIT9313. With strong sequence homology in the leader sequences but 'hyper-variability' in the core peptides, a single lanthionine synthetase encoded in the genome was shown to differentially dehydrate and form lanthionine bridges in all 29 unique precursors. A high throughput screen was combined with saturation mutagenesis libraries of the lasso peptide microcin J25. This method created nearly 100 variants with the four-gene cassette, where half harbored some level of antibacterial activity.[217] A more comprehensive library was designed and screened for the trunkamide pathway heterologously expressed in *E. coli*.[218] Over 300 unique cyclic peptides were observed in up to quadruple-saturating mutant libraries, revealing a highly plastic three-enzyme modifying pathway synthesizing the cyclic cyanobactin RiPPs. Moderate substrate tolerance was observed by a fourth modifying enzyme, the prenyltransferase TruF, which interestingly recognizes the cyclic peptides and not the leader sequence. This work was continued with this system and other cyanobactins by Schmidt and coworkers,[179] where unnatural derivatives were created *in vitro* with up to five different post-translational modifications in a combinatorial fashion. A highlight of this work was the production of different-sized macrocycles of up to 22 amino acids by utilizing PatG from the patellamide pathway. These papers provide exciting insight into the potential for full-pathway promiscuity in RiPP biosynthesis and the strengths of having the peptide scaffold directly encoded in the genome. While other classes of natural products not mentioned in this chapter have been subject to natural combinatorial biosynthesis, it should be noted that terpenoids also display an enormous natural diversity and thus are also attractive targets for combinatorially made (non)natural products. Peters and coworkers[219,220] showed the amazing substrate flexibility of bacterial- and plant-derived diterpene synthases, cyclases, and cytochrome P450s by using a modular metabolic engineering approach[221] for production of a large number of labdane scaffolds in *E. coli*.

11.9 Concluding Remarks

This chapter has aimed to survey the current status and challenges facing the field of natural products research and its continuing merger with synthetic biology. With the exponential growth of sequenced yet uncharacterized biosynthetic gene clusters and a pressing need for new therapeutics due to microbial resistance, we are currently limited in our abilities to quickly identify, *in silico* dereplicate, metabolomically prioritize, and heterologously produce and engineer new metabolites. The last decade has seen an explosion of DNA sequencing and synthesis technologies that have given us unparalleled access to Nature's blueprints of its impressive metabolic diversity. High throughput screening methods and DNA assembly strategies have begun to tackle Big Data issues, and further innovation of different synthetic parts and approaches is key to advancing the field of natural products. How will refactoring rationales evolve? Will subtly engineered polycistronic operons be tailored for optimal macromolecular assembly in specialized heterologous hosts? Will microbial hosts themselves be refactored to optimally control temporally regulated metabolic output? Can synthetic cell technology be engineered to afford complex cofactor assembly or mammalian post-translational machinery? Will microbial communities be engineered to coordinate and metabolically partition biosynthesis of a desired metabolite? The coming decade will test our creativity, intuition, and adaptability to effectively assimilate new scientific and technological breakthroughs constantly emerging from the exciting fields of synthetic biology and natural product biosynthesis.

References

1. M. S. Butler, *J. Nat. Prod.*, 2004, **67**, 2141.
2. D. Mastropaolo, A. Camerman, Y. Luo, G. D. Brayer and N. Camerman, *Proc. Natl. Acad. Sci. U. S. A.*, 1995, **92**, 6920.
3. M. Mackay and D. C. Hodgkin, *J. Chem. Soc.*, 1955, 3261.
4. J. S. S. Damsté, M. Strous, W. I. C. Rijpstra, E. C. Hopmans, J. A. J. Geenevasen, A. C. T. van Duin, L. A. van Niftrik and M. S. M. Jetten, *Nature*, 2002, **419**, 708.
5. Y. Shimizu, H. N. Chou, H. Bando, G. Van Duyne and J. Clardy, *J. Am. Chem. Soc.*, 1986, **108**, 514.

6. H.-W. Fehlhaber, M. Girg, G. Seibert, K. Hobert, P. Welzel, Y. Van Heijenoort and J. Van Heijenoort, *Tetrahedron*, 1990, **46**, 1557.
7. P. F. Wiley, K. Gerzon, E. H. Flynn, M. V. Sigal, O. Weaver, U. C. Quarck, R. R. Chauvette and R. Monahan, *J. Am. Chem. Soc.*, 1957, **79**, 6062.
8. T. Matsumoto, M. Yanagiya, S. Maeno and S. Yasuda, *Tetrahedron Lett.*, 1968, **9**, 6297–6300.
9. G. J. Pitt, *Acta Crystallogr.*, 1952, **5**, 770.
10. T. Hamada, S. Matsunaga, G. Yano and N. Fusetani, *J. Am. Chem. Soc.*, 2005, **127**, 110.
11. E. P. Abraham, E. Chain, C. M. Fletcher, H. W. Florey, A. D. Gardner, N. G. Heatley and M. A. Jennings, *Eur. J. Clin. Pharmacol.*, 1992, **42**, 3.
12. J. W.-H. Li and J. C. Vederas, *Science*, 2009, **325**, 161.
13. S. D. Bentley, K. F. Chater, A.-M. Cerdeño-Tárraga, G. L. Challis, N. R. Thomson, K. D. James, D. E. Harris, M. A. Quail, H. Kieser, D. Harper, A. Bateman, S. Brown, G. Chandra, C. W. Chen, M. Collins, A. Cronin, A. Fraser, A. Goble, J. Hidalgo, T. Hornsby, S. Howarth, C.-H. Huang, T. Kieser, L. Larke, L. Murphy, K. Oliver, S. O'Neil, E. Rabbinowitsch, M.-A. Rajandream, K. Rutherford, S. Rutter, K. Seeger, D. Saunders, S. Sharp, R. Squares, S. Squares, K. Taylor, T. Warren, A. Wietzorrek, J. Woodward, B. G. Barrell, J. Parkhill and D. A. Hopwood, *Nature*, 2002, **417**, 141.
14. N. Zaburannyi, M. Rabyk, B. Ostash, V. Fedorenko and A. Luzhetskyy, *BMC Genomics*, 2014, **15**, 97.
15. M. F. Freeman, C. Gurgui, M. J. Helf, B. I. Morinaka, A. R. Uria, N. J. Oldham, H.-G. Sahl, S. Matsunaga and J. Piel, *Science*, 2012, **338**, 387.
16. M. C. Wilson, T. Mori, C. Rückert, A. R. Uria, M. J. Helf, K. Takada, C. Gernert, U. A. E. Steffens, N. Heycke, S. Schmitt, C. Rinke, E. J. N. Helfrich, A. O. Brachmann, C. Gurgui, T. Wakimoto, M. Kracht, M. Crüsemann, U. Hentschel, I. Abe, S. Matsunaga, J. Kalinowski, H. Takeyama and J. Piel, *Nature*, 2014, **506**, 58.
17. J. C. Way, J. J. Collins, J. D. Keasling and P. A. Silver, *Cell*, 2014, **157**, 151.
18. B. A. Rudd and D. A. Hopwood, *J. Gen. Microbiol.*, 1979, **114**, 35.
19. K. Tahlan, H. U. Park and S. E. Jensen, *Can. J. Microbiol.*, 2004, **50**, 803.
20. J. Piel, *Proc. Natl. Acad. Sci. U. S. A.*, 2002, **99**, 14002.
21. J. Piel, G. Wen, M. Platzer and D. Hui, *ChemBioChem*, 2004, **5**, 93.
22. H.-W. Nützmann and A. Osbourn, *Curr. Opin. Biotechnol.*, 2014, **26**, 91.
23. D. J. Kliebenstein, *Annu. Rev. Phytopathol.*, 2012, **50**, 155.
24. U. Heinig, M. Gutensohn, N. Dudareva and A. Aharoni, *Curr. Opin. Biotechnol.*, 2013, **24**, 239–246.
25. M. Komatsu, T. Uchiyama, S. Omura, D. E. Cane and H. Ikeda, *Proc. Natl. Acad. Sci. U. S. A.*, 2010, **107**, 2646.
26. J. P. Gomez-Escribano and M. J. Bibb, *Microb. Biotechnol.*, 2011, **4**, 207.

27. Y.-M. Chiang, C. E. Oakley, M. Ahuja, R. Entwistle, A. Schultz, S.-L. Chang, C. T. Sung, C. C. C. Wang and B. R. Oakley, *J. Am. Chem. Soc.*, 2013, **135**, 7720.
28. V. Kolisnychenko, G. Plunkett, C. D. Herring, T. Fehér, J. Pósfai, F. R. Blattner and G. Pósfai, *Genome Res.*, 2002, **12**, 640.
29. G. Pósfai, G. Plunkett, T. Fehér, D. Frisch, G. M. Keil, K. Umenhoffer, V. Kolisnychenko, B. Stahl, S. S. Sharma, M. de Arruda, V. Burland, S. W. Harcum and F. R. Blattner, *Science*, 2006, **312**, 1044.
30. B. Csörgo, T. Fehér, E. Tímár, F. R. Blattner and G. Pósfai, *Microb. Cell Fact.*, 2012, **11**, 11.
31. F. J. Isaacs, P. A. Carr, H. H. Wang, M. J. Lajoie, B. Sterling, L. Kraal, A. C. Tolonen, T. A. Gianoulis, D. B. Goodman, N. B. Reppas, C. J. Emig, D. Bang, S. J. Hwang, M. C. Jewett, J. M. Jacobson and G. M. Church, *Science*, 2011, **333**, 348.
32. C. C. Liu and P. G. Schultz, *Annu. Rev. Biochem.*, 2010, **79**, 413.
33. A. E. Osbourn and B. Field, *Cell. Mol. Life Sci.*, 2009, **66**, 3755.
34. M. J. Bibb, *Curr. Opin. Microbiol.*, 2005, **8**, 208.
35. G. Liu, K. F. Chater, G. Chandra, G. Niu and H. Tan, *Microbiol. Mol. Biol. Rev.*, 2013, 77, 112.
36. E. J. Lawlor, H. A. Baylis and K. F. Chater, *Genes Dev.*, 1987, **1**, 1305.
37. M. A. Fernández-Moreno, J. Caballero, D. A. Hopwood and F. Malpartida, *Cell*, 1991, **66**, 769.
38. I. Höfer, M. Crüsemann, M. Radzom, B. Geers, D. Flachshaar, X. Cai, A. Zeeck and J. Piel, *Chem. Biol.*, 2011, **18**, 381.
39. X. Cai, R. Teta, C. Kohlhaas, M. Crüsemann, R. Ueoka, A. Mangoni, M. F. Freeman and J. Piel, *Chem. Biol.*, 2013, **20**, 839.
40. A. Espah Borujeni, A. S. Channarasappa and H. M. Salis, *Nucleic Acids Res.*, 2014, **42**, 2646.
41. H. M. Salis, E. A. Mirsky and C. A. Voigt, *Nat. Biotechnol.*, 2009, **27**, 946.
42. M. J. Bibb, G. R. Janssen and J. M. Ward, *Gene*, 1985, **38**, 215.
43. T. Murakami, T. G. Holt and C. J. Thompson, *J. Bacteriol.*, 1989, **171**, 1459.
44. T. Schmitt-John and J. W. Engels, *Appl. Microbiol. Biotechnol.*, 1992, **36**, 493.
45. N. Seghezzi, P. Amar, B. Koebmann, P. R. Jensen and M. J. Virolle, *Appl. Microbiol. Biotechnol.*, 2011, **90**, 615.
46. S. M. Wu, C. Feng, J. Zhong and L. D. Huan, *World J. Microbiol. Biotechnol.*, 2011, **27**, 99.
47. V. A. Rhodius, T. H. Segall-Shapiro, B. D. Sharon, A. Ghodasara, E. Orlova, H. Tabakh, D. H. Burkhardt, K. Clancy, T. C. Peterson, C. A. Gross and C. A. Voigt, *Mol. Syst. Biol.*, 2014, **9**, 702.
48. T. Ellis, X. Wang and J. J. Collins, *Nat. Biotechnol.*, 2009, **27**, 465.
49. K. J. Verstrepen and F. M. Klis, *Mol. Microbiol.*, 2006, **60**, 5.
50. P. J. Rutledge and G. L. Challis, *Nat. Rev. Microbiol.*, 2015, **13**, 509.
51. M. R. Seyedsayamdost, *Proc. Natl. Acad. Sci. U. S. A.*, 2014, **111**, 7266.

52. T. Lincke, S. Behnken, K. Ishida, M. Roth and C. Hertweck, *Angew. Chem., Int. Ed. Engl.*, 2010, **49**, 2011–2013.
53. M. Bibb and A. Hesketh, *Methods Enzymol.*, 2009, **458**, 93.
54. M. J. Smanski, H. Zhou, J. Claesen, B. Shen, M. A. Fischbach and C. A. Voigt, *Nat. Rev. Microbiol.*, 2016, **14**, 135.
55. T. Wakimoto, Y. Egami, Y. Nakashima, Y. Wakimoto, T. Mori, T. Awakawa, T. Ito, H. Kenmoku, Y. Asakawa, J. Piel and I. Abe, *Nat. Chem. Biol.*, 2014, **10**, 648.
56. M. Welch, S. Govindarajan, J. E. Ness, A. Villalobos, A. Gurney, J. Minshull and C. Gustafsson, *PLoS One*, 2009, **4**, e7002.
57. G. Boel, R. Letso, H. Neely, W. N. Price, K.-H. Wong, M. Su, J. D. Luff, M. Valecha, J. K. Everett, T. B. Acton, R. Xiao, G. T. Montelione, D. P. Aalberts and J. F. Hunt, *Nature*, 2016, **529**, 358–363.
58. H. Dong, L. Nilsson and C. G. Kurland, *J. Mol. Biol.*, 1996, **260**, 649.
59. A. Fuglsang, *Protein Expression Purif.*, 2003, **31**, 247.
60. P. M. Sharp and W. H. Li, *Nucleic Acids Res.*, 1987, **15**, 1281.
61. C. Kimchi-Sarfaty, J. M. Oh, I.-W. Kim, Z. E. Sauna, A. M. Calcagno, S. V. Ambudkar and M. M. Gottesman, *Science*, 2007, **315**, 525.
62. Y.-W. Shieh, P. Minguez, P. Bork, J. J. Auburger, D. L. Guilbride, G. Kramer and B. Bukau, *Science*, 2015, **350**, 678.
63. J. N. Wells, L. T. Bergendahl and J. A. Marsh, *Cell Rep.*, 2016, **14**, 679.
64. V. K. Mutalik, J. C. Guimaraes, G. Cambray, C. Lam, M. J. Christoffersen, Q.-A. Mai, A. B. Tran, M. Paull, J. D. Keasling, A. P. Arkin and D. Endy, *Nat. Methods*, 2013, **10**, 354.
65. C. Lou, B. Stanton, Y.-J. Chen, B. Munsky and C. A. Voigt, *Nat. Biotechnol.*, 2012, **30**, 1137–1142.
66. C. Bai, Y. Zhang, X. Zhao, Y. Hu, S. Xiang, J. Miao, C. Lou and L. Zhang, *Proc. Natl. Acad. Sci. U. S. A.*, 2015, **112**, 12181.
67. D. E. Cameron and J. J. Collins, *Nat. Biotechnol.*, 2014, **32**, 1276.
68. J. W. Tobias, T. E. Shrader, G. Rocap and A. Varshavsky, *Science*, 1991, **254**, 1374.
69. J. Fernandez-Rodriguez and C. A. Voigt, *Nucleic Acids Res.*, 2016, **44**, 6493.
70. J. T. Fitzgerald, L. K. Charkoudian, K. R. Watts and C. Khosla, *J. Am. Chem. Soc.*, 2013, **135**, 3752.
71. Z. Shao, G. Rao, C. Li, Z. Abil, Y. Luo and H. Zhao, *ACS Synth. Biol.*, 2013, **2**, 662.
72. K. Yamanaka, K. A. Reynolds, R. D. Kersten, K. S. Ryan, D. J. Gonzalez, V. Nizet, P. C. Dorrestein and B. S. Moore, *Proc. Natl. Acad. Sci. U. S. A.*, 2014, **111**, 1957.
73. G. Zhang, Q. Cao, J. Liu, B. Liu, J. Li and C. Li, *AIChE J.*, 2015, **61**, 3172.
74. K. Temme, D. Zhao and C. A. Voigt, *Proc. Natl. Acad. Sci. U. S. A.*, 2012, **109**, 7085.
75. M. J. Smanski, S. Bhatia, D. Zhao, Y. Park, L. B. A. Woodruff, G. Giannoukos, D. Ciulla, M. Busby, J. Calderon, R. Nicol, D. B. Gordon, D. Densmore and C. A. Voigt, *Nat. Biotechnol.*, 2014, **32**, 1241.

76. S. E. Unkles, V. Valiante, D. J. Mattern and A. A. Brakhage, *Chem. Biol.*, 2014, **21**, 502.
77. T. Ellis, T. Adie and G. S. Baldwin, *Integr. Biol.*, 2011, **3**, 109.
78. R. J. Roberts, *Proc. Natl. Acad. Sci. U. S. A.*, 2005, **102**, 5905.
79. C. D. Smolke, *Nat. Biotechnol.*, 2009, **27**, 1099.
80. C. Engler, R. Kandzia and S. Marillonnet, *PLoS One*, 2008, **3**, e3647.
81. D. G. Gibson, L. Young, R.-Y. Chuang, J. C. Venter, C. A. Hutchison and H. O. Smith, *Nat. Methods*, 2009, **6**, 343.
82. T. Gaj, C. A. Gersbach and C. F. Barbas III, *Trends Biotechnol.*, 2013, **31**, 397.
83. P. Horvath and R. Barrangou, *Science*, 2010, **327**, 167.
84. P. Mali, L. Yang, K. M. Esvelt, J. Aach, M. Guell, J. E. Dicarlo, J. E. Norville and G. M. Church, *Science*, 2013, **339**, 823.
85. R. Carlson, *Nat. Biotechnol.*, 2009, **27**, 1091.
86. S. F. Brady, *Nat. Protoc.*, 2007, **2**, 1297.
87. M. R. Rondon, P. R. August, A. D. Bettermann, S. F. Brady, T. H. Grossman, M. R. Liles, K. A. Loiacono, B. A. Lynch, I. A. MacNeil, C. Minor, C. L. Tiong, M. Gilman, M. S. Osburne, J. Clardy, J. Handelsman and R. M. Goodman, *Appl. Environ. Microbiol.*, 2000, **66**, 2541.
88. S. F. Brady and J. Clardy, *J. Am. Chem. Soc.*, 2000, **122**, 12903.
89. S. Hrvatin and J. Piel, *J. Microbiol. Methods*, 2007, **68**, 434.
90. K. M. Fisch, C. Gurgui, N. Heycke, S. A. van der Sar, S. A. Anderson, V. L. Webb, S. Taudien, M. Platzer, B. K. Rubio, S. J. Robinson, P. Crews and J. Piel, *Nat. Chem. Biol.*, 2009, **5**, 494.
91. B. H. Kvitko, I. A. McMillan and H. P. Schweizer, *Appl. Environ. Microbiol.*, 2013, **79**, 4869.
92. D. Du, L. Wang, Y. Tian, H. Liu, H. Tan and G. Niu, *Sci. Rep.*, 2015, **5**, 8740.
93. J. Fu, X. Bian, S. Hu, H. Wang, F. Huang, P. M. Seibert, A. Plaza, L. Xia, R. Müller, A. F. Stewart and Y. Zhang, *Nat. Biotechnol.*, 2012, **30**, 440.
94. Q. Liu, Q. Shen, X. Bian, H. Chen, J. Fu, H. Wang, P. Lei, Z. Guo, W. Chen, D. Li and Y. Zhang, *Sci. Rep.*, 2016, **6**, 34623.
95. J. Yin, M. Hoffmann, X. Bian, Q. Tu, F. Yan, L. Xia, X. Ding, A. F. Stewart, R. Müller, J. Fu and Y. Zhang, *Sci. Rep.*, 2015, **5**, 15081.
96. S. Hu, Z. Liu, X. Zhang, G. Zhang, Y. Xie, X. Ding, X. Mo, A. F. Stewart, J. Fu, Y. Zhang and L. Xia, *Sci. Rep.*, 2016, **6**, 29087.
97. Z. Shao, H. Zhao and H. Zhao, *Nucleic Acids Res.*, 2009, **37**, 1.
98. N. Kouprina and V. Larionov, *Nat. Rev. Genet.*, 2006, **7**, 805.
99. A. C. Ross, L. E. S. Gulland, P. C. Dorrestein and B. S. Moore, *ACS Synth. Biol.*, 2015, **4**, 414.
100. D. Montiel, H. Kang, F. Chang, Z. Charlop-Powers and S. F. Brady, *Proc. Natl. Acad. Sci. U. S. A.*, 2015, **112**, 8953.
101. H.-S. Kang, Z. Charlop-Powers and S. F. Brady, *ACS Synth. Biol.*, 2016, **5**, 1002.

102. W. Jiang, X. Zhao, T. Gabrieli, C. Lou, Y. Ebenstein and T. F. Zhu, *Nat. Commun.*, 2015, **6**, 8101.
103. C. T. Walsh and M. A. Fischbach, *J. Am. Chem. Soc.*, 2010, **132**, 2469.
104. T. Weber, K. Blin, S. Duddela, D. Krug, H. U. Kim, R. Bruccoleri, S. Y. Lee, M. A. Fischbach, R. Müller, W. Wohlleben, R. Breitling, E. Takano and M. H. Medema, *Nucleic Acids Res.*, 2015, **43**, W237.
105. P. Cimermancic, M. H. Medema, J. Claesen, K. Kurita, L. C. Wieland Brown, K. Mavrommatis, A. Pati, P. A. Godfrey, M. Koehrsen, J. Clardy, B. W. Birren, E. Takano, A. Sali, R. G. Linington and M. A. Fischbach, *Cell*, 2014, **158**, 412.
106. M. H. Medema and M. A. Fischbach, *Nat. Chem. Biol.*, 2015, **11**, 639.
107. M. T. Henke and N. L. Kelleher, *Nat. Prod. Rep.*, 2016, **33**, 942.
108. R. D. Kersten, Y.-L. Yang, Y. Xu, P. Cimermancic, S.-J. Nam, W. Fenical, M. A. Fischbach, B. S. Moore and P. C. Dorrestein, *Nat. Chem. Biol.*, 2011, **7**, 794.
109. R. D. Kersten, N. Ziemert, D. J. Gonzalez, B. M. Duggan, V. Nizet, P. C. Dorrestein and B. S. Moore, *Proc. Natl. Acad. Sci. U. S. A.*, 2013, **110**, E4407.
110. M. I. Vizcaino and J. M. Crawford, *Nat. Chem.*, 2015, **7**, 411.
111. D. D. Nguyen, C.-H. Wu, W. J. Moree, A. Lamsa, M. H. Medema, X. Zhao, R. G. Gavilan, M. Aparicio, L. Atencio, C. Jackson, J. Ballesteros, J. Sanchez, J. D. Watrous, V. V. Phelan, C. van de Wiel, R. D. Kersten, S. Mehnaz, R. De Mot, E. A. Shank, P. Charusanti, H. Nagarajan, B. M. Duggan, B. S. Moore, N. Bandeira, B. Ø. Palsson, K. Pogliano, M. Gutiérrez and P. C. Dorrestein, *Proc. Natl. Acad. Sci. U. S. A.*, 2013, **110**, E2611.
112. M. Wang, J. J. Carver, V. V. Phelan, L. M. Sanchez, N. Garg, Y. Peng, D. D. Nguyen, J. Watrous, C. A. Kapono, T. Luzzatto-Knaan, C. Porto, A. Bouslimani, A. V. Melnik, M. J. Meehan, W.-T. Liu, M. Crüsemann, P. D. Boudreau, E. Esquenazi, M. Sandoval-Calderón, R. D. Kersten, L. A. Pace, R. A. Quinn, K. R. Duncan, C.-C. Hsu, D. J. Floros, R. G. Gavilan, K. Kleigrewe, T. Northen, R. J. Dutton, D. Parrot, E. E. Carlson, B. Aigle, C. F. Michelsen, L. Jelsbak, C. Sohlenkamp, P. Pevzner, A. Edlund, J. McLean, J. Piel, B. T. Murphy, L. Gerwick, C.-C. Liaw, Y.-L. Yang, H.-U. Humpf, M. Maansson, R. A. Keyzers, A. C. Sims, A. R. Johnson, A. M. Sidebottom, B. E. Sedio, A. Klitgaard, C. B. Larson, C. A. Boya, P. D. Torres-Mendoza, D. J. Gonzalez, D. B. Silva, L. M. Marques, D. P. Demarque, E. Pociute, E. C. O'Neill, E. Briand, E. J. N. Helfrich, E. A. Granatosky, E. Glukhov, F. Ryffel, H. Houson, H. Mohimani, J. J. Kharbush, Y. Zeng, J. A. Vorholt, K. L. Kurita, P. Charusanti, K. L. McPhail, K. F. Nielsen, L. Vuong, M. Elfeki, M. F. Traxler, N. Engene, N. Koyama, O. B. Vining, R. Baric, R. R. Silva, S. J. Mascuch, S. Tomasi, S. Jenkins, V. Macherla, T. Hoffman, V. Agarwal, P. G. Williams, J. Dai, R. Neupane, J. Gurr, A. M. C. Rodríguez, A. Lamsa, C. Zhang, K. Dorrestein, B. M. Duggan, J. Almaliti, P.-M. Allard, P. Phapale, L.-F. Nothias, T. Alexandrov,

M. Litaudon, J.-L. Wolfender, J. E. Kyle, T. O. Metz, T. Peryea, D.-T. Nguyen, D. VanLeer, P. Shinn, A. Jadhav, R. Müller, K. M. Waters, W. Shi, X. Liu, L. Zhang, R. Knight, P. R. Jensen, B. Ø. Palsson, K. Pogliano, R. G. Linington, M. Gutiérrez, N. P. Lopes, W. H. Gerwick, B. S. Moore, P. C. Dorrestein and N. Bandeira, *Nat. Biotechnol.*, 2016, **34**, 828.

113. K. C. Peach, W. M. Bray, D. Winslow, P. F. Linington and R. G. Linington, *Mol. Biosyst.*, 2013, **9**, 1837.
114. K. L. Kurita, E. Glassey and R. G. Linington, *Proc. Natl. Acad. Sci. U. S. A.*, 2015, **112**, 11999.
115. T. Roemer and C. Boone, *Nat. Chem. Biol.*, 2013, **9**, 222.
116. A. B. Parsons, R. L. Brost, H. Ding, Z. Li, C. Zhang, B. Sheikh, G. W. Brown, P. M. Kane, T. R. Hughes and C. Boone, *Nat. Biotechnol.*, 2004, **22**, 62.
117. M. Costanzo, A. Baryshnikova, J. Bellay, Y. Kim, E. D. Spear, C. S. Sevier, H. Ding, J. L. Y. Koh, K. Toufighi, S. Mostafavi, J. Prinz, R. P. St Onge, B. VanderSluis, T. Makhnevych, F. J. Vizeacoumar, S. Alizadeh, S. Bahr, R. L. Brost, Y. Chen, M. Cokol, R. Deshpande, Z. Li, Z.-Y. Lin, W. Liang, M. Marback, J. Paw, B.-J. San Luis, E. Shuteriqi, A. H. Y. Tong, N. van Dyk, I. M. Wallace, J. A. Whitney, M. T. Weirauch, G. Zhong, H. Zhu, W. A. Houry, M. Brudno, S. Ragibizadeh, B. Papp, C. Pál, F. P. Roth, G. Giaever, C. Nislow, O. G. Troyanskaya, H. Bussey, G. D. Bader, A.-C. Gingras, Q. D. Morris, P. M. Kim, C. A. Kaiser, C. L. Myers, B. J. Andrews and C. Boone, *Science*, 2010, **327**, 425.
118. M. Costanzo, B. VanderSluis, E. N. Koch, A. Baryshnikova, C. Pons, G. Tan, W. Wang, M. Usaj, J. Hanchard, S. D. Lee, V. Pelechano, E. B. Styles, M. Billmann, J. van Leeuwen, N. van Dyk, Z.-Y. Lin, E. Kuzmin, J. Nelson, J. S. Piotrowski, T. Srikumar, S. Bahr, Y. Chen, R. Deshpande, C. F. Kurat, S. C. Li, Z. Li, M. M. Usaj, H. Okada, N. Pascoe, B.-J. San Luis, S. Sharifpoor, E. Shuteriqi, S. W. Simpkins, J. Snider, H. G. Suresh, Y. Tan, H. Zhu, N. Malod-Dognin, V. Janjic, N. Przulj, O. G. Troyanskaya, I. Stagljar, T. Xia, Y. Ohya, A.-C. Gingras, B. Raught, M. Boutros, L. M. Steinmetz, C. L. Moore, A. P. Rosebrock, A. A. Caudy, C. L. Myers, B. Andrews and C. Boone, *Science*, 2016, **353**, aaf1420.
119. C. W. Johnston, M. A. Skinnider, C. A. Dejong, P. N. Rees, G. M. Chen, C. G. Walker, S. French, E. D. Brown, J. Bérdy, D. Y. Liu and N. A. Magarvey, *Nat. Chem. Biol.*, 2016, **12**, 233.
120. S. Mo, D. H. Kim, J. H. Lee, J. W. Park, D. B. Basnet, Y. H. Ban, Y. J. Yoo, S. Chen, S. R. Park, E. A. Choi, E. Kim, Y.-Y. Jin, S.-K. Lee, J. Y. Park, Y. Liu, M. O. Lee, K. S. Lee, S. J. Kim, D. Kim, B. C. Park, S. Lee, H. J. Kwon, J.-W. Suh, B. S. Moore, S.-K. Lim and Y. J. Yoon, *J. Am. Chem. Soc.*, 2011, **133**, 976.
121. J. Cortes, S. F. Haydock, G. A. Roberts, D. J. Bevitt and P. F. Leadlay, *Nature*, 1990, **348**, 176.
122. G. A. Roberts, J. Staunton and P. F. Leadlay, *Eur. J. Biochem.*, 1993, **214**, 305.

123. C. M. Kao, L. Katz and C. Khosla, *Science*, 1994, **265**, 509.
124. B. A. Pfeifer, S. J. Admiraal, H. Gramajo, D. E. Cane and C. Khosla, *Science*, 2001, **291**, 1790.
125. M. Jiang and B. A. Pfeifer, *Metab. Eng.*, 2013, **19**, 42.
126. Y. Yang, Y. Lin, L. Li, R. J. Linhardt and Y. Yan, *Metab. Eng.*, 2015, **29**, 217.
127. J. H. An and Y. S. Kim, *Eur. J. Biochem.*, 1998, **257**, 395.
128. N. L. Pohl, M. Hans, H. Y. Lee, Y. S. Kim, D. E. Cane and C. Khosla, *J. Am. Chem. Soc.*, 2001, **123**, 5822.
129. A. J. Hughes and A. Keatinge-Clay, *Chem. Biol.*, 2011, **18**, 165.
130. M. C. Walker, B. W. Thuronyi, L. K. Charkoudian, B. Lowry, C. Khosla and M. C. Y. Chang, *Science*, 2013, **341**, 1089.
131. E. M. Musiol-Kroll, F. Zubeil, T. Schafhauser, T. Härtner, A. Kulik, J. McArthur, I. Koryakina, W. Wohlleben, S. Grond, G. J. Williams, S. Y. Lee and T. Weber, *ACS Synth. Biol.*, 2017, **6**, 421.
132. I. Koryakina, C. Kasey, J. B. McArthur, A. N. Lowell, J. A. Chemler, S. Li, D. A. Hansen, D. H. Sherman and G. J. Williams, *ACS Chem. Biol.*, 2017, **12**, 114.
133. Y. Liu, C. Hazzard, A. S. Eustáquio, K. A. Reynolds and B. S. Moore, *J. Am. Chem. Soc.*, 2009, **131**, 10376.
134. A. S. Eustáquio, R. P. McGlinchey, Y. Liu, C. Hazzard, L. L. Beer, G. Florova, M. M. Alhamadsheh, A. Lechner, A. J. Kale, Y. Kobayashi, K. A. Reynolds and B. S. Moore, *Proc. Natl. Acad. Sci. U. S. A.*, 2009, **106**, 12295.
135. M. C. Wilson and B. S. Moore, *Nat. Prod. Rep.*, 2012, **29**, 72.
136. C. T. Walsh, R. V. O'Brien and C. Khosla, *Angew. Chem., Int. Ed. Engl.*, 2013, **52**, 7098.
137. M. Winn, J. K. Fyans, Y. Zhuo and J. Micklefield, *Nat. Prod. Rep.*, 2016, **33**, 317.
138. T. Stachelhaus, H. D. Mootz and M. A. Marahiel, *Chem. Biol.*, 1999, **6**, 493.
139. G. L. Challis, J. Ravel and C. A. Townsend, *Chem. Biol.*, 2000, **7**, 211.
140. M. Crüsemann, C. Kohlhaas and J. Piel, *Chem. Sci.*, 2013, **4**, 1041.
141. H. Kries, D. L. Niquille and D. Hilvert, *Chem. Biol.*, 2015, **22**, 640.
142. A. Converti, B. Aliakbarian, J. M. Domínguez, G. Bustos Vázquez and P. Perego, *Braz. J. Microbiol.*, 2010, **41**, 519.
143. C. Fleige, F. Meyer and A. Steinbüchel, *Appl. Environ. Microbiol.*, 2016, **82**, 3410.
144. E. H. Hansen, B. L. Møller, G. R. Kock, C. M. Bünner, C. Kristensen, O. R. Jensen, F. T. Okkels, C. E. Olsen, M. S. Motawia and J. Hansen, *Appl. Environ. Microbiol.*, 2009, **75**, 2765.
145. J. Ni, F. Tao, H. Du and P. Xu, *Sci. Rep.*, 2015, **5**, 13670.
146. D. L. Klayman, A. J. Lin, N. Acton, J. P. Scovill, J. M. Hoch, W. K. Milhous, A. D. Theoharides and A. S. Dobek, *J. Nat. Prod.*, 1984, **47**, 715.
147. T. E. Wallaart, N. Pras, A. C. Beekman and W. J. Quax, *Planta Med.*, 2000, **66**, 57.

148. T. Kuzuyama and H. Seto, *Proc. Jpn. Acad., Ser. B*, 2012, **88**, 41.
149. S. W. Kim and J. D. Keasling, *Biotechnol. Bioeng.*, 2001, **72**, 408.
150. V. J. J. Martin, D. J. Pitera, S. T. Withers, J. D. Newman and J. D. Keasling, *Nat. Biotechnol.*, 2003, **21**, 796.
151. J. D. Newman, J. Marshall, M. Chang, F. Nowroozi, E. Paradise, D. Pitera, K. L. Newman and J. D. Keasling, *Biotechnol. Bioeng.*, 2006, **95**, 684.
152. D. J. Pitera, C. J. Paddon, J. D. Newman and J. D. Keasling, *Metab. Eng.*, 2007, **9**, 193.
153. H. Tsuruta, C. J. Paddon, D. Eng, J. R. Lenihan, T. Horning, L. C. Anthony, R. Regentin, J. D. Keasling, N. S. Renninger and J. D. Newman, *PLoS One*, 2009, **4**, e4489.
154. D. Ro, E. M. Paradise, M. Ouellet, K. J. Fisher, K. L. Newman, J. M. Ndungu, K. A. Ho, R. A. Eachus, T. S. Ham, J. Kirby, M. C. Y. Chang, S. T. Withers, Y. Shiba, R. Sarpong and J. D. Keasling, *Nature*, 2006, **440**, 940.
155. P. J. Westfall, D. J. Pitera, J. R. Lenihan, D. Eng, F. X. Woolard, R. Regentin, T. Horning, H. Tsuruta, D. J. Melis, A. Owens, S. Fickes, D. Diola, K. R. Benjamin, J. D. Keasling, M. D. Leavell, D. J. McPhee, N. S. Renninger, J. D. Newman and C. J. Paddon, *Proc. Natl. Acad. Sci. U. S. A.*, 2012, **109**, E111.
156. C. J. Paddon, P. J. Westfall, D. J. Pitera, K. Benjamin, K. Fisher, D. McPhee, M. D. Leavell, A. Tai, A. Main, D. Eng, D. R. Polichuk, K. H. Teoh, D. W. Reed, T. Treynor, J. Lenihan, M. Fleck, S. Bajad, G. Dang, D. Dengrove, D. Diola, G. Dorin, K. W. Ellens, S. Fickes, J. Galazzo, S. P. Gaucher, T. Geistlinger, R. Henry, M. Hepp, T. Horning, T. Iqbal, H. Jiang, L. Kizer, B. Lieu, D. Melis, N. Moss, R. Regentin, S. Secrest, H. Tsuruta, R. Vazquez, L. F. Westblade, L. Xu, M. Yu, Y. Zhang, L. Zhao, J. Lievense, P. S. Covello, J. D. Keasling, K. K. Reiling, N. S. Renninger and J. D. Newman, *Nature*, 2013, **496**, 528.
157. Q. Huang, C. A. Roessner, R. Croteau and A. I. Scott, *Bioorg. Med. Chem.*, 2001, **9**, 2237.
158. J. M. Dejong, Y. Liu, A. P. Bollon, R. M. Long, S. Jennewein, D. Williams and R. B. Croteau, *Biotechnol. Bioeng.*, 2006, **93**, 212.
159. B. Engels, P. Dahm and S. Jennewein, *Metab. Eng.*, 2008, **10**, 201.
160. P. K. Ajikumar, W.-H. Xiao, K. E. J. Tyo, Y. Wang, F. Simeon, E. Leonard, O. Mucha, T. H. Phon, B. Pfeifer and G. Stephanopoulos, *Science*, 2010, **330**, 70.
161. K. Zhou, K. Qiao, S. Edgar and G. Stephanopoulos, *Nat. Biotechnol.*, 2015, **33**, 377.
162. H. Renault, J.-E. Bassard, B. Hamberger and D. Werck-Reichhart, *Curr. Opin. Plant Biol.*, 2014, **19**, 27.
163. H. Ichinose and H. Wariishi, *Biochem. Biophys. Res. Commun.*, 2013, **438**, 289.
164. H. Minami, J.-S. Kim, N. Ikezawa, T. Takemura, T. Katayama, H. Kumagai and F. Sato, *Proc. Natl. Acad. Sci. U. S. A.*, 2008, **105**, 7393.

165. A. Z. Nielsen, B. Ziersen, K. Jensen, L. M. Lassen, C. E. Olsen, B. L. Møller and P. E. Jensen, *ACS Synth. Biol.*, 2013, **2**, 308.
166. G. Goñi, A. Zöllner, M. Lisurek, A. Velázquez-Campoy, S. Pinto, C. Gómez-Moreno, F. Hannemann, R. Bernhardt and M. Medina, *Biochim. Biophys. Acta*, 2009, **1794**, 1635.
167. M. F. Freeman, M. J. Helf, A. Bhushan, B. I. Morinaka and J. Piel, *Nat. Chem.*, 2016, **9**, 387.
168. H. Kuwahara, M. Alazmi, X. Cui and X. Gao, *Nucleic Acids Res.*, 2016, **44**, W217.
169. A. Chessher, R. Breitling and E. Takano, *ACS Biomater. Sci. Eng.*, 2015, **1**, 345.
170. W. Bonacci, P. K. Teng, B. Afonso, H. Niederholtmeyer, P. Grob, P. A. Silver and D. F. Savage, *Proc. Natl. Acad. Sci. U. S. A.*, 2012, **109**, 478.
171. J. B. Parsons, S. D. Dinesh, E. Deery, H. K. Leech, A. A. Brindley, D. Heldt, S. Frank, C. M. Smales, H. Lünsdorf, A. Rambach, M. H. Gass, A. Bleloch, K. J. McClean, A. W. Munro, S. E. J. Rigby, M. J. Warren and M. B. Prentice, *J. Biol. Chem.*, 2008, **283**, 14366.
172. A. D. Lawrence, S. Frank, S. Newnham, M. J. Lee, I. R. Brown, W.-F. Xue, M. L. Rowe, D. P. Mulvihill, M. B. Prentice, M. J. Howard and M. J. Warren, *ACS Synth. Biol.*, 2014, **3**, 454.
173. M. B. Quin, S. A. Perdue, S.-Y. Hsu and C. Schmidt-Dannert, *Appl. Microbiol. Biotechnol.*, 2016, **100**, 9187.
174. S. Tippmann, J. Anfelt, F. David, J. M. Rand, V. Siewers, M. Uhlén, J. Nielsen and E. P. Hudson, *ACS Synth. Biol.*, 2017, **6**, 19.
175. J.-K. Guterl and V. Sieber, *Eng. Life Sci.*, 2013, **13**, 4.
176. Q. Cheng, L. Xiang, M. Izumikawa, D. Meluzzi and B. S. Moore, *Nat. Chem. Biol.*, 2007, **3**, 557.
177. J. A. Kalaitzis, Q. Cheng, P. M. Thomas, N. L. Kelleher and B. S. Moore, *J. Nat. Prod.*, 2009, **72**, 469.
178. D. A. Hansen, C. M. Rath, E. B. Eisman, A. R. H. Narayan, J. D. Kittendorf, J. D. Mortison, Y. J. Yoon and D. H. Sherman, *J. Am. Chem. Soc.*, 2013, **135**, 11232.
179. D. Sardar, Z. Lin and E. W. Schmidt, *Chem. Biol.*, 2015, **22**, 907.
180. R. Maini, S. Umemoto and H. Suga, *Curr. Opin. Chem. Biol.*, 2016, **34**, 44.
181. H. H. Wang, P.-Y. Huang, G. Xu, W. Haas, A. Marblestone, J. Li, S. P. Gygi, A. C. Forster, M. C. Jewett and G. M. Church, *ACS Synth. Biol.*, 2012, **1**, 43.
182. A. Zemella, L. Thoring, C. Hoffmeister and S. Kubick, *ChemBioChem*, 2015, **16**, 2420.
183. H. Murakami, H. Saito and H. Suga, *Chem. Biol.*, 2003, **10**, 655.
184. C. J. Hipolito and H. Suga, *Curr. Opin. Chem. Biol.*, 2012, **16**, 196.
185. K. P. Adamala, D. A. Martin-Alarcon, K. R. Guthrie-Honea and E. S. Boyden, *Nat. Chem.*, 2017, **1**, 431–439.
186. C. J. Tsoi and C. Khosla, *Chem. Biol.*, 1995, **2**, 355.
187. J. M. Crawford, B. C. R. Dancy, E. A. Hill, D. W. Udwary and C. A. Townsend, *Proc. Natl. Acad. Sci. U. S. A.*, 2006, **103**, 16728.

188. J. M. Crawford, P. M. Thomas, J. R. Scheerer, A. L. Vagstad, N. L. Kelleher and C. A. Townsend, *Science*, 2008, **320**, 243.
189. J. M. Crawford, T. P. Korman, J. W. Labonte, A. L. Vagstad, E. A. Hill, O. Kamari-Bidkorpeh, S.-C. Tsai and C. A. Townsend, *Nature*, 2009, **461**, 1139.
190. A. L. Vagstad, A. G. Newman, P. A. Storm, K. Belecki, J. M. Crawford and C. A. Townsend, *Angew. Chem., Int. Ed. Engl.*, 2013, **52**, 1718.
191. A. G. Newman, A. L. Vagstad, P. A. Storm and C. A. Townsend, *J. Am. Chem. Soc.*, 2014, **136**, 7348.
192. Y. Xu, T. Zhou, S. Zhang, P. Espinosa-Artiles, L. Wang, W. Zhang, M. Lin, A. A. L. Gunatilaka, J. Zhan and I. Molnár, *Proc. Natl. Acad. Sci. U. S. A.*, 2014, **111**, 12354.
193. R. McDaniel, A. Thamchaipenet, C. Gustafsson, H. Fu, M. Betlach and G. Ashley, *Proc. Natl. Acad. Sci. U. S. A.*, 1999, **96**, 1846.
194. H. G. Menzella, J. R. Carney and D. V. Santi, *Chem. Biol.*, 2007, **14**, 143.
195. R. Ueoka, A. R. Uria, S. Reiter, T. Mori, P. Karbaum, E. E. Peters, E. J. N. Helfrich, B. I. Morinaka, M. Gugger, H. Takeyama, S. Matsunaga and J. Piel, *Nat. Chem. Biol.*, 2015, **11**, 705.
196. F. Lipmann, *Science*, 1971, **173**, 875.
197. K. J. Weissman and R. Müller, *ChemBioChem*, 2008, **9**, 826.
198. M. Hahn and T. Stachelhaus, *Proc. Natl. Acad. Sci. U. S. A.*, 2004, **101**, 15585.
199. M. Hahn and T. Stachelhaus, *Proc. Natl. Acad. Sci. U. S. A.*, 2006, **103**, 275.
200. H. G. Menzella, R. Reid, J. R. Carney, S. S. Chandran, S. J. Reisinger, K. G. Patel, D. A. Hopwood and D. V. Santi, *Nat. Biotechnol.*, 2005, **23**, 1171.
201. T. B. Kakule, Z. Lin and E. W. Schmidt, *J. Am. Chem. Soc.*, 2014, **136**, 17882.
202. M. L. Nielsen, T. Isbrandt, L. M. Petersen, U. H. Mortensen, M. R. Andersen, J. B. Hoof and T. O. Larsen, *PLoS One*, 2016, **11**, e0161199.
203. R. H. Baltz, P. Brian, V. Miao and S. K. Wrigley, *J. Ind. Microbiol. Biotechnol.*, 2006, **33**, 66.
204. K. T. Nguyen, D. Ritz, J.-Q. Gu, D. Alexander, M. Chu, V. Miao, P. Brian and R. H. Baltz, *Proc. Natl. Acad. Sci. U. S. A.*, 2006, **103**, 17462.
205. V. Miao, M.-F. Coëffet-Le Gal, K. Nguyen, P. Brian, J. Penn, A. Whiting, J. Steele, D. Kau, S. Martin, R. Ford, T. Gibson, M. Bouchard, S. K. Wrigley and R. H. Baltz, *Chem. Biol.*, 2006, **13**, 269.
206. X. Cai, S. Nowak, F. Wesche, I. Bischoff, M. Kaiser, R. Fürst and H. B. Bode, *Nat. Chem.*, 2016, **9**, 379.
207. D. Reimer, K. N. Cowles, A. Proschak, F. I. Nollmann, A. J. Dowling, M. Kaiser, R. Ffrench-Constant, H. Goodrich-Blair and H. B. Bode, *ChemBioChem*, 2013, **14**, 1991.
208. D. Reimer, F. I. Nollmann, K. Schultz, M. Kaiser and H. B. Bode, *J. Nat. Prod.*, 2014, 77, 1976.

209. R. W. Gantt, P. Peltier-Pain and J. S. Thorson, *Nat. Prod. Rep.*, 2011, **28**, 1811.
210. C. Zhang, B. R. Griffith, Q. Fu, C. Albermann, X. Fu, I.-K. Lee, L. Li and J. S. Thorson, *Science*, 2006, **313**, 1291.
211. H. Yu and X. Chen, *Org. Biomol. Chem.*, 2016, **14**, 2809.
212. R. M. Kohli, C. T. Walsh and M. D. Burkart, *Nature*, 2002, **418**, 658.
213. X. Vila-Farres, J. Chu, D. Inoyama, M. A. Ternei, C. Lemetre, L. J. Cohen, W. Cho, B. V. B. Reddy, H. A. Zebroski, J. S. Freundlich, D. S. Perlin and S. F. Brady, *J. Am. Chem. Soc.*, 2017, **139**, 1404.
214. S. R. Woodward, L. J. Cruz, B. M. Olivera and D. R. Hillyard, *EMBO J.*, 1990, **9**, 1015–1020.
215. M. S. Donia, J. Ravel and E. W. Schmidt, *Nat. Chem. Biol.*, 2008, **4**, 341–343.
216. B. Li, D. Sher, L. Kelly, Y. Shi, K. Huang, P. J. Knerr, I. Joewono, D. Rusch, S. W. Chisholm and W. A. van der Donk, *Proc. Natl. Acad. Sci. U. S. A.*, 2010, **107**, 10430.
217. S. J. Pan and A. J. Link, *J. Am. Chem. Soc.*, 2011, **133**, 5016.
218. D. E. Ruffner, E. W. Schmidt and J. R. Heemstra, *ACS Synth. Biol.*, 2015, **4**, 482.
219. S. Mafu, M. Jia, J. Zi, D. Morrone, Y. Wu, M. Xu, M. L. Hillwig and R. J. Peters, *Proc. Natl. Acad. Sci. U. S. A.*, 2016, **113**, 2526.
220. M. Jia, K. C. Potter and R. J. Peters, *Metab. Eng.*, 2016, **37**, 24.
221. A. Cyr, P. R. Wilderman, M. Determan and R. J. Peters, *J. Am. Chem. Soc.*, 2007, **129**, 6684.

Section IV: Biocatalysis for Modern Synthesis

CHAPTER 12

Self-contained Biocatalysts

JON D. STEWART*

University of Florida, Department of Chemistry, P.O. Box 117200, Gainesville, FL 32611, USA
Email: jds2@chem.ufl.edu

12.1 Introduction

While hydrolases, mutases, epimerases and isomerases generally operate without exogenous cofactors, many synthetically important conversions involve redox reactions that demand either a source or destination for electrons. This is particularly the case for nicotinamide-dependent reactions, which demand a continuing supply of $NAD(P)^+$ or NAD(P)H to support substrate oxidations or reductions, respectively. Due to their high costs, nicotinamides are almost never used stoichiometrically; instead, provision is made for their *in situ* regeneration from a sacrificial substrate or by direct coupling to an electrode. For many practical reasons, the former remains much more common, and will be the focus of this review. Before tackling the problem of self-contained redox biocatalysis, the three main strategies for cofactor regeneration are outlined below. These are illustrated schematically, using ketone reduction as an example of a biocatalytic conversion, in Figure 12.1.

12.1.1 The "One-enzyme" Solution

The simplest approach to nicotinamide regeneration requires only a second enzyme-catalyzed reaction that occurs in the complementary redox direction. For those alcohol dehydrogenases (also known as ketoreductases or

Catalysis Series No. 32
Modern Biocatalysis: Advances Towards Synthetic Biological Systems
Edited by Gavin Williams and Mélanie Hall

Published by the Royal Society of Chemistry, www.rsc.org

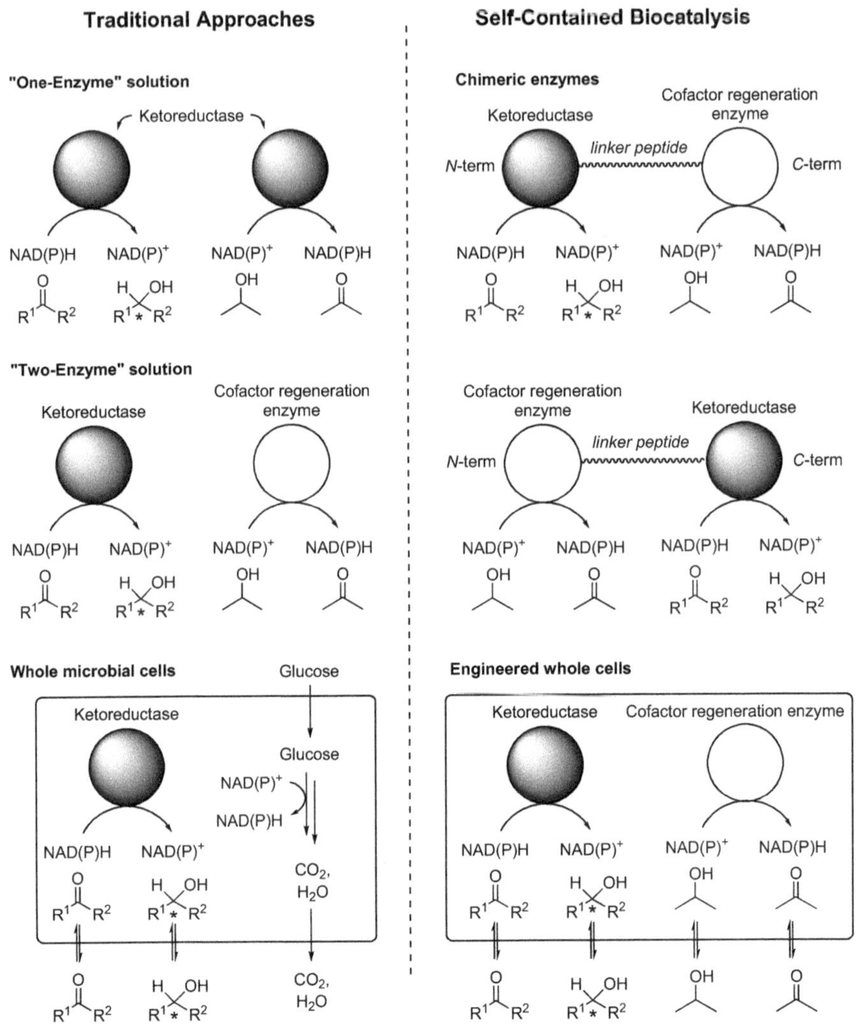

Figure 12.1 Schematic illustrations of cofactor regeneration strategies.

KREDs) that accept isopropanol/acetone as well as the substrate of synthetic interest, a single enzyme can accomplish both the desired transformation as well as cofactor regeneration (illustrated for a carbonyl reduction in Scheme 12.1). This is a biochemical equivalent to the Meerwein–Pondorf–Verley reduction/Oppenauer oxidation. In addition to requiring only a single biocatalyst, this approach also allows the highly polar nicotinamide cofactor to remain bound to the enzyme active site throughout the reaction, opening the possibility of carrying out these reactions in non-aqueous solvents. Moreover, because acetone is usually the most volatile reaction component, its removal by vacuum or pervaporation can be used to drive the equilibrium position toward the desired direction.

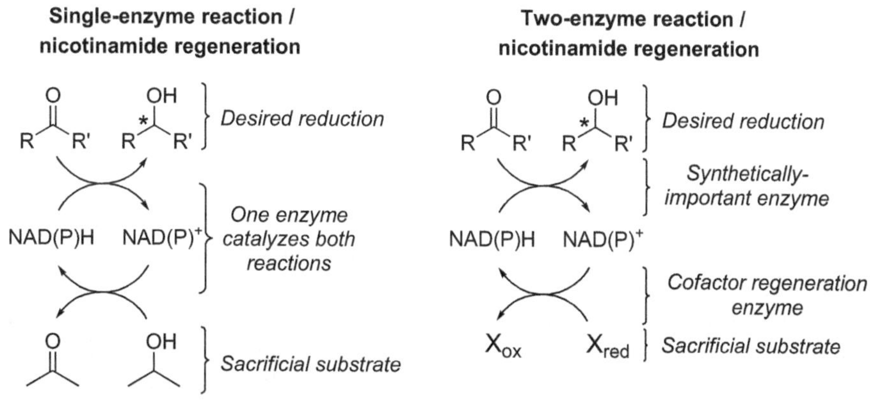

Scheme 12.1 One- and two-enzyme strategies for NAD(P)H regeneration.

One disadvantage of the "one-enzyme" solution to nicotinamide regeneration is that it is only applicable to the subset of $NAD(P)^+$/NAD(P)H-dependent reactions that are readily reversible. While carbonyl reductions/alcohol oxidations fall within this scope, Baeyer–Villiger oxidations, cytochrome P450-mediated oxidations, alkene reductions,[†] *etc.*, cannot be readily accommodated by this strategy. The other major drawback to the "one-enzyme" approach is that the equilibrium for the desired reaction depends on the thermodynamics and reactant concentration of the sacrificial substrate (usually *i*-PrOH/acetone). A large excess of the second substrate not only detracts from the economics of the process but can also result in enzyme deactivation.

12.1.2 The "Two-enzyme Solution"

Using a second enzyme—one tailored for nicotinamide cofactor regeneration—in addition to the one catalyzing the desired synthetic reaction eliminates the drawbacks of the "one-enzyme" approach (Scheme 12.1). Because the two enzymes catalyze different reactions, there is no requirement that their substrate specificities overlap. This also allows for choosing sacrificial substrates based on their costs, reduction potentials (to drive the overall reaction equilibrium in the desired direction without resorting to large molar excesses) and lack of detrimental effects on enzyme activity and stability. In addition, because a single cofactor regeneration enzyme/substrate pair can be coupled to many different synthetic transformations, process optimization (see Chapter 17) can be simplified since reaction conditions for only one of the two reactions must be developed. The two major drawbacks of the "two-enzyme" strategy are the additional costs of the second (cofactor regeneration) enzyme and the requirement that it has

[†]Cases of alkene reduction in nicotinamide-free system using flavin-dependent ene-reductases have been reported and rely on concomittant dehydrogenation of a suitable cosubstrate.[1–3]

good activity and stability under the reaction conditions demanded by the synthetically important enzyme. In practice, these requirements have been easily met by a variety of cofactor regeneration enzymes.

12.1.3 Whole Microbial Cells

Whole cells (usually microbial) are a logical extension of the "two-enzyme" strategy for cofactor regeneration (for recent reviews, see ref. 4–7). In favorable cases, the synthetically important enzyme is produced directly by the microbial cells; otherwise, cloning and molecular biology strategies are used for its heterologous expression in a suitable microbial host. In practice, even when native cells produce the enzyme of interest, they are re-engineered to increase its relative concentration, which often increases the volumetric productivity of the process. When whole microbial cells are used as the biocatalyst, a sacrificial substrate is nearly always included whose metabolism provides a continuous supply of the cofactor required by the synthetically important enzyme.[8] In most cases, the sacrificial substrate is a carbon source, *e.g.*, glucose, sucrose, glycerol, *etc.*, that also supports growth of the organism. In this scheme, the "cofactor regeneration enzyme" actually encompasses all of the enzyme-catalyzed, catabolic processes involved in carbon source utilization such as glycolysis and the citric acid cycle.[‡] There are two key advantages to using whole cells as the "cofactor regeneration enzyme." First, the enzyme that catalyzes the synthetically useful transformation, the cofactor regeneration system and the required cofactor are all supplied by the cells, which minimizes their costs. Second, because entire metabolic processes are available for carbon source metabolism, the sacrificial substrate can be used completely. For example, glucose oxidation by whole cells yields 10 NAD(P)H when consumed by glycolysis and the citric acid cycle as compared to 1 NAD(P)H when oxidized by glucose dehydrogenase. These advantages, however, must be balanced against some important drawbacks. First, because entire metabolic networks must be preserved during the biotransformation reaction, it is almost always essential that whole cells remain intact. High substrate concentrations desirable for synthetic purposes often cause cell membrane leakage and/or destruction. In addition, the presence of many enzymes in the reaction can lead to side-products and/or undesired stereoisomers. It is for this reason that the synthetically important enzyme is often overproduced so that its catalytic activity dominates, even when competing with all others in the cell. In some cases, this may make cofactor regeneration rate-limiting. Membrane transport of the substrate and/or product can also be partly or completely rate-limiting. Finally, the larger quantity of extraneous biomass inherent in using whole cells can complicate downstream processing and

---

[‡]Optimization of methanol oxidation pathway (dissimilation) in *Pichia pastoris* engineered strains has allowed the use of such strain for recombinant technology and self-sufficient whole-cell biocatalysis.[9,10]

product isolation (see Chapter 17). For all of these reasons, the use of unmodified whole cells for redox biocatalysis has declined in popularity as other cofactor regeneration schemes have become better developed.

One additional difficulty with using whole cells is that their growth phase can have major impacts on the efficiency of the biocatalytic process. Specifically, it is commonly observed that actively growing (log phase) cells perform well, but stationary phase cells have little or no activity. This seriously impacts reaction productivity because low cell densities have high catalytic efficiencies on a per-cell basis, but low volumetric productivities. High cell densities lead to stationary phase and poor catalytic efficiencies on a per-cell basis. One solution is to grow cells under optimal conditions that yield high levels of the desired enzyme, then carry out the biotransformation in a medium that supports high metabolism but not cell division. We have found that a high-glucose medium that lacks a nitrogen source can be very useful for this purpose when engineered *Escherichia coli* cells are used as the biocatalyst.[11]

12.1.4 Self-contained Biocatalysis

The "one-enzyme" solution to cofactor regeneration is an example of a self-contained biocatalytic process. Aside from the substrate and cofactor, a single enzyme provides everything needed for the desired reaction. The goal of the efforts reviewed here is to bring the "two-enzyme" and whole cell solutions closer to this ideal. One approach explored recently is to prepare artificial fusion proteins that encompass both the biocatalytic enzyme as well as a cofactor regeneration partner. Such a fusion guarantees that the two proteins will always be in a 1 : 1 molar ratio (although their relative catalytic efficiencies may differ). Several examples of this approach will be detailed below.

The second major effort in creating self-contained biocatalysts involves the engineering of whole cells to optimize their ability to supply redox cofactors. While this is a natural consequence of normal catabolism, these processes are not always sufficient to keep up with demand by an overexpressed redox enzyme. By overexpressing both the biocatalytic enzyme as well as one or more cofactor regeneration and/or biosynthetic enzymes, the per-cell productivity can be increased significantly. The goal is to maintain the advantages of whole microbial cells (inexpensive, simple to employ and supplying all needed cofactors) while overcoming their principle deficiency (low volumetric productivity).

This chapter focuses almost exclusively on biocatalytic processes involving an externally added substrate, rather than one derived from the host cell's metabolite pool using metabolic engineering strategies (one exception was included that was particularly relevant). While guaranteeing cofactor regeneration is essential to the success of metabolic engineering studies involving redox enzymes, their major focus is usually on optimizing the starting material supply level and eliminating losses to competing metabolic fates.

12.2 Chimeric Enzymes

In mythology, a chimera had the body and head of a lion, the head of a goat arising from its back, and a tail ending in a snake head. In the context of self-contained biocatalysis, chimeric proteins combine a redox enzyme that catalyzes the reaction of synthetic interest with a second enzyme that returns the redox cofactor (usually NAD(P)H) to the form needed for the next turnover. One key advantage of this strategy is that only a single protein must be isolated, and the two catalytic activities can never become separated during purification or during the biocatalytic process. In addition, because the two active sites are always physically near one another, diffusion times for the redox cofactor are minimized. The main drawback to this strategy (apart from the time necessary to carry out the molecular cloning steps required to construct the fusion protein gene) is that the catalytic efficiencies of the two partners can be compromised when their N- or C-termini are connected to a second protein. This can require optimization of the connecting linker peptide and empirical screening to determine the best order for the two partners.

For NAD(P)H, three enzymes have been used most commonly as regeneration partners: formate dehydrogenase, glucose dehydrogenase and phosphite dehydrogenase. These partners have been fused to cytochrome P450's, Baeyer–Villiger monooxygenases, a ketoreductase and two amino acid dehydrogenases. Each is discussed separately below.

12.2.1 Cytochrome P450's

Cytochrome P450 monooxygenases catalyze a variety of oxidative transformations including unactivated C–H hydroxylations, *O*-, *N*- and *S*-dealkylations, epoxidations and heteroatom oxidations. In addition to the heme-containing active site that actually carries out substrate oxidation, these enzymes also require one or more electron transport proteins that shuttle a pair of electrons from a nicotinamide to the Fe center in tight coordination with O_2 binding and activation. In a few cases, such as the well-studied *Bacillus megaterium* P450$_{BM3}$ fatty acid hydroxylase, the oxidizing hemoprotein domain is naturally fused with an NADH reductase domain. This makes electron transport highly efficient and dramatically simplifies the use of this enzyme for biocatalysis. The Arnold group in particular has used P450$_{BM3}$ as the basis for many protein engineering efforts that have improved its catalytic efficiency and extended its reaction range significantly (see ref. 12 and references therein).

In contrast to the intrinsically self-contained P450$_{BM3}$, the vast majority of P450's depend upon one or more exogenous electron transport domains, normally found on separate proteins. Cytochrome P450's are classified into families (I–X) by the number and arrangement of these electron transport partners. In many cases, one or more of these proteins are membrane-bound, which further increases the difficulty of using these proteins for preparative synthesis. For this reason, the majority of P450-mediated biocatalysis utilizes intact whole cells of the producing organism (often fungi). Unfortunately, because many microorganisms produce a range of P450's with overlapping substrate ranges, mixtures of products often result, which both lowers the yield and complicates product isolation.

Inspired by native fusion P450's such as $P450_{BM3}$, many groups have created artificial fusion proteins containing all components linked into a single entity. Because this field has been well-reviewed recently,[13] only a few recent examples are detailed below.

Rua *et al.* used a "molecular Lego" strategy to fuse the P450 2C20 with the reductase segment of $P450_{BM3}$ to create a self-sufficient chimera. This macaque P450 is closely related to human P450 2C8, which metabolizes around 10% of prescribed drugs.[14] The goal was to create a simple-to-employ protein for drug metabolism studies that eliminated the need for animal testing. The chimeric P450 had similar kinetic properties for two representative drugs (paclitaxel and amodiaquine) as the reconstituted wild-type enzyme.

The Kim group focused on the C_6 hydroxylation of daidzen **1**, catalyzed by P450 102D1 (Scheme 12.2).[15] Oxidized derivatives of this isoflavone have good therapeutic potential, but their chemical synthesis is difficult and the compounds occur at very low concentrations in fermented soybean. The authors fused the P450 daidzen C_6 hydroxylase with the electron transport domains of the self-sufficient CYP102D1 and compared the performance of the chimera with the reconstituted single protein mixture. The chimera had a k_{cat} value approximately 6-fold higher than the reconstituted system, demonstrating the success of the design. While the chimera could be overexpressed well in *E. coli*, its production depleted the host cells of NADPH. This led the authors to overexpress the chimera in *Streptomyces avermilitis*, an industrially used strain that produces avermectin antibiotics. This host was chosen because it maintained a higher $NADPH/NADP^+$ ratio and whole cells yielded a higher titer of product **2** than the corresponding whole *E. coli* cells.

For many years, the soluble camphor-hydroxylating P450 ($P450_{cam}$) served as the most-studied model system for this class of enzymes.[16] The native enzyme catalyzes the 5-*exo*-hydroxylation of camphor, but engineered variants with greater substrate range have been created, *e.g.*, the Y96F/V247L mutant oxidizes (−)-limonene. The need for separate electron transport proteins is the major drawback of $P450_{cam}$, which led Xu and co-workers to

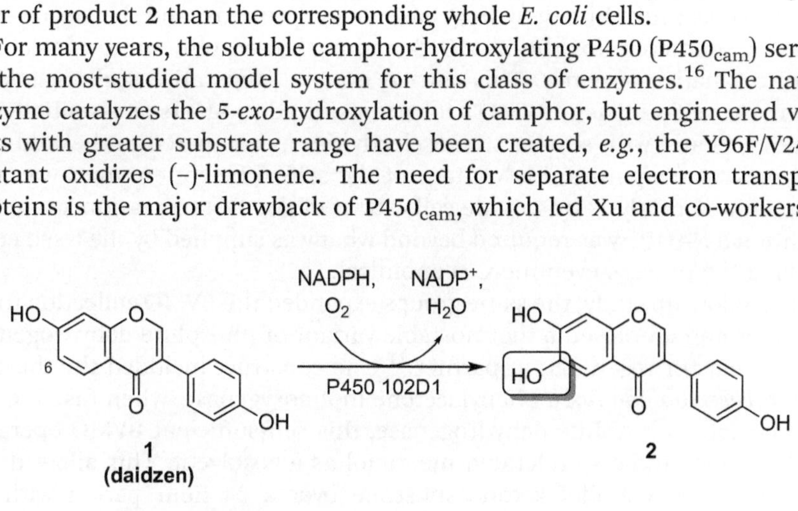

Scheme 12.2 Daidzen hydroxylation.

fuse this double mutant hemoprotein with the reductase and ferredoxin domains of the self-sufficient $P450_{SMO}$.[17] Four different linkers were explored to determine the best construct. Chimeras incorporating two of the linkers showed no catalytic activity; of the remaining two, one gave better efficiency than the other. Reactions in engineered *E. coli* cells were further improved by co-expressing glucose dehydrogenase, presumably because cellular NADPH levels were increased.

12.2.2 Baeyer–Villiger Monooxygenases

These flavoproteins catalyze the biochemical equivalent of the Criegee mechanism for ketone oxidations, using O_2 as the oxidant and yielding water as the only by-product (Scheme 12.3). This avoids the need for potentially explosive peracids, peroxides, *etc.* Moreover, because the key bond migration occurs in the chiral environment of a protein, these enzymes usually afford high regio- and stereoselectivities. A lack of protein stability for some Baeyer–Villiger monooxygenases (BVMO's) and a requirement for NADPH are the major reason for their (so far) modest impact on organic synthesis. The discovery and characterization of new BVMO's with greater stability has overcome the first problem. In the 1990s and early 2000s, whole cell methods were developed to solve the NADPH supply problem.[11,18,19] While effective, these whole-cell methods required that the *E. coli* cells remained intact throughout the reaction, which limited the usable substrate concentration and imposed membrane transport rate limitations.

The Mihovilovic and Fraaije groups developed an elegant solution to the NADPH supply problem by covalently linking several BVMO's with phosphite reductase.[20] The $NADP^+$-mediated oxidation of phosphite strongly favors P^V formation on thermodynamic grounds and the by-product phosphate is highly innocuous. Three of the four targeted BVMO's were efficiently overexpressed as phosphite dehydrogenase fusion proteins, and the steady-state kinetic parameters for the BVMO domains showed only minor deviations from the native enzymes. This carried over to preparative scale reactions, where the self-sufficient BVMO's formed oxidized products at essentially the same rates and with similar stereoselectivities as compared to the wild-type enzymes. One significant advantage of the self-sufficient enzymes was that they allowed catalysis by a crude cell extract of the overexpression strain. No additional $NADP^+$ was required beyond what was supplied by the lysed cells, making the process even more economical.

In a follow-up study, the same groups expanded the BVMO collection to six enzymes and employed a thermostable variant of phosphite dehydrogenase as the cofactor regeneration partner.[21] One construct included the thermostable *Thermobifida fusca* phenylacetone monooxygenase; when fused to the thermostable phosphite dehydrogenase, this self-sufficient BVMO operated well at 50 °C and also tolerated methanol as a cosolvent. This allowed full conversion of a model ketone substrate over a 24 hour period with no interference from enzyme inactivation.

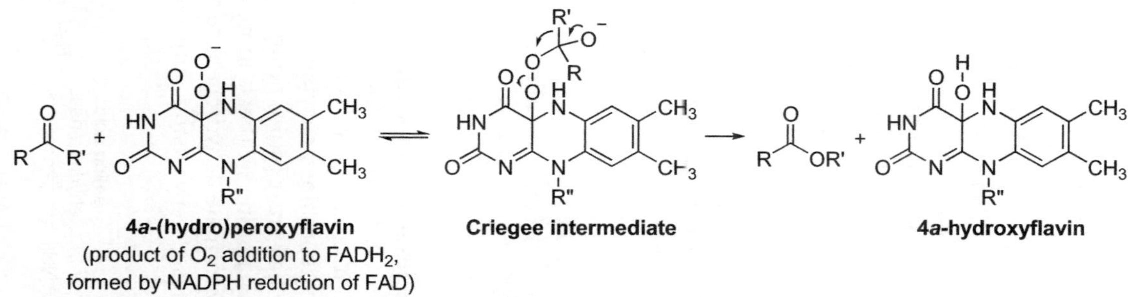

Scheme 12.3 Flavoprotein Baeyer–Villiger oxidation mechanism.

12.2.3 Amino Acid Dehydrogenases

Amino acid dehydrogenases catalyze the NADH-dependent reductive amination of α-keto acids, yielding the corresponding α-amino acids (Scheme 12.4). Because the starting materials are achiral, this stereoselective reduction can provide an economical route to a wide variety of amino acids, both natural and non-natural. Nearly all amino acid dehydrogenases are highly selective for NADH (over NADPH) and providing this cofactor in a cost-competitive manner has been a key driver of work in this area.

The most common NADH regeneration strategy involves formate oxidation by formate dehydrogenase (Scheme 12.4). Formate is inexpensive and the by-product (CO_2) is a gas whose loss to the atmosphere pulls the equilibrium strongly toward NADH production. The large-scale production of L-*tert*-leucine using leucine dehydrogenase and *Candida boidinii* formate dehydrogenase in a membrane reactor was one of the early success stories of industrial biocatalysis[22] and this compound is still an important chiral building block.

Lu *et al.* recently explored the possibility of assembling leucine dehydrogenase and formate dehydrogenase in order to increase the process efficiency.[23] In this case, the two proteins were not joined covalently, but assembled by non-covalent interactions in an assembly inspired by the cellulosome.[24] Specifically, the large protein cohesin forms a scaffold onto which dockerin modules bind with very high affinities in the presence of Ca^{2+} ($K_D \approx 10^{-9}$–10^{-12} M). Both leucine dehydrogenase and formate dehydrogenase were fused to dockerin modules and mixed with a synthetic scaffold containing two different cohesins domains to ensure a 1:1:1 stoichiometry between the three proteins as well as a defined spatial relationship. The scaffold also contained a cellulose-binding domain, which allowed the assembled complex to be immobilized on an insoluble cellulose support. The final complex was indeed more efficient than the corresponding free proteins and it could be recycled, albeit with some loss of activity. This strategy, which combines enzyme complex formation, immobilization and purification, has great future promise.

In an effort to make phenylalanine synthesis more cost-efficient, Jiang and Fang fused *Candida boidinii* formate dehydrogenase with *Bacillus halodurans* phenylalanine dehydrogenase.[25] By using ammonium formate as the

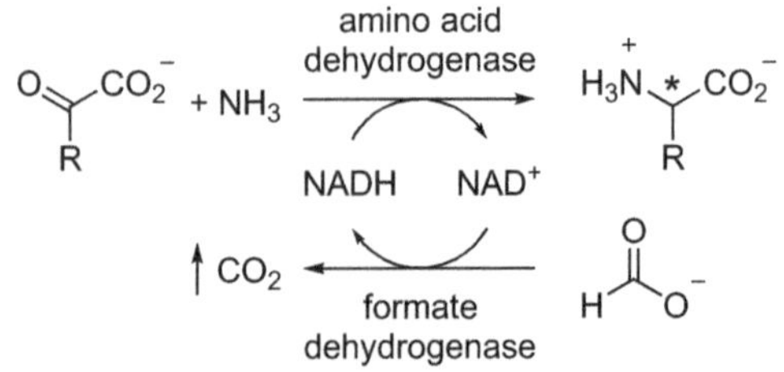

Scheme 12.4 Amino acid dehydrogenase route to α-amino acids.

co-substrate along with phenylpyruvate, all the necessary reaction components were included. The fusion protein showed somewhat better thermal stability than the separate enzymes as well as a moderate improvement in catalytic efficiency (1.5-fold).

12.2.4 3-Ketoacyl-carrier-protein Reductase

Fatty acid biosynthesis proceeds by an iterative process involving sequential Claisen condensation, β-keto ester reduction, dehydration and alkene reduction. This process continues until the C_{16} (palmitate) stage, at which point a thioesterase hydrolytically liberates the free fatty acid. These and related modules from polyketide synthases have proven useful as biocatalysts for asymmetric synthesis (for a recent example, see ref. 26). One drawback, however, is that some modules—such as the 3-ketoacyl-carrier-protein reductase from a cyanobacterium (*Synechococcus* sp.)—are membrane-bound, which makes them difficult to use for preparative synthesis.

Hölsch and Weuster-Botz devised a general strategy based on self-sufficient fusion proteins to provide the reduced nicotinamide needed by the reductase that also enhances its aqueous solubility.[27] Previous work had demonstrated that fusing the ketoreductase to the maltose-binding protein yielded a soluble fusion protein; unfortunately, the affinity tag was designed only to enhance purification. This was improved by fusing the ketoreductase to an NAD^+-dependent formate dehydrogenase (FDH), which both increased ketoreductase solubility and also supplied its essential cofactor.

Fusion proteins were constructed with the ketoreductase preceding FDH and *vice versa*. Unfortunately, both showed a propensity to form inclusion bodies when overexpressed in *E. coli*, although it was possible to obtain some soluble protein for characterization. The construct with FDH at the N-terminus catalyzed both formate oxidation and carbonyl reduction; the converse arrangement afforded only ketoreductase activity. Even in the former case, however, the efficiency of the FDH was reduced as compared to the free enzyme. A small library of linkers with varying lengths was evaluated. The results showed that linker length was more critical than linker sequence. The best fusion construct provided FDH activity that was 80% that of the free enzyme. When compared to combining the individual ketoreductase and FDH proteins in solution, the optimized fusion construct gave essentially the same activity.

12.3 Engineered Whole Cells

Given that whole cells already regenerate cofactors by normal metabolism, why incorporate additional enzymes? There are several reasons:

- *By incorporating cofactor regeneration enzymes into whole cells, one can increase the potential flux for NAD(P)H regeneration beyond what can be supplied by native metabolism.* For example, when a highly efficient cofactor-requiring enzyme is highly overproduced, its cofactor demand can easily outstrip the cell's ability to supply to cofactor.

- *Cofactor regeneration by native metabolism involves long sequential chains of reactions, all of which must retain function for the cell to maintain its ability to supply cofactors.* For example, when a carbon source such as glucose is used for *E. coli* cells, major sources of NADH are step 6 of glycolysis (glyceraldehyde-3-phosphate dehydrogenase) and the citric acid cycle. The major sources of NADPH are the pentose phosphate pathway and the citric acid cycle under aerobic conditions.[28] This scheme is therefore vulnerable to the loss of any single enzyme in these pathways.
- *Most native pathways that supply cofactors rely on an intact cell membrane.* As noted above, high substrate loadings (desirable for high space–time yields) usually damage or even dissolve cell membranes, leading to dispersion of cellular contents throughout the reaction mixture. Under these conditions, native cofactor regeneration pathways are essentially non-functional because the necessary enzyme components are much more dilute.
- *Cells with efficient cofactor regeneration can be employed as crude lysates, rather than as whole cells.* Intact cell membranes introduce membrane transport issues for both substrates and products, and these rates can limit the biocatalytic performance of whole cells. When a simple and efficient cofactor regeneration scheme is combined with a highly overexpressed biocatalytic enzyme, the dilution that occurs from dispersing cellular contents into the reaction mixture still leaves the two enzymes at sufficiently high concentrations to carry on the desired reaction with good space–time yields. This completely avoids any mass transport limitations imposed by cell membranes.

Two enzymes have dominated cofactor regeneration strategies in whole cells: formate dehydrogenase (NADH) and glucose dehydrogenase (NADH and NADPH). In both cases, the "metabolic pathway" required for cofactor regeneration involves only a single enzyme, which can be overexpressed at high levels. This solves the cofactor flux problem as well as the requirement for multi-enzyme pathways.

Several examples of whole-cell self-sufficient biocatalysts are described below, organized by the synthetic reaction of interest. In some cases, the whole cells remain intact during the bioprocess; in others, cells are deliberately lysed either prior to the reaction or *in situ*, by the addition of high substrate loadings. The latter is particularly useful since it avoids all the steps required for cell lysis and also allows the biocatalytic cells to be accumulated and stored prior to the reaction (usually by freezing). Decoupling of the biocatalyst preparation from its use is particularly valuable for large-scale industrial-scale processes.

Because multiple proteins must be co-expressed when creating self-sufficient whole cells, balancing the catalytic activities of the relevant enzymes is an important consideration. When fusion proteins are employed, there is a guaranteed 1 : 1 molar ratio between the protein components; this cannot be assumed when individual proteins are co-expressed in a single cell. It is therefore necessary to balance both protein expression levels, as modified by the relative specific activities in order to create the optimum whole-cell biocatalyst.

12.3.1 Cytochrome P450's

Lu and Mei constructed a self-sufficient whole cell *E. coli* strain for indigo production based on a mutant of $P450_{BM3}$.[29] Because native $P450_{BM3}$ is already a self-sufficient enzyme, Lu and Mei focused on NADPH supply to improve the overall biocatalytic process. Both the $P450_{BM3}$ and a glucose dehydrogenase were co-expressed from a single plasmid. One problem was that the glucose dehydrogenase was largely produced in insoluble inclusion bodies. After extensive optimization studies, it was possible to increase the specific activities of $P450_{BM3}$ and glucose dehydrogenase to 8200 and 0.045 U mg^{-1}. While these conditions yielded 2.9 mM indigo from an initial indole concentration of 5.0 mM, it is clear that there is still potential for better matching of the two specific activities.

More recently, Li and co-workers constructed a more complex self-sufficient *E. coli* strain designed to hydroxylate three representative heterocycles (Scheme 12.5). The system was based on a triple mutant of P450pyr (known as $P450_{TM}$) since this variant provided higher enantioselectivity toward these substrates as compared to the wild-type enzyme.[30] This hemoprotein also requires two additional proteins (ferredoxin and ferredoxin

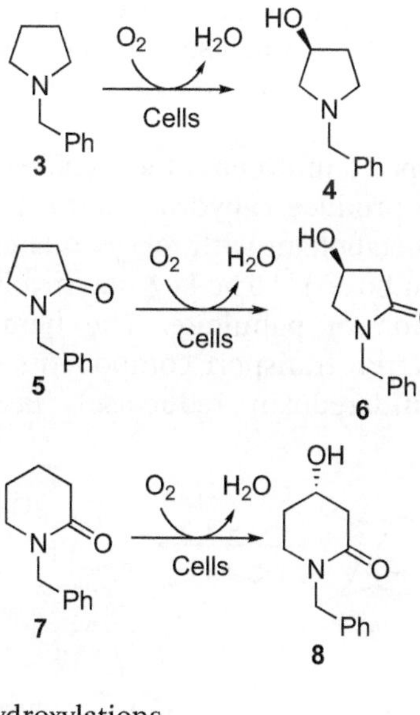

Scheme 12.5 $P450_{TM}$ hydroxylations.

reductase). Glucose dehydrogenase was chosen as the NAD(P)H regeneration catalyst. In total, four separate proteins were co-expressed in a single *E. coli* strain ($P450_{TM}$, ferredoxin, ferredoxin reductase and glucose dehydrogenase). This was an ambitious undertaking and a number of molecular biology approaches were explored before the best protein production system was identified (11.7 U g^{-1} cdw and 140 U g^{-1} protein for $P450_{TM}$ and glucose dehydrogenase, respectively). When induced cells were suspended in a phosphate buffer lacking a nitrogen source (to prevent entry into stationary phase) and glucose was provided, the target compounds were hydroxylated with high efficiencies and stereoselectivities. By comparing the reaction time courses for the four-protein strain *versus* an otherwise identical strain lacking glucose dehydrogenase, it was clear that *in situ* cofactor regeneration was critical in prolonging the reaction time beyond the initial 6 h.

Kino and co-workers used $P450_{cam}$ to hydroxylate 2-adamantanone stereoselectively at the 5-position (Scheme 12.6). Like the previous example, $P450_{cam}$ requires both a ferredoxin and ferredoxin reductase (putidaredoxin and putidaredoxin reductase) in addition to the hemoprotein component. These three proteins were co-expressed in a single *E. coli* strain, which relied on host metabolism to supply the required NADH. A next-generation strain was then prepared in which glucose oxidation by glucose dehydrogenase directly supplied NADH in a single step and the efficiencies of the two strains were compared. When used directly after growth, cells lacking glucose dehydrogenase afforded 5.4 mM **10** after 8 hours with no added glucose; the same cells yielded 19 mM **10** after 8 hours when glucose was added to the reaction medium, demonstrating that host cell metabolism was capable of supplying NADH. Interestingly, when the strain co-expressing glucose dehydrogenase was substituted, nearly the same concentration of **10** was formed after 8 hours (17 mM), although glucose was metabolized at a much faster rate. By adding solid **9** portionwise, the rate of substrate consumption increased by *ca.* 1.3-fold and a slightly higher yield of **10** was obtained (90% conversion).

The Kim group developed a multi-enzyme cascade supported by a cofactor regeneration enzyme to produce ω-hydroxypalmitic acid by diverting and supplementing *E. coli* metabolism with exogenous enzymes (Scheme 12.7; added enzymes shown in boxes).[31] The key step involved the P450-catalyzed intracellular hydroxylation of palmitate. The hemoprotein of CYP153A was paired with the electron transport components of the $P450_{cam}$ system (putidaredoxin and putidaredoxin reductase). Because ω-hydroxylation

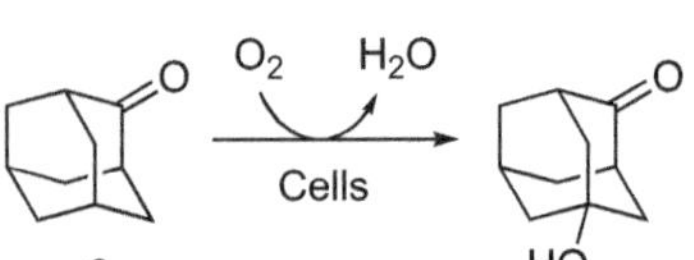

9 **10**

Scheme 12.6 P450-catalyzed adamantanone hydroxylation.

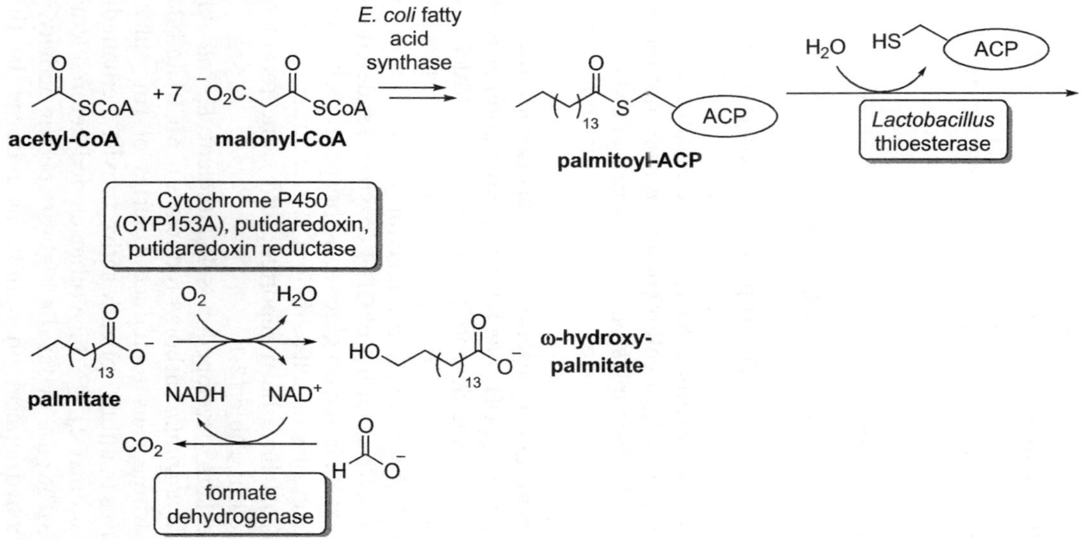

Scheme 12.7 Production of ω-hydroxy fatty acids using P450 oxidation.

required NADH, *C. boidinii* formate dehydrogenase was co-expressed and its substrate (formate) was included in the reaction mixtures. To increase the intracellular palmitate level, additional copies of some of its biosynthetic enzymes were also co-expressed (*Lactobacillus* thioesterase and *E. coli* β-hydroxyacetyl-ACP dehydratase; not shown). A number of gene knockouts were also incorporated to prevent diversion of biosynthetic precursors to other, non-palmitate pathways. Finally, a disabling key *E. coli* alcohol dehydrogenase (AdhE) eliminated competition for NADH and yielded a 50% higher NADH/NAD$^+$ ratio as compared to wild-type. The optimized strain yielded 611 mg L^{-1} ω-hydroxypalmitate from glucose, the highest level reported in shake flasks.

12.3.2 Ketoreductases

Pairing a ketoreductase (alcohol dehydrogenase) with a cofactor regeneration enzyme forms an extremely simple metabolic pathway composed of only two catalysts (an NAD(P)H-producing enzyme and an NAD(P)H-consuming enzyme). A key advantage is that even if cells lyse under the reaction conditions, nicotinamide cofactor regeneration can easily continue since the complexities of host cell metabolism need not remain functional. As will be seen in the examples below, this can allow very high substrate loadings (≥1.0 M), which makes these processes more attractive from a preparative standpoint.

Because many synthetically useful ketoreductases are NADPH-dependent, glucose dehydrogenase (GDH) has been the most popular cofactor regeneration partner. The enzyme accepts both NAD$^+$ and NADP$^+$, the substrate (glucose) is inexpensive and biocompatible and the product gluconolactone spontaneously hydrolyzes under the reaction conditions, strongly favoring NAD(P)H formation. The cloning of GDH enzymes rendered former NADPH regeneration strategies based on glucose-6-phosphate dehydrogenase largely obsolete. Many examples of self-sufficient *E. coli* strains combining a ketoreductase with GDH have been reported, and some recent representative examples are listed in Table 12.1.

Zhu and coworkers developed a self-sufficient *E. coli* strain that co-expresses *Candida magnolia* carbonyl reductase (CMCR) and *Bacillus subtilis* GDH and used whole cells to reduce a series of ring-substituted acetophenone derivatives to afford optically active 1-aryl-1,2-ethanediols with high stereoselectivities (entry 1).[32,33] Five other cloned ketoreductases were also evaluated, but CMCR gave by far the highest specific activity. Impressively, one substrate (α-hydroxyacetophenone) was reduced by the lyophilized whole cells (50 mg mL^{-1}) at a concentration of 1.0 M within 16 hour. For best results, 1.0 mM NADP$^+$ was also added to the reaction mixture. At such high substrate loading, it is possible that this reaction takes place from a crude lysate formed *in situ* during the reaction, and for this reason, the cellular NADP$^+$ became too dilute and efficient ketone reduction required additional NADP$^+$.

Table 12.1 Recent examples of carbonyl reductions by engineered self-sufficient *E. coli* cells.

Entry	Substrates	Products	Ketoreductase	Cofactor regeneration enzyme	Ref.
1	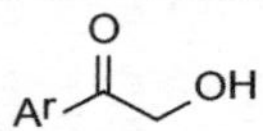	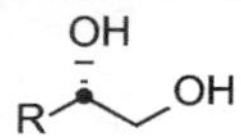	*Candida magnolia* carbonyl reductase	*B. subtilis* GDH	32
	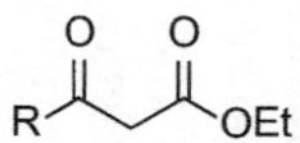	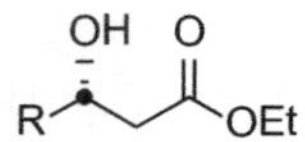			
2	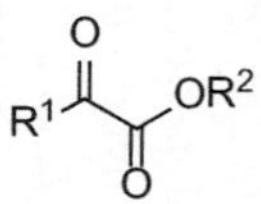	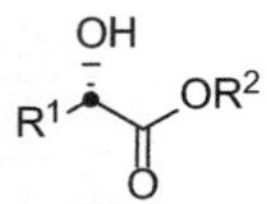	*Bacillus* sp. ECU0013 carbonyl reductase	*B. subtilis* GDH	34
	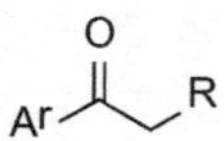	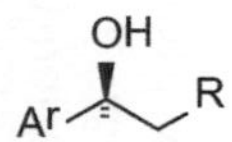			
3	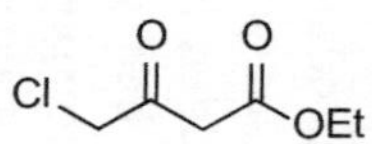	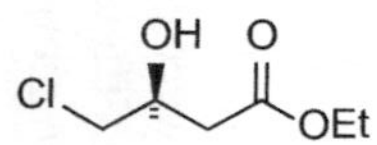	*Pichia stipitis* carbonyl reductase	*B. megaterium* GDH	35
4	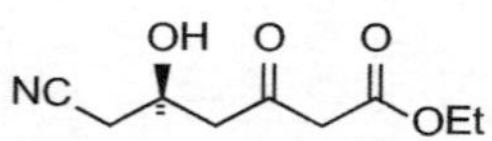	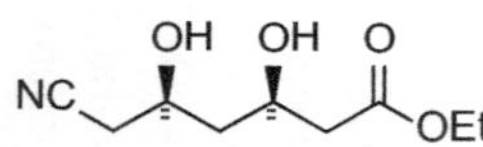	*Saccharomyces cerevisiae* carbonyl reductase (YDL124w)	*B. megaterium* GDH	36
5	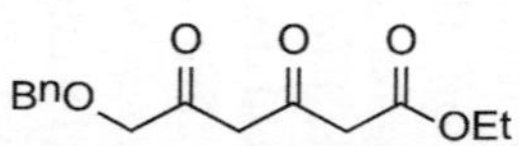	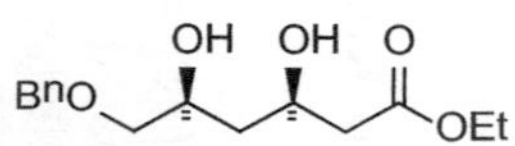	*Acinetobacter baylyi* diketoreductase	*B. megaterium* GDH	38

Xu, Li and co-workers profiled a *Bacillus* sp. ketoreductase against a variety of representative β-keto esters, α-keto esters and substituted acetophenones (entry 2).[34] The NADPH required by the ketoreductase was supplied by a co-expressed GDH, which allowed whole cells to be used in the presence of glucose as the biocatalytic reagent. For both β-keto esters and substituted acetophenones, alcohols were formed with >99% ee; the enzyme proved less stereoselective toward α-keto esters. In some cases, however, specific activities were very low (<0.1 U mg^{-1} protein). The engineered whole cells were used in a biphasic water–toluene mixture to reduce 1.3 M ethyl 4-chloro-3-oxobutyrate to the corresponding (*R*)-alcohol in 92% yield and with 99.6% ee. To avoid enzyme deactivation, the substrate was added in three equal portions. More glucose was also added at the same time, and as in the previous example, a small amount of exogenous NADP$^+$ had to be supplied (likely because of *in situ* cell lysis).

The Ying group also studied the asymmetric reduction of ethyl 4-chloro-3-oxobutyrate since the alcohol can be used as a chiral building block, *e.g.*, for statin drugs such as Lipitor® (entry 3).[35] In earlier studies, the group had identified and cloned an (*S*)-selective reductase from *Pichia stipitis* that was NADH-dependent. This was combined in a self-sufficient *E. coli* strain overexpressing GDH that provided an NADH supply. After preliminary studies, biphasic conditions (water–*n*-BuOAc or water–ethyl caprylate) yielded *ca.* 1.0 and 1.4 M of the target alcohol in the organic phase in the presence of glucose and 0.10 mM NAD$^+$.

Biocatalytic production of a more advanced statin building block was explored by Chen and co-workers (entry 4).[36] The self-sufficient *E. coli* strain overexpressed *Saccharomyces cerevisiae* YDL124w[37] along with GDH to supply the requisite NADPH. After optimization, the substrate was reduced with high diastereoselectivity at a concentration of 0.15 M after 7 h.

The same group tackled the more ambitious double reduction of a statin precursor (entry 5).[38] The NADPH-dependent diketoreductase from *Acinetobacter baylyi* and GDH were co-expressed in *E. coli*. One unusual observation was that the catalytic efficiency of ketone reduction was limited by the intracellular cofactor concentration in *E. coli*.

The Zheng group very recently reported an interesting deracemization process for 2-hydroxy acids (Scheme 12.8).[39] Unlike a traditional kinetic resolution that is burdened by a maximum product yield of 50%, deracemizations allow 100% of an initially racemic starting material to be converted to a single product enantiomer. A number of (*S*)-selective 2-hydroxyacid dehydrogenases were identified by genome mining (see Chapter 1) and evaluated; this yielded an FMN-dependent enzyme from *Pseudomonas aeruginosa* NSUT that had good catalytic efficiency, very high (*S*)-stereoselectivity and high overexpression in *E. coli*. The ultimate electron acceptor for this FMN-dependent enzyme was not determined and O_2, a flavocytochrome b_2 or ubiquinone, are all possibilities. All would be readily accessible in a whole-cell bioprocess. An NADH-dependent *Leuconostoc mesenteroides* (*R*)-2-ketoacid dehydrogenase was co-expressed in the same *E. coli* strain.

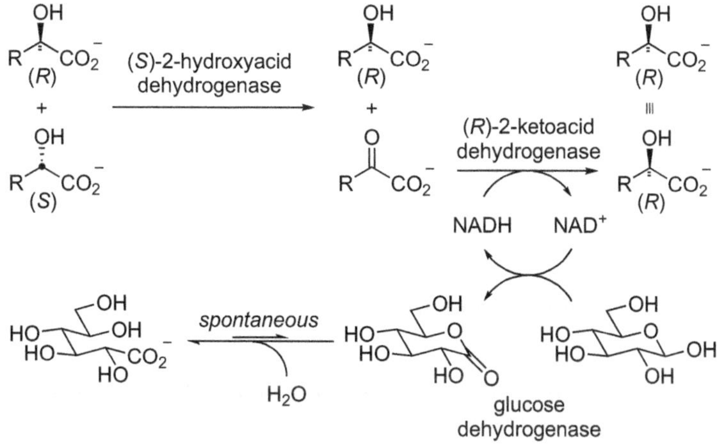

Scheme 12.8 2-Hydroxyacid deracemization.

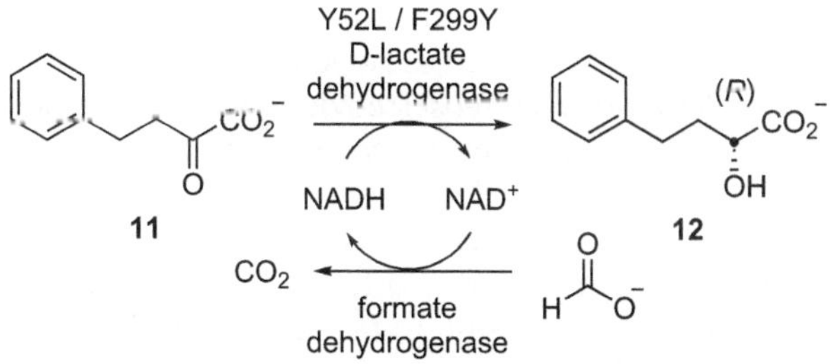

Scheme 12.9 2-Hydroxyacid synthesis by a lactate dehydrogenase mutant.

Finally, *Exiguobacterium sibiricum* GDH was introduced to regenerate the NADH needed by the ketoreductase. Whole cells of the final *E. coli* strain were used to deracemize 19 substrates successfully in 2–6 hours, with final optical purifies of >99% ee and a substrate loading of 20 mM. An important key to the success of this strategy is that the oxidation step does *not* rely on nicotinamide cofactors, which avoids the need to maintain separate nucleotide pools, one favoring NAD(P)H and the other $NAD(P)^+$ so that both oxidation and reduction can occur simultaneously.

As noted above, most synthetically useful ketoreductases rely on NADPH, but some are NADH-selective. Most glucose dehydrogenases show dual specificity,[40] but it is also possible to co-express an NAD^+/NADH cofactor regeneration enzyme, *e.g.*, formate dehydrogenase. The final two examples in this section illustrate this strategy.

Chiral 2-hydroxyacids such as **12** are important pharmaceutical building blocks (Scheme 12.9).[41] A double mutant of *Lactobacillus bulgaricus* D-lactate dehydrogenase was developed to allow 2-keto acids with large groups at C_3 to be substrates for the enzyme. Because the dehydrogenase is NADH-dependent, *C. boidinii* formate dehydrogenase was co-expressed, then the

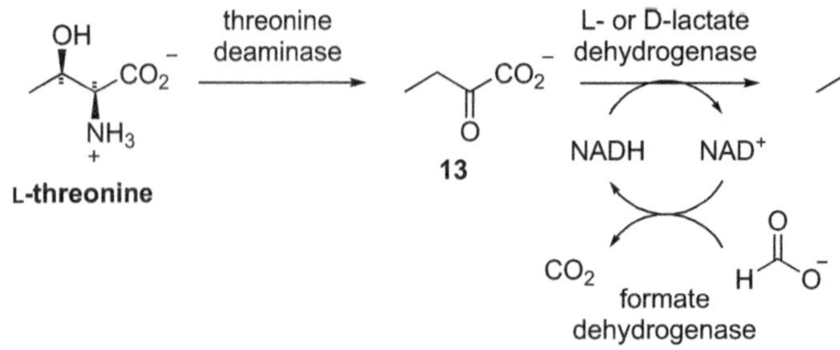

Scheme 12.10 2-Hydroxybutyrate synthesis from L-threonine.

resulting whole *E. coli* cells were used to reduce **11**. After optimizing the reaction conditions, 72 mM **12** could be formed with >99% ee with a productivity of 48 mM h^{-1}.

The Zhu group developed an ingenious route to both enantiomers of 2-hydroxybutyrate **14** that used L-threonine as the starting material (Scheme 12.10).[42] This solved the substrate supply problem since **13** is not commercially available. Two NADH-dependent lactate dehydrogenases (LDH's) with complementary stereoselectivities were identified (*Oryctolagus cuniculus* L-LDH for (*S*)-**14** and *Staphylococcus epidermis* D-LDH for (*R*)-**14**). Formate dehydrogenase was used to regenerate intracellular NADH at the expense of added formate. Two *E. coli* strains were created. Each overexpressed threonine deaminase and FDH along with one of the two LDH's. These were used to convert 750 mM L-threonine to (*R*)- and (*S*)-**14** in yields of 93% and 97%, respectively, and with optical purities of >99% ee. Because threonine deaminase is a pyridoxal phosphate-dependent enzyme, adding exogenous pyridoxal phosphate allowed the conversions to occur somewhat more rapidly.

12.3.3 Amino Acid Dehydrogenase

As described previously, leucine dehydrogenase can be used to synthesize analogs, especially L-*tert*-leucine, which is a valuable chiral building block for pharmaceutical synthesis. Xu and co-workers identified and cloned a novel leucine dehydrogenase from *Exiguobacterium sibiricum* that gave greater activity, even down to 0 °C. The same enzyme also accepted a variety of other α-keto acids, which prompted the development of a self-sufficient biocatalyst system (Scheme 12.11).[43] This used glucose dehydrogenase for NADH regeneration at the expense of glucose. Whole cells of the self-sufficient *E. coli* were used to prepare 78 g of L-*tert*-leucine from a 1.0 L reaction in 5.5 hour and with an 80% yield.

12.3.4 NAD^+ Regeneration

Most biocatalytic processes involve NAD(P)H-dependent reductions, since these processes convert prochiral sp^2 carbons into chiral sp^3 stereocenters.

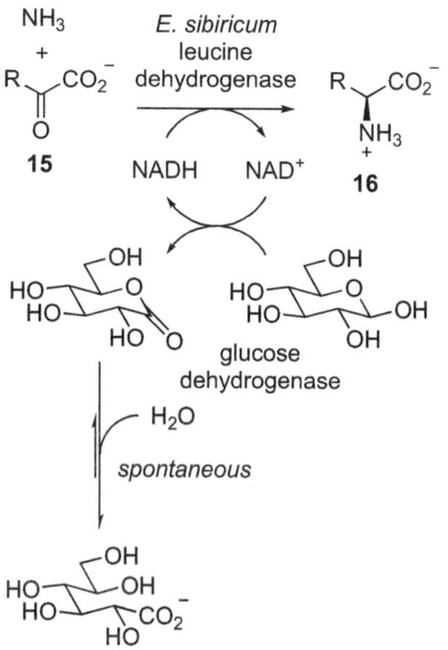

Scheme 12.11 Leucine dehydrogenase-catalyzed synthesis of α-amino acids.

However, the converse reaction, functional group oxidation at the expense of $NAD(P)^+$, does find important applications in preparative synthesis. For these reasons, some self-sufficient whole-cell systems have been developed to supply nicotinamides predominantly in their oxidized forms. The key to this approach is to identify a highly exothermic reaction of NAD(P)H that yields an innocuous by-product. To date, the best enzymes for this purpose are water-forming NAD(P)H oxidases, first popularized by the Bommarius group more than a decade ago.[44] One advantage of coupling synthetically useful reactions with an NAD(P)H oxidase is that no co-substrate need be added as long as the reaction mixture is in contact with air; moreover, the cofactor regeneration by-product (water) poses no separation problem for downstream processing.

The groups of Ma and Xu co-expressed the water-forming *Lactobacillus brevis* NADH oxidase along with an NADH-dependent (2*R*,3*R*)-butanediol dehydrogenase.[45] The stereoselectivity of the dehydrogenase allowed conversion of various butanediol feedstocks to chiral acetoin. For example, 43 $g L^{-1}$ *meso*-2,3-butanediol was converted to 37 $g L^{-1}$ of (3*S*)-acetoin in 12 hours. When 43 $g L^{-1}$ (2*R*,3*R*)-2,3-butanediol was used as the feedstock, 42 $g L^{-1}$ of (3*R*)-acetoin were formed. Finally, the same *E. coli* cells could be used to produce optically pure (2*S*,3*S*)-2,3-butanediol by the kinetic resolution of a mixture of 2,3-butanediol diastereomers. Unfortunately, the final yield was relatively poor because the desired (2*S*,3*S*)-enantiomer made up only a small fraction of the input substrate mixture.

The growth of the biodiesel industry has led to a glut of glycerol, the by-product of transesterification of triacylglycerides with short-chain alcohols such as methanol or ethanol. Zhang and co-workers devised a self-sufficient whole-cell strategy to valorize glycerol by selective C_2-oxidation to yield dihydroxyacetone (Scheme 12.12).[46] NAD^+-dependent *E. coli* glycerol dehydrogenase catalyzed the desired oxidation, and the water-forming NADH oxidase from *Lactobacillus pentosus* returned the nicotinamide to its oxidized form for the next round of turnover. An *E. coli* strain that overexpressed both proteins could be used in the form of whole cells, which disappointingly yielded only 0.2 $g\,L^{-1}$ of dihydroxyacetone, even when additional NAD^+ was supplied to the reaction. Much better results were obtained from a crude lysate of the same strain, which yielded 4 $g\,L^{-1}$ of dihydroxyacetone with no need for additional NAD^+. While no explanation for the performance difference was provided, one can speculate that membrane transport of the highly polar glycerol and dihydroxyacetone molecules may have been at least partly rate-limiting. It is also possible that additional metabolic pathways may have consumed much of the dihydroxyacetone product when intact cells were employed.

The Li group combined several enzymes in an ingenious conversion of *meso*- or racemic epoxides into (*R*)-α-hydroxy ketones (Scheme 12.13).[47] A highly active epoxide hydrolase from *Sphingomonas* sp. HXN-200 catalyzed the initial ring-opening of **17**, which provided the (*R*,*R*)-alcohols with very high stereoselectivities. Butanediol dehydrogenase from *B. subtilis* catalyzed the NAD^+-dependent oxidation of **18** to afford the corresponding α-hydroxy ketones **19**. *L. brevis* water-forming NADH oxidase regenerated the NAD^+. After preliminary studies with separate strains for the three enzymes, a triple

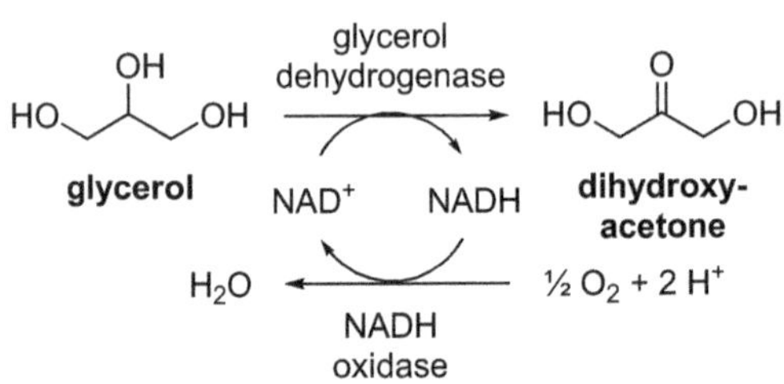

Scheme 12.12 Oxidation of glycerol to dihydroxyacetone.

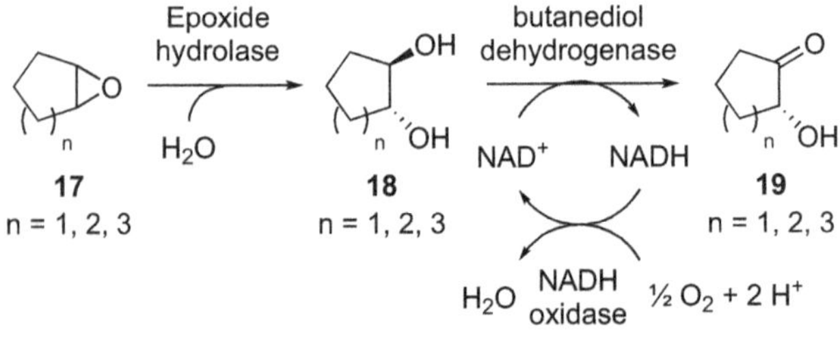

Scheme 12.13 Epoxide desymmetrization.

overexpression strain that co-produced all three enzymes was created. Whole cells were used in a two-phase system (aqueous buffer–hexadecane) with epoxides present at 100 or 200 mM. These provided 52%–71% conversion (depending on the substrate) with 98% or 99% ee. The same system was also applied to the kinetic resolution of a racemic epoxide.

More recently, the Li group described several highly ambitious, multi-step conversions of alkenes to α-hydroxy acids, α-hydroxy primary amines and α-amino acids.[48] Because the longest reaction sequence required six different conversions, four modules were designed for interconnections in various ways to generate all of the final products (Scheme 12.14). α-Hydroxyaldehyde **23** was the key intermediate. This was produced from substituted styrenes by initial enantioselective epoxidation by a well-characterized styrene monooxygenase, hydrolysis by the same epoxide hydrolase as in the previous example, and finally, an FAD-coupled oxidation of the primary alcohol by AlkJ to give (*S*)-**23**. The AlkJ alcohol dehydrogenase could use several terminal electron acceptors to regenerate its FAD cofactor. A transamination yielded amino alcohol **24**. The alanine co-substrate was regenerated by reductive amination using alanine dehydrogenase. The pathway to the final two products involved an NAD^+-mediated oxidation of **23** to the corresponding (*S*)-α-hydroxy acid **25**. Further oxidation yielded the corresponding α-keto acid **26**, which was transaminated to phenylglycine **27**. In this case, the transaminase was glutamate-dependent, and the α-ketoglutarate co-product (shown as α-KG in Scheme 12.14) was converted back to glutamate by an NAD^+-dependent glutamate dehydrogenase.

E. coli strains that co-expressed all of the required proteins for production of **24**, **25** and **27** were created and their performance evaluated. A variety of ring-substituted styrenes were used as feedstocks. Good conversions (69%–99%) were found for whole-cell conversions of styrenes to **25** at substrate loadings of 20 mM. A two phase-system was employed (1 : 1 aqueous buffer–hexadecane) and glucose was included to support cofactor regeneration by host–cell pathways. Similar conditions were used to produce amino alcohols **24**, although conversions were somewhat more variable. In this case, additional glucose and ammonia were included to support the alanine dehydrogenase. Similar results were found in the production of phenylglycines **27**.

Interestingly, no special provision was made for regenerating nicotinamides; despite this, the artificial pathways performed well. Whether including additional cofactor regeneration modules would increase the volumetric productivity of these already-impressive systems remains to be seen.

12.3.5 Non-*E. coli* Systems

The vast majority of self-sufficient biocatalysts are based on *E. coli*. This is unsurprising in light of the highly developed tools for protein overproduction that exist for this organism and its fast and simple growth on

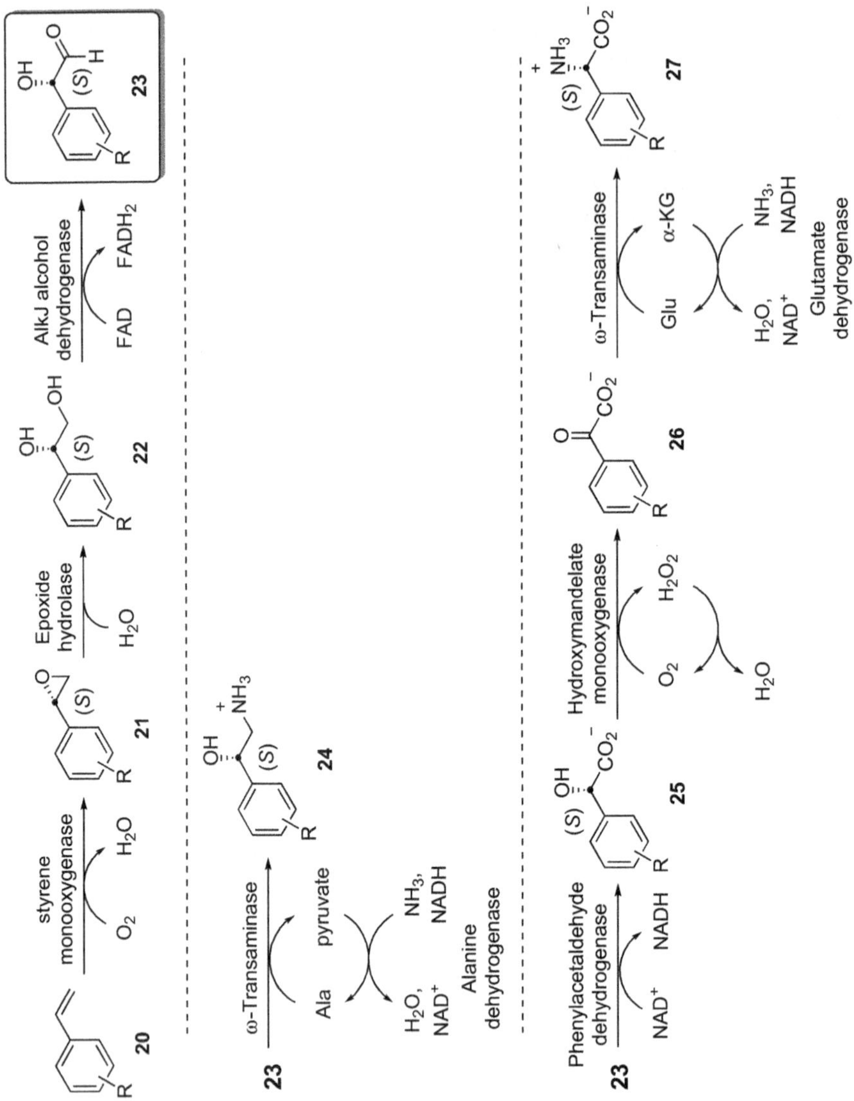

Scheme 12.14 Modular alkene conversions.

inexpensive media. For certain reactions, however, *E. coli* may not be the optimum host, and there are a few examples that use other organisms as hosts. Two representative cases are described below.

A whole-cell route to xylitol based on engineered *Gluconobacter oxydans* was reported a few years ago.[49] This valuable sugar substitute can be produced from D-arabitol by *G. oxydans*, although at an undesirably low efficiency. A membrane-bound *G. oxydans* arabitol dehydrogenase oxidizes D-arabitol in the periplasm to D-xylulose, which is imported into the cytoplasm. Here D-xylulose is converted to xylitol by xylose reductase in an NADH-dependent reaction. Because the native *G. oxydans* xylose reductase is relatively slow, Zhang *et al.* overexpressed this protein to achieve higher flux through the rate-limiting step in xylitol synthesis. Glucose dehydrogenase from *B. subtilis* was also co-expressed to increase the available NADH pool. Using whole cells of the engineered *G. oxydans* strain, 30 g L^{-1} D-arabitol yielded 12 g L^{-1} xylitol (the wild-type organism gave >2-fold lower yield).

The Kunze group used *Arxula adeninivorans* as the basis for a self-sufficient biocatalytic route to 1-(*R*)-phenylethanol **29** (Scheme 12.15).[50] This host organism was chosen since it efficiently expresses proteins and it can use a number of carbon and nitrogen sources for growth. In a previous study, the same group devised a self-sufficient *A. adeninivorans* strain that reduced acetophenone **28** to 1-(*S*)-phenylethanol **29**.[51] Here, the NADPH-dependent, (*R*)-selective *Lactobacillus brevis* alcohol dehydrogenase was overexpressed in *A. adeninivorans* along with one of two NADPH-regenerating enzymes (glucose dehydrogenase or glucose-6-phosphate dehydrogenase). The resulting strains were evaluated in a number of different forms: crude extract,

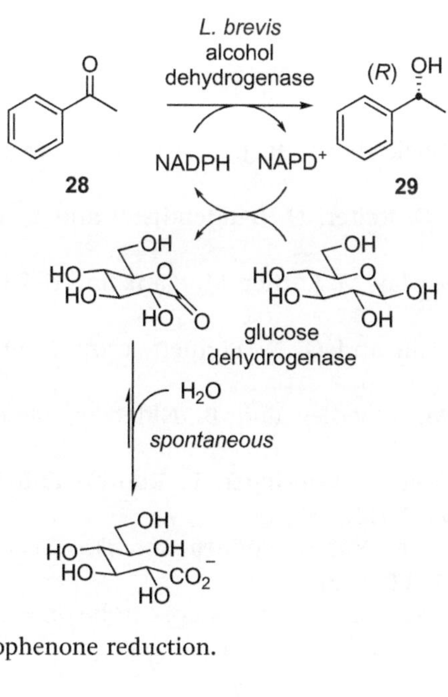

Scheme 12.15 Acetophenone reduction.

an immobilized crude extract, permeabilized whole cells and immobilized, permeabilized whole cells. In general, glucose-6-phosphate dehydrogenase was a more efficient NADPH regeneration catalyst than glucose dehydrogenase. In line with previous observations, additional $NADP^+$ was required when permeabilized whole cells were employed. Because only aqueous conditions were used, substrate loading was restricted by the solubility limit of **28** and **29** (*ca.* 40 mM). The long-term stability of the process was probed by incorporating immobilized, permeabilized cells co-expressing glucose dehydrogenase and the alcohol dehydrogenase into a semi-continuous flow system. The feed solution contained 40 mM **28**, 200 mM glucose and 0.25 mM $NADP^+$ (along with other buffer components) and the system was run for 300 hour. It gave a space–time yield of 9.7 $g\,L^{-1}\,day^{-1}$, which, under the process conditions, corresponded to 0.41 $kg\,kg^{-1}\,cdw\,day^{-1}$. Only the desired (*R*)-alcohol was observed. The authors note that the current process productivity is not at an industrially suitable level, and that two-phase systems will likely be required to overcome the low substrate concentrations imposed by its limited aqueous solubility.

12.4 Conclusions

The value of self-contained biocatalysts for preparative synthesis is well established. By overproducing all of the necessary enzyme activities, the system as a whole can achieve much better performance. Balancing catalytic activities of multiple enzymes remains a challenge, however. This is particularly true for chimeric self-sufficient enzymes since a 1 : 1 molar ratio is inescapable. Despite these challenges, it seems clear that self-sufficient biocatalysis will remain an important topic for the foreseeable future.

References

1. Y. S. Niino, S. Chakraborty, B. J. Brown and V. Massey, *J. Biol. Chem.*, 1995, **270**, 1983.
2. C. Stueckler, T. C. Reiter, N. Baudendistel and K. Faber, *Tetrahedron*, 2010, **66**, 663.
3. C. K. Winkler, D. Clay, M. Entner, M. Plank and K. Faber, *Chem. - Eur. J.*, 2014, **20**, 1403.
4. Y. Wang, K.-Y. San and G. N. Bennett, *Curr. Opin. Biotechnol.*, 2013, **24**, 994.
5. R. Kratzer, J. M. Woodley and B. Nidetzky, *Biotechnol. Adv.*, 2015, **33**, 1641.
6. T. Bayer, S. Milker, T. Wiesinger, F. Rudroff and M. D. Mihovilovic, *Adv. Synth. Catal.*, 2015, **357**, 1587.
7. M. Lakshmanan, K. Yu, L. Koduru and D.-Y. Lee, *J. Ind. Microbiol. Biotechnol.*, 2015, **42**, 1401.
8. Bakers' yeast (*Saccharomyces cerevisiae*) can be an exception to this rule. Commercially available dried yeast was grown on high levels of

molasses, and retained stored carbohydrates that can temporarily support metabolism in the absence of exogenously-added sugars. For longer-term reactions, however, glucose or sucrose is usually added to the reaction medium.
9. K. Schroer, K. P. Luef, F. S. Hartner, A. Glieder and B. Pscheidt, *Metab. Eng.*, 2010, **12**, 8.
10. M. Geier, C. Brandner, G. A. Strohmeier, M. Hall, F. S. Hartner and A. Glieder, *Beilstein J. Org. Chem.*, 2015, **11**, 1741.
11. A. Z. Walton and J. D. Stewart, *Biotechnol. Prog.*, 2004, **20**, 403.
12. J. A. McIntosh, C. C. Farwell and F. H. Arnold, *Curr. Opin. Chem. Biol.*, 2014, **19**, 126.
13. S. J. Sadeghi and G. Gilardi, *Biotechnol. Appl. Biochem.*, 2013, **60**, 102.
14. F. Rua, S. J. Sadeghi, S. Castrignanò, G. Di Nardo and G. Gilardi, *J. Inorg. Biochem.*, 2012, **117**, 277.
15. K.-Y. Choi, E. Jung, H. Yun, Y.-H. Yang and B.-G. Kim, *Appl. Microbiol. Biotechnol.*, 2014, **98**, 8191.
16. *Cytochrome P450*, ed. P. R. Ortiz de Montellano, Plenum Press, New York, 2nd edn, 1995.
17. Z.-J. Luan, Y.-C. Yin, A.-T. Li, H.-L. Yu and J.-H. Xu, *J. Mol. Catal. B: Enzym.*, 2015, **116**, 78.
18. J. D. Stewart, K. W. Reed, C. A. Martinez, J. Zhu, G. Chen and M. M. Kayser, *J. Am. Chem. Soc.*, 1998, **120**, 3541.
19. A. Z. Walton and J. D. Stewart, *Biotechnol. Prog.*, 2002, **18**, 262.
20. D. E. Torres Pazmiño, R. Snajdrova, B.-J. Baas, M. Ghobrial, M. D. Mihovilovic and M. W. Fraaije, *Angew. Chem., Int. Ed.*, 2008, **47**, 2275.
21. D. E. Torres Pazmiño, A. Riebel, J. de Lange, F. Rudroff, M. D. Mihovilovic and M. W. Fraaije, *ChemBioChem*, 2009, **10**, 2595.
22. J. Wöltinger, A. S. Bommarius, K. Drauz and C. Wandrey, *Org. Process Res. Dev.*, 2001, **5**, 241.
23. J. Lu, Y. Zhang, D. Sun, W. Jiang, S. Wang and B. Fang, *Appl. Biochem. Biotechnol.*, 2016, **180**, 1180.
24. This consortium of enzymes assembled on a three-dimensional scaffold cooperatively degrades cellulose into soluble carbohydrates that can be utilized as a carbon and energy source.
25. W. Jiang and B.-S. Fang, *J. Ind. Microbiol. Biotechnol.*, 2016, **43**, 577.
26. C. B. Bailey, M. E. Pasman and A. T. Keatinge-Clay, *Chem. Commun.*, 2016, **52**, 792.
27. K. Hölsch and D. Weuster-Botz, *Biotechnol. Appl. Biochem.*, 2010, **56**, 131.
28. Unlike its eukaryotic counterparts that use NAD^+, *E. coli* isocitrate dehydrogenase utilizes only $NADP^+$.
29. Y. Lu and L. Mei, *J. Ind. Microbiol. Biotechnol.*, 2007, **34**, 247.
30. S. Q. Pham, G. Pompidor, J. Liu, X. D. Li and Z. Li, *Chem. Commun.*, 2012, **48**, 4618.
31. C. Sung, E. Jung, K.-Y. Choi, J.-H. Bae, M. Kim, J. Kim, E.-J. Kim, P. I. Kim and B.-G. Kim, *Appl. Microbiol. Biotechnol.*, 2015, **99**, 6667.

32. X. Chen, T. Mei, Y. Cui, Q. Chen, X. Liu, J. Feng, Q. Wu and D. Zhu, *ChemistryOpen*, 2015, **4**, 483.
33. In addition to the 9 acetophenone derivatives investigated, 1-hydroxy-2-pentanone was also studied.
34. Y. Ni, C.-X. Li, L.-J. Wang, J. Zhang and J.-H. Xu, *Org. Biomol. Chem.*, 2011, **9**, 5463.
35. Q. Ye, H. Cao, G. Zang, L. Mi, M. Yan, Y. Wang, Y. Zhang, X. Li, J. Li, L. Xu, J. Xiong, P. Ouyang and H. Ying, *Appl. Microbiol. Biotechnol.*, 2010, **88**, 1277.
36. X. Wu, X. Gou and Y. Chen, *Process Biochem.*, 2015, **50**, 104.
37. Referred to as carbonyl reductase CR1 by the authors.
38. X. Wu, J. Jiang and Y. Chen, *ACS Catal.*, 2011, **1**, 1661.
39. Y.-P. Xue, H. Zeng, X.-L. Jin, Z.-Q. Liu and Y.-G. Zheng, *Microb. Cell Fact.*, 2016, **15**, 162.
40. "Dual specificity" refers to enzymes that accept both NAD($^+$/H) and NADP($^+$/H).
41. B. Sheng, Z. Zheng, M. Lv, H. Zhang, T. Qin, C. Gao, C. Ma and P. Xu, *PLoS One*, 2014, **9**, e104204.
42. P. Yao, Y. Cui, S. Yu, Y. Du, J. Feng, Q. Wu and D. Zhu, *Adv. Synth. Catal.*, 2016, **358**, 2923.
43. J. Li, J. Pan, J. Zhang and J.-H. Xu, *J. Mol. Catal. B: Enzym.*, 2014, **105**, 11.
44. B. R. Riebel, P. R. Gibbs, W. B. Wellborn and A. S. Bommarius, *Adv. Synth. Catal.*, 2002, **344**, 1156.
45. Z. Xiao, C. Lv, C. Gao, J. Qin, C. Ma, Z. Liu, P. Liu, L. Li and P. Xu, *PLoS One*, 2010, **5**, e8860.
46. J. Zhang, Z. Cui, H. Chang, X. Fan, Q. Zhao and W. Wei, *Biotechnol. Lett.*, 2016, **38**, 1559.
47. J. Zhang, S. Wu, J. Wu and Z. Li, *ACS Catal.*, 2014, **5**, 51.
48. S. Wu, Y. Zhou, T. Wang, H.-P. Too, D. I. C. Wang and Z. Li, *Nat. Commun.*, 2016, 7, 11917.
49. J. Zhang, S. Li, H. Xu, P. Zhou, L. Zhang and P. Ouyang, *J. Agric. Food Chem.*, 2013, **61**, 2861.
50. M. Rauter, A. Prokoph, J. Kasprzak, K. Becker, K. Baronian, R. Bode, G. Kunze and H.-M. Vorbrodt, *Appl. Microbiol. Biotechnol.*, 2015, **99**, 4723.
51. M. Rauter, J. Kasprzak, K. Becker, K. Baronian, R. Bode, G. Kunze and H. M. Vorbrodt, *J. Mol. Catal. B: Enzym.*, 2014, **104**, 8.

CHAPTER 13

Designing Multi-enzymatic Systems for the Preparation of Optically Active Molecules

LÍA MARTÍNEZ-MONTERO AND IVÁN LAVANDERA*

Department of Organic and Inorganic Chemistry, Biotechnology Institute of Asturias, University of Oviedo, Avenida Julián Clavería, Oviedo 33006, Spain

*Email: lavanderaivan@uniovi.es

13.1 Introduction

Living organisms have evolved into highly efficient systems assembling metabolic networks that allow them to grow and interact with the natural medium. In natural processes, the maximisation of resources and energy saving are two of the main evolutionary pressures. As a strategy to improve metabolic efficiency, living systems make use of different catalysts working together with an exquisite selectivity to achieve these complex biochemical pathways. In those nets, the catalysts usually are cross-regulated and the product of one reaction is the substrate of the following one.[1]

Nowadays, the need of (chiral) precursors for fine chemicals in the, *e.g.*, agrochemical, flavor and fragrance, and pharmaceutical industry is increasing. Nonetheless, this necessity must also go in hand with the design of sustainable processes according to the more restrictive legislative regulations concerning the quality and safety in the chemical industry.

Catalysis Series No. 32
Modern Biocatalysis: Advances Towards Synthetic Biological Systems
Edited by Gavin Williams and Mélanie Hall

Published by the Royal Society of Chemistry, www.rsc.org

Also, the limitation of resources undoubtedly drives industry towards cost-effective, shorter time-consuming, and energy-saving chemical methods. Currently, the "process efficiency" concept is not only related to a high chemical yield of a pure compound, but also to a minimum use of harmful organic solvents and reagents to obtain the minimum amount of waste and/or pollutants in the overall process.[2] In this sense, catalytic methodologies replacing stoichiometric amounts of reagents have consequently emerged as very popular and useful strategies. Particularly in the last decades, the use of enzymes as natural and environmentally friendly catalysts has provided to the organic chemists a powerful tool for tackling the current challenges previously mentioned.[3–5]

Moreover, recent advances in gene manipulation have contributed to the preparation of new biocatalysts with improved stability and activity properties (see Chapter 7) in large amounts at reasonable costs. These studies provide tailored enzymes with broader substrate spectra and higher (or new) catalytic activities, allowing transformations that a few years ago were unthinkable.[6,7] Due to the joint effort of several scientific disciplines, novel strategies and methods are applied nowadays in the field of biocatalysis, being seriously envisaged by industry as an alternative to prepare chemicals not only on the laboratory scale (milligram to gram), but also on industrial scale (kilogram to ton).[8,9]

In this scenario, the combination of multiple biocatalysts might be very plausible, since enzymes usually operate under similar reaction conditions concerning pH, temperature and also the solvent medium (*i.e.*, water).[10–15] Also, the high specificity frequently shown by enzymes leads to a high control of reactivities, allowing the presence of different substrates, cofactors, intermediates and catalysts in the same medium. On the contrary, these high specificities also lead to a limited substrate scope, which obviously limits the application of multi-enzymatic protocols. In this sense, the use of molecular biology is addressing, at least partially, this drawback by discovering new biocatalysts with a broader substrate scope (see Chapters 1 and 2). Likewise, while not the focus of this chapter, it is worth mentioning the great efforts that have also been made in order to combine enzymes with other chemical catalysts (see Chapter 15) such as organo- and metal-based ones,[16] opening the door to possible applications of biocatalysts to completely unexplored transformations.

In these multi-enzymatic processes, in many cases the product of one biocatalytic reaction acts as the substrate for the next one. This allows rapid complexity generation, the shifting of reaction equilibria,[17] the elimination of inhibition problems or the minimisation of the decomposition of unstable intermediates. These protocols can be implemented *via* the use of isolated enzymes in an homogeneous phase or by employing whole-cell systems containing the biocatalysts of interest to create artificial pathways (also called artificial metabolisms).[18] In nature, the existence of multifunctional biocatalysts or enzyme complexes is very common. This approach can also be mimicked by crosslinking or immobilising several proteins in a support (see Chapter 16), by fusing genes encoding different enzymes or by modular assembling of proteins on synthetic scaffolds.[19]

From an operational point of view, the application of these methodologies presents several advantages over the classic step-by-step design. As they are performed in the same vessel ('one-pot'), there is no need for intermediate purification, and therefore costs, operating time and waste are reduced, improving the atom economy and the overall process yield.[20,21] In this regard, these systems can be considered as 'greener' methods in comparison with the more classical synthetic methodologies.

Among the different examples that can be found in the literature concerning multi-enzymatic protocols, the term 'cascade' recurrently appears associated with them. While its definition has been clarified by some authors,[22,23] there is no commonly accepted meaning of this term, nor of related descriptors such as 'domino' or 'tandem' – neither in the context of the chemical sciences in general, nor within the particular field of biocatalysis. Therefore, many authors still use these three terms rather interchangeably, and, particularly within the biocatalysis research community, 'cascade' seems to have become the descriptor of choice for almost any type of one-pot multi-step process. For the purpose of this chapter, we have chosen to adopt a fairly broad conception of multi-enzymatic cascades, including concurrent or sequential biocatalytic one-pot systems, even if some of the steps involve chemocatalysis and/or spontaneous (non-catalysed) transformations. Particularly, we will focus on reaction designs which provide the final products in enantioenriched form, although a few non-stereoselective examples will also be reported due to their outstanding synthetic relevance and applicability.

The majority of the multi-enzymatic systems described can be assigned to one out of four designs as recently described by Schrittwieser *et al.* (Scheme 13.1),[24] namely linear-, parallel-, orthogonal-, and cyclic cascades. Some other examples will cover, *e.g.*, dynamic and enantioconvergent transformations.

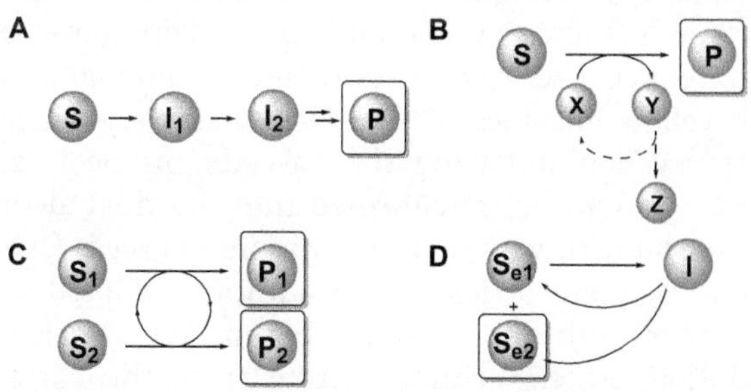

Scheme 13.1 Main types of cascade protocols. (A) Linear; (B) orthogonal; (C) parallel; and (D) cyclic. S = substrate, S_e = substrate enantiomer; I = intermediate; P = product; X, Y, and Z = co-substrate and co-product(s).

13.2 Multi-enzymatic Linear Cascades

The most common system consists of consecutive transformations in one-pot (Scheme 13.1A) where a starting material or substrate (*S*) is converted *via* one or several (reactive) intermediates (I_n), which shall not be isolated, to form the final product (*P*). This type of cascade has a practical advantage as the storage and handling of toxic or unstable intermediates can be minimised, and also the displacement of unfavourable equilibria can be achieved. For more synthetic examples, see Chapter 14.

13.2.1 Synthesis of Chiral Hydroxy-functionalised Compounds

Several biocatalytic methods to synthesise enantiopure secondary alcohols have been developed in the last few years aiming to fulfil the increasing demand of this type of highly valuable compounds. Among them, stereoselective reduction of ketones and enantioselective oxidation of racemic *sec*-alcohols using alcohol dehydrogenases (ADHs), have recently gained increasing relevance.[25] These enzymes need a nicotinamide cofactor [NAD(P)H/$^+$] to perform the desired transformations, and due to their high costs, efficient recycling methodologies have been designed in order to utilise them in catalytic amounts (see Chapter 12).[26,27] Another family of enzymes able to perform redox processes are ene-reductases (EREDs), which can stereoselectively reduce double bonds.[28] These biocatalysts are flavin-dependent and they also need a nicotinamide cofactor to obtain the reduced form of the flavin. The only necessary requisite is that the double bond must be conjugated to an electron-withdrawing group. Therefore, one could envisage the combination of an ERED and an ADH to both reduce an α,β-unsaturated carbonylic derivative, and introduce chirality in up to three different carbon centres at the same time (Scheme 13.2A). The main problem of these linear cascades is the competition for the ADH between the unsaturated substrate and the saturated one, which can provide as by-product the unsaturated alcohol that cannot be reduced by the ERED.

In this sense, several examples can be found in the literature using this methodology. For instance, Gatti and co-workers have employed this methodology to reduce a series of α-substituted cinnamaldehyde derivatives combining old yellow enzymes (OYEs) from Baker's yeast and horse liver ADH (HLADH).[29] When following the cascade protocol, α,β-unsaturated aldehydes (2 $g L^{-1}$) could be transformed into the final alcohol derivatives in high yield (>80%) with very high enantiomeric excess (>90%) after 12 h at 30 °C. In a subsequent article, these authors obtained several interesting precursors of tetralin and chroman-based drugs such as Robalzotan, Ebalzotan and Rotigotine.[30] Thus, starting from the corresponding α,β-unsaturated aldehydes or ketones (5 $g L^{-1}$), and combining OYEs with HLADH, *Candida parapsilosis* ADH (CPADH), *Rhodococcus erythropolis* ADH (READH), or *Deinococcus radiodurans* ADH (DRADH), the final alcohols could

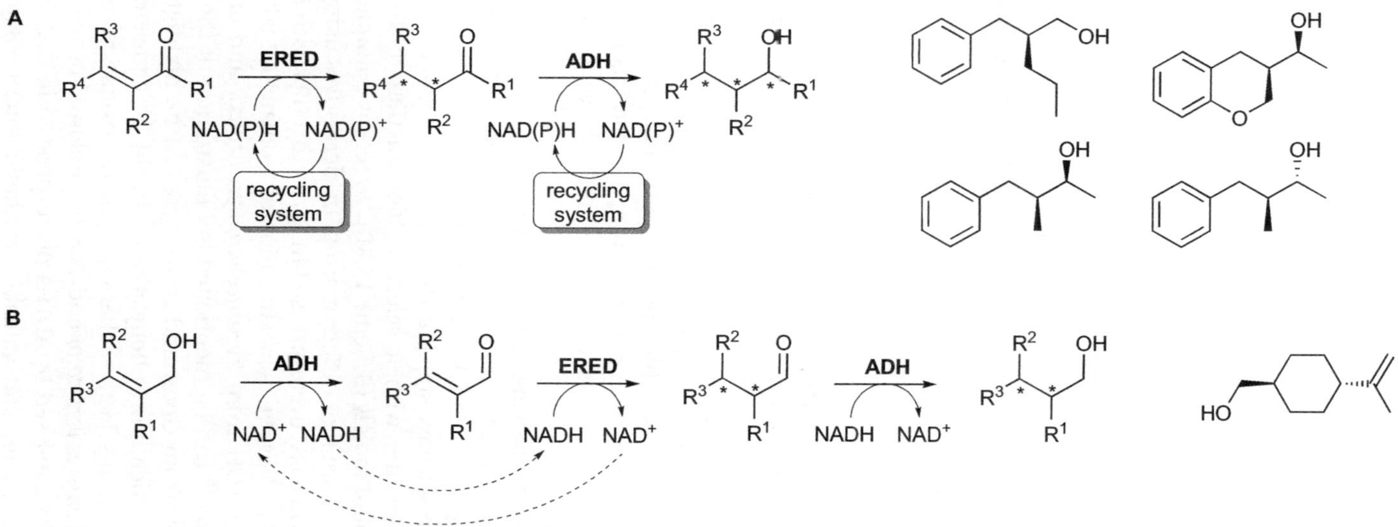

Scheme 13.2 (A) Combination of an ERED and an ADH to obtain chiral alcohols starting from α,β-unsaturated carbonylic compounds. (B) Combination of an ADH and an ERED starting from allylic alcohols.

be attained in usually very high conversions (>85%) and selectivities (ee > 90%) after 12 h at 30 °C and pH 7. In a recent example, the cascade was applied to obtain two out of four isomers of odorant Muguesia®.[31] In this case, an α,β-unsaturated ketone (0.8 g L^{-1}) was selected as the suitable substrate to synthesise the chiral alcohols with perfect enantio- and diastereoselectivities (>99%) in high yields (75–78%) after 24 h at 30 °C and pH 7. The enzymes of choice in this case were OYE3 from Baker's yeast and READH or *Parvibaculum lavamentivorans* ADH (PLADH). In all these examples, the system glucose–glucose dehydrogenase (GDH) was selected to recycle the nicotinamide cofactor employed by both catalysts (the ERED and the ADH).

A similar approach was described by Hauer *et al.*, using as starting material a series of allylic primary alcohols (Scheme 13.2B).[32] In this case, the nicotinamide cofactor could be internally recycled, as the first step was the oxidation of the unsaturated alcohol (2 mM) into the corresponding aldehyde catalysed by the ADH (requiring the nicotinamide cofactor in its oxidised form, NAD^+), which in a subsequent step was reduced by the ERED (requiring the nicotinamide cofactor in its reduced form, NADH) into the saturated aldehyde. Playing with the proportion of the reduced and oxidised forms of the NAD-cofactor, the saturated alcohols could be obtained at high extent when an excess of NADH was used. By combining different wild-type and structurally modified EREDs with HLADH, after 24 h at 30 °C and pH 7.5, the final products could be attained in excellent conversions (>95%). The best results were provided with OYE1 from *Saccharomyces pastorianus*.

More attractive (and more challenging) methods include deracemisation protocols. Thus, a highly valuable optically pure product with a theoretical 100% yield and 100% ee can be obtained from a cheap racemic substrate in a concurrent 'one-pot' process combining multiple catalysts.[33] Recently, efficient systems have appeared for the deracemisation (or stereoinversion) of secondary alcohols, employing isolated enzymes or whole-cell systems.[34] These processes can be performed by combination of two stereoselective enzymes, *i.e.* two enantiocomplementary ADHs (Scheme 13.3A), or by arrangement of a non-selective step (*i.e.* the oxidation of both enantiomers into the ketone intermediate) with a selective bioreduction to afford the enantioenriched alcohol moiety (Scheme 13.3B). As recent examples of the first set-up, the deracemisation of different aromatic derivatives using ADH from *Rhodococcus ruber* (ADH-A) overexpressed in *E. coli* together with *Lactobacillus brevis* ADH (LBADH),[35] or the deracemisation of 1-phenyl-1,2-ethanediol due to the action of an ADH from *Zygosaccharomyces rouxii* and an ADH from *Candida parapsilosis*,[36] can be mentioned. As illustration of the second system, a contribution from Gotor and co-workers can be highlighted. In this particular case, 2,2-dichloro-1-phenylethanol could be deracemised by the combination of a non-selective chemoenzymatic oxidation mediated by the laccase from *Trametes versicolor* and TEMPO, followed by reduction of the ketone intermediate catalysed by ADH-A overexpressed in *E. coli*.[37]

Vicinal diols are useful and valuable synthetic intermediates for many pharmaceuticals, bioactive compounds, and chiral reagents. Li and co-workers

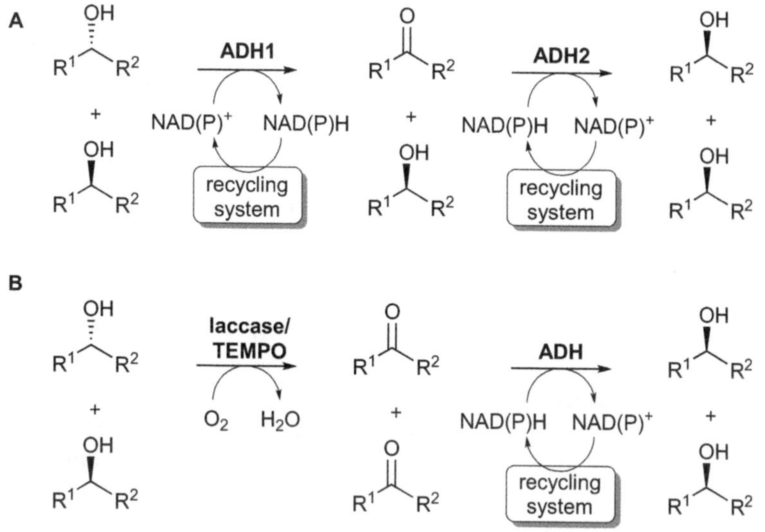

Scheme 13.3 Deracemisation of *sec*-alcohols combining: (A) Two stereoselective and enantiocomplementary biocatalysts; and (B) one non-selective oxidation followed by a selective bioreduction.

envisaged a linear cascade to synthesise various aromatic derivatives overexpressing in the same *E. coli* host a monooxygenase (MO) and an epoxide hydrolase (EH).[38] Thus, combining styrene monooxygenase (SMO) with epoxide hydrolase from *Sphingomonas* sp. HXN-200 (SpEH) or *Solanum tuberosum* (StEH), and starting from the corresponding styrene compounds (10 mM), the final (*S*)- or (*R*)-diols, respectively, could be attained with usually high yields (>80%) and selectivities (>90% ee). Selective epoxidation mediated by SMO afforded the (*S*)-epoxides, and later their opening at the β- (catalysed by SpEH) or α-position (catalysed by StEH), determined the final configuration. The reactions were carried out in a biphasic system at 30 °C for 8 h.

A second approach to obtain chiral diols consists of the consecutive use of a carboligase and an ADH. In this case, two aldehydes selectively react in the presence of a biocatalyst able to perform the aldol addition, affording a chiral hydroxy ketone intermediate that is subsequently reduced by the ADH into the final derivative. Rother and co-workers have shown the application of this system to synthesise (1*R*,2*R*)-1-phenylpropane-1,2-diol with excellent conversion and diastereomeric excess (>99%), using benzaldehyde lyase from *Pseudomonas fluorescens* (BAL) and ADH from *Ralstonia* sp. (RasADH) in aqueous medium. A modularised approach was achieved by introducing the lyophilised *E. coli* cells overexpressing both biocatalysts in 'teabags' of polyvinylidene fluoride membrane.[39,40] Thus, this 'one-pot two-step' transformation could be achieved in methyl *tert*-butyl ether (MTBE) in the presence of a small quantity of buffer pH 9. After performing the ligation reaction, 'teabags' containing BAL were removed and changed by the ones enclosing RasADH. In order to recycle the cofactor used in the second step, cyclohexanol was added as co-substrate. After 17 h at 30 °C, a remarkable concentration of 339 mM of the final product was obtained. This system was also employed in a 'micro-aqueous' environment in which the lyophilised cells of both biocatalysts were directly added in MTBE sequentially or simultaneously. Again, high concentrations of the final product were detected (363–440 mM).[41] Finally, in a very recent contribution, the authors have demonstrated the application of this system to obtain the four possible stereoisomers of 1-phenylpropane-1,2-diol *via* combination of two carboligases (BAL or a benzoylformate decarboxylase variant from *Pseudomonas putida*) and two ADHs (RasADH or LBADH) in organic medium at high substrate concentrations.[42] In this case, the use of 1,5-pentanediol as 'smart' co-substrate[43] for the cofactor recycling was also notable.

Domínguez de María *et al.* have also described a cascade system applicable to carboligation reactions between two aldehydes.[44] In this case, due to the fact that small aldehydes can be harmful for enzymes and also very volatile, these molecules were obtained after oxidation of the corresponding primary alcohols in a process catalysed by an alcohol oxidase. Then, they reacted with a second aldehyde (6 mM) present in the reaction media by the action of the lyase enzyme. Using alcohol oxidase from *Hansenula* sp. and BAL, good conversions (>90%) and excellent selectivities of different chiral α-hydroxy ketones were attained after 16 h. Another approach for the synthesis of enantioenriched α-hydroxy ketones has been developed by Li *et al.* Thus, epoxides reacted with EHs to form the corresponding diols, which in a second step were regioselectively oxidised by the action of an ADH.[45] In this case, different *meso*-epoxides (100–200 mM) were employed as starting material, and SpEH and butanediol dehydrogenase from *Bacillus subtilis* (BDHA) overexpressed in *E. coli* cells were used as catalysts. Lactate dehydrogenase or NADH oxidase were utilised as cofactor recycling enzymes, and the final derivatives were obtained in moderate to high conversions (50–85%) and excellent ee (>98%) at 30 °C after 6–12 h.

Ethyl (*R*)-3-hydroxyglutarate is a key intermediate for the synthesis of Rosuvastatin. In a recent contribution, Wu, Zhu and co-workers have designed a cascade synthesis of this derivative using a halohydrin dehalogenase (HHDH) and a nitrilase (Scheme 13.4A).[46] (*S*)-4-Chloro-3-hydroxybutyrate was employed as starting material and the biotransformation was made in the presence of *E. coli* cells overexpressing the genes of HHDH from *Agrobacterium radiobacter* AD1 and nitrilase from *Arabidopsis thaliana*. HHDH was responsible for the epoxide formation *via* intramolecular nucleophilic substitution and for the subsequent addition of the cyanide anion present in the reaction mixture, affording the chiral β-hydroxy nitrile intermediate. This compound was hydrolysed by the nitrilase to form the final derivative. Due to inhibition effects, better results were observed by sequential addition of the nitrilase to the reaction medium. Following this protocol, high concentrations (up to 1.2 M) of the desired product were attained after 7 h at 30 °C and pH 8–9.

α-Hydroxy acids are found in numerous natural products, as well as in pharmaceutical and plant-protection agents. A biocatalytic approach to

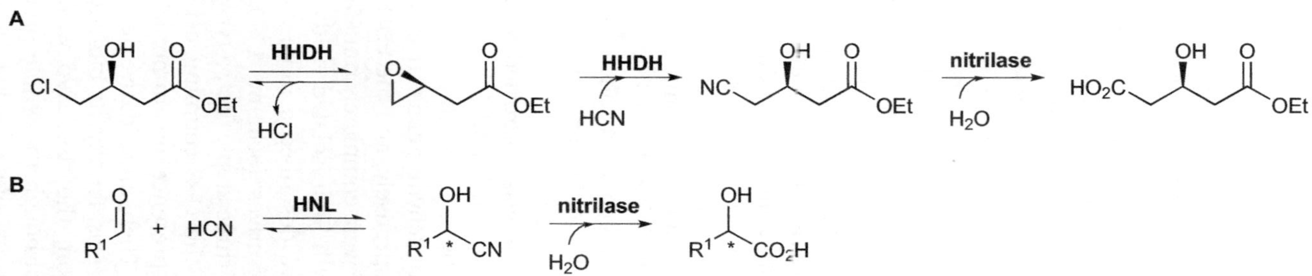

Scheme 13.4 Synthesis of chiral hydroxy acids: (A) starting from a halohydrin combining the use of a halohydrin dehalogenase and a nitrilase; and (B) starting from an aldehyde combining the use of a hydroxynitrile lyase and a nitrilase.

synthesise them makes use of a lyase and a nitrilase (Scheme 13.4B).[47] The nucleophilic addition of cyanide to an aldehyde is reversibly catalysed by a hydroxynitrile lyase (HNL), thus providing the chiral cyanohydrin intermediate that can be hydrolysed in a second step to produce the α-hydroxy acid. Stolz, Sheldon and co-workers have exploited this protocol to obtain (*S*)-mandelic acid using HNL from *Manihot esculenta* and nitrilase from *Pseudomonas fluorescens* EBC 191. In some cases, these biocatalysts were applied as cross-linked enzymatic aggregates (CLEAs).[48] As a side reaction, (*S*)-mandeloamide was also observed, therefore the application of an amidase to improve the yield of mandelic acid was also investigated.[49] These transformations are usually performed under acidic pH (4.5–5.5) in a biphasic system at 25 °C. Finally, the application of these enzymes overexpressed in the same *E. coli* host was also reported, combined with the use of organic co-solvents[50] or ionic liquids.[51]

Due to the natural occurrence of α-amino acids, a direct approach to obtain the corresponding α-hydroxy acids would be the transformation of the amino group into the hydroxylated derivative. Thus, starting from a series of natural amino acids and combining an amino acid deaminase (AAD), to obtain the α-keto acid intermediate, with an ADH, the final chiral compounds were synthesised (Scheme 13.5A). Kroutil *et al.* applied this system employing AAD from *Proteus myxofaciens* together with L- and D-isocaproate reductases (L- and D-Hic) from *Lactobacillus paracasei* and *Lactobacillus confuses*, respectively, to get access to both enantiomers of a series of aliphatic and aromatic derivatives after 7 h at 21 °C and pH 7 in high yields (>75%) and excellent enantioselectivity (>99%).[52] To deaminate the substrates, it was necessary to employ an oxygen atmosphere. For the recycling of the cofactor needed in the bioreduction, formate and FDH were used. In another contribution, the co-expression of these three biocatalysts (AAD, ADH and FDH) in a single vector in *E. coli* was optimised to perform this cascade.[53] Following a similar methodology, (*R*)- and (*S*)-2-hydroxybutyric acids were synthesised from L-threonine by co-expressing in *E. coli* L-threonine deaminase from *E. coli*, L- or D-lactate dehydrogenase from *Oryctolagus cuniculus* or *Staphylococcus epidermidis*, respectively, and the FDH from *Candida boidinii*. Up to 750 mM of the substrate were completely transformed to the product in optically pure form (>99%) with high isolated yields.[54]

Kroutil and co-workers have described a more complex system starting from different phenolic derivatives (Scheme 13.5B).[55] In this case, the cascade protocol consisted of three steps: first, a tyrosine phenol lyase from *Citrobacter freundii* catalysed the C–C coupling between the phenol compound and pyruvate in the presence of ammonia, affording the corresponding L-tyrosine derivative. Then, an oxidative deamination was mediated by L-AAD from *Proteus myxofaciens*. Finally, its bioreduction was catalysed by L- or D-Hic. Due to inhibition, the first step was performed separately from the second and third, although the whole process was performed in 'one-pot'. Preparative transformations of substituted phenols (23–96 mM) afforded the corresponding (*R*)- and (*S*)-*p*-hydroxyphenyl lactic acids in moderate to high yields (58–85%) and enantiopure form (>97% ee).

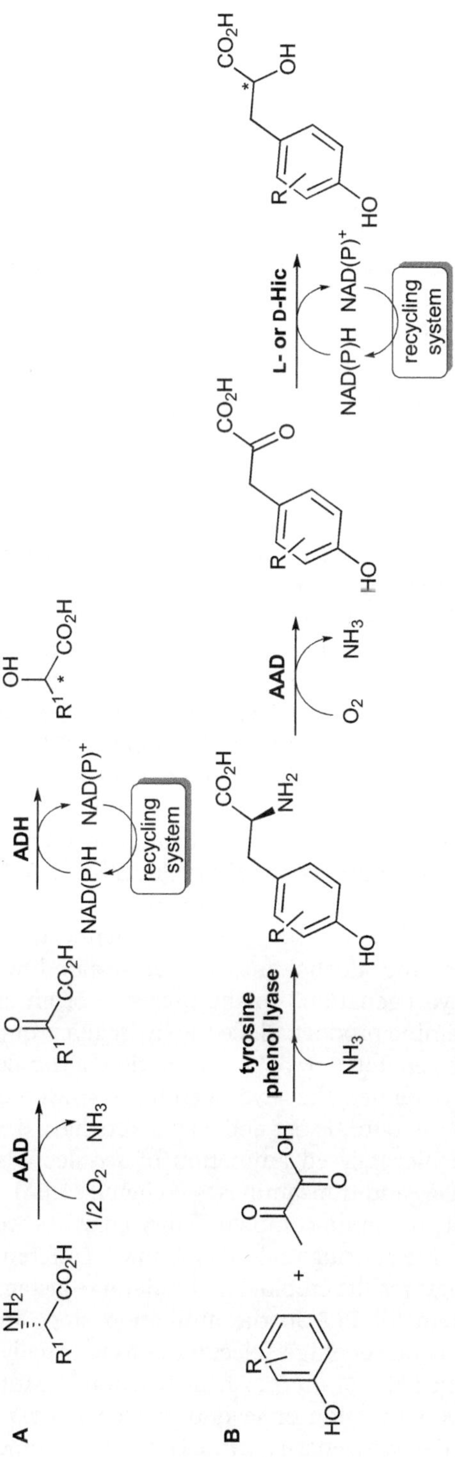

Scheme 13.5 Synthesis of chiral α-hydroxy acids: (A) starting from L-amino acids combining the use of α-amino acid deaminase and an ADH; and (B) combining the use of a tyrosine phenol lyase, an amino acid deaminase and L- or D-Hic.

López *et al.* described the synthesis of a polyalcohol starting from benzyloxycarbonyl (Cbz) *N*-protected ethanolamine.[56] After oxidation of the alcohol into the aldehyde mediated by chloroperoxidase from *Caldaryomices fumago*, the addition of dihydroxyacetone phosphate catalysed by rhamnulose-1-phosphate aldolase afforded the final derivative. The use of immobilised biocatalysts permitted yields of approximately 70% in the stirred-tank and packed-bed reactor modes. Furthermore, the presence of dioxane as co-solvent allowed a higher product formation (31 $g L^{-1}$). Lemaire and co-workers have designed several 'one-pot' cascade processes applied to the synthesis of functionalised sugars using aldolases. Derivatives of D-fructose 6-phosphate,[57] D-arabinose 5-phosphate,[57] and nitrocyclitol (see Chapter 14)[58] can be mentioned.

13.2.2 Synthesis of Chiral Amino-functionalised Compounds

Optically active amines are valuable building blocks in the synthesis of pharmaceuticals and agrochemicals, also used in multiple applications as resolving agents, chiral auxiliaries, and organocatalytic reagents. In the last decades, biocatalytic methods have appeared as an alternative to conventional chemical methodologies to synthesise these derivatives. While lipases have been historically the first enzymes of choice to get access to these compounds through kinetic resolutions,[59] other enzymes have appeared in recent years as very appealing tools. In this sense, of special relevance are the transaminases (TAs), which can provide the stereoselective amination of carbonylic groups or the enantioselective deamination of racemic amines.[60] TAs are dependent on pyridoxal 5′-phosphate (PLP) that is usually employed in catalytic amounts as a result of the development of very efficient recycling systems. In fact, the aminating species is pyridoxamine 5′-phosphate (PMP). These enzymes work under similar conditions compared to oxidoreductases, therefore, many of the cascades already mentioned in this chapter could be applicable.

The amination of alcohols can be performed chemically through catalytic "dehydrogenation" of the alcohol into the corresponding carbonyl compound, and then "hydrogenation" in the presence of an ammonia source leads to the desired amine product. Since the hydrogen required for the final hydrogenation step is generated by dehydrogenation of the alcohol in the first step, this approach is named the 'hydrogen-borrowing' methodology (also called 'redox-neutral'). Kroutil, Faber and co-workers have developed the first example of selective biocatalysed amination of *sec*-alcohols by combining alcohol dehydrogenases and transaminases (Scheme 13.6A). Due to the high selectivity of ADHs, two enantiocomplementary enzymes were necessary to quantitatively oxidise the starting material (50 mM). Different protocols were tried, obtaining the best results employing the alanine–lactate dehydrogenase (LDH) recycling system for PLP in the amination step. Moderate to high conversions (30–91%) and very high selectivities were usually observed (up to 96% ee) at 30 °C and pH 7.5 after 24 h of reaction.[61] Mutti, Turner *et al.* achieved the selective amination of secondary alcohols *via* arrangement of ADHs with amine dehydrogenases (AmDHs). These biocatalysts reduce

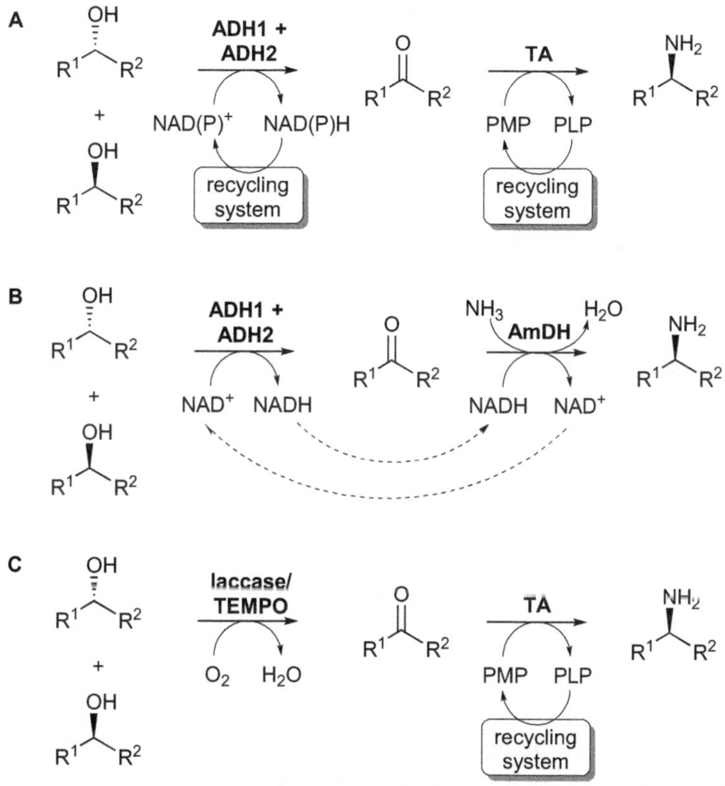

Scheme 13.6 Amination of alcohols through multi-enzymatic designs combining: (A) two ADHs and a TA; (B) two ADHs and an AmDH; and (C) the laccase–TEMPO system with a TA.

carbonylic compounds into amines in the presence of ammonia and the nicotinamide cofactor. Therefore, it was envisaged that by coupling both transformations, NADH could be internally recycled in a closed loop, leading to chiral amines from primary or racemic secondary alcohols (Scheme 13.6B). When starting from racemic derivatives, two stereocomplementary ADHs (LBADH and ADH from *Aromatoleum aromaticum*) were necessary to achieve quantitatively the oxidation step. Aliphatic substrates were better accepted than aromatic ones by the AmDH from *B. badius*. The cascade process was performed at pH 8.7 and 30 °C for 48 h. Under these conditions, the substrates (20 mM) were converted in moderate to high conversions with excellent enantioselectivity.[62] A very recent contribution in this field has been reported by Gotor-Fernández, Lavandera and co-workers, who envisaged the use of a non-selective chemoenzymatic oxidation to achieve this cascade. For this purpose, the laccase from *Trametes versicolor* and TEMPO system was utilised to oxidise a series of aromatic racemic *sec*-alcohols (Scheme 13.6C). Combining it in a sequential mode with a TA, alcohols (25 mM) were transformed into the enantiopure amines in high conversions (>80%) after 40 h at 30 °C. In this case, to recycle the cofactor needed by the TA, isopropylamine or *cis*-but-2-ene-1,4-diamine[63] were selected as suitable co-substrates.[64]

The deracemisation of primary amines on secondary centres has been demonstrated by Turner *et al.* by the joined action of a monoamine oxidase (MAO) and a transaminase (Scheme 13.7A).[65] MAO from *Aspergillus niger* (MAO-N) was able to selectively oxidise the (*S*)-enantiomers of a series of amines, providing a mixture of the prochiral ketone intermediates and the untouched (*R*)-antipodes. In a subsequent step mediated by a commercial TA with (*R*)-selectivity, the prochiral ketones were transformed into the (*R*)-derivatives. High conversions (>80%) and selectivities were achieved under mild reaction conditions. As hydrogen peroxide was generated in the oxidation step, a catalase was added to the reaction medium. Alanine–LDH–GDH was utilised as PLP recycling system for the amination step.

The amination of non-activated positions has also been shown possible *via* the combination of several biocatalysts. Bühler and co-workers have designed a whole-cell system with all necessary fused enzymes to obtain the ω-functionalisation of methyl dodecanoate.[66] By coupling an alkane monooxygenase (MO) from *Pseudomonas putida* and a transaminase (TA) from *Chromobacterium violaceum*, the substrate was converted with absolute regiospecificity to the terminal amine *via* two sequential oxidations performed by the MO followed by the amination step. While the effective concentration of the final derivative was not high (0.13 mM), this concept demonstrated that the concurrent reactivity of enzymes can achieve transformations that in other chemical ways would be highly impeded. Flitsch *et al.* showed the selective amination of ethylbenzenes (1 mM) at the benzylic position by overexpressing

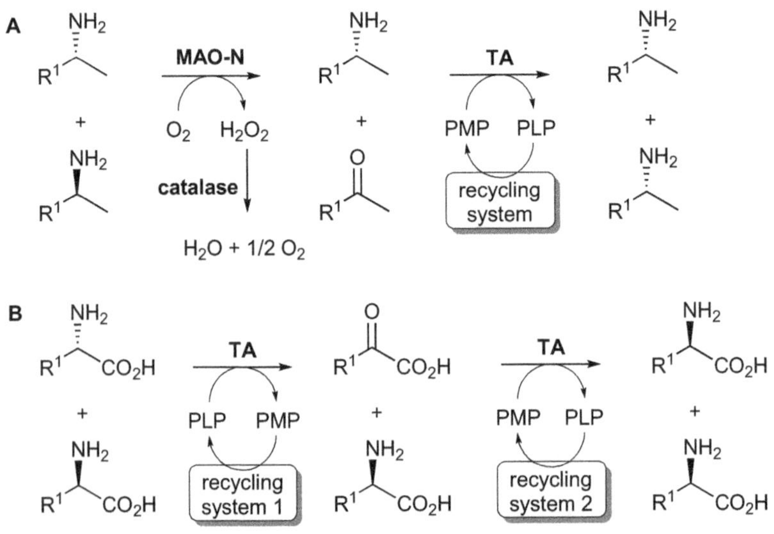

Scheme 13.7 Deracemisation of: (A) primary amines combining a monoamine oxidase and a transaminase; and (B) α-amino acids using two TAs with opposite stereoselectivity.

in the same host a P450 monooxygenase, two ADHs with opposite selectivity (LBADH and ADH from *Rhodococcus erythropolis*) and TA from *Arthrobacter* sp. Both ADHs were necessary as the first oxidation mediated by a P450 monooxygenase proceeded with very low selectivity. Although different (*R*)-amines with excellent ee (>97%) were obtained only at low conversions (up to 26%), the potential of this methodology was clearly demonstrated.[67]

Chiral substituted secondary amines are important architectures that can be found in many biologically active natural products and pharmaceuticals. A synthesis of enantiomerically pure mono- and disubstituted piperidines and pyrrolidines was achieved using, *e.g.* a biocatalytic cascade involving carboxylic acid reductase (CAR), TA and imine reductase (IRED) enzymes (Scheme 13.8).[68] IRED enzymes are emerging as very useful biocatalysts due to their excellent stereoselectivities in the reduction of imine derivatives at the expense of a nicotinamide cofactor. Hence, starting from δ-keto acids, the final compounds were accessed with high enantio- and diastereoselectivities by: (1) reduction into the corresponding δ-keto aldehyde catalysed by CAR; (2) regioselective amination mediated by the TA into the amino ketone that spontaneously cyclised into the cyclic imine; and (3) reduction of the imine catalysed by IRED. Using 5 mM substrate concentration, different 2-substituted piperidines were obtained with excellent conversions (>90%) and high selectivities (>90% ee) after 24 h at pH 7 and 30 °C. Also, a panel of 2-substituted piperidines and pyrrolidines (5 mM) were deracemised to afford the (*S*)-amines in high conversions (>95%) and ee values (>97%) using 6-hydroxy-D-nicotine oxidase (an amine oxidase) with an IRED from *Streptomyces* sp. after 24 h at 30 °C and pH 7.4.[69] On the other hand, the combination of TAs with synthases in 'one-pot' has provided cyclic amines such as alkaloid[70] or strictosidine[71] derivatives (see Chapter 14).

Transferases and TAs have also been combined to synthesise different amino alcohol compounds. Rother and co-workers described the synthesis of (1*R*,2*R*)-norpseudoephedrine and (1*R*,2*S*)-norephedrine in a 'one-pot two-step' synthesis starting from benzaldehyde and pyruvate. In the first step, the thiamine diphosphate-dependent acetohydroxyacid synthase I was responsible for the decarboxylation and ligation of pyruvate to benzaldehyde producing (*R*)-phenylacetylcarbinol, which in the second step reacted with a TA to give the final derivatives in high yields and excellent enantio- and diastereoselectivities. Two TAs with opposite stereoselectivity afforded both diastereoisomers. The use of alanine as amine donor was remarkable, since the co-product obtained (pyruvate) was one of the substrates needed in the reaction, therefore internal recycling was feasible (Scheme 13.9A).[72] Ward *et al.* showed two different biocatalytic strategies for the synthesis of (2*S*,3*R*)-2-amino-1,3,4-butanetriol (ABT). Firstly, a sequential cascade of three enzymatic steps (two TAs and one transketolase) starting from serine, pyruvate and glycolaldehyde as substrates. Secondly, a two-step recycling cascade was developed for the simultaneous synthesis of ABT and hydroxypyruvate mediated by a TA using serine as donor to aminate erythrulose (previously obtained by the action of a transketolase) (Scheme 13.9B). This last system

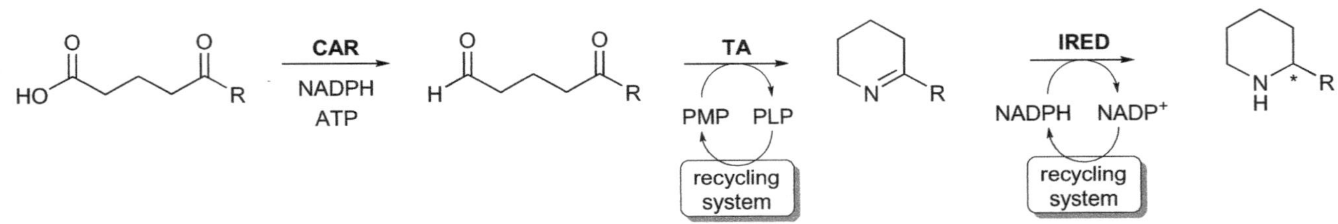

Scheme 13.8 Synthesis of 2-substituted piperidines by combined action of a carboxylic acid reductase, a TA and an imine reductase.

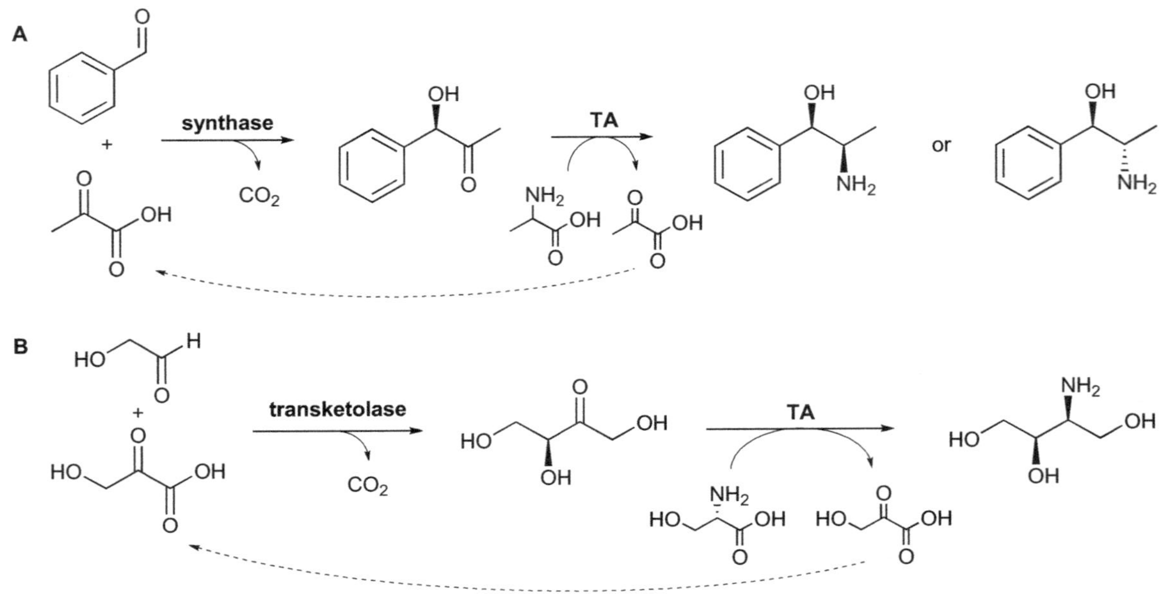

Scheme 13.9 Synthesis of amino alcohol derivatives combining transferases and TAs.

appeared as the most promising, and although conversion was low (9%), the product was obtained with excellent stereoselectivity (ee > 95%).[73]

Amino acids appear as essential chiral building blocks of diverse pharmaceutical drugs, agrochemicals and chiral ligands. Park and Shin showed the deracemisation of α-amino acids coupling two TAs with opposite stereoselectivity and different substrate pattern, so the necessary amine donor and amine acceptor in both steps were accepted by only one of the two transaminases (Scheme 13.7B). Depending on the enzymes employed, L- or D-amino acids were accessible in enantiopure form and in many cases with very high conversions. Moreover, in a process run under very mild conditions (pH 7 and 37 °C), the substrate concentration utilised was remarkably high (100 mM).[74] Another recent example of deracemisation of arylalanines into the D-antipodes made use of two stereocomplementary amino acid dehydrogenases in 'one-pot'.[75]

An impressive example of multi-enzymatic catalysis to produce D-amino acids, among other derivatives, has been shown by Li and co-workers.[76] All cascades described in this contribution used styrene compounds as starting material (10 mM). By overexpressing simultaneously different biocatalysts in the same *E. coli* host, a great number of aromatic 1,2-amino alcohols, α-hydroxy acids and α-amino acids were obtained in moderate to excellent yields and very high enantiomeric excess. In all cases, the biotransformations were done in a biphasic system using hexadecane as co-solvent. The initial steps were common to all cascades: (1) epoxidation using SMO from *Pseudomonas* sp.; (2) hydrolysis with EH from *Sphingomonas* sp. HXN-200 to obtain the chiral diols; and (3) oxidation of the primary alcohol to the hydroxy aldehyde catalysed by ADH from *Pseudomonas putida* GPo1. Then, the overexpression of CV-TA or phenylacetaldehyde dehydrogenase from *E. coli* afforded the corresponding 1,2-amino alcohols or α-hydroxy acids, respectively. Even more complex was the pathway to obtain α-amino acids. In addition to the enzymes previously mentioned to synthesise the α-hydroxy acids, two more biocatalysts were overexpressed: hydroxymandelate oxidase from *Streptomyces coelicolor*, responsible for the formation of the α-keto acid intermediate, and branch chain amino acid transaminase from *E. coli* to form the final products (Scheme 13.10). Also a catalase (to eliminate H_2O_2 obtained as by-product in the oxidation of the α-hydroxy acid) and glutamate dehydrogenase (to recycle the amine donor in the last amination step) were expressed in *E. coli.*

Kroutil, Faber and co-workers have also described a 'one-pot two-step' cascade process to synthesise L-tyrosine analogues starting from benzene analogues, *via ortho*-hydroxylation of the phenyl ring mediated by a variant of monooxygenase P450 BM3 from *Bacillus megaterium*, and subsequent reaction with tyrosine phenol lyase in the presence of ammonia. Due to the formation of some by-products, *e.g.* other phenolic derivatives, the yields were usually low or moderate (6–49%), but the final compounds were attained with excellent enantioselectivity (>97% ee).[77] This group has also designed a multi-enzymatic protocol to achieve the synthesis of 6-aminohexanoic acid from cyclohexanol (Scheme 13.11). Due to reaction condition

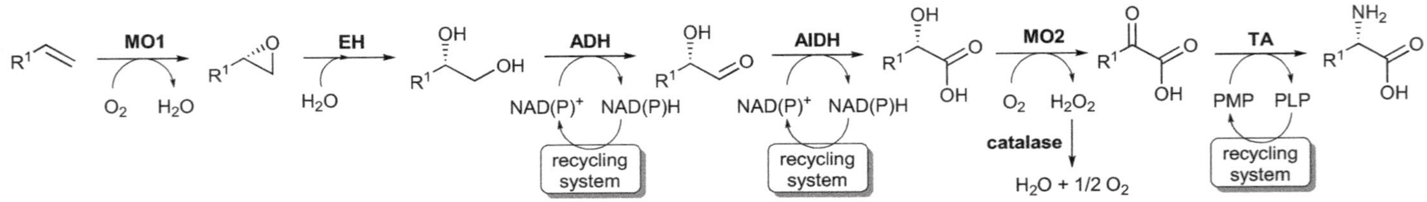

Scheme 13.10 Multi-enzymatic cascade system to obtain α-amino acids from alkene derivatives.

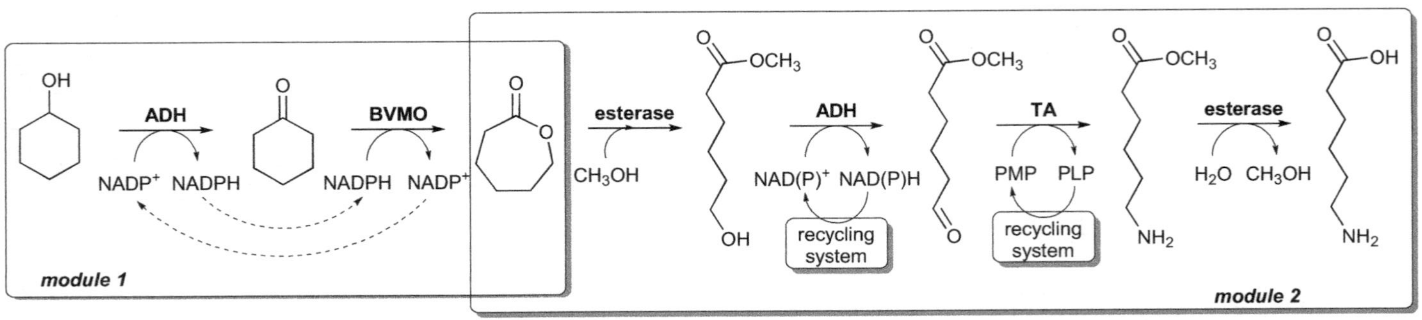

Scheme 13.11 Two-module six-step process to obtain 6-aminohexanoic acid from cyclohexanol.

incompatibilities, the system was divided into two modules: (1) oxidation of cyclohexanol into ε-caprolactone by coupling an ADH and a Baeyer–Villiger monooxygenase (BVMO) with internal cofactor regeneration; and (2) methanolysis of the lactone by an esterase to form the ω-hydroxy ester, followed by its oxidation mediated by a second ADH to form the terminal aldehyde. Finally, this compound was aminated by a TA to produce the ω-amino ester, and as last step the ester moiety was hydrolysed by an esterase rendering the final product. The key step was the use of methanol to open the lactone, as the undesired formation of the ω-hydroxy acid intermediate *via* hydrolysis ended the cascade, since this derivative was not accepted by the second ADH. Under optimum conditions, 6-aminohexanoic acid was obtained in 75% conversion from ε-caprolactone (50 mM).[78]

13.2.3 Synthesis of Chiral Cyclic Carboxylic Acid Derivatives[79]

Agudo and Reetz coupled a mutant from monooxygenase P450 BM3 with different variants of ERED YqjM from the Old Yellow Enzyme family, to obtain both enantiomers of methyl 3-oxocyclohexane-1-carboxylate from methyl cyclohex-1-enecarboxylate (Scheme 13.12). The first enzyme was responsible for a double oxidation to produce the cyclohexenone derivative, which was then reduced by the ERED to form the final compound. Different *E. coli* preparations were tested for overexpressing one or both biocatalysts in the same host, reaching, after 75 min, yields up to 70% of the enantiopure compound.[80]

Likewise, the preparation of lactones has attracted increasing attention. These compounds are important because of their structural implications as basic chemicals, but also due to the fact that they are valuable building blocks for polymers and natural product synthesis. Furthermore, they display a broad biological profile and are also important flavour and aroma constituents. Among the biocatalytic pathways to synthesise enantioenriched derivatives, different multi-enzymatic cascade protocols can be mentioned. Beginning from linear α,β-unsaturated keto esters, the sequential selective bioreduction of the double bond mediated by an ERED, followed by the transformation of the ketone into the hydroxy ester by an ADH action, was envisaged. For γ- or δ-derivatives, the spontaneous (or acid-catalysed) intramolecular attack of the alcohol to the ester is highly favoured, thus providing the corresponding lactones (Scheme 13.13A). Using this synthetic route, Pietruszka *et al.* obtained a family of 3,4-, 3,5-, and 4,5-disubstituted γ-butyrolactones,[81] and Gatti *et al.* synthesised all isomers of *Nicotiana tabacum* lactone.[82] Another cascade approach can be visualised starting from cyclic derivatives such as cycloalkenones, using an ERED and a BVMO that can provide a lactone with several chiral centres (Scheme 13.13B). Liu and Li described the synthesis of 6-substituted δ-lactones by the employment of this protocol. To improve the ee of the products, an additional hydrolase-catalysed step was performed, selectively hydrolysing the minor enantiomer produced into the hydroxy acid.[83] Using cyclic alcohols as substrates, the

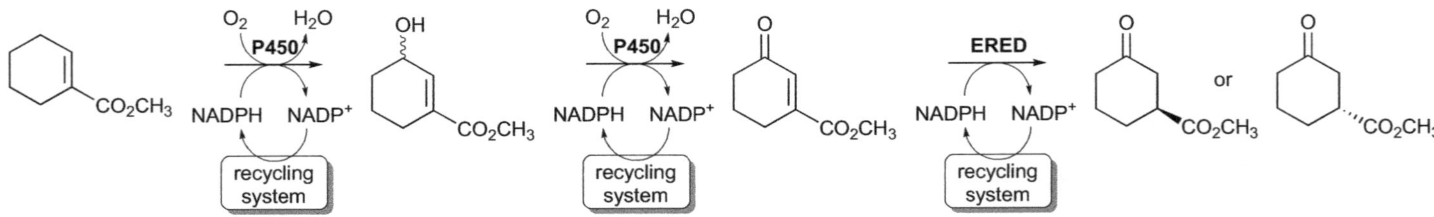

Scheme 13.12 Synthesis of a chiral carboxylic acid derivative combining the use of a monooxygenase and an ERED.

A

B

C

D

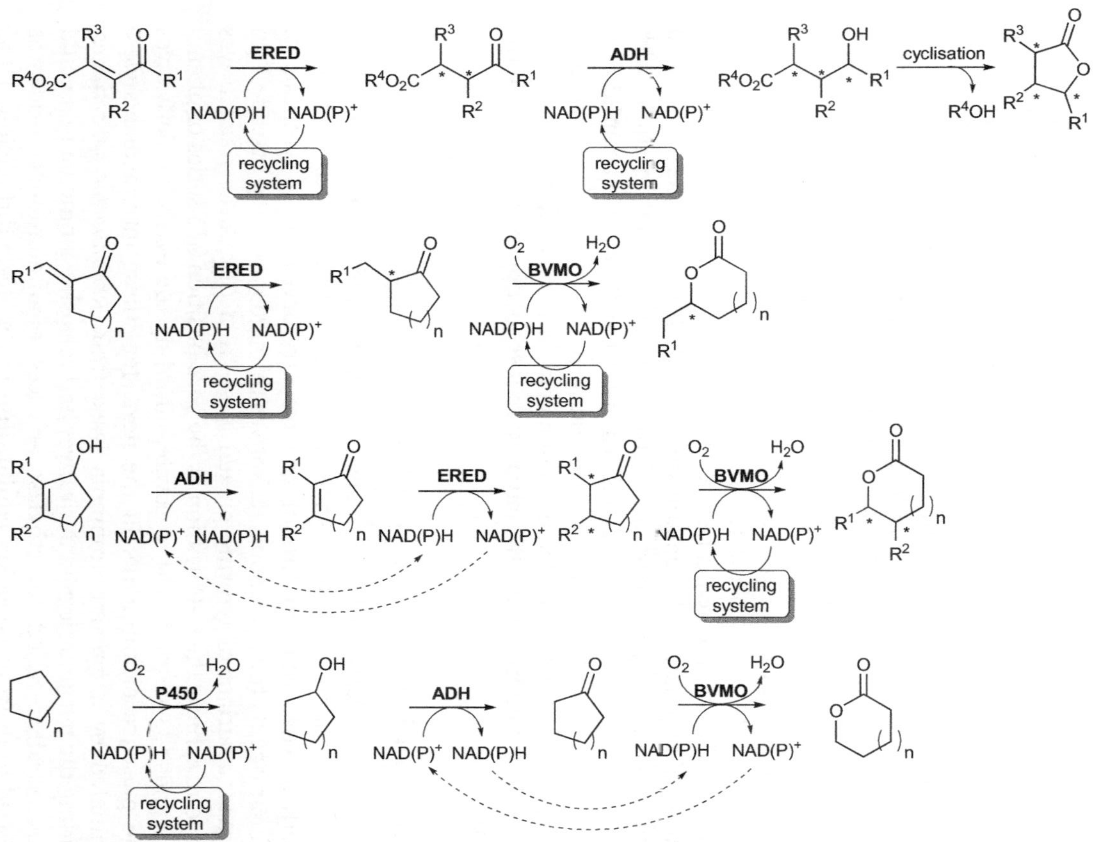

Scheme 13.13 Synthesis of (chiral) lactones through multi-enzymatic cascades starting from: (A) linear keto esters; (B) cyclic α,β-unsaturated ketones; (C) cyclic alcohols; and (D) cycloalkanes.

straightforward synthesis of the corresponding lactones can be envisaged *via* combination of an oxidative step catalysed by an ADH followed by a second oxidation mediated by a BVMO. In the case that a double bond would be present, an ERED could also be included to generate more stereogenic centres (Scheme 13.13C). An advantage of this cascade is that the nicotinamide cofactor needed by ADH and BVMO can be internally recycled if both catalysts present the same cofactor preference. In recent contributions, the Rudroff, Mihovilovic and Bornscheuer groups have exploited this cascade to obtain valuable lactones starting from easily accessible materials with engineered whole-cell systems co-expressing all the biocatalysts in the same *E. coli* host. Using prochiral compounds, desymmetrisation processes were possible, thus affording the enantioenriched lactones in high yields and ee values.[84–86] The application of these transformations in order to get access to polymers obtained by lipase-mediated oligomerisation has also been shown.[87,88] Moreover, the synthesis of lactones from cycloalkanes has been recently reported by a 'one-pot' sequential action of a P450 monooxygenase, an ADH and a BVMO (Scheme 13.13D).[89]

The use of an engineered galactose oxidase (GOase) variant applied to the oxidation of aliphatic and aromatic amino alcohols allowed the synthesis of lactams such as 2-pyrrolidone or valerolactam in 'one-pot' reactions with xanthine dehydrogenase or aldehyde oxidase from *E. coli*. Under mild conditions (pH 7.4 and 30 °C), high conversions (up to 85%) could be attained.[90] After oxidation of the alcohol and spontaneous intramolecular cyclisation to form the imine intermediate, the latter was subsequently oxidised to yield the lactam.

13.2.4 Synthesis of Natural Product Derivatives

α,β-Unsaturated carboxylic acids such as ferulic and coumaric acids are natural compounds that have shown, among other activities, anticancer properties. The group of Kroutil has recently developed a synthesis of *p*-coumaric acids from phenolic derivatives through a two-step cascade.[91] In this case, a variant of tyrosine phenol lyase from *Citrobacter freundii* was selected to achieve the C–C ligation at the *para*-position of a series of phenol compounds (23–46 mM) with pyruvate, affording the L-tyrosine derivatives in the presence of ammonia. In the next step, these intermediates were deaminated by a tyrosine ammonia lyase from *Rhodobacter sphaeroides*, providing the *p*-coumaric acid derivatives and releasing ammonia that could be theoretically recycled. Nonetheless, an excess of ammonium chloride was employed to drive the reaction equilibrium in the desired direction. Thus, after 24 h at pH 10 and 30 °C in the presence of diethyl ether (5% v/v) as co-solvent, the final compounds were obtained in high yields (>80%). Also, the co-expression of both enzymes in the same *E. coli* host was achieved to obtain a bi-functional catalyst. In an extension of this work, different vinyl arenes were synthesised from phenol derivatives through a three-step cascade involving the two previously mentioned steps followed by the decarboxylation of the coumaric acids in the presence of a ferulic acid decarboxylase from *Enterobacter* sp.[92]

As examples of multi-enzymatic linear cascades making use of other types of biocatalysts, the following contributions can be underlined: the synthesis of vanillin from ferulic acid employing a decarboxylase and an oxygenase,[93] the conversion of flavanones into their respective dihydrochalcones mediated by chalcone isomerase and an ene-reductase,[94] the synthesis of pinorenisol from eugenol combining vanillyl-alcohol oxidase and a laccase,[95] the regioselective *O*-methylation of catechols mediated by an adenosyltransferase, an *O*-methyltransferase, and a nucleosidase,[96] and the simultaneous reaction of an uridine phosphorylase and a purine nucleoside phosphorylase to obtain vidarabine.[97]

13.3 Multi-enzymatic Orthogonal Cascades

Orthogonal designs, in contrast to linear systems, comprise only a single synthetically relevant transformation, which is coupled to secondary 'auxiliary' reactions that promote the formation of the desired product. The regeneration of cofactors or co-substrates, or the removal of troublesome co-products can be ascribed to these systems (Scheme 13.1B). While the description of cofactor regeneration systems as biocatalytic cascades may be questionable, there are examples of complex auxiliary reaction set-ups that do undoubtedly deserve to be termed orthogonal cascades.

Yang *et al.* described the synthesis of L-2-aminobutyric acid catalysed by a TA. As an amine donor, L-aspartic acid was employed, releasing 2-oxosuccinic acid as co-product. This compound spontaneously decarboxylated, affording pyruvate that could also act as substrate for the TA forming L-alanine, complicating the purification of the target amino acid. To overcome this problem, two coupled reactions were simultaneously used: (1) acetolactate synthase was introduced to remove the pyruvate intermediate producing 3-hydroxybutan-2-one; and (2) addition of alanine racemase and D-amino acid oxidase to remove L-alanine by racemisation, followed by the alanine oxidation re-forming pyruvate (Scheme 13.14).[98] Another recent example of this type of cascade was reported by Kroutil and co-workers using L-alanine as amine donor for TAs that possessed (*R*)-selectivity. These biocatalysts, due to their stereochemical preference, just accept D-amino acids as donors. Since L-amino acids are more accessible than their enantiomers, this research group combined the use of (*R*)-selective TAs with L-alanine by introduction of an intermediate racemisation step that is catalysed by an alanine racemase. Thus, D-alanine was formed in the reaction medium, and then acted as substrate for the (*R*)-selective TAs affording the desired amines with high conversions.[99]

13.4 Multi-enzymatic Parallel Cascades

In parallel cascade designs, two synthetically relevant reactions are coupled simultaneously to afford various derivatives. The link of both transformations

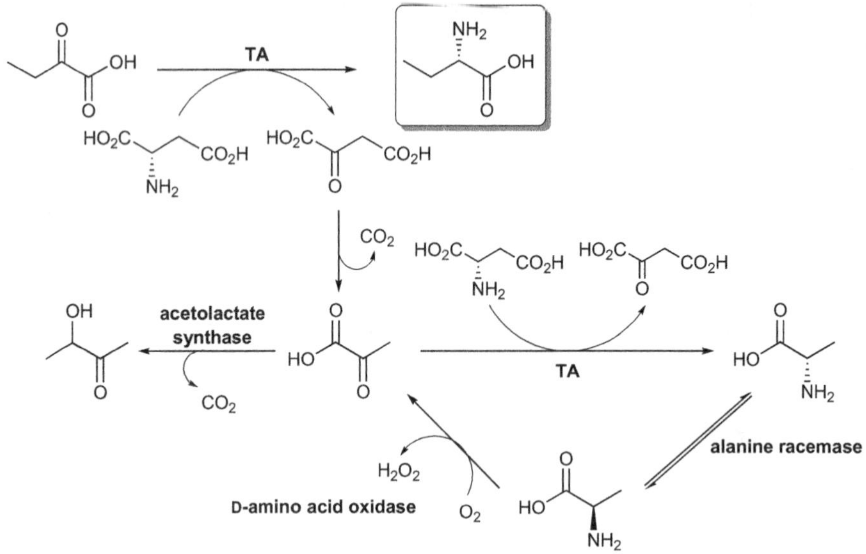

Scheme 13.14 Orthogonal multi-enzymatic cascade applied to the synthesis of L-2-aminobutyric acid.

is the cofactor regeneration that perfectly matches in each reaction (Scheme 13.1C). Multi-enzymatic examples of this methodology have been reported, *e.g.* the stereoselective reduction of a ketone catalysed by an ADH coupled to the oxidative kinetic resolution of a chiral ketone,[100] or coupled to the stereoselective oxidation of sulphides[101] mediated by a BVMO in the so-called Parallel Kinetic Asymmetric Transformations. While these designs ensured the efficient use of the involved cofactors, they presented as main disadvantage the formation of several products, which required complicated separation protocols.

This drawback was smartly overcome by coupling two enzymatic reactions that afforded exactly the same product in a convergent manner. Kara *et al.* have recently described the production of ε-caprolactone, an important monomer for biodegradable polymers, by coupling the oxidation of cyclohexanone mediated by cyclohexanone monooxygenase from *Acinetobacter* sp., with the oxidation of 1,6-hexanediol catalysed by ADH from *Thermoanaerobacter ethanolicus* (TeSADH). Using 2 equivalents of the ketone and one of the diol, conversions up to 20 mM of the final product were attained after 18 h.[102] As an update of this system, dodecane was used as organic co-solvent, constituting a second non-miscible organic phase that allowed lactone conversions higher than 50 mM.[103]

13.5 Multi-enzymatic Cyclic Cascades

These protocols are usually employed to deracemise derivatives such as amines, α-amino acids, *sec*-alcohols, or α-hydroxy acids, and combine a non-selective process (usually a reductive reaction) with a highly selective one (usually an oxidation step).[104] Therefore, one enantiomer of the racemic substrate is selectively transformed into a prochiral intermediate (*e.g.*, an imine or a ketone), which is non-selectively converted back to the racemic starting material (Scheme 13.1D). Over several cycles, the more slowly reacting enantiomer in the selective transformation accumulates.

These protocols usually encompass an enzymatic transformation with a conventional chemical reaction. Therefore, most of the multi-enzymatic cyclic cascades correspond to the insertion of one of these chemical methods in a linear cascade system in the presence of other biocatalysts. The group of Turner has been especially active in the field, and has recently reported a cascade to synthesise 2,5-disubstituted pyrrolidines.[105] Starting from different 1,4-diketones, after regioselective amination catalysed by a commercial TA, the amino ketone compound spontaneously cyclised forming a pyrroline intermediate. The latter was then stereoselectively transformed into the final derivatives *via* cyclic deracemisation combining a variant of monoamine oxidase from *Aspergillus niger* (MAO-N) and ammonia–borane complex as chemical non-selective reducing agent (Scheme 13.15A). Under optimised conditions, different enantiopure pyrrolidines were obtained in a concentration up to 25 mM in a quantitative manner. Likewise, D- and L-phenylalanine derivatives were accessed from the corresponding cinnamic acids by combination of a phenyl ammonia lyase (PAL) variant from *Anabaena variabilis* and an L- or D-amino acid deaminase (AAD), together with the non-selective ammonia–borane reducing complex (Scheme 13.15B).[106] Depending on the selectivity of the AAD, both antipodes could be synthesised. The cascade protocol was carried out at 37 °C and pH 9.6, obtaining the products with moderate to high conversions (>60%) and excellent selectivities (>98% ee). Moreover, in collaboration with the group of Kroutil, the cyclic deracemisation of benzylisoquinolines was achieved. This process was mediated by a variant of MAO-N in the presence of the morpholine–borane complex, followed by the reaction of the accumulated (*S*)-enantiomers with the oxidative berberine bridge enzyme (BBE), forming the final berbines in a kinetic resolution process.[107] Using DMSO as co-solvent, some derivatives were attained in high yields (>80%) and in enantiopure fashion.

Until recently, cyclic cascades were limited to chemo-enzymatic systems. Nonetheless, Ward, Turner, Hollmann and co-workers demonstrated that the combination of a monoamine oxidase and an artificial transfer hydrogenase (ATH; *i.e.*, a biotinylated iridium complex bound to streptavidin, see Chapter 3) can carry out the same transformations with high efficiency and minimal interference of the catalysts.[108] Further extension of this work has been reported for the deracemisation of a cyclic aromatic amine by coupling the MAO with the ATH. The reduced form of the iridium complex was recycled by NADPH, and therefore, glucose–GDH was employed to recover the reduced form of the nicotinamide cofactor.[109]

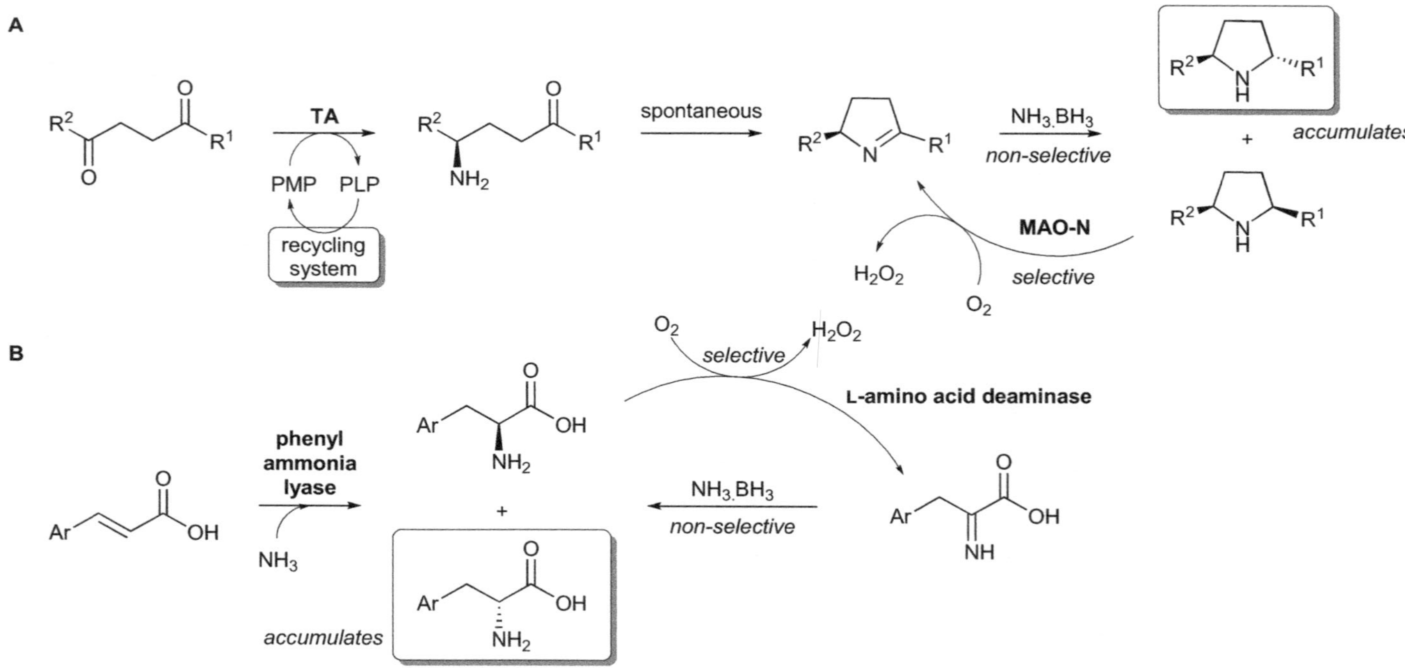

Scheme 13.15 Cyclic cascades applied in multi-enzymatic protocols combining: (A) a TA and MAO-N with the ammonia–borane complex and (B) a phenyl ammonia lyase and an L-amino acid deaminase with the ammonia–borane complex.

13.6 Other Multi-enzymatic Systems

13.6.1 Enantioconvergent Processes

These strategies involve transformations in which both enantiomers of a racemic mixture are converted into the same product enantiomer, or into a homochiral mixture of products through different reaction pathways, one *via* retention of configuration and the second one through inversion of configuration (Scheme 13.16A).[110] Therefore, for this strategy, an inverting enzyme must be used.

Among the different biocatalysts able to perform these transformations, most of the examples have been shown by epoxide hydrolases due to the fact that they can perform the opening of substituted epoxides at the α- or β-position. Starting from a racemic mixture, if one EH can hydrolyse one enantiomer introducing the molecule of water at the α-position (*via* inversion of configuration) and the second one catalyses the nucleophilic attack on the other enantiomer at the β-position (*via* retention of configuration), an enantioconvergent process can be obtained. This has been shown using whole-cell preparations where a multitude of different enzymes with opposite selectivities can be found, but also procedures employing two isolated EHs have provided good results. For instance, Lee and co-workers have developed this system to synthesise chiral aromatic diol derivatives.[111,112]

Faber *et al.* have described an enantioconvergent process of different *sec*-alcohols in the presence of two stereocomplementary sulfatases (Scheme 13.16B).[113] These enzymes are able to hydrolyse sulfate esters with retention or inversion of configuration. Arylsulfatase from *Pseudomonas aeruginosa* (PAS), an *R*-selective enzyme hydrolysing with retention of configuration, was combined with *sec*-alkylsulfatase Pisa1 that presents *S*-preference with inversion of configuration. Using this protocol, sulfate esters afforded the corresponding unprotected (*R*)-alcohols with high conversions (>80%) and ee (>90%).

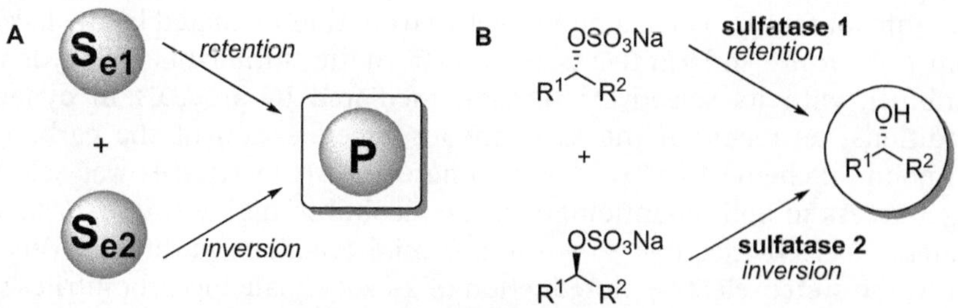

Scheme 13.16 (A) General scheme for enantioconvergent processes; and (B) synthesis of chiral *sec*-alcohols combining two sulfatases with opposite selectivity using this strategy.

13.6.2 Dynamic Kinetic Resolutions (DKRs)

This approach relies on a racemisation step of the substrate enantiomers combined with a selective transformation over one single enantiomer through a kinetic resolution process.[114] In theory, this would afford the final (enantiopure) product in quantitative yield. Most of the DKR examples found in the literature are related to the discrimination of racemic alcohols, amines, carbonylic and carboxylic acid derivatives.

Focusing on multi-enzymatic DKR protocols, in many cases those make use of racemases as the agent in charge of the racemisation process. The application of these biocatalysts is still limited due to their narrow substrate specificity and low stability in organic solvents. In fact, they are traditionally employed in processes involving amino acid or hydroxy acid derivatives. For instance, the combination of an amino acid amidase and a racemase co-expressed in *E. coli* was envisaged in order to synthesise several phenylalanine derivatives from the corresponding racemic phenylalanine amides,[115] but also the combination of a D-acylase with *N*-acetyl amino acid racemase to afford D-allylglycine,[116] or the combination of L-*N*-carbamoylase and *N*-succinyl-amino acid racemase immobilised on different supports for the production of different optically pure natural and unnatural L-α-amino acids starting from the corresponding racemic *N*-formyl- and *N*-carbamoyl-amino acids.[117] On the other hand, ADHs have been successfully used to accomplish racemisation of enantiopure alcohols,[118] so the combination of ADHs with lipase-catalysed asymmetric transformations to develop DKRs can be envisioned. However, the low stability of oxidoreductases in net organic solvents has hampered this approach. Musa *et al.* have overcome this drawback by immobilisation of a mutant from TeSADH in a xerogel.[119] When coupling this preparation with *Candida antarctica* lipase B (CAL-B) in hexane at 50 °C for 24 h, the (*R*)-ester derivatives of a series of *sec*-alcohols were obtained with conversions up to 75% in moderate ee. Although the racemisation process was not fast enough for real synthetic purposes, it is a promising preliminary study that presents great possibilities for the future.

Other multi-enzymatic protocols make use of a chemoenzymatic DKR as one step of a cascade process. Díaz-Rodríguez, Gotor *et al.* have envisaged a 'one-pot' sequential system to deracemise 2-phenyl-1-propanol, a key core of the profenol family. Thus, a non-selective oxidation mediated by the laccase from *T. versicolor* and TEMPO system to afford the chiral labile aldehyde was combined with its selective reduction mediated by an ADH in dynamic conditions, as result of the spontaneous racemisation of the carbonylic compound (Scheme 13.17A).[120] A commercial ADH or HLADH was selected to get access to both enantiomers of the alcohol in high yields (>70%) and moderate selectivities (>82% ee) under mild reaction conditions. Very recently, the stereoselective bioreduction of 2-oxocycloalkanecarbonitriles was concurrently coupled to a whole cell-catalysed nitrile hydrolysis in 'one-pot'. The first step, mediated by ADHs, involved the spontaneous racemisation of the substrate leading, through a DKR process, to the cyclic β-hydroxy nitriles

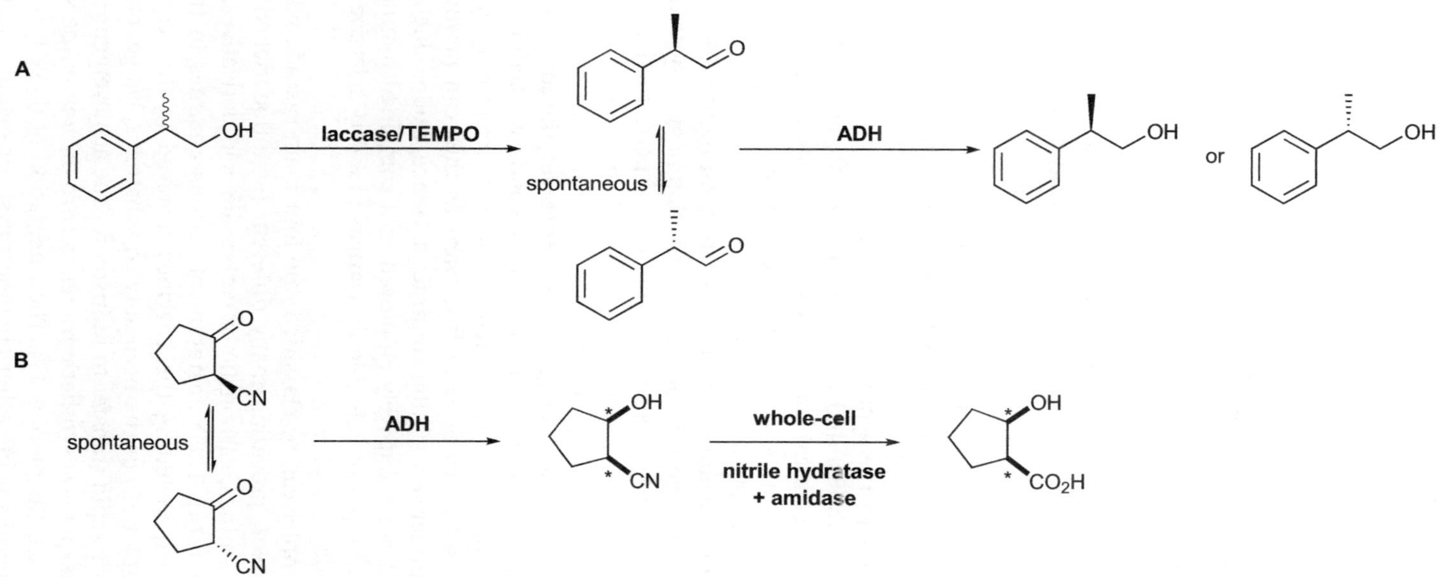

Scheme 13.17 Examples of multi-enzymatic cascade protocols including a DKR step: (A) synthesis of a profenol core by combination of the laccase/TEMPO system and an ADH; and (B) synthesis of cyclic β-hydroxy acids coupling an ADH and a whole-cell system containing a nitrile hydratase and an amidase.

in very high enantio- and diastereomeric ratios (Scheme 13.17B).[121] Then, the reaction with the nitrile hydratase and amidase present in the whole cells of *Rhodococcus rhodochrous* provided the final 2-hydroxycycloalkanecarboxylic acids with excellent overall yield (>95%) and optical purity (de >98%, ee >99%) after 24 h at 30 °C and pH 7.

13.7 Summary and Outlook

The application of biocatalytic methods in academic and industrial laboratories has gained importance in the last decades. The first studies mainly focused on the application of whole-cell systems or isolated enzymes in desymmetrisation or kinetic resolution processes. Later, biocatalysts were applied in other systems that could afford enantioenriched products in quantitative yields starting from racemates, such as dynamic kinetic resolutions, deracemisations or enantioconvergent processes. This methodology expansion has also come in hand with tremendous advances in protein engineering (see Chapter 7), DNA sequencing and computational design (see Chapter 4), conferring impressive novel capabilities to enzymes and also higher productivities in the synthesis of the biocatalysts. This has led in recent years to more robust and elegant strategies applied to the efficient production of pharmaceutical compounds and fine chemicals at industrial scale under more environmentally benign conditions, as required by the more restrictive governmental legislations.

The need for more sustainable processes based on highly efficient transformations is turning attention towards the development of systems where the combination of several (bio)catalytic steps can be performed in 'one-pot'. These strategies provide multi-step catalytic cascade processes that avoid the isolation of the corresponding intermediates, minimise the reaction times, the energy required and the waste generated. Moreover, the concentration of unstable intermediates and the formation of secondary reactions are also decreased in these protocols. In particular, the concurrent employment of different enzymes seems to be straightforward as they can perform their natural reactions under very similar conditions. Hence, oxidoreductases and transferases have been extensively employed in a combined manner to afford many different types of valuable derivatives in a very selective fashion with excellent yields.

Most of the examples can be classified into four main cascade systems or combinations thereof, providing many different (enantioenriched) targets such as alcohols, amines, carbonylic or carboxylic acid derivatives, among others (see Chapter 14). The integration of various enzymes in the same microorganism host, including the enzymes involved in the recycling of cofactors (see Chapter 12) or the necessary co-substrates, *via* genetic modification, will allow a rapid transfer to industry for the achievement of highly productive and inexpensive synthetic routes *in vitro* or *in vivo* (the so-called synthetic biology, see Chapters 8–11). The integration of these protocols in more complex networks using other chemocatalysts or metalloenzymes will also expand their potential applications (see Chapter 15). However, to get access to novel enzymes (see Chapters 1–4) with new activities and/or improved stabilities at higher expression levels in reduced timeframes, new contributions from different multidisciplinary and interdependent areas will be required.

Acknowledgements

Financial support from MICINN (Project CTQ2013-44153-P) and the Principado de Asturias (Project FC-15-GRUPIN14-002) are gratefully acknowledged. L.M.-M. thanks the Principado de Asturias for her predoctoral fellowship Severo Ochoa.

References

1. D. L. Nelson and M. M. Cox, *Lehninger: Principles of Biochemistry*, W. H. Freeman, New York, 2004.
2. P. Anastas and N. Eghbali, *Chem. Soc. Rev.*, 2010, **39**, 301.
3. K. Faber, *Biotransformations in Organic Chemistry*, Springer-Verlag, Berlin, 6th edn, 2011.
4. *Enzyme Catalysis in Organic Synthesis*, ed. K. Drauz, H. Gröger and O. May, Wiley-VCH, Weinheim, 3rd edn, 2012.
5. *Science of Synthesis, Biocatalysis in Organic Synthesis*, ed. K. Faber, W.-D. Fessner and N. J. Turner, Georg Thieme Verlag, Stuttgart, 2015.
6. G. A. Behrens, A. Hummel, S. K. Padhi, S. Schätzle and U. T. Bornscheuer, *Adv. Synth. Catal.*, 2011, **353**, 2191.
7. T. Davids, M. Schmidt, D. Boettcher and U. T. Bornscheuer, *Curr. Opin. Chem. Biol.*, 2013, **17**, 215.
8. B. M. Nestl, B. A. Nebel and B. Hauer, *Curr. Opin. Chem. Biol.*, 2011, **15**, 187.
9. D. Muñoz Solano, P. Hoyos, M. J. Hernáiz, A. R. Alcántara and J. M. Sánchez-Montero, *Bioresour. Technol.*, 2012, **115**, 196.
10. B. T. Ueberbacher, M. Hall and K. Faber, *Nat. Prod. Rep.*, 2012, **29**, 337.
11. R. Sigrist, B. Z. da Costa, A. J. Marsaioli and L. G. de Oliveira, *Biotechnol. Adv.*, 2015, **33**, 394.
12. V. Köhler and N. J. Turner, *Chem. Commun.*, 2015, **51**, 450.
13. J. Muschiol, C. Peters, N. Oberleitner, M. D. Mihovilovic, U. T. Bornscheuer and F. Rudroff, *Chem. Commun.*, 2015, **51**, 5798.
14. E. García-Junceda, I. Lavandera, D. Rother and J. H. Schrittwieser, *J. Mol. Catal. B: Enzym.*, 2015, **114**, 1.
15. Q. Ji, B. Wang, J. Tan, L. Zhu and L. Li, *Process Biochem.*, 2016, **51**, 1193.
16. H. Gröger and W. Hummel, *Curr. Opin. Chem. Biol.*, 2014, **19**, 171.
17. R. Abu and J. M. Woodley, *ChemCatChem*, 2015, **7**, 3094.
18. W.-D. Fessner and C. Walter, *Angew. Chem., Int. Ed.*, 1992, **31**, 614.

19. D. E. Torres Pazmiño, R. Snajdrova, B.-J. Baas, M. Ghobrial, M. D. Mihovilovic and M. W. Fraaije, *Angew. Chem., Int. Ed.*, 2008, **47**, 2275.
20. R. A. Sheldon, *Green Chem.*, 2007, **9**, 1273.
21. Y. Ni, D. Holtmann and F. Hollmann, *ChemCatChem*, 2014, **6**, 930.
22. S. F. Mayer, W. Kroutil and K. Faber, *Chem. Soc. Rev.*, 2001, **30**, 332.
23. J.-C. Wasilke, S. J. Obrey, R. T. Baker and G. C. Bazan, *Chem. Rev.*, 2005, **105**, 1001.
24. E. Ricca, B. Brucher and J. H. Schrittwieser, *Adv. Synth. Catal.*, 2011, **353**, 2239.
25. T. S. Moody, S. Mix, G. Brown and D. Beecher in *Science of Synthesis, Biocatalysis in Organic Synthesis*, ed. K. Faber, W.-D. Fessner and N. J. Turner, Georg Thieme Verlag, Stuttgart, 2015, p. 421.
26. C. Rodríguez, I. Lavandera and V. Gotor, *Curr. Org. Chem.*, 2012, **16**, 2525.
27. S. Kara, J. H. Schrittwieser, F. Hollmann and M. B. Ansorge-Schumacher, *Appl. Microbiol. Biotechnol.*, 2014, **98**, 1517.
28. R. Stuermer, B. Hauer, M. Hall and K. Faber, *Curr. Opin. Chem. Biol.*, 2007, **11**, 203.
29. E. Brenna, F. G. Gatti, D. Monti, F. Parmeggiani and A. Sacchetti, *ChemCatChem*, 2012, **4**, 653.
30. E. Brenna, F. G. Gatti, L. Malpezzi, D. Monti, F. Parmeggiani and A. Sacchetti, *J. Org. Chem.*, 2013, **78**, 4811.
31. E. Brenna, M. Crotti, F. G. Gatti, D. Monti, F. Parmeggiani, A. Pugliese and S. Santangelo, *J. Mol. Catal. B: Enzym.*, 2015, **114**, 37.
32. S. Reich, B. M. Nestl and B. Hauer, *ChemBioChem*, 2016, **17**, 561.
33. A. Díaz-Rodríguez, I. Lavandera and V. Gotor, *Curr. Green Chem.*, 2015, **2**, 192.
34. C. V. Voss, C. C. Gruber and W. Kroutil, *Synlett*, 2010, 991.
35. C. E. Paul, I. Lavandera, V. Gotor-Fernández, W. Kroutil and V. Gotor, *ChemCatChem*, 2013, **5**, 3875.
36. B. Lia, Y. Nie, X. Q. Mu and Y. Xu, *J. Mol. Catal. B: Enzym.*, 2016, **129**, 21.
37. K. Kędziora, A. Díaz-Rodríguez, I. Lavandera, V. Gotor-Fernández and V. Gotor, *Green Chem.*, 2014, **16**, 2448.
38. S. Wu, Y. Chen, Y. Xu, A. Li, Q. Xu, A. Glieder and Z. Li, *ACS Catal.*, 2014, **4**, 409.
39. J. Wachtmeister, A. Jakoblinnert, J. Kulig, H. Offermann and D. Rother, *ChemCatChem*, 2014, **6**, 1051.
40. J. Wachtmeister, P. Mennicken, A. Hunold and D. Rother, *ChemCatChem*, 2016, **8**, 607.
41. A. Jakoblinnert and D. Rother, *Green Chem.*, 2014, **16**, 3472.
42. J. Wachtmeister, A. Jakoblinnert and D. Rother, *Org. Process Res. Dev.*, 2016, **20**, 1744.
43. S. Kara, D. Spickermann, J. H. Schrittwieser, C. Leggewie, W. J. H. van Berkel, I. W. C. E. Arends and F. Hollmann, *Green Chem.*, 2013, **15**, 330.

44. M. Pérez-Sánchez, C. R. Müller and P. Domínguez de María, *ChemCatChem*, 2013, **5**, 2512.
45. J. Zhang, S. Wu, J. Wu and Z. Li, *ACS Catal.*, 2015, **5**, 51.
46. P. Yao, L. Wang, J. Yuan, L. Cheng, R. Jia, M. Xie, J. Feng, M. Wang, Q. Wu and D. Zhu, *ChemCatChem*, 2015, **7**, 1438.
47. F. van Rantwijk and A. Stolz, *J. Mol. Catal. B: Enzym.*, 2015, **114**, 25.
48. C. Mateo, A. Chmura, S. Rustler, F. van Rantwijk, A. Stolz and R. A. Sheldon, *Tetrahedron: Asymmetry*, 2006, **17**, 320.
49. A. Chmura, S. Rustler, M. Paravidino, F. van Rantwijk, A. Stolz and R. A. Sheldon, *Tetrahedron: Asymmetry*, 2013, **24**, 1225.
50. O. Sosedov, K. Matzer, S. Bürger, C. Kiziak, S. Baum, J. Altenbuchner, A. Chmura, F. van Rantwijk and A. Stolz, *Adv. Synth. Catal.*, 2009, **351**, 1531.
51. S. Baum, F. van Rantwijk and A. Stolz, *Adv. Synth. Catal.*, 2012, **354**, 113.
52. E. Busto, N. Richter, B. Grischek and W. Kroutil, *Chem. Eur. J.*, 2014, **20**, 11225.
53. G. Gourinchas, E. Busto, M. Killinger, N. Richter, B. Wiltschi and W. Kroutil, *Chem. Commun.*, 2015, **51**, 2828.
54. P. Yao, Y. Cui, S. Yu, Y. Du, J. Feng, Q. Wu and D. Zhu, *Adv. Synth. Catal.*, 2016, **358**, 2923.
55. E. Busto, R. C. Simon, N. Richter and W. Kroutil, *ACS Catal.*, 2016, **6**, 2393.
56. M. Pešić, C. López, J. López-Santín and G. Álvaro, *Appl. Microbiol. Biotechnol.*, 2013, **97**, 7173.
57. I. Sánchez-Moreno, V. Hélaine, N. Poupard, F. Charmantray, B. Légeret, L. Hecquet, E. García-Junceda, R. Wohlgemuth, C. Guérard-Hélaine and M. Lemaire, *Adv. Synth. Catal.*, 2012, **354**, 1725.
58. F. Camps Bres, C. Guérard-Hélaine, V. Hélaine, C. Fernandes, I. Sánchez-Moreno, M. Traïkia, E. García-Junceda and M. Lemaire, *J. Mol. Catal. B: Enzym.*, 2015, **114**, 50.
59. E. Busto, V. Gotor-Fernández and V. Gotor, *Chem. Rev.*, 2011, **111**, 3998.
60. M. Fuchs, J. E. Farnberger and W. Kroutil, *Eur. J. Org. Chem.*, 2015, 6965.
61. K. Tauber, M. Fuchs, J. H. Sattler, J. Pitzer, D. Pressnitz, D. Koszelewski, K. Faber, J. Pfeffer, T. Haas and W. Kroutil, *Chem. Eur. J.*, 2013, **19**, 4030.
62. F. G. Mutti, T. Knaus, N. S. Scrutton, M. Breuer and N. J. Turner, *Science*, 2015, **349**, 1525.
63. L. Martínez-Montero, V. Gotor, V. Gotor-Fernández and I. Lavandera, *Adv. Synth. Catal.*, 2016, **358**, 1618.
64. L. Martínez-Montero, V. Gotor, V. Gotor-Fernández and I. Lavandera, *Green Chem.*, 2017, **19**, 474.
65. E. O'Reilly, C. Iglesias and N. J. Turner, *ChemCatChem*, 2014, **6**, 992.
66. M. Schrewe, N. Ladkau, B. Bühler and A. Schmid, *Adv. Synth. Catal.*, 2013, **355**, 1693.
67. P. Both, H. Busch, P. P. Kelly, F. G. Mutti, N. J. Turner and S. L. Flitsch, *Angew. Chem., Int. Ed.*, 2016, **55**, 1511.

68. S. P. France, S. Hussain, A. M. Hill, L. J. Hepworth, R. M. Howard, K. R. Mulholland, S. L. Flitsch and N. J. Turner, *ACS Catal.*, 2016, **6**, 3753.
69. R. S. Heath, M. Pontini, S. Hussain and N. J. Turner, *ChemCatChem*, 2016, **8**, 117.
70. B. R. Lichman, E. D. Lamming, T. Pesnot, J. M. Smith, H. C. Hailes and J. M. Ward, *Green Chem.*, 2015, **17**, 852.
71. E.-M. Fischereder, D. Pressnitz and W. Kroutil, *ACS Catal.*, 2016, **6**, 23.
72. T. Sehl, H. C. Hailes, J. M. Ward, R. Wardenga, E. von Lieres, H. Offermann, R. Westphal, M. Pohl and D. Rother, *Angew. Chem., Int. Ed.*, 2013, **52**, 6772.
73. M. F. Villegas-Torres, R. J. Martinez-Torres, A. Cázares-Körner, H. Hailes, F. Baganz and J. Ward, *Enzyme Microb. Technol.*, 2015, **81**, 23.
74. E.-S. Park and J.-S. Shin, *Adv. Synth. Catal.*, 2014, **356**, 3505.
75. F. Parmeggiani, S. T. Ahmed, M. P. Thompson, N. J. Weise, J. L. Galman, D. Gahloth, M. S. Dunstan, D. Leys and N. J. Turner, *Adv. Synth. Catal.*, 2016, **358**, 3298.
76. S. Wu, Y. Zhou, T. Wang, H.-P. Too, D. I. C. Wang and Z. Li, *Nature Commun.*, 2016, **7**, 11917.
77. A. Dennig, E. Busto, W. Kroutil and K. Faber, *ACS Catal.*, 2015, **5**, 7503.
78. J. H. Sattler, M. Fuchs, F. G. Mutti, B. Grischek, P. Engel, J. Pfeffer, J. M. Woodley and W. Kroutil, *Angew. Chem., Int. Ed.*, 2014, **53**, 14153.
79. For non-cyclic derivatives, see Chapter 14.
80. R. Agudo and M. T. Reetz, *Chem. Commun.*, 2013, **49**, 10914.
81. T. Classen, M. Korpak, M. Schölzel and J. Pietruszka, *ACS Catal.*, 2014, **4**, 1321.
82. E. Brenna, F. G. Gatti, D. Monti, F. Parmeggiani, A. Sacchetti and J. Valoti, *J. Mol. Catal. B: Enzym.*, 2015, **114**, 77.
83. J. Liu and Z. Li, *ACS Catal.*, 2013, **3**, 908.
84. N. Oberleitner, C. Peters, J. Muschiol, M. Kadow, S. Saß, T. Bayer, P. Schaaf, N. Iqbal, F. Rudroff, M. D. Mihovilovic and U. T. Bornscheuer, *ChemCatChem*, 2013, **5**, 3524.
85. N. Oberleitner, C. Peters, F. Rudroff, U. T. Bornscheuer and M. D. Mihovilovic, *J. Biotechnol.*, 2014, **192**, 393.
86. C. Peters, R. Kölzsch, M. Kadow, L. Skalden, F. Rudroff, M. D. Mihovilovic and U. T. Bornscheuer, *ChemCatChem*, 2014, **6**, 1021.
87. S. Schmidt, C. Scherkus, J. Muschiol, U. Menyes, T. Winkler, W. Hummel, H. Gröger, A. Liese, H.-G. Herz and U. T. Bornscheuer, *Angew. Chem., Int. Ed.*, 2015, **54**, 2784.
88. S. Schmidt, H. C. Büchsenschütz, C. Scherkus, A. Liese, H. Gröger and U. T. Bornscheuer, *ChemCatChem*, 2015, **7**, 3951.
89. A. Pennec, F. Hollmann, M. S. Smit and D. J. Opperman, *ChemCatChem*, 2015, **7**, 236.
90. S. Herter, S. M. McKenna, A. R. Frazer, S. Leimkühler, A. J. Carnell and N. J. Turner, *ChemCatChem*, 2015, **7**, 2313.

91. E. Busto, M. Gerstmann, F. Tobola, E. Dittmann, B. Wiltschi and W. Kroutil, *Catal. Sci. Technol.*, 2016, **6**, 8098.
92. E. Busto, R. C. Simon and W. Kroutil, *Angew. Chem., Int. Ed.*, 2015, **54**, 10899.
93. T. Furuya, M. Miura and K. Kino, *ChemBioChem*, 2014, **15**, 2248.
94. M. Gall, M. Thomsen, C. Peters, I. V. Pavlidis, P. Jonczyk, P. P. Grünert, S. Beutel, T. Scheper, E. Gross, M. Backes, T. Geißler, J. P. Ley, J.-M. Hilmer, G. Krammer, G. J. Palm, W. Hinrichs and U. T. Bornscheuer, *Angew. Chem., Int. Ed.*, 2014, **53**, 1439.
95. E. Ricklefs, M. Girhard, K. Koschorreck, M. S. Smit and V. B. Urlacher, *ChemCatChem*, 2015, **7**, 1857.
96. J. Siegrist, S. Aschwanden, S. Mordhorst, L. Thöny-Meyer, M. Richter and J. N. Andexer, *ChemBioChem*, 2015, **16**, 2576.
97. I. Serra, S. Daly, A. R. Alcantara, D. Bianchi, M. Terreni and D. Ubiali, *RSC Adv.*, 2015, **5**, 23569.
98. L. Zhu, R. Tao, Y. Wang, Y. Jiang, X. Lin, Y. Yang, H. Zheng, W. Jiang and S. Yang, *Appl. Microbiol. Biotechnol.*, 2011, **90**, 903.
99. N. Richter, J. E. Farnberger, D. Pressnitz, H. Lechner, F. Zepeck and W. Kroutil, *Green Chem.*, 2015, **17**, 2952.
100. A. Rioz-Martínez, F. R. Bisogno, C. Rodríguez, G. de Gonzalo, I. Lavandera, D. E. Torres Pazmiño, M. W. Fraaije and V. Gotor, *Org. Biomol. Chem.*, 2010, **8**, 1431.
101. F. R. Bisogno, A. Rioz-Martínez, C. Rodríguez, I. Lavandera, G. de Gonzalo, D. E. Torres Pazmiño, M. W. Fraaije and V. Gotor, *ChemCatChem*, 2010, **2**, 946.
102. A. Bornadel, R. Hatti-Kaul, F. Hollmann and S. Kara, *ChemCatChem*, 2015, **7**, 2442.
103. A. Bornadel, R. Hatti-Kaul, F. Hollmann and S. Kara, *Tetrahedron*, 2016, **72**, 7222.
104. M. Rachwalski, N. Vermue and F. P. J. T. Rutjes, *Chem. Soc. Rev.*, 2013, **42**, 9268.
105. E. O'Reilly, C. Iglesias, D. Ghislieri, J. Hopwood, J. L. Galman, R. C. Lloyd and N. J. Turner, *Angew. Chem., Int. Ed.*, 2014, **53**, 2447.
106. F. Parmeggiani, S. L. Lovelock, N. J. Weise, S. T. Ahmed and N. J. Turner, *Angew. Chem., Int. Ed.*, 2015, **54**, 4608.
107. J. H. Schrittwieser, B. Groenendaal, V. Resch, D. Ghislieri, S. Wallner, E.-M. Fischereder, E. Fuchs, B. Grischek, J. H. Sattler, P. Macheroux, N. J. Turner and W. Kroutil, *Angew. Chem., Int. Ed.*, 2014, **53**, 3731.
108. V. Köhler, Y. M. Wilson, M. Dürrenberger, D. Ghislieri, E. Churakova, T. Quinto, L. Knörr, D. Häussinger, F. Hollmann, N. J. Turner and T. R. Ward, *Nature Chem.*, 2013, **5**, 93.
109. Y. Okamoto, V. Köhler and T. R. Ward, *J. Am. Chem. Soc.*, 2016, **138**, 5781.
110. S. R. Wallner, M. Pogorevc, H. Trauthwein and K. Faber, *Eng. Life Sci.*, 2004, **4**, 512.

111. S. Hwang, C. Y. Choi and E. Y. Lee, *Biotechnol. Bioprocess Eng.*, 2008, **13**, 453.
112. J. Y. Min and E. Y. Lee, *J. Ind. Eng. Chem.*, 2012, **18**, 160.
113. M. Schober, M. Toesch, T. Knaus, G. A. Strohmeier, B. van Loo, M. Fuchs, F. Hollfelder, P. Macheroux and K. Faber, *Angew. Chem., Int. Ed.*, 2013, **52**, 3277.
114. O. Verho and J.-E. Bäckvall, *J. Am. Chem. Soc.*, 2015, **137**, 3996.
115. K. Yasukawa and Y. Asano, *Adv. Synth. Catal.*, 2012, **354**, 3327.
116. S. Baxter, S. Royer, G. Grogan, F. Brown, K. E. Holt-Tiffin, I. N. Taylor, I. G. Fotheringham and D. J. Campopiano, *J. Am. Chem. Soc.*, 2012, **134**, 19310.
117. P. Soriano-Maldonado, F. J. Las Heras-Vazquez, J. M. Clemente-Jimenez, F. Rodriguez-Vico and S. Martínez-Rodríguez, *Appl. Microbiol. Biotechnol.*, 2015, **99**, 283.
118. C. C. Gruber, B. M. Nestl, J. Gross, P. Hildebrant, U. T. Bornscheuer, K. Faber and W. Kroutil, *Chem. Eur. J.*, 2007, **13**, 8271.
119. I. Karume, M. M. Musa, O. Bsharat, M. Takahashi, S. M. Hamdan and B. El Ali, *RSC Adv.*, 2016, **6**, 96616.
120. A. Díaz-Rodríguez, N. Ríos-Lombardía, J. H. Sattler, I. Lavandera, V. Gotor-Fernández, W. Kroutil and V. Gotor, *Catal. Sci. Technol.*, 2015, **5**, 1443.
121. E. Liardo, N. Ríos-Lombardía, F. Morís, J. González-Sabín and F. Rebolledo, *Org. Lett.*, 2016, **18**, 3366.

CHAPTER 14

Artificial Biocatalytic Cascades to Alcohols and Amines

JOERG H. SCHRITTWIESER,*[a] STEFAN VELIKOGNE[b] AND WOLFGANG KROUTIL*[a,b]

[a] Institute of Chemistry, Organic and Bioorganic Chemistry, University of Graz, NAWI Graz, BioTechMed Graz, Heinrichstrasse 28, 8010 Graz, Austria ; [b] ACIB GmbH, c/o Department of Chemistry, University of Graz, Heinrichstrasse 28, 8010 Graz, Austria
*Email: joerg.schrittwieser@uni-graz.at; wolfgang.kroutil@uni-graz.at

14.1 Introduction

A cascade reaction is the combination of several chemical steps in a single reaction vessel without isolation of the intermediates.[1] In the case of a biocatalytic cascade, at least one biocatalyst (for instance an isolated enzyme, or whole microbial cells) is required for at least one step in the sequence of transformations. This definition includes linear combinations of multiple biotransformations, which are also referred to as biocatalytic tandem reactions, as well as domino reactions, which are spontaneous multi-step reactions that are merely initiated by an enzymatic transformation; thus, only the first step is enzyme catalysed while the following steps proceed spontaneously.[2] The latter processes are only scarcely found in recent literature. The biocatalytic cascades discussed in this review comprise at least two linear steps that each require one biocatalyst, which means that at least two enzymes are involved in the overall process.

Cascade reactions offer the advantage of circumventing the isolation of reaction intermediates, which not only saves resources, reagents and time,

Catalysis Series No. 32
Modern Biocatalysis: Advances Towards Synthetic Biological Systems
Edited by Gavin Williams and Mélanie Hall

Published by the Royal Society of Chemistry, www.rsc.org

but is also useful when unstable intermediates, which are formed *in situ* and are directly consumed in the next step, are involved in the reaction sequence. Consequently, the cascade approach can lead to higher yields compared to a classical sequence of single-step transformations, while at the same time increasing the synthetic efficiency by saving operational work-up steps. In fact, artificial biocatalytic cascades can be seen as an extension of biosynthesis in organisms producing the molecules of life.[3]

The present book chapter focuses on artificial linear cascades of the last five years involving biocatalysts only and does not discuss cascades that combine enzyme- and chemo-catalysed steps (see Chapters 13 and 15);[4–7] cascades employing non-natural metalloproteins are at the borderline (see Chapter 3).[8] Furthermore, co-immobilization strategies of cascade enzymes will not be considered, and metabolic engineering of natural pathways is out of scope of this review.

In this context, 'artificial cascade' means that enzymes from different organisms are combined for a reaction sequence that does not occur in nature for the specific substrates studied. Linear cascade means that the product of a first step is the substrate of a second subsequent step (Scheme 14.1A), which is in contrast to enzymatic cofactor recycling systems[9–12] (Scheme 14.1B) or enzymatic parallel cascades (see Chapters 12 and 13).[13,14] The latter have also been termed 'parallel interconnected transformations' and are based on the same general reaction scheme as the cofactor recycling systems.[13]

Biocatalytic one-pot reactions can be performed in a simultaneous or sequential mode. In the case of a simultaneous cascade, all reagents and

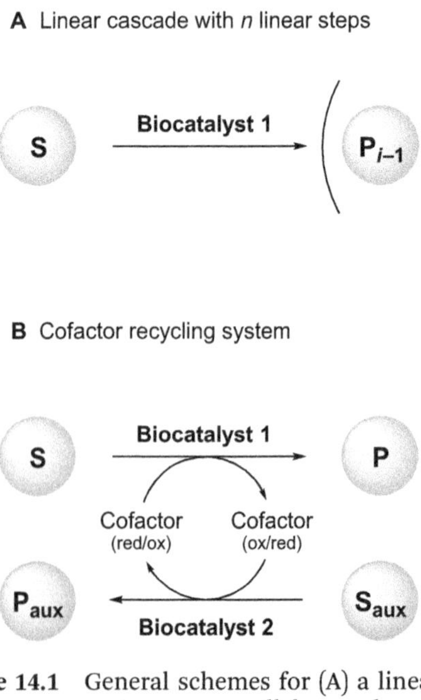

Scheme 14.1 General schemes for (A) a linear cascade and (B) a cofactor recycling system or a parallel cascade.

biocatalysts are present from the outset of the reaction, while in sequential one-pot reactions, some reagents and/or biocatalysts are added to the reaction mixture at a later point in time. For instance, an enzyme required for the second reaction step might be added after the first step has run to completion. A sequential strategy can be required when the reaction conditions of two steps are not compatible with each other, due to *e.g.* different pH optima, inhibition of the first step by reagents/products of the second step, insufficient chemoselectivity of the second step or other reasons.

Cascades have been reviewed frequently,[3,15–22] in some cases focusing on particular techniques such as deracemisation[23–29] or on specific product or enzyme classes.[30–33] Some recent reviews have referred to *in vitro* cascades by other names, for instance 'cell-free metabolic engineering',[34] 'cell-free enzymatic reaction cascades'[35] or 'systems biocatalysis'.[36,37] Other related reviews deal with *de novo* redox pathways in the host *Escherichia coli*,[38] thermodynamic constraints,[39] aspects of process technology,[40] multi-enzymatic synthesis in cells,[41] designer microbes,[42] and continuous flow processes.[43]

The present literature review is structured according to the functional group formed in the last enzymatic step and therein sorted according to the last enzyme in the linear sequence. The functional groups considered in this review are alcohols and amines.

14.2 Alcohols

Alcohols and their derivatives play an important role not only in nature but also in the chemical and pharmaceutical sector, since various hydroxylated compounds and especially vicinal diols are used as valuable synthetic intermediates for many pharmaceuticals, bioactive compounds and chiral agents.[44–46]

In the examples discussed in this section, the alcohol moiety is formed in the last reaction step by an alcohol dehydrogenase, an aldolase or an epoxide hydrolase.

14.2.1 Alcohol Dehydrogenases

14.2.1.1 C–C Bond Formation and Ketone Reduction

Thiamine diphosphate (ThDP) is nature's cofactor of choice for a variety of carbonyl 'umpolung' processes, including C–C bond forming reactions that afford α-hydroxy ketones as products.[47–49] ThDP is used by numerous enzymes of the C–C lyase class (EC 4.1), of which benzaldehyde lyase (BAL), benzoylformate decarboxylase (BFD) and pyruvate decarboxylase (PDC) are most frequently employed in biocatalysis.[50] ThDP-dependent enzymes have recently been combined with alcohol dehydrogenases (ADHs) for accessing chiral vicinal diols. For instance, the diol (1*R*,2*R*)-**4** was prepared using a two-step-one-pot cascade involving benzaldehyde lyase (BAL) from *Pseudomonas fluorescens* and an ADH from a *Ralstonia* sp. (Scheme 14.2).[51]

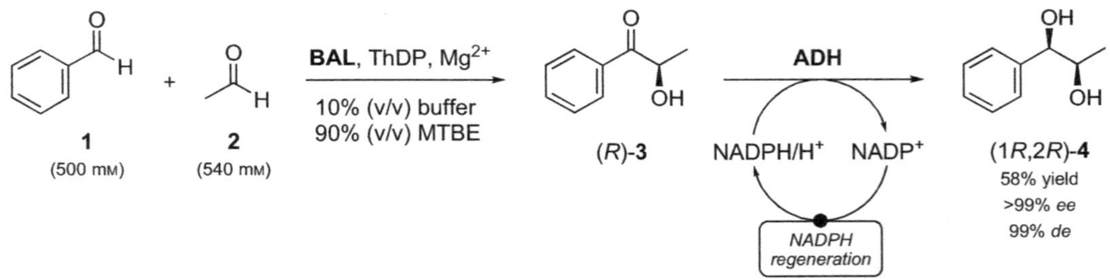

Scheme 14.2 Biocatalytic synthesis of vicinal diol (1*R*,2*R*)-**4** using a two-step one-pot cascade. Enzyme abbreviations: BAL, benzaldehyde lyase; ADH, alcohol dehydrogenase.

The reaction was performed in a micro-aqueous system containing 10% (v/v) water or buffer and 90% (v/v) of methyl *tert*-butyl ether (MTBE) at a concentration of 500 mM benzaldehyde. The advantage of this system is that the concentration of benzaldehyde can be elevated to levels beyond its solubility limit in aqueous media. Comparing the simultaneous cascade with the same cascade performed in sequential mode (adding the ADH after completion of the BAL reaction), it was found that the latter gave the better results in terms of conversion (93% *vs.* 75%), while the ee and de were the same in both cases (>99% ee; 99.6% and 99.8% de, respectively). This was attributed mainly to the competing reduction of benzaldehyde by *Ralstonia* ADH that was observed in the simultaneous cascade, which led to the accumulation of benzyl alcohol (39 mM). Performing the two-step reaction on a preparative scale (1.06 g benzaldehyde), product (1*R*,2*R*)-**4** was obtained with 58% isolated yield and excellent optical purity (>99% ee, 99% de). In a follow-up study, the same cascade was investigated using whole cells placed in a polyvinylidene fluoride (PVDF) teabag, aiming at a better recyclability of the biocatalysts by simply removing the bag from the reaction vessel.[52] This new reaction setup was tested in (i) sequential mode, adding the ADH after completion of the BAL reaction, (ii) simultaneous mode, adding all enzymes and reagents at the start of the reaction, and (iii) mixed cascade mode, adding the ADH after 3 h, when benzaldehyde had been largely but not completely consumed by BAL. The highest levels of product formation (339 mM from 500 mM benzaldehyde) were attained in the mixed mode, while the simultaneous cascade led to the highest space–time yield (3.25 $g L^{-1} h^{-1}$) due to the shorter time required for the reaction to reach completion.

Coupling a ThDP-dependent acetylacetoin synthase (AAS) from *Bacillus licheniformis* to the NADH-dependent acetylacetoin reductase (AAR) from the same organism allowed to access *syn*-α-alkyl-α,β-dihydroxyketones **8** with high enantioselectivities (>95% ee; Scheme 14.3) and good isolated yields (60–70%).[53,54] Interestingly, the corresponding symmetrical diketones

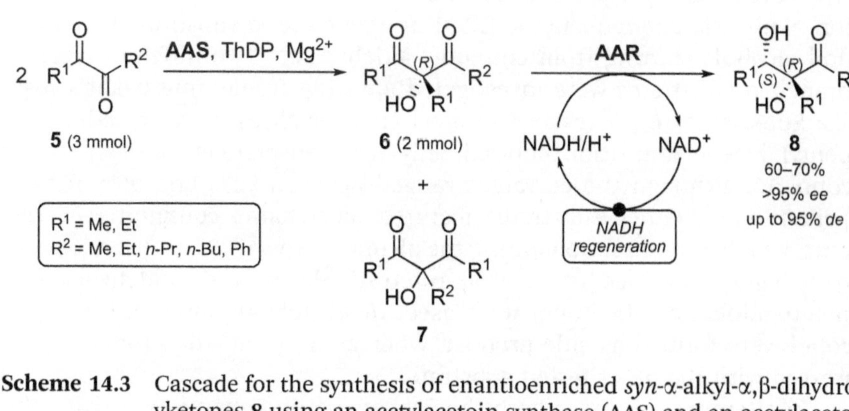

Scheme 14.3 Cascade for the synthesis of enantioenriched *syn*-α-alkyl-α,β-dihydroxyketones **8** using an acetylacetoin synthase (AAS) and an acetylacetoin reductase (AAR).

7 formed as by-products were not reduced by the strictly (*S*)-stereospecific AAR and could be separated from the final products by flash chromatography. The reaction sequence was also employed for the synthesis of the chiral green tea flavour compound 3-hydroxy-3-methylnonane-2,4-dione.[55]

14.2.1.2 Deracemisation of Secondary Alcohols

For the deracemisation of racemic *sec*-alcohols to the optically pure compounds,[23–29] a two-step cascade was established involving two stereo-complementary alcohol dehydrogenases (ADHs) from *Rhodococcus ruber* and *Lactobacillus brevis* (Scheme 14.4).[56] In this process, one enantiomer of the racemic alcohol is oxidized to the ketone, leaving the desired stereoisomer behind. By using a second stereo-complementary ADH, the intermediate ketone is reduced to the desired alcohol enantiomer. In contrast to a previous protocol based on the same principle,[57] regeneration of the oxidised nicotinamide cofactor NAD^+ was achieved *via* the oxidative metabolism of the *E. coli* expression host. Using this method, the model substrate *rac*-1-phenylethanol (**9**) was converted into its optically pure (*R*)-form within 30 min. A preparative-scale reaction (50 mg) with substrate **9** afforded enantiopure (*R*)-**9** with an isolated yield of 82% after purification. Furthermore, various other alcohols **10–15** were successfully converted, reaching 59% to >99% ee, whereby in most cases the ee was above 90%.

14.2.1.3 Double Reduction of α,β-Unsaturated Carbonyl Compounds

The combination of C=C-reduction of α,β-unsaturated ketones catalysed by ene-reductases with C=O-reduction of the resulting saturated carbonyl compounds provides access to chiral, branched alcohols (Scheme 14.5). Several recent studies have developed variations of this biocatalytic cascade sequence.

The constitutively expressed ADHs of baker's yeast were used in combination with its ene-reductases (ERs) to generate α-substituted primary chiral alcohols starting from cinnamic aldehydes.[58] Two different types of α-branched substrates were investigated applying fermenting baker's yeast cells: Substrates **16** possessing an alkyl chain such as 2-propyl, *n*-butyl or *n*-pentyl in the α-position showed only low conversions (8–14%) to the alcohol **17**, although the ee values ranged between 92% and 96% for the (*S*)-enantiomer. Other substrates featured side-chains containing hetero-atoms, which allowed for obtaining the alcohol with 34–39% conversion and mostly higher ee values (up to 99%). Interestingly, when the aldehydes with a hetero-aliphatic substituent were used, detectable amounts of the allylic alcohol were formed as side product, whereas no such side products were observed with the other tested substrates.

In an attempt to improve both the conversion and enantioselectivity, the baker's yeast mediated setup was replaced by a cascade using the purified

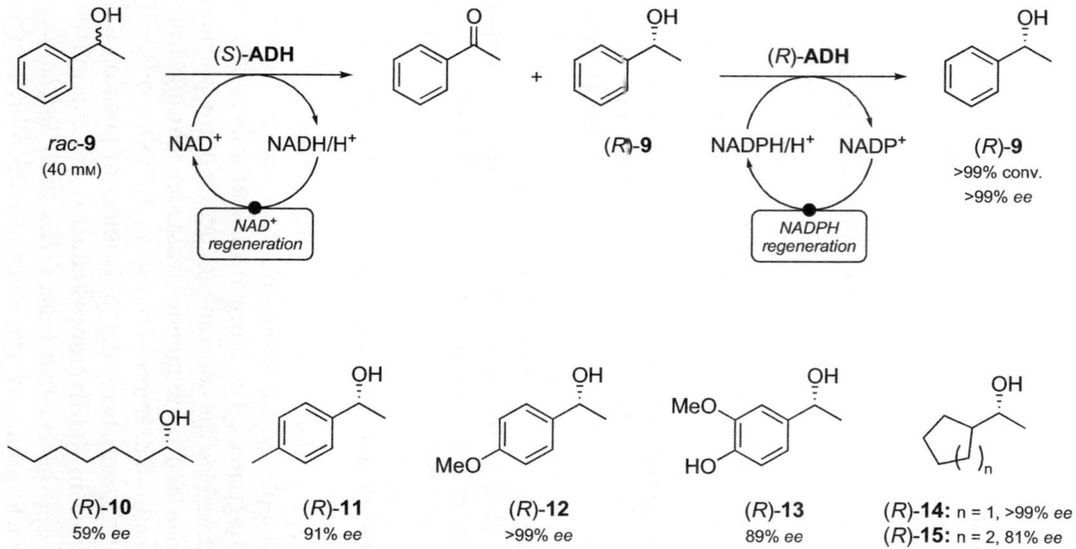

Scheme 14.4 Deracemisation of secondary alcohols **9–15** using a pair of stereocomplementary alcohol dehydrogenases (ADHs).

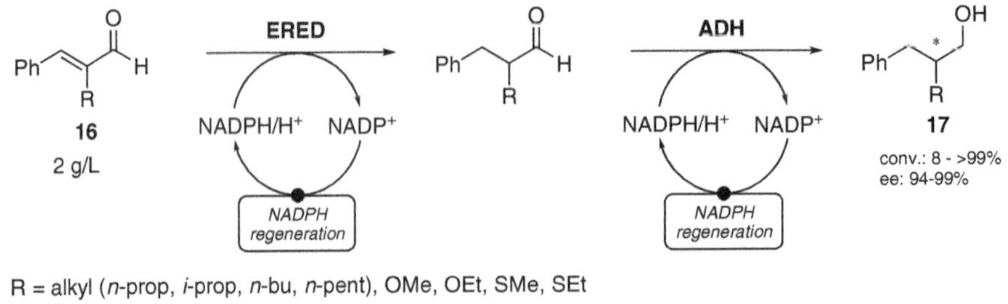

Scheme 14.5 Biocatalytic synthesis of α-substituted alcohols *via* double reduction using an ene-reductase (ERED) and an alcohol dehydrogenase (ADH).

ene-reductases OYE2 and OYE3 in combination with horse liver alcohol dehydrogenase (HL-ADH). Moreover, the substrates were adsorbed on polymeric adsorbent beads to allow for *in situ* substrate feeding and product removal. With this modified system, conversions increased significantly reaching up to >99%, while the optical purities were comparable to those obtained in the yeast reactions.

The above-mentioned cascade has then been investigated with bicyclic substrates **18a–c** and **20**, whose reduction products may serve as synthetic precursors for the pharmaceuticals Robalzotan, Ebalzotan and Rotigotine.[59] Reductions were performed with purified ene-reductase OYE2 and two different ADHs, originating from horse liver (HL-ADH) or *Candida parapsilosis* (*Cp*-ADH), depending on the substrate (see Figure 14.1). Depending on the substrate and enzyme combination, the product alcohols were obtained with high conversion and good to perfect enantioselectivity in almost all cases. Furthermore, alcohols **19a–c** and **21** were synthesised on a preparative scale (1.25 mmol, 200–238 mg) with high isolated yields (83–96%) and optical purities (91–99% ee).

Besides their application in pharmaceutical chemistry, various alcohols such as 3-methyl-4-phenyl-2-butanol (**22**; trade name: Muguesia®) are used as flavours and fragrances. The biocatalytic synthesis of selected isomers of **22** was demonstrated *via* the concomitant action of the previously described enzymes in a two-step simultaneous cascade, starting from the corresponding α,β-unsaturated ketone (Figure 14.2).[60] Out of the four possible stereoisomers of **22**, only two—the (*S*)-isomers in position C3—display the floral notes desired in the fragrance industry. For the introduction of the chiral centre at C3, the ene-reductase OYE3 from *Saccharomyces cerevisiae* was chosen, which produced the intermediate α-branched ketone with 98% ee for the (*S*)-enantiomer. The desired product alcohols were obtained by using two different alcohol dehydrogenases (ADHs), namely *Re*-ADH from *Rhodococcus erythropolis* to produce (2*S*,3*S*)-**22** and *Pl*-ADH from *Parvibaculum lavamentivorans* for the synthesis of the (2*R*,3*S*)-product. A preparative synthesis with 5 g L^{-1} substrate loading showed 90% and 95% conversion after 48 h for the products (2*S*,3*S*)-**22** and (2*R*,3*S*)-**22**, respectively.

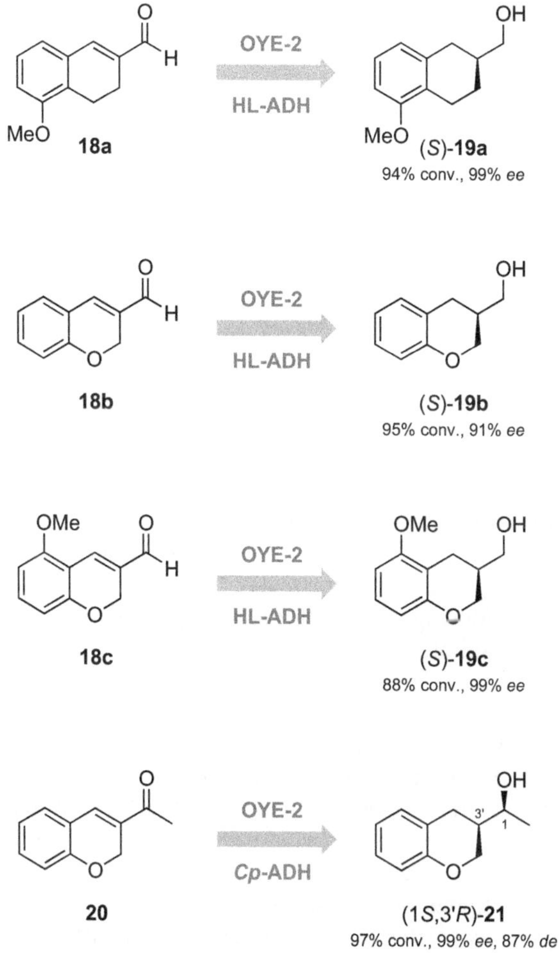

Figure 14.1 Production of pharmaceutical precursors *via* double reduction of α,β-unsaturated carbonyl compounds **18a–c** and **20**.

14.2.2 Phosphatases and Epoxide Hydrolases

Several stereoselective multi-enzymatic cascades have been developed for the synthesis of nitrocyclitols. Most of them require at least two enzymes, namely a DHAP-dependent aldolase and a phytase or phosphatase for phosphate hydrolysis (Scheme 14.6).[61–63] However, reaction systems based on D-fructose-6-phosphate aldolase from *E. coli* (FSA), which accepts non-phosphorylated aldol donors such as dihydroxyacetone (**25**), have also been reported.[64] In all cases, the cascade sequence consists of an enzyme-catalysed aldol reaction followed by a spontaneous, stereoselective Henry reaction of the aldol adduct **26**, yielding the final nitrocyclitol **27**. In the reactions based on DHAP-dependent aldolases, the use of chemically

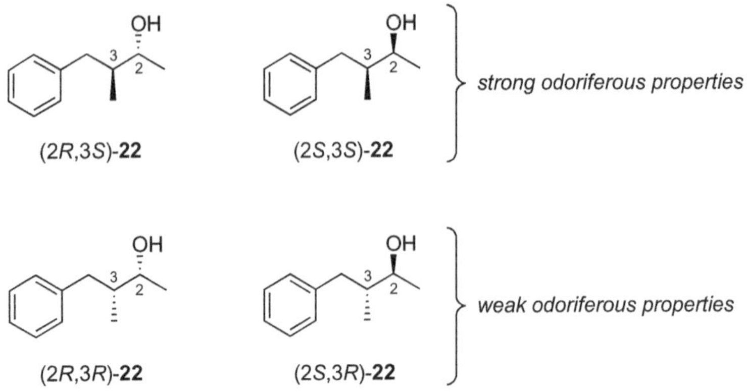

Figure 14.2 The four stereoisomers of the fragrance compound 3-methyl-4-phenyl-2-butanol.

synthesized dihydroxyacetone phosphate (DHAP, **24**) was avoided by preparing **24** *in situ* from **25** employing an ATP-dependent dihydroxyacetone kinase (DHAK) together with acetate kinase for ATP regeneration (Scheme 14.6).[65] Choosing either L-fuculose-1-phosphate aldolase (FucA) or L-rhamnulose-1-phosphate aldolase (RhuA) for the enzymatic carboligation step, nitrocyclitols with different absolute configurations at the C2- and C3-carbons were obtained. While FucA provided the expected (2*R*,3*S*)-configuration in most cases, RhuA was much less reliable for the expected (2*R*,3*R*)-configuration, hence leading to product mixtures. The Henry reaction, however, was highly stereoselective at the C1- and C6-carbons resulting in a *trans* relative configuration for the hydroxymethyl moiety and the nitro group.

Styrene and its derivatives have been shown to be suitable substrates for the biocatalytic synthesis of vicinal diols with pharmaceutical applications *via* multi-enzyme cascades.[66] For instance, (*S*)-1-phenylethane-1,2-diol (**30a**, R = H) is a synthetic precursor of the antidepressant (*R*)-fluoxetine. Employing a styrene monooxygenase (SMO) from *Sphingomonas* sp. and two different epoxide hydrolases (EHs), (*S*)- as well as (*R*)-configured diols **30** were produced with good conversions and high optical purity from the corresponding styrene derivatives **28** (Scheme 14.7). Access to both enantiomers of **30** from the same intermediate (*S*)-**29** was realised using epoxide hydrolases with complementary regioselectivity: opening of the epoxide **29** by attack of water at the α-carbon atom leads to inversion of configuration and hence affords the (*R*)-product, while epoxide hydrolysis *via* attack at the β-carbon proceeds with stereochemical retention to give (*S*)-**30**. The biotransformations were performed in a two-phase system consisting of potassium phosphate buffer and *n*-hexadecane in a 1 : 1 mixture, using *E. coli* whole cells coexpressing the two enzymes of the cascade. Regeneration of the nicotinamide cofactor required for the monooxygenation step was accounted for by cell metabolism, with D-glucose being added as sacrificial co-substrate.

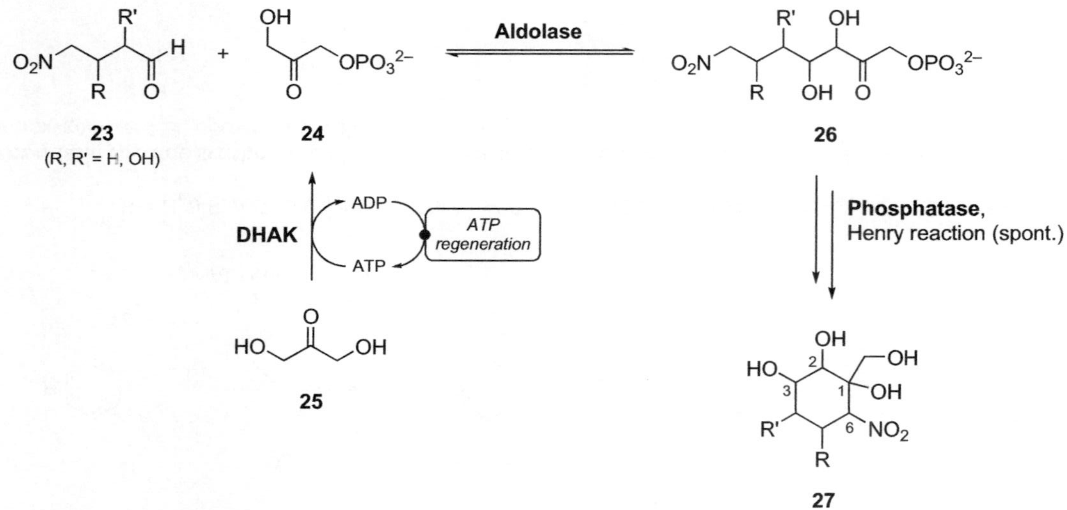

Scheme 14.6 Stereoselective chemoenzymatic synthesis of nitrocyclitols **27** with *in situ* formation of dihydroxyacetone (**24**) catalysed by ATP-dependent dihydroxyacetone kinase (DHAK).

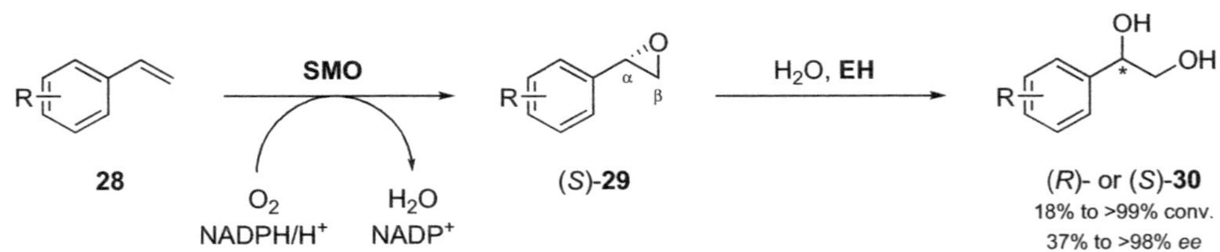

28 (*S*)-**29** (*R*)- or (*S*)-**30**

18% to >99% conv.

37% to >98% *ee*

R = H, *p*-F, *p*-Cl, *p*-Br, *p*-Me, *p*-MeO, *p*-CF_3, *m*-F, *m*-Cl, *m*-Br, *m*-Me, *m*-MeO, *o*-F, *o*-Cl, *o*-Me

Scheme 14.7 Bi-enzymatic cascade synthesis of vicinal diols **30** starting from styrene derivatives **28**. Enzyme abbreviations: SMO, styrene monooxygenase; EH, epoxide hydrolase.

14.3 Amines

Amines are important building blocks for several areas of the chemical industry: chiral amines and nitrogen heterocycles, for instance, are frequently encountered in bioactive molecules such as alkaloids or active pharmaceutical ingredients and are hence prime targets for medicinal chemistry.[67–69] Linear diamines, to name another example, are produced in million ton per year quantities for use as monomers in polyamide formation.

Over the course of the last decade, substantial efforts have been devoted to the development of efficient biocatalytic methods for amine synthesis, and many of these methods involve multi-enzyme reaction systems.[69,70]

14.3.1 Transaminases

Transaminases (EC 2.6.1) are a family of transferase enzymes that use pyridoxal 5′-phosphate (PLP) as cofactor for transferring an amino group from a donor substrate (*e.g.*, an α-amino acid, such as alanine, or isopropylamine) onto a carbonyl compound, and typically show a high stereoselectivity in this process. The broad substrate scope of these enzymes, their excellent stereoselectivity and the availability of a large number of (*R*)- as well as (*S*)-selective enzymes have helped transaminases become the most important biocatalysts for the synthesis of amines.[33,70–73] Nevertheless, shifting the reaction equilibrium has been shown to be challenging, particularly when alkyl or aryl ketones are used as acceptor substrates with alanine as the amino donor.[74] In such cases, measures for shifting the equilibrium of the reaction to the product side need to be taken if (near-)complete conversion is to be attained. In this context, the use of one or more auxiliary enzymes that remove the reaction co-product pyruvate is a commonly employed strategy. For instance, pyruvate can be removed by reduction using lactate dehydrogenase (LDH), or recycled to L-alanine using L-alanine dehydrogenase (AlaDH) and ammonia.[75,76] Acetone, which is the co-product formed when isopropylamine is the donor, can be reduced by an alcohol dehydrogenase (ADH), provided that this enzyme is not active with the carbonyl compound that is to be aminated.[77] Scheme 14.8 provides an overview of multi-enzyme systems that have been developed for driving the equilibrium of transaminase reactions to the amine product when starting from the ketone. Most of these 'orthogonal cascades' are well-established and have already been extensively reviewed elsewhere;[20,33,70,72] therefore, they will not be discussed in detail here, since the focus is on linear cascades. However, two systems that were reported only recently deserve to be mentioned: In the first, an NAD^+-reducing [NiFe]-hydrogenase from *Ralstonia eutropha* H16 was used for NADH regeneration in the well-known L-alanine dehydrogenase system (Scheme 14.8B), thereby enabling transamination reactions at the expense of ammonia and molecular hydrogen.[78] In the second, the AlaDH system was complemented with an alanine racemase, rendering possible the use of the cheaper L-alanine instead of its D-enantiomer for transaminations using

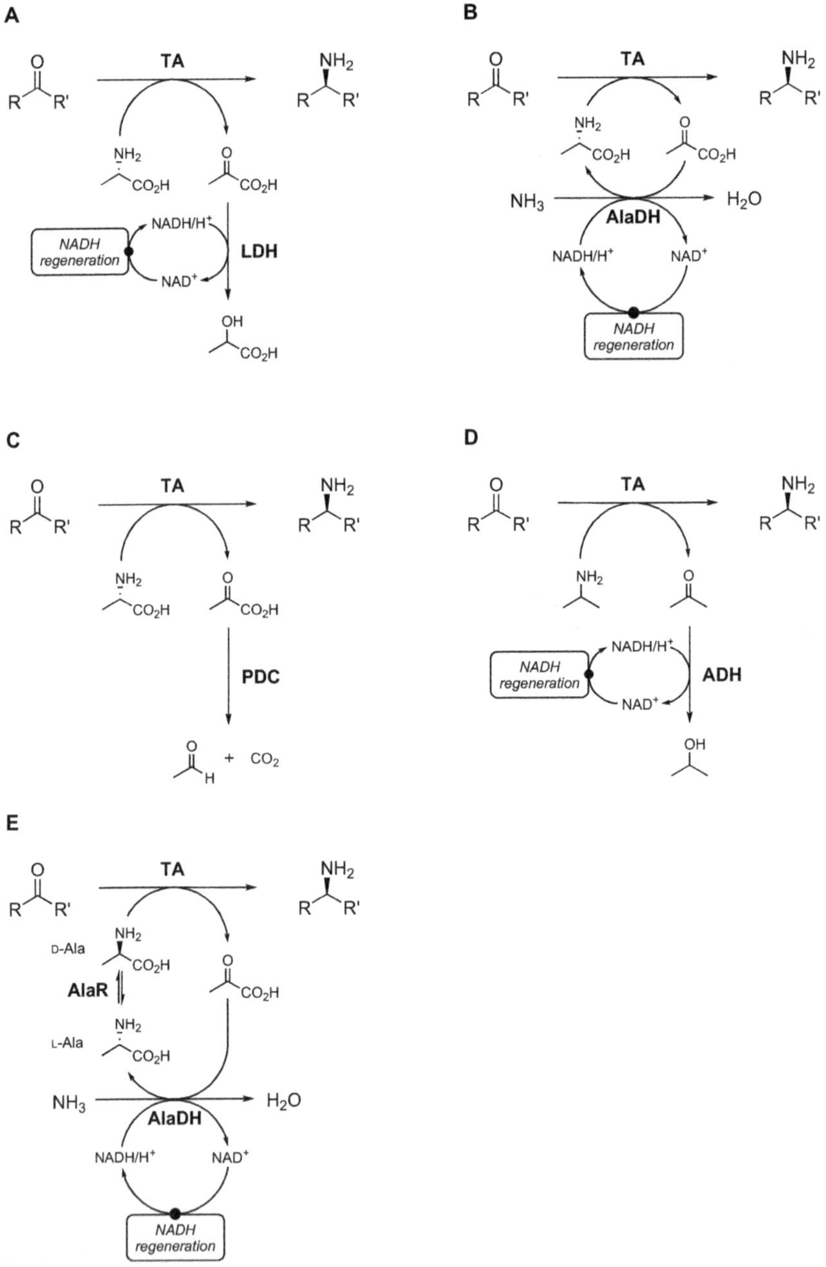

Scheme 14.8 Common multi-enzyme systems for shifting the equilibrium of transaminase reactions to the product side. Enzyme abbreviations: TA, transaminase; LDH, lactate dehydrogenase; AlaDH, L-alanine dehydrogenase; PDC, pyruvate decarboxylase; ADH, alcohol dehydrogenase; AlaR, alanine racemase.

(*R*)-selective transaminases (Scheme 14.8E).[79] Under optimised conditions, comparable levels of conversion were achieved using 2 equivalents of L-alanine in the presence of alanine racemase as have been attained with the traditional AlaDH setup and five equivalents of D-alanine. As a consequence, the cost contribution of the amine donor to the overall process is reduced by more than 90%.

14.3.1.1 Alcohol-to-amine Cascades

Alcohols are attractive synthetic intermediates because they can be accessed by a variety of reliable methods; additionally, the alcohol functionality is broadly distributed in renewable raw materials. Substitution of an alcohol moiety by a nucleophile (*e.g.*, an amine) requires some form of activation to render the hydroxyl group amenable to nucleophilic displacement. Alternatively, oxidation of the alcohol affords a carbonyl compound that can undergo condensation with an amine or ammonia to form an imine, which then can be reduced to the target amine. In principle, the redox equivalents (formally, a molecule of hydrogen, H_2) liberated in the alcohol oxidation step can be re-used for the imine reduction, resulting in an overall redox-neutral transformation. This so-called 'borrowing hydrogen' concept is a current focus of research in transition metal catalysis,[80–86] and in recent years it has been applied in biocatalysis as well, using transaminases or amine dehydrogenases in the reductive amination step (for systems using the latter class of enzymes, see Section 14.3.2).

Two basic concepts of biocatalytic alcohol-to-amine interconversion systems involving transaminases have been published, which differ in the type of enzyme used in the oxidation step and—as a consequence—in the overall redox balance: The first of these systems (Scheme 14.9A) uses an alcohol oxidase for oxidation—a reaction that requires molecular oxygen, which in the process is reduced to hydrogen peroxide. The redox equivalents obtained from alcohol oxidation are thereby 'lost' and hence this system is not suitable for the 'borrowing hydrogen' concept. If, in contrast, the oxidation step is catalysed by an alcohol dehydrogenase, which transfers the redox equivalents onto $NAD(P)^+$ to give NAD(P)H, a closed redox cycle can be established. However, a mediating enzyme is needed, since the transaminase responsible for the amination step does not consume NAD(P)H but *e.g.* alanine. This mediating function can be fulfilled by L-alanine dehydrogenase (AlaDH), which regenerates the transaminase co-substrate L-alanine from pyruvate and at the same time consumes the reduced NADH cofactor to provide NAD^+ for the oxidation step (Scheme 14.9B). The only reagent that is required in stoichiometric amounts in this system is ammonia, added for instance in the form of ammonium chloride.

A system with two redox-independent steps (Scheme 14.9A) was reported using a variant[87] of galactose oxidase (GalOx) from *Fusarium* sp. NRRL 2903 for alcohol oxidation and transaminases from *Vibrio fluvialis* or *Paracoccus denitrificans* for reductive amination.[88] In addition, horseradish peroxidase

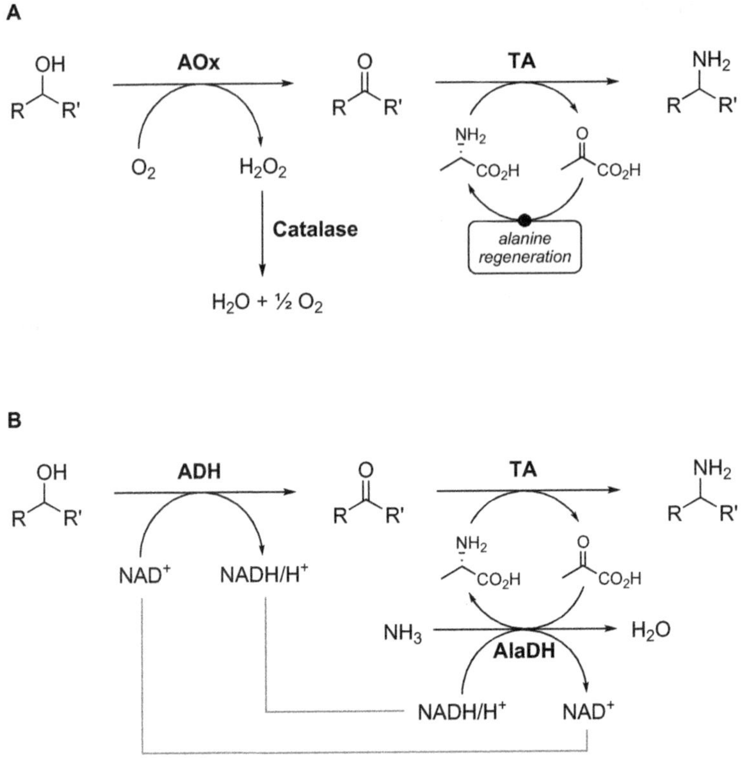

Scheme 14.9 Two multi-enzyme systems for alcohol-to-amine interconversion: **(A)** Combination of alcohol oxidase (AOx) and transaminase (TA), **(B)** 'hydrogen-borrowing' biocatalyst network comprising an alcohol dehydrogenase (ADH), transaminase (TA) and L-alanine dehydrogenase (AlaDH).

and 2,2′-azino-bis(3-ethylbenzothiazoline-6-sulfonic acid) (ABTS) were added for decomposition of H_2O_2, while L-alanine dehydrogenase (AlaDH) and glucose dehydrogenase (GDH) were used for regenerating the transaminase co-substrate, L-alanine. Since the overall reaction rate turned out to be limited by the oxidation step, the biotransformations were carried out in a pressurised chamber under 4 bar of oxygen. Due to limitations in the substrate scope of the applied oxidase, the cascade reaction was applicable to benzylic and cinnamic primary alcohols only (Figure 14.3A); in the best cases (benzyl alcohol and *p*-methylbenzyl alcohol), 50 mM of substrate were quantitatively aminated within 20 h at room temperature. Cinnamyl alcohol was also efficiently converted into the corresponding amine, (*E*)-3-phenyl-2-propen-1-amine (**39**), using the enzymatic cascade, and the synthetic utility of this reaction was demonstrated by further conversion of **39** into the antifungal agent naftifine (**47**). The target compound was obtained in 51% overall yield *via* four steps from cinnamyl alcohol.

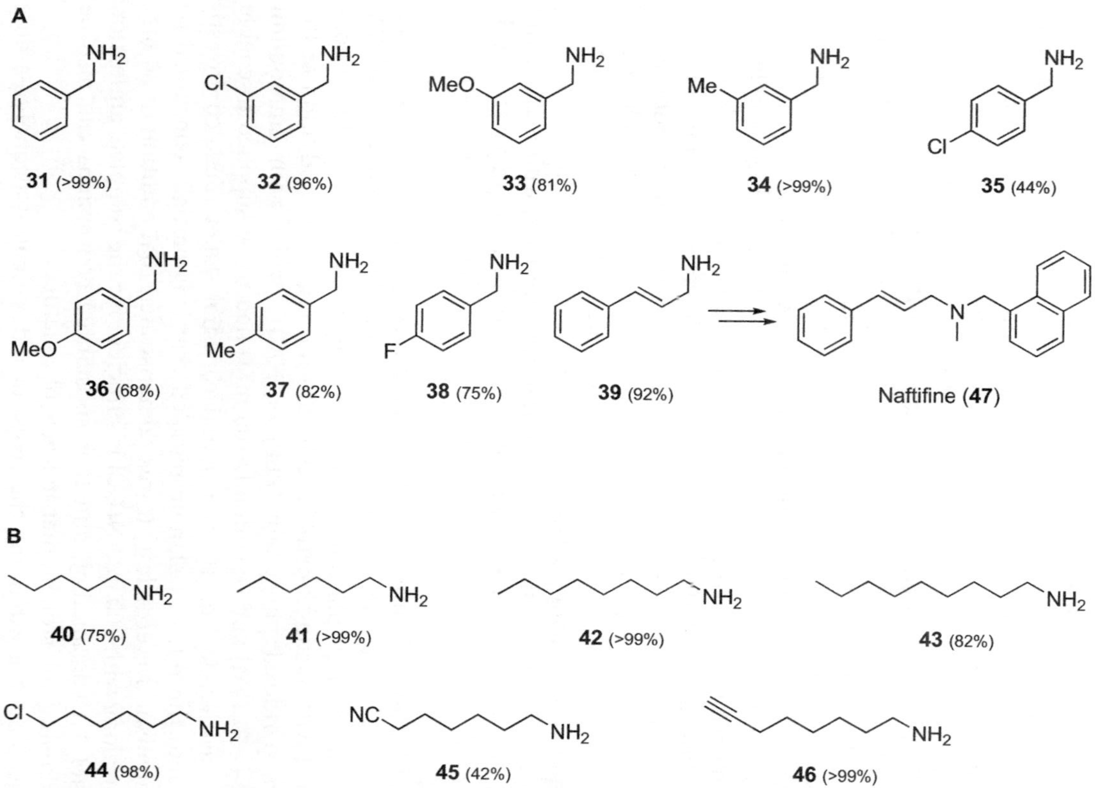

Figure 14.3 Examples of amines obtained from the corresponding primary alcohols using a biocatalytic cascade comprising transaminases and (A) galactose oxidase from *Fusarium* sp. NRRL 2903 or (B) long-chain alcohol oxidase from *Aspergillus fumigatus* according to Scheme 14.9A. Values in parentheses represent conversions into product within 20 h of reaction time.

The amination method was later extended to aliphatic primary alcohols of medium chain-length (C_4–C_{11}) by replacing galactose oxidase with a long-chain alcohol oxidase (LCAO) from *Aspergillus fumigatus*.[89] Quantitative conversion to the amine was observed for several substrates (*e.g.*, 1-hexanol, 1-octanol, 7-octyne-1-ol; Figure 14.3B); however, the reactions were run at a substantially lower substrate concentration (10 mM) than in the original cascade system using GalOx (50 mM).

The redox–neutral transformation of alcohols into amines using a combination of alcohol dehydrogenase, transaminase, and alanine dehydrogenase (Scheme 14.9B) has been intensively investigated by several research groups over the past five years. The feasibility of this reaction system was first demonstrated in a study that mainly focused on linear primary alcohols and terminal diols as substrates (Figure 14.4).[90] The thermostable ADH from *Bacillus stearothermophilus*, transaminase from *Chromobacterium violaceum*, and L-alanine dehydrogenase from *Bacillus subtilis* were used as biocatalysts. At a substrate concentration of 50 mM, the amination of 1-hexanol reached completion within 24 h, but with increasing chain length lower conversions were observed (*e.g.*, 50% with 1-octanol). The addition of 1,2-dimethoxyethane (DME, 10% v/v) as organic cosolvent improved conversions substantially in some cases, hinting at solubility limitations as the reason for the slower turnover of longer-chain substrates. The addition of DME also proved beneficial in the double amination of terminal diols, which was demonstrated using 1,8-octanediol and 1,10-decanediol as starting materials. Both compounds were quantitatively aminated within 20 h, and the di-amination of 1,10-decanediol was also performed on preparative scale (174 mg of substrate), affording the terminal diamine **50**—a valuable building block for polyamide formation—in 70% isolated yield.

The amination of similar alcohol and diol substrates using an *E. coli* whole-cell biocatalyst co-expressing the responsible enzymes (ADH, TA, AlaDH) from a single plasmid has also been reported.[91] Biotransformations performed with resting microbial cells did not require the addition of the cofactors pyridoxal phosphate (PLP) or NAD^+, while both ammonium chloride (≥50 mM) and L-alanine (≥100 mM) had to be supplied to enable complete conversion of 10 mM 1,10-decanediol. Later, detailed investigations into the role of L-alanine revealed that it is partially consumed by cell catabolism. Nevertheless, it was demonstrated that addition of 20 mM L-alanine along with 100 mM NH_4Cl is sufficient for the complete amination of 10 mM 1,10-decanediol, and that substantial conversions can also be achieved using sub-stoichiometric levels of L-alanine.[92]

Studies aimed at extending the scope of redox–neutral alcohol-to-amine interconversion to secondary alcohols revealed some additional complications:[93] First, if starting materials and products are chiral, the stereoselectivity of the involved enzymes has a profound influence on the overall reaction system. For instance, the oxidation of a racemic alcohol may require the use of two stereocomplementary alcohol dehydrogenases, and indeed the authors have used a combination of (*S*)-selective ADH-A from

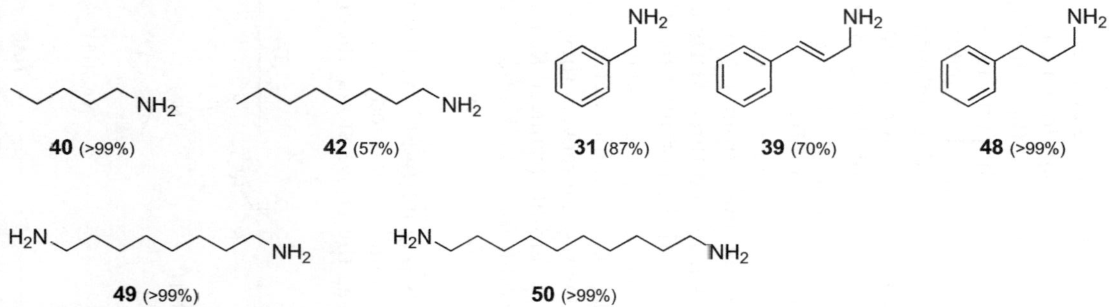

Figure 14.4 Examples of amines that have been produced from the corresponding primary alcohols *via* biocatalytic alcohol-to-amine interconversion using the ‘borrowing hydrogen’ concept according to Scheme 14.9B. Values in parentheses represent conversions into product within 24 h of reaction time.

Rhodococcus ruber DSM 44541 and (*R*)-selective ADH-007 from Codexis in selected cases. Second, the reactions were also affected by thermodynamic limitations—in contrast to the primary alcohols used in the initial study, the secondary alcohols could not be converted quantitatively, and amine formation did not exceed 64%. Moreover, significant amounts of the intermediate ketones accumulated in most reactions, which the authors attributed to the presence of NADH oxidase activity in the *E. coli* whole-cell biocatalysts. Lastly, a decrease in the enantiomeric excess of the amine was observed over time. To overcome some of these limitations, a modified reaction system was designed, in which pyruvate—the coproduct of the transamination step—is reduced to lactate by a lactate dehydrogenase (LDH) (Scheme 14.10) rather than being recycled to L-alanine (Scheme 14.8A). This modified redox-neutral cascade, when implemented using purified enzymes, led to higher levels of amine formation and significantly improved optical purities (*e.g.*, 91% of cyclohexylamine from cyclohexanol; 64% of (*S*)-octan-2-amine in 96% ee from (*S*)-2-octanol).

In a related study, the biocatalytic amination of isosorbide (**61**), a bicyclic secondary diol that can be obtained from D-glucose, was investigated (Scheme 14.11).[94] The authors screened three ADHs and 27 transaminases for their activity on isosorbide and the potential reaction intermediates before choosing levodione reductase from *Leifsonia aquatica*, transaminase from *Paracoccus denitrificans*, and alanine dehydrogenase from *Bacillus subtilis* as best-suited biocatalysts. Moreover, they engineered the transaminase for increased substrate affinity (reduced K_M) using a site-saturation approach. Under optimised conditions, the biotransformation afforded the mono-amination product (2*S*,5*S*)-**62** in 7% conversion (20 mM from 300 mM isosorbide).

Very recently, ether alcohols (Figure 14.5) have been aminated using the ADH/transaminase/AlaDH-system.[95] Thermodynamic limitations became apparent also in this study, as the conversions under a variety of initially

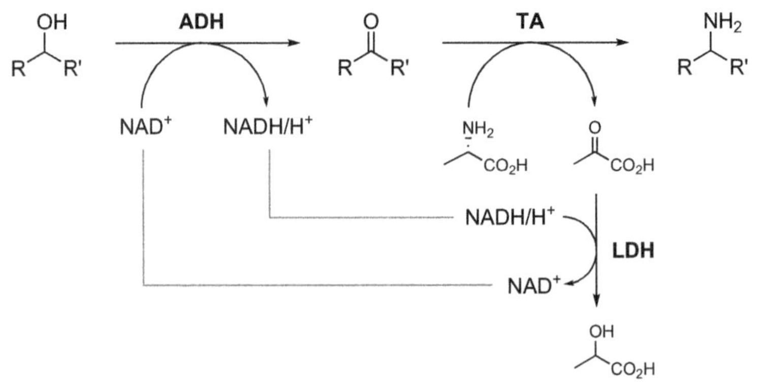

Scheme 14.10 Modified redox-neutral multi-enzyme system for alcohol-to-amine interconversion. Enzyme abbreviations: ADH, alcohol dehydrogenase; TA, transaminase; LDH, lactate dehydrogenase.

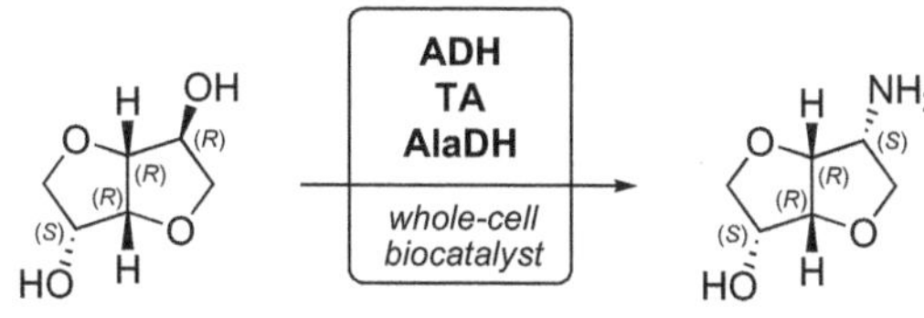

Isosorbide (**61**)

62
up to 7% conv.

Scheme 14.11 Amination of isosorbide (**61**) using a whole-cell amination biocatalyst. Enzyme abbreviations: ADH, alcohol dehydrogenase; TA, transaminase; AlaDH, L-alanine dehydrogenase.

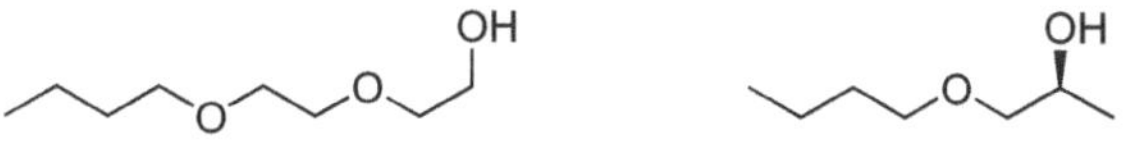

63 **64**

Figure 14.5 Ether alcohols studied as substrates for biocatalytic alcohol-to-amine interconversion.[95]

tested reaction conditions did not exceed 40%. An in-depth analysis of reaction equilibria combined with mathematical modelling of the cascade process suggested that an ammonia concentration of at least 700 mM would be required for near-complete conversion (>95%) of 10 mM of substrate. Since the employed enzymes were not stable under these conditions, a step-wise addition of a total 280 mM ammonia was used instead, pushing conversion to 60%.

Encouraged by the success of biocatalytic alcohol-to-amine interconversion systems, in particular those using primary alcohols as starting materials, researchers have also incorporated this functional group interconversion into more complex biocatalytic cascades. Above all, the production of amines directly from hydrocarbons *via* hydroxylation, alcohol oxidation and transamination has been investigated. For instance, the alkane monooxygenase AlkBGT from *Pseudomonas putida* GPo1 was co-expressed with a transaminase from *Chromobacterium violaceum* in an *E. coli* host, and resting cells of the recombinant biocatalyst were used for the synthesis of 12-aminododecanoic acid methyl ester (**68**; Scheme 14.12).[96] The monooxygenase AlkBGT catalyses two productive steps—hydroxylation of **65** and oxidation of **66** to **67**—in this system, but is also responsible for an undesired side reaction, namely the oxidation of **67** to the diacid monoester **69**.[97] The cofactor required by the monooxygenase, NADH, was provided *via* cell metabolism, which also accounted for the removal of the transamination byproduct, pyruvate. However, the amine donor for the transaminase, L-alanine, had to be supplied externally in a concentration of 50 mM. Using 1.4 g L^{-1} (cell dry-weight) of the biocatalyst, 0.13 mM (4.5%) of ω-amino ester **68** were produced from 2.9 mM of methyl dodecanoate **65** within 90 min,

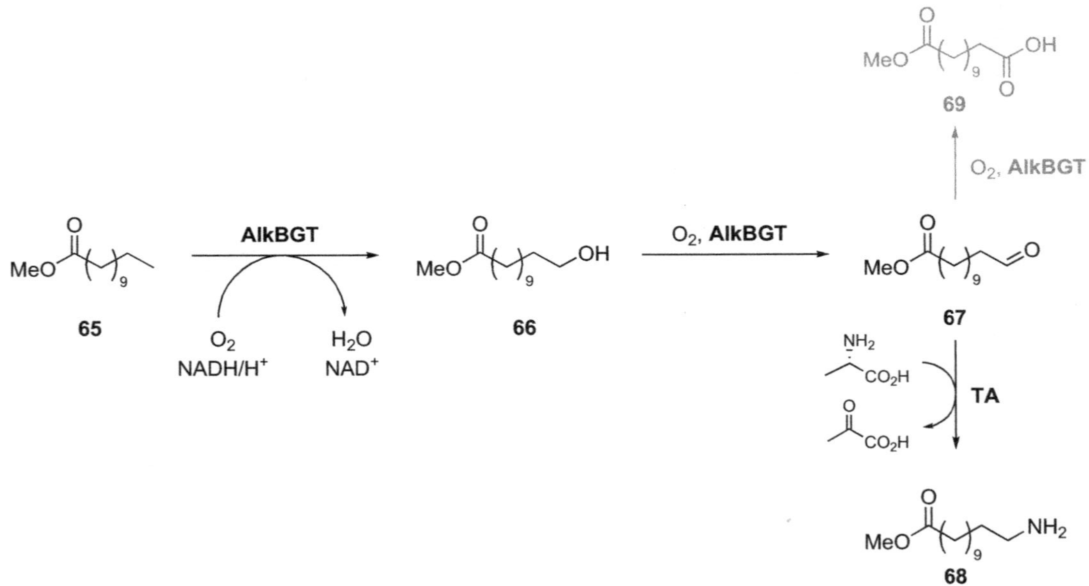

Scheme 14.12 Biocatalytic conversion of dodecanoic acid methyl ester (**65**) into 12-aminododecanoic acid methyl ester (**68**) *via* hydroxylation and alcohol-to-amine interconversion. Enzyme abbreviations: AlkBGT, alkane monooxygenase from *Pseudomonas putida* GPo1; TA, transaminase.

along with equal levels of the diacid monoester **69**. Extension of the substrate scope to non-functionalised hydrocarbons was also possible, as demonstrated by the production of 1-octylamine (0.22 mM) from *n*-octane (1.4 mM).[98] The authors later increased the productivity of the whole-cell biocatalyst substantially by following three optimisation strategies: (i) Introduction of the outer membrane protein AlkL from *P. putida* Gpo1 led to improved substrate uptake by the cells, which in turn increased the reaction rate of all productive biotransformations 6.8–8.1-fold; (ii) co-expression of L-alanine dehydrogenase from *Bacillus subtilis* enabled the biocatalyst to use NH_4Cl as terminal nitrogen source for the transamination, although the reactions in this case were slower compared to those carried out in the presence of externally added L-alanine; and (iii) complementing the multi-enzyme pathway with alcohol dehydrogenase AlkJ from *P. putida* Gpo1, which is linked to the electron transport chain and hence catalyses irreversible alcohol oxidation, increased the intracellular concentration of the intermediate oxo-ester **67** and thereby made the transamination step considerably faster. The combined effects of these three adjustments increased the formation rate of product **68** 10-fold and its analytical yield per unit biomass 3.5-fold.

A whole-cell biocatalyst capable of converting ethylbenzene derivatives **70** into 1-phenylethylamines **73** has been developed along similar lines (Scheme 14.13).[99] In this system, benzylic hydroxylation is catalysed by a

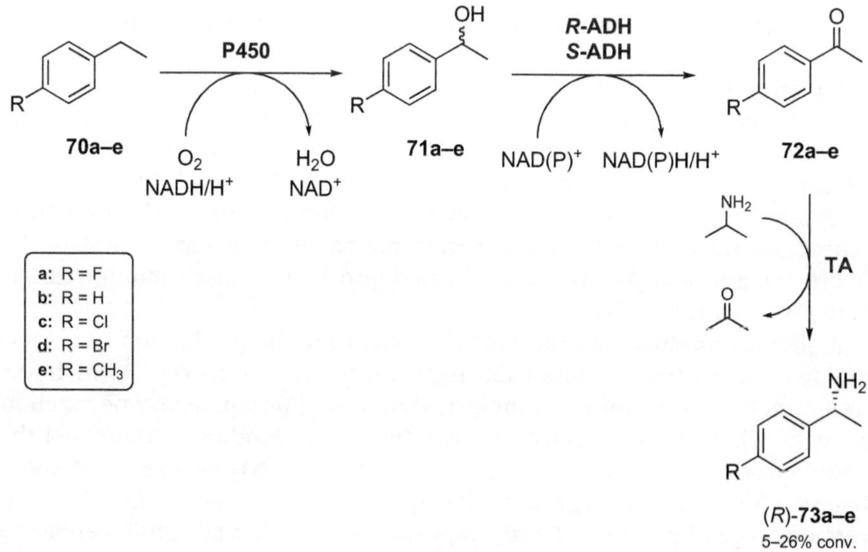

Scheme 14.13 Biocatalytic conversion of ethylbenzene derivatives **70** into (*R*)-1-phenylethylamine derivatives **73** *via* hydroxylation and alcohol-to-amine interconversion. Enzyme abbreviations: P450, P450 monooxygenase; *R*-ADH, (*R*)-selective alcohol dehydrogenase; *S*-ADH, (*S*)-selective alcohol dehydrogenase; TA, transaminase.

self-sufficient P450 monooxygenase variant, P450cam Y96F,[100] while a pair of stereocomplementary alcohol dehydrogenases—the (*R*)-selective ADH from *Lactobacillus brevis* and the (*S*)-selective ADH from *Rhodococcus erythropolis*—are responsible for oxidation of the intermediate chiral alcohol **71** to ketone **72**. The final transamination step is catalysed by an (*R*)-selective transaminase from *Arthrobacter* sp. Isopropylamine (200 mM) was used as the amino donor, but since it inhibited the P450 reaction, the authors chose to add it only after 24 h, once the conversion of the ethylbenzene derivative **70** (1 mM initial concentration) into ketone **72** was completed. After another 24 h, the (*R*)-amines **73** were obtained with 5–26% conversion and in excellent enantiomeric purity (ee for **73a**: 97.5%).

Recently, a biocatalytic cascade for the conversion of fatty acids into ω-hydroxycarboxylic acids has been developed.[101] While these molecules are interesting as such—for instance, as building blocks for polyesters—they can also be converted further into ω-aminocarboxylic acids, which in turn can serve as monomers for polyamides. The alcohol-to-amine interconversion was realised using an *E. coli* whole-cell biocatalyst coexpressing the membrane-associated alcohol dehydrogenase AlkJ from *Pseudomonas putida* GPo1 and a transaminase from *Silicibacter pomeroyi* (Scheme 14.14, Whole-cell biocatalyst 2).[102] Using 7.2 g L^{-1} (cell dry-weight) of this biocatalyst and 10 mM of L-alanine as amino donor, 4 mM of 12-hydroxydodecanoic acid were transformed into 2.6 mM (65%) of 12-aminododecanoic acid within 2 h. 1,12-Dodecanedioic acid (0.8 mM) was formed as a side product, which was attributed to an overoxidation activity of the alcohol dehydrogenase AlkJ. A one-pot coupling of hydroxycarboxylic acid production and alcohol-to-amine interconversion was also shown to be possible. To this end, *sec*-hydroxy fatty acids were first incubated with an *E. coli* biocatalyst that transformed the substrates near-quantitatively into the corresponding ω-hydroxycarboxylic acids *via* alcohol oxidation, Baeyer–Villiger monooxygenation, and ester hydrolysis (Scheme 14.14, Whole-cell biocatalyst 1). After adjusting the pH of the reaction medium from 8.0 to 9.5, the amination biocatalyst and 10 mM L-alanine were added. In this way, 5 mM of 12-hydroxystearic acid (**74**) were transformed into 2.5 mM 11-aminoundecanoic acid (**79**) within less than 3 h.

A similar, modularised biocatalytic system for the production of a polyamide monomer was reported starting from cyclohexanol (**57**),[103] which was transformed into 6-aminohexanoic acid (**85**) *via* a 6-step, 6-enzyme reaction sequence that was subdivided into two 'modules': Module 1 comprised the double oxidation of cyclohexanol to ε-caprolactone (**81**) by a *sec*-ADH and a Baeyer–Villiger monooxygenase (Scheme 14.15). Module 1, *i.e.* the conversion of cyclohexanol (**57**) into ε-caprolactone (**81**), has been independently reported by two research groups already in 2013.[104,105] Module 2 was initially planned to involve the hydrolysis of the lactone and conversion of the resulting 6-hydroxyhexanoic acid into **85** by the well-established *prim*-ADH/TA/AlaDH-system. The cascade stopped at the hydroxyacid stage, however, as all tested alcohol dehydrogenases turned out to be unable to oxidise

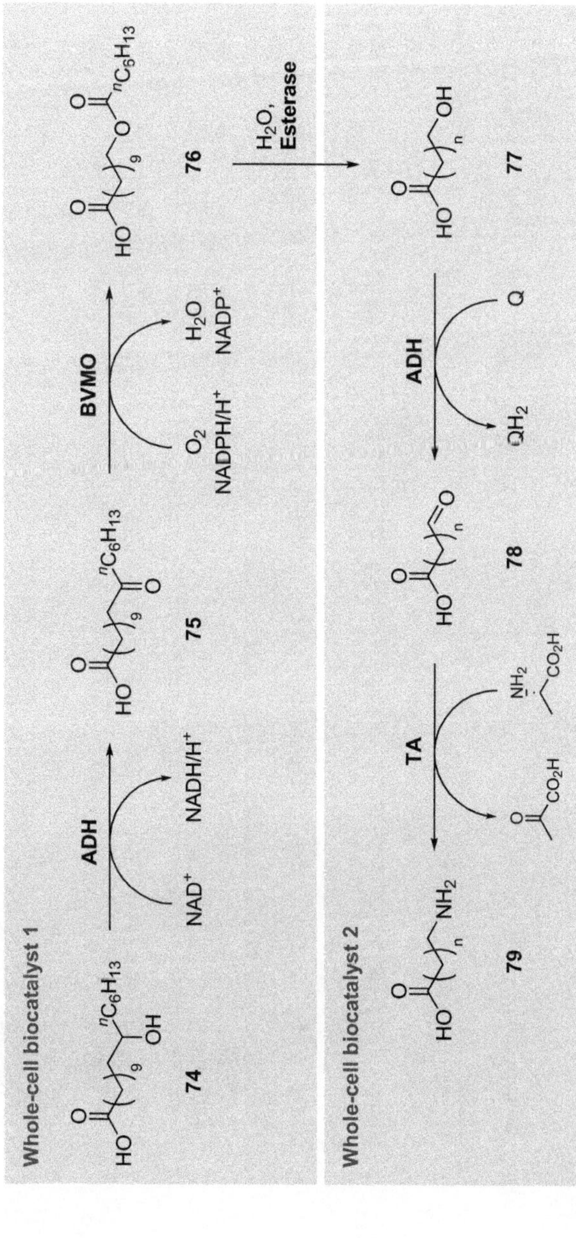

Scheme 14.14 Oxidative cleavage of hydroxyfatty acids **74** and further conversion into ω-aminocarboxylic acids **79** by two multi-enzyme whole-cell biocatalysts. Abbreviations: ADH, alcohol dehydrogenase; BVMO, Baeyer–Villiger monooxygenase; TA, transaminase; Q, ubiquinone.

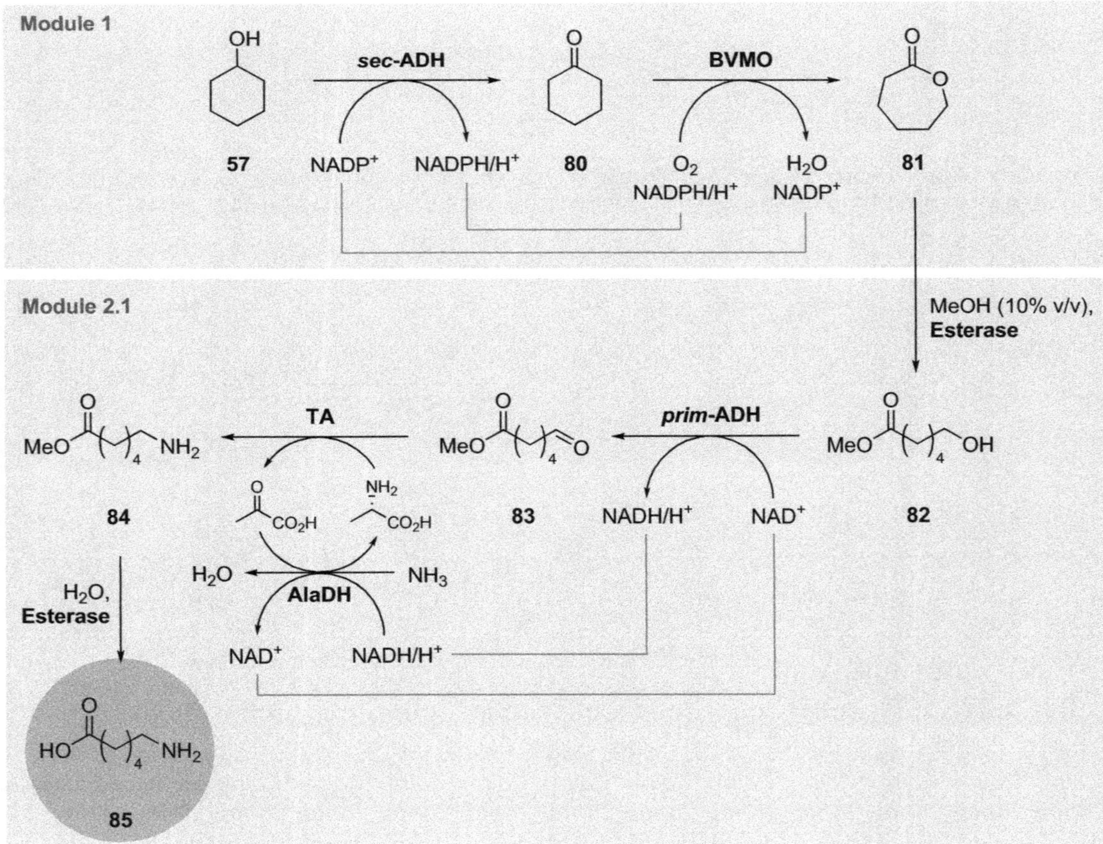

Scheme 14.15 Modularised multi-enzyme system for the production of 6-aminohexanoic acid (**85**) from hexanol (**57**). Enzyme abbreviations: *sec*-ADH, alcohol dehydrogenase specific for secondary alcohols; BVMO, Baeyer–Villiger monooxygenase; *prim*-ADH, alcohol dehydrogenase specific for primary alcohols; TA, transaminase; AlaDH, L-alanine dehydrogenase.

6-hydroxyhexanoic acid to the corresponding oxo-acid. The key to overcoming this limitation was an *in situ* 'capping' of the carboxylic acid moiety by conversion of ε-caprolactone into the corresponding methyl ester. This transformation appears challenging in an aqueous environment, as the competing hydrolysis of the lactone is thermodynamically favoured. Still, a screening of 30 hydrolases revealed that horse liver esterase displayed a sufficient kinetic preference for alcoholysis over hydrolysis to allow the conversion of **81** into >90% of methyl ester **82** in the presence of only 10% v/v of methanol. The following conversion of **82** into the amino ester **84** proceeded smoothly, and hydrolysis of **84** finally afforded the target compound, 6-aminohexanoic acid (**85**). The modified second module (termed Module 2.1 by the authors) produced **85** in 75% conversion from ε-caprolactone (50 mM), while running both modules simultaneously resulted in the formation of 24% of **85** from 50 mM cyclohexanol.

A final example of biocatalytic alcohol-to-amine interconversion using transaminases involves again a modularised multi-enzyme system: *E. coli* whole-cell biocatalysts were constructed that perform the 1,2-functionalisation of terminal alkenes leading to α-hydroxy acids, α-amino acids, or 1,2-amino alcohols as products.[106] For all final products, the reaction sequence starts with the conversion of the alkene **28** into the vicinal diol **30** *via* epoxidation by styrene monooxygenase and hydrolysis of the intermediate epoxide **29** by an epoxide hydrolase (Module 1, Scheme 14.16; *cf.* Section 14.2.2). Conversion of **30** into the 1,2-amino alcohols **88** was then achieved by the already mentioned combination of ADH, transaminase, and AlaDH (Module 3). Multi-enzyme sequences for the transformation of diols **30** into α-hydroxy acids (Module 2) and of the α-hydroxy acids into α-amino acids (Module 4) were also developed (see Chapter 13). For convenience, each multi-enzyme module was encoded on a separate plasmid, so that *E. coli* cells could be transformed with two or more plasmids as needed. An *E. coli* strain containing Modules 1 and 3 converted 60 mM of styrene (**28**, R = H) into 42 mM (70%) of the corresponding amino alcohol (*S*)-**88** (R = H) with an ee of 98%. A range of ring-substituted styrene derivatives was transformed by the same whole-cell biocatalyst with good conversions (16–81%) and excellent optical purities (91% to >99% ee) at 10 mM substrate concentration.

14.3.1.2 C–C Bond Formation and Transamination Cascade

In recent years, ThDP-dependent lyases (*cf.* Section 14.2.1.1) have been combined with transaminases in cascade systems for the production of 1,2-amino alcohols, a class of compounds with outstanding importance in medicinal chemistry.[107] The feasibility of such a cascade of C–C coupling and transamination has first been demonstrated by combining a transketolase from *E. coli* and β-alanine:pyruvate aminotransferase from *Pseudomonas aeruginosa* PAO2 co-expressed in a single *E. coli* host, for the production of 3-amino-1,2,4-butanetriol (**94**) from achiral starting materials

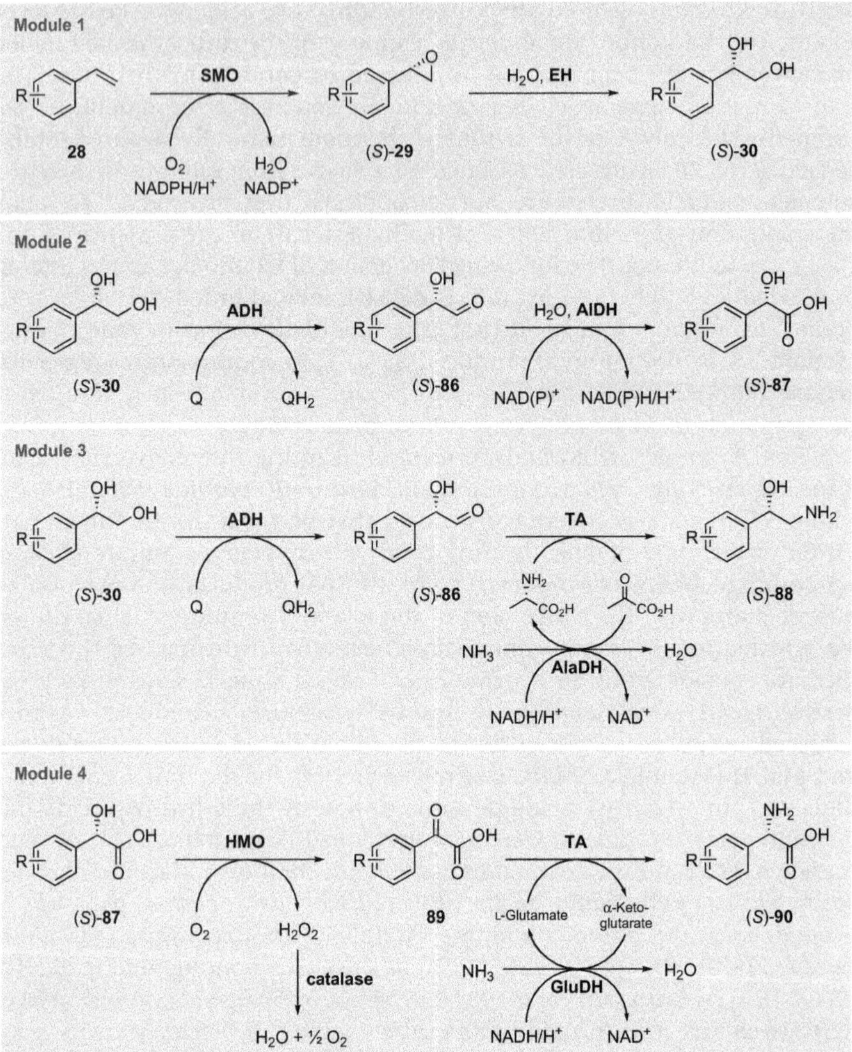

Scheme 14.16 Modularised multi-enzyme system for the production of 1,2-diols (*S*)-**30**, α-hydroxy acids (*S*)-**87**, 1,2-amino alcohols (*S*)-**88** or α-amino acids (*S*)-**90** from styrene derivatives **28**. Abbreviations: SMO, styrene monooxygenase; EH, epoxide hydrolase; ADH, alcohol dehydrogenase; TA, transaminase; AlaDH, L-alanine dehydrogenase; HMO, hydroxymandelate oxidase; GluDH, glutamate dehydrogenase; Q, ubiquinone.

hydroxypyruvate (**92**) and glycolaldehyde (**91**), using (*S*)-1-phenylethylamine as amino donor (Scheme 14.17).[108] Both enzymes displayed excellent stereoselectivity, forming (2*R*,3*S*)-**94** as a single stereoisomer,[50] but their activities were misbalanced: the transketolase was more active than the transaminase by three orders of magnitude and thus the first step was complete within 30 min, while more than four days were required for the second step to reach 21% conversion. Moreover, product **94** was found to be slowly degrading under the reaction conditions. These limitations were addressed in follow-up studies by switching to a different transaminase—the well-known enzyme from *Chromobacterium violaceum*—and by carefully optimising the fermentation conditions to achieve more balanced expression levels of the two enzymes.[109–111] These adjustments, in combination with mathematical reaction modelling and model-based process optimisation, formed the basis for an improved cascade run at 100 mM initial substrate concentration with continuous (*S*)-1-phenylethylamine feeding, which afforded (2*S*,3*R*)-**94** in 62% conversion after two-and-a-half days. A process for the synthesis of (2*S*,3*S*)-2-amino-1,3-pentanetriol has been developed and optimised in a similar manner, in this case using isopropylamine as amino donor.[111,112]

The same general principle—combining biocatalytic carbon–carbon bond formation and transamination in one pot—has been applied in the asymmetric synthesis of norephedrine and norpseudoephedrine from pyruvate and benzaldehyde (Scheme 14.18A).[113] For the C–C coupling step, the ThDP-dependent acetohydroxyacid synthase I (AHAS-I) from *E. coli* was used, which produced intermediate (*R*)-**96** in >98% optical purity. This intermediate was reductively aminated by the (*S*)-selective transaminase from *Chromobacterium violaceum* to afford (1*R*,2*S*)-norephedrine **97** or by the (*R*)-selective transaminase from *Aspergillus terreus* to give (1*R*,2*R*)-norpseudoephedrine **98**, both compounds being obtained as single stereoisomers. L-Alanine was chosen as amino donor, affording pyruvate as coproduct of the transamination step. As a consequence, the second reaction in the cascade replenishes the substrate of the first, and the authors have coined the term 'recycling cascade' for processes of this kind.[114] However, a simultaneous coupling of both biotransformation steps turned out to be problematic when the transaminase from *C. violaceum* was used, since this enzyme aminated benzaldehyde (**1**) much more efficiently than (*R*)-**96**. The cascade process hence accumulated benzylamine as an undesired side product while forming **97** in only 2% conversion. A sequential one-pot setup (*i.e.*, adding the transaminase after 1 h, once benzaldehyde was completely consumed)

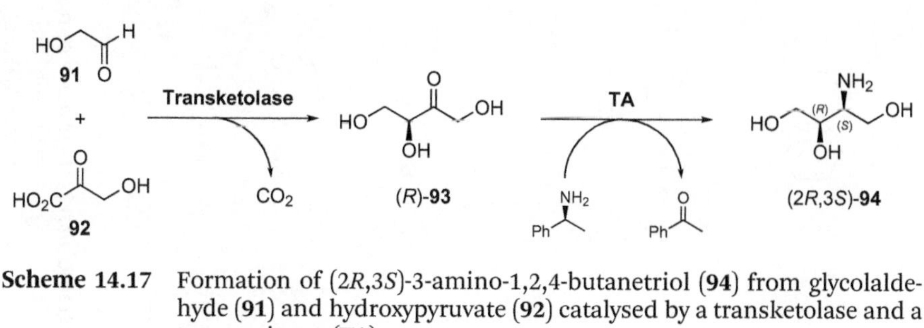

Scheme 14.17 Formation of (2*R*,3*S*)-3-amino-1,2,4-butanetriol (**94**) from glycolaldehyde (**91**) and hydroxypyruvate (**92**) catalysed by a transketolase and a transaminase (TA).

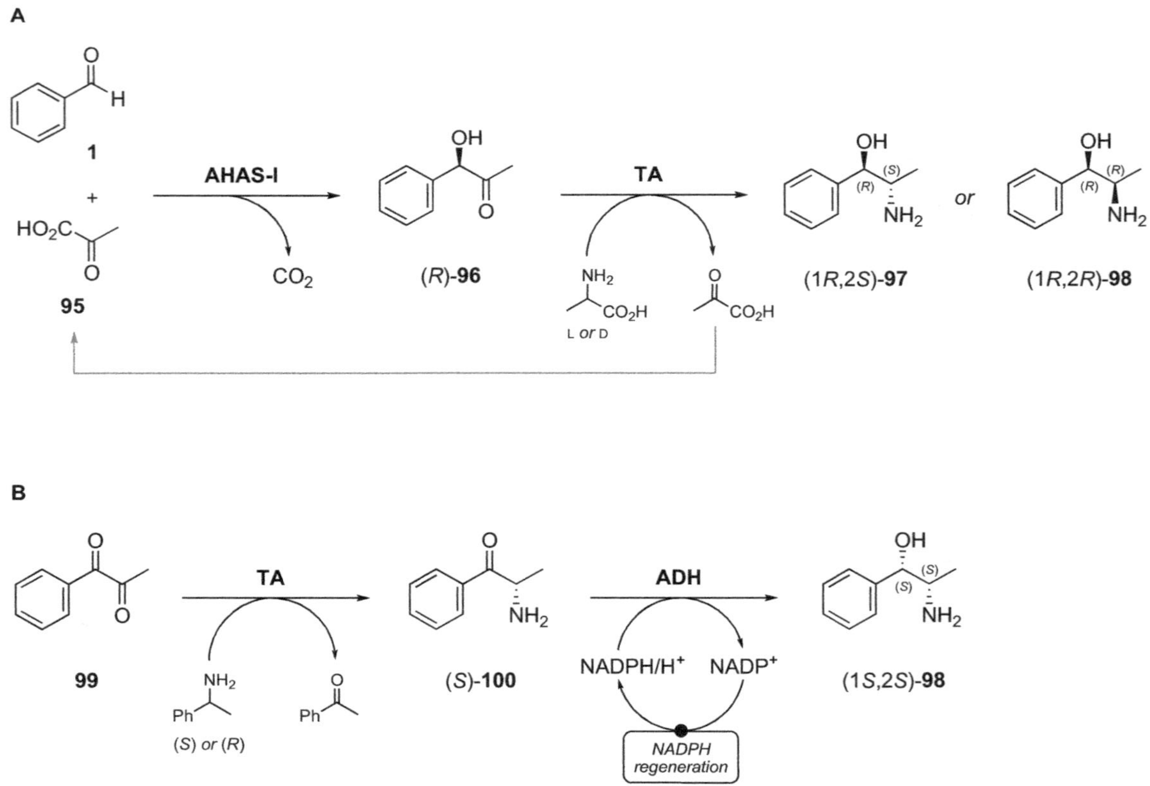

Scheme 14.18 Synthesis of norephedrine (**97**) and norpseudoephedrine (**98**) isomers *via* (A) C–C coupling and reductive amination or (B) reductive amination and carbonyl reduction. Enzyme abbreviations: AHAS-I, acetohydroxyacid synthase I; TA, transaminase; ADH, alcohol dehydrogenase.

worked much better, giving **97** in 78% conversion.[115] The reaction using the *A. terreus* transaminase even produced 96% of **98**. In both cases, the subsequent addition of another equivalent of benzaldehyde and fresh AHAS-I led to further product formation, demonstrating that pyruvate recycling was indeed taking place. In further work, the authors not only increased the space-time yield of their process by raising the initial benzaldehyde concentration from 10 mM to 100 mM but also sought to access the opposite enantiomers of both products by employing an (*S*)-selective variant of pyruvate decarboxylase from *Acetobacter pasteurianus* (*Ap*PDC-E469G) for the C–C coupling step. Unfortunately, the stereoselectivity of *Ap*PDC-E469G proved insufficient under the investigated reaction conditions, as (1*S*,2*S*)-**97** and (1*S*,2*R*)-**98** were obtained in only 70% diastereomeric excess.[116] However, one of these stereoisomers was finally obtained in high optical purity *via* an alternative two-step one-pot process in which 1-phenyl-1,2-propanedione (**99**) was subjected to reductive amination by suitable transaminase and subsequent carbonyl reduction catalysed by alcohol dehydrogenase from *Lactobacillus brevis* (Scheme 14.18B). This alternative cascade required pH-induced deactivation of the transaminase after completion of the first step to minimise side-product formation, and afforded (1*S*,2*S*)-**98** in 80% conversion, >99% ee and >98% de.[116]

14.3.1.3 *Amination and Amine Oxidation or Ene Reduction*

Transaminases have been combined with engineered variants of monoamine oxidase from *Aspergillus niger* (MAO-N variants D5 and D9) and ammonia–borane as reducing agent for the enantio- and diastereoselective synthesis of 2,5-disubstituted pyrrolidines (Scheme 14.19).[117] Thereby, prochiral 1,4-diketones **101** were first selectively mono-aminated by commercial transaminases ATA-113 or ATA-117 (both available from Codexis) at the sterically more accessible carbonyl group. The resulting aminoketones **102** underwent spontaneous cyclisation to imines **103**, which in a second step were converted into the target pyrrolidines **104** by the combined action of MAO-N and $NH_3 \cdot BH_3$. This formal asymmetric C=N-reduction is based on the principle that borane reduces **103** to a mixture of *cis*- and *trans*-**104** (with a slight excess of the *cis*-diastereomer), while MAO-N oxidises exclusively the (5*S*)-isomer of **104**. Hence, the (5*R*)-isomer accumulates in the reaction mixture over several reduction/oxidation cycles. No interference between the two biocatalysts was observed, as the transaminases exclusively acted on the less hindered ketone moiety of **101**, whereas MAO-N shows a strong preference for oxidation at the benzylic centre. Nevertheless, a simultaneous bi-enzymatic cascade was not possible, since $NH_3 \cdot BH_3$ reduced the starting material, diketone **101**. The sequential combination of the two steps worked well though, providing access to both *cis*- and *trans*-pyrrolidines **104** in >99% de and with quantitative conversion. A preparative-scale reaction using 1-phenyl-1,4-pentanedione (**101**, R = Me, R′ = H; 220 mg) as starting

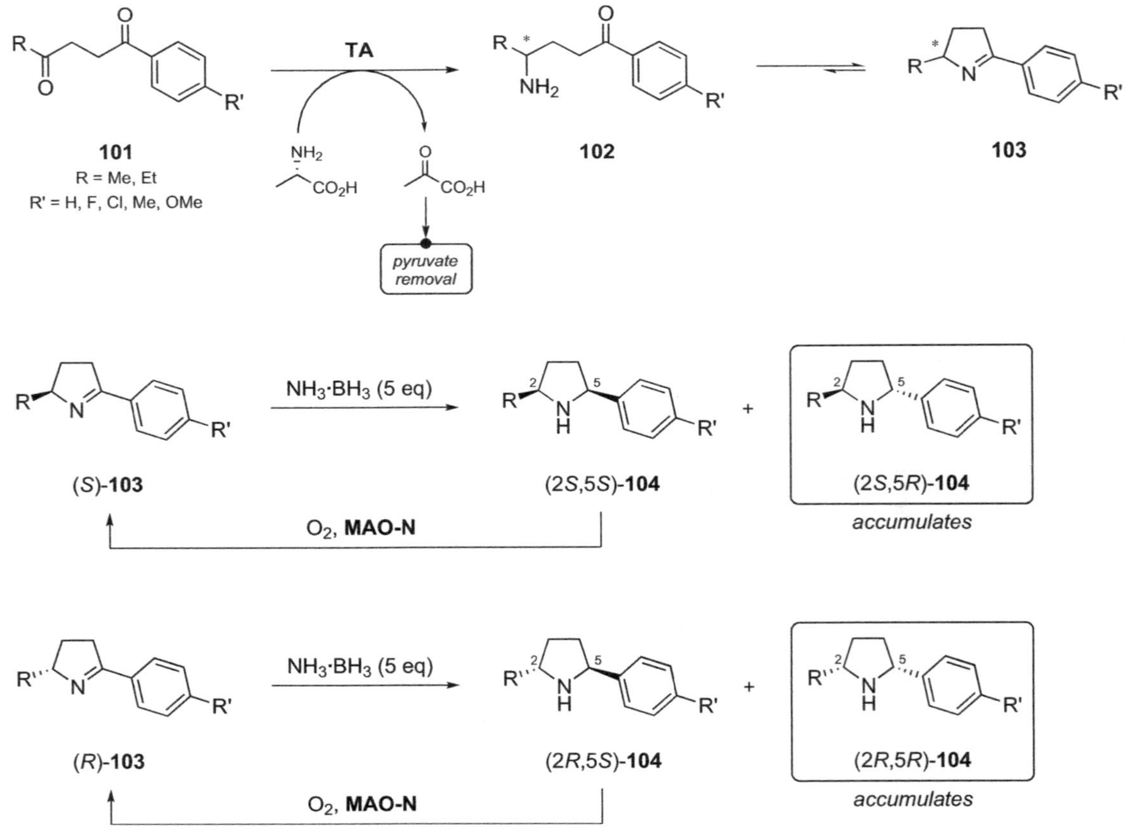

Scheme 14.19 Enantio- and diastereoselective synthesis of 2,5-disubstituted pyrrolidines *via* sequential biotransformation with transaminases (TAs) and monoamine oxidase (MAO-N).

material and transaminase ATA-113 for the first step afforded (2*S*,5*R*)-**104** (R = Me, R′ = H) in 82% isolated yield and stereoisomerically pure form.

Recently, the combination of ene-reductases (ERs) and transaminases for the preparation of amines containing two stereogenic centres from the corresponding enones has been reported. In a first proof-of-principle study, the conversion of 3-methylcyclohex-2-en-1-one (**105**) into 3-methylcyclohexanamine **107** was investigated using old yellow enzyme (OYE1) from *Saccharomyces pastorianus* and transaminase from *Vibrio fluvialis* (Scheme 14.20).[118] While the first step—the asymmetric C=C-reduction of **105** to give (*S*)-**106**—proceeded with excellent stereoselectivity (ee > 99%), the subsequent reductive amination using wild-type *Vf*-TA furnished a mixture of diastereomers with only a slight excess (14% de) of the (1*R*,3*S*)-product. Using molecular modelling and multiple-sequence analysis based on a 3DM database, Leu56 was identified as a promising site for mutagenesis aimed at improving the stereoselectivity. Indeed, variant L56V provided product (1*R*,3*S*)-**107** in a significantly increased diastereomeric purity of 65% de, while the stereoselectivity was inverted for variant L56I, which gave (1*S*,3*S*)-**107** in 70% de. The optical purity of the (1*R*,3*S*)-isomer was further improved to 89% de by performing the biotransformation in the presence of 30% (v/v) DMSO. In a second report, the authors used an analogous cascade process for accessing the (1*R*,3*R*)-isomer of **107** in 97% de using ene-reductase variant YqjM C26D/I69T and *Vf*-TA variant L56I.[119] In all these systems, a simultaneous combination of the two biocatalytic steps was possible without undesired side reactions, because the investigated transaminases were unreactive towards the starting enone while they readily converted the saturated ketone **106**. The same chemoselectivity was also observed in a related study where three enones (Figure 14.6) were reduced with old yellow enzymes from *S. cerevisiae* or *S. pastorianus* and a set of 24 commercially available transaminases from Codexis was screened for reduction of the corresponding saturated ketones.[120] Using this enzyme 'toolbox', two of the four possible stereoisomers of each target amine were obtained in high diastereomeric excess, independent of whether the biotransformation was carried out as sequential one-pot process or simultaneous cascade.

14.3.2 Amine Dehydrogenases and Imine Reductases

The reduction of imines and the reductive amination of ketones using ammonia are well-established reactions in synthetic organic chemistry, but biocatalysts for these transformations have long been unavailable. Only recently, imine-reducing enzymes with a synthetically relevant scope have been developed by protein engineering and identified from natural sources.[121] These enzymes include artificial transfer hydrogenases, which reduce imines at the expense of formate by virtue of an artificial organometallic cofactor (see Chapter 3),[122,123] amine dehydrogenases,[124–126] which catalyse the reductive amination of ketones with ammonia, and imine reductases,[121,127]

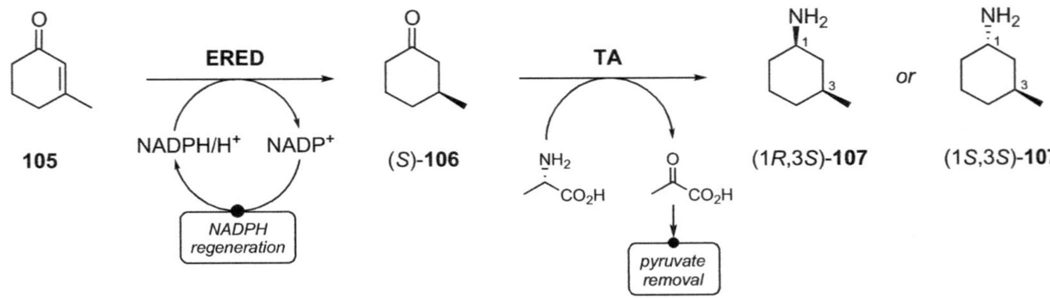

Scheme 14.20 Combination of an ene-reductase (ERED) and a transaminase (TA) for the asymmetric synthesis of 3-methylcyclohexanamine (**107**).

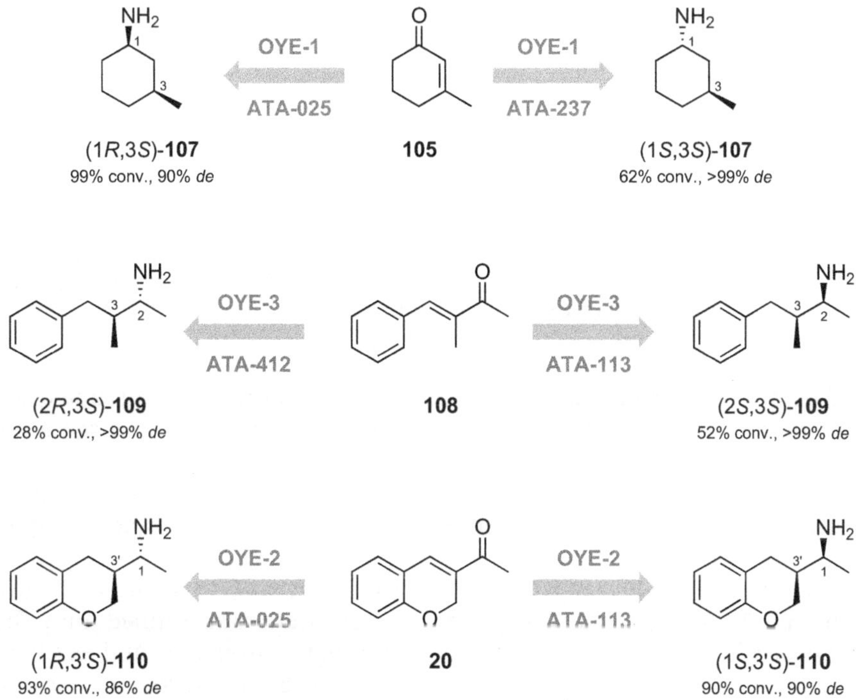

Figure 14.6 Substrates investigated and products obtained *via* biocatalytic C=C-reduction and transamination.[120]

which mainly reduce cyclic imines and iminium ions. Although these enzymes have only been employed in biocatalysis during the last five years, all of them have already been integrated into cascade processes.

14.3.2.1 *Alcohol-to-amine Interconversion Using Alcohol Dehydrogenases and Amine Dehydrogenases*

In contrast to the alcohol-to-amine interconversion systems based on transaminases described above (Section 14.3.1.1), a combination of alcohol dehydrogenases and amine dehydrogenases allows a direct coupling of the alcohol oxidation and reductive amination steps *via* internal cofactor recycling (Scheme 14.21).

The amination of a broad range of secondary and primary alcohols based on this reaction system has been reported,[128] whereby an amine dehydrogenase engineered from *Bacillus badius* phenylalanine dehydrogenase (Ph-AmDH),[124] as well as a chimeric amine dehydrogenase (Ch1-AmDH)[125] were chosen as biocatalysts for the reductive amination step. Alcohol dehydrogenases from *Lactobacillus brevis* (*Lb*-ADH) and *Aromatoleum aromaticum* (*Aa*-ADH) were employed for alcohol oxidation. Preliminary experiments using crude cell-free extracts revealed an interference from

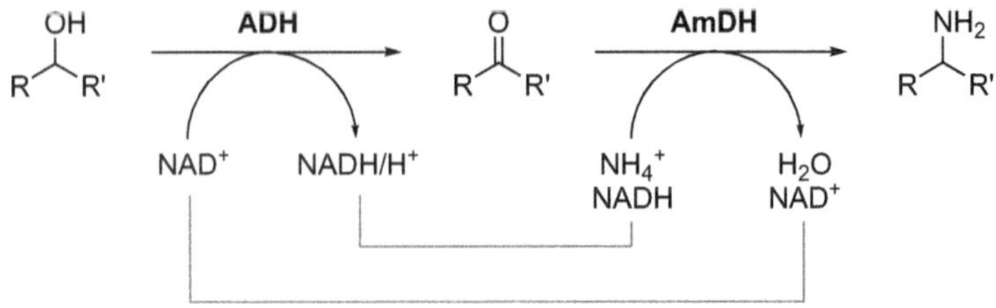

Scheme 14.21 Conversion of alcohols into amines *via* a 'borrowing hydrogen' cascade comprising an alcohol dehydrogenase (ADH) and an amine dehydrogenase (AmDH).

NADH oxidases present in the *E. coli* expression host, which disrupted the closed redox cycle and hence led to low overall conversions and the accumulation of the intermediate ketone. Moreover, the presence of a His-tag on Ph-AmDH was found to destabilise the *Lb*-ADH, which also prevented an efficient biotransformation. Hence, the bi-enzymatic cascade required purified enzymes from which the His-tag had been cleaved. Employing these highly pure enzyme preparations, a wide range of (*R*)-configured, (*S*)-configured, and racemic secondary alcohols were transformed into the corresponding (*R*)-amines with moderate to high conversions and excellent optical purities (Table 14.1, entries 1–8). Since both investigated alcohol dehydrogenases are highly enantioselective (*Lb*-ADH: *R*-selective, *Aa*-ADH: *S*-selective), the amination of racemic alcohols required the presence of both ADHs. The amination of primary alcohols was also investigated and proceeded smoothly for aliphatic substrates with a chain length smaller or equal to C_6 (Table 14.1, entries 9–11). In an independent study published only two months later, similar results were obtained with the same concept but different enzymes.[129] The amine dehydrogenase used in this work was created by protein engineering of a leucine dehydrogenase from *Exiguobacterium sibiricum* while the ADH originated from *Streptomyces coelicolor*.[130] The latter enzyme offers the advantage of oxidising a variety of *sec*-alcohols with low enantioselectivity, hence allowing the complete conversion of racemic substrates without the need for a second ADH. In contrast to the first report,[128] purification of the enzymes was not required, as reactions performed with crude cell-free extracts proceeded efficiently without accumulation of the intermediate ketones. A selection of small aliphatic secondary alcohols (acyclic and cyclic) was aminated, providing the corresponding amines in up to 97% conversion and (where applicable) >99% ee of the (*R*)-enantiomer (Table 14.1, entries 12–16).

Interestingly, thermodynamic limitations, which were observed in the transaminase-based amination of *sec*-alcohols, did not become apparent in either of the two studies, as evidenced by the near-complete conversion achieved in many reactions. Most likely, the large surplus of ammonia (2 M for reactions containing 20 mM substrate in one case and 50 mM in the other) helped push the equilibrium to the product side.

Table 14.1 Selected results of the redox-neutral conversion of alcohols into amines using alcohol dehydrogenases and amine dehydrogenases.[a]

Entry	Substrate	Product	ADH	AmDH	Conv. [%]	ee [%]	Ref.
1	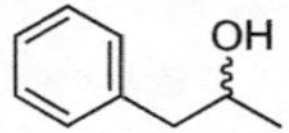 *rac*-**111**	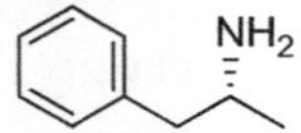 (*R*)-**112**	*Aa*-ADH, *Lb*-ADH	Ph-AmDH	81	>99	128
2	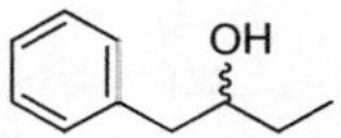 *rac*-**113**	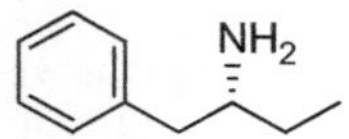 (*R*)-**114**	*Aa*-ADH, *Lb*-ADH	Ph-AmDH	87	>99	128
3	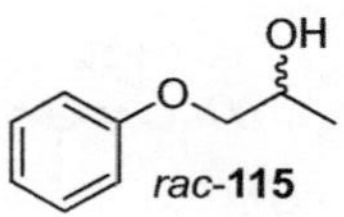 *rac*-**115**	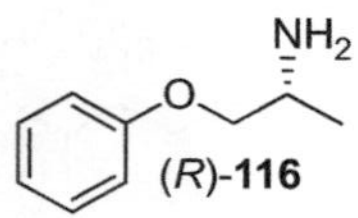 (*R*)-**116**	*Aa*-ADH, *Lb*-ADH	Ch1-AmDH	84	>99	128
4	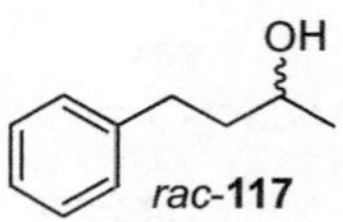 *rac*-**117**	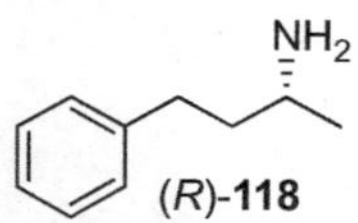 (*R*)-**118**	*Aa*-ADH, *Lb*-ADH	Ch1-AmDH	92	83	128
5	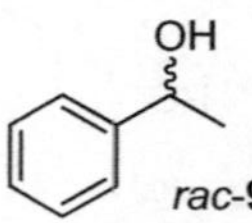 *rac*-**9**	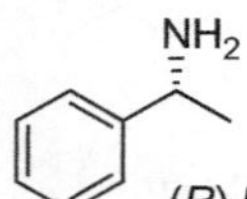 (*R*)-**56**	*Aa*-ADH, *Lb*-ADH	Ch1-AmDH	16	>99	128
6	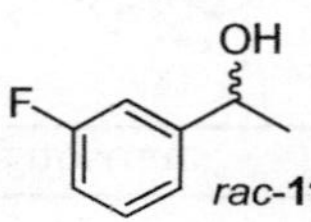 *rac*-**119**	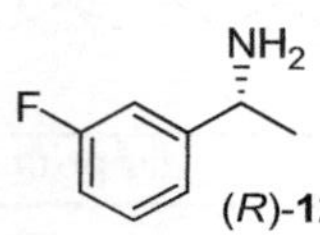 (*R*)-**120**	*Aa*-ADH, *Lb*-ADH	Ch1-AmDH	20	>99	128

Table 14.1 (*Continued*)

Entry	Substrate	Product	ADH	AmDH	Conv. [%]	ee [%]	Ref.
7	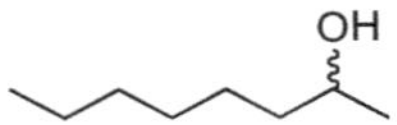 *rac*-**10**	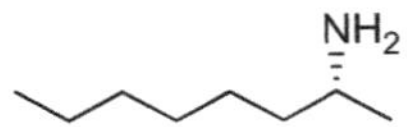 (*R*)-**53**	*Aa*-ADH, *Lb*-ADH	Ch1-AmDH	93	99	128
8	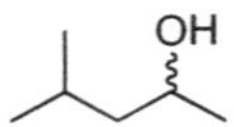 *rac*-**121**	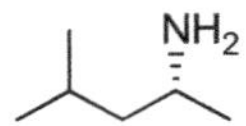 (*R*)-**122**	*Aa*-ADH, *Lb*-ADH	Ch1-AmDH	80	>99	128
9	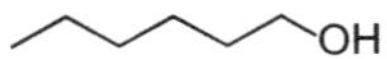 **123**	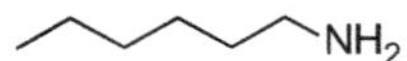 **40**	hT-ADH	Ch1-AmDH	99	*Na*	128
10	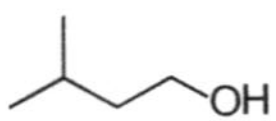 **124**	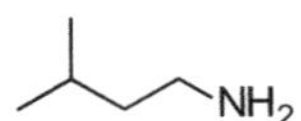 **125**	hT-ADH	Ch1-AmDH	99	*Na*	128
11	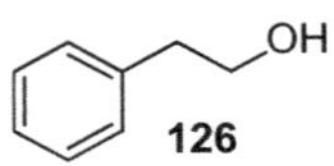 **126**	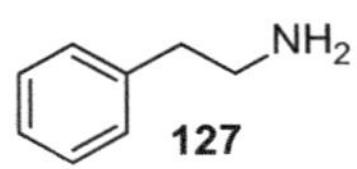 **127**	hT-ADH	Ph-AmDH	10	*Na*	128
12	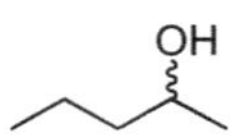 *rac*-**128**	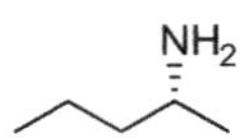 (*R*)-**129**	*Sc*-CR	*Es*-LeuDH-DM	94	>99	129
13	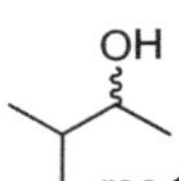 *rac*-**130**	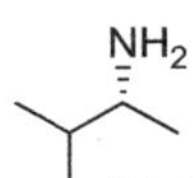 (*R*)-**131**	*Sc*-CR	*Es*-LeuDH-DM	93	>99	129

14	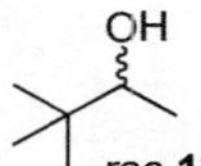 *rac*-**132**	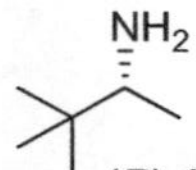 (*R*)-**133**	*Sc*-CR	*Es*-LeuDH-DM	29	>99	129
15	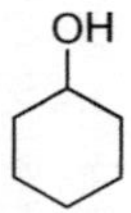 **57**	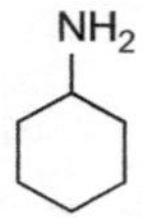 **58**	*Sc*-CR	*Es*-LeuDH-DM	97	*Na*	129
16	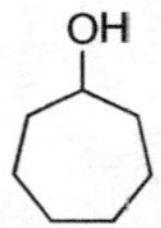 **134**	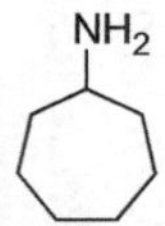 **135**	*Sc*-CR	*Es*-LeuDH-DM	35	*Na*	129

[a]*Reaction conditions:* Entries 1–3 and 7–8: substrate (20 mM), NAD^+ (1 mM), *Aa*-ADH (purified enzyme, 1 mg mL^{-1}), *Lb*-ADH (purified enzyme, 1 mg mL^{-1}), AmDH (purified enzyme, 40 U mL^{-1}), ammonium formate buffer (2 M, pH 8.5), 30 °C, 48 h; entries 4–6: substrate (20 mM), NAD^+ (1 mM), *Aa*-ADH (purified enzyme, 2.2 mg mL^{-1}), *Lb*-ADH (purified enzyme, 2.6 mg mL^{-1}), AmDH (purified enzyme, 80 U mL^{-1}), ammonium formate buffer (2 M, pH 8.5), 30 °C, 48 h; entries 9–11: substrate (20 mM), NAD^+ (1 mM), hT-ADH (purified enzyme, 3.7 mg mL^{-1}), AmDH (purified enzyme, 40 U mL^{-1}, except entry 11: 80 U mL^{-1}), ammonium chloride buffer (2 M, pH 8.7), 30 °C, 48 h; entries 12–16: substrate (50 mM), NAD^+ (1 mM), *Sc*-CR (lyophilised cell-free extract, 10 mg mL^{-1}), *Es*-LeuDH-DM (lyophilised cell-free extract, 10 mg mL^{-1}), ammonium chloride buffer (2 M, pH 9.5), 30 °C, 24 h. *Aa*-ADH = (*S*)-selective alcohol dehydrogenase from *Aromatoleum aromaticum*; *Lb*-ADH = (*R*)-selective ADH *Lactobacillus brevis* (NADH-dependent variant); hT-ADH = alcohol dehydrogenase from *Bacillus stearothermophilus*; Ph-AmDH = amine dehydrogenase engineered from *Bacillus badius* phenylalanine dehydrogenase;[124] Ch1-AmDH = chimeric amine dehydrogenase;[125] *Sc*-CR = alcohol dehydrogenase from *Streptomyces coelicolor*; *Es*-LeuDH-DM = amine dehydrogenase engineered from *Exiguobacterium sibiricum* leucine dehydrogenase. *Na* = not applicable.

14.3.2.2 Deracemisation of Amines

The combined enantioselective oxidation of *rac*-amines by monoamine oxidase from *Aspergillus niger* (MAO-N) and non-stereoselective reduction of the resulting imine or iminium ion by ammonia–borane is a well-established method for the deracemisation of structurally diverse primary, secondary and tertiary amines (Scheme 14.22A).[22,25,26] The enantiomeric purity of the product depends on the formal number of redox cycles the reacting enantiomer undergoes, and for large cycle numbers it is limited by the enantioselectivity of the amine oxidase.[131] With the advent of artificial transfer hydrogenases (see Chapter 3) and imine reductases, it has become possible to perform the deracemisation of α-chiral amines *via* a bi-enzymatic cascade (Scheme 14.22B and C), in which both biocatalysts contribute to the overall stereoselectivity of the process. This was first demonstrated by combining engineered variants of MAO-N with an artificial transfer hydrogenase based on streptavidin ligated with the biotinylated iridium(III) complex **136** (Scheme 14.22B).[132] While the free complex and MAO-N suffered from mutual inactivation, the complex embedded in the streptavidin binding site was fully compatible with the oxidase. Deracemisation of two chiral secondary amines was achieved as well as the stereoinversion of a third model compound (Figure 14.7A); in all three cases the (*R*)-enantiomers were obtained with quantitative conversion and excellent optical purity (ee > 99%). The (*S*)-enantiomers of structurally related amines (Figure 14.7B) were recently obtained using a deracemisation system that combined a variant of 6-hydroxy-D-nicotine oxidase (6-HDNO) from *Arthrobacter nicotinovorans* and an imine reductase (IRED) from *Streptomyces* sp. GF3587.[133] Cooperation of the two enzymes with respect to the overall stereoselectivity was apparent in this system, as in several cases the deracemisation afforded the product amine in significantly higher enantiomeric excess (95–99%) than that obtained from the reduction of the corresponding imine using the IRED alone (37–89%).

14.3.2.3 Synthesis of Disubstituted N-Heterocycles Using Transaminases and Imine Reductases

Several enzymatic cascade systems have recently been developed for the synthesis of disubstituted piperidines and pyrrolidines.[134] All these systems make use of the regioselective mono-amination of diketones (*cf.* Scheme 14.19) or keto aldehydes catalysed by transaminases and the subsequent stereoselective reduction of the generated cyclic imines by imine reductases (Scheme 14.23). Several of the keto aldehydes were prepared *in situ* from the corresponding keto acids by biocatalytic reduction employing a carboxylic acid reductase from *Mycobacterium marinum*, thus extending the cascade by one upstream step. A broad range of substrates featuring various substitution patterns was investigated, and some interesting observations regarding the enantio- and diastereoselectivity of imine reductases were made: (i) A stereogenic centre in α-position of the aldehyde

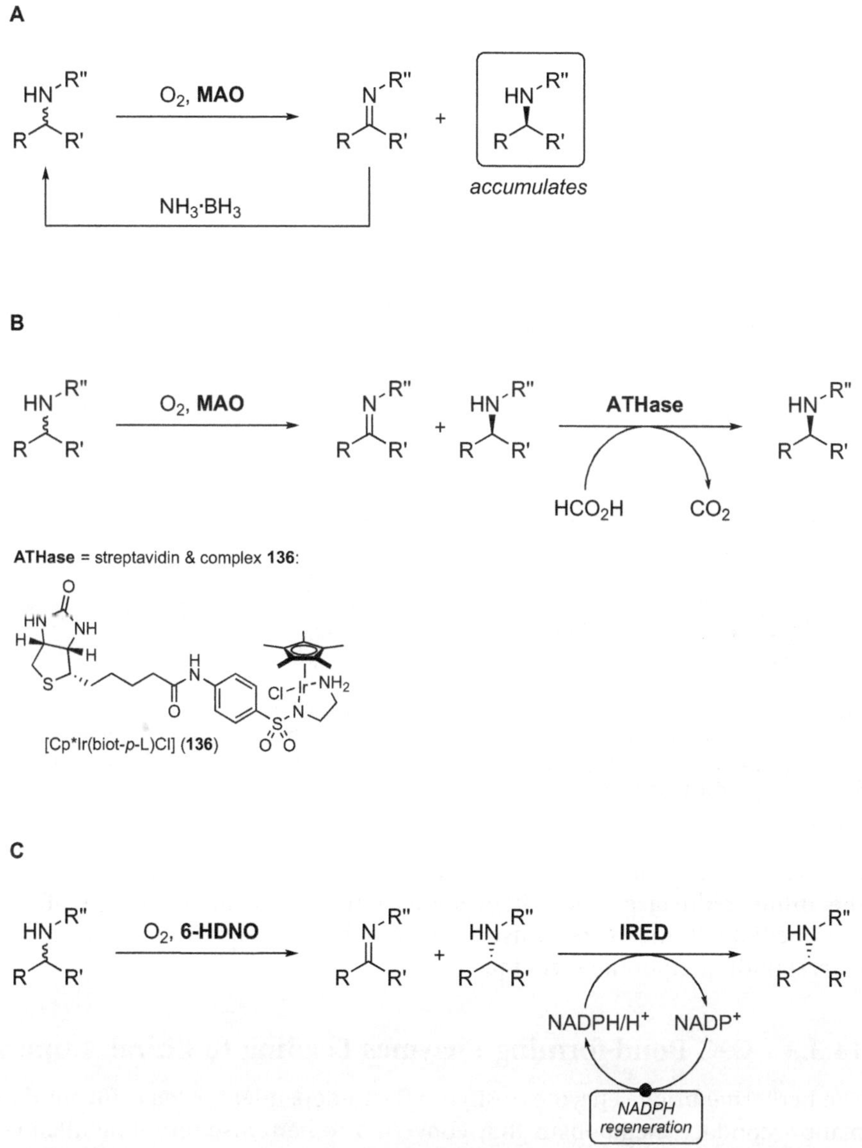

Scheme 14.22 Deracemisation of α-chiral amines: Combination of enantioselective oxidation by monoamine oxidase (MAO) and (A) a non-stereoselective chemical reduction, or (B) asymmetric reduction using an Ir-dependent artificial transfer hydrogenase (ATHase); (C) combination of enantioselective oxidation by (*R*)-6-hydroxy-D-nicotine oxidase (6-HDNO) and asymmetric reduction using an imine reductase (IRED).

is largely retained under the reaction conditions and has no influence on the stereopreference of the imine reductase (Scheme 14.23B); (ii) if the final reaction product is a 2,6- or 2,4-disubstituted piperidine, the *trans*-isomers

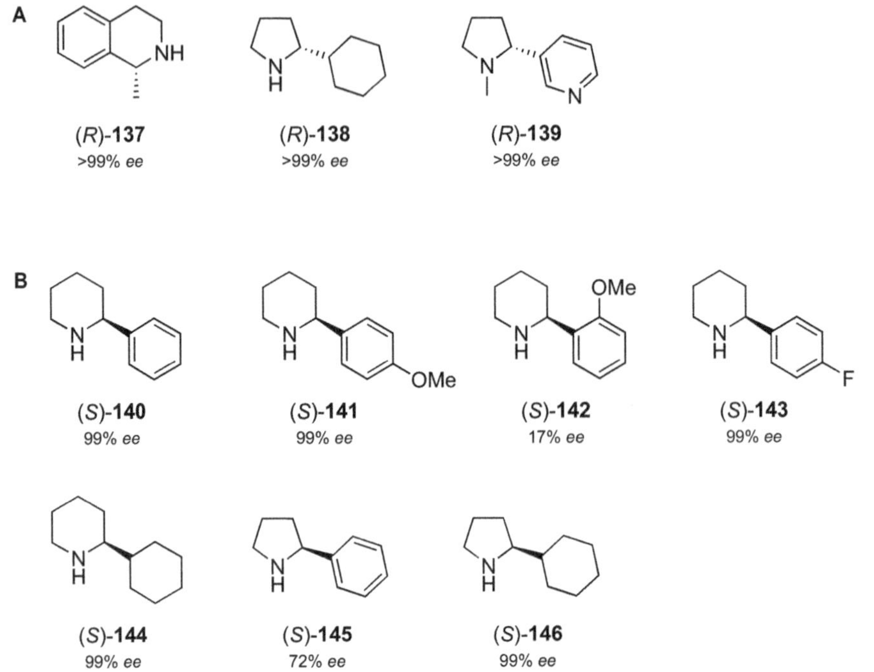

Figure 14.7 Examples of amines obtained by deracemisation using (A) monoamine oxidase variants and an artificial transfer hydrogenase (Scheme 14.22B) or (B) an (*R*)-6-hydroxy-D-nicotine oxidase variant and an imine reductase (Scheme 14.22C).

are preferentially formed, irrespective of the inherent stereopreference of the imine reductase; and (iii) a stereogenic centre in α-position of the intermediate imine is racemised, presumably *via* equilibration with the corresponding enamine (Scheme 14.23C).

14.3.3 C–C Bond-forming Enzymes Leading to Chiral Amines

The berberine bridge enzyme (BBE) is a flavin-dependent oxidase involved in plant secondary metabolism that converts the benzylisoquinoline alkaloid (*S*)-reticuline (**157b**) into the berbine alkaloid (*S*)-scoulerine (**159b**) *via* oxidative C–C coupling. The substrate scope of BBE has been shown to extend to a range of benzylisoquinoline derivatives carrying different substituents on the aromatic rings, and the enzyme catalysed the kinetic resolution of racemic substrates with excellent enantioselectivity, affording optically pure (*S*)-berbines and (*R*)-benzylisoquinolines as products.[135–138] Based on this kinetic resolution, a chemo-bienzymatic cascade process was developed that transforms racemic benzylisoquinolines **157** into (*S*)-berbines **159** with near-quantitative conversion and excellent enantiomeric excess (Scheme 14.24A).[139] In this cascade, a variant of monoamine oxidase from

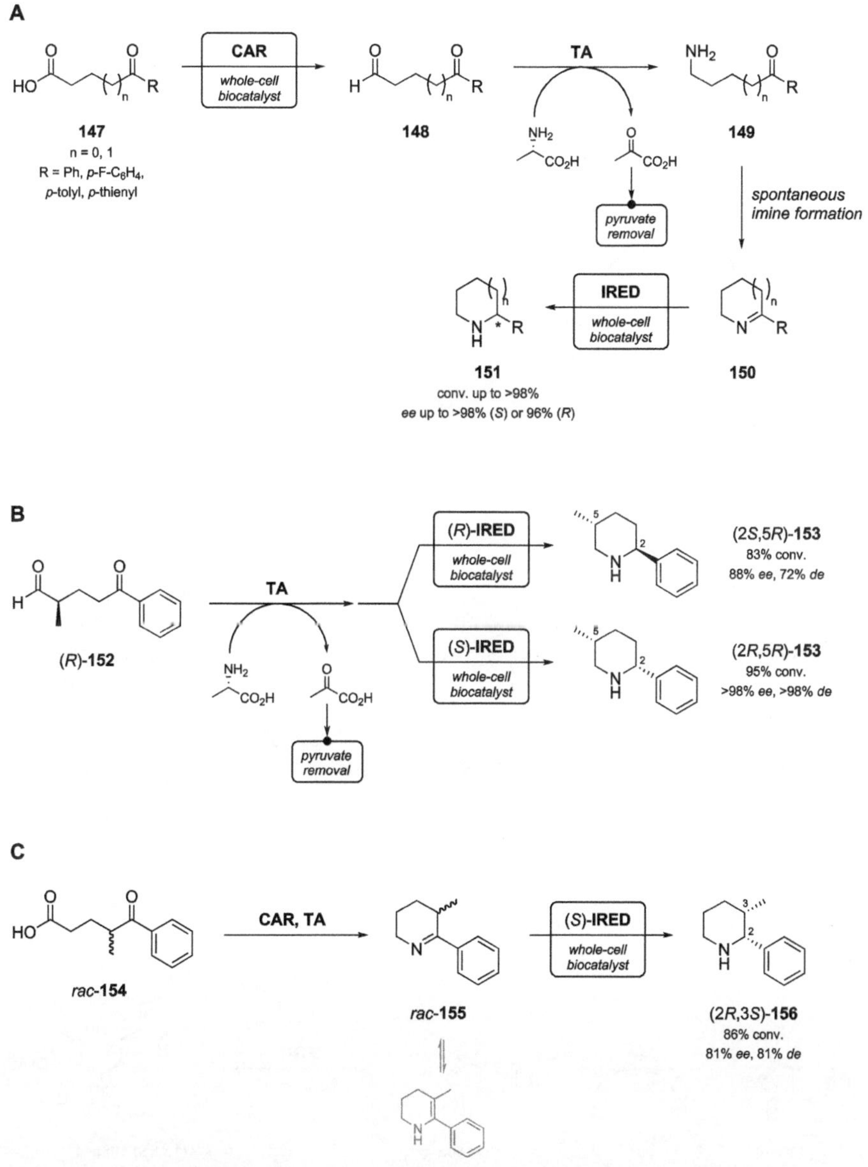

Scheme 14.23 Asymmetric synthesis of substituted piperidines and pyrrolidines *via* multi-enzyme cascade systems comprising carboxylic acid reductases (CARs), transaminases (TAs) and imine reductases (IREDs).

Aspergillus niger engineered for conversion of sterically demanding substrates (MAO-N D11) together with morpholine-borane as reducing agent enables the deracemisation of *rac*-benzylisoquinolines to give optically pure (*S*)-isomers (*cf.* Scheme 14.22A). The latter are oxidatively cyclised by

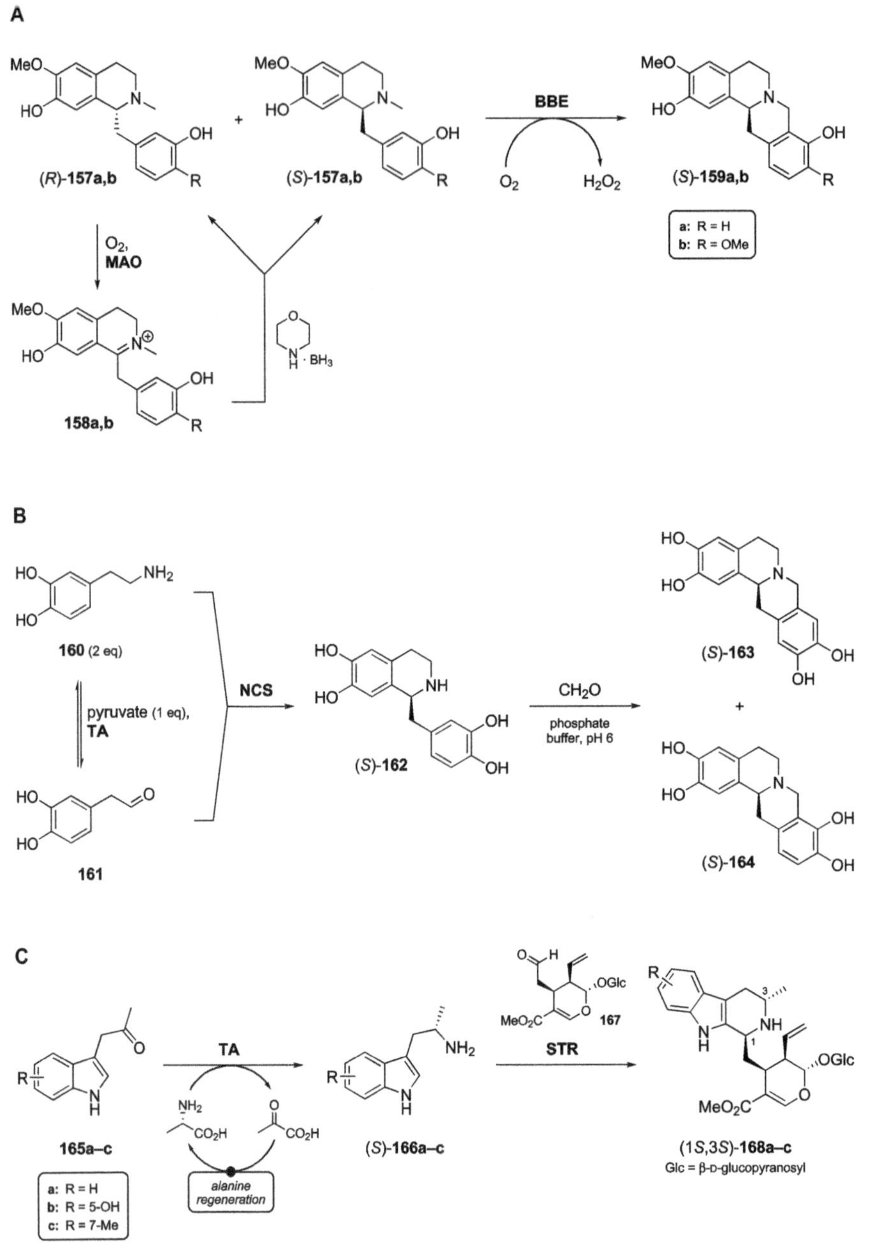

Scheme 14.24 Biocatalytic cascades for the asymmetric synthesis of alkaloids: (A) Transformation of *rac*-benzylisoquinolines **157** into (*S*)-berbines **159** by monoamine oxidase (MAO-N), berberine bridge enzyme (BBE) and morpholine–borane, (B) condensation of two molecules of dopamine (**160**) to (*S*)-norlaudanosoline (**162**) using a transaminase (TA) and a norcoclaurine synthase (NCS), and (C) preparation of C3-methyl-substituted strictosidine analogues **168** using a transaminase (TA) and a strictosidine synthase (STR).

BBE, affording the corresponding (*S*)-berbines as final products. The cascade was carried out on preparative scale (0.5 mmol) with two substrates (*rac*-**157a** and *rac*-**157b**), and the corresponding products (*S*)-**159a** and (*S*)-**159b** were isolated in 88% and 80% yield, respectively, and in optically pure form (ee > 97%).

An alternative biocatalytic cascade approach to berbine alkaloids, which uses transaminase from *Chromobacterium violaceum* and norcoclaurine synthase (NCS) from *Thalictrum flavum* as biocatalysts, was recently reported:[140] In this 'triangular cascade', as the authors called it, the transaminase catalysed the conversion of dopamine (**160**) into a 1 : 1 mixture of **160** and 3,4-dihydroxyphenylacetaldehyde (**161**) at the expense of pyruvate as amino group acceptor (Scheme 14.24B). Norcoclaurine synthase couples **160** and **161** in an asymmetric biocatalytic Pictet–Spengler reaction, affording the benzylisoquinoline alkaloid (*S*)-norlaudanosoline (**162**) in high conversion (up to 87%) and excellent enantioselectivity (99% ee). Addition of formaldehyde to the reaction mixture then led to a second, this time spontaneous, Pictet–Spengler-reaction, which provided the optically pure (*S*)-berbine **163** along with its regioisomer **164** in a 7 : 1 ratio. Carrying out the three-step reaction on preparative scale (0.5 mmol of dopamine), (*S*) **163** was obtained in 12% overall isolated yield and in >95% ee.

Another Pictet–Spenglerase next to norcoclaurine synthase is strictosidine synthase (STR), a key C–C-bond forming enzyme of the monoterpenoid indole alkaloid biosynthetic pathway. This enzyme, too, has been combined with transaminases in a cascade system, in this case in a linear sequence in which both biocatalysts control the formation of a stereogenic centre (Scheme 14.24C):[141] transaminases from *Arthrobacter* sp. and *Silicibacter pomeroyi* were used to generate the (*R*)- and (*S*)-enantiomers, respectively, of a series of α-methyltryptamine derivatives **166** by reductive amination of the corresponding ketones **165**. The amines **166** were then subjected to the enzymatic Pictet–Spengler reaction catalysed by STR from *Ophiorrhiza pumila*, using the natural aldehyde substrate secologanin (**167**) as second reaction partner. Thus, the (1*S*,3*R*)- and (1*S*,3*S*)-isomers of C3-methyl-substituted strictosidine derivatives **168** were obtained in stereoisomerically pure form. The reaction sequence was performed both in a simultaneous fashion—with all enzymes and starting materials present from the outset—and as sequential steps in one pot, whereby the transamination was run to completion and the transaminase whole-cell biocatalyst was removed before addition of the strictosidine synthase. The latter option provided better overall conversions than the former (56% to >99% *vs.* 31–97%), which was attributed to a potential side reaction of secologanin with the transaminase. The optical purity, on the other hand, was excellent in both cases, as all products were obtained as single stereoisomers. The isolation and purification of the products required column chromatography and preparative HPLC, but afforded products **168** in 70–80% isolated yield.

14.4 Conclusion

A broad variety of biocatalytic cascade processes have been developed over the past five years. All these multi-enzyme systems share the common advantages that arise from avoiding intermediate isolation and purification steps: reduced consumption of chemicals, time and energy. The additional benefits that these cascades offer are as multi-faceted as the systems themselves: some allow the control of multiple stereogenic centres, some enable functional group transformations that have no known single-step counterpart and some avoid the problems associated with unstable reaction intermediates. Although most of the examples discussed in this chapter are two- or three-enzyme sequences, more complex systems are becoming increasingly common, and efficient and high-yielding cascades comprising five or more enzymes have been shown to be possible. In the light of recent developments in the areas of enzyme discovery (see Chapter 1), protein engineering (see Chapter 7), cloning and heterologous protein expression, and of the ever-growing toolbox of enzymes used in biocatalysis, we believe that the development of multi-enzyme cascade reactions will continue at a still increasing pace.

Acknowledgements

S.V. has been supported by the Austrian BMWFJ, BMVIT, SFG, Standortagentur Tirol and ZIT through the Austrian FFG-COMET-Funding Program. COST Action CM1303 "Systems Biocatalysis" is acknowledged.

References

1. W. Kroutil and M. Rueping, *ACS Catal.*, 2014, **4**, 2086.
2. S. F. Mayer, W. Kroutil and K. Faber, *Chem. Soc. Rev.*, 2001, **30**, 332.
3. E. García-Junceda, I. Lavandera, D. Rother and J. H. Schrittwieser, *J. Mol. Catal. B: Enzym.*, 2015, **114**, 1.
4. B. Bechi, S. Herter, S. McKenna, C. Riley, S. Leimkuhler, N. J. Turner and A. J. Carnell, *Green Chem.*, 2014, **16**, 4524.
5. C. A. Denard, J. F. Hartwig and H. Zhao, *ACS Catal.*, 2013, **3**, 2856.
6. H. Gröger and W. Hummel, *Curr. Opin. Chem. Biol.*, 2014, **19**, 171.
7. A. C. Marr and S. Liu, *Trends Biotechnol.*, 2011, **29**, 199.
8. A. J. Carnell, *ChemCatChem*, 2014, **6**, 958.
9. V. Uppada, S. Bhaduri and S. B. Noronha, *Curr. Sci.*, 2014, **106**, 946.
10. S. Kara, J. H. Schrittwieser, F. Hollmann and M. B. Ansorge-Schumacher, *Appl. Microbiol. Biotechnol.*, 2013, **98**, 1517.
11. C. Rodriguez, I. Lavandera and V. Gotor, *Curr. Org. Chem.*, 2012, **16**, 2525.
12. F. Hollmann, I. W. C. E. Arends and K. Buehler, *ChemCatChem*, 2010, **2**, 762.

13. A. Rioz-Martínez, F. R. Bisogno, C. Rodríguez, G. de Gonzalo, I. Lavandera, D. E. Torres Pazmiño, M. W. Fraaije and V. Gotor, *Org. Biomol. Chem.*, 2010, **8**, 1431.
14. W. F. Liu and P. Wang, *Biotechnol. Adv.*, 2007, **25**, 369.
15. R. Sigrist, B. Z. D. Costa, A. J. Marsaioli and L. G. de Oliveira, *Biotechnol. Adv.*, 2015, **33**, 394.
16. V. Köhler and N. J. Turner, *Chem. Commun.*, 2015, **51**, 450.
17. J. Muschiol, C. Peters, N. Oberleitner, M. D. Mihovilovic, U. T. Bornscheuer and F. Rudroff, *Chem. Commun.*, 2015, **51**, 5798.
18. K. B. Otte and B. Hauer, *Curr. Opin. Biotechnol.*, 2015, **35**, 16.
19. I. Oroz-Guinea and E. García-Junceda, *Curr. Opin. Chem. Biol.*, 2013, **17**, 236.
20. E. Ricca, B. Brucher and J. H. Schrittwieser, *Adv. Synth. Catal.*, 2011, **353**, 2239.
21. F. R. Bisogno, I. Lavandera and V. Gotor, in *Kirk-Othmer Encyclopedia of Chemical Technology*, 2011, pp. 1–20.
22. J. H. Schrittwieser, J. Sattler, V. Resch, F. G. Mutti and W. Kroutil, *Curr. Opin. Chem. Biol.*, 2011, **15**, 249.
23. A. Díaz-Rodríguez, I. Lavandera and V. Gotor, *Curr. Green Chem.*, 2015, **2**, 192.
24. W. G. Birolli, I. M. Ferreira, N. Alvarenga, D. D. Santos, I. L. de Matos, J. V. Comasseto and A. L. M. Porto, *Biotechnol. Adv.*, 2015, **33**, 481.
25. N. J. Turner, *Chem Rev*, 2011, **111**, 4073.
26. N. J. Turner, *Curr. Opin. Chem. Biol.*, 2010, **14**, 115.
27. W. Kroutil, C. Voss and C. Gruber, *Synlett*, 2010, **2010**, 991.
28. S. Servi, D. Tessaro and G. Pedrocchi-Fantoni, *Coord. Chem. Rev.*, 2008, **252**, 715.
29. C. C. Gruber, I. Lavandera, K. Faber and W. Kroutil, *Adv. Synth. Catal.*, 2006, **348**, 1789.
30. E. O'Reilly and N. J. Turner, *Perspect. Sci.*, 2015, **4**, 55.
31. E. Brenna, M. Crotti, F. G. Gatti, F. Parmeggiani, A. Pugliese and S. Santangelo, in *Importance of Chirality to Flavor Compounds*, ed. K.-H. Engel and G. Takeoka, American Chemical Society, 2016, pp. 59–75.
32. J.-H. Seo, S.-M. Lee, J. Lee and J.-B. Park, *J. Biotechnol.*, 2015, **216**, 158.
33. R. C. Simon, N. Richter, E. Busto and W. Kroutil, *ACS Catal.*, 2014, **4**, 129.
34. Q. M. Dudley, A. S. Karim and M. C. Jewett, *Biotechnol. J.*, 2015, **10**, 69.
35. J.-K. Guterl and V. Sieber, *Eng. Life Sci.*, 2013, **13**, 4.
36. W.-D. Fessner, *New Biotechnol.*, 2015, **32**, 658.
37. S. Servi, R. Wohlgemuth and W.-D. Fessner, CMST COST Action CM1303: Systems Biocatalysis, http://www.cost.eu/domains_actions/cmst/Actions/CM1303, Accessed 15.09.2016.
38. T. Bayer, S. Milker, T. Wiesinger, F. Rudroff and M. D. Mihovilovic, *Adv. Synth. Catal.*, 2015, **357**, 1587.
39. R. Abu and J. M. Woodley, *ChemCatChem*, 2015, 7, 3094.
40. R. Xue and J. M. Woodley, *Bioresour. Technol.*, 2012, **115**, 183.

41. F. Lopez-Gallego and C. Schmidt-Dannert, *Curr. Opin. Chem. Biol.*, 2010, **14**, 174.
42. M. B. Quin and C. Schmidt-Dannert, *Curr. Opin. Biotechnol.*, 2014, **29**, 55.
43. R. Yuryev, S. Strompen and A. Liese, *Beilstein J. Org. Chem.*, 2011, **7**, 1449.
44. R. K. Pandey, R. A. Fernandes and P. Kumar, *Tetrahedron Lett.*, 2002, **43**, 4425.
45. J. H. Kim, H. Yang, J. Park and G. J. Boons, *J. Am. Chem. Soc.*, 2005, **127**, 12090.
46. G. B. Jones, M. Guzel and S. B. Heaton, *Tetrahedron: Asymmetry*, 2000, **11**, 4303.
47. M. Müller, G. A. Sprenger and M. Pohl, *Curr. Opin. Chem. Biol.*, 2013, **17**, 261.
48. M. Müller, *Adv. Synth. Catal.*, 2012, **354**, 3161.
49. M. Pohl, B. Lingen and M. Müller, *Chem. Eur. J.*, 2002, **8**, 5288.
50. Please note that in the original publication the carbon chain of **94** is inversely numbered and the compound hence named 2-amino-1,3,4-butanetriol. According to this nomenclature (which is not in agreement with the current version of the IUPAC nomenclature rules), the obtained stereoisomer has the absolute configuration (2*S*,3*R*).
51. A. Jakoblinnert and D. Rother, *Green Chem.*, 2014, **16**, 3472.
52. J. Wachtmeister, A. Jakoblinnert, J. Kulig, H. Offermann and D. Rother, *ChemCatChem*, 2014, **6**, 1051.
53. P. P. Giovannini, G. Fantin, A. Massi, V. Venturi and P. Pedrini, *Org. Biomol. Chem.*, 2011, **9**, 8038.
54. P. P. Giovannini, M. Mantovani, A. Grandini, A. Medici and P. Pedrini, *J. Mol. Catal. B: Enzym.*, 2011, **69**, 15.
55. O. Bortolini, P. P. Giovannini, S. Maietti, A. Massi, P. Pedrini, G. Sacchetti and V. Venturi, *J. Mol. Catal. B: Enzym.*, 2013, **85–86**, 93.
56. C. E. Paul, I. Lavandera, V. Gotor-Fernández, W. Kroutil and V. Gotor, *ChemCatChem*, 2013, **5**, 3875.
57. C. V. Voss, C. C. Gruber, K. Faber, T. Knaus, P. Macheroux and W. Kroutil, *J. Am. Chem. Soc.*, 2008, **130**, 13969.
58. E. Brenna, F. G. Gatti, D. Monti, F. Parmeggiani and A. Sacchetti, *ChemCatChem*, 2012, **4**, 653.
59. E. Brenna, F. G. Gatti, L. Malpezzi, D. Monti, F. Parmeggiani and A. Sacchetti, *J. Org. Chem.*, 2013, **78**, 4811.
60. E. Brenna, M. Crotti, F. G. Gatti, D. Monti, F. Parmeggiani, A. Pugliese and S. Santangelo, *J. Mol. Catal. B: Enzym.*, 2015, **114**, 37.
61. L. El Blidi, Z. Assaf, F. Camps Bres, H. Veschambre, V. Théry, J. Bolte and M. Lemaire, *ChemCatChem*, 2009, **1**, 463.
62. L. El Blidi, M. Ahbala, J. Bolte and M. Lemaire, *Tetrahedron: Asymmetry*, 2006, **17**, 2684.
63. L. El Blidi, D. Crestia, E. Gallienne, C. Demuynck, J. Bolte and M. Lemaire, *Tetrahedron: Asymmetry*, 2004, **15**, 2951.

64. J. A. Castillo, C. Guérard Hélaine, M. Gutiérrez, X. Garrabou, M. Sancelme, M. Schürmann, T. Inoue, V. Hélaine, F. Charmantray, T. Gefflaut, L. Hecquet, J. Joglar, P. Clapés, G. A. Sprenger and M. Lemaire, *Adv. Synth. Catal.*, 2010, **352**, 1039.
65. F. Camps Bres, C. Guérard-Hélaine, V. Hélaine, C. Fernandes, I. Sánchez-Moreno, M. Traïkia, E. García-Junceda and M. Lemaire, *J. Mol. Catal. B: Enzym.*, 2015, **114**, 50.
66. S. Wu, Y. Chen, Y. Xu, A. Li, Q. Xu, A. Glieder and Z. Li, *ACS Catal.*, 2014, **4**, 409.
67. T. C. Nugent, ed., *Chiral Amine Synthesis*, Wiley-VCH, Weinheim, 2010.
68. T. C. Nugent and M. El-Shazly, *Adv. Synth. Catal.*, 2010, **352**, 753.
69. D. Ghislieri and N. Turner, *Top. Catal.*, 2014, **57**, 284.
70. M. Höhne and U. T. Bornscheuer, *ChemCatChem*, 2009, **1**, 42.
71. H. Kohls, F. Steffen-Munsberg and M. Höhne, *Curr. Opin. Chem. Biol.*, 2014, **19**, 180.
72. D. Koszelewski, K. Tauber, K. Faber and W. Kroutil, *Trends Biotechnol.*, 2010, **28**, 324.
73. S. Mathew and H. Yun, *ACS Catal.*, 2012, **2**, 993.
74. P. Tufvesson, J. S. Jensen, W. Kroutil and J. M. Woodley, *Biotechnol. Bioeng.*, 2012, **109**, 2159.
75. J.-S. Shin and B.-G. Kim, *Biotechnol. Bioeng.*, 1999, **65**, 206.
76. D. Koszelewski, I. Lavandera, D. Clay, G. M. Guebitz, D. Rozzell and W. Kroutil, *Angew. Chem., Int. Ed.*, 2008, **47**, 9337.
77. K. E. Cassimjee, C. Branneby, V. Abedi, A. Wells and P. Berglund, *Chem. Commun.*, 2010, **46**, 5569.
78. A. K. Holzer, K. Hiebler, F. G. Mutti, R. C. Simon, L. Lauterbach, O. Lenz and W. Kroutil, *Org. Lett.*, 2015, **17**, 2431.
79. N. Richter, J. E. Farnberger, D. Pressnitz, H. Lechner, F. Zepeck and W. Kroutil, *Green Chem.*, 2015, **17**, 2952.
80. M. H. S. A. Hamid, P. A. Slatford and J. M. J. Williams, *Adv. Synth. Catal.*, 2007, **349**, 1555.
81. T. D. Nixon, M. K. Whittlesey and J. M. J. Williams, *Dalton Trans.*, 2009, 753.
82. G. E. Dobereiner and R. H. Crabtree, *Chem. Rev.*, 2010, **110**, 681.
83. G. Guillena, D. J. Ramón and M. Yus, *Chem. Rev.*, 2010, **110**, 1611.
84. A. J. A. Watson and J. M. J. Williams, *Science*, 2010, **329**, 635.
85. S. Bähn, S. Imm, L. Neubert, M. Zhang, H. Neumann and M. Beller, *ChemCatChem*, 2011, **3**, 1853.
86. C. Gunanathan and D. Milstein, *Science*, 2013, **341**, 1229712.
87. L. Sun, I. P. Petrounia, M. Yagasaki, G. Bandara and F. H. Arnold, *Protein Eng.*, 2001, **14**, 699.
88. M. Fuchs, K. Tauber, J. Sattler, H. Lechner, J. Pfeffer, W. Kroutil and K. Faber, *RSC Adv.*, 2012, **2**, 6262.
89. M. Pickl, M. Fuchs, S. M. Glueck and K. Faber, *ChemCatChem*, 2015, **7**, 3121.
90. J. H. Sattler, M. Fuchs, K. Tauber, F. G. Mutti, K. Faber, J. Pfeffer, T. Haas and W. Kroutil, *Angew. Chem., Int. Ed.*, 2012, **51**, 9156.

91. S. Klatte and V. F. Wendisch, *Bioorg. Med. Chem.*, 2014, **22**, 5578.
92. S. Klatte and V. F. Wendisch, *Microb. Cell Fact.*, 2015, **14**, 1.
93. K. Tauber, M. Fuchs, J. H. Sattler, J. Pitzer, D. Pressnitz, D. Koszelewski, K. Faber, J. Pfeffer, T. Haas and W. Kroutil, *Chem. – Eur. J.*, 2013, **19**, 4030.
94. A. Lerchner, S. Achatz, C. Rausch, T. Haas and A. Skerra, *ChemCatChem*, 2013, **5**, 3374.
95. C. M. Palacio, C. G. Crismaru, S. Bartsch, V. Navickas, K. Ditrich, M. Breuer, R. Abu, J. Woodley, K. Baldenius, B. Wu and D. B. Janssen, *Biotechnol. Bioeng.*, 2016, **113**, 1853.
96. M. Schrewe, N. Ladkau, B. Bühler and A. Schmid, *Adv. Synth. Catal.*, 2013, **355**, 1693.
97. Engineering of the whole-cell catalyst to make the diacid monoester the main product has also been reported. See: M. Schrewe, M. K. Julsing, K. Lange, E. Czarnotta, A. Schmid and B. Bühler, *Biotechnol. Bioeng.*, 2014, **111**, 1820.
98. N. Ladkau, M. Assmann, M. Schrewe, M. K. Julsing, A. Schmid and B. Bühler, *Metab. Eng.*, 2016, **36**, 1.
99. P. Both, H. Busch, P. P. Kelly, F. G. Mutti, N. J. Turner and S. L. Flitsch, *Angew. Chem., Int. Ed.*, 2016, **55**, 1511.
100. A. Robin, V. Köhler, A. Jones, A. Ali, P. P. Kelly, E. O'Reilly, N. J. Turner and S. L. Flitsch, *Beilstein J. Org. Chem.*, 2011, **7**, 1494.
101. J. W. Song, E. Y. Jeon, D. H. Song, H. Y. Jang, U. T. Bornscheuer, D. K. Oh and J. B. Park, *Angew. Chem., Int. Ed.*, 2013, **52**, 2534.
102. J.-W. Song, J.-H. Lee, U. T. Bornscheuer and J.-B. Park, *Adv. Synth. Catal.*, 2014, **356**, 1782.
103. J. H. Sattler, M. Fuchs, F. G. Mutti, B. Grischek, P. Engel, J. Pfeffer, J. M. Woodley and W. Kroutil, *Angew. Chem., Int. Ed.*, 2014, **53**, 14153.
104. S. Staudt, U. T. Bornscheuer, U. Menyes, W. Hummel and H. Groger, *Enzyme Microb. Technol.*, 2013, **53**, 288.
105. H. Mallin, H. Wulf and U. T. Bornscheuer, *Enzyme Microb. Technol.*, 2013, **53**, 283.
106. S. Wu, Y. Zhou, T. Wang, H.-P. Too, D. I. C. Wang and Z. Li, *Nat. Commun.*, 2016, **7**, 11917.
107. T. Sehl, Z. Maugeri and D. Rother, *J. Mol. Catal. B: Enzym.*, 2015, **114**, 65.
108. C. U. Ingram, M. Bommer, M. E. B. Smith, P. A. Dalby, J. M. Ward, H. C. Hailes and G. J. Lye, *Biotechnol. Bioeng.*, 2007, **96**, 559.
109. L. Rios-Solis, N. Bayir, M. Halim, C. Du, J. M. Ward, F. Baganz and G. J. Lye, *Biochem. Eng. J.*, 2013, **73**, 38.
110. L. Rios-Solis, M. Halim, A. Cázares, P. Morris, J. M. Ward, H. C. Hailes, P. A. Dalby, F. Baganz and G. J. Lye, *Biocatal. Biotransform.*, 2011, **29**, 192.
111. L. Rios-Solis, P. Morris, C. Grant, A. O. O. Odeleye, H. C. Hailes, J. M. Ward, P. A. Dalby, F. Baganz and G. J. Lye, *Chem. Eng. Sci.*, 2015, **122**, 360.

112. M. E. B. Smith, B. H. Chen, E. G. Hibbert, U. Kaulmann, K. Smithies, J. L. Galman, F. Baganz, P. A. Dalby, H. C. Hailes, G. J. Lye, J. M. Ward, J. M. Woodley and M. Micheletti, *Org. Process Res. Dev.*, 2010, **14**, 99.
113. T. Sehl, H. C. Hailes, J. M. Ward, R. Wardenga, E. von Lieres, H. Offermann, R. Westphal, M. Pohl and D. Rother, *Angew. Chem., Int. Ed.*, 2013, **52**, 6772.
114. An analogous 'recycling cascade' has been developed for the production of (2*R*,3*S*)-3-amino-1,2,4-butanetriol; however, only limited conversions were obtained. See: M. F. Villegas-Torres, R. J. Martinez-Torres, A. Cázares-Körner, H. Hailes, F. Baganz and J. Ward, *Enzyme Microb. Technol.*, 2015, **81**, 23.
115. S. Wu, Y. Chen, Y. Xu, A. Li, Q. Xu, A. Glieder and Z. Li, *ACS Catal.*, 2014, **4**, 409.
116. T. Sehl, H. C. Hailes, J. M. Ward, U. Menyes, M. Pohl and D. Rother, *Green Chem.*, 2014, **16**, 3341.
117. E. O'Reilly, C. Iglesias, D. Ghislieri, J. Hopwood, J. L. Galman, R. C. Lloyd and N. J. Turner, *Angew. Chem., Int. Ed.*, 2014, **53**, 2447.
118. L. Skalden, C. Peters, J. Dickerhoff, A. Nobili, H. J. Joosten, K. Weisz, M. Hohne and U. T. Bornscheuer, *ChemBioChem*, 2015, **16**, 1041.
119. L. Skalden, C. Peters, L. Ratz and U. T. Bornscheuer, *Tetrahedron*, 2016, **72**, 7207.
120. D. Monti, M. C. Forchin, M. Crotti, F. Parmeggiani, F. G. Gatti, E. Brenna and S. Riva, *ChemCatChem*, 2015, **7**, 3106.
121. J. H. Schrittwieser, S. Velikogne and W. Kroutil, *Adv. Synth. Catal.*, 2015, **357**, 1655.
122. D. Gamenara and P. Dominguez de Maria, *Org. Biomol. Chem.*, 2014, **12**, 2989.
123. O. Pàmies, M. Diéguez and J. E. Bäckvall, *Adv. Synth. Catal.*, 2015, **357**, 1567.
124. M. J. Abrahamson, J. W. Wong and A. S. Bommarius, *Adv. Synth. Catal.*, 2013, **355**, 1780.
125. B. R. Bommarius, M. Schurmann and A. S. Bommarius, *Chem. Commun.*, 2014, **50**, 14953.
126. A. S. Bommarius and S. K. Au, in *Science of Synthesis: Biocatalysis in Organic Synthesis 2*, ed. K. Faber, W.-D. Fessner and N. J. Turner, Thieme, Stuttgart, 2015, pp. 335–358.
127. G. Grogan and N. J. Turner, *Chem. Eur. J.*, 2016, **22**, 1900.
128. F. G. Mutti, T. Knaus, N. S. Scrutton, M. Breuer and N. J. Turner, *Science*, 2015, **349**, 1525.
129. F.-F. Chen, Y.-Y. Liu, G.-W. Zheng and J.-H. Xu, *ChemCatChem*, 2015, **7**, 3838.
130. L.-J. Wang, C.-X. Li, Y. Ni, J. Zhang, X. Liu and J.-H. Xu, *Bioresour. Technol.*, 2011, **102**, 7023.
131. W. Kroutil and K. Faber, *Tetrahedron: Asymmetry*, 1998, **9**, 2901.

132. V. Köhler, Y. M. Wilson, M. Dürrenberger, D. Ghislieri, E. Churakova, T. Quinto, L. Knörr, D. Häussinger, F. Hollmann, N. J. Turner and T. R. Ward, *Nat. Chem.*, 2013, **5**, 93.
133. R. S. Heath, M. Pontini, S. Hussain and N. J. Turner, *ChemCatChem*, 2016, **8**, 117.
134. S. P. France, S. Hussain, A. M. Hill, L. J. Hepworth, R. M. Howard, K. R. Mulholland, S. L. Flitsch and N. J. Turner, *ACS Catal.*, 2016, **6**, 3753.
135. V. Resch, H. Lechner, J. H. Schrittwieser, S. Wallner, K. Gruber, P. Macheroux and W. Kroutil, *Chem. Eur. J.*, 2012, **18**, 13173.
136. V. Resch, J. H. Schrittwieser, S. Wallner, P. Macheroux and W. Kroutil, *Adv. Synth. Catal.*, 2011, **353**, 2377.
137. J. H. Schrittwieser, V. Resch, J. H. Sattler, W.-D. Lienhart, K. Durchschein, A. Winkler, K. Gruber, P. Macheroux and W. Kroutil, *Angew. Chem., Int. Ed.*, 2011, **50**, 1068.
138. J. H. Schrittwieser, V. Resch, S. Wallner, W.-D. Lienhart, J. H. Sattler, J. Resch, P. Macheroux and W. Kroutil, *J. Org. Chem.*, 2011, **76**, 6703.
139. J. H. Schrittwieser, B. Groenendaal, V. Resch, D. Ghislieri, S. Wallner, E. M. Fischereder, E. Fuchs, B. Grischek, J. H. Sattler, P. Macheroux, N. J. Turner and W. Kroutil, *Angew. Chem., Int. Ed.*, 2014, **53**, 3731.
140. B. R. Lichman, E. D. Lamming, T. Pesnot, J. M. Smith, H. C. Hailes and J. M. Ward, *Green Chem.*, 2015, **17**, 852.
141. E.-M. Fischereder, D. Pressnitz and W. Kroutil, *ACS Catal.*, 2016, **6**, 23.

CHAPTER 15

Emerging Fields in One-pot Multi-step Synthesis with Combined Chemo- and Bio-catalysts: Sequential- and Domino-type Process Concepts as well as Compartmentation Strategies

HARALD GRÖGER

Chair of Organic Chemistry I, Faculty of Chemistry, Bielefeld University, Universitätsstr. 25, 33615 Bielefeld, Germany
Email: harald.groeger@uni-bielefeld.de

15.1 Introduction and Current Status Overview

Organic-synthetic multi-step cascade processes running in a one-pot fashion offer many advantages and have raised a broad interest in the chemocatalysis community.[1] Many successful examples have been reported and a key prerequisite is achieving compatibility of the individual steps (which typically run in organic media). The integration of chemical steps towards cascades without the need for work-up of intermediates is also of interest for the chemical industry, and the 'second generation process' for the drug

Catalysis Series No. 32
Modern Biocatalysis: Advances Towards Synthetic Biological Systems
Edited by Gavin Williams and Mélanie Hall

Published by the Royal Society of Chemistry, www.rsc.org

sertraline is a very successful example of such an achievement (being awarded with the US Presidential Green Chemistry Award).[2] In biotechnology, naturally occurring cascades are of interest since, *e.g.* any type of fermentation process can be regarded as a tandem process in a whole cell as 'biological reactor'. Again, compatibility of the individual reaction steps is a key prerequisite which is typically fulfilled since enzymes operate under comparable (physiological) conditions in water. In contrast to studying 'pure' chemocatalytic cascades, biocatalytic cascades such as microbial fermentations (synthetic biology),[3] and multi-step enzymatic cascades[4] (see Chapters 13 and 14), for a long time merging these two different 'worlds' of catalysis—chemocatalysis and biocatalysis—was studied to a much less extent in spite of tremendous synthetic diversity, which could result from such a combination. One reason for this young field of chemoenzymatic one-pot process development might be the simple assumption that chemocatalysts and biocatalysts are hardly compatible, in particular when considering metal components as chemocatalysts of choice (which enable a broad portfolio of reactions in organic chemistry). However, in recent years, several proofs of concept for such one-pot combinations of chemo- and biocatalysts were demonstrated. These chemoenzymatic one-pot processes comprise combinations of metal catalysts as well as organocatalysts with enzymes. Furthermore, organic as well as aqueous media have been found to be suitable solvent systems for such processes. For conducting such chemoenzymatic one-pot processes, in general two strategies are conceivable, which are graphically summarized in Scheme 15.1. Taking biosynthetic pathways as a 'blue print', a first option are domino-type one-pot processes, in which both reaction steps proceed concurrently in the presence of the catalysts being present in the reaction mixture from the beginning on (Scheme 15.1, Part (A)). Such an approach requires that all catalysts involved are compatible with one another and that they can operate under the defined 'process window'. The second one-pot option consists of an initial reaction carried out with the first catalyst, *e.g.* a chemocatalyst, followed by addition of the second catalyst, *e.g.* a biocatalyst, after completion of the first reaction step (Scheme 15.1, Part (B)). Both concepts show advantages and disadvantages, and case by case the one or the other concept turns out to be more suitable. For example, in the case of non-stable intermediates, the one-pot domino-type concept is more recommendable, whereas in the case of cross-inhibition of the catalyst used in the second step by a substrate from the first step, the one-pot sequential-mode appears to be more advantageous.

15.2 Overview of Selected Current Emerging Fields

Based on recently published reviews about state of the art in one-pot processes with combined chemo- and biocatalysts,[5,6] this contribution provides an overview of emerging fields in chemoenzymatic synthesis and selected major achievements in this field, which (at least for most examples)

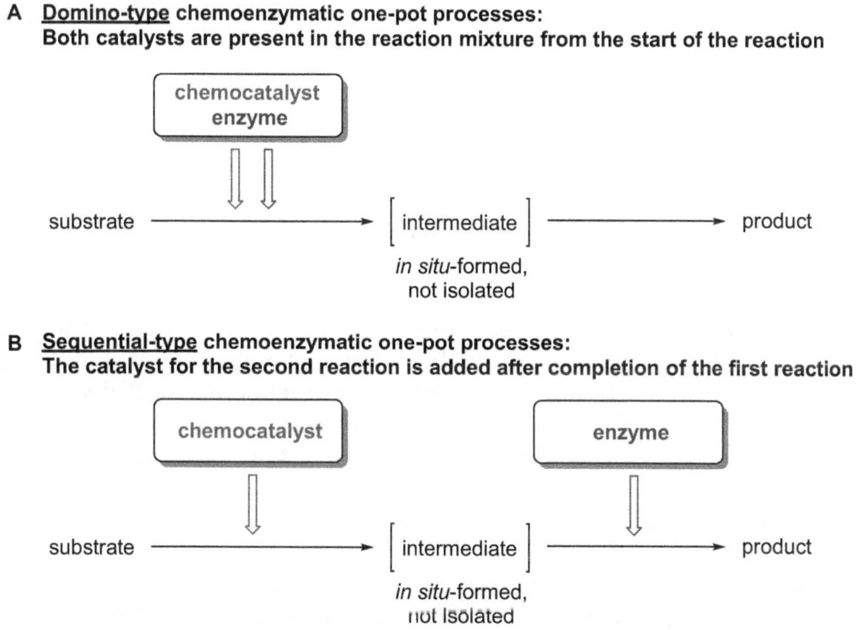

Scheme 15.1 Strategies for chemoenzymatic one-pot processes under combination of chemo- and biocatalysts.

were reported only recently. In the past few years, besides the expansion of sequential chemoenzymatic one-pot synthesis (with addition of a catalyst for the second step after completion of the initial step), a further major challenge has been the development of domino-type tandem reactions, in which chemo- and biocatalytic reactions proceed simultaneously. In parallel, a further challenge consisted of finding solutions for one-pot processes under combination of chemo- and biocatalysts, which turned out not to be compatible with each other. A major concept being used for such cases are compartmentation of the catalysts within one reactor. It is noteworthy that the size of such compartments can broadly vary ranging from the molecular to the nano- and macroscopic scale. A further strategy for combination of non-compatible catalysts is compartmentation in different reactors. Although not being a pure 'one-pot process', the same goal, namely combination of reactions steps without the need for isolation of the intermediates, can be achieved. Thus, in the following, the focus will be on recent developments in these three fields (i) sequential-type one-pot multi-step processes, (ii) domino-type one-pot multi-step processes, and (iii) compartmentation of catalysts within either one reactor or within multi-reactor systems. In addition, a further focus will be on such combinations of chemo- and biocatalysts, which are based on the use of such types of chemocatalysts and enzyme classes not having been combined earlier in multi-step chemoenzymatic one-pot processes.

15.3 New Sequential-type Chemoenzymatic One-pot Syntheses

15.3.1 Introduction and Overview

In recent years, major achievements for one-pot processes running in a sequential mode—in which the catalysts for the second step were added after completion of the first step—have been made. It is noteworthy that a broad range of chemo- as well as biocatalysts can be already combined towards efficient one-pot processes.

One notable fact is that a broad range of metal complexes have turned out to be suitable for combination with enzymes in sequential-type two-step one-pot processes, running *e.g.* in water, thus enabling combination of metal-catalyzed reactions such as cross-coupling reactions, hydrogenation and olefin metathesis, with biotransformations.[7] Furthermore, organocatalysts have been proven to be tolerated by biocatalysts. Although less studied, several types of one-pot multi-step processes based on sequential organo- and biocatalytic reactions have been developed.[8]

Studies from various groups have underlined that numerous biocatalysts are quite robust and tolerant against the presence of complex reaction mixtures from previous reaction steps within such one-pot processes, as well as compatible with many metal catalysts and organocatalysts. The success of this field is also the subject of several extensive reviews, describing the state of the art and the progress in this field.[5,6]

15.3.2 Emerging Fields and Selected Recent Examples of Sequential-type Chemoenzymatic One-pot Syntheses

Referring to comprehensive reviews summarizing earlier examples and pioneer work in this field,[5,6] some selected examples of very recent developments in this field will be presented in this section.

One of the most recent examples of such a sequential multi-step one-pot processes reported by Monti and Guidotti *et al.* underlines the suitability to combine heterogeneous catalysts with enzymes as well as the opportunity to link chemocatalytic reactions under exclusion of water with enzymatic steps running in water.[9] In detail, an epoxidation of limonene with a heterogeneous, titanium-grafted silica catalyst in acetonitrile or α,α,α-trifluorotoluene as the organic medium was combined with a limonene epoxide hydrolase being subsequently added as a buffered solution, thus leading to a formal dihydroxylation of (+)- and (−)-limonene without the need for purification of the intermediate (Scheme 15.2). For example, when combining such a regio- and diastereoselective epoxidation of (−)-limonene, **1**, in the presence of the heterogeneous titanium catalyst Ti/SiO_2, with the subsequent enzymatic epoxide opening reaction in a one-pot process, high conversion of the individual steps together with high compatibility of the catalytic systems and a yield of 72% for the formed (*R*,*R*,*S*)-enantiomer **5** were observed.

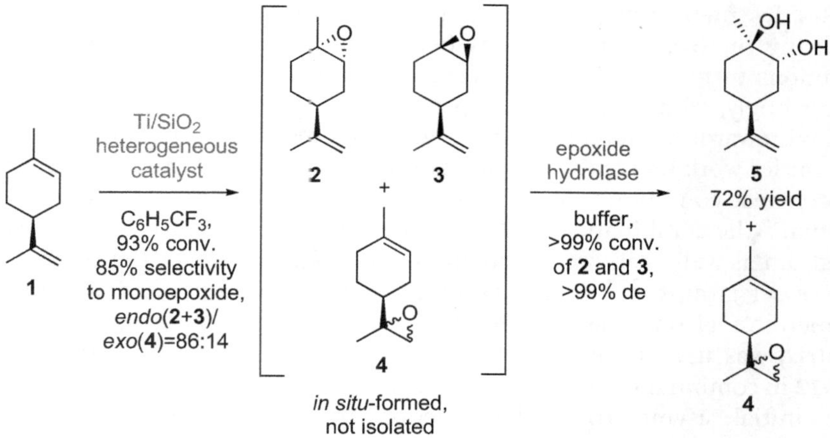

Scheme 15.2 Combination of a heterogeneous metal-catalyzed epoxidation with an enzyme-catalyzed hydrolytic ring-opening in a one-pot process running in a sequential-mode.

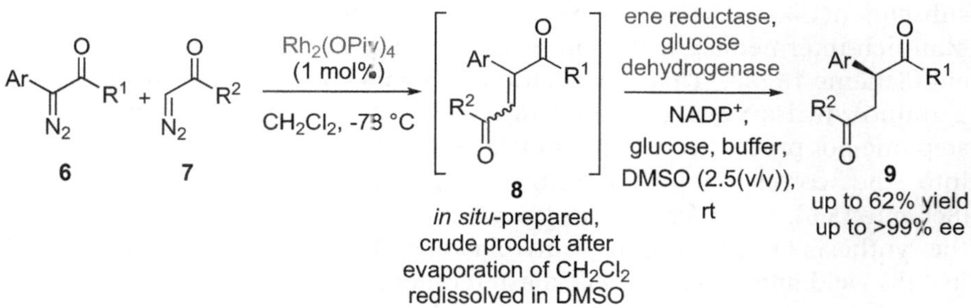

Scheme 15.3 Combination of rhodium-catalyzed alkene formation and enzymatic C=C double bond reduction.

A novel retrosynthetic pathway towards chiral 1,4-dicarbonyl compounds **9** by means of a chemoenzymatic two-step one-pot process based on a metal-catalyzed alkene formation and subsequent enantioselective enzymatic ene reduction has been disclosed by the Zhao group (Scheme 15.3).[10] In an initial step, a rhodium-catalyzed heterocoupling of two diazoesters **6** and **7** was carried out in methylene chloride, delivering the corresponding 1,4-dicarbonyl alkenes **8**. A solvent switch from methylene chloride to DMSO (through evaporation of methylene chloride and dissolving the crude product in DMSO) and direct subsequent enzymatic reduction of the C=C double bond in **8** then furnished the desired chiral 2-aryl-substituted succinate derivatives **9** in up to 62% yield over two steps and with >99% ee in most cases. For the biotransformation, ene-reductases have been utilized as biocatalyst in combination with *in situ*-cofactor regeneration with a glucose dehydrogenase and D-glucose as reducing agent being used in stoichiometric amount.

Besides metal catalysis, in recent years organocatalysis also gained increasing interest in the catalysis field, and a broad range of asymmetric reactions turned out to be catalyzed efficiently by chiral organocatalysts.[11] Accordingly, with respect to one-pot processes utilizing enzymes, their potential towards combination with organocatalysts has been studied.[8] Based on earlier work on combining an asymmetric organocatalytic aldol reaction under neat conditions with a subsequent biocatalytic reaction in aqueous media,[8b] the combination of these two stereoselective reactions in aqueous medium as well as in organic medium has been realized by the Gröger and Berkessel groups.[12,13] In the latter example, an immobilized biocatalyst, namely an alcohol dehydrogenase being entrapped in a superabsorber matrix, was used in an organic solvent for the reduction of intermediate (*R*)-**12** in combination with a dissolved chiral organocatalyst, which catalyzed the initial asymmetric aldol reaction of aldehyde **10** with acetone (Scheme 15.4).[13] This one-pot process running in a sequential mode led to the desired 1,3-diol (1*R*,3*S*)-**13** with >99% ee and a diastereomeric ratio exceeding >35 : 1. In addition, high conversions were achieved for both steps.

The Faber, Kroutil and Pietruszka group jointly reported the successful combination of an asymmetric organocatalytic Mannich-type reaction and a subsequent diastereoselective biocatalytic reduction of the keto moiety in the Mannich-intermediate with isopropanol serving as reducing agent and solvent (Scheme 15.5).[14] This combination of organo- and biocatalysis led to the α-amino-γ-hydroxy acid derivative **16**, which was transformed in a further two-step one-pot process consisting of transesterification and lactone formation into the corresponding α-amino-γ-butyrolactone in *N*-protected form (Scheme 15.5). This overall four-step two-pot process has been exemplified for the synthesis of the α-amino-γ-butyrolactone (3*R*,5*S*)-**18**, which was obtained in 51% yield and with excellent diastereomeric ratio (*d.r.* = 99 : 1) and enantiomeric excess (>99% ee). Optionally, initial aldimine substrate formation can be included in the one-pot process, thus enabling combination of three reactions in a one-pot synthesis. However, enantioselectivity of the proline-catalyzed aldol reaction was lower in this case. Since suitable organo- and biocatalysts have been identified for the selective formation of both stereogenic centers, this approach enables a highly enantio- and diastereoselective access to all four stereoisomers of the target molecule, which were obtained in high yield and with excellent diastereo- and enantioselectivity. Furthermore, conversion of (3*R*,5*S*)-butyrolactone (3*R*,5*S*)-**18** into natural, non-proteinogenic α-amino acid (3*R*,5*S*)-γ-hydroxynorvaline was achieved.

15.4 New Tandem-type (Domino-type) Chemoenzymatic One-pot Syntheses

15.4.1 Introduction and Overview

For a long time, chemoenzymatic tandem-type (domino-type) one-pot processes have been rare and this field still represents a great challenge

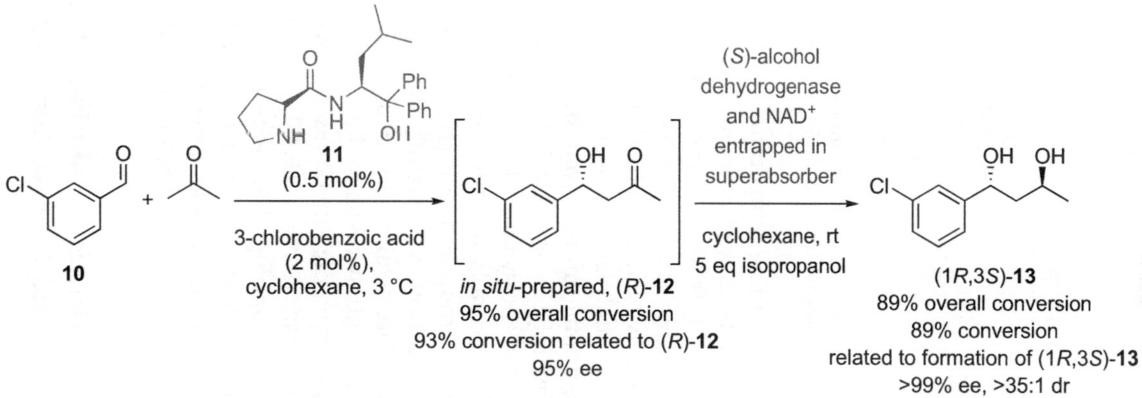

Scheme 15.4 Combination of an enantioselective aldol reaction with an enzymatic reduction in a one-pot process running in an organic medium.

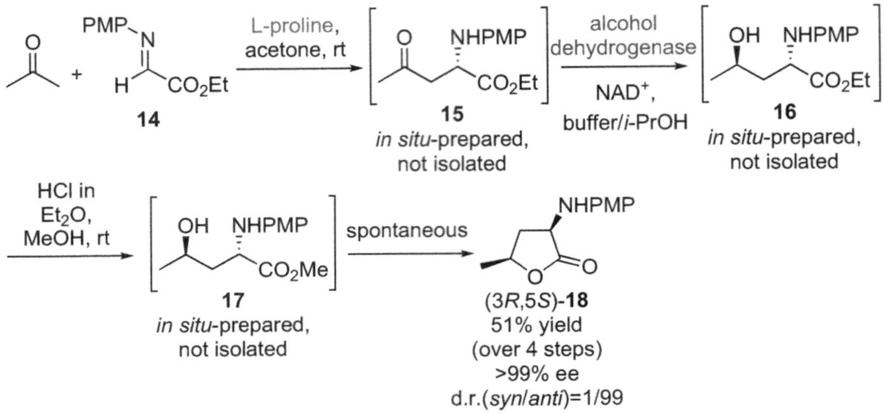

Scheme 15.5 Combination of an asymmetric organocatalytic Mannich-type reaction and an enzymatic reduction in a one-pot process.

since the process window (including temperature, solvent and so on) has to match with the chemocatalysts as well as enzyme catalysts, and undesired interactions between compounds of the various steps have to be avoided. Examples for early stage domino one-pot processes include the pioneer work by the van Bekkum group, which combined an enzymatic carbohydrate isomerization with a metal-catalyzed hydrogenation,[15] the proof of concept for dynamic kinetic resolutions through combined metal-catalyzed racemization and enzymatic hydrolysis or acylation reactions by the Williams group,[16] and the highly efficient dynamic kinetic resolution processes with lipases and Ru-catalysts for alcohols and amines developed for a broad range of examples by the Bäckvall group and Kim group, respectively.[17]

Recently, the range of reactions was significantly extended and selected examples are described in more detail in the following sections. A major part of such contributions can be subdivided into the following research fields: (i) extension of the concept of dynamic kinetic resolutions, in particular those running in organic media (see Section 15.4.2), (ii) domino-type cascade processes with concurrently running chemo- and biocatalytic steps, favorably running in water to enable the use of a broad range of enzymes (see Section 15.4.3), and (iii) related one-pot processes with simultaneously interacting chemo- and biocatalysts (see Section 15.4.4).

15.4.2 Novel Dynamic Kinetic Resolutions: Expanded Substrate Scope and Process Concepts

Chemoenzymatic dynamic kinetic resolutions (DKRs) represent the probably most intensively studied type of one-pot processes under combination of chemo- and biocatalysts. In addition, this type of process has been successfully developed towards a mature synthetic technology with wide application scope, and impressive technical feasibility thereof has been demonstrated.[17] Among others, earlier impressive breakthrough contributions in this field were made by the Williams group,[16] Bäckvall group,[17,18] Kim group,[19] and Berkessel group,[20] in which a lipase was applied for a resolution *via* acylation of an alcohol or an amine substrate to an ester or amide in combination with a metal-catalyzed racemization of the substrate.

A major advantage of this type of DKRs utilizing immobilized lipases from an organic chemist's perspective is that the process can be conducted in organic medium, which solubilizes the substrate and product and enables a simple work-up of the reaction mixture *via* separation of the product mixture from the heterogenous biocatalyst by, *e.g.* filtration.

This field of dynamic kinetic resolutions through combination of a chemocatalytic racemization and a biocatalytic resolution has been already addressed in various comprehensive reviews.[17,21] Thus, the focus of this chapter will be on selected recent developments.

A goal gaining wide interest in organic synthesis is still to identify further applications of chemoenzymatic DKRs particularly in the synthesis of chiral life science molecules such as pharmaceuticals and agrochemicals. Accordingly, recent examples comprise applications in natural product chemistry such as total synthesis of the small ring macrolide xyolide, for which a required *O*-acylated allylic alcohol building block was prepared through chemoenzymatic dynamic kinetic resolution utilizing the lipase CAL-B for the resolution and a ruthenium racemization catalyst.[22] In addition, analogous application of such type of DKR towards racemic amines was applied by the You and Jia groups as a key step within a 7-step synthesis of the drug rasagiline mesylate, which was obtained in 25% overall yield and with 99.9% ee.[23] Combination of CAL-B with a palladium nanocatalyst for racemization of indan-1-amine at a substrate loading of 200 $g\,L^{-1}$ gave the desired amide in high yield exceeding >90% and excellent enantiomeric excess of >99% ee.

Such a type of DKR of racemic amines was further integrated in a multi-step one-pot cascade process by the Córdova group. Therein, the racemic amines are formed *in situ via* a chemocatalytic reductive amination reaction, followed by addition of lipase and racemization chemocatalyst for the DKR process.[24]

Besides broadening the substrate scope of this type of chemoenzymatic DKR towards target molecules serving as intermediates in the synthesis of natural products and pharmaceuticals, challenges with respect to the use of other metal complexes based on, *e.g.* iron or titanium, enabling novel types of DKR reactions, were studied. So far combinations of iron-catalysts with enzymes in one-pot processes are still rare. A combination of an iron–porphyrin complex for *in situ*-cofactor regeneration with a glucose dehydrogenase-catalyzed oxidation process for various carbohydrates was reported by the Gröger group (see Section 15.4.4).[25] On the other hand, iron is a highly attractive metal from various perspectives such as economical access, abundant availability and non-toxicity. Furthermore, a broad range

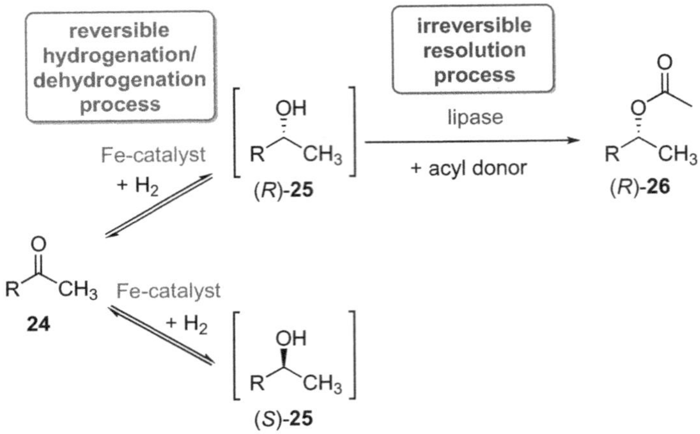

Scheme 15.6 Iron-complex for hydrogenation of ketones and racemization of alcohols enabling a DKR process in combination with enzymatic resolution.

of iron complexes are known to be suitable for redox processes. Making use of the benefits of iron complexes, the Rueping group combined an iron carbonyl-complex with a lipase for a one-pot transformation of ketones into enantiomerically enriched acetates (Scheme 15.6).[26] In detail, the individual reactions consist of an initial hydrogenation of a ketone, **24**, thus forming the racemic alcohol *rac*-**25** under consumption of molecular hydrogen used as reducing agent. Subsequently, a dynamic kinetic resolution of *rac*-**25** by means of a lipase-catalyzed acylation of the alcohol with ethyl acetate as acyl donor and iron-catalyzed racemization of the alcohol *rac*-**25** through an oxidation–reduction sequence then gives the desired ester products (*R*)-**26**. In analogy to previous chemoenzymatic DKRs of alcohols,[17] the iron catalysts turned out to be compatible with the biocatalyst.[26] The substrate scope comprises aliphatic, benzylic, aromatic as well as heteroaromatic ketones, and the resulting acetate products (*R*)-**26** were obtained in high yields and with high enantioselectivities.

A chemoenzymatic dynamic kinetic resolution process recently developed by Bäckvall *et al.* is the combination of a dimeric palladium complex with a heterocyclic carbene ligand as racemization catalyst for axially chiral allenyl alcohols, with a lipase for their acylation towards the synthesis of the corresponding esters in good yields and with high enantioselectivities.[27]

A conceptual alternative concept for a chemoenzymatic DKR based on lipase-catalyzed resolution and metal-catalyzed racemization of alcohols was disclosed by the Akai group (Scheme 15.7).[28,29] Whereas in the previously developed DKRs by the Bäckvall, Kim and Berkessel groups the racemization of the racemic alcohol moiety is based on a redox process through a prochiral ketone intermediate (utilizing, *e.g.* the so-called Shvo catalyst as shown by the Bäckvall group),[17–20] the Akai group applied various vanadium-based catalysts, *e.g.* of type **30**, **31** and **32**, for racemization through reversible

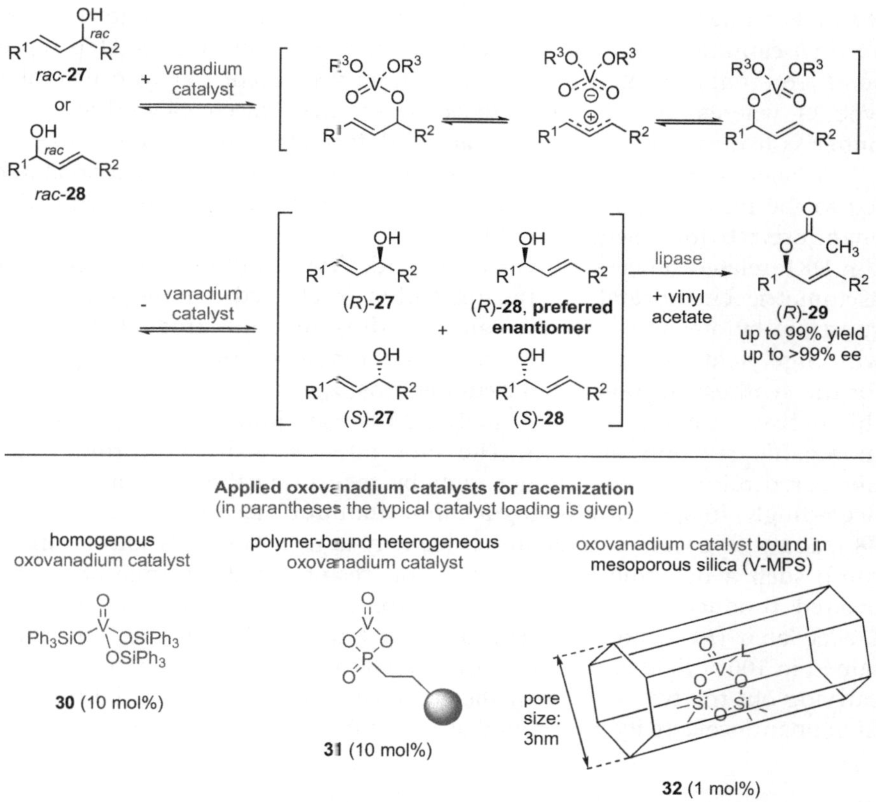

Scheme 15.7 Dynamic kinetic resolution of allylic alcohols through combination of a lipase-catalyzed resolution and vanadium-based racemization catalysts.

'formal removal' of the hydroxy group and, thus, formation of a prochiral cationic intermediate (Scheme 15.7).[28,29] This racemization works particularly well for alcohols which form stable carbocations. This is the case, for example, when utilizing allylic alcohols of type *rac*-**27** and *rac*-**28**. Combination of vanadates with a lipase for acylation of the alcohol moiety (thus forming esters which are resistant against racemization) proceeds efficiently leading to the desired ester products (*R*)-**29** with impressive yields of up to 99% and excellent enantiomeric excesses of up to >99% ee (Scheme 15.7).[28,29] After studying a range of vanadate catalysts, the mesoporous silica catalyst **32** proved very suitable, allowing its use in low catalytic loading of 1 mol% and showing a very low leaching of vanadium with <0.0003%.[28c,29]

Besides metal-based catalysts, racemization in chemoenzymatic DKR processes can also be conducted through the use of a heterogeneous, resin-based Brønsted acid. This has been demonstrated by Deska and Bäckvall *et al.*,[30] who achieved compatibility of the chemo- and biocatalytic reactions through the application of an easy-to-use 'tea bag' in combination with a two-phase solvent system (consisting of isooctane and water) as well as vinyl caprate as a lipophilic acyl donor. The chemocatalytic step represents a reversible isomerization/racemization reaction of an allylic alcohol of type **33**, whereas the enantioselective step is the lipase-catalyzed acylation under consumption of vinyl caprate and formation of the desired ester (*R*)-**35**. Such a process set-up also turned out to prevent side-reactions and led to the formation of the desired products in high yields and with high enantioselectivities (Scheme 15.8).

A DKR-related process, in which the undesired enantiomer formed in an asymmetric chemocatalytic transformation is cleaved and recycled to its substrate by means of enzyme catalysis, thus enabling product formation with high yield and enantioselectivity, was developed by the Moberg group for the synthesis of *O*-acylated cyanohydrins, *e.g.* (*R*)-**39** (Scheme 15.9).[31] In this process, a chiral titanium-catalyst **38** is used jointly with lipase CAL-B in organic reaction medium. The biocatalyst selectively hydrolyzes the undesired minor (*S*)-enantiomer, thereby re-forming the substrate, *e.g.* **36**. Accordingly, in spite of utilizing a chiral titanium-based Lewis acid catalyst **38** leading to a moderate enantioselectivity (*e.g.* 24% ee at room temperature), such a dynamic kinetic resolution results in the formation of the desired products with excellent enantiomeric excess. For example, the *O*-acetylated (*R*)-cyanohydrin (*R*)-**39** formed from (*E*)-2-butenal, **36**, is obtained in 100% yield and with >99% ee (Scheme 15.9). In comparison, when carrying out the two steps with the same catalysts in a sequential mode, a high enantioselectivity is obtained as well, albeit at the expense of the yield.

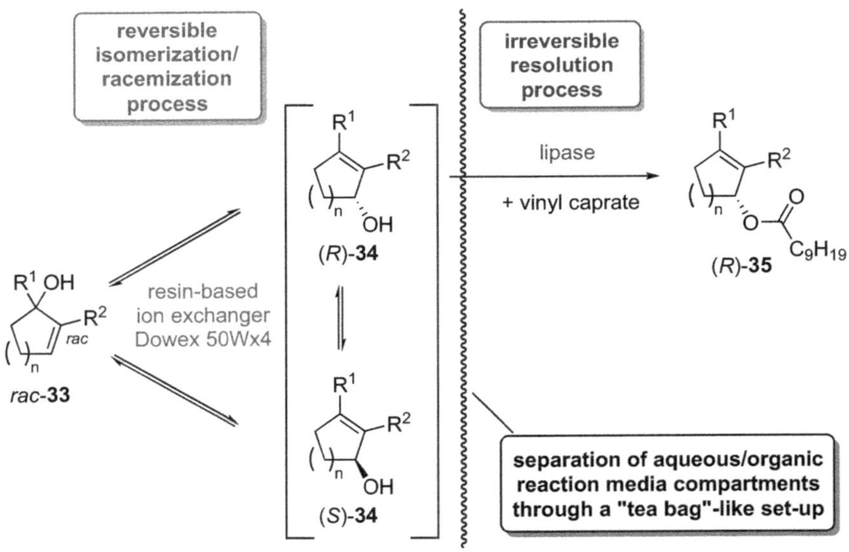

Scheme 15.8 Chemoenzymatic DKR of allylic alcohols *via* combination of a reversible substrate isomerization/racemization with an enzymatic acylation in a 'tea bag'-like system.

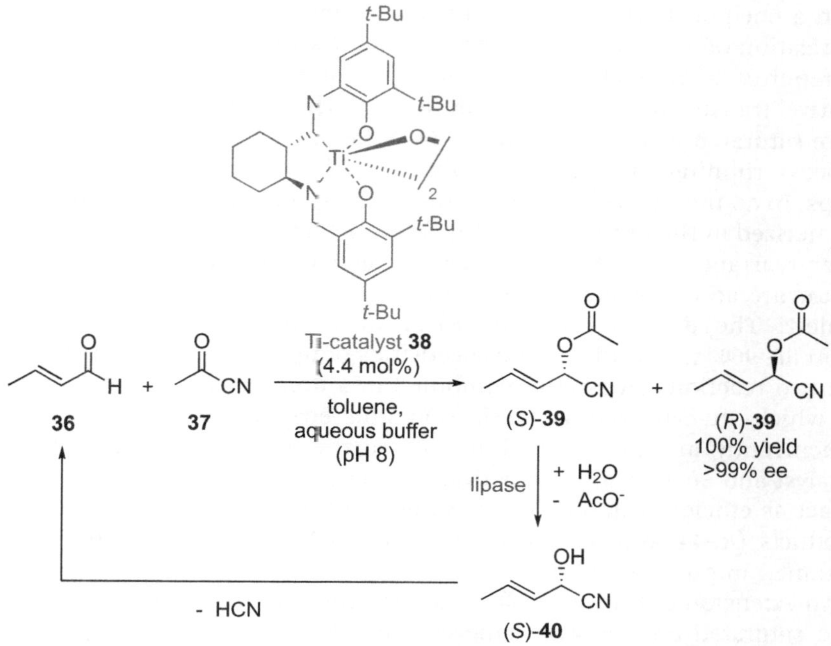

Scheme 15.9 Synthesis of *O*-acylated cyanohydrins in high ee *via* minor enantiomer recycling by means of an enzyme in combination with a chemocatalyst with modest enantioselectivity.

This '*in situ*-minor enantiomer recycling'-concept is complementary to an earlier DKR pioneer work towards *O*-acylated cyanohydrins, which was developed successfully by the Hanefeld group,[32] and based on a reversible cyanohydrin formation in the presence of an Amberlite-type resin (OH^--form). This enabled an efficient racemization, in combination with an efficient lipase-catalyzed enantioselective acylation, leading to the desired *O*-acylated cyanohydrins with excellent conversion and enantioselectivity.

In an extension of this concept, the Moberg group reported a transformation of the minor product enantiomer resulting from the titanium-catalyzed reaction to the major enantiomer by means of cyanoformate as an acylation reagent in combination with a lipase and a carbonic acid anhydrase.[33] The latter enzyme is of importance for removal of carbon dioxide after lipase-catalyzed hydrolysis of the minor enantiomer of the acylated cyanohydrin, thus shifting the reaction in the desired direction.

15.4.3 Domino-type Cascade Processes with Concurrently Running Chemo- and Biocatalytic Steps

Besides racemization with chemocatalysts (see Section 15.4.2), other types of chemocatalytic transformations have been utilized as well for combination with biotransformations towards one-pot processes.

In a chemocatalytic rearrangement reaction, Ru-based catalysts for isomerization of allylic alcohols were combined with alcohol dehydrogenases, as reported by Entrechem researchers. The system was used for the quantitative transformation of racemic allylic alcohols into enantiomerically pure saturated alcohols of type (*R*)-**44** (Scheme 15.10).[34] First, this one-pot process running in aqueous medium was based on sequential reaction steps. In an initial chemocatalytic transformation, allylic alcohols *rac*-**41** are isomerized in the presence of a Ru(IV)-catalyst **42**, thus leading to enones **43** after rearrangement. Subsequently, an alcohol dehydrogenase and a cofactor are added, which enable an asymmetric reduction of the carbonyl moiety. The desired products (*R*)-**44** were obtained in yields ranging from 85–90% and with enantioselectivities of up to >99% ee. Furthermore, the two reactions steps were combined in a domino-type one-pot process, in which Ru-catalyzed isomerization and enzymatic reduction proceed concurrently in aqueous medium. These results demonstrate that metal catalyst and enzyme are compatible with each other and that they are able to act as efficient catalysts in the same 'process window'. The yield of the products (*R*)-**44** were in the range of 60–86%, and these products were obtained in excellent >99% ee.

An extension of this technology towards the conversion of allylic alcohols into saturated amines was achieved recently by the same group through utilization of the efficient Ru-based isomerization catalyst in combination with a transaminase,[35] which runs in a one-pot tandem process in aqueous

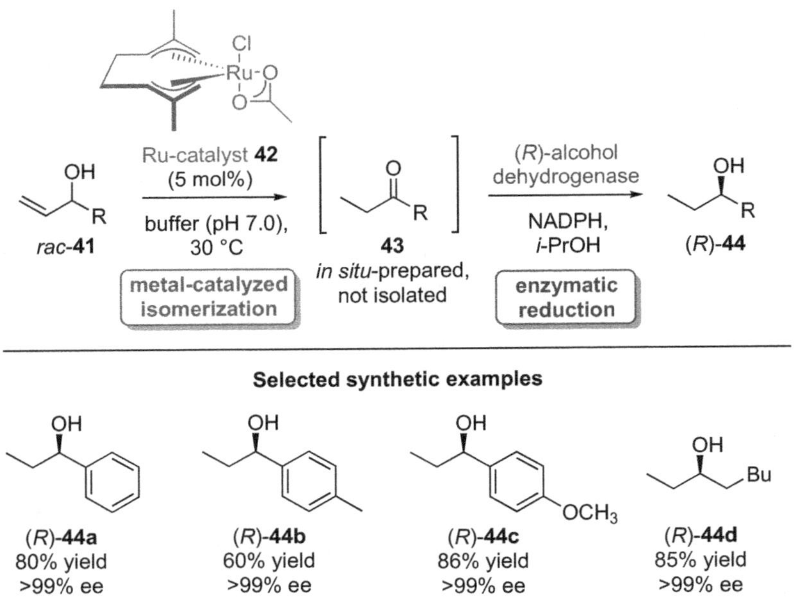

Scheme 15.10 Combined metal-catalyzed isomerization of allylic alcohols with an enzymatic reduction in a domino-type one-pot two-step process.

medium. The resulting amines were obtained in high yields and, in case of chiral products, with excellent enantioselectivities.

Besides isomerization, metal-catalyzed C=C-bond formations have been combined with enzymes in enzymatic domino-type one-pot processes. Such a process has been developed by the Zhao and Hartwig groups, and enabled a synthetic access to epoxy-functionalized fatty acids, *e.g.* **49**, by combining an olefin metathesis with an enzymatic epoxidation reaction (Scheme 15.11).[7e] This one-pot process consists of initial metathesis reaction of an unsaturated fatty acid such as **45** with an alkene (*e.g.* **46**) in the presence of a water-compatible ruthenium catalyst **47**, thus forming an unsaturated fatty acid (*e.g.* **48**), which then undergoes enzymatic epoxidation by means of a P450-monooxygenase. Since the first (metathesis) step is reversible, the combination with the subsequent, irreversible monooxygenase-catalyzed reaction shifts the equilibrium permanently. Accordingly, in this domino-type one-pot process, higher conversion can be obtained compared to a sequential two-step process with work-up of the alkene intermediate. For example, epoxide **49** was obtained in 48% yield in the domino-type one-pot process (as shown in Scheme 15.11) corresponding to a 1.6-fold higher yield compared to the (hypothetical) yield, which was calculated from the two separated reaction steps.

As mentioned in Section 15.3.2, besides metal catalysts, a major focus in recent years has also been on organocatalysts. This is in particular true for the field of asymmetric organocatalysis,[11] and a broad range of reactions turned out to be suitable by means of such methodologies. Accordingly, there has also been a great interest in designing one-pot processes under combination of organo- and biocatalysts. Besides sequential two-step one-pot processes, also combinations of organo- and biocatalysts within domino-type one-pot processes have been established. An early work is the dynamic kinetic resolution of α-amino acid ester based on a successful combination of an aldehyde-catalyzed racemization with an enzymatic hydrolysis.[36] Very recently, the Pietruszka group reported the suitability of laccases as redox enzymes for combination with a chiral organocatalyst in various processes, enabling a one-pot process for the α-arylation of aldehydes (Scheme 15.12).[37] In detail, starting from 1,4-dihydroxybenzene, **50**, oxidation in the presence of a laccase and air leads to the formation of quinone, **51**, which then undergoes an asymmetric organocatalytic Michael reaction with aldehydes. After rearomatization and a final cyclization step, the resulting 2,3-dihydrobenzofuran-2,5-diol products of type (*R*,*R*)-**54** are formed in yields ranging from 52–98% and in most cases with good to very high enantioselectivities of up to 96% ee (Scheme 15.12).

A combination of two asymmetric transformations, consisting of an enantioselective organocatalytic aldol reaction with a diastereoselective biocatalytic reduction in aqueous medium, was reported by Gröger and Berkessel *et al.*, exemplified for the synthesis of the 1,3-diol (1*R*,3*S*)-**58** (Scheme 15.13).[38] The combination of these two types of reaction has been demonstrated earlier to work well within a one-pot process under sequential

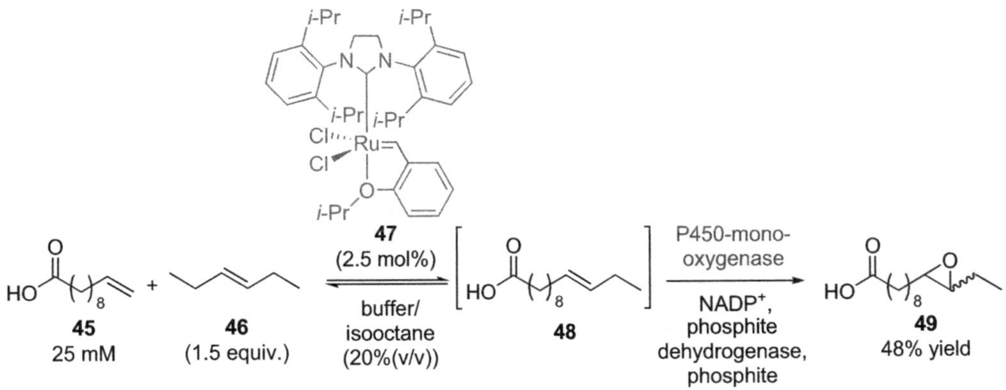

Scheme 15.11 Combination of olefin metathesis and enzymatic epoxidation in a domino-type one-pot process.

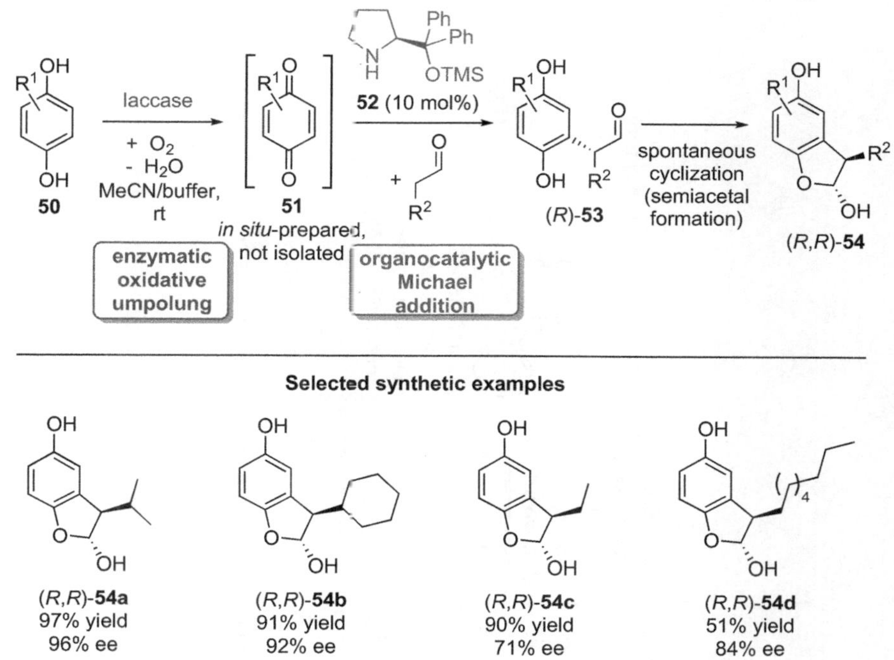

Scheme 15.12 Domino-type one-pot process combining a laccase-catalyzed oxidative umpolung with an asymmetric organocatalytic Michael addition and subsequent cyclization.

conduction of the two reaction steps.[8b,12] This one-pot synthesis also demonstrates the compatibility of the enzyme (used in the second step) with the organocatalyst and other components of the initial aldol reaction. With respect to a domino-type tandem process, however, major tasks consisted of suppressing the unwanted reduction of aldehyde **55** (used in the first step) by the enzyme (utilized for the second step) and to adjust the acetone concentration in a way so that the aldol reaction still proceeds but without leading to undesired enzymatic oxidation of the final 1,3-diol product (1*R*,3*S*)-**58** (with acetone as oxidation agent). A detailed process development work was carried out, leading to the desired combination of asymmetric organocatalytic aldol reaction of **55** with acetone and enzymatic reduction of the intermediate (*R*)-**57** in a domino-type one-pot process in aqueous medium. The process was operated at high substrate concentration of 500 mM of aldehyde **55** and 9 equivalents of acetone.

15.4.4 Related One-pot Processes with Simultaneously Interacting Chemo- and Biocatalysts

While not representing multi-step domino processes, some work in the field of artificial cofactor-regeneration should be mentioned here since it

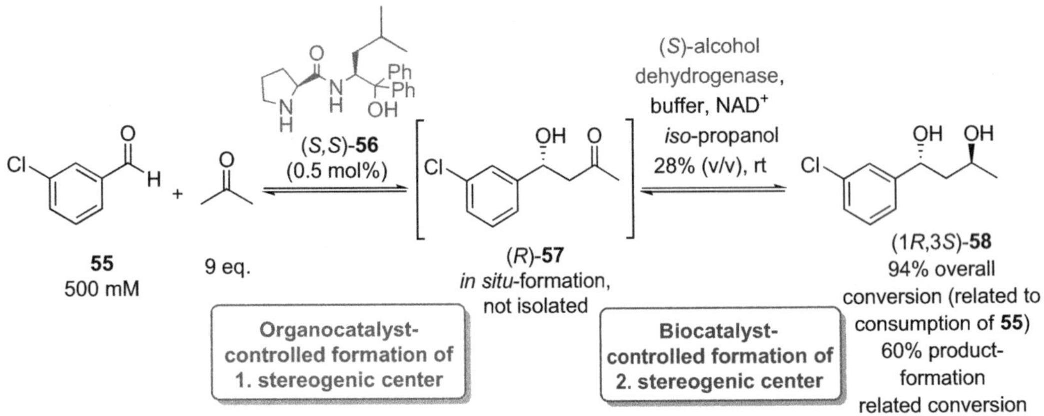

Scheme 15.13 Domino-type one-pot process based on asymmetric organocatalytic aldol reaction and diastereoselective enzymatic reduction of the aldol intermediate.

underlines that metal complexes and enzymes can be combined efficiently, operating simultaneously in the same 'process window'. Such examples include *in situ*-recycling of cofactors consumed in a biotransformation utilizing transition metal complexes based on ruthenium and iron.[25,39] Different types of such processes utilizing redox enzymes for the desired oxidative or reductive transformation in combination with 'artificial' cofactor-regeneration with a chemocatalyst have been developed over the years, and these contributions have been reviewed elsewhere already.[40] In the following, some recent contributions are briefly summarized.

A related process is the laccase-mediated regeneration of palladium(0) to palladium(II). This combination as a 'catalytic system' enabled the oxidation of primary alcohols towards aldehydes under consumption of molecular oxygen and formation of water.[41] In detail, this process, which has been exemplified for the oxidation of veratryl alcohol, consists of an oxidation of the alcohol moiety while a water-soluble Pd(II) component is transformed into a palladium(0) species. In turn, the Pd(II) is *in situ* regenerated through oxidation with a laccase. The reduced form of the laccase is then re-oxidized by means of a four-electron reduction of dioxygen into water. Such integration of the laccase in the oxidation system and its combination with the palladium complex gave a 7-fold higher catalytic efficiency. However, a challenge for future work is to increase turnover number and turnover frequency, which did not exceed TON = 33 and TOF = 1.4 h^{-1}.

It should be added that combination of organo- and biocatalysis not only turned out to be suitable for preparative organic synthesis but also for applications in the field of electrochemistry and electricity generation. This has been shown by Sigman and Minteer *et al.*, who developed a hybrid catalytic system, consisting of 4-amino-TEMPO as an organocatalyst and an oxalate oxidase as enzyme component for a complete electrochemical oxidation of the biofuel glycerol.[42] In this electrochemical oxidation under full transformation of glycerol into carbon dioxide, as many as 16 electrons per molecule of glycerol have been collected.

Furthermore, besides aiming at efficient syntheses of complex organic molecules, a combination of enzymes with chemocatalysts towards domino-type one-pot processes was recently applied in the field of biofuel production, as exemplified by the so-called methanol reforming process running at room temperature.[43] In such a process, methanol is converted into molecular hydrogen, and the Deska and Prechtl groups jointly reported such a process based on an initial enzyme-catalyzed conversion of methanol into formaldehyde hydrate utilizing an oxidase (and a catalase for removal of hydrogen peroxide formed in the oxidation process), followed by a metal-catalyzed transformation under formation of molecular hydrogen in the presence of an air-stable ruthenium complex.

15.5 The Concept of Compartmentation for Combination of Chemo- and Biocatalytic Steps Towards One-pot Syntheses

15.5.1 Introduction and Overview: Compartmentation Strategies in Chemoenzymatic One-pot Syntheses and Size Scale of Compartments

Although a range of one-pot processes with a surprisingly large number of chemo- and biocatalysts tolerant to each other have been developed, a further task consisted of identifying solutions for combination of enzymes and chemocatalysts, which are 'non-tolerant' to each other. One general strategy is compartmentation of such catalysts in order to avoid a direct contact with each other. Interestingly, the size of such compartments can vary broadly along the 'size scale' from microscopic to macroscopic (Scheme 15.14).

15.5.2 Nanoscale Reactor Compartmentation

Taking eukaryotic cells as a 'blue print', the groups of van Hest and Lecommandoux designed an artificial compartment for combining incompatible enzymatic reactions, which require different reaction conditions.[44] The compartment itself is a polymersome-type particle and represents a so-called 'functional cell mimic' (Scheme 15.15). The major difference to previous other work with polymersomes is that this polymersome scaffold contains sub-compartments. In these 'artificial organelles' with one enzyme in each of the separated subunits and the 'cytosolic-like

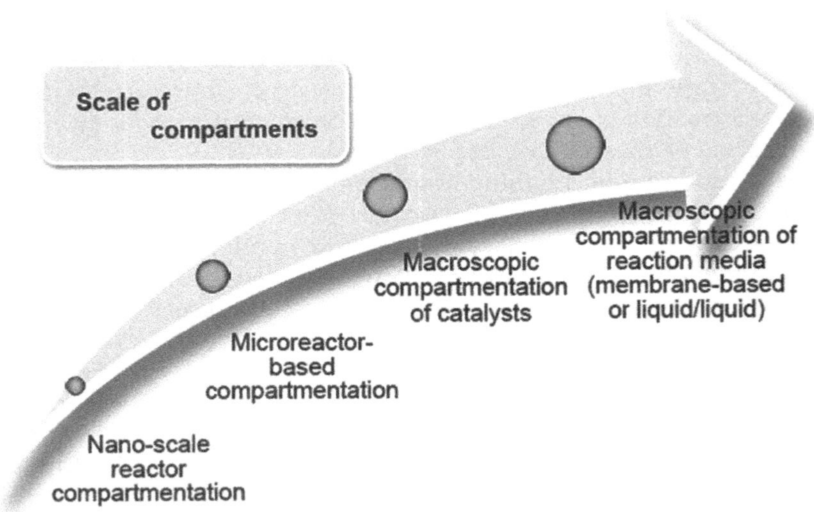

Scheme 15.14 Type of compartments depending on their size.

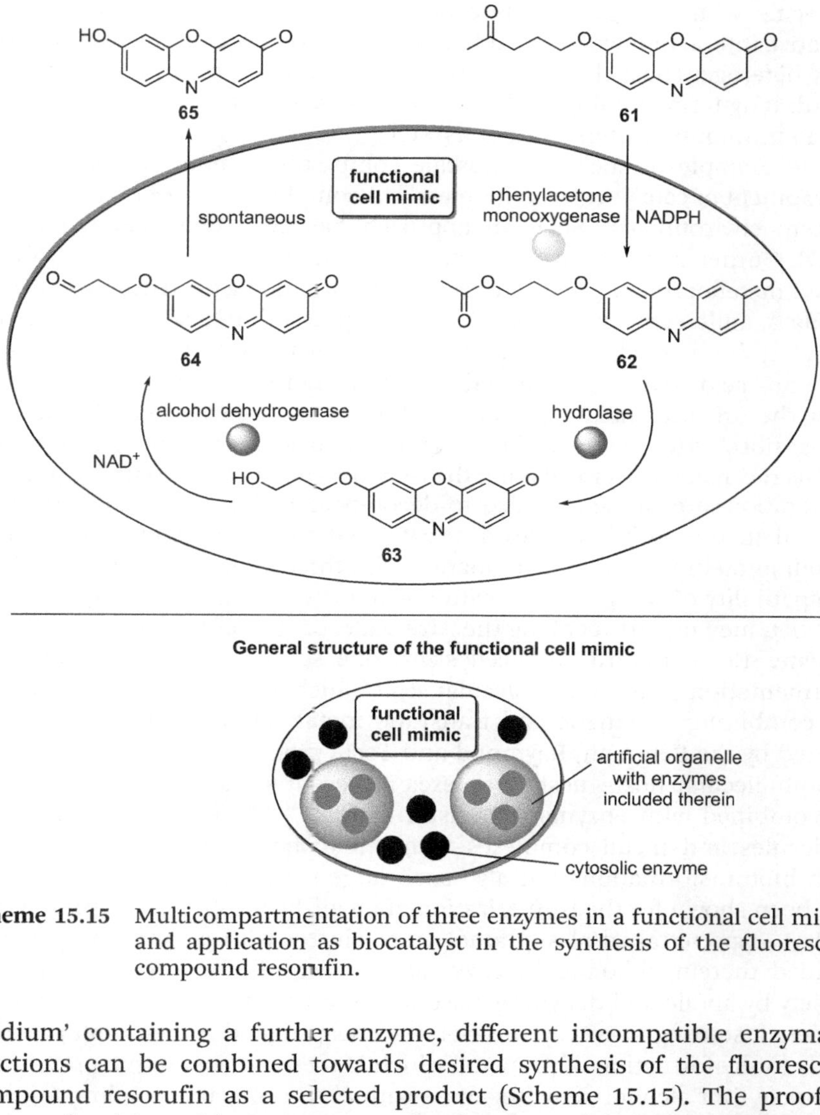

Scheme 15.15 Multicompartmentation of three enzymes in a functional cell mimic and application as biocatalyst in the synthesis of the fluorescent compound resorufin.

medium' containing a further enzyme, different incompatible enzymatic reactions can be combined towards desired synthesis of the fluorescent compound resorufin as a selected product (Scheme 15.15). The proof of concept for this multi-compartmentation has been demonstrated by means of three enzymes in a cascade, namely a phenylacetone monooxygenase, a hydrolase (which is lipase B from *Candida antarctica* or alcalase) and an alcohol dehydrogenase. Although a classic chemocatalyst is not involved in this cascade process, one might regard the enzymes embedded in separate compartments as 'artificial biocatalyst systems'.

It is noteworthy that not only the biocatalyst might be integrated into an immobilized form in a defined compartment, but also a metal component utilized as chemocatalyst. Several tasks are addressed by means of this approach. First, a direct contact of the chemocatalyst with the enzyme (as a molecule with a large size) is avoided, which is beneficial in case of a negative interaction with each other. Second, separation of the immobilized, thus heterogeneous, chemocatalyst from the reaction mixture is simplified. Third, long-term stability of the chemocatalyst might be improved through such an immobilization strategy as well.

One example of such a 'nanoscale compartmentation' is based on the entrapment of catalytically active metals or complexes thereof in a 'molecular protein environment'. Such an approach has been demonstrated by the Ward, Turner and Hollmann groups, and the resulting artificial iridium-based metalloprotein (see Chapter 3) was combined with redox enzymes towards a multi-step one-pot synthesis of L-pipecolinic acid.[45] In the initial step, L-lysine is oxidized to the corresponding imino acid by means of an L-amino acid oxidase, followed by non-enantioselective hydrogenation with the artificial metalloprotein leading to the racemic α-amino acid. A D-enantioselective amino acid oxidase then re-oxidizes only the D-enantiomer of this racemate, thus furnishing the L-enantiomer as the remaining product. In addition, a catalase is added to decompose hydrogen peroxide which is formed in the amino acid oxidase-catalyzed transformation. The artificial metalloprotein is used in combination with three other enzymes, and a high compatibility of such catalysts with one another was demonstrated.

When moving further along the 'size scale' of compartments (according to Scheme 15.14) towards reduced sizes, one strategy consists of the compartmentation on a 'supramolecular scale'. Such a supramolecular approach for combining enzymatic and transition metal catalysis has been demonstrated by the Bergman, Raymond and Toste groups, who demonstrate that supramolecular host–guest complexes represent suitable catalysts that can be combined with enzymatic transformations.[46] Thus, not only did small molecules and metal complexes turn out to be suited for combinations with biotransformations but also such large supramolecular clusters. This has been shown for the isomerization of an allylic alcohol in the presence of a Ga_4L_6-type tetrahedral supramolecular cluster with a Ru(II)-complex embedded therein, followed by enzymatic reduction of the formed carbonyl moiety by an alcohol dehydrogenase within a domino-type one-pot process. In detail, starting from allyl alcohol, the encapsulated ruthenium(II) catalyst leads to the formation of propenal which is then reduced in the presence of an horse liver alcohol dehydrogenase with *in situ*-cofactor recycling utilizing a formate dehydrogenase, leading to 61% of the desired alcohol over the two-step one-pot process. It should be added that the same groups also reported other one-pot domino-type processes with supramolecular host–guest complexes and enzymes, such as the combination of a lipase-catalyzed enantioselective hydrolysis of racemic allenic acetates and subsequent diastereoselective cyclization of the formed allene intermediates through hydroalkoxylation by means of an Au(I)-encapsulated Ga_4L_6-type supramolecular cluster.[46]

Another strategy for immobilization of the chemocatalyst component for applications in chemoenzymatic one-pot processes has been developed by

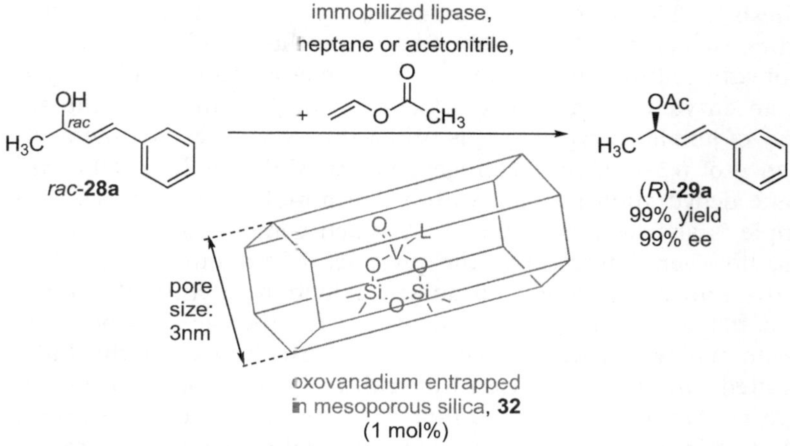

Scheme 15.16 Dynamic kinetic resolution of allylic alcohols through combination of a heterogeneous vanadium-type mesoporous catalyst with an immobilized lipase.

the Akai group by combining a heterogeneous vanadium-type mesoporous catalyst with an immobilized lipase towards a dynamic kinetic resolution of allylic alcohols.[28c,29] A selected example is shown in Scheme 15.16. In detail, an oxovanadium complex was embedded into a mesoporous silica matrix furnishing the heterogeneous catalyst **32** (see also Section 15.4.2). The size of this matrix enables the substrate to reach the integrated vanadium catalyst, while avoiding direct contact with the lipase and thus, a deactivation of the enzyme caused by interaction with the vanadium complex. Combining chemo- and biocatalyst under those compartmentation conditions led to a very efficient resolution process operating at low catalyst loading of 1 mol%. Furthermore, the heterogenized chemocatalyst was successfully recycled, giving yields >99% and enantioselectivities of 99% ee over 6 reaction cycles, whereas yield slightly decreased to 85% in the 7th reaction cycle.[28c,29]

15.5.3 Process Engineering Strategies in Chemoenzymatic One-pot Syntheses: Use of One Liquid Medium and Compartmentation in Different Reactor Segments in a Flow Process

A review highlighting the state of the art in the field of chemoenzymatic synthesis in continuously running flow processes has been published by the Liese group.[47] In detail, examples demonstrating the suitability of the concepts '*in vitro*-type single-reactor processes' and '*in vitro*-type cascade-reactor processes' for combination of chemocatalytic or classic chemical reactions with biotransformations are given therein.

The concept of catalyst compartmentation within a flow chemistry-process has been applied successfully by the Sieber group for a straightforward synthesis of 2-keto-3-deoxy sugar acids, *e.g.*, **74**, starting from the corresponding sugars (*e.g.*, **72**), which is based on the transformation of suitable carbohydrates through combination of a heterogeneous gold-catalyzed step with an enzymatic reaction step (Scheme 15.17).[48] In detail, the aldehyde moiety of an aldose-type sugar is oxidized to the corresponding acid in the presence of 0.5% of the heterogeneous Au/Al_2O_3 catalyst, followed by enzymatic dehydration reaction with a thermostable dehydratase. A selected example is shown in Scheme 15.17. When combining these two reaction steps, however, interestingly the presence of protein buffer salts had a negative impact on the chemocatalytic oxidation reaction. It is noteworthy that a range of reasons were identified for the lack of incompatibility between the two reaction types: the negative impact of the buffer and undesired binding of released iron–sulfur-cluster or other low-molecular-weight contaminants from the protein as well as the enzyme itself on the catalyst surface. *Vice versa*, the side-product hydrogen peroxide formed in the chemocatalytic reaction deactivates the enzyme. In addition, process windows are quite different in terms of pH and temperature for both catalysts. Since a one-pot process turned out to be not possible due to such incompatibility issues, as a solution a compartmented continuously running flow process was designed. First, chemo- and biocatalysts were located in separated reactor compartments, and in between the filtrate from the gold-catalyzed oxidation process was collected in a second reaction vessel, which contained the catalase in order to decompose the hydrogen peroxide formed in the initial reaction. The use of different reactor compartments also enabled adjustment of the reaction parameters (pH and temperature) for each reaction individually. By means of this combined fed-batch and continuous-flow strategy, the desired products were formed efficiently. For example, L-arabinose (**72**) was oxidized with a conversion of about 90% to L-arabonate **73**. The subsequent dehydration then gave the desired product 2-keto-3-deoxy arabonate (**74**) in 58% yield (Scheme 15.17).

An extension of the dynamic kinetic resolution concept of Akai, consisting of a lipase-catalyzed resolution in combination with a vanadium-catalyzed racemization, towards a continuously running process with the two catalysts immobilized in different compartments, was reported by the de Souza group (Scheme 15.18).[49] This flow-process was exemplified for the DKR-related

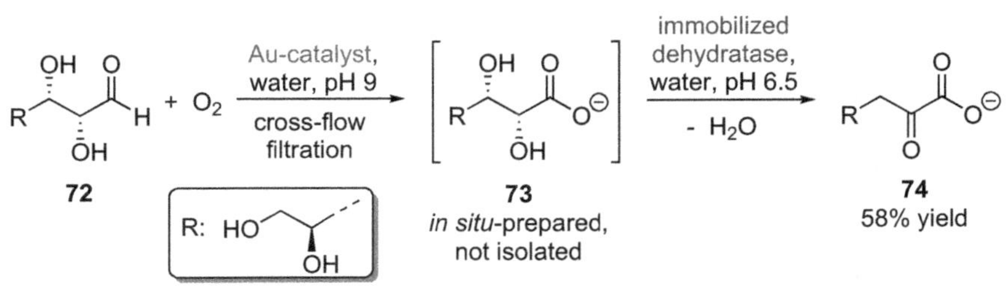

Scheme 15.17 Combination of gold-catalysis and enzymatic hydrolysis starting from sugar substrates under catalysts compartmentation.

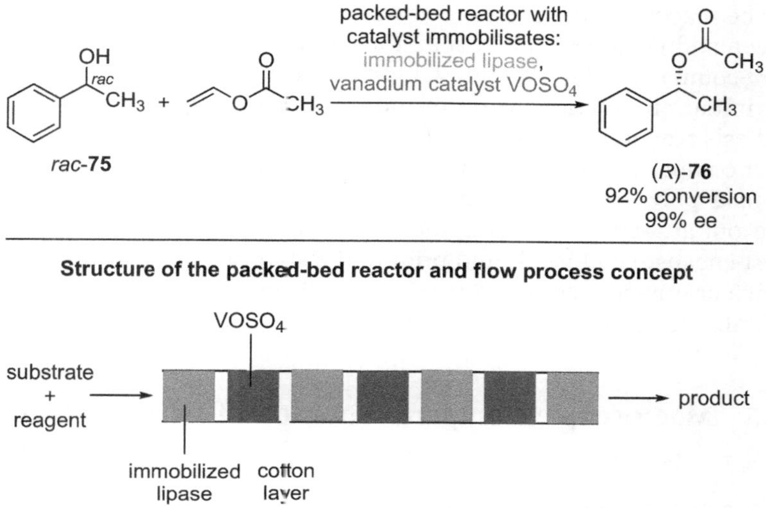

Scheme 15.18 Flow DKR process with combined chemocatalyst and enzyme embedded in an alternate mode within a packed-bed reactor.

esterification of racemic 1-phenylethanol (*rac*-**75**) and is based on the use of a single packed-bed reactor, in which both catalysts, namely the lipase from *Candida antarctica* B (CAL-B) and the racemization catalyst $VOSO_4$ are included. A major task has been to overcome the incompatibility of the lipase CAL-B with the vanadyl oxide $VOSO_4$. Accordingly, the construction of the packed-bed reactor is based on four layers of immobilized lipase and three layers of $VOSO_4$, which were alternately disposed along a column and separated by a thin (cotton) layer from each other. By means of this strategy, in which alternately kinetic resolution and racemization steps are carried out in the four and three separated layers, respectively, the resulting ester (*R*)-**76** was obtained with both high conversion (92%) and enantioselectivity (99% ee), at a selectivity of 68% and a residence time of 5.2 min when utilizing vinyl acetate as an acyl donor.

15.5.4 Microscopic Compartmentation of Catalysts in Polymer Beads/Particles

In a further compartmention strategy, polymer balls with entrapped water phase were used. These compartments can be regarded as microscopic compartments showing a decreased size of the compartment volumes compared with those of the thimble technology described below. In such a set-up, the reaction compartments are separated by a membrane and can now differ with respect to the type of liquid phase. For example, whereas the major liquid phase is organic, the polymer balls might contain an aqueous phase. Thus, enzymes can be ideally immobilized in such polymer balls and separated from chemocatalysts, which themselves require an organic phase and might be inactive in the presence of water. Such a complementary challenge has been addressed by Kourist *et al.* for the two-step one-pot transformation of *para*-coumaric acid (**77**) into 4,4′-dihydroxy-*trans*-stilbene (**79**) by means of an initial enzymatic decarboxylation and a subsequent metal-catalyzed metathesis reaction (Scheme 15.19).[50] As the metathesis catalysts turned out to be incompatible with an aqueous phase, encapsulation of the decarboxylase in a polyvinylalcohol/polyethylene glycol (PVA/PEG) cryogel particles turned out attractive. By drying the organic phase after removal of the biocatalyst-encapsulated beads and prior to the start of the metathesis reaction, an efficient one-pot cascade process under formation of the desired substituted stilbene **79** was realized in 90% yield (Scheme 15.19).

15.5.5 Macroscopic Compartmentation of Reaction Media

15.5.5.1 Membrane-based Solutions

A strategy through macroscopic compartmentation, which has been recently broadly used in the field of chemoenzymatic one-pot synthesis, is based on the use of so-called polydimethylsiloxane (PDMS)-thimbles for separating different reaction phases from each other. The first example of chemoenzymatic one-pot synthesis with such a technique was reported by the Gröger group, combining a Wacker oxidation of styrene utilizing a Pd/Cu-based catalyst system with a subsequent enzymatic reduction in order to transform styrenes in a one-pot two-step process enantioselectively into 1-phenylethanols (Scheme 15.20).[51] A compartmentation turned out to be necessary due to the strong negative impact of the copper catalyst components (for the initial Wacker oxidation) on the enzyme utilized in the second reaction step. As a solution, a one-pot process based on a compartmentation of the palladium/copper-based catalyst system (needed for the Wacker oxidation) and an alcohol dehydrogenase for the subsequent reduction enabled an efficient formation of the desired target products (*R*)-**82** with high conversion and enantioselectivities in many cases (Scheme 15.20). In detail, the Wacker oxidation was carried out in the interior of a PDMS thimble and after passing the polydimethylsiloxane (PDMS)-membrane, the formed ketone intermediate was enzymatically reduced towards the desired chiral alcohols. Both reactions proceed in an aqueous medium, but only the hydrophobic components such as ketones **81** can pass the membrane to the exterior solution, whereas the metal components are kept in the interior phase. In addition, recycling of the palladium and copper components remaining in the interior phase was demonstrated with 15 reaction cycles.

This technology for separation of reaction compartments has been subsequently applied by other groups. A one-pot combination of a halogenase-catalyzed aryl halogenation with a subsequent palladium-catalyzed functionalization through Suzuki cross-coupling reaction was described by Micklefield and Greaney *et al.*, thus allowing selectively transformation of one of two similar unactivated C–H positions.[52] This chemoenzymatic

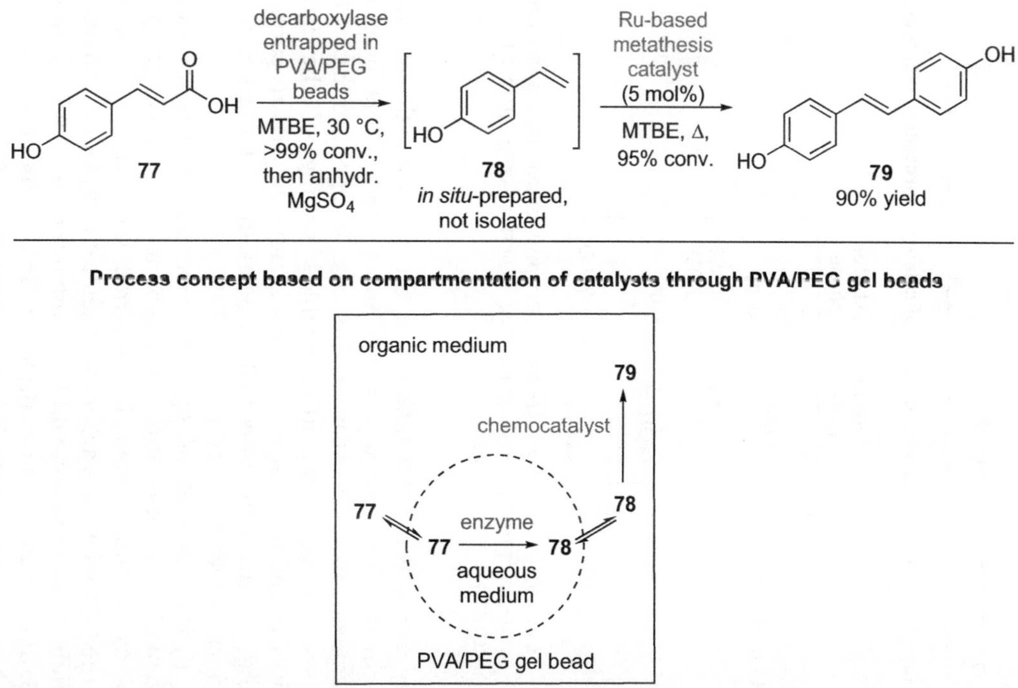

Scheme 15.19 Compartmentation of a decarboxylase *via* encapsulation in PVA/PEG cryogel beads and combination with a metathesis reaction for the synthesis of a substituted stilbene.

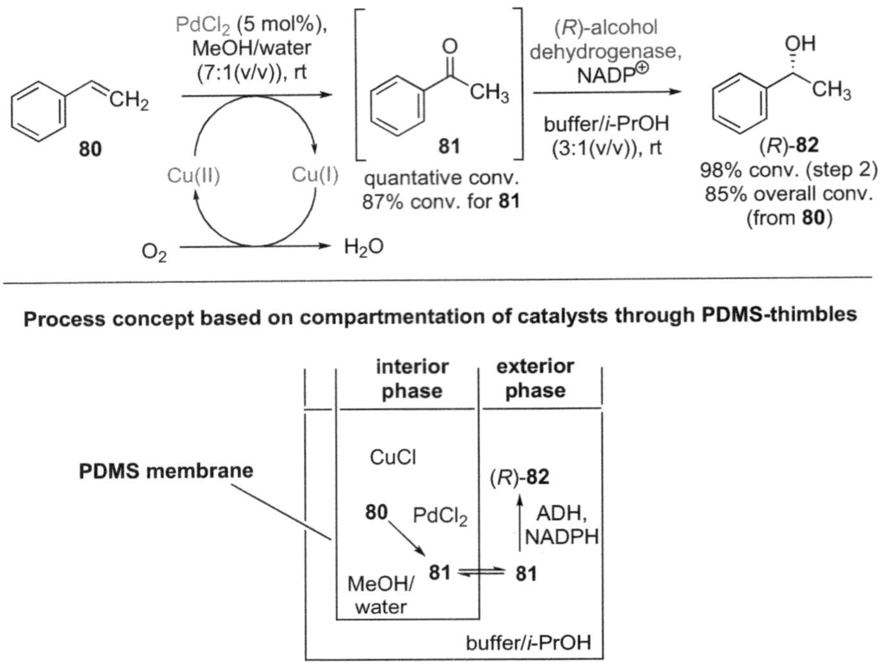

Scheme 15.20 Combination of Wacker-oxidation and biocatalytic reduction in a one-pot process with separated reaction compartments using a 'thimble technology'.

methodology under compartmentation of the two catalytic systems was applied in the synthesis of several aryl-substituted indole heterocycles running in aqueous medium. Again in this case the application of a PDMS-membrane-based compartmentation enabled solving of the issue of incompatibility of the biocatalytic and palladium-catalyzed reaction steps, thus avoiding the need for intermediate work-up and purification. The desired products where obtained in up to 74% yield.

A formal hydroamination of styrene, consisting of Wacker oxidation and an enzymatic transamination, was reported by the Gröger group.[53] In this process, at first styrene is converted into acetophenone in the presence of the Pd/Cu-based Wacker oxidation catalyst, followed by an *in situ* transformation of acetophenone under formation of 1-phenylethylamine in the presence of the transaminase. Under optimized conditions, the desired product is obtained with up to 93% conversion and 99% ee.

In addition, as described in Section 15.4.2, the application of an easy-to-use 'tea bag' for separation of chemo- and biocatalytic reactions (which can be also considered as a membrane-like compartmentation) in combination with a two-phase solvent system consisting of isooctane and water has been reported by Deska and Bäckvall *et al.*, and led to an efficient chemoenzymatic DKR process.[30]

15.5.5.2 Liquid/Liquid Solutions

Liquid compartments can also be considered as compartmentation tools for realizing challenging chemoenzymatic one-pot processes without the need for intermediate isolation and purification. This has been demonstrated by the Liese, Kara and Faber groups for the two-step transformation of ferulic acid into the fragrance compound 4-ethylguaiacol under extraction of the intermediate and direct derivatization thereof in a further reactor (Scheme 15.21).[54] The initial step represents an enzymatic oxidation of ferulic acid (**83**) by means of a phenolic acid decarboxylase, leading to 4-vinylguaiacol (**84**) as an intermediate. Subsequent metal-catalyzed hydrogenation with Pd/C furnishes the desired target compound **85.** Since, however, the applied enzyme faces product inhibition and deactivation by the formed product **84**, continuous product removal by means of a second organic phase was carried out. In such an organic phase, the hydrogenation of **84** can also be conducted. Thus, reaction engineering was carried out based on extraction of product **84** formed in the biotransformation, and direct hydrogenation of **84** being present in the organic layer after phase separation. Note that this organic layer contains the intermediate vinylguaiacol as a substrate for hydrogenation but not the polar hydrophilic ferulic acid. In order to minimize the amount of intermediate **84** in the aqueous phase, the extracted organic phase containing **84** is continuously pumped into a second reactor, where the organic solvent is distilled off. Whereas the distilled organic solvent is re-used for a next extraction process, the intermediate **84** remains in the second reactor as a crude product. By means of this strategy, 92% conversion at a substrate loading of 33 $g\,L^{-1}$ of ferulic acid (170 mM) was achieved. Subsequent hydrogenation furnishes the desired product **85**. By combining these two reactions in such a two-step

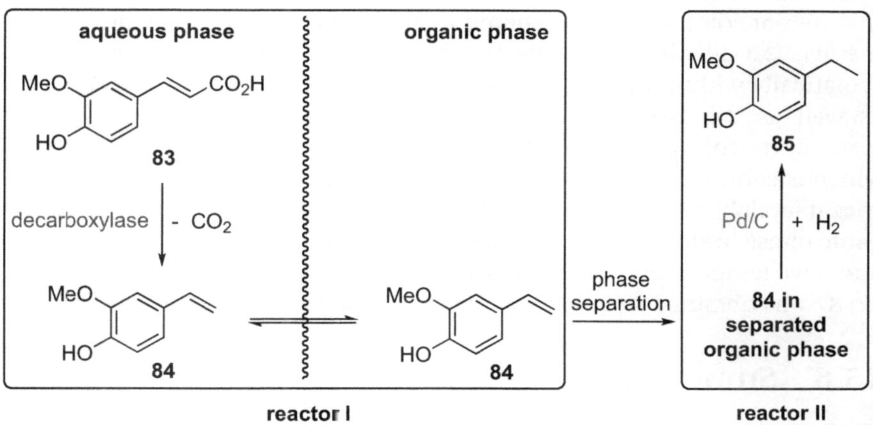

Scheme 15.21 Two-phase systems as a concept for compartmentation in chemoenzymatic synthesis without intermediate isolation.

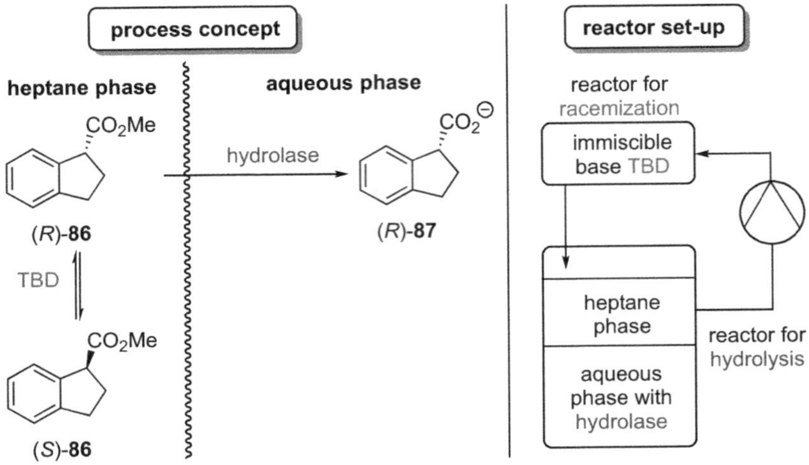

Scheme 15.22 Compartmentation of a TBD-based racemization in an organic phase and an enzymatic hydrolysis in a two-phase-system for a DKR-related carboxylate synthesis (TBD: 1,5,7-triazabicyclo[4.4.0]dec-5-ene).

two-reactor process strategy without intermediate purification, the desired product **85** was obtained in 70% overall yield.

Whereas a broad diversity of dynamic kinetic resolutions for alcohols and amines through combination of chemo- and biocatalysts exists (see Section 15.4.2), analogous processes for racemic carboxylates as substrates are rare. This is due, among other reasons, to the fact that base-catalyzed racemization at high pH is hardly compatible with the preferred reaction conditions of most hydrolases. A strategy to circumvent this problem has been demonstrated by the Pietruszka group through development of a formal dynamic kinetic resolution, in which the two catalytic reactions run in different phases and in which the catalytic components are separated in different compartments (Scheme 15.22).[55] In detail, the racemic substrate methyl 2,3-dihydro-1*H*-indene-1-carboxylate (*rac*-**86**) is hydrolyzed enzymatically within a two phase system and the resulting carboxylate (*R*)-**87** as well as the enzyme are kept in aqueous medium of this two-phase system. In contrast, the ester substrate remains in the organic phase, which is continuously pumped through a syringe packed with the base 1,5,7-triazabicyclo[4.4.0]dec-5-ene (TBD), which remains immobilized in the organic phase under the chosen reaction conditions, such as hexane as solvent and low temperature. By means of this process, the desired carboxylate (*R*)-**87** was obtained in 95% yield and high enantiomeric excess of >96% ee.

15.6 Summary

In recent years the combination of chemo- and biocatalysts towards multistep chemoenzymatic one-pot syntheses has gained increasing interest due to a range of benefits. For example, by direct transformation of the formed intermediates without purification steps, a number of time-consuming process unit operations steps can be avoided and overall amount of solvent consumption can be significantly reduced. A major challenge in developing such processes is to achieve compatibility of the different reaction steps with one another, which often run preferably under different 'process windows'. However, promising proof of concept work demonstrated the suitability of combining chemo- and bio-catalysts towards one-pot syntheses as an attractive process concept. In addition, several emerging fields in this research area arose and have been addressed by numerous groups, such as the development of domino-type tandem reactions, in which chemo- and biocatalytic reactions proceed simultaneously, as well as the development of solutions for one-pot processes under combination of chemo- and bio-catalysts, which turned out to be incompatible. A major concept being used for such cases is compartmentation of the catalysts within one reactor, and it is noteworthy that the size of such compartments can vary broadly from the molecular to the nano and macroscopic scale. Thus, synthetically attractive solutions have allowed chemoenzymatic one-pot processes based on (i) sequential-type one-pot multi-step processes, (ii) domino-type one-pot multi-step processes, and (iii) compartmentation of catalysts within either one reactor or multi-reactor systems.

References

1. L. F. Tietze, G. Brasche and K. M. Gericke, *Domino Reactions in Organic Synthesis*, Wiley-VCH, Weinheim, 2006.
2. G. P. Taber, D. M. Pfisterer and J. C. Colberg, *Org. Process Res. Dev.*, 2004, **8**, 38.
3. *Synthetic Biology: Industrial and Environmental Applications*, ed. M. Schmidt, Wiley-VCH, Weinheim, 2012.
4. S. P. France, L. J. Hepworth, N. J. Turner and S. L. Flitsch, *ACS Catal.*, 2017, 7, 710.
5. (a) C. A. Denard, J. F. Hartwig and H. Zhao, *ACS Catal.*, 2013, **3**, 2856; (b) J. H. Schrittwieser, S. Velikogne, M. Hall and W. Kroutil, *Chem. Rev.*, 2018, **118**, 270; (c) F. Rudroff, M. D. Mihovilovic, H. Gröger, R. Snajdrova, H. Iding and U. T. Bornscheuer, *Nature Catal.*, 2018, **1**, 12.
6. (a) H. Gröger and W. Hummel, *Curr. Opin. Chem. Biol.*, 2014, **19**, 171; (b) H. Gröger and W. Hummel, in *Cascade Biocatalysis*, ed. S. Riva and W.-D. Fessner, Wiley-VCH, Weinheim, 2014, chapter 19, p. 427; (c) H. Gröger, in *Cooperative Catalysis*, ed. R. Peters, Wiley-VCH, Weinheim, 2015, chapter 11, p. 325; (d) H. Gröger and W. Hummel, in *Science of Synthesis, Biocatalysis in Organic Synthesis*, ed. K. Faber, W.-D. Fessner, and N. J. Turner, vol. 3, Thieme, Stuttgart, 2015, chapter 3.8.2, p. 491.
7. Selected examples: (a) C. Simons, U. Hanefeld, I. W. C. E. Arends, T. Maschmeyer and R. A. Sheldon, *Top. Catal.*, 2006, **40**, 35; (b) E. Burda, W. Hummel and H. Gröger, *Angew. Chem.*, 2008, **120**, 9693; *Angew.*

Chem., Int. Ed., 2008, **47**, 9551; (c) A. Prastaro, P. Ceci, E. Chiancone, A. Boffi, R. Cirilli, M. Colone, G. Fabrizi, A. Stringaro and S. Cacchi, *Green Chem.*, 2009, **11**, 1929; (d) K. Tenbrink, M. Seßler, J. Schatz and H. Gröger, *Adv. Synth. Catal.*, 2011, **353**, 2363; (e) C. A. Denard, H. Huang, M. J. Bartlett, L. Lu, Y. Tan, H. Zhao and J. F. Hartwig, *Angew. Chem.*, 2014, **126**, 475; *Angew. Chem., Int. Ed.*, 2014, **53**, 465.

8. Selected examples: (a) M. Edin, J.-E. Bäckvall and A. Córdova, *Tetrahedron Lett.*, 2004, **45**, 7697; (b) K. Baer, M. Kraußer, E. Burda, W. Hummel, A. Berkessel and H. Gröger, *Angew. Chem.*, 2009, **121**, 9519; *Angew. Chem., Int. Ed.*, 2009, **48**, 9355; (c) A. Berkessel, I. Jurkiewicz and R. Mohan, *ChemCatChem*, 2011, **3**, 319.
9. C. Palumbo, E. E. Ferrandi, C. Marchesi, D. Monti, S. Riva, R. Psaro and M. Guidotti, *ChemistrySelect*, 2016, **1**, 1795.
10. Y. Wang, M. J. Bartlett, C. A. Denard, J. F. Hartwig and H. Zhao, *ACS Catal.*, 2017, 7, 2548.
11. (a) A. Berkessel and H. Gröger, *Asymmetric Organocatalysis-from Biomimetic Concepts to Powerful Methods for Asymmetric Synthesis*, Wiley-VCH, Weinheim, 2005; (b) H. Pellissier, *Tetrahedron*, 2007, **63**, 9267; (c) *Science of Synthesis: Asymmetric Organocatalysis*, ed. B. List and K. Maruoka, vol. 1–2, Thieme, Stuttgart, 2012.
12. G. Rulli, N. Duangdee, K. Baer, W. Hummel, A. Berkessel and H. Gröger, *Angew. Chem.*, 2011, **123**, 8092; *Angew. Chem., Int. Ed.*, 2011, **50**, 7944.
13. M. Heidlindemann, G. Rulli, A. Berkessel, W. Hummel and H. Gröger, *ACS Catal.*, 2014, **4**, 1099.
14. R. C. Simon, E. Busto, J. H. Schrittwieser, J. H. Sattler, J. Pietruszka, K. Faber and W. Kroutil, *Chem. Commun.*, 2014, **50**, 15669.
15. (a) M. Makkee, A. P. G. Kieboom, H. van Bekkum and J. A. Roels, *J. Chem. Soc., Chem. Commun.*, 1980, 930; (b) M. Makkee, A. P. G. Kieboom and H. van Bekkum, *Carbohydr. Res.*, 1985, **138**, 237.
16. (a) J. V. Allen and J. M. J. Williams, *Tetrahedron Lett.*, 1996, **37**, 1859; (b) P. M. Dinh, J. A. Howarth, A. R. Hudnott, J. M. J. Williams and W. Harris, *Tetrahedron Lett.*, 1996, **37**, 7623.
17. Reviews: (a) O. Pàmies and J.-E. Bäckvall, *Chem. Rev.*, 2003, **103**, 3247; (b) I. Hussain and J.-E. Bäckvall, in *Enzyme Catalysis in Organic Synthesis*, ed. K. Drauz, H. Gröger and O. May, 3rd edn, vol. 2, Wiley-VCH, Weinheim, 2012, chapter 43, p. 1777.
18. Selected examples: L. E. Larsson, B. A. Persson and J. E. Bäckvall, *Angew. Chem.*, 1997, **109**, 1256; *Angew. Chem., Int. Ed. Engl.*, 1997, **36**, 1211; (b) B. A. Persson, A. L. E. Larsson, M. Le Ray and J. E. Bäckvall, *J. Am. Chem. Soc.*, 1999, **121**, 1645.
19. Selected example: J. H. Choi, Y. H. Kim, S. H. Nam, S. T. Shin, M. J. Kim and J. Park, *Angew. Chem.*, 2002, **114**, 2479; *Angew. Chem., Int. Ed.* 2002, **41**, 2373.
20. A. Berkessel, M. L. Sebastian-Ibarz and T. N. Müller, *Angew. Chem.*, 2006, **118**, 6717; *Angew. Chem., Int. Ed.*, 2006, **45**, 6567.

21. (a) M. Ahmed, T. Kelly and A. Ghanem, *Tetrahedron*, 2012, **68**, 6781; (b) P. Hoyos, V. Pace and A. R. Alcántara, *Adv. Synth. Catal.*, 2012, **354**, 2585.
22. R. K. Rej, A. Jana and S. Nanda, *Tetrahedron*, 2014, **70**, 2634.
23. G. Ma, Z. Xu, P. Zhang, J. Liu, X. Hao, J. Ouyang, P. Liang, S. You and X. Jia, *Org. Process Res. Dev.*, 2014, **18**, 1169.
24. C. Palo-Nieto, S. Afewerki, M. Anderson, C.-W. Tai, P. Berglund and A. Córdova, *ACS Catal.*, 2016, **6**, 3932.
25. H. Maid, P. Böhm, S. M. Huber, W. Bauer, W. Hummel, N. Jux and H. Gröger, *Angew. Chem.*, 2011, **123**, 2445; *Angew. Chem., Int. Ed.*, 2011, **50**, 2397.
26. O. El-Sepelgy, A. Brzozowska and M. Rueping, *ChemSusChem*, 2017, **10**, 1664.
27. J. Deska, C. del Pozo, Ochoa and J.-E. Bäckvall, *Chem. – Eur. J.*, 2010, **16**, 4447.
28. S. Akai, K. Tanimoto, Y. Kanao, M. Egi, T. Yamamoto and Y. Kita, *Angew. Chem.*, 2006, **118**, 2654; *Angew. Chem., Int. Ed.*, 2006, **45**, 2592; (b) S. Akai, R. Hanada, N. Fujiwara, Y. Kita and M. Egi, *Org. Lett.*, 2010, **12**, 4900; (c) M. Egi, K. Sugiyama, M. Saneto, R. Hanada, K. Kato and S. Akai, *Angew. Chem.*, 2013, **125**, 3742; *Angew. Chem. Int. Ed.*, 2013, **52**, 3654.
29. Review: S. Akai, *Chem. Lett.*, 2014, **43**, 746
30. C. M. Sapu, T. Görbe, R. Lihammar, J.-E. Bäckvall and J. Deska, *Org. Lett.*, 2014, **16**, 5952.
31. A. Laurell and C. Moberg, *Eur. J. Org. Chem.*, 2011, 3980.
32. (a) Y.-K. Li, A. J. Stratthof and U. Hanefeld, *Tetrahedron: Asymmetry*, 2002, **13**, 739; (b) L. Veum and U. Hanefeld, *Tetrahedron: Asymmetry*, 2004, **15**, 3707.
33. A. Laurell Nash, K. Widyan and C. Moberg, *ChemCatChem*, 2014, **6**, 3314.
34. N. Rios-Lombardia, C. Vidal, E. Liardo, F. Moris, J. Garcia-Alvarez and J. Gonzalez-Sabin, *Angew. Chem.*, 2016, **128**, 8833; *Angew. Chem., Int. Ed.* 2016, **55**, 8691.
35. N. Rios-Lombardia, C. Vidal, M. Cocina, F. Moris, J. Garcia-Alvarez and J. Gonzalez-Sabin, *Chem. Commun.*, 2015, **51**, 10937.
36. (a) S.-T. Chen, W.-H. Huang and K.-T. Wang, *J. Org. Chem.*, 1994, **59**, 7580; (b) T. Riermeier, U. Dingerdissen, P. Groß, W. Holla, M. Beller and D. Schichl, *DE Pat. Appl. DE 19955283*, 2001; (c) V. Zimmermann, M. Beller and U. Kragl, *Org. Process Res. Dev.*, 2005, **10**, 622.
37. S. Suljic, J. Pietruszka and D. Worgulla, *Adv. Synth. Catal.*, 2015, **357**, 1822.
38. G. Rulli, N. Duangdee, W. Hummel, A. Berkessel and H. Gröger, *Eur. J. Org. Chem.*, 2017, 812–817.
39. G. Hilt, B. Lewall, G. Montero, J. H. P. Utley and E. Steckhan, *Liebigs Ann.*, 1997, 2289.
40. (a) J. C. Lewis, *ACS Catal.*, 2013, **3**, 2954; (b) T. Heinisch and T. R. Ward, *Acc. Chem. Res.*, 2016, **49**, 1711.

41. Y. Mekmouche, L. Schneider, P. Rousselot-Pailley, B. Faure, A. J. Simaan, C. Bochot, M. Réglier and T. Tron, *Chem. Sci.*, 2015, **6**, 1247.
42. D. P. Hickey, M. S. McCammant, F. Giroud, M. S. Sigman and S. D. Minteer, *J. Am. Chem. Soc.*, 2014, **136**, 15917.
43. L. E. Heim, D. Thiel, C. Gedig, J. Deska and M. H. G. Prechtl, *Angew. Chem.*, 2015, **127**, 10447; *Angew. Chem., Int. Ed.* 2015, **54**, 10308.
44. R. J. R. W. Peters, M. Marguet, S. Marais, M. W. Fraaije, J. C. M. van Hest and S. Lecommandoux, *Angew. Chem.*, 2014, **126**, 150; *Angew. Chem., Int. Ed.*, 2014, **53**, 146.
45. V. Köhler, Y. M. Wilson, M. Dürrenberger, D. Gishlieri, E. Churakova, T. Quinto, L. Knörr, D. Häussinger, F. Hollmann, N. J. Turner and T. R. Ward, *Nat. Chem.*, 2013, **5**, 93.
46. Z. J. Wang, K. N. Clary, R. G. Bergman, K. N. Raymond and F. D. Toste, *Nat. Chem.*, 2013, **5**, 100.
47. R. Yuryev, S. Strompen and A. Liese, *Beilstein J. Org. Chem.*, 2011, **7**, 1449.
48. J. M. Sperl, J. M. Carsten, J.-K. Guterl, P. Lommes and V. Sieber, *ACS Catal.*, 2016, **6**, 6329.
49. A. S. de Miranda, M. V. de, M. Silva, F. C. Dias, S. P. de Souza, R. A. C. Leão and R. O. M. A. de Souza, *React. Chem. Eng.*, 2017, **2**, 375.
50. A. Gómez Baraibar, D. Reichert, C. Mügge, S. Seger, H. Gröger and R. Kourist, *Angew. Chem.*, 2016, **128**, 15043; *Angew. Chem., Int. Ed.* 2016, **55**, 14823.
51. H. Sato, W. Hummel and H. Gröger, *Angew. Chem.*, 2015, **127**, 4570; *Angew. Chem., Int. Ed.* 2015, **54**, 4488.
52. J. Latham, J.-M. Henry, H. H. Sharif, B. R. K. Menon, S. A. Shepherd, M. F. Greaney and J. Micklefield, *Nat. Commun.*, 2016, **7**, 11873.
53. F. Uthoff, H. Sato and H. Gröger, *ChemCatChem*, 2017, **9**, 555.
54. L. Pesci, M. Baydar, S. Glueck, K. Faber, A. Liese and S. Kara, *Org. Process Res. Dev.*, 2017, **21**, 85.
55. J. Pietruszka, R. C. Simon, F. Kruska and M. Braun, *Eur. J. Org. Chem.*, 2009, 6217.

Section V: Applied Biocatalysis

CHAPTER 16

Technical Biocatalysis

ANDRÉS ILLANES,* LORENA WILSON AND CARLOS VERA

School of Biochemical Engineering, Pontificia Universidad Católica de Valparaíso, Avenida Brasil 2085, 2340000 Valparaíso, Chile
*Email: andres.illanes@pucv.cl

16.1 Enzyme Catalysis: The Route from Degradation to Synthesis

Enzymes have been used as process catalysts for more than a century.[1] Early processes were based on simple degradation reactions catalyzed by robust hydrolases acting in soluble form in aqueous media. Enzymes from those early days were mostly extracted from animal and plant tissues and fluids, but the impressive development of industrial microbiology and fermentation technology since the middle of the past century produced a radical change and microbial enzymes came vigorously into play. By 1970 most enzymes were already of microbial origin[2] and this trend was reinforced with the advances in the fields of genetic and protein engineering that occurred in the following decades, allowing the cloning and expression of improved genes from different origins into suitable microbial hosts.[3] Conventional uses of enzymes at that time were almost entirely based on hydrolytic reactions and fields of applications were mostly related to food (amylases, proteases, glycosidases), detergents (initially proteases, but later on amylases, lipases and cellulases) and leather manufacturing (proteases). Most of these conventional uses were done under poorly controlled and non-optimized conditions with respect to enzyme performance.[4]

However, in the 1960s a process based entirely on enzyme technology was developed for the large-scale production of glucose from cornstarch

Catalysis Series No. 32
Modern Biocatalysis: Advances Towards Synthetic Biological Systems
Edited by Gavin Williams and Mélanie Hall

Published by the Royal Society of Chemistry, www.rsc.org

(Staley process) using two sequential enzyme-catalyzed reactions (liquefaction of gelatinized starch with α-amylase and saccharification of the resulting dextrins with glucoamylase) under optimized conditions.[5] A decade later, a third enzymatic step was added to convert glucose into fructose in a highly controlled and optimized continuous operation in column reactors packed with immobilized glucose isomerase.[6] The production of high-fructose syrup (HFS) departed from the classical hydrolase-catalyzed reactions, using a glucose (xylose) isomerase instead, which was an intracellular and more unstable protein that required stabilization for making the process viable. This was certainly a milestone in enzyme catalysis for being the first large-scale industrial process using an immobilized enzyme which, by improving the catalyst efficiency of use (mass of product produced per unit mass of catalyst[†]), was an absolute requirement for making the process economically viable. After more than forty years, HFS is still the largest industrial-scale product from enzyme catalysis. Immobilization technology has developed vigorously since then and immobilized enzymes were used beyond the food industry in the pharmaceutical sector, where the process for producing the semi-synthetic antibiotic intermediate 6-amino penicillanic acid was established using penicillin acylase in a cost-effective and environmentally sound process that soon displaced the former chemical one;[7,8] soon after, a process using immobilized bacterial cells containing a non-hydrolytic enzyme, nitrile hydratase, was developed for the industrial production of acrylamide and nicotinamide on a large scale, introducing biocatalysis in the chemical industry.[9]

About 80% of the ongoing enzymatic processes are still based on degradation reactions catalyzed by hydrolases in aqueous medium,[10] where a potentially low added-value is implicit. However, their use as catalysts in high added-value processes of organic synthesis has been actively pursued in the last three decades.[11] This was a formidable task, since enzymes involved in biosynthetic routes are complex, intracellular, unstable co-enzyme-requiring proteins, which are unsuitable as process catalysts. A major breakthrough came from the pioneering work of Klibanov and others in the early 1980s on enzyme catalysis in non-aqueous media.[12] The paradigm of enzymes as water catalysts was challenged, showing that, under proper conditions, enzymes could perform in low water organic media, and in such media, hydrolases can catalyze reverse reactions of synthesis, *i.e.* bond formation instead of breakage.[13] This was a remarkably appealing strategy allowing the use of well-conditioned, robust and technologically established hydrolases as catalysts for reactions of organic synthesis of potentially high added-value. The road was paved for enzymes as catalysts for organic synthesis. Use of enzymes in organic synthesis has come a long way since then.

Figure 16.1 shows the evolution of enzyme catalysis, highlighting those milestones corresponding to breakthroughs in enzyme technology.[14] As can be observed, the development of industrial biocatalysis was promoted by

---

[†]Also referred to as turnover number (TON).

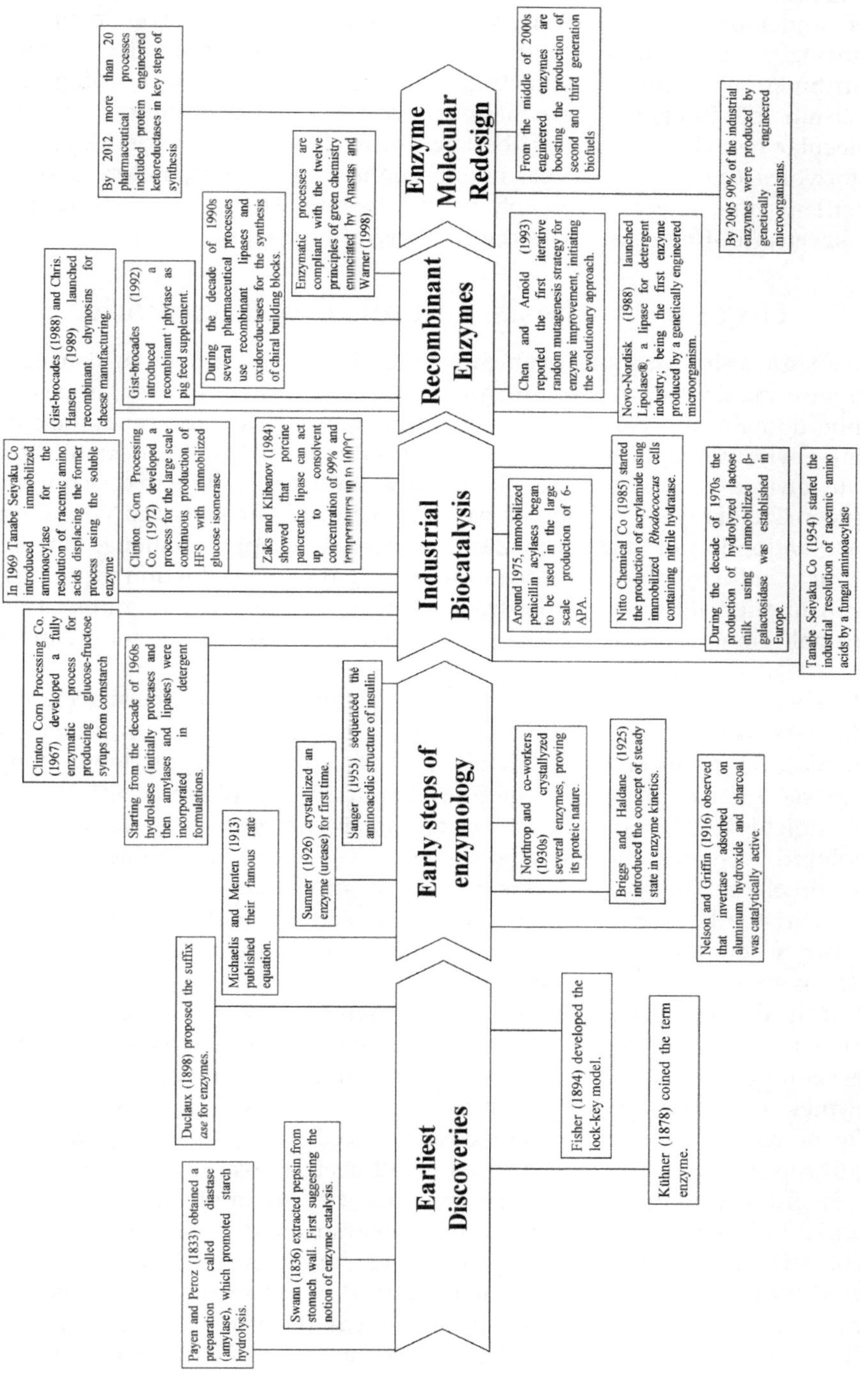

Figure 16.1 Evolution of biocatalysis.

immobilization techniques producing robust catalysts that were efficiently used under process conditions, with the consequent reduction in manufacturing process costs.[15,16] In the early 1990s, recombinant enzymes contributed to a further reduction in production cost, broadening the spectrum of industrial enzymatic processes.[3,17] Furthermore, protein molecular redesign starting in the late 1990s allowed for the development of enzymes that effectively act on non-natural substrates, tailoring their specificity and selectivity, which contributed to even further increasing the spectrum of biocatalysis in the industry of organic synthesis.[18,19]

16.2 Organic Synthesis: Chemical or Enzymatic?

The chemical industry has been threatened in recent decades by concepts like sustainable growth and environmental care, which are now issues of public domain and concern. The chemical industry has the reputation of being environmentally aggressive, so that guidelines have been established for an environmentally sustainable chemical industry, leading to the concept of green chemistry enunciated in twelve principles by Anastas and Warner in 1998.[20] Adoption of such principles has been a complex challenge for the chemical industry, since it involves the redesigning of production processes, which implies additional expenses with uncertain return. Programs to reduce pollution by industry self-regulation have been considered for the most part still unsuccessful.[21]

This situation certainly applies to the fine chemical and pharmaceutical industries where, beyond environmental reasons, hazardous materials and reaction conditions are not allowed due to safety regulations. Many processes for the production of fine chemicals and pharmaceuticals are still stoichiometric, and if catalyzed, mostly homogeneous catalysis is employed.[22] Heterogeneous chemical catalysis is more acceptable in terms of green chemistry, but problems associated with catalyst leaching from the matrix are hard to solve and this is particularly critical in the pharmaceutical industry where allowable levels of metal catalysts in the product are on the order of a few parts per million or less.[23]

Within this scenario, biocatalysis has emerged as a sound technological option for the synthesis of fine chemicals and pharmaceuticals. Chemical precision and activity under moderate conditions are salient properties of enzymes as catalysts that have profound technological implications: significant reduction in equipment cost, reduced energy consumption and significant reduction in downstream operations and cost. However, enzymes are far from ideal process catalysts, because they are labile and costly, and reputed to be effective only on natural substrates and in aqueous environments, which are quite different conditions compared to those usually required for organic synthesis. As analyzed in the next section, the latter two comments are certainly arguable, but in general terms these considerations still prevail and have certainly hampered the entry of biocatalysis in the organic synthesis industry. This situation is certainly changing at a sustained pace, particularly in the pharmaceutical area where chemical precision is a major asset; in fact, it is estimated that more than 50% of the chiral pharmaceuticals produced at industrial scale are synthesized at least partially by biocatalysis with the intended purpose of producing bioactive molecules with high enantiomeric excess, thereby avoiding the formation of a diastereomer which is inactive or, even worse, deleterious.[24] But beyond strictly economic considerations, biocatalysis is in full compliance with green chemistry principles, so this is a potent driving force for its consolidation as a sound technological option for industrial organic synthesis.[25] Pharmaceuticals and fine chemicals are mostly complex molecules whose syntheses involve many reaction steps, so that biocatalysis can be envisaged within the context of chemoenzymatic processes where some key steps requiring high molecular precision (regio-, chemo- and stereoselectivity) are enzyme-catalyzed, while others are conducted by conventional chemical catalysis.[26] The key role of enzyme catalysis in the development of greener large scale processes can be observed in Table 16.1, which summarizes the winning processes and technologies of the Presidential Green Chemistry Challenge between 2000 and 2015.

Table 16.1 Award of the Presidential Green Chemistry Challenge to biocatalytic processes between 2000 and 2015.

Year	Winner company	Brief description of the winning technology
2012	Buckman International, Inc.	Buckman's Maximyze® enzymes (mostly cellulases) increased the number of fibrils that bind the wood fibers to each other, thus making paper with improved strength and quality without additional chemicals or energy. Buckman's process also allowed papermaking with less wood fiber and higher percentages of recycled paper.
2012	Codexis, Inc.; Professor Yi Tang, University of California, Los Angeles	Professor Tang developed a manufacturing process for simvastatin synthesis using a low-cost acyl donor and LovD, a highly regioselective acyltransferase from *E. coli* obtained by directed evolution.
2010	Merck & Co and Codexis, Inc.	A (*R*)-selective transaminase obtained by directed evolution was employed for the production of sitagliptin by converting a ketone to a chiral amine. The enzymatic process eliminated the high-pressure hydrogenation, all metals (rhodium and iron), and the wasteful chiral purification step.
2009	Eastman Chemical Company	Eastman's new method used immobilized lipases to make esters, saving energy and avoiding both strong acids and organic solvents. Esters are an important class of cosmetic ingredients, comprising emollients, emulsifiers and specialty performance ingredients.

Table 16.1 (*Continued*)

Year	Winner company	Brief description of the winning technology
2006	Codexis, Inc.	Codexis designed a green process for manufacturing atorvastatin using three enzymes engineered by directed evolution. This biocatalytic process involved less unit operations than earlier processes, did not require fractional distillation of the product, and reduced the formation of byproducts, waste generation and use of solvents.
2005	Archer Daniels Midland (ADM) and Novozymes	Novozymes and ADM developed Lipozyme®, an immobilized enzyme that interesterifies triglycerides to produce low trans edibles fats and oils. Novozymes developed a cost-effective immobilized enzyme, and ADM developed a process to stabilize the immobilized enzyme for successful commercial production.
2004	Buckman Laboratories International, Inc.	Paper mills traditionally use hazardous solvents, such as mineral spirits, to remove sticky contaminants. Optimyze® technology provided a novel enzyme for removing stickiness from paper products prior to recycling, increasing the percentage of paper that can be recycled. Major components of the sticky contaminants in paper are polyvinyl acetate and similar materials. Optimyze® contains an esterase enzyme that catalyzed the hydrolysis of this type of polymers to polyvinyl alcohol, which is not sticky and water-soluble.
2003	Professor Richard A. Gross, Rensselaer Polytechnic Institute	Professor Gross developed a transesterification process using immobilized *Candida antarctica* Lipase B for manufacturing polyesters. Lipase catalyzes intrachain exchange reactions between polymer chains as well as transesterification reactions between a monomer and a polymer. This technique required less energy and produced less toxic substances than conventional methods that typically use heavy metal catalysts and hazardous solvents. This innovation also made it possible to manufacture new types of polyesters.
2001	Novozymes North America, Inc.	Novozymes North America developed BioPreparation™, a technology to separate natural waxes, oils, and contaminants from cotton before fabric manufacturing. This technology used enzymes instead of corrosive chemicals and saved 7–12 billion gallons of water per year. A bacterial pectate lyase was the main scouring agent that degraded pectin to release the entangled waxes and other components from the cotton surface.

Table 16.1 (*Continued*)

Year	Winner company	Brief description of the winning technology
2000	Professor Chi-Huey Wong, The Scripps Research Institute	Professor Wong developed effective enzymes, novel substrates and processes for large-scale organic synthesis in order to replace traditional reactions requiring toxic metals and hazardous solvents. His methods used enzymes, environmentally acceptable solvents, and mild reaction conditions.

16.3 Enzyme Catalysis in Organic Synthesis. SWOT Analysis

Enzyme catalysis is rapidly evolving from conventional degradation processes in aqueous media to more complex processes of organic synthesis whose intrinsic added value is much higher, representing nowadays the most promising and challenging area for enzyme catalysis. On the one hand, robust industrial enzymes like lipases, proteases and glycosidases can be used as catalysts for organic synthesis in properly designed reaction media,[27] while more complex coenzyme-requiring labile enzymes are beginning to emerge along with technological developments in enzyme stabilization, cofactor retention and regeneration.[28,29] According to a recent market survey, 45% of the enzymes used in organic synthesis are hydrolases, 30% are oxidoreductases, 17% are lyases, 6% are isomerases and 2% are transferases. This picture indicates that hydrolases (acting properly as hydrolytic enzymes or in reverse reactions of synthesis) are still predominant, but oxidoreductases and lyases represent already a significant share of biocatalysis in organic synthesis.

Introduction of biocatalysis in the organic synthesis industry has been however troublesome because it used to be an industry not acquainted to dealing with bioactive compounds, so that some biases existed: enzymes are too expensive, they are only effective on their natural substrates and in their natural habitats, validation is difficult because of batch-to-batch differences in enzyme quality, and they are too labile and prone to inhibition. Arguments against these comments can certainly be put forward:

- Enzymes are indeed expensive, but the question is what is the impact of the enzyme cost on the production cost of the process using it, so the real question here is enzyme efficiency of use (see Chapter 17). Besides, enzyme prices have been continuously decreasing because of intense competition and technological advances in enzyme production. Furthermore, inexpensive, coenzyme-independent and robust hydrolases can be used in properly engineered reaction conditions, and significant technological advances in coenzyme regeneration and recycling are allowing the use of other classes of enzymes, at least in high added-value processes where high molecular precision is mandatory.[30]

- It is now well established that enzymes can act efficiently in non-conventional media and on non-natural substrates; both protein, genetic and media engineering are powerful tools for these achievements.[31] Enzyme catalysis in non-aqueous environments has come a long way since the early developments of biocatalysis in organic solvents and can be considered now as a mature technology.[32] On the other hand, hydrolases are considered now quite promiscuous catalysts with a rather broad substrate specificity and able to catalyze diverse reactions of organic synthesis.[33]
- The use of substances of biological origin indeed adds in complexity for process validation (see Chapter 17). However, substitution of petroleum-derived raw materials by materials coming from natural resources generates the same problem, so this is a concern that the manufacturing industry is necessarily facing.[34] With respect to biocatalysis, enzyme production by fermentation has come to be a highly developed technology so that tight production monitoring and control warrant an acceptable batch-to-batch variation.[35]
- Enzymes are indeed quite labile catalysts not well endowed for effective performance at conditions far from physiological. This is certainly a major weak spot of enzymes as process catalysts when compared to chemical catalysts, but it is equally true that most technological advances experienced in biocatalysis in recent decades have been precisely focused on tackling this problem. Impressive advances with respect to enzyme activity and stability under process conditions have been obtained by screening from biotic[36] and metagenomic pools,[37] by genetic manipulations: *i.e.* site-directed mutagenesis[38] and directed evolution,[39] by biocatalyst engineering: immobilization,[40] auto-aggregation,[41] chemical modification,[42] and medium engineering[31] (see next section and Chapters 1, 2 and 7).
- Product (and sometimes substrate) inhibition is a rather common feature in enzyme catalysis that may represent a problem, which is less common in chemical organic synthesis.[43] This is indeed a drawback of enzymes and strategies for avoiding or reducing it have been proposed. Product inhibition can be effectively reduced using heterogeneous systems where the inhibitor can be partitioned out from the enzyme-containing phase;[44] on the other hand, protein engineering strategies have been used to reduce inhibition[45] and significant reduction in inhibition has been obtained by enzyme immobilization.[40]

An unbiased appraisal of the potential of enzymes as catalysts for organic synthesis should balance their strengths and opportunities derived from them with their weaknesses and threats. Undoubtedly, the exquisite specificity and selectivity of enzymes is a major advantage for organic synthesis, especially for the production of pharmaceuticals and fine chemicals where high molecular precision is demanded for their proper function, being a distinctive advantage over conventional chemical processes. The case of aspartame (*N*-(L-α-aspartyl)-L-phenylalanine 1-methyl ester), a leading non-caloric sweetener, illustrates this well: the chemical synthesis of this dipeptide ester requires several protection and deprotection steps, while in the enzymatic synthesis with the metalloprotease thermolysin, only the amino group in aspartic acid needs to be protected, since the enzyme is regioselective for the carboxyl group proximal to the amino group and enantioselective for the L-phenylalanine methyl ester, so that the racemate can be used as substrate.[46] Enzymes are well-endowed catalysts for asymmetric organic synthesis.[47] This is particularly relevant in the pharmaceutical field where a significant fraction of the active principles in commercial drugs are chiral compounds and there is a sustained tendency to replace racemates by enantiomerically pure compounds because of safety, legal and economic considerations.[48–50] Enzymes are chiral molecules themselves so that substrate chirality is recognized upon the formation of the enzyme–substrate complex at the active site and in this way, both enantiomers of the racemate will react at quite different rates affording kinetic resolution.[51] Asymmetric synthesis with enzymes also allows for the production of pure enantiomers from pro-chiral substrates.[52] These are outstanding opportunities for enzymes as selective catalysts for organic synthesis.

Enzymes are highly evolved catalysts for effective performance under the mild conditions of cell physiology. This is an outstanding property of enzymes, which, as stated before, has profound technological consequences in terms of energy demand, product purification and waste management. An illustrative example of this strength and underlying opportunity is the industrial production of acrylamide. The chemical process involves harsh chemical catalysts and is energy-intensive, requiring high temperatures which impair product quality. A biocatalytic process was developed with nitrile hydratase-containing bacterial cells operating at low temperature with a product yield of 100% so that energy, downstream operation and waste management costs were significantly reduced.[53,54] Nowadays, the biocatalytic process shares a significant fraction of the total production of acrylamide with an estimate of over 50 000 tons per year.

Enzymes are considered versatile catalysts acting not only on their natural substrates; in fact, enzymes are considered now rather promiscuous meaning that they can catalyze reactions barely related to their physiological function.[55,56] This is quite important in the context of organic synthesis where substrates are in many cases synthetic molecules to which enzymes certainly did not evolve to act upon, so that this broadened scope of action represents a clear opportunity for enzymes.

Sustainability has been considered inherent to biotechnology.[57] Concepts like the E-factor (mass ratio of waste to product) and the atom economy (ratio of molecular weight of the product to the sum of molecular weights of all substances involved in the reaction) are now commonplace metrics for assessing process sustainability.[58] In this context, enzymes are appealing catalyst from an environmental perspective; as stated before, enzymatic processes comply with green chemistry principles, a feature that is being increasingly appreciated by the chemical industry and the society as a whole.[59,60] The industrial production of cephalexin with penicillin acylase clearly illustrates the opportunity derived from this strength: a completely enzyme-catalyzed process is now in operation with a claimed reduction in E-factor from 50 to 15,[61] thereby rendering this process competitive with the chemical semi-synthesis of the antibiotic.[62]

It seems clear that there are now new and challenging opportunities for biocatalysis in organic chemistry, particularly in the synthesis of fine chemicals, pharmaceuticals and intermediates, where it will be a player of increasing importance.[63]

Conditions for organic synthesis are in most cases far from those at which enzymes perform their role, so it is undoubtedly that enzyme activity and stability at such conditions is a major hurdle for biocatalysis. However, as stated before, significant advances in different fields are contributing to solve this problem whose negative impact is expected to decrease in time. On the other hand, the ranges of operational conditions are rather narrow when working with enzymes, which can be a limitation in certain cases; also product inhibition can be a severe limitation. This problem can be tackled by several strategies like molecular redesign of enzymes,[64,65] enzyme immobilization[58] and *in-situ* product removal.[66,67] Enzyme prices are high though the observed tendency is a decline in price as production technologies progress and competition gets tougher. These weaknesses that represent a threat to the extended use of enzymes in organic chemistry are consistently being overcome by the above mentioned progress experienced by biocatalysis in recent years.

Reluctance of the chemical industry to adopt biocatalysis is understandable and catalytic processes, especially biocatalytic processes, should offer a clear-cut advantage to justify the replacement of conventional processes of organic synthesis, which is seldom the case. This is certainly a major threat for biocatalysis; however, environmental regulations are a driving force for moving from stoichiometric to catalytic processes and from homogeneous to heterogeneous catalysis. Resource efficiency of use and waste minimization are now considered essential parameters in green chemistry, which further highlights the increasing importance of biocatalysis within the chemical industry.[68] Chemical catalysis is also moving ahead and the development of catalytic processes within the framework of green chemistry is also a powerful driving force within the chemical industry, as illustrated by the case of the catalytic synthesis of polyisobutenes, for which a process much in compliance with the principles of green chemistry has been developed.[69] In fact, green chemistry is becoming a paradigm of the chemical industry, whose most important aspect relates to the design of processes and products, which means a statement of human intention.[70] Green chemistry "is not simply good process chemistry; it is the highest efficiency potential that exists for each chemical process, serving as both an inspiration for and a measure of the best process chemistry".[71] Within this scenario, biocatalysis will compete with chemical catalysis, and the competitive edge should be based on clear advantages. Biocatalysis will be a probable choice when (i) no chemical counterpart exists, (ii) high selectivity or specificity is mandatory, (iii) mild process conditions are required or (iv) stringent environmental regulations are enforced. Catalyst stability, catalyst cost and product yield are probably the major weak spots of biocatalysis that should be acknowledged in each case.

16.4 Enzyme Catalysis in Non-conventional Media

Until well into the 20th century, enzymes were used almost exclusively in aqueous media in which hydrolytic enzymes, namely proteases, glycosidases and lipases, were catalyzing degradation reactions conducted at high water activity mostly in homogeneous catalysis. Such reactions have an intrinsically low added-value, so the use of enzymes as catalyst for high added-value reactions of organic synthesis started to be considered as a key issue in biocatalysis about four decades ago. As stated in Section 16.1, hydrolases could be used as catalysts for reactions of synthesis provided its hydrolytic capacity was arrested. This required the use of non-conventional reaction media where water activity was low enough to reduce their hydrolytic potential. Besides, substrates (and products) of organic synthesis were frequently poorly soluble or insoluble in water. This was a powerful driving force for exploring enzyme behavior in non-conventional reaction media.[32]

16.4.1 Enzyme Catalysis in Organic Solvents

A major impulse came from the work of enzyme catalysis in organic solvents which was vigorously developed in the 1980s. Enzyme performance in such media was usually poor,[72,73] but an impressive amount of information soon accumulated on how to deal with such a problem.[74] More recently, the strategy of optimizing reaction conditions to suit the available biocatalyst has been replaced by the reverse strategy of designing the biocatalyst to suit the reaction conditions, so that now enzymes can be engineered to withstand the harsh conditions imposed by an organic medium.[75] Among hydrolytic enzymes, lipases are an exception for being structurally conditioned to act in hydrophobic media and in water–oil interfaces, so a significant part of the development of organic synthesis in organic solvents has been conducted with lipases.[76–78] A list of illustrative examples of synthesis with the use of lipases in organic solvents is presented in Table 16.2. Most of these lipases were used in immobilized form to allow higher stability and reusability. Use of enzymes in organic solvents is certainly not restricted to lipases and important applications of other hydrolytic enzymes, like proteases,[79–82] acylases[83–85] and glycosidases[86–89] have been reported. Non-hydrolytic enzymes are becoming increasingly important in organic synthesis and important applications of oxidoreductses,[90,91] lyases[92] and transferases[93a] requiring organic media have been reported.

For some time, biocatalysis in organic synthesis was closely tied to biocatalysis in organic solvents. There are in fact several potential advantages of using organic solvents as reaction media for biocatalysis: robust

Table 16.2 Processes of organic synthesis using lipases in organic solvents.

Lipase source	Solvent	Product	Ref.
Alcaligenes sp.	Methanol	Biodiesel	94
Aspergillus oryzae	Toluene	Flurbiprofen	95
Candida antarctica			
Candida antarctica	*tert*-Amyl alcohol	Ascorbyl palmitate	96
Candida antarctica	2-Methyl-2-butanol/DMSO	Fructose monolaurate	97
Candida antarctica	DMSO	Xylitol monoester	98
Candida antarctica	*tert*-Butanol	Ascorbyl palmitate	99
Candida antarctica	Isooctane–butanol	Biodiesel	100
Candida rugosa	Isooctane	Ibuprofen	101
Mucor miehei	Heptane	Interesterified soybean oil	102
Pseudomonas flurescens	Hexane	Biodiesel	103
Pseudomonas stutzeri	*tert*-Butanol	Ascorbyl palmitate	104
Rhizomucor miehei	Methanol, ethanol	Biodiesel	105
Rhizomucor miehei	Isooctane	Ibuprofen	106
Thermomyces lanuginosus	2-Methyl-2-butanol	Ascorbyl oletae	107
Thermomyces lanuginosus	Hexane	β-Sitosterol ester	108

hydrolases can be used in reverse reactions, thermodynamically unfavorable reactions in water can be performed, the use of poorly water soluble or insoluble substrates becomes possible, thermal stability may be increased, product inhibition may be reduced and specificity and selectivity can be tuned.[32] There is however a pitfall: the use of organic solvents in biocatalysis moves it away from environmental benignity, which is an alleged advantage of biocatalysis. Organic solvents are obviously different with this respect:[109] some of them, like ethyl lactate, have been claimed to be quite innocuous;[110] this is also the case for glycerol which is now a plentiful output from biodiesel production[111] although its high viscosity is a limiting factor.[112] Evaluation of organic solvents according to their environmental impact is a complex task that should consider life cycle analysis.[113,114] A tentative classification of some of the most used organic solvents in the chemical industry is presented in Table 16.3.

Organic solvents represent no less than 80% of the mass utilized in the pharmaceutical and fine chemical synthetic processes; most of them are environmentally offensive and even toxic[49] and a thorough classification of organic solvents from this perspective has been reported.[115] Concern about the use of organic solvents for organic synthesis led to an intensive search for alternative solvents more in compliance with green chemistry[116] and the concept of green solvents was soon established,[93] a desirable green solvent being natural, nontoxic, cheap, and readily available.[117] In the particular case of biocatalysis, beyond the restrictions imposed for good sustainable chemistry, the solvent must be compatible with enzyme structure and function, which limits the choice further.[118] Indeed, the best solvent for biocatalysis is no solvent at all and many reactions have been conducted in the absence of added solvent and using one of the substrates as such.[119–121]

Table 16.3 Organic solvents used in the chemical industry according to a green chemistry perspective.

Acceptable	Less desirable	Undesirable
Water	Cyclohexane	Pentane
Acetone	Methyl cyclohexane	Hexane
Ethanol	Heptane	Benzene
Methanol	Isooctane	Dichloromethane
1-Propanol	Toluene	Dichloroethane
2-Propanol	Xylenes	Dimethoxyethane
1-Butanol	Acetonitrile	Dioxane
tert-Butanol	Acetic acid	Diethyl ether
Glycerol	Ethyleneglycol	Diisopropyl ether
Ethyl acetate	Methyl *tert*-butyl ether	Chloroform
Isopropyl acetate	Tetrahydrofuran	Carbon tetrachloride
Methyl ethyl ketone	Methyl tetrahydrofuran	Pyridine
Ethyl lactate	Dimethyl sulfoxide	Dimethyl formamide

However, a high excess of the substrate acting as solvent is required, so the handling of unreacted substrate remains a problem; increased viscosity may also be a problem when working in solvent-free media.

Organic solvents for biocatalysis can be used as cosolvents in water-miscible monophasic mixtures, in water-immiscible biphasic systems and in close to anhydrous systems.[122] Water-miscible cosolvents are not the best medium for biocatalysis, since such solvents tend to strip away the essential water surrounding the enzyme, thereby distorting its active configuration;[123,124] enzymes are in this case mostly used in immobilized form.[125] However, solvents like glycols, polyols and glymes have shown to be good cosolvents for enzyme performance.[126–128] Use of cosolvents may be a particularly good choice when substrates are poorly soluble in more hydrophobic solvents.[129] Enzymatic synthesis in hydrophobic solvents has been thoroughly studied; such systems can be considered microheterogeneous since water is tightly bound to the enzyme and no water phase is observable to the naked eye. A small amount of water should always be present to provide the enzyme with a minimal hydration level that allows activity expression; that amount varies from one enzyme to another and can be as low as one monolayer of water molecules surrounding the enzyme molecule.[130] Enzymes are insoluble in such media and in many occasions exhibit an outstanding thermal stability,[131,132] so that immobilization is not strictly necessary even though it may help by favoring enzyme–substrate contact and precluding the direct contact of the enzyme molecules with the hydrophobic solvent.[133] Activity and stability of enzymes in hydrophobic solvents can be severely hampered[64] so that several strategies of biocatalyst and medium engineering have been used to improve enzyme performance under such conditions;[134] a comprehensive review on this topic has been published recently.[135] Lipases are a unique case among enzymes for being structurally conditioned to non-aqueous environments,[136] so that some of them perform even better in hydrophobic solvents than in aqueous medium, thus lipases constitute the most relevant class of enzymes for biocatalysis in organic medium.[137] Another microheterogeneous system is reverse micelles, which are formed when a tiny amount of water in which the enzyme is dissolved is added to a hydrophobic solvent in the presence of a surfactant. Potential advantages of this system are high mass transfer rates, increased stability and even hyperactivation in some cases;[138] however, reverse micelles are mechanically weak, unstable, and surfactants may be deleterious for many enzymes, so their potential as process biocatalysts is limited.[79] Aqueous–organic biphasic systems are those in which phases are distinguishable by the naked eye. Even though mass transfer limitations may be a problem, this system is appealing in the case of substrates and products that are poorly soluble in water or when products are strong enzyme inhibitors.[139]

Among the many different systems of biocatalysis in organic media, probably the microheterogeneous system using strong hydrophobic solvents is the one in which the benefit of non-aqueous biocatalysis is more clearly perceived. Careful selection of the designed enzyme, the solvent and the reaction conditions are necessary to obtain acceptable levels of activity that approach those obtained in aqueous medium.[140]

16.4.2 Enzyme Catalysis in Other Non-conventional Media

Considerable effort has been devoted to develop greener solvents for reactions of organic synthesis;[141] this is particularly relevant in biocatalysis for conserving its environmentally benign character, so major effort was raised to the use of organic solvents as reaction media.[67] The subject of alternative reaction systems has been thoroughly reviewed in a recent book by Kerton and Marriott.[142] Within the field of biocatalysis, several non-conventional media have been studied; worth-mentioning are the use of gases and supercritical fluids, ionic liquids and deep eutectic solvents, and high-solid systems. The goal of these systems is the compliance with green chemistry principles and their adequacy for efficient enzyme performance.

16.4.2.1 Gases

Biocatalysis in gaseous media was proven feasible and some advantages like high mass transfer rates, removal of volatile inhibitors, higher stability than in aqueous medium and asepsis were claimed.[143] The system is restricted to gaseous or highly volatile compounds and the control of water activity, reactor design and operation are complex so that few applications, such as the selective removal of specific highly toxic pollutants from industrial gas effluents, are envisaged.[144]

16.4.2.2 Supercritical Fluids (SCF)

Fluids above their critical pressure and critical temperature are called supercritical fluids (dense gases). At such conditions, fluid behavior can be considered intermediate between gases and liquids, which can be advantageous as a medium for biocatalysis. In fact, SCF densities are similar to that of liquids, while viscosities are similar to that of gases. The former allows an acceptable solubility for many compounds, very much as in liquids, while the latter implies very high mass transfer rates, very much as in gases. Other advantages of SCF refer to product recovery, since the solvent can be simply eliminated by pressure relief under green chemistry conditions.[145] Compatibility of enzyme performance and supercritical conditions is the main restriction of SCF in biocatalysis; in fact, there are only a few acceptable SCF, namely ethane, ethylene, fluoroform, sulfur hexafluoride and carbon dioxide.[146,147] Carbon dioxide, despite not being the optimal SCF for biocatalysis, has been by far the most used due to safety, environmental and economic considerations.[148] Other SCF like sulfur hexafluoride compare favorably to carbon dioxide in terms of reaction rates and enzyme stability.[149]

Carbon dioxide's critical pressure of 7.3 MPa can be considered rather high, but the critical temperature of 31 °C leaves an ample margin of work even for moderately mesophilic enzymes. Supercritical carbon dioxide ($SCCO_2$) provides a non-oxidizing environment which is important for avoiding oxidation reactions during product recovery, and as a highly non-polar solvent it is well suited for hydrophobic but not for hydrophilic substrates; in the latter case, the addition of a more polar entrainer has been proposed.[150] $SCCO_2$ has been profusely used as medium in many synthetic reactions, such as the oxidation of cholesterol with reverse micelles containing cholesterol oxidase,[151] the synthesis of pyruvic acid from acetaldehyde and carbon dioxide with pyruvate decarboxylase,[152] the reduction of aldehydes to alcohols with NADH-dependent alcohol dehydrogenases,[153] the kinetic resolution of racemic alcohols with free and immobilized lipases,[154] the ring-opening polymerization of lactones with immobilized lipase,[155] the esterification of acids and alcohols with immobilized lipase,[156] the enantioselective acylation of racemic alcohols with different lipases,[157] the synthesis of biodiesel with immobilized lipase,[158] the synthesis of sugar fatty acid esters with immobilized lipase,[159] the ring-opening polymerization of L-lactide with immobilized lipase[160] and the synthesis of cocoa butter analog with immobilized lipase.[161] As seen, the enzymes mostly used in $SCCO_2$ are lipases and in most cases in an immobilized form; stringent reaction conditions and the hydrophobicity of the reaction medium are the reasons behind this. Temperature and pressure (above critical) are key variables in determining enzyme behavior in terms of kinetics, stability and selectivity. Activity may increase over the critical temperature within a limited range, but activity usually decreases with pressure above critical because of the increase in mass transfer limitations.[150] Stability of enzymes at the stringent pressure and temperatures of SCF are likely to be low; however, enzymes are more rigid than in aqueous systems, which may be favorable in terms of stability, so this is not a general statement and some enzymes have shown acceptable stability in such media; notwithstanding, enzyme stability is of major concern so that all strategies envisaged for increasing enzyme stability are welcome here.

Water activity is a key variable for biocatalysis in SCF, and is difficult to control; a very low water content is required for proper enzyme performance so this is also a challenge for biocatalysis in SCF. In the case of $SCCO_2$, carbamate may be formed upon reaction with the lysine residues on the enzyme surface, and in the presence of high water activity, carbonic acid may also be formed that will reduce enzyme activity.[162] Selectivity of the enzyme can be tuned by manipulating the pressure and temperature, which is a very interesting option for asymmetric catalysis, as illustrated by the enantioselective acetylation of racemic 1-(*p*-chlorophenyl)-2,2,2-trifluoroethanol with vinyl acetate catalyzed by an immobilized lipase in $SCCO_2$.[157] $SCCO_2$ has been used in conjunction with ionic liquids (see Section 16.4.2.3) in biphasic systems, which allows coupling reaction and product extraction in a single tentatively environmentally benign operation.[163] Reactor design and operation for SCF biocatalysis is challenging mostly because of the high pressures required; in batch systems, mandatory pressurization–depressurization cycles are detrimental for enzyme activity, so the use of continuous SCF enzymatic processes has been proposed to reduce such damage and increase productivity.[164] Despite its limitations and complexities for scale-up to productive levels, biocatalysis in $SCCO_2$ is an interesting option for the synthesis of valuable compounds under the green chemistry concept.

16.4.2.3 Ionic Liquids

Ionic liquids (ILs) have been extensively used in recent decades as reaction media for both chemical and enzymatic organic synthesis.[165] The use of organic solvents as reaction media for biocatalysis has been seriously threatened since it contradicts the concept of clean technology, which is a salient attribute of biocatalysis. Greener solvents have been actively pursued in recent decades and among them ILs clearly stand out. ILs are simply molten salts that exist in liquid stage at around room temperature. Favorable properties of ILs are an extremely low vapor pressure, a wide liquid range, high ionic and thermal conductivity, high thermal and chemical stability, low flammability and high dissolution capacity of a wide range of substrates. They are considered green solvents in the sense that they can be contained and recovered after removal of products from the reacted medium. ILs compatible with enzyme activity (room temperature ILs) are composed of a rather complex organic cation (*i.e.* 1,3-dialkylimidazolium, *N*-alkylpiridinium, tetraalkylammonium) and a rather simple, frequently inorganic, anion (*i.e.* tetrafluoroborate, hexafluoroantimony, hexafluorophosphate, dicyanamide). The proper selection of cations and anions allows access to water-miscible, partially water-miscible and water-immiscible ILs; this is a remarkable advantage since the solvent can be tuned to match the properties of the enzyme in terms of catalytic behavior and operational stability[166] and also in terms of enantioselectivity in asymmetric organic synthesis.[167,168] Most of the enzymes used in ILs are lipases and proteases. Lipases have been used in transesterification reactions,[169,170] in the synthesis of sugar esters[171] and alkyl esters,[172] in the acylation of flavonoids[173] and in the production of biodiesel.[174,175]

Proteases are in general not well endowed for performing in organic solvents; however, they are active and stable in some ILs where proteolysis is avoided; they have been used in the ligation of peptides and protein fragments,[176] in transesterification reactions,[177] in the enzymatic resolution of racemic esters[178] and in peptide synthesis.[179,180] Oxidases,[181] lyases[182] and glycosidases[183] have also been reported to catalyze reactions in ILs.

Even though enzymes are in many cases active, stable and selective in ILs, a major hurdle is their low solubility in such media so that most active enzymes remain suspended in powder form and if dissolved, conformational changes are induced that result in inactivation.[184] Stabilization is then a major issue in biocatalysis in ILs[185] that can be achieved through immobilization[186] and microemulsification.[187]

Despite the obvious potential of ILs in biocatalysis, application is constrained by major drawbacks. High prices and complex removal of products from the spent reaction medium are indeed limitations but, beyond that, the question has been raised about how green these solvents really are.[188] The claim of ILs as green solvents may not be valid from a life cycle analysis perspective, which has been considered a requirement before validating such claim; ILs are synthetic products so such warning seems well sustained.[189] These considerations underline the search in recent years of novel and greener neoteric solvents. Other biotechnological applications of ILs, such as cellulose dissolution to provide it as a plentiful and cheap raw material, their use in biphasic systems with $SCCO_2$, and as components of bioelectrochemical energy cells, seem more promising technologies than their use as a bulk solvent for biocatalysis.[190]

16.4.2.4 Deep Eutectic Solvents

Deep eutectic solvents (DES) have been recently introduced as a new class of room temperature liquids which represent a promising alternative to ILs in terms of cost effectiveness and most certainly from a life cycle analysis consideration, since they are formed mostly from compounds derived from natural products. DES are mixtures of compounds whose melting point is below the ones of their components, which is explained by the formation of intermolecular hydrogen bonds.[191] Commonly used natural DES are choline chloride-based, namely, choline chloride–alcohols,[192] choline chloride–organic acids,[193] choline chloride–glucose,[194] choline chloride–fructose,[195] choline chloride–urea,[196] choline chloride–glycerol,[197] and choline chloride–1,4-butanediol mixtures;[198] in some cases, choline–acetate has been a preferred DES for biocatalysis with lipases following considerations about enzyme activity and stability.[199]

DES find applications in many different areas, such as metal processing and CO_2 adsorption,[200] separation processes, extraction of bioactive compounds from natural products,[201,202] stabilization of natural pigments,[203] drug delivery vehicles,[204] chemical catalysis and biocatalysis.[205] Application of DES as reaction media for biocatalysis is rather recent and restrained by the fact that some of them, like urea, are strong hydrogen bond donors that are protein denaturants, and some have melting points higher than temperatures compatible with enzyme activity;[206] however, some enzymes are remarkably stable and active in choline chloride–urea[207] and choline chloride–glycerol DES,[208] displaying a better performance and higher protein stability compared to ILs. Lipases have been the most-used enzymes for exploring biocatalysis in DES, and have been applied to the synthesis of biodiesel,[209,210] the transesterification of esters with alcohols[211] and the synthesis of esters.[212]

DES are gaining increasing technological relevance in many fields. Their use as green solvents for biocatalysis is an attractive alternative to ILs; however, the technology is still in an early stage of development so it is premature to forecast its impact in the field.

16.4.2.5 High-solids Systems

Semi-solid and nearly-solid systems refer to those media where a small amount of liquid exists in which the substrate becomes saturated and converted by the enzyme into a product that precipitates out as soon as its solubility limit is exceeded.[213] In principle, this is an attractive system for reactions of synthesis with hydrolases since water activity is low enough (actually much lower than in other non-conventional reaction media) to suppress the reverse or competing reactions of hydrolysis. Other potential advantages are the very high volumetric productivity that can be attained because of the absence of obnoxious solvents and the facilitated downstream operations after reaction because most of the reactor volume will be occupied by the precipitated product;[214,215] however, mass transfer limitations are likely to be severe, which results in very low reaction rates, and mixing can be a serious problem when scaling-up to production level.[216]

Reactions at very high substrate concentrations, approaching saturation and beyond, will also reduce water activity, thereby being a sound alternative to other non-conventional systems and particularly useful when substrates are poorly soluble in other non-aqueous solvents.[217] Except for the case of substrate inhibition, this strategy will also be favorable from kinetic considerations. Increased viscosity may be problematic and also enzyme removal from the reaction medium can be difficult. Even so, this strategy has proved to be successful for carrying out synthesis reactions with highly water-soluble substrates, as illustrated by the case of the synthesis of semi-synthetic β-lactam antibiotics with penicillin acylase[218,219] and the synthesis of lactose-derived prebiotic oligosaccharides with β-glycosidases.[220,221]

16.4.3 Strategies for Building-up Enzyme Biocatalysts for Organic Synthesis

Organic synthesis reactions are usually conducted at stringent conditions, usually in non-aqueous media, so biocatalysts capable of performing efficiently under such conditions are required.[222] As can be observed in Figure 16.2, several strategies of biocatalyst engineering have been proposed for such purpose, namely: addition of protective additives,[223] chemical modification of enzyme structure,[134,224] immobilization in solid carriers[40] and by aggregation,[225,226] and protein engineering.[227–230]

16.4.3.1 Protective Additives

Protective additives have been massively used to increase enzyme activity and stability. Some of them have been specifically used to improve enzyme performance in organic solvents, namely: inorganic salts such as KCl,[231,232] polyethylene glycol,[233] glycerol, sorbitol, trehalose,[234] crown ethers, complex

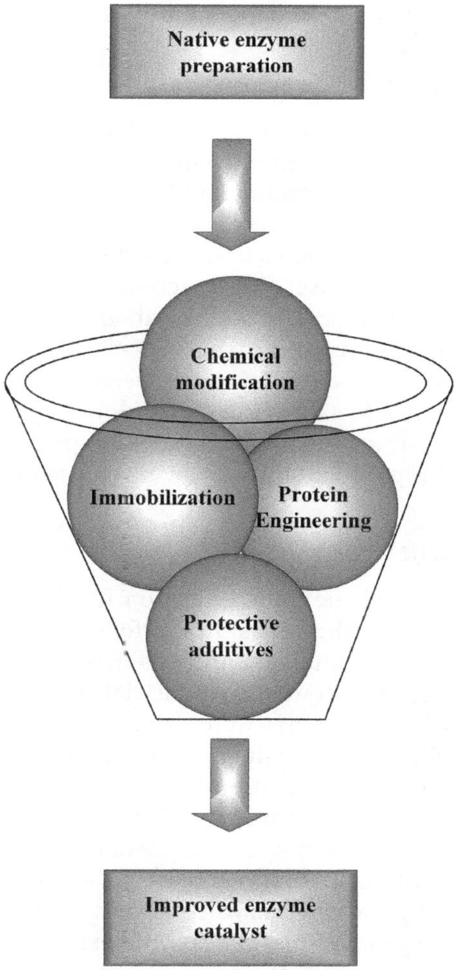

Figure 16.2 Strategies of biocatalyst engineering.

carbohydrates, surfactants[235,236] and cyclodextrins.[237] An illustrative example is the use of methyl-β-cyclodextrin in preparing lyophilized α-chymotrypsin and subtilisin, with which much higher activities compared to the free counterparts were obtained in pure organic solvents.[238,239] Additives have also been used successfully for improving lipase performance in terms not only of activity and stability but also selectivity, which is of paramount importance in reactions of organic synthesis.[236] The effect of additives on enzyme properties is not fully understood yet; most of them are used on an empirical basis and few systematic studies on this aspect have been published.[240,241] This strategy is quite straightforward and many candidate stabilizers are available, but they should prove their compatibility with the reaction system and comply with environmental and safety regulations; besides, separation of the additive from the final product is an additional operation that has to be taken into consideration.[135]

16.4.3.2 Chemical Modification

Chemical modification of the amino acid residues of the enzyme protein is a rather simple and straightforward procedure that does not require previous knowledge of the enzyme structure and has been used frequently for improving protein stability (for incorporation of non-canonical amino acids, see Chapter 6). It may also increase the solubility of the enzyme in non-aqueous media and increase the reaction rate. Monomethoxy polyethyleneglycol was used in several studies about structure modification of enzymes for performing in organic medium;[242,243] significant improvements in activity, stability and selectivity have been reported,[244,245] although mechanisms for such effects have been seldom reported.[246,247] A major disadvantage of this strategy is that it has to be performed each time the enzyme is used and reproducibility has been reported to be low.[135]

16.4.3.3 Immobilization

First developments of enzyme immobilization for the production of industrial catalysts date back to the 1960s. In the following decades, enzyme immobilization research was intense and considered as the most promising output of biotechnology. However, industrial processes adopting such technology, although important, were few. In recent years, significant advances in immobilization strategies and support design, as well as rational approaches of immobilization based on enzyme configuration and topography, have opened opportunities of using immobilized enzymes in synthetic reactions in non-conventional media. After all these years, an impressive amount of information has been gathered on support structure, protein surface configuration, enzyme–support interactions, catalyst recovery and support regeneration after enzyme exhaustion; some of this information is focused on a production process perspective. Notwithstanding, enzyme immobilization is still facing many challenges, and opportunities keep on emerging, especially in relation to industrial synthetic processes and also in biomedical applications.

An ample palette of immobilization methodologies is now available;[248–250] however, the one to be chosen will depend on each particular case and many variables will be involved in optimizing the immobilization process, which is a multi-variable, multi-parameter system. Immobilization methods can be broadly classified as carrier-bound or carrier-free (see Figure 16.3).

Carrier-bound: This is the most reported technique for enzyme immobilization and consists in the attachment or containment of the enzyme to or into a solid support. In most cases, a significant increase in enzyme stability is

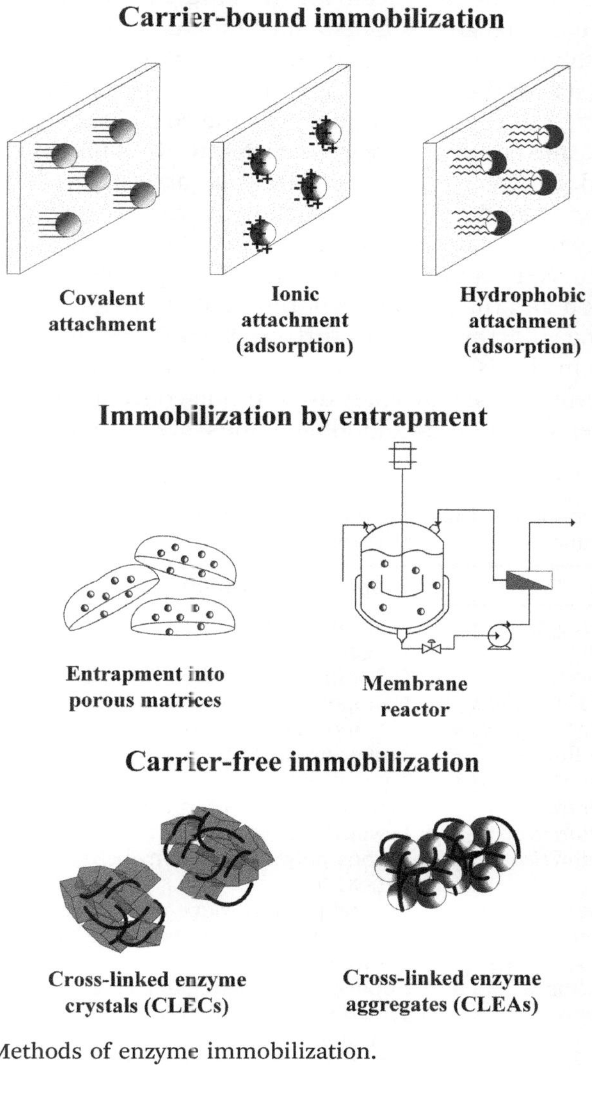

Figure 16.3 Methods of enzyme immobilization.

obtained and the spent catalyst can be easily removed and recovered from the reaction medium. Many studies about enzyme immobilization by covalent and non-covalent binding to solid supports and encapsulation have been developed.[251–253] At present, the trend is a more rational approach to the immobilization of enzymes to supports, due to the advances in enzyme three-dimensional structure elucidation and support microstructure determination that allows for determining the protein distribution within the support.[250,254] The properties of the support are key determinants not only of the immobilization process but also of the catalytic properties of the immobilized enzyme. In the case of using the enzyme in a non-conventional medium, special attention should be made to the compatibility of the support with it. Support properties that must be considered are the surface area and permeability, the presence of functional groups for enzyme binding, the chemical, thermal and mechanical stability, the resistance to microbial attack, the compatibility with the intended use of the enzyme, the availability, the preparation procedure and cost.[255–257] With respect to the latter, the costs of the support material and reagents, and the technology associated to the immobilization have to be considered. Different supports offer a wide variation in their physical and chemical properties for enzyme attachment (*e.g.* particle size, pore size, hydrophilic/hydrophobic balance and surface chemistry). The differences in morphological characteristics and physical properties can affect enzyme immobilization and the properties of the catalyst due to the close proximity of the enzyme and the support surface. Stability of the support in non-conventional media is a pre-requisite for use in synthetic reactions. Some supports of immobilized enzymes used in non-conventional media (organic solvents, supercritical fluids and ionic liquids) are presented in Table 16.4.

Table 16.4 Supports of immobilized enzymes used in non-conventional media: organic solvents, supercritical carbon dioxide ($SCCO_2$), and ionic liquids (IL).

Enzyme/source	Support/non-conventional medium	Ref.
Alcohol dehydrogenase/*H. volcanii*	Metal-derivatized-epoxy Sepabeads/ organic solvent	258
α-Amylase/*A. oryzae*	Ca-alginate beads and NaY zeolite/$SCCO_2$	259
Chloroperoxidase/*C. fumago*	Nanoporous silica glass/organic solvent	260
Chloroperoxidase/*C. fumago*	Mesoporous silica/organic co-solvents	261
Commercial cellulase	Amberlite XAD4/IL	262
β-Galactosidase/*B. circulans*	Agarose gel/IL	263
Lipase B/*C. antarctica*	Lewatit/organic solvent/IL	264
Lipase B/*C. antarctica*	Lewatit/$SCCO_2$	265–267
Lipase B/*C. antarctica*	Porous polystyrene divinylbenzene/ IL-$SCCO_2$	268
Lipase/*P. cepacia*	Accurel polypropylene/organic solvent	269
Lipase *P. stutzeri*	Epoxy-functionalized polymers/organic solvent	270
T. lanuginosus lipase	Eupergit C, silica/organic co-solvents	271
Y. lipolytica lipase	Functionalized carbon nanotubes/organic solvent	272

Entrapment consists of the retention of the catalyst within the inner cavities of a solid porous matrix usually formed by organic polymers, like polyacrylamide, collagen, alginate, resins, and polyvinyl alcohol (PVA). Entrapment has been mostly used as a technique for cell immobilization; in the case of enzyme immobilization, leakage from the matrix may be severe due to their relatively low molecular size; leakage can be avoided if the enzyme is first precipitated and cross-linked to form a protein aggregate of a size similar or higher than a yeast cell. Entrapment within semipermeable membranes can be considered in this category, even though the enzyme is simply contained by the membrane but not necessarily attached to it.[273] An illustrative example is the entrapment of *Lactobacillus kefir* alcohol dehydrogenase together with its coenzyme NADPH in polyvinyl alcohol (PVA) gel beads. The immobilized preparation transformed a number of hydrophobic ketones to the corresponding enantiomerically pure (*R*)-alcohols in pure hexane medium. The PVA matrix protected both the enzyme and its cofactor from the deleterious effects of the organic solvent; however, due to the low diffusivity of the cofactor within the gel, a high concentration of the cofactor was needed.[274] Another example is the entrapment of cross-linked enzyme aggregates of penicillin acylase in PVA gel beads; the immobilized enzymes were much more stable than the free enzyme in 75% (v/v) dioxane, showing that the PVA matrix provided a highly hydrophilic environment that protected the enzyme from the aggressive organic solvent; in fact, partition experiments showed that the concentration of dioxane inside the matrix was considerably lower than in the reaction medium.[275]

Carrier-free: In this case, the enzyme protein constitutes its own matrix so that there is no inert support and much higher (one order of magnitude or more) specific activities than in carrier-bound enzymes can be obtained.[276] Cross-linked enzyme crystals (CLECs) are produced by cross-linking of highly purified enzyme crystals with bifunctional reagents, such as glutaraldehyde.[277] Cross-linked enzyme aggregates (CLEAs) are produced by cross-linking enzyme protein aggregates, where enzymes are precipitated by conventional non-denaturing protein precipitation techniques and cross-linked with bifunctional reagents. Both carrier-free types of catalysts exhibit increased resistance toward organic solvents after the formation of water-insoluble particles. CLEAs have the advantage over CLECs of not requiring a purified enzyme as starting material, which is a major asset in terms of cost;[41,278] also, several conventional techniques used for protein purification are available that allow high recovery yields of enzyme activity. Major drawbacks of CLEAs with respect to CLECs are a lower mechanical stability and the difficulty of obtaining particles of controlled and regular size. However, because of low cost and simplicity, CLEAs appear as a promising technology for producing enzyme catalysts for industrial processes. In fact, a large number of technologically important enzymes have been produced in this format, and appropriate reactor configurations have been developed for their proper handling and recovery.[279] CLEAs are particularly well-suited systems for the immobilization of multimeric enzymes, since cross-linking of the subunits is promoted.[280] This strategy also allows the co-immobilization of enzymes and polymers; dextran sulfate and polyethyleneimine coated CLEAs of penicillin acylase were prepared by adding polymers of different sizes before the precipitation stage. Experiments showed that CLEAs coated with polymers have an increased stability in organic media, which was attributed to the creation of a hydrophilic microenvironment for the enzyme rather than a conformational stiffening effect.[281] A wide variety of CLEAs of hydrolases, oxidoreductases, lyases, and transferases have been used for producing alcohols, amines, carboxylic acids and derivatives, carbohydrates, amino acids, peptides, and polyamides.[282] Worthwhile mentioning are the *in-situ* formation of peracetic acid with perhydrolase CLEAs used for the Baeyer–Villiger oxidation of cyclohexanone, yielding 63% caprolactone,[283] and the synthesis of valuable dipeptides in organic media with alcalase CLEAs.[284]

Immobilized enzymes are generally more stable than their soluble counterparts and this represents their most important advantage; however, expressed activity is in most cases reduced because of mass transfer limitations and conformational changes. Immobilization of enzymes in nanomaterials is an interesting option for producing stable and active (devoid of mass transfer limitations) enzyme catalysts.[285] Such nanobiocatalysts have been mostly studied for medical applications;[286] their potential as process catalysts is still to be assessed, their efficient recovery and reuse still remaining an issue to be addressed. New flower-like nanostructures have been used for enzyme immobilization, and increases both in enzyme stability and activity were reported; these new generation of nanobiocatalysts are potential candidates as process catalysts.[287]

16.4.3.4 Protein Engineering

The first reports on the use of protein engineering strategies (see Chapter 7) for improving enzyme activity and stability in organic solvents were published by Arnold and co-workers in the 1990s.[288] Three main protein engineering approaches have been used for improving enzyme properties:

- Rational design: Based on an in-depth knowledge of the enzyme sequence, structure, function and catalytic mechanism, computational design strategies are used to identify substitutions of key residues in order to improve a desired feature of the enzyme. Rational design has the advantage of a simpler screening; however, its major drawback is the difficulty in predicting the effect of a point mutation. Its application requires knowledge of the enzyme structure or a homology model, knowledge of the structure–function relationships and computational design.[43,289]
- Directed evolution: This implies an iterative two-step protocol. Firstly, a library containing a molecular diversity is generated by random mutagenesis and *in-vitro* recombination. Secondly, the library members

showing improvement in a desired feature are identified by high-throughput screening or selection. The selected library members are used as starting point for the next round. Its application requires a gene template and a functional assay for selection or screening of improved variants.[291–293]

- Semi-rational design: This corresponds to a wise mixture of the two strategies outlined above. Smart or knowledge-based libraries are designed using information on protein sequence, structure–function relationships, as well as computational predictive algorithms to pre-select promising target sites of modification and limiting the amino acid diversity for protein engineering. Its application requires knowledge of the enzyme structure or a homology model, knowledge of structure–function relationships, computational design, and a functional assay for selection or screening of improved variants, or direct characterization of a number of mutants.[290]

Enzyme stability in non-aqueous media has seldom been improved by rational design because of a lack of understanding of the enzyme–solvent molecular interactions. Progress in the understanding of protein stability based on advances in structural biology, bioinformatics, and molecular modeling are necessary for improving the success rate of rational design for tailoring enzyme stability in non-aqueous media.[229,290] The rapidly expanding databases of new enzyme sequences are now allowing the fast evaluation of natural protein variation and the incorporation of this information into both rational and semi-rational enzyme engineering programs. In the long term, this will lead to improved rules for the rational design of proteins and allow a rapid development of new improved enzymes for industrial applications in organic synthesis.[228]

Many interesting studies on the improvement of enzyme activity, stability and selectivity have been reported using directed evolution, thereby mimicking natural evolution for selection of improved traits. This strategy is allowing the improvement and discovery of enzyme activities that can be projected to industrial applications in organic synthesis.[230] An illustrative example is the protein engineering system developed for a ligninolytic enzyme consortium.[294]

Protein engineering strategies are undoubtedly very potent tools for enzyme improvement; however, they are more complex than others and may require considerable time and effort for achieving success, so that cost–benefit analysis should be taken into consideration when using this strategy for developing enzymes for industrial applications.

Strategies described above, with their pros and cons, are allowing the production of improved enzymes and new activities which are already changing the scenario of industrial biocatalysis. The combined use of more than one of the strategies presented is likely to produce some synergy in obtaining improved catalysts. For example, chemical modification of the enzyme has been considered as a tool for improving enzyme immobilization by introducing additional functional groups in the enzyme surface for a better interaction with the support;[42] protein engineering tools have also been used for the same purpose.[295,296]

Whatever the strategy or strategies used for improving enzyme performance in synthetic reactions, it has to be kept in mind that for industrial purposes, they should be as simple as possible, scalable to production level and hopefully not representing a significant part of the enzyme production cost (see Chapter 17).

16.4.4 Industrial Perspective of Enzyme Catalysis: Reality, Challenges and Opportunities

Enzymes have been used inadvertently since ancient times, but the rational use of enzymes as industrial catalysts can be traced back to the late 19th century, gaining some relevance in the first half of the 20th century.[14] As previously mentioned, enzymes were then mostly crude hydrolase preparations from plant or animal origin, acting in soluble form in aqueous media, usually catalyzing reactions of depolymerization with a limited added-value and under poorly controlled conditions. A strong impulse for enzyme catalysis came after World War II as a consequence of the development of industrial microbiology and biotechnology that allowed the massive production of microbial enzymes under controlled conditions; this was a significant step forward since intensive and reliable production of a wider spectrum of enzymes was then possible. Enzymatic processes under controlled and optimized conditions for enzyme performance were developed in the 1960s, and later on, the development of immobilized enzymes widened the scope of biocatalysis allowing the use of less stable, non-hydrolytic and complex intracellular enzymes, as illustrated in Section 16.1. In parallel, the impressive developments of genetic engineering since 1970 allowed the expression of foreign genes into suitable microbial hosts, thereby expanding the scope of biocatalysis further; some time later, the development of protein engineering tools allowed the fine-tuning of the enzyme structure for better performance, and high-throughput-screening and metagenomic tools were developed taking full advantage of the microbial diversity as an endless enzyme reservoir (see Chapters 1 and 2).[122] Until the 1980s, most of the industrial enzyme use still referred to low added-value processes using mostly hydrolases in conventional aqueous media. These are now for the most part well-established technologies still representing the major share of the enzyme market; however, there are significant improvements still ongoing, especially related to food, feed and biofuels. The latter deserves a special mention since second generation bioethanol and derived products are demanding highly active and robust hydrolases (mostly cellulases and hemicellulases) effective at high solid concentrations and under harsh conditions, which is a challenge being already assumed;[297] new non-hydrolytic enzymes capable of efficient degradation of cellulose from complex lignocellulosic matrices are also coming into play.[298] It is to be expected that second generation biofuels will represent a breakthrough in terms of enzyme market as new high-tonnage plants of biofuels from biomass will come into operation in the next few years, within the concept of biorefinery.[299,300] However, the extent to which biofuels can be sustained in the short term will be very much dependent on the volatility of oil prices.[301]

The intrinsically high added-value of processes of organic synthesis was a powerful driving force to explore enzymes as biocatalysts for such processes. This is undoubtedly the leading edge for the development of industrial biocatalysis. There is now a very impressive number of synthetic reactions that have been conducted with enzymes and enzyme-containing cells;[302] some of them have reached industrial status and many more are on the verge of acquiring it.

Lipases play a central role in biocatalysis for organic synthesis because they perform well in non-aqueous environments and are robust non-coenzyme-requiring hydrolases. Their use in large-scale processes, like biodiesel production, has been up to now restricted by the high incidence of the enzyme cost, being for the moment not competitive with ongoing production by chemical synthesis.[303,304] But smaller scale high added-value processes using lipases are now in commercial operation, mostly for the production of biopharmaceuticals, taking advantage of their exquisite enantioselectivity and enantiospecificity. Some processes worth-mentioning are the industrial production of (*S*)-ibuprofen with *Candida rugosa* lipase, the oral hypoglycemic drug saxagliptin with immobilized *Candida antarctica* lipase B (CALB), a precursor of the anticancer drug paclitaxel with *Pseudomonas cepacia* lipase, the antihypertensive drug diltiazem with *Serratia marcescens* lipase and the HIV inhibitor carbovir with *Pseudomonas fluorescens* lipase.[305,306] Lipases have also been considered as key enzymes for the cosmetics industry.[307]

Oxidoreductases are powerful catalysts for producing chiral building blocks for the synthesis of pharmaceuticals and other bioactive molecules. However, they are complex coenzyme-requiring enzymes, so their use as process biocatalysts is challenged by stabilization and coenzyme regeneration strategies. Despite such complexities, oxidoreductases are being increasingly used in the synthesis of high added-value pharmaceuticals. This is illustrated by the case of the synthesis of the lateral chain of the anticholesterolemic drug atorvastatin in which one step is catalyzed by a NADPH-dependent engineered ketoreductase (glucose dehydrogenase used for coenzyme regeneration) and one step by a halohydrin dehalogenase, by one step in the synthesis of amphetamines intermediates catalyzed by a *Rhodococcus erythropolis* alcohol dehydrogenase (formate dehydrogenase used for coenzyme regeneration), and by one step in the synthesis of the leukotriene receptor antagonist montelukast, used in the treatment of asthma, catalyzed by a NADPH-dependent engineered ketoreductase (glucose dehydrogenase used for coenzyme regeneration). In all such reactions, advantage is taken of the high stereoselectivity of those enzymes. The use of whole cells containing these poorly stable coenzyme-dependent enzymes appears attractive and in fact it is estimated that commercial processes with oxidoreductases are carried out more frequently in whole-cell bioconversions (see Chapter 12), despite their potential drawbacks due to cell inhibition by substrates or products of reaction and their degradation by cell metabolism, and also because of environmental considerations. Some examples of commercial processes using oxidoreductases in whole-cell format are the use of *Rhodococcus erythropolis* in one step of the synthesis of the HIV inhibitor atazanivir, and the use of *Escherichia coli* and *Geotrichum candidum* cells in one step of the synthesis of the lateral chain of atorvastatin.

The commercial use of other enzymes, such as transaminases and aldolases (lyases), is still incipient. An engineered transaminase has been used in the stereoselective amination of prositagliptin to produce sitagliptin,[308] a compound used as a hypoglycemic drug, as an alternative to the chemocatalytic route. Aldolases are also starting to be used in the synthesis of valuable precursor molecules with multiple chiral centers, taking advantage of their strict diastereoselectivity, as exemplified by the synthesis of intermediates of statins which are used as anticholesterolemic drugs.[306]

Up to now, enzymes have been mostly used as single entities in one-step processes. However, a growing trend in biocatalysis is the use of multi-enzymatic reaction systems (see Chapters 13 and 14). Several types of such systems have been envisaged, including the use of auxiliary enzymes for cofactor regeneration already analyzed. Multi-enzyme systems include cascade reactions in which enzymes act in separate reactors, the product of one reaction being the substrate of the next one, and in one-pot processes in which all enzymes involved are acting together in a single reactor. The former system has the obvious advantage that each enzyme is acting at its optimum conditions and should be a preferred strategy when significant compromise between operational variables (*i.e.* temperature, pH) exist or when reaction rates are quite different in magnitude. One-pot systems, on the contrary, will work at compromised operational conditions, which means that each enzyme will be acting at sub-optimal conditions; this may or may not be important depending on how distant such optimum conditions are. One-pot systems are quite flexible systems in which the enzymes can act in sequence, in parallel or in network; for sequential reactions, the concentration of the intermediates is low, which can be detrimental for the following enzymes but beneficial if product and/or substrate inhibitions are strong; they also allow unfavorable equilibria to be driven to completion. A distinctive advantage of one-pot systems is that separation and purification of intermediates is avoided, so that considerable process improvement can be obtained by significant saving in downstream operations. Optimally balancing the conditions of operation is quite challenging and the use of computational tools has been proposed for process simulation under different conditions.[309] A thorough analysis of the options for multi-enzyme processes in terms of process configuration, reactor types, modes of operation and biocatalyst format has been presented by Xue and Woodley (see Chapter 17).[310] Enzymes within a one-pot system can be co-immobilized within the same matrix for performing sequential reactions, which may be advantageous to reduce the impact of diffusional limitations.[311] Of particular interest in this respect is the co-immobilization of enzymes in CLEAs (combi-CLEAs), offering a simple system with the advantage of the close proximity of the enzymes.[312] Potential applications of combi-CLEAs are numerous. A triple combi-CLEA of α-amylase, glucoamylase and pullulanase was used for the one-pot, one-catalyst hydrolysis of starch.[313] We have recently used combi-CLEAs of arabinofuranosidase and β-glucosidase for enhancing aroma release in wine and obtained a better performance compared to the separate enzymes in free form.[314]

It may be premature to assess the magnitude of the impact of enzyme catalysis in the production of high added-value bioactive compounds. However, it is clear that enzyme catalysis is gaining a prominent role in industrial biotechnology as applied to food supply,[315] health,[63] environmental purposes (see Chapter 18)[316] and sustainable fuel production,[317] which are major issues in modern society. Sustainability is an enforced task for technology, and biotechnology is a powerful driving-force for sustainability; within that framework, biocatalysis will play a part especially in those areas, like the pharmaceutical and fine chemical industries, where chemical technology still prevails.[57]

Present limitations of biocatalysis mostly refer to the poor stability of enzymes under process conditions and their high cost. The former is being overtaken by the impressive advances in genetic and protein engineering,[318–320] biocatalyst and medium engineering.[32,296] Enzyme cost has been in fact a major hurdle for industrial biocatalysis; however, few insights on cost analysis of biocatalysts for industrial processes exist, even though some guidelines have been proposed for an early assessment of the impact of catalyst cost on biocatalytic processes.[321] Impact of enzyme cost on production cost will be less significant for high added-value processes; in the case of large scale, more conventional processes, prices for enzymes are expected to keep on declining as production technology develops to high standards of efficiency and competition for supply gets tougher.

References

1. J. E. Puskas, J. Kantor and G. Shrikhande, *AIChE J.*, 2016, **63**, 266.
2. P. W. Lambert and J. L. Meers, *Philos. Trans. R. Soc. B*, 1983, **300**, 263.
3. A. L. Demain and P. Vaishnav, *Biotechnol. Adv.*, 2009, **27**, 297.
4. O. Kirk, T. Vedel Borchert and C. Crone Fuglsang, *Curr. Opin. Biotechnol.*, 2002, **13**, 345.
5. P. Michael-Sinclair, *Chem. Eng.*, 1965, **72**, 90.
6. S. H. Bhosale, M. B. Rao and V. V. Deshpande, *Microbiol. Rev.*, 1996, **60**, 300.
7. A. Illanes and L. Wilson, in *Enzyme Biocatalysis: Principles and Applications*, ed. A. Illanes, Springer, London, 2008, p. 273.
8. K. Buchholz, *Appl. Microbiol. Biotechnol.*, 2016, **100**, 3825.

9. A. Liese, K. Seelbach, A. Buchholz and J. Haberland, in *Industrial Biotransformations*, ed. A. Liese, K. Seelbach and C. Wandrey, Wiley-VCH Verlag, Weinheim, 2nd edn, 2006, p. 447.
10. S. Sanchez and A. L. Demain, *Org. Process Res. Dev.*, 2011, **15**, 224.
11. E. García-Junceda, J. F. García-García and A. Bastida, *Bioorg. Med. Chem.*, 2004, **12**, 1817.
12. A. M. Klibanov, *CHEMTECH*, 1986, **6**, 354.
13. A. M. Klibanov, *Nature*, 2001, **409**, 241.
14. S. L. Neidleman, in *Biocatalysts for Industry*, ed. J. S. Dordick, Plenum Press, New York, 1991, pp. 21–33.
15. A. Schmid, J. S. Dordick, B. Hauer, A. Kiener, M. Wubbolts and B. Witholt, *Nature*, 2001, **409**, 258.
16. J. M. Guisan, in *Immobilization of Enzymes and Cells*, Springer Dordrecht, 2006, pp. 1–14.
17. S. Panke, M. Held and M. Wubbolts, *Curr. Opin. Biotechnol.*, 2004, **15**, 272.
18. O. Kuchner and F. H. Arnold, *Trends Biotechnol.*, 1997, **15**, 523.
19. R. J. Kazlauskas and U. T. Bornscheuer, *Nat. Chem. Biol.*, 2009, **5**, 526.
20. P. T. Anastas and M. M. Kirchhoff, *Acc. Chem. Res.*, 2002, **35**, 686.
21. S. Gamper-Rabindran and S. R. Finger, *J. Regul. Econ.*, 2013, **43**, 1.
22. I. Delidovich and R. Palkovits, *Green Chem.*, 2016, **18**, 590.
23. D. R. Abernethy, A. J. DeStefano, T. L. Cecil, K. Zaidiand and R. L. Williams, *Pharm. Res.*, 2010, **27**, 750.
24. R. Ciriminna and M. Pagliaro, *Org. Process Res. Dev.*, 2013, **17**, 1479.
25. R. Wohlgemuth, *Curr. Opin. Biotechnol.*, 2010, **21**, 713.
26. J. Tao, L. Zhao and N. Ran, *Org. Process Res. Dev.*, 2007, **11**, 259.
27. U. T. Bornscheuer and R. J. Kazlauskas, *Hydrolases in Organic Synthesis*, Wiley VCH, Weinheim, 1999, 336 p.
28. W. A. van der Donk and H. Zhao, *Curr. Opin. Biotechnol.*, 2003, **14**, 421.
29. W. Liu and P. Wang, *Biotechnol. Adv.*, 2007, **25**, 369.
30. B. M. Nestl, S. C. Hammer, B. A. Nebel and B. Hauer, *Angew. Chem., Int. Ed.*, 2014, **53**, 3070.
31. P. V. Iyer and L. Ananthanaraya, *Process Biochem.*, 2008, **43**, 1019.
32. A. Illanes, in *White Biotechnology for Sustainable Chemistry*, ed. M. A. Z. Coelho and B. D. Ribeiro, The Royal Society of Chemistry, Cambridge, 2016, ch. 3, pp. 36–51.
33. E. Busto, V. Gotor-Fernández and V. Gotor, *Chem. Soc. Rev.*, 2010, **39**, 4504.
34. F. Cherubini, *Energy Convers. Manage.*, 2010, **51**, 1412.
35. B. H. Junker and H. Y. Wang, *Biotechnol. Bioeng.*, 2006, **95**, 226.
36. M. Ferrer, O. Golyshina, A. Beloqui and P. N. Golyshin, *Curr. Opin. Microbiol.*, 2007, **10**, 207.
37. H. L. Steele, K.-E. Jaeger, R. Daniel and W. R. Streit, *J. Mol. Microbiol. Biotechnol.*, 2009, **16**, 25.
38. K. Hernandez and R. Fernandez-Lafuente, *Enzyme Microb. Technol.*, 2011, **48**, 107.

39. P. A. Dalby, *Curr. Opin. Struct. Biol.*, 2011, **21**, 473.
40. C. Mateo, J. M. Palomo, G. Fernandez-Lorente, J. M. Guisan and R. Fernandez-Lafuente, *Enzyme Microb. Technol.*, 2007, **40**, 1451.
41. R. Sheldon, *Appl. Microbiol. Biotechnol.*, 2011, **92**, 467.
42. D. A. Cowan and R. Fernandez-Lafuente, *Enzyme Microb. Technol.*, 2011, **49**, 326.
43. K. M. Koeller and C.-H. Wong, *Nature*, 2001, **409**, 232.
44. A. C. Spieß, W. Eberhard, M. Peters, M. F. Eckstein, L. Greiner and J. Büchs, *Chem. Eng. Process.*, 2008, **47**, 1034.
45. C. Garcia-Galan, A. Berenguer-Murcia, R. Fernandez-Lafuente and R. C. Rodrigues, *Adv. Synth. Catal.*, 2011, **353**, 2285.
46. M. Erbeldinger, A. J. Mesiano and A. J. Russell, *Biotechnol. Prog.*, 2000, **16**, 1129.
47. V. Gotor, I. Alfonso and E. García-Urdiales, *Asymmetric Organic Synthesis with Enzymes*, Wiley-VCH, Weinheim, 2008, 324 pp.
48. H. Caner, E. Groner and L. Levy, *Drug Discovery Today*, 2004, **9**, 105.
49. J. M. Woodley, *Trends Biotechnol.*, 2008, **26**, 321.
50. M. C. Núñez, M. E. García-Rubiño, A. Conejo-García, O. Cruz-López, M. Kimatrai, M. A. Gallo, A. Espinosa and J. M. Campos, *Curr. Med. Chem.*, 2009, **16**, 2064.
51. H. Pellissier, *Tetrahedron*, 2008, **64**, 1563.
52. A. J. J. Straathof, S. Panke and A. Schmid, *Curr. Opin. Biotechnol.*, 2002, **13**, 548.
53. S. M. Thomas, R. DiCosimo and V. Nagarajan, *Trends Biotechnol.*, 2002, **20**, 238.
54. N. Ran, L. Zhao, Z. Chen and J. Tao, *Green Chem.*, 2008, **10**, 361.
55. U. T. Bornscheuer and R. J. Kazlauskas, *Angew. Chem., Int. Ed.*, 2004, **43**, 6032.
56. P. Gatti-Lafranconi and F. Hollfelder, *ChemBioChem*, 2013, **14**, 285.
57. H.-P. Meyer, *Org. Process Res. Dev.*, 2011, **15**, 180.
58. R. A. Sheldon, *Chem. Soc. Rev.*, 2012, **41**, 1437.
59. S. Wenda, S. Illner, A. Mell and U. Kragl, *Green Chem.*, 2011, **13**, 3007.
60. R. Ciriminna and M. Pagliaro, *Org. Process Res. Dev.*, 2013, **17**, 14794.
61. M. Gavrilescu and Y. Chisti, *Biotechnol. Adv.*, 2005, **23**, 471.
62. M. H. A. Janssen, http://repository.tudelft.nl/file/200584/169747, 2006, ISBN:90-9020754-6.
63. G. W. Huisman and S. J. Collier, *Curr. Opin. Chem. Biol.*, 2013, **17**, 284.
64. J. W. Woodley, *Curr. Opin. Chem. Biol.*, 2013, **17**, 310.
65. Y. Li and P. C. Cirino, *Biotechnol. Bioeng.*, 2014, **111**, 1273.
66. R. C. Rodrigues, C. Ortiz, A. Berenguer-Murcia, R. Torres and R. Fernández-Lafuente, *Chem. Soc. Rev.*, 2013, **42**, 6290.
67. D. J. Pollard and J. M. Woodley, *Trends Biotechnol.*, 2007, **25**, 66.
68. R. A. Sheldon, *Green Chem.*, 2016, **18**, 3180.
69. S. F. Rach and F. E. Kühn, *Sustainability*, 2009, **1**, 35.
70. P. Anastas and N. Eghbali, *Chem. Soc. Rev.*, 2010, **39**, 301.
71. J. L. Tucker, *Org. Process Res. Dev.*, 2006, **10**, 315.

72. A. M. Klibanov, *Trends Biotechnol.*, 1997, **15**, 97.
73. V. Bansal, Y. Delgado, M. D. Legault and G. Barletta, *Molecules*, 2012, **17**, 1870.
74. A. M. P. Koskinen and A. M. Klibanov, *Enzymatic Reactions in Organic Media*, Blackie Academic & Professional, 1996, 314 pp.
75. M. N. Gupta and I. Roy, *Eur. J. Biochem.*, 2004, **271**, 2575.
76. V. Gotor-Fernández, R. Brieva and V. Gotor, *J. Mol. Catal. B: Enzym.*, 2006, **40**, 111.
77. F. Hasan, A. Ali Shah and A. Hameed, *Enzyme Microb. Technol.*, 2006, **39**, 235.
78. P. Adlercreutz, *Chem. Soc. Rev.*, 2013, **42**, 6406.
79. F. Bordusa, *Chem. Rev.*, 2002, **102**, 4817.
80. D. Kumar and T. C. Bhalla, *Appl. Microbiol. Biotechnol.*, 2005, **68**, 726.
81. E. Quiroga, N. Priolo, D. Obregón, J. M. Marchese and S. Barberis, *Biochem. Eng. J.*, 2008, **39**, 115.
82. R. A. Abusham, R. N. R. Rahman, A. B. Salleh and M. Basri, *Microb. Cell Fact.*, 2009, **8**, 20.
83. L. M. van Langen, N. H. P. Oosthoek, D. T. Guranda, F. van Rantwijk, V. Švedas and R. A. Sheldon, *Tetrahedron: Asymmetry*, 2000, **11**, 4593.
84. A. Illanes, L. Wilson, C. Altamirano, Z. Cabrera, L. Alvarez and C. Aguirre, *Enzyme Microb. Technol.*, 2007, **40**, 195.
85. O. Romero, J. Vergara, R. Fernández-Lafuente, J. M. Guisán, A. Illanes and L. Wilson, *Biotechnol. Bioeng.*, 2009, **103**, 472.
86. F. van Rantwijk, M. Woudenberg-van Oosterom and R. A. Sheldon, *J. Mol. Catal. B: Enzym.*, 1999, **6**, 511.
87. M. Faijes and A. Planas, *Carbohydr. Res.*, 2007, **342**, 1581.
88. I. Mladenoska, E. Winkelhausen and S. Kuzmanova, *Food Technol. Biotechnol.*, 2008, **46**, 311.
89. M. Y. Rather and S. Mishra, *Sustainable Chem. Processes*, 2013, **1**, 7.
90. E. García-Urdiales, I. Alfonso and V. Gotor, *Chem. Rev.*, 2005, **105**, 313.
91. M. M. Musa, K. I. Ziegelmann-Fjeld, C. Vieille, J. G. Zeikus and R. Phillips, *Angew. Chem., Int. Ed.*, 2007, **46**, 3091.
92. I. Oroz-Guinea, K. Hernández, F. Camps Bres, C. Guérard-Hélaine, M. Lemaire, P. Clapés and E. García-Junceda, *Adv. Synth. Catal.*, 2015, **357**, 1951.
93. H.-P. Meyer, E. Eichhorn, S. Hanlon, S. Lütz, M. Schürmann, R. Wohlgemuth and R. Coppolecchia, *Catal. Sci. Technol.*, 2013, **3**, 29.
94. L. Soler, A. Illanes and L. Wilson, *Catal. Today*, 2015, **259**, 177.
95. P. Spizzo, A. Basso, C. Ebert, L. Gardossi, V. Ferrario, D. Romano and F. Molinari, *Tetrahedron*, 2007, **63**, 11005.
96. I. C. R. Costa, F. K. Sutili, G. V. V. da Silva, S. G. F. Leite, L. S. M. Miranda and R. O. M. A. de Souza, *J. Mol. Catal. B: Enzym.*, 2014, **102**, 127.
97. Z. Jin, S. Liang, X. Zhang, S. Han, C. Ren, Y. Lin and S. Zheng, *Biotechnol. Bioprocess Eng.*, 2013, **18**, 365.

98. E. Castillo, F. Pezzotti, A. Navarro and A. López-Munguía, *J. Biotechnol.*, 2003, **102**, 251.
99. L. A. Lerin, A. Richetti, R. Dallago, H. Treichel, M. A. Mazutti, J. V. Oliveira, O. A. C. Antunes, E. G. Oestreicher and D. de Oliveira, *Food Bioprocess Technol.*, 2012, **5**, 1068.
100. E. Su and D. Wei, *J. Mol. Catal. B: Enzym.*, 2008, **55**, 118.
101. L. S. Yon, F. N. Gonawan, A. H. Kamaruddin and M. Hekarl Uzir, *Ind. Eng. Chem. Res.*, 2013, **52**, 9441.
102. M. Akimoto, M. Izawa, K. Hoshino, K.-I. Abe and H. Takahashi, *Appl. Biochem. Biotechnol.*, 2003, **104**, 105.
103. M. M. Soumanou and U. T. Bornscheuer, *Eur. J. Lipid Sci. Technol.*, 2003, **105**, 656.
104. L. Santibáñez, L. Wilson and A. Illanes, *J. Am. Oil Chem. Soc.*, 2014, **91**, 405.
105. R. Costa Rodrigues, G. Volpato, K. Wada and M. A. Záchia Ayub, *J. Am. Oil Chem. Soc.*, 2008, **85**, 925.
106. A. Sánchez, F. Valero, J. Lafuente and C. Sola, *Enzyme Microb. Technol.*, 2000, **27**, 157.
107. D. Reyes-Duarte, N. Lopez-Cortes, P. Torres, F. Comelles, J. L. Parra, S. Peña, A. V. Ugidos, A. Ballesteros and F. J. Plou, *J. Am. Oil Chem. Soc.*, 2011, **88**, 57.
108. A. Sengupta and M. Ghosh, *Eur. J. Lipid Sci. Technol.*, 2011, **113**, 763.
109. C. Capello, U. Fischer and K. Hungerbühler, *Green Chem.*, 2007, **9**, 927.
110. S. Aparicio and R. Alcalde, *Green Chem.*, 2009, **11**, 65.
111. A. E. Díaz-Álvarez, J. Francos, B. Lastra-Barreira, P. Crochet and V. Cadierno, *Chem. Commun.*, 2011, **47**, 6208.
112. Y. Gu and F. Jérôme, *Green Chem.*, 2010, **12**, 1127.
113. C. Jiménez-González, A. D. Curzons, D. J. C. Constable and V. L. Cunningham, *Clean Technol. Environ. Policy*, 2005, **7**, 42.
114. R. K. Henderson, C. Jiménez-González, D. J. C. Constable, S. R. Alston, G. G. A. Inglis, G. Fisher, J. Sherwood, S. P. Binks and A. D. Curzons, *Green Chem.*, 2011, **13**, 854.
115. D. Prat, J. Hayler and A. Wells, *Green Chem.*, 2014, **16**, 4546.
116. P. G. Jessop, *Green Chem.*, 2011, **13**, 1391.
117. C.-J. Li and B. M. Trost, *Proc. Natl. Acad. Sci. U. S. A.*, 2008, **105**, 13197.
118. K. M. Polizzi, A. S. Bommarius, J. M. Broering and J. F. Chaparro-Riggers, *Curr. Opin. Chem. Biol.*, 2007, **11**, 220.
119. A. K. Prasad, M. Husain, B. K. Singh, R. K. Gupta, V. K. Manchanda, C. E. Olsend and V. S. Parmar, *Tetrahedron Lett.*, 2005, **46**, 4511.
120. S. Shah and M. N. Gupta, *Process Biochem.*, 2007, **42**, 409.
121. H.-H. Li, Y.-H. He, Y. Yuan and Z. Guan, *Green Chem.*, 2011, **13**, 185.
122. A. Illanes, A. Cauerhff, L. Wilson and G. R. Castro, *Bioresour. Technol.*, 2012, **115**, 48.
123. S. Torres and G. Castro, *Food Technol. Biotechnol.*, 2004, **42**, 271.
124. N. Doukyu and H. Ogino, *Biochem. Eng. J.*, 2010, **48**, 270.

125. T. Montes, V. Grazú, I. Manso, B. Galán, F. López-Gallego, R. González, J. A. Hermoso, J. L. García, J. M. Guisán and R. Fernández-Lafuente, *Adv. Synth. Catal.*, 2007, **349**, 459.
126. M. Arroyo, R. Torres-Guzmán, I. de la Mata, M. Castillón and C. Acebal, *Biotechnol. Prog.*, 2000, **16**, 368.
127. C. Aguirre, M. Toledo, V. Medina and A. Illanes, *Process Biochem.*, 2002, **38**, 351.
128. S. Tang, C. L. Jones and H. Zhao, *Bioresour. Technol.*, 2013, **129**, 667.
129. A. Illanes and A. Fajardo, *J. Mol. Catal. B: Enzym.*, 2001, **11**, 587.
130. D. S. Clark, *Philos. Trans. R. Soc. London B*, 2004, **359**, 1299.
131. N. A. Turner and E. N. Vulfson, *Enzyme Microb. Technol.*, 2000, **27**, 108.
132. J. Mansfeld and R. Ulbrich-Hofmann, *Biotechnol. Bioeng.*, 2007, **97**, 672.
133. H. Ogino and H. Ishikawa, *J. Biosci. Bioeng.*, 2001, **91**, 109.
134. E. P. Hudson, R. K. Eppler and D. S. Clark, *Curr. Opin. Biotechnol.*, 2005, **16**, 6373.
135. V. Stepankova, S. Bidmanova, T. Koudelakova, Z. Prokop, R. Chaloupkova and J. Damborsky, *ACS Catal.*, 2013, **3**, 2823.
136. F. Secundo, G. Carrea, C. Tarabiono, P. Gatti-Lafranconi, S. Brocca, M. Lotti, K. E. Jaeger, M. Puls and T. Eggert, *J. Mol. Catal. B: Enzym.*, 2006, **39**, 166.
137. M. Petkar, A. Lali, P. Caimi and M. Daminati, *J. Mol. Catal. B: Enzym.*, 2006, **39**, 83.
138. B. Orlich and R. Schomäcker, *Adv. Biochem. Eng./Biotechnol.*, 2002, **75**, 185.
139. C. Aguirre, I. Concha, J. Vergara, R. Riveros and A. Illanes, *Process Biochem.*, 2010, **45**, 1163.
140. P. Adlercreutz, *Biocatalysis in Organic Synthesis with Enzymes in Non-Aqueous Media*, ed. G. Carrea and S. Riva, Wiley-VCH, Weinheim, 2008, pp. 3–23.
141. M. P. C. Marques, N. M. T. Lourenço, P. Fernandes and C. C. C. R. de Carvalho, in *Green Solvents I. Properties and Applications in Chemistry*, ed. A. Mohammad and Inamuddin, Springer, Dordrecht, 2012, pp. 147–173.
142. F. M. Kerton and F. Marriott, *Alternative Solvents for Green Chemistry*, RSC Publishing, Cambridge, UK, 2nd edn, 2013, 350 pp.
143. E. Barzana, A. Klibanov and M. Karel, *Appl. Biochem. Biotechnol.*, 1987, **15**, 25.
144. B. C. Dravis, K. E. LeJeune, A. D. Hetro and A. J. Russell, *Biotechnol. Bioeng.*, 2000, **69**, 235.
145. Ž. Knez and M. Habulin, *J. Supercrit. Fluids*, 2002, **23**, 29.
146. S. V. Kamat, B. Iwaskewycz, E. J. Beckman and A. J. Russell, *Proc. Natl. Acad. Sci. U. S. A.*, 1993, **90**, 2940.
147. S. V. Kamat, E. J. Beckman and A. J. Russell, *Crit. Rev. Biotechnol.*, 1995, **15**, 41.
148. E. Ramsey, Q. Sun, Z. Zhang, C. Zhang and W. Gou, *J. Environ. Sci.*, 2009, **21**, 720.

149. E. Celia, E. Cernia, C. Palocci, S. Soro and T. Turchet, *J. Supercrit. Fluids*, 2005, **33**, 193.
150. K. Rezaei, F. Temelli and E. Jenab, *Biotechnol. Adv.*, 2007, **25**, 272.
151. M. A. Kane, G. A. Baker, S. Pandey and F. V. Bright, *Langmuir*, 2000, **16**, 4901.
152. M. Miyazaki, M. Shibue, K. Ogino, H. Nakamura and H. Maeda, *Chem. Commun.*, 2001, **18**, 1800.
153. T. Matsuda, *J. Biosci. Bioeng.*, 2013, **115**, 233.
154. P. Lozano, T. de Diego, D. Carrié, M. Vaultier and J. L. Iborra, *Biotechnol. Prog.*, 2003, **19**, 380.
155. F. C. Loeker, C. J. Duxbury, R. Kumar, W. Gao, R. A. Gross and S. M. Howdle, *Macromolecules*, 2004, **37**, 2450.
156. M. D. Romero, L. Calvo, C. Alba, M. Habulin, M. Primožič and Ž. Knez, *J. Supercrit. Fluids*, 2005, **33**, 77.
157. T. Matsuda, T. Harada, K. Nakamura and T. Ikariya, *Tetrahedron: Asymmetry*, 2005, **16**, 909.
158. V. Rathore and G. Madras, *Fuel*, 2007, **86**, 2650.
159. M. Habulin, S. Sabeder and Ž. Knez, *J. Supercrit. Fluids*, 2008, **45**, 338.
160. R. García-Arrazola, D. A. López-Guerrero, M. Gimeno and E. Bárzana, *J. Supercrit. Fluids*, 2009, **51**, 197.
161. H. Shekarchizadeh, M. Kadivar, H. S. Ghaziaskar and M. Rezayat, *J. Supercrit. Fluids*, 2009, **49**, 209.
162. Ž. Knez, *J. Supercrit. Fluids*, 2009, **47**, 357.
163. P. Lozano, T. de Diego and J. L. Iborra, *Chem. Today*, 2007, **25**, 76.
164. P. Lozano, E. Garcia-Verdugo, S. V. Luis, M. Pucheault and M. Vaultier, *Curr. Org. Synth.*, 2011, **8**, 810.
165. S. V. Malhotra, V. Kumar and V. S. Virinder, *Curr. Org. Synth.*, 2007, **4**, 370.
166. H. Zhao, *J. Chem. Technol. Biotechnol.*, 2010, **85**, 891.
167. Z. Yang and W. Pan, *Enzyme Microb. Technol.*, 2005, **37**, 19.
168. J. Durand, E. Teuma and M. Gómez, *C. R. Chim.*, 2007, **10**, 152.
169. F. van Rantwijk, F. Secundo and R. A. Sheldon, *Green Chem.*, 2006, **8**, 282.
170. Y. Abe, K. Kude, S. Hayase, M. Kawatsura, K. Tsunashima and T. Itoh, *J. Mol. Catal. B: Enzym.*, 2008, **51**, 81.
171. F. Ganske and U. T. Bornscheuer, *Org. Lett.*, 2005, 7, 3097.
172. T. De Diego, P. Lozano, M. A. Abad, K. Steffensky, M. Vaultier and J. L. Iborra, *J. Biotechnol.*, 2009, **140**, 234.
173. M. H. Katsoura, A. C. Polydera, L. Tsironis, A. D. Tselepis and H. Stamatis, *J. Biotechnol.*, 2006, **123**, 491.
174. S. H. Ha, M. N. Lan, S. H. Lee, S. M. Hwang and Y.-M. Koo, *Enzyme Microb. Technol.*, 2007, **41**, 480.
175. N. I. Ruzich and A. S. Bassi, *Can. J. Chem. Eng.*, 2010, **88**, 277.
176. N. Wehofsky, C. Wespe, V. Cerovsky, A. Pech, E. Hoess, R. Rudolph and F. Bordusa, *ChemBioChem*, 2008, **9**, 1493.
177. J. A. Laszlo and D. L. Compton, *Biotechnol. Bioeng.*, 2001, **75**, 181.

178. J.-C. Plaquevent, J. Levillain, F. Guillen, C. Malhiac and A.-C. Gaumont, *Chem. Rev.*, 2008, **108**, 5035.
179. H. Vallette, L. Ferron, G. Coquerel, A.-C. Gaumont and J.-C. Plaquevent, *Tetrahedron Lett.*, 2004, **45**, 1617.
180. H. Noritomi, K. Suzuki, M. Kikuta and S. Kato, *Biochem. Eng. J.*, 2009, **47**, 27.
181. K. Okrasa, E. Guibé-Jampel and M. Therisod, *Tetrahedron: Asymmetry*, 2003, **14**, 2487.
182. W.-Y. Lou, R. Xu and M.-H. Zong, *Biotechnol. Lett.*, 2005, **27**, 1387.
183. M. C. Galan, R. A. Jones and A.-T. Tran, *Carbohydr. Res.*, 2013, **375**, 35.
184. M. Moniruzzaman, K. Nakashima, N. Kamiya and M. Goto, *Biochem. Eng. J.*, 2010, **48**, 295.
185. E. Fehér, B. Major, K. Bélafi-Bakó and L. Gubicza, *Biochem. Soc. Trans.*, 2007, **35**, 1624.
186. J. Gorke, F. Srienc and R. Kazlauskas, *Biotechnol. Bioproc. Eng.*, 2010, **15**, 40.
187. M. Moniruzzaman, N. Kamiya, K. Nakashima and M. Goto, *Green Chem.*, 2008, **10**, 497.
188. J. H. Clark and S. J. Tavener, *Org. Process Res. Dev.*, 2007, **11**, 149.
189. Y. Zhang, B. Bhavik, R. Bakshi and E. Sahle Demessie, *Environ. Sci. Technol.*, 2008, **42**, 1724.
190. M. K. Potdar, G. F. Kelso, L. Schwarz, C. Zhang and M. T. W. Hearn, *Molecules*, 2015, **20**, 16788.
191. Y. Dai, J. van Spronsen, G.-J. Witkamp, R. Verpoorte and Y. H. Choi, *Anal. Chim. Acta*, 2013, **766**, 61.
192. W. Bi, M. Tian and K. H. Row, *J. Chromatogr. A*, 2013, **1285**, 22.
193. A. P. Abbott, D. Boothby, G. Capper, D. L. Davies and R. K. Rasheed, *J. Am. Chem. Soc.*, 2004, **126**, 9142.
194. A. Hayyan, F. S. Mjalli, I. M. Al-Nashef, Y. M. Al-Wahaibi, T. Al-Wahaibi and M. A. Hashim, *J. Mol. Liq.*, 2013, **178**, 137.
195. A. Hayyan, F. S. Mjalli, I. M. Al-Nashef, Y. M. Al-Wahaibi, T. Al-Wahaibi and M. A. Hashim, *Thermochim. Acta*, 2012, **541**, 70.
196. C. D'Agostino, R. C. Harris, A. P. Abbott, L. F. Gladden and M. D. Mantle, *Phys. Chem. Chem. Phys.*, 2011, **13**, 21383.
197. A. P. Abbott, R. C. Harris, K. S. Ryder, C. D'Agostino, L. F. Gladden and M. Mantle, *Green Chem.*, 2011, **13**, 82.
198. B. Tang and K. H. Row, *Monatsh. Chem.*, 2013, **144**, 1427.
199. Z.-L. Huang, B.-P. Wu, Q. Wen, T.-X. Yang and Z. Yang, *J. Chem. Technol. Biotechnol.*, 2014, **89**, 1975.
200. E. L. Smith, A. P. Abbott and K. S. Ryder, *Chem. Rev.*, 2014, **114**, 11060.
201. X.-L. Qi, X. Peng, Y.-Y. Huang, L. Li, Z.-F. Wei, Y.-G. Zu and Y.-J. Fu, *Ind. Crops Prod.*, 2015, **70**, 142.
202. B.-Y. Zhao, P. Xu, F.-X. Yang, H. Wu, M.-H. Zong and W.-Y. Lou, *ACS Sustainable Chem. Eng.*, 2015, **3**, 2746.
203. Y. Dai, R. Verpoorte and Y. H. Choi, *Food Chem.*, 2014, **159**, 116.

204. H. G. Morrison, C. C. Sun and S. Neervannan, *Int. J. Pharm.*, 2009, **378**, 136.
205. Q. Zhang, K. De Oliveira Vigier, S. Royer and F. Jérôme, *Chem. Soc. Rev.*, 2012, **41**, 7108.
206. Z. Maugeri and P. Domínguez de María, *RSC Adv.*, 2012, **2**, 421.
207. J. T. Gorke, F. Srienc and R. J. Kazlauskas, *Chem. Commun.*, 2008, **10**, 1235.
208. S. H. Kim, S. Park, H. Yu, J. H. Kim, H. J. Kim, Y.-H. Yang, Y. H. Kim, K. J. Kim, E. Kand and S. H. Lee, *J. Mol. Catal. B: Enzym.*, 2016, **128**, 65.
209. H. Zhao, C. Zhang and T. D. Crittle, *J. Mol. Catal. B: Enzym.*, 2013, **85–86**, 243.
210. Y. Zhang, X. Xia, M. Duan, Y. Han, J. Liu, M. Luo, C. Zhao, Y. Zu and Y. Fu, *J. Mol. Catal. B: Enzym.*, 2016, **123**, 35.
211. E. Durand, J. Lecomte and P. Villeneuve, *Eur. J. Lipid Sci. Technol.*, 2013, **115**, 379.
212. M. C. Bubalo, A. J. Tušek, M. Vinković, K. Radošsević, V. Gaurina SrLek and I. R. Redovniković, *J. Mol. Catal. B: Enzym.*, 2015, **122**, 188.
213. P. Chaiwut, P. Kanasawud and P. J. Halling, *Enzyme Microb. Technol.*, 2007, **40**, 954.
214. M. I. Youshko and V. K. Švedas, *Adv. Synth. Catal.*, 2002, **344**, 894.
215. A. P. Basso, P. Spizzo and M. Toniutti, *J. Mol. Catal. B: Enzym.*, 2006, **39**, 105.
216. M. Erbeldinger, X. Ni and P. J. Halling, *Biotechnol. Bioeng.*, 1998, **59**, 68.
217. M. I. Youshko, H. Moody, A. Bukhanov, W. H. J. Boosten and V. K. Švedas, *Biotechnol. Bioeng.*, 2004, **85**, 323.
218. M. I. Youshko, L. M. van Langen, E. de Vroom, F. van Rantwijk, R. A. Sheldon and V. K. Švedas, *Biotechnol. Bioeng.*, 2001, **73**, 426.
219. A. Illanes, L. Wilson, O. Corrotea, L. Tavernini, F. Zamorano and C. Aguirre, *J. Mol. Catal. B: Enzym.*, 2007, **47**, 72.
220. C. Vera, C. Guerrero, R. Conejeros and A. Illanes, *Enzyme Microb. Technol.*, 2012, **50**, 188.
221. C. Vera, C. Guerrero, A. Illanes and R. Conejeros, *Biotechnol. Prog.*, 2014, **30**, 59.
222. S. Hari Krishna, *Biotechnol. Adv.*, 2002, **20**, 239.
223. M.-Y. Lee and J. S. Dordick, *Curr. Opin. Biotechnol.*, 2002, **13**, 376.
224. U. Akbar, C. D. Aschenbrenner, M. R. Harper, H. R. Johnson, J. S. Dordick and D. S. Clark, *Biotechnol. Bioeng.*, 2007, **96**, 1030.
225. L. Cao, L. van Langen and R. A. Sheldon, *Curr. Opin. Biotechnol.*, 2003, **14**, 387.
226. R. A. Sheldon, *Adv. Synth. Catal.*, 2007, **349**, 1289.
227. H. E. Schoemaker, D. Mink and M. G. Wubbolts, *Science*, 2003, **299**, 1694.
228. J. R. Cherry and A. L. Fidantsef, *Curr. Opin. Chem.*, 2003, **14**, 438.
229. H. J. Park, J. C. Joo, K. Park, Y. H. Kim and Y. J. Yoo, *J. Biotechnol.*, 2013, **163**, 346.

230. H. Renata, Z. J. Wang and F. H. Arnold, *Angew. Chem., Int. Ed.*, 2015, **54**, 3351.
231. A. P. Borole and B. H. Davison, *Appl. Biochem. Biotechnol.*, 2008, **146**, 215.
232. A. L. Serdakowski and J. S. Dordick, *Trends Biotechnol.*, 2008, **26**, 48.
233. K. Dabulis and A. M. Klibanov, *Biotechnol. Bioeng.*, 1993, **41**, 566.
234. M. Pazhang, K. Khajeh, B. Ranjbar and S. Hosseinkhani, *J. Biotechnol.*, 2006, **127**, 45.
235. D. Das, S. Roy, S. Debnath and P. K. Das, *Chem. – Eur. J.*, 2010, **16**, 4911.
236. Y.-R. Liang, Q. Wu and X.-F. Lin, *Chem. Rec.*, 2017, **17**, 90.
237. P. R. Santagapita, L. G. Brizuela, M. F. Mazzobre, H. L. Ramírez, H. R. Corti, R. V. Santana and M. P. Buera, *Carbohydr. Polym.*, 2011, **83**, 203.
238. K. Griebenow, Y. D. Laureano, A. M. Santos, I. M. Clemente, L. Rodríguez, M. W. Vidal and G. J. Barletta, *J. Am. Chem. Soc.*, 1999, **121**, 8157.
239. Y. Ooe, S. Yamamoto, M. Kobayashi and H. Kise, *Biotechnol. Lett.*, 1999, **21**, 385.
240. D. J. van Unen, J. F. J. Engbersen and D. N. Reinhoudt, *Biotechnol. Bioeng.*, 2002, 77, 248.
241. R. K. Eppler, R. S. Komor, J. Huynh, J. S. Dordick, J. A. Reimer and D. S. Clark, *Proc. Natl. Acad. Sci. U. S. A.*, 2006, **103**, 5706.
242. F. M. Veronese, R. Largajolli, E. Boccu, C. A. Benassi and O. Schiavon, *Appl. Biochem. Biotechnol.*, 1985, **11**, 141.
243. Y. Inada, K. Takahashi, T. Yoshimoto, A. Ajima, A. Matsushima and Y. Saito, *Trends Biotechnol.*, 1986, **4**, 190.
244. I. J. Castellanos, W. Al-Azzam and K. Griebenow, *J. Pharm. Sci.*, 2005, **94**, 327.
245. B. Castillo, R. J. Sola, A. Ferrer, G. Barletta and K. Griebenow, *Biotechnol. Bioeng.*, 2008, **99**, 9.
246. B. G. Davis, *Curr. Opin. Biotechnol.*, 2003, **14**, 379.
247. A. Díaz-Rodríguez and B. G. Davis, *Curr. Opin. Chem. Biol.*, 2011, **15**, 211.
248. C. Mateo, V. Grazu, J. M. Palomo, F. Lopez-Gallego, R. Fernandez-Lafuente and J. M. Guisan, *Nat. Protoc.*, 2007, **2**, 1022.
249. L. Betancor and H. R. Luckarift, *Trends Biotechnol.*, 2008, **26**, 566.
250. T. Jesionowski, J. Zdarta and B. Krajewska, *Adsorption*, 2014, **20**, 801.
251. J. M. Guisán, *Enzyme Microb. Technol.*, 1998, **10**, 3752.
252. M. N. Gupta, *Biotechnol. Appl. Biochem.*, 1991, **4**, 1.
253. C. Mateo, O. Abián, R. Fernández-Lafuente and J. M. Guisán, *Enzyme Microb. Technol.*, 2000, **26**, 509.
254. J. M. Bolivar, I. Eisl and B. Nidetzky, *Catal. Today*, 2016, **259**, 66.
255. S. Cantone, V. Ferrario, L. Corici, C. Ebert, D. Fattor, P. Spizzo and L. Gardossi, *Chem. Soc. Rev.*, 2013, **42**, 6262.
256. S. Datta, L. R. Christena and Y. R. S. Rajaram, *Biotechnology*, 2013, **3**, 1.

257. P. Zucca and E. Sanjust, *Molecules*, 2014, **19**, 14139.
258. D. Alsafadi and F. Paradisi, *Mol. Biotechnol.*, 2014, **56**, 240.
259. D. Senyay-Oncel and O. Yesil-Celiktas, *J. Mol. Catal. B: Enzym.*, 2013, **91**, 72.
260. A. Borole, S. Dai, C. L. Cheng, M. Rodriguez, Jr. and B. H. Davison, *Appl. Biochem. Biotechnol.*, 2004, **113**, 273.
261. F. A. Muñoz-Guerrero, S. Águila, R. Vazquez-Duhalt and J. B. Alderete, *J. Mol. Catal. B: Enzym.*, 2015, **116**, 1.
262. P. Lozano, B. Bernal, J. M. Bernal, M. Pucheault and M. Vaultier, *Green Chem.*, 2011, **13**, 1406.
263. S. Gómez, C. Bayón, S. Navarrete, J. M. Guisán and M. J. Hernáiz, *RSC Adv.*, 2016, **6**, 79554.
264. P. Hara, U. Hanefeld and L. T. Kanerva, *Green Chem.*, 2009, **11**, 250.
265. T. Matsuda, R. Kanamaru, K. Watanabe, T. Kamitanaka, T. Harada and K. Nakamura, *Tetrahedron: Asymmetry*, 2003, **14**, 2087.
266. J. Zhou, S. Villarroya, W. Wang, M. F. Wyatt, C. J. Duxbury, K. J. Thurecht and S. M. Howdle, *Macromolecules*, 2006, **39**, 5352.
267. S. Villarroya, J. Zhou, K. J. Thurecht and S. M. Howdle, *Macromolecules*, 2006, **39**, 9080.
268. P. Lozano, E. García-Verdugo, J. M. Bernal, D. F. Izquierdo, M. I. Burguete, G. Sánchez-Gómez and S. V. Luis, *ChemSusChem*, 2012, **5**, 790.
269. R. N. Patel, A. Banerjee, Y. R. Pendri, J. Liang, C. P. Chen and R. Mueller, *Tetrahedron: Asymmetry*, 2006, **17**, 175.
270. A. Aires-Trapote, P. Hoyos, A. R. Alcántara, A. Tamayo, J. Rubio, A. Rumbero and M. J. Hernáiz, *Org. Process Res. Dev.*, 2015, **19**, 687.
271. F. J. Plou, M. A. Cruces, M. Ferrer, G. Fuentes, E. Pastor, M. Bernabé, M. Christensen, F. Comelles, J. L. Parra and A. Ballesteros, *J. Biotechnol.*, 2002, **96**, 55.
272. W. Feng, X. Sun and P. Ji, *Soft Matter*, 2012, **8**, 7143.
273. P. Jochems, Y. Satyawali, L. Diels and W. Dejonghe, *Green Chem.*, 2011, **13**, 1609.
274. D. M.-R. De Temiño, W. Hartmeier and M. B. Ansorge-Schumacher, *Enzyme Microb. Technol.*, 2005, **36**, 3.
275. L. Wilson, A. Illanes, B. C. C. Pessela, O. Abian, R. Fernández-Lafuente and J. M. Guisán, *Biotechnol. Bioeng.*, 2004, **86**, 558.
276. L. Cao, F. Van Rantwijk and R. A. Sheldon, *Org. Lett.*, 2000, **2**, 1361.
277. J. D. Vaghjiani, T. S. Lee, G. J. Lye and M. K. Turner, *Biocatal. Biotransform.*, 2000, **18**, 151.
278. J. D. Cui and S. R. Jia, *Crit. Rev. Biotechnol.*, 2015, **35**, 15.
279. R. A. Sheldon, *Org. Process Res. Dev.*, 2011, **15**, 213.
280. L. Wilson, L. Betancor, G. Fernández-Lorente, M. Fuentes, A. Hidalgo, J. M. Guisán, B. Pessela and R. Fernández-Lafuente, *Biomacromolecules*, 2004, **5**, 814.
281. L. Wilson, A. Illanes, O. Abián, B. C. C. Pessela, R. Fernández-Lafuente and J. M. Guisán, *Biomacromolecules*, 2004, **5**, , 852.

282. R. A. Sheldon, S. van Pelt, S. Kanbak-Aksu, J. A. Rasmussen and M. H. A. Janssen, *Aldrichim. Acta*, 2013, **46**, 81.
283. G. Chávez, J. A. Rasmussen, M. Janssen, G. Mamo, R. Hatti-Kaul and R. A. Sheldon, *Top. Catal.*, 2014, **57**, 349.
284. P. Vossenberg, H. H. Beeftink, T. Nuijens, M. A. Cohen Stuart and J. Tramper, *J. Mol. Catal. B: Enzym.*, 2012, **75**, 43.
285. P. Wang, *Appl. Biochem. Biotechnol.*, 2008, **152**, 343.
286. S. A. Ansari and Q. Husain, Potential applications of enzymes immobilized on/in nano materials: a review, *Biotechnol. Adv.*, 2012, **30**, 512.
287. C. Altinkaynak, S. Tavlasoglu, N. Özdemir and I. Ocsoy, *Enzyme Microb. Technol.*, 2016, **93**, 105.
288. K. Chen and F. H. Arnold, *Nat. Biotechnol.*, 1991, **9**, 1073.
289. M. Adamczak and S. Hari Krishna, *Food Technol. Biotechnol.*, 2004, **42**, 251.
290. S. Lutz, *Curr. Opin. Biotechnol.*, 2010, **21**, 734.
291. K.-E. Jaeger and M. T. Reetz, *Curr. Opin. Chem. Biol.*, 2000, **4**, 68.
292. U. T. Bornscheuer and M. Pohl, *Curr. Opin. Chem. Biol.*, 2001, **5**, 137.
293. M. T. Reetz, *Chem. Rec.*, 2016, **16**, 2449.
294. M. Alcalde, *Trends Biotechnol.*, 2015, **33**, 155.
295. E. S. Redeker, D. T. Ta, D. Cortens, B. Billen, W. Guedens and P. Adriaensens, *Bioconjugate Chem.*, 2013, **24**, 1761.
296. R. K. Singh, M. K. Tiwari, R. Singh and J.-K. Lee, *Int. J. Mol. Sci.*, 2013, **14**, 1232.
297. A. A. Modenbach and S. E. Nokes, *Biomass Bioenergy*, 2013, **56**, 5264.
298. S. J. Horn, G. Vaaje-Kolstad, B. Westereng and V. G. H. Eijsink, *Biotechnol. Biofuels*, 2012, **5**, 45.
299. A. Limayem and S. C. Ricke, *Prog. Energy Combust. Sci.*, 2012, **38**, 449.
300. M. O. S. Dias, T. L. Junqueira, O. Cavalett, L. G. Pavanello, M. P. Cunha, C. D. F. Jesus, R. Maciel Filho and A. Bonomi, *Appl. Energy*, 2013, **109**, 72.
301. S. C. Albers, A. M. Berklund and G. D. Graff, *Nat. Biotechnol.*, 2016, **34**, 814.
302. K. Drauz, H. Gröger and O. May, *Enzyme Catalysis in Organic Synthesis*, Wiley-VCH, Weinheim, Germany, 3rd edn, 2012,1985 pp.
303. S. V. Ranganathan, S. L. Narasimhan and K. Muthukumar, *Bioresour. Technol.*, 2008, **99**, 3975.
304. L. Fjerbaek, K. V. Christensen and B. Norddahl, *Biotechnol. Bioeng.*, 2009, **102**, 1298.
305. U. T. Bornscheuer, G. W. Huisman, R. J. Kazlauskas, S. Lutz, J. C. Moore and K. Robins, *Nature*, 2012, **485**, 185.
306. D. Muñoz Solano, P. Hoyos, M. J. Hernáiz, A. R. Alcántara and J. M. Sánchez-Montero, *Bioresour. Technol.*, 2012, **115**, 196.
307. M. B. Ansorge-Schumacher and O. Thum, *Chem. Soc. Rev.*, 2013, **42**, 6475.

308. C. K. Savile, J. M. Janey, E. C. Mundorff, J. C. Moore, S. Tam, W. R. Jarvis, J. C. Colbeck, A. Krebber, F. J. Fleitz, J. Brands, P. N. Devine, G. W. Huisman and G. J. Hughes, *Science*, 2010, **329**, 305.
309. P. A. Santacoloma, G. Sin, K. V. Gernaey and J. M. Woodley, *Org. Process Res. Dev.*, 2011, **15**, 203.
310. R. Xue and J. M. Woodley, *Bioresour. Technol.*, 2012, **115**, 183.
311. D. N. Tran and K. J. Balkus, *ACS Catal.*, 2011, **1**, 956.
312. R. A. Sheldon and S. van Pelt, *Chem. Soc. Rev.*, 2013, **42**, 6223.
313. S. Talekar, A. Pandharbale, M. Ladole, S. Nadar, M. Mulla, K. Japhalekar, K. Pattankude and D. Arage, *Bioresour. Technol.*, 2013, **147**, 269.
314. K. Ahumada, P. Urrutia, A. Illanes and L. Wilson, *Food Bioprod. Process.*, 2015, **94**, 555.
315. P. Fernandes, *Enzyme Res.*, 2010, DOI: 10.4061/2010/862537.
316. M. A. Rao, R. Scelza, F. Acevedo, M. C. Diez and L. Gianfreda, *Chemosphere*, 2014, **107**, 145.
317. S. N. Naik, V. V. Goud, P. K. Rout and A. K. Dalai, *Renewable Sustainable Energy Rev.*, 2010, **14**, 578.
318. A. S. Bommarius, J. K. Blum and M. J. Abrahamson, *Curr. Opin. Chem. Biol.*, 2011, **15**, 194.
319. M. Wang, T. Si and H. Zhao, *Bioresour. Technol.*, 2012, **115**, 117.
320. T. Davids, M. Schmidt, D. Böttcher and U. T. Bornscheuer, *Curr. Opin. Chem. Biol.*, 2013, **17**, 215.
321. P. Tufvesson, J. Lima-Ramos, M. Nordblad and J. M. Woodley, *Org. Process Res. Dev.*, 2011, **15**, 266.

CHAPTER 17

Biocatalytic Process Engineering

JOHN M. WOODLEY

Department of Chemical and Biochemical Engineering, Technical University of Denmark (DTU), DK-2800 Lyngby, Denmark
Email: jw@kt.dtu.dk

17.1 Introduction

Much has been written about the opportunities and possibilities of either replacing existing chemical manufacturing reaction steps with biocatalytic process steps, or entering new markets with entirely new products made *via* biocatalysis, mostly in combination with fermentation and/or other catalytic steps.[1–7] Many of the other chapters in this book are testament to those opportunities. Indeed, there can be little doubt about the superb possibilities provided by the unrivalled selective chemistry which biocatalysis can offer, in general carried out under mild reaction conditions (ambient temperature, pressure and neutral pH). For these reasons alone, biocatalysis appears as particularly attractive tool to the synthetic chemist. In addition to the synthetic rationale, biocatalytic reactions offer also the possibility of economically improved processes, as well as a number of potential environmental benefits.[8–12] Today, excellent application examples can be found in numerous industrial sectors. In this chapter, examples will not be presented in detail since the aim is to explain the bioprocess engineering principles and concepts which underlie the development, scale-up and eventual implementation of a given biocatalytic reaction scheme. Indeed,

Catalysis Series No. 32
Modern Biocatalysis: Advances Towards Synthetic Biological Systems
Edited by Gavin Williams and Mélanie Hall

Published by the Royal Society of Chemistry, www.rsc.org

there are many excellent examples in the other chapters of this book as well as other texts with detailed descriptions of relevant examples.[13–15]

Some of the reasons most often cited for interest in implementing a biocatalytic process (rather than conventional chemical process) in industry are outlined in Table 17.1. The list of reasons presented in this table clearly underlines the fact that first and foremost implementation and commercialization are dependent upon the product value and the specific requirements of a given industry sector. It is interesting to note that the majority, by number, of processes implemented to date have focused on the pharmaceutical sector, for the most part due to the high turnover of target molecules, as a result of high attrition rates as well as patent-limited product lifetimes. This all contributes to the need for new chemistries capable of synthesizing non-natural molecules. Likewise, the target molecules themselves (often containing multiple chiral centers), as well as the synthetic routes to make such molecules, have become increasingly complex, which tends to favor highly selective biocatalytic methods (avoiding protection and subsequent de-protection steps).

For chiral molecules, the move towards optically pure entities, since the FDA guidelines[16] were introduced in 1992 (necessitating proof of the pharmacology of each of the potential enantiomers, independently), has also proven an important driver. Nevertheless, biocatalytic synthesis has today developed to such an extent that many new opportunities are forthcoming, and can increasingly be found in the fine chemicals sector (in particular in the synthesis of products in the price range around 50–100 $\$\,kg^{-1}$).

The development of an expanded portfolio of synthetic targets will also place increasing emphasis on bioprocess engineering and process development, since many of the new opportunities are in the synthesis of lower value products, compared to pharmaceuticals. Nevertheless, regardless of the industry sector, in order to transfer a given biocatalytic reaction from a synthetically useful laboratory tool into a commercially viable industrial process, bioprocess engineering and process implementation need to be

Table 17.1 Rationale for introduction of biocatalytic processes from a synthetic perspective.

Industry sector	Rationale for introduction
Pharmaceuticals	• Avoidance of protection and de-protection steps in syntheses • Syntheses of optically pure molecules containing multiple chiral centers
Fine chemicals	• Replacement of hazardous and difficult syntheses • 'Natural' syntheses
Bulk chemicals	• Use of alternative 'greener' raw materials • Extension of existing syntheses to incorporate flexibility for cheaper raw materials (reduced purity) and to produce higher value products (increased purity)

carried out in an effective and systematic manner. Compared to chemical processes, there has to date been relatively little attention paid to this aspect of biocatalysis.[17] Additionally, the required bioprocess engineering and process development need to be carried out within a regulatory and economic context appropriate for the given product value and specification.

17.2 Types of Biocatalytic Process

Growth-associated whole-cell bioconversion is usually referred to as 'fermentation', and will not be discussed further here, except to say that in such cases the substrate is used for production of the biocatalyst, as well as the reaction of interest, and furthermore the productivity is limited by the growth rate of the cells (which can be low in aerobic fermentations). In contrast, biocatalytic processes can be defined as those where catalyst production and the conversion step operate independently. Figure 17.1 classifies such biocatalytic processes into three groups dependent upon the format of the biocatalyst used for the conversion: (a) Resting whole-cell biocatalysis; (b) Isolated enzyme biocatalysis; (c) Immobilized enzyme biocatalysis. The format of the biocatalyst is of utmost importance in developing a suitable process structure, and ultimately the process flowsheet.

The first case, resting whole-cell biocatalysis, is schematically shown in Figure 17.1a and represents those cases where the growth of the biocatalyst and the conversion of the substrate to product can be separated, while still operating in whole-cell format. Compared to fermentation, this brings several advantages, including the potential to use alternative reaction media (such as water without the nutrients and salts necessary for fermentation), with potential benefits for downstream processing. Likewise an extra degree

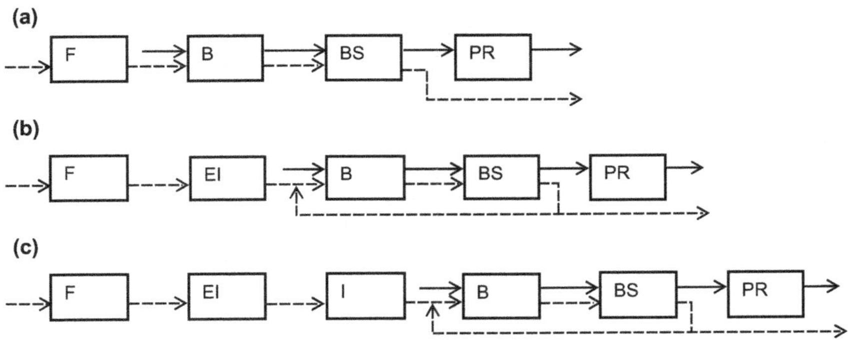

Figure 17.1 Examples of typical biocatalytic process flowsheets, dependent upon the format of the biocatalyst: (a) Resting whole-cell biocatalysis; (b) Isolated enzyme biocatalysis; (c) Immobilized enzyme biocatalysis. Continuous lines represent substrate/product process streams, while dashed lines represent the biocatalyst stream. Unit operations are identified as follows: F – fermentation; B – biocatalysis; EI – enzyme isolation; I – immobilization; BS – biocatalyst separation; PR – product recovery.

of freedom is introduced *via* the option to increase cell concentration after the fermentation. In this way the productivity (space–time yield) of the reaction can be increased many-fold over a conventional aerobic fermentation. This also implies that each step, growth of the biocatalyst and use of the biocatalyst, can operate optimally. In principle, the cells could be recycled at the end of the conversion step in order to reduce the cost contribution of the fermentation (biocatalyst production) to the final product, although this is rarely justified in practice. The second group, isolated enzyme biocatalysis, is schematically shown in Figure 17.1b and represents those reactions where the enzyme is used outside the cell in which it was produced. In such cases, preferably the enzyme has been secreted during the fermentation using an extracellular expression system. Alternatively, the cells need to be disrupted following fermentation (and the cell debris removed), in order to release the soluble, although unpurified, enzyme. During this process some loss of enzyme (activity) can be expected. Using an unpurified enzyme avoids the added cost of protein purification. The additional protein present in such an unpurified mixture will, provided it does not contain proteases (which can degrade the enzyme of interest), usually be beneficial to the desired enzyme stability. Such a scheme also has the advantage of overcoming potential substrate diffusional limitations into cells which can limit the conversion rate of non-natural substrates, since no active transport mechanism into the cell is available. Unsurprisingly, the effect is more pronounced when converting bulky substrates. Conditions for the use of such an enzyme catalyst usually include the requirement that the enzyme can be easily reused and recycled (*e.g.* by ultrafiltration) in order to keep the cost contribution of the biocatalyst low enough. Ideally such systems could be operated in a single batch, disposing of the enzyme after a single use. However, in many cases the cost of enzyme isolation is such that this cannot be justified and then it is necessary to recycle the enzyme. This can be done by isolating the enzyme from the reaction medium (usually water) by ultrafiltration. However, perhaps more effective is to consider the third scheme, immobilized enzyme biocatalysis, schematically shown in Figure 17.1c. Here the enzyme is immobilized, usually on a spherical solid or more likely within a porous support, to facilitate removal from the product stream, *via* simple (micro)-filtration, and thereby enables subsequent recycling at sufficient flux and with minimal losses. Many processes operate with multiple recycles (in some cases over 100) to reduce the cost contribution of the biocatalyst to the final product. An additional benefit of immobilization may be the added stability conferred to the enzyme by supporting it in a more stable conformation (see Chapter 16).

17.3 Requirements for Implementing a Scalable Industrial Process

Regardless of the biocatalyst format, resting whole-cells, isolated enzyme or immobilized enzyme, and the associated flowsheet (Figure 17.1a–c), it is necessary to meet certain economic requirements or targets in order to design a commercially viable industrial process. These targets become the drivers of the necessary bioprocess engineering to develop the process. The required bioprocess engineering is focused primarily on reaction enhancement and process intensification.

The implementation of any new process in industry is of course first and foremost dependent upon a suitable economic return on investment. This means that even at an early stage of process development (in the laboratory) it is essential to consider the costs of potential manufacturing. Clearly, this is also necessary to consider when implementing a new biocatalytic process, and can be used to establish a first set of targets for scalability. However, for biocatalytic processes, the conditions used in a manufacturing plant (*e.g.* high concentration of reagents in the M range) are far away from those found in nature (*e.g.* low concentration of reagents in the μM to mM range). This is not at all surprising since enzymes and microbial cells have been evolved over millennia to operate very efficiently at low concentrations of substrate. However, the consequence is that for any new biocatalytic reaction it is necessary to develop the biocatalyst and adapt the conditions so that a process can be developed which matches the requirements of the relevant industry sector. This means that not only must the potential manufacturing cost of a product be considered but also the cost of biocatalyst and process development. Indeed, it is highly advantageous at an early stage to consider if such extensive development can be justified. In some industry sectors, the timeline required to deliver a process (to launch the product) may also determine the outcome of such a decision (*e.g.* the pharmaceutical sector).

One of the challenges with assessing potential development and manufacturing costs at an early stage of process development is that an estimate of the potential for process improvement needs to be made. For biocatalytic processes this is particularly difficult, because not only are there many process technologies available to improve the process, and overcome biocatalyst limitations, but additionally a vast array of biological techniques are available for improvement of the biocatalyst (see Chapter 7). In itself, the multitude of techniques makes it difficult to plan a suitable research and development program that will lead to an improved process in a given time period. For example, genetic engineering (resulting in improved expression of the enzyme of interest or changing the host organism) as well as protein engineering (resulting in improved properties to better match industrial reaction conditions, see Chapter 7) offer the opportunity to improve the biocatalyst and its associated production process. This implies that an assessment of the feasibility of a potential biocatalytic process is more about assessing the feasibility of the required development program, than assessing the optimal scaled-up production process. Furthermore, in comparison with chemical processes, there is also rather limited information available concerning the scale-up and development of industrial biocatalytic processes. This is in part because biocatalytic processes, although several hundred of them have been implemented to date, represent a rather limited variety of cases. For the future it will be important to understand more about the optimal development procedure on a wider variety of chemistries and cases. In addition to the particular focus on biocatalyst modification and improvement, there are some basic economic requirements which also need to be met in order to enable implementation. All these economic goals are best described in terms of process metrics, and can very helpfully be used to provide targets associated with given economic demands. In this way, rather than defining the feasibility of a process at an early stage, instead the metrics can be used to help define a suitable development strategy.

The structure of the process, as defined in the previous section, sets the framework for identification of suitable process metrics as potential targets to meet given economic demands. The metrics are defined in terms of the main economic cost drivers and will be described in the following sections.

17.3.1 Reaction Yield and Process Yield

For the process itself to be economically feasible, it is a primary requirement that the reaction step adds value to the starting material (substrate). The simplest way to calculate this is to ensure that the value of the product exceeds the value of the substrate and other raw materials. However, while this is a necessary condition, it is not sufficient because the value added by the reaction is also dependent on the reaction yield. For the purpose of economic analysis it is most helpful to describe this as the mass of product produced relative to the mass of the substrate consumed (or key-substrate, if there is more than one), over the course of the reaction. For low-value products where the cost of a production is often determined by the cost of the substrate (up to 70% of the production cost), the yield needs to be high enough to achieve an economic process. In some cases it may also be necessary to include product recovery steps in the determination of the product yield on the key substrate, and this gives rise to the definition of a second metric, the process yield, since some product can be lost in recovery due to step inefficiencies. Nevertheless, as an initial metric, the reaction yield is perhaps adequate, since process yield estimates require significant assumptions at an early stage (prior to the flowsheet being defined). For high-value products (*e.g.* pharmaceuticals), the required minimum value of the reaction yield for an economic process will be determined by the downstream, product recovery, requirements (rather than the relative value of substrate and product). In some cases, the maximum reaction yield will be limited by the thermodynamic equilibrium, emphasizing also the importance of determining this, especially under the conditions to be used in the process. The value added by the reaction will ultimately determine the maximum allowable cost contribution of the biocatalyst, reactor and downstream processing. The metrics used to assess these individual contributions are summarized in the following sections.

17.3.2 Biocatalyst Yield

An essential issue in evaluation of a new biocatalytic process concerns the cost contribution of the biocatalyst.[18] Despite the enormous improvements in fermentation technology in the last two decades, coupled with enhanced expression as a result of recombinant DNA technology, it is nevertheless essential to assess the cost contribution of the biocatalyst to the product. The so-called 'biocatalyst yield' is the metric which measures this. It is defined as the mass of product produced relative to the mass of biocatalyst used, for the production of that product. In many processes the operational lifetime of the biocatalyst is such that reuse is possible in multiple batches, meaning that the best measure of biocatalyst yield is not just over the lifetime of one batch, but rather the entire operational lifetime of the biocatalyst. The necessity to measure this value over the whole lifetime of the biocatalyst clearly makes practical measurement and estimation difficult. However, this is often essential, since expensive biocatalysts (*i.e.* enzymes requiring recovery and isolation following fermentation, linked with immobilization) will require recycle and reuse in order to make a low enough contribution to the overall cost of making a given product. No matter the format of the biocatalyst, development is required in order to express sufficient activity for a given reaction.

Most enzymes are assayed, by scientists in the field, for their initial specific activity (activity per mass of biocatalyst). Usually this is defined in terms of enzyme units (*i.e.* amount of biocatalyst required to convert a specified amount of a standard substrate per minute, under standard conditions). Frequently this is expressed in micromoles of substrate and enzyme, but translating this into a useful metric to assess the potential feasibility of a commercial biocatalytic process requires using units such as g product per g biocatalyst per hour. Clearly, to achieve this requires knowledge of the mass and purity of the biocatalyst. This is dependent upon the biocatalyst format and will therefore determine the minimum value of the biocatalyst productivity required. In principle the biocatalyst productivity should (dependent upon the rate law) be constant throughout a reaction, but in practice it may be masked by time-dependent mass-transfer limitations or the biocatalyst may lose activity due to inactivation over time. Hence, rather than the initial biocatalyst productivity, it is more useful to evaluate the average productivity over the total reaction period. For this reason, the initial activity, often used as the basis for comparing biocatalyst performance, should be treated with particular caution. Provided the conditions of the process (*e.g.* pH, temperature, substrate concentration) are known, the biocatalyst yield turns out to be a much better way of making comparisons between biocatalysts.

Larger product markets imply larger biocatalyst requirements and can justify higher investment in biocatalyst development, reducing the cost of biocatalyst production. Nevertheless, dependent upon format, 'protein' can be produced at a given scale for a given cost. The dependence on scale is

Table 17.2 Examples of required biocatalyst yields for immobilized enzymes.

Product market	Product value/$ kg^{-1}	Biocatalyst yield (kg product per kg biocatalyst)
High value (*e.g.* pharmaceuticals)	500	50
Medium value (*e.g.* fine chemicals)	100	500
Low value (*e.g.* biofuels)	1	10 000

significant, favoring larger markets. The fact that this cost is fixed for a given scale is often cited as an advantage for the application of biocatalysts over the use of metal catalysts for example, where the cost may fluctuate according to the market. Since, in the case of biocatalysis, the biocatalyst is contributing as a catalytic entity alone, its cost should be evaluated in relation to the value of the product, and more specifically to the cost difference between the product and the substrate. Hence, knowledge about the specific activity of a biocatalyst is insufficient. Biocatalyst stability, the dynamic change of reaction rate with respect to time on account of denaturation, must also be known. Together in a single batch, the amount of product that can be produced for a given charge of biocatalyst can be calculated. It is also necessary to account for enzyme recovery or immobilization, dependent upon the biocatalyst format. Based upon acceptable contributions to the final product cost, it is possible to calculate a reasonable biocatalyst yield (kg product per kg biocatalyst) for a given product class. Some examples for immobilized enzymes implemented in processes are given in Table 17.2. Such information is invaluable, since it sets a target for biocatalyst specific activity and stability improvement.

This information also allows an estimation of the total amount of biocatalyst required for a given application and also, even more critically, the fraction of total enzyme (required to achieve the necessary biocatalyst yield) in a single batch. Interestingly, if the cost of biocatalyst in a single batch is very high, implying multiple recycles are necessary to give a high enough biocatalyst yield for the overall process, it becomes important to consider the risks that might be entailed by loss of that batch, due to a processing problem. For example, if there is a risk of over-feeding of an inhibitory substrate to the biocatalyst, in error, it could be hard to justify the use of an immobilized enzyme, where multiple reuses (recycles) would be required.

17.3.3 Productivity

Productivity (referred to as space–time yield by chemical engineers) is defined either as the mass of product produced per time per unit volume of reactor or per unit mass of biocatalyst. It is normal to use the averaged space–time yield, since the rate of production will not always be constant, either due to a decrease in substrate concentration and/or biocatalyst inactivation (see previous section).

The productivity, in combination with knowledge of the required demand for a given product (market) in a specified time period (*e.g.* per year), will give information about the required plant capacity. Frequently it is necessary to use existing process plant for biocatalytic processes, since building dedicated purpose-built plant is usually hard to justify when production volumes are relatively small (*e.g.* less than 10 000 tons per year). Nevertheless, knowledge about the productivity is of vital importance in assessing a given conversion, since it will form a first basis for knowledge of plant sizing. For biocatalytic reactions, the productivity can always be increased by adding more biocatalyst. Although limits may result from reactor capacity for the biocatalyst, especially with immobilized enzymes, perhaps even more important is that the biocatalyst yield will often be compromised when using very high biocatalyst concentration. In such cases, calculation of productivity can give a target for the minimum specific activity required of the biocatalyst.

17.3.4 Product Concentration

The product concentration leaving the reactor will determine the scale of the downstream process which will be sized according to the recovery of a given mass of product from a given fermentation volume. In general, since biocatalytic reactions are carried out at ambient temperature in an aqueous medium, the energy required to evaporate water is significant. In this way, the product concentration also determines the operating cost of the downstream process. In conventional chemical processes, it would be more usual to use the purity of the product leaving the reactor as a metric, but in the case of biocatalysis, the aqueous reaction medium means that the product concentration dominates. Hence, a high product concentration leaving the reactor is particularly important for low-value products. As discussed previously, achieving such a high concentration would not be usual in the natural environment of biocatalysts, making this a particularly challenging development target. Table 17.3 gives typical examples of required product concentrations for some enzyme-catalyzed processes. Clearly, the minimum allowable concentration is also dependent upon the ease of product recovery, which is itself dependent upon the type (and properties) of the molecule produced (and any by-products).

Table 17.3 Examples of required product concentrations for enzyme catalyzed processes.

Product type	Product value/$ kg^{-1}	Product concentration leaving reactor/g L^{-1}
High value	500	50
Medium value	100	150
Low value	1	300

17.4 Biocatalytic Process Technology

In principle, conventional chemical engineering provides the necessary equipment and unit operations for all biocatalytic processing. However, there are some special considerations required with biocatalysis and, in particular, the integration of the unit operations to create a complete process requires specialist knowledge. The following sections outline the features which require particular attention in the context of reactors and downstream processing. The improvement of the biocatalyst and process to achieve the required process metrics for implementation are discussed in the section 'Process development'.

17.4.1 Biocatalytic Reactor Options

As stated previously, in many cases the scale of production using biocatalysis is relatively small and this makes it hard to justify designing and installing a dedicated purpose-built reactor for a specific reaction. Nevertheless, in all cases (whether using a multi-purpose or dedicated process plant) it is necessary to match the specific features and characteristics of a biocatalyst and the reaction to a given reactor. In principle, the same types of reactors available for chemical conversions are also available for biocatalytic reactions, and all offer particular benefits or drawbacks dependent on the specific characteristics of the biocatalyst and reaction (see Figure 17.2). The characteristics that need to be considered include productivity in the biocatalytic process – which determines the mass transfer requirements and the amount of biocatalyst required, biocatalyst yield – which determines the biocatalyst recycle requirements, the presence of more than one phase – which determines the mixing required, the need to add reagents (*e.g.* for pH control) – which also determines the mixing required, as well as the biocatalyst kinetics and reaction thermodynamics. When using biocatalysis it is rare to find issues with heating and cooling, although ease of operation and format of the biocatalyst (resting whole-cells, isolated enzyme or immobilized enzyme) are also important.[19]

Many of the criteria by which reactors might be selected were established some years ago[20–22] when the cost of the biocatalyst was a paramount consideration. This early work has guided several large-scale processes, including the production of High Fructose Corn Syrup (HFCS) using immobilized glucose isomerase.[23] In recent years, technological developments have subsequently widened the options for biocatalytic reactors (*e.g.* fluidized bed reactors or membrane reactors). Likewise, improvements in rDNA technology, as well as protein expression, mean that less emphasis is placed on biocatalyst yield, and as the value of synthetic targets decreases more emphasis must be placed on reaction yield. Nevertheless, the economic requirements for the individual process metrics need to be met.

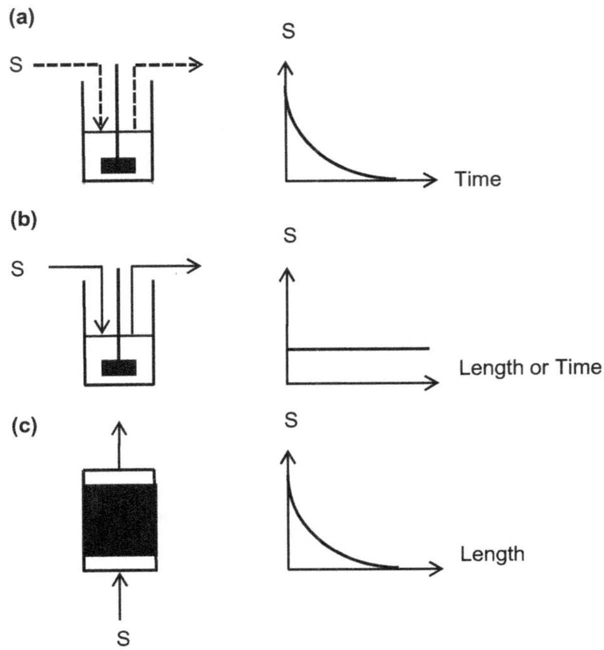

Figure 17.2 Conventional reactor types available for biocatalytic reactions. (a) (Fed)-batch stirred tank reactor; (b) Continuous stirred tank reactor; (c) Continuous packed-bed reactor, operating in plug-flow mode. On the left of the figure, a schematic representation of the reactor is shown, and on the right of the figure, the substrate concentration profile as a function of length (through the reactor) or time in the reactor is shown.

17.4.2 Downstream Processing

No matter how effective the synthesis step to produce the desired product, the effective recovery and purification of the product is an obvious necessity to scale-up and implement a process effectively at industrial scale. Frequently this is a step overlooked until late in process development, and only limited scientific studies have been reported.[24,25] The extent of purification will be dependent upon the market and final application of the product, but the proceeding steps of concentration and isolation require special attention because the products are usually present in water at relatively low concentrations and are heavily influenced by the biocatalytic reaction step.

Following the reaction, it is important to remove the biocatalyst first and then concentrate the product stream so as to reduce the scale and operating cost of the following product recovery operations.[25] Clearly, the higher the concentration of product coming from the reaction step, the better, but often some further concentration step is required. Since most biocatalytic reactions are carried out in water, concentration by evaporation is expensive, so usually (at large scale) the product is first extracted into a lower-boiling point organic solvent in order to reduce the cost of subsequent evaporation. In general, biocatalysts operate under mild conditions such that the process is in line with 'green' process concepts but also, critically, because the highly functionalized and complex molecules that are frequently the product of an enzyme-catalyzed reaction are usually unstable under other conditions. For this reason, distillation under reduced pressure, solvent extraction, or crystallization are the most common product recovery operations. Chromatography is used solely for high value products.

17.4.3 Special Cases Deserving of Particular Attention

Some special downstream processing cases which require specific process technology solutions that need to be considered are outlined in the following sections.

17.4.3.1 Inhibitory or Toxic Products

The need to leave the reactor at a high concentration of product places particular demands upon the reaction and the biocatalyst. Biocatalysts usually operate effectively only at low concentrations of substrate and therefore product. In cases where the biocatalyst is sufficiently stable, there can be a good argument for continuous stirred-tank reactor operation. Regardless of the reactor type, at higher concentrations of product, enzymes frequently show product inhibition (reversible loss of enzyme activity) or toxicity (irreversible loss of enzyme activity). For this reason it can be beneficial to remove the product during the course of the reaction using so-called *in-situ* product removal (ISPR, see Section 17.5.4.1).

17.4.3.2 Multi-substrate Reactions

In nature, enzymes can be found that catalyze the cleavage ($A \rightarrow B + C$) or formation ($A + B \rightarrow C$) of molecules. Both approaches can be useful in the context of synthetic biocatalytic schemes (or pathways). Additionally, enzymes can be found that transfer functional groups. Such enzymes, known as transferases, use two substrates, and swap functional groups to create two new products ($A + B \rightarrow C + D$). These enzymes are particularly useful, for instance for the synthesis of chiral centers. For practical operation, it is advantageous that one of the substrates is in stoichiometric excess over the other ($A > B$ or $B > A$), for example to drive the equilibrium towards the product. Consequently, the isolation of the product will also need to take into account removal of the excess substrate (ideally for recycling, dependent upon separation costs). Since the excess substrate will be at a higher concentration than the product in the outlet stream, it is important to consider at an early stage the relative ease of separation of the two alternative substrates (and potential recycle of the substrate in excess). An interesting example discussing some of the issues involved in such transferase type reactions concerns the use of ω-transaminase, a transferase that swaps amino groups between ketones.[26] Such reactions are very important in the pharmaceutical industry for the synthesis of optically pure chiral amines.[27,28]

17.5 Technology Toolbox for the Development of Biocatalytic Processes

A toolbox of different technologies is available for development of processes in order to meet the required economic metrics, and some of the key options are mentioned in the following sections.

17.5.1 Biocatalyst Engineering

Biocatalyst engineering, involving either protein engineering or modification of the host organism containing the biocatalyst (genetic engineering), is enormously powerful. Many excellent texts cover the field,[29–37] and the reader is directed to these and similar texts. Most 'protein' enhancements are focused on improvement (or alteration) of reaction selectivity, reaction rate and/or enzyme substrate scope, by 'swapping' amino acids in the protein structure. Random approaches yield significant improvements for biocatalysis, although they require high-throughput methodologies. Hence the field is now developing a whole range of new technologies including the introduction of more computer-based and semi-rational approaches (see Chapters 4, 5 and 7).[38] In the context of this chapter, it is interesting also to see that a start has been made to biocatalyst modifications focused on process needs[39,40] and more recently the potential and limitations of developing microbial tolerance towards inhibitory or toxic products has been discussed.[41]

17.5.2 Biocatalyst Immobilization

A reaction catalyzed by an isolated enzyme (Figure 17.1b) is easier to implement than a whole-cell catalyzed reaction (Figure 17.1a) due to its simplicity (*e.g.* fewer side reactions, resulting in simpler downstream processing; ease of biocatalyst separation, resulting in effective recycle). The price to be paid for this is higher upstream cost (on account of protein purification and the associated potential losses) and therefore reuse of the enzyme is often essential, potentially *via* immobilization (Figure 17.1c), to keep costs down. As a rule-of-thumb, the crudest possible form of the enzyme should be used while maintaining product quality.[42,43]

Immobilization is often the key to improving the operational performance of an enzyme.[44,45] Especially for use in non-aqueous media, such as an organic solvent, it is difficult to use biocatalysts without immobilization or cross-linking since the enzymes are prone to aggregation. Other potential benefits are enhanced stability, the possibility of use in a packed bed (or even fluidized bed) and prevention of protein contamination in the product stream.[46] This can be particularly important in the pharmaceutical sector. Likewise, fast and easy separation of the biocatalyst from the reaction medium is sometimes a key factor for enzymatic resolution of racemic chiral molecules, where the reaction has to be stopped at a certain conversion in order to preserve the high enantiomeric excess (ee) of the product. Simpler downstream processing and easy biocatalyst recovery for immobilized enzymes also lead to an improved process economy, which may be essential for developing a competitive process. Enzyme immobilization is a particularly important intensification method for obtaining adequate productivity based on the biocatalyst. In principle, improvements in activity, stability and selectivity are all potentially possible,[47,48] although improved stability and separation are the usual arguments for immobilization. A general method of immobilization that can be applied to any enzyme is not available and therefore the typical approach is trial and error. Efficient immobilization protocols should take into account the physicochemical properties of the carrier (or matrix) (*e.g.* hydrophobicity/hydrophilicity) as well as the enzyme, to obtain the best compromise between stability, activity, handling and cost.[49,50] For large scale production of immobilized biocatalysts, the procedures should be quick, robust, scalable and reproducible and the enzyme should be stable during each step of immobilization. In addition, working environment issues such as the handling of cross-linking chemicals and dust-producing materials should be considered.[51] Some examples of industrially used immobilized enzymes at the multi-ton scale are glucose isomerase (on an inorganic carrier) for production of high fructose corn syrup, penicillin G acylase (covalently attached to a polyacrylate support) for the production of semi-synthetic penicillins, lactase (on ion-exchange resins) for producing low lactose milk, TL lipase (on silica) for fat modification, and lipase B from *Candida antarctica* (NZ435) ('CalB') (on polyacrylate supports) for chiral resolutions, for example, in the manufacture of pharmaceutical intermediates. By selecting the appropriate immobilization technique, savings can be made in terms of the added cost of the biocatalyst to the process. A further factor is the selection of an appropriate enzyme loading on the resin.[52] Optimal cost effectiveness will depend on the cost of both enzyme and immobilization matrix and the proportions in which they are mixed. To make 1 kg of immobilized enzyme, a loading of about 50–100 g of protein is usually used. Carrier costs vary between 25 and 250 $\$\,kg^{-1}$ and the cost of the enzyme lies in the range 400–3000 $\$\,kg^{-1}$ for standard enzymes such as lipases and proteases (specialist enzymes can cost between 15 and 60 000 $\$\,kg^{-1}$).

It is important to keep in mind that the cost of the immobilized enzyme does not say much on its own. For an effective industrial process of a low-value product, a productivity of around 10 tons per kg immobilized enzyme is required. This normally requires considerable biocatalyst stability. For example, immobilized glucose isomerase used to produce high fructose corn syrup has an operating life time of about 1 year. Depending on the number of recycles of the enzyme (up to 200 may be required), the cost contribution to the produced product varies between a few hundred dollars per kg (for pharmaceuticals) down to a few cents per kg (for bulk chemicals), but is most normally in the range 0.15–15 $\$ \, kg^{-1}$.[18,53]

17.5.3 Reaction Engineering

Reaction engineering refers to the ability to alter conditions in the reaction such that optimal use can be made of the biocatalyst and substrate, and in some cases help to achieve the required economic metrics to enable implementation. The concentration of substrate(s) and product(s) will be determined by the reaction thermodynamics, in a given media and under specified reaction conditions. Likewise, the concentration of biocatalyst as well as substrate(s) to be used is of course dependent upon the reaction kinetics. For many reactions, the choice of media to be used also makes a very important difference to the reaction performance. It has become well established that addition of an organic solvent to an otherwise aqueous medium can be used to great effect in biocatalytic reactions. Some of the advantages and disadvantages of using different types of media, such as aqueous, organic[54] and aqueous–organic biphasic,[55] are briefly listed in Table 17.4. Today, other suggested and successfully tested biocatalytic reaction media such a $scCO_2$, ionic liquids, deep-eutectic solvents and vapor phase reactions remain primarily for application in the laboratory, rather than at a large scale, although many show considerable promise (see Chapter 16).

Table 17.4 Features of aqueous, organic, and aqueous–organic biocatalytic reaction media.

Media	Phases	Features
Aqueous	1	Good enzyme compatibility Poorly water-soluble substrates only present at low concentrations
Organic	1	May change the equilibrium compared with aqueous phase Selection of solvents required to ensure enzyme compatibility Use of immobilization *via* adsorption
Aqueous–organic	2	Beneficial conversion of poorly-water soluble substrates Beneficial for ISPR of poorly water-soluble products Selection of solvents required to ensure enzyme compatibility Potential mass transfer limitation for rate Need to determine minimum requirements for mixing Potential emulsification of phases making separation difficult Potential interfacial denaturation of biocatalysts

17.5.4 Process Engineering

Process engineering based improvements to help in achieving the required economic metrics can also come from some specific tailor-made solutions for bioprocesses, some of which are highlighted in the following sections.

17.5.4.1 In-situ *Product Removal (ISPR)*

Throughout the development of a process, it is essential to examine not only the biocatalytic reaction, but also the preceding and following reaction steps as well as the necessary product recovery and impurity removal steps. Changes to pH, temperature, pressure and solvent should be minimized. However, the need to address the problems of inhibitory or toxic products is so common it deserves special mention here.[24,41,56–58] A common approach is to use *in-situ* product removal (ISPR).[59–61] ISPR has been widely reported and studied, although more in-depth engineering studies are still required to provide guidelines for implementation.

The basic concept as implemented in industrial processes uses a recycle loop, where biocatalyst is retained in the reactor (Figure 17.3).

The recycle loop passes a product removal unit ('ISPR') in which product is selectively removed, for example into (or onto) a second phase. Common operations for the removal include extraction into a water-immiscible organic solvent, where the product selectively partitions into the organic solvent, or the use of resins, where the product selectively absorbs in the resin, or adsorbs on the surface.[62] Some further details are given in the sections which follow. The system can also be linked to chromatography for products that justify the costs.[63] In principle, ISPR can also be used to shift the equilibrium of a thermodynamically unfavorable reaction by means of selectively removing product over substrate. Some examples have been reported in the literature where this proved effective,[64] but it is clear it will not work in all cases, despite the many reported claims.

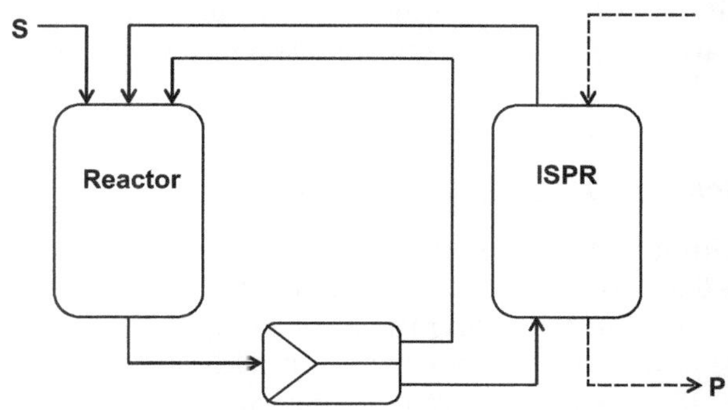

Figure 17.3 Typical ISPR reactor configuration with external loop for product removal (and recycling).

17.5.4.2 Aqueous–Organic Biphasic Systems

The addition of an organic water-miscible solvent can help solubilize poorly water-soluble compounds. However, the solubility increases are limited and the water-miscible solvents are often harmful to the enzymes themselves. In other cases, two-liquid phase biocatalysis may be used to enhance solubility. In addition, an established way to run some batch processes with biocatalysts is to avoid the toxic and inhibitory concentrations by supplying the substrate *via* a second liquid phase. In this case, the solvent acts as a substrate 'reservoir'. Order-of-magnitude improvements in product concentration can be achieved.[65,66] In general, conventional equipment can be used although the extent of mixing, and ease of subsequent phase separation, is critical. Extra parameters to be evaluated are the phase ratio of the two phases, mass transfer, and the partition of the substrate and product between the phases. It remains difficult to select the organic solvent and this is the subject of ongoing research, although much has been learnt from whole-cell biocatalysis studies.[66] Likewise, kinetic data in such processes need to be evaluated[67] in order to avoid operating under mass transfer limited conditions. Emulsification can also be a problem, although the use of membrane technology is an interesting potential solution.[68] Several processes have been scaled-up using this technology, and indeed the use of a second liquid phase is a common method of ISPR.[69]

17.5.4.3 Use of Resins

An alternative to organic solvents is to use a resin to supply the substrate or remove the product.[70] Such an approach has several advantages since it avoids the difficulties of organic solvent selection, both from an environmental perspective and due to the problems frequently faced by contact of enzyme with organic solvents (either dissolved solvent effects, stripping essential water, or liquid–liquid interfacial effects). Similar parameters need to be investigated as with two-liquid phase biocatalysis, but it is clear that capacity of the resin is particularly important. Indeed the option of a recycle stream containing the resin to either supply substrate or remove product looks particularly attractive. A few processes have been scaled-up using this technology.[71]

17.5.4.4 Benchmarking Improvements

The extent of improvement possible is of course very important in justifying the necessary development time to implement such technology. Table 17.5 presents three examples detailing the improvements achieved as a result of implementing process technology based on addition of resins, a second liquid phase or ISPR. The improvements are significant and in many cases can be achieved relatively simply using rules-of-thumb and examples of past precedent in the scientific literature.

Table 17.5 Examples of improvement as a result of implementing alternative process technologies.

Technology	Resins[71,72]	ISPR[64]	TLPB[73]
Additional reagents	Addition of Amberlite XAD 7 in recycle loop	Use of reduced pressure (350 mbar)	Addition of solvents (1-pentanol; 1-hexanol; isobutanol)
Reaction	Carbonyl reductase	Carbonyl reductase	Pyruvate decarboxylase
Biocatalyst	Whole-cell	Whole-cell	Whole-cell
Rationale	Overcome substrate and product inhibition	Overcome equilibrium limitation	Overcome toxic substrate
Improvement	Productivity 12.5-Fold	Reaction yield 1.4-Fold	Biocatalyst yield 9.2-Fold

It is clear that documenting more such examples in the future is required so that process development strategies can be effectively benchmarked.

17.6 Systematic and Accelerated Process Development

The development of a biocatalytic process from laboratory scale to achieve suitable economic metrics is an essential step in scale-up and implementation. A number of technologies are available to achieve this objective, as discussed in the previous sections. Nevertheless, the investment each requires, and the gains in terms of productivity, have still not been fully benchmarked even though the technologies mentioned here have all been implemented in large-scale industrial processes.

Clearly, for many reactions it is essential to intensify a reaction or process in order that it can be applied in an industrial context. This is not surprising given that the conditions required in a biocatalytic reactor are very different to those in nature. Protein engineering offers the engineer involved in scale-up and implementation of a biocatalytic process a powerful tool.[40] Although it appears as an efficient approach to search for the ideal biocatalyst,[74] in reality, it is a combined effort of protein and process engineering that is required to implement the process.

It requires significant experimental work to determine the laboratory process metrics for a potential reaction of interest, and partly for this reason they are not often reported in the scientific literature. Nevertheless, having determined the laboratory process metrics for a reaction of interest, it is also subsequently necessary to evaluate how to improve the metrics to meet the necessary 'target' threshold values to implement the process at large-scale. Improving the metrics requires process intensification, and indeed this has been the subject of considerable bioprocess engineering efforts in the last 15–20 years. Substrates will most normally be added in batch mode at the start of the reaction. However, in cases where the solubility is limited, alternative methods must be used. Substrates can be used above saturation concentration in the form of a slurry reactor or two-liquid phase reactor (as discussed in the previous section). In other cases, if the substrates are toxic or inhibitory, it can be essential to feed the substrate to the reactor, which can be done directly, if the solubility in the reaction medium is high, or *via* another phase (organic solvent or resin). The approach discussed here can be complemented by protein and enzyme engineering efforts. In many cases, this is an essential requirement alongside process engineering. Following process intensification, the process can be scaled-up. The philosophy of intensifying a process prior to scale-up helps to provide confidence that the process metrics necessary for economic implementation will ultimately be achieved. This then forms the basis for a systematic methodology for process development.

While a systematic method for process development is advantageous to improve the chances of implementing potentially beneficial processes, the speed at which the development is done is of paramount importance in many industrial sectors, especially for pharmaceutical processes. Collection of the necessary data (such as kinetics) to characterize a new biocatalyst needs to be central in accelerating process development. For simple reactions, models such as the well-known Michaelis–Menten kinetics can be used, but at the concentrations required for a scalable process this is usually inadequate because inhibitory effects and toxic effects may dominate. This also means, especially in reactions involving more than one substrate or more than one product (*e.g.* transferase reactions), that the kinetic expressions become very complicated and it may therefore prove difficult to accurately estimate the parameters in the expression. In all cases, it is important to determine the rate limiting step (*e.g.* substrate conversion, oxygen mass transfer, product removal) in order to optimize the reaction system.[75–77]

It is clear that to collect suitable kinetic (and other characterization) data about the biocatalyst will require considerable effort and there is no doubt that much can be made of the various new technologies available for automation and miniaturization, in principle allowing savings in resources (both in reagents and enzyme), as well as in time, by running reactions in parallel. For example, significant progress has been made in using micro-titer plates to test reaction conditions.[78–80] Most recently, the data collected using micro-titer plates has started to be integrated with modelling, such that parameters can be collected to build models in a very short time.[81] An alternative and complementary approach is to use micro-flow reactors which can be operated in continuous mode for the collection of data (also in parallel). Here the flow in the reactor is usually laminar and the effects of diffusion can also be studied. This also allows evaluation of alternative configurations of downstream unit operations, such as liquid–liquid extraction.[82] Both systems, micro-titer plates and micro-flow reactors, have been used with enzymatic systems as part of a move towards more efficient and accelerated development procedures.

17.7 Future Perspectives

In order to capitalize upon the unique properties of enzymes as catalysts, it is becoming evident that future biocatalysis will not just use single steps but become part of the wider synthetic approach towards multi-step and cascade catalysis (see Chapters 13–15). Two types of multi-step process can be envisaged for the future. The first concerns those reactions where an enzymatic step is coupled with the preceding or following chemical step. Such two-step chemo-enzymatic processes can be operated in a single reactor (so-called one-pot) or in two separate reactors.[83–86] Additionally the one-pot option can be separated (in space or time) within the reactor.[87] Multiple benefits result from such integration, although clearly, targets for improvement include the need to match the rates and concentrations, as well as conditions (pH, temperature and pressure). New bioprocess engineering concepts will be required to address the challenges here. The second type of multi-step process concerns those enzymatic reactions coupled with a preceding, following or parallel, enzymatic step.[88–91] Likewise here new bioprocess engineering concepts will be required, in particular to address the challenges of cross-activity between enzyme reactions on other substrates(s) and products(s) in the synthetic pathway, and to establish guidelines for combining enzymes together. Additional possibilities exist for developing alternative formats of the biocatalyst (containing multiple enzymes) and cloning both enzymes into a single microbial catalyst.

The implementation and development of new biocatalytic processes requires considerable investment in a range of technologies from ISPR to protein engineering. Nevertheless it is clear that as knowledge is built in each of these areas (in particular to identify limits for application), and a suitable methodology for process development, scale-up and implementation is built, the necessary tools will become available. This will be important not only to implement single step biocatalytic reactions, but for the future also to help address multi-step biocatalytic reactions in order that processes can be designed in such a way as to capitalize upon the best features of both conventional catalysis as well as biological catalysis.

References

1. A. Schmid, J. S. Dordick, B. Hauer, A. Kiener, M. Wubbolts and B. Witholt, *Nature*, 2001, **409**, 258.
2. S. M. Thomas, R. DiCosimo and V. Nagarajan, *Trends Biotechnol.*, 2002, **20**, 238.
3. A. Schmid, F. Hollmann, B. J. Park and B. Bühler, *Curr. Opin. Biotechnol.*, 2002, **13**, 359.
4. A. J. J. Straathof, S. Panke and A. Schmid, *Curr. Opin. Biotechnol.*, 2002, **13**, 548.
5. J. M. Woodley, M. Breuer and D. Mink, *Chem. Eng. Res. Des.*, 2013, **91**, 2029.

6. B. Thompson, T. S. Moon and D. R. Nielsen, *Curr. Opin. Biotechnol.*, 2014, **30**, 17.
7. A. J. J. Straathof, *Chem. Rev.*, 2014, **114**, 1871.
8. R. A. Sheldon and F. Van Rantwijk, *Aust. J. Chem.*, 2004, **57**, 281.
9. J. M. Woodley, *Trends Biotechnol.*, 2008, **26**, 321.
10. K. Seungdo, C. Jimenez-Gonzalez and B. F. Dale, *Int. J. Life Cycle Assess.*, 2009, **14**, 392.
11. K. R. Jegannathan and P. H. Nielsen, *J. Cleaner Prod.*, 2013, **42**, 228.
12. S. Illner, R. Plagemann, P. Saling and U. Kragl, *J. Mol. Catal. B*, 2014, **102**, 106.
13. *Industrial Biotransformations*, ed. A. Liese, K. Seelbach and C. Wandrey, Wiley-VCH Verlag GmbH and Co KGaA, Weinheim, 2nd edn, 2006.
14. *Asymmetric Catalysis on Industrial Scale*, ed. H.-U. Blaser and H.-J. Federsel, Wiley-VCH Verlag GmbH and Co KGaA, Weinheim, 2nd edn, 2010.
15. *Enzyme Catalysis in Organic Synthesis*, ed. K. Drauz, H. Gröger and O. May, Wiley-VCH Verlag GmbH and Co KGaA, Weinheim, 3rd edn, 2012.
16. FDA, *Chirality*, 1992, **4**, 338.
17. P. Tufvesson, J. Lima-Ramos, N. Al-Haque, K. V. Gernaey and J. M. Woodley, *Org. Process Res. Dev.*, 2013, **17**, 1233.
18. P. Tufvesson, J. Lima-Ramos, M. Nordblad and J. M. Woodley, *Org. Process Res. Dev.*, 2011, **15**, 266.
19. J. M. Woodley, in *Enzyme Catalysis in Organic Synthesis*, ed. K.-H. Drauz, H. Gröger and O. May, Wiley VCH, Weinheim, 2012, p. 217.
20. M. D. Lilly and P. Dunnill, *Process Biochem.*, 1971, 29–32.
21. M. D. Lilly and P. Dunnill, in *Methods in Enzymology*, ed. K. Mosbach, Academic Press, New York, Vol. XLIV, 1976, p. 717.
22. J. M. Woodley and M. D. Lilly, in *Applied Biocatalysis*, ed. J. M. S. Cabral, D. Best and J. Tramper, Harwood Academic, Chur, 1994, p. 371.
23. S. Pedersen, in *Industrial Application of Immobilized Biocatalysts*, ed. A. Tanaka, T. Tosa and T. Kobayashi, Marcel Dekker, New York, 1993, p. 185.
24. D. R. Yazbeck, C. A. Martinez, S. Hu and J. Tao, *Tetrahedron: Asymmetry*, 2004, **15**, 2757.
25. A. Oudshoorn, C. Van den Berg, C. P. M. Roelands, A. J. J. Straathof and L. A. M. van der Wielen, *Process Biochem.*, 2010, **45**, 1605.
26. P. Tufvesson, J. Lima-Ramos, J. S. Jensen, N. Al-Haque, W. Neto and J. M. Woodley, *Biotechnol. Bioeng.*, 2011, **108**, 1479.
27. C. K. Savile, J. M. Janey, E. C. Mundorff, J. C. Moore, S. Tam, W. R. Jarvis, J. C. Colbeck, A. Krebber, F. J. Fleitz, J. Brands, P. N. Devine, G. W. Huisman and G. J. Hughes, *Science*, 2010, **329**, 305.
28. M. D. Truppo, H. Strotman and G. Hughes, *ChemCatChem*, 2012, **4**, 1071.
29. F. H. Arnold, *Nature*, 2001, **409**, 253.
30. R. Chen, *Trends Biotechnol.*, 2001, **18**, 13.
31. N. J. Turner, *Trends Biotechnol.*, 2003, **21**, 474.
32. N. J. Turner, *Nat. Chem. Biol.*, 2009, **5**, 567.

33. A. S. Bommarius, J. K. Blum and M. J. Abrahamson, *Curr. Opin. Chem. Biol.*, 2011, **15**, 194.
34. G. A. Strohmeier, H. Pichler, O. May and M. Gruber-Khadjawi, *Chem. Rev.*, 2011, **111**, 4141.
35. U. T. Bornscheuer, G. W. Huisman, R. J. Kazlauskas, S. Lutz, J. C. Moore and K. Robins, *Nature*, 2012, **485**, 185.
36. R. E. Cobb, R. Chao and H. Zhao, *AIChE J.*, 2013, **59**, 1432.
37. Y. Li and P. C. Cirono, *Biotechnol. Bioeng.*, 2014, **111**, 1273.
38. S. Lutz, *Curr. Opin. Biotechnol.*, 2010, **21**, 734.
39. G. W. Huisman, J. Liang and A. Krebber, *Curr. Opin. Chem. Biol.*, 2010, **14**, 122.
40. J. M. Woodley, *Curr. Opin. Chem. Biol.*, 2013, **17**, 310.
41. H. Ling, W. Teo, B. Chen, S. S. J. Leong and M. W. Chang, *Curr. Opin. Biotechnol.*, 2014, **29**, 99.
42. D. J. Pollard and J. M. Woodley, *Trends Biotechnol.*, 2007, **25**, 66.
43. A. Liese and L. Hilterhaus, *Chem. Soc. Rev.*, 2013, **42**, 6236.
44. R. A. Sheldon, *Adv. Synth. Catal.*, 2007, **349**, 1289.
45. R. DiCosimo, J. McAuliffe, A. J. Poulose and G. Bohlmann, *Chem. Soc. Rev.*, 2013, **42**, 6437.
46. U. T. Bornscheuer, *Angew. Chem., Int. Ed.*, 2003, **43**, 3336.
47. C. Mateo, J. M. Palomo, G. Fernandez-Lorente, J. M. Guisan and R. Fenandez-Lafuente, *Enzyme Microb. Technol.*, 2007, **40**, 1451.
48. R. C. Rodrigues, A. Berenguer-Murcia and R. Fernandez-Lafuente, *Adv. Synth. Catal.*, 2011, **353**, 2216.
49. U. Hanefeld, L. Gardossi and E. Magner, *Chem. Soc. Rev.*, 2009, **38**, 453.
50. L. Cao, L. van Langen and R. Sheldon, *Curr. Opin. Biotechnol.*, 2003, **14**, 387.
51. O. Kirk and M. W. Christensen, *Org. Process Res. Dev.*, 2002, **6**, 446.
52. J. A. Bosley and A. D. Pielow, *J. Am. Oil Chem. Soc.*, 1997, **74**, 107.
53. D. J. Rozzell, *Bioorg. Med. Chem.*, 1999, **7**, 2253.
54. A. M. Klibanov, *Chemtech*, 1986, 354.
55. J.-S. Shin and B.-G. Kim, *Biotechnol. Bioeng.*, 1997, **55**, 348.
56. L. J. Bruce and A. J. Daugulis, *Biotechnol. Prog.*, 1991, **7**, 116.
57. T.-M. Lo, W. S. Teo, H. Ling, B. Chen, A. Kang and M. Wook Chang, *Biotechnol. Adv.*, 2013, **31**, 903.
58. C. Willrodt, R. Karande, A. Schmid and M. K. Julsing, *Curr. Opin. Biotechnol.*, 2015, **35**, 52.
59. R. P. Chauhan and J. M. Woodley, *CHEMTECH*, 1997, **27**, 26.
60. G. J. Lye and J. M. Woodley, *Trends Biotechnol.*, 1999, **17**, 395.
61. J. M. Woodley, M. Bisschops, A. J. J. Straathof and M. Ottens, *J. Chem. Technol. Biotechnol.*, 2008, **83**, 121.
62. J. T. Dafoe and A. J. Daugulis, *Biotechnol. Lett.*, 2014, **36**, 443.
63. N. Wagner, M. Fuereder, A. Bosshart, S. Panke and M. Bechtold, *Org. Process Res. Dev.*, 2012, **16**, 323.
64. S. J. Calvin, D. Mangan, I. Miskelly, T. S. Moody and P. J. Stevenson, *Org. Process Res. Dev.*, 2012, **16**, 82.

65. G. J. Lye and J. M. Woodley, in *Multiphase Bioreactor Design*, ed. J. M. S. Cabral, M. Mota and J. Tramper, Taylor and Francis, London, 2001, p. 115.
66. A. J. J. Straathof, *Biotechnol. Prog.*, 2003, **19**, 755.
67. J. M. Woodley, A. J. Brazier and M. D. Lilly, *Biotechnol Bioeng.*, 1991, **37**, 133.
68. C. G. P. H. Schroën and J. M. Woodley, *Biotechnol. Prog.*, 1997, **13**, 276.
69. J. M. Woodley and M. D. Lilly, *Chem. Eng. Sci.*, 1990, **45**, 2391.
70. P.-Y. Kim, D. J. Pollard and J. M. Woodley, *Biotechnol. Prog.*, 2001, **23**, 74.
71. J. T. Vicenzi, M. J. Zmijewski, M. R. Reinhard, B. E. Landen, W. Muth and P. G. Marler, *Enzyme Microb. Technol.*, 1997, **20**, 494.
72. B. A. Anderson, M. M. Hansen, A. R. Harkness, C. L. Henry, J. T. Vicenzi and M. J. Zmijewski, *J. Am. Chem. Soc.*, 1995, **117**, 12358.
73. B. Rosche, M. Breuer, B. Hauer and P. L. Rogers, *Biotechnol. Bioeng.*, 2004, **86**, 788.
74. S. G. Burton, D. A. Cowan and J. M. Woodley, *Nat. Biotechnol.*, 2002, **20**, 37.
75. G. Maria, *Chem. Biochem. Eng. Q.*, 2004, **18**, 195.
76. G. Caminal, J. Lafuente, J. Lopez-Santin, M. Poch and C. Sola, *Biotechnol. Bioeng.*, 1987, **29**, 366.
77. D. Vasic-Racki, Z. Findrik and A. V. Presecki, *Appl. Microbiol. Biotechnol.*, 2011, **91**, 845.
78. S. D. Doig, S. C. R. Pickering, G. J. Lye and J. M. Woodley, *Biotechnol. Bioeng.*, 2002, **80**, 42.
79. G. J. Lye, P. A. Dalby and J. M. Woodley, *Org. Process Res. Dev.*, 2002, **6**, 434.
80. G. J. Lye, P. Ayazi-Shamlou, F. Baganz, P. A. Dalby and J. M. Woodley, *Trends Biotechnol.*, 2003, **21**, 29.
81. B. H. Chen, M. Micheletti, F. Baganz, J. M. Woodley and G. J. Lye, *Chem. Eng. Sci.*, 2009, **64**, 403.
82. U. Kruhne, S. Heintz, R. Ringborg, I. P. Rosinha, P. Tufvesson, K. V. Gernaey and J. M. Woodley, *Green Process Synth.*, 2014, **3**, 23.
83. K. P. J. Gustafson, R. Lihammar, O. Verho, K. Engström and J.-E. Bäckvall, *J. Org. Chem.*, 2014, **79**, 3747.
84. S. Wallace and E. P. Balskus, *Curr. Opin. Biotechnol.*, 2014, **30**, 1.
85. H. C. Hailes, P. A. Dalby and J. M. Woodley, *J. Chem. Technol. Biotechnol.*, 2007, **82**, 1063.
86. P. N. R. Vennestrøm, C. H. Christensen, S. Pedersen, J.-D. Grunwaldt and J. M. Woodley, *ChemCatChem*, 2010, **2**, 249.
87. E. Burda, W. Bauer, W. Hummel and H. Gröger, *ChemCatChem*, 2010, **2**, 67.
88. P. A. Santacoloma, G. Sin, K. V. Gernaey and J. M. Woodley, *Org. Process Res. Dev.*, 2011, **15**, 203.
89. R. Xue and J. M. Woodley, *Bioresour. Technol.*, 2012, **115**, 183.
90. R.-C. Zheng, Z.-Y. Yang, C.-C. Li, Y.-G. Zheng and Y.-C. Shen, *J. Mol. Catal. B*, 2014, **102**, 161.
91. L. Cantarella, A. Gallifuoco, A. Malandra, L. Martikova, A. Spera and M. Cantarella, *Enzyme Microb. Technol.*, 2011, **48**, 345.

CHAPTER 18

Enzymes for Detection and Decontamination of Chemical Warfare Agents

ZBYNEK PROKOP,*[a,b] TANA KOUDELAKOVA,[a] SARKA BIDMANOVA[a] AND JIRI DAMBORSKY[a,b]

[a] International Clinical Research Center, St. Anne's University Hospital Brno, Pekarska 53, 656 91 Brno, Czech Republic; [b] Enantis, s.r.o., Kamenice 34, 625 00 Brno, Czech Republic
*Email: zbynek@chemi.muni.cz

18.1 Introduction

Nature equips us with numerous biological tools for the improvement of a wide range of production processes, which reduce energy and raw material consumption, while generating less waste and toxic side-products. As a result, modern biotechnology has had a significant impact on environmental applications, such as the production of renewable energies and bioremediation.[1] Compared with traditional physicochemical methods, bioremediation is a cost-effective, safe and nature-friendly biotechnology.[1–6] The major approach in the biodegradation of organic pollutants employs whole microorganisms, such as indigenous or introduced bacteria and fungi. Most bioremediation systems operate under aerobic conditions, but anaerobic environments may also permit microbial degradation of recalcitrant molecules.[3,5] In order to be biodegraded, contaminants must

Catalysis Series No. 32
Modern Biocatalysis: Advances Towards Synthetic Biological Systems
Edited by Gavin Williams and Mélanie Hall

Published by the Royal Society of Chemistry, www.rsc.org

interact with enzymatic systems in the degrading organisms. Both bacteria and fungi use various intracellular and extracellular enzymes often assembled in the biochemical pathways. More sophisticated systems involving genetically modified microorganisms are typically restricted to laboratory demonstrations of proof-of-principle, while there are only a few examples of the application in the field. Whether the organisms are genetically modified or not, microbial bioremediation relies on generally slow microbial growth that is limited by numerous factors such as mass transfer and contaminant bioavailability, the level of oxygen, pH, temperature, the nutrient level at contaminant sites and the presence of other toxic or growth inhibiting factors.[1,2,5] This has led to an interest in the use of formulated enzymes rather than live microorganisms as bioremediation agents.

Unlike many microbes, extracellular or isolated enzymes can remain effective in a wide range of pH and temperature ranges, particularly if they are immobilized on some carrier (see Chapter 16).[3] They are effective at low pollutant concentrations and are active in the presence of microbial predators or antagonists. Enzymes are more mobile than microorganisms because of their smaller size.[7] Employment of enzymes has the advantage of using the power of enzyme engineering to maximize enzyme efficiency (see Chapter 7), yet the use of the final product does not involve the release of a genetically engineered organism. Advances in bioinformatics and molecular modeling are helpful in the development of advanced semi-rational engineering strategies targeting various enzyme properties. The discovery of novel enzymes with desired technological and economical value is accelerated also by exploring vast natural diversity in genomic databases or by isolation of novel variants (*e.g.*, thermophilic or cold-adapted enzymes) from metagenomes of extreme environments (see Chapters 1 and 2). The gap between the vast information in genetic databases and a limited amount of functional data can be closed by advances in experimental screening methods, *e.g.*, lab on a chip, by the development of the database mining and by the application of advanced technological formulations and novel materials, *e.g.*, next generation co-solvents or nanomaterials.

Enzymatic bioremediation is particularly suited to situations where rapid remediation is needed in specific cases in time scale as short as a few minutes. Enzymes developed for remediation of specific toxic compounds can also be applied as antidotes for poisonings and as prophylaxis for people at risk of exposure.[2] These characteristics make enzymes attractive biocatalysts for applications in the neutralization of chemical weapons, which will be the major focus of this chapter. Benign biocatalytic methods are recommended for decontamination of warfare chemicals because enzymatic preparations are not corrosive, flammable or harmful to a user. Potentially, an enzyme-based decontaminant is benign enough to be directly applied to the skin of exposed personnel.[8–10] Unlike most chemical catalysts, enzymes with different properties and specificities can be mixed together in a single versatile formulation since they generally function under similar mild condition.[1,8,9]

18.2 Chemical Warfare Agents

Chemical weapons have been utilized for debilitation of enemies since prehistory with evidence from ancient times and the Middle Ages.[11] The natural chemical weapons used at those times were mainly based on smoke, animal and insect venom, poisonous plants, and later on toxic elements.[11–13] The synthesis of the first toxic chemicals in the 18th century and the great flourishing of chemistry in the 19th and 20th century together with the development of delivery systems started the modern chemical warfare era and turned chemical weapons into means of mass destruction, mass murders and genocide.[11,13] The first massive large-scale use of chemical weapons occurred during World War I, which is therefore known also as the "Chemist's War".[13,14] In total, over 113 000 tons of chemical warfare agents (mainly chlorine gas, phosgene, and mustard gas) was released killing around 90 000 people and causing over 1.2 million casualties.[12,14] Not even the horror of such a war has prevented the further exploration and use of chemical weapons.[12,13] The deadly organophosphorus nerve agents (tabun, originally an insecticide, and its derivatives sarin and soman) were discovered, produced, and stockpiled in heavily industrialized Germany before and during World War II.[11,13] Sulfur mustard, sarin, soman, and newly developed V-type nerve gases (VX and Russian VX) were produced and/or stockpiled by both sides also during the Cold War.[11,12]

There is evidence that chemical warfare agents have been perceived as an immoral means of combat and a taboo throughout history.[15] The employment of chemical warfare agents can be "invisible" for some time because of the delayed effect. Moreover, the chemicals can be unconsciously spread further from the contaminated area and/or by affected people.[13,15] The psychological impact on survivors is devastating. Therefore, the usage of chemical weapons should have been prevented by several treaties made in history.[11,13,15] The Geneva Protocol of 1925 prohibits the use of asphyxiating, poisonous and other gases in a war.[15] The Chemical Weapons Convention of 1993, which has 192 state-parties (in October 2016) extends the Geneva Protocol and prohibits development, production, stockpiling, and use of chemical weapons.[15,16] However, numerous confirmed and countless alleged applications of chemical warfare agents in localized conflicts (mainly in the Middle East) in 20th and 21st centuries have been described.[11–13,17–21] Apart from combat, and incidents related to abandoned stockpiles, chemical weapons have harmed people in several terrorist attacks.[12,13,17,22–24] A couple of chemical warfare agents (*e.g.*, sarin) were manufactured and misused by a Japanese religious sect Aum Shinrikyo in the 1990s.[13,25,26] Iraqi Al Qaeda used bombs and chlorine to intimidate people in Iraq between 2006 and 2007.[12,24] These cases underlined that chemical warfare agents could currently represent a threat mainly to unprotected civilian population as a mean of terrorism or 'New Wars' where no borders between a war and an organized crime can be recognized.[13,15,27,28]

18.2.1 Nerve Agents

Organophosphate chemicals are the deadliest group of the modern chemical warfare agents.[29] Despite being called nerve gases, these agents are liquid, stable, easily dispersed and highly toxic.[29–31] The group of nerve agents comprises (i) volatile G-type compounds (tabun, GA; sarin, GB; soman, GD; ethylsarin, GE; cyclosarin, GF), and (ii) more persistent V-type compounds (VG, VX and Russian VX) (Scheme 18.1). The first organophosphates of each type were originally intended for the agricultural use of pesticides.

All these organophosphate chemicals contain essential structural features (Scheme 18.1): the phosphorus atom is bonded directly to oxygen or sulfur forming a phosphoryl- or thiophosphoryl-group, R^1 and R^2 are defined as an alkoxy-, alkyl-, *N*-alkylamino- or *N,N*-dialkylamino-groups, and the leaving group X is typically a halogen- or pseudohalogen-, aryloxy-, thioxy-, or phosphate-group.[29] The nature of the leaving group influences phosphorylation ability of the compound and its stability in the aqueous medium. All nerve agents are chiral compounds, and the individual enantiomers differ in their toxicity.[32,33]

Known nerve agents rapidly and fully inhibit the activity of the human enzyme acetylcholinesterase (AChE) by irreversible phosphorylation of the essential serine residue in the AChE active site.[29,31] Phosphorylated AChE cannot catalyze the hydrolysis of the neurotransmitter acetylcholine to choline at neural chemical synapses and neuromuscular junctions of the cholinergic type. Based on the method of exposure and the dose, the peripheral nervous system or central nervous system can be affected and muscarinic and/or nicotinic receptors stimulated.[31,34] As long as acetylcholine is not removed by AChE, the signal remains triggered. Characteristic signs and symptoms of nerve agent poisoning can be apparent within seconds.[30] An exposed person can evince constricted pupils (miosis), headache, nausea, vomiting, difficulties in breathing, chest oppression, muscular weakness and twitching, cough, rhinorrhea, drooling, excessive sweating, abdominal cramps, and anxiety.[30,31] After exposure to a lethal dose, an untreated person can die within minutes.[30,34] The standard clinically available antidote is the combination of anticholinergic atropine and pralidoxime, which can reactivate the inhibited enzyme.[35] Reactivation of inhibited AChE, however, becomes impossible after aging of the nerve agent–enzyme complex, a change probably caused by a loss of an alkyl- or alkoxy-group, which is relevant mainly to soman poisoning.[31,35]

18.2.2 Blister Agents

Blister agents or vesicants are named based on their ability to form blisters and vesicles on the skin and mucous membranes.[29,36] The mode of action of these cytotoxic alkylating agents is more complex. Chemical weapons based on blister agents were intended to cause injury rather than to kill.[30] The group of blister agents comprises: mustards (the most prominent member sulfur mustard and its derivatives, Scheme 18.2), systemic arsenical poisons

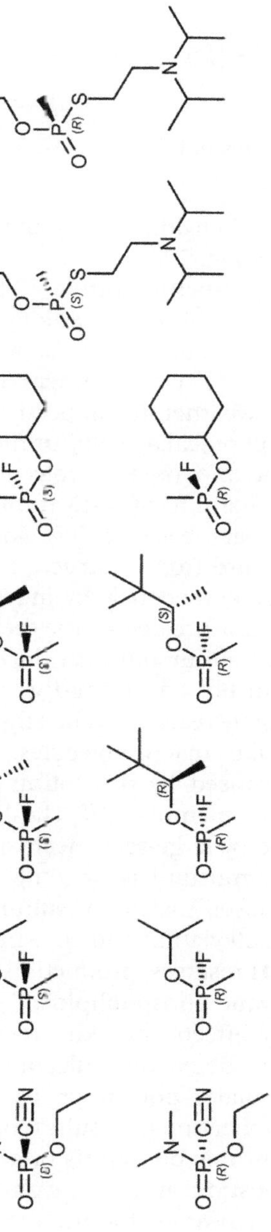

tabun sarin soman cyclosarin VX

Scheme 18.1 Chemical structures of nerve agents.

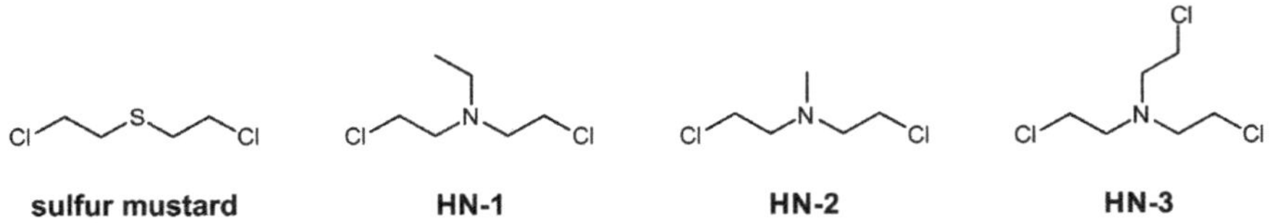

Scheme 18.2 Chemical structures of blister agents.

(lewisite, mustard–lewisite, phenyldichloroarsine), and a systemic poison and an urticant phosgene oxime.[29,36]

Sulfur mustard (bis(2-chloroethyl)sulfide), known also as Yperite, LOST, S-Lost, Yellow Cross, is a bifunctional alkylating agent.[36] It is a pale yellow oily liquid, which easily vaporizes in the warm dry environment and penetrates ordinary clothes.[17,29,36] On the other hand, sulfur mustard is very persistent in cold and damp weather. It can persist for two weeks sprayed on the soil or for years as a result of leakage. Sulfur mustard is sparingly soluble in water but soluble in fat and organic solvents. While sulfur mustard dissolved in water exhibits half-life of 4–15 minutes, bulk sulfur mustard may persist in water for years because the skin of hydrolysis products protects the remaining mustard from destruction.[29,36]

Intoxication is induced percutaneously by ingestion or inhalation.[29] The length and intensity of the exposure determine the local or systematic nature of the damage.[36] Toxicity of sulfur mustard depends on temperature and humidity; it is more toxic in hot and humid weather. Hydrolysis of sulfur mustard produces an extremely reactive cyclic ethylene sulfonium ion which alkylates a variety of cellular macromolecules.[17,36,37] Blistering between epidermis and dermis is caused by disruption of the interface between the epidermis and basement membrane.[36] The formation of large blisters filled with lymph is followed by large-scale necrosis.[29] The understanding of the delayed effect of sulfur mustard is ongoing.[36] The generally accepted hypotheses suggest that delayed toxicity of sulfur mustard is connected to different molecular actions: alkylation and crosslinking of DNA, inactivation of proteins, depletion of GSH reserves, production of reactive oxygen species, and peroxidation of membrane phospholipids.[36,38]

Sulfur mustard primarily affects the skin, respiratory tract, and eyes.[36] Acute symptoms comprise edema and ulceration necrosis of epithelial tissue. Symptoms of systematic poisoning are nausea, vomiting, fever, and malaise. Delayed complications of sulfur mustard poisoning can be conjunctivitis, blindness, chronic bronchitis or cancer. Lethality, caused by inhaling the aerosol or ingestion, is 2–5%, significantly lower than in the case of nerve agents.[11,13,17] However, the number of injuries together with psychological impact on victims is high. As no specific antidote is known, the supportive treatment is the only option.[17] The potential of antioxidants as a countermeasure of sulfur mustard has been intensively studied.[38,39]

Nitrogen mustards represent other bifunctional alkylating agents with similar vesicant, cytotoxic and mutagenic effects as sulfur mustard.[17,36]

The group of nitrogen mustards produced as potential military agents comprises HN-1, bis(2-chloroethyl)ethylamine, HN-2, bis(2-chloroethyl)-methylamine, and HN-3, tris(2-chloroethyl)amine. These tertiary bis(2-chloroethyl)amines are oily liquids sparingly soluble or insoluble in water, but miscible with organic solvents. Their relative potency for vesicant effects slightly decreases as follows: HN-3 > HN-1 > HN-2.[36] The dermal effect is enhanced by moisture. Nitrogen mustards alkylate nucleophilic parts of cellular macromolecules through the instable cyclic immonium cation which originates from hydrolysis. Similarly, as in the case of sulfur mustard, the mainly affected parts of the body are eyes, skin, the respiratory tract and the gastrointestinal tract.[17] The length of the latency period of the poisoning varies for several hours and days for eye and skin damage, respectively.[36] No specific antidote is known.[17]

18.2.3 Other Agents

Other significant warfare chemicals belong to choking and blood agents. Although their military importance has vanished, their misuse by terrorists against unprotected civilian population still represent a real threat.[29] Choking, or pulmonary, agents can cause lung irritation, damage, and pulmonary edema.[38] Choking agents comprise: chlorine gas (Cl_2) – a choking agent with strong oxidizing potential, chloropicrin (PS, Aquinite, Klop) – a strong lacrimator and lung irritant inhibiting mitochondrial enzymes, phosgene (CG) and diphosgene (DP) – pulmonary agents causing an acute lung injury, and phosphine – a pesticide interacting with a mitochondrial cytochrome oxidase.[17,29,38] Blood agents, or asphyxiants, target blood cells, and/or cellular respiration (oxygen transport or utilization) and thus cause tissue hypoxia.[17,29,30,38] Blood agents include: arsine, a highly toxic form of arsenic acting as a hemolytic agent; hydrogen cyanide, a potent inhibitor of mitochondrial respiration; and inflammable cyanogen chloride (Mauguinite), a reversible inhibitor of a cytochrome oxidase.[17,38] The specific antidotes (hydroxocobalamine, dicobalt edentate, 4-dimethylaminophenol, amyl nitrite/sodium nitrite/sodium thiosulfate) are available only in the case of hydrogen cyanide or cyanogen chloride poisoning.[17,40]

18.2.4 Decontamination of Chemical Warfare Agents

Parallel to the use and preparation of novel chemical weapons, there has naturally always been an interest in the protection against those toxic chemicals. Protective gear and equipment for early detection and decontamination of warfare agents have been developed.[11] The complete Personal Protective Equipment (PPE) includes a face mask connected to an oxygen source or a canister with a high-efficiency particulate aerosol (HEPA) filter, a vapor-impermeable/air-permeable suit, and liquid-impermeable butyl rubber gloves and boots.[41] Self-detoxifying materials (*e.g.*, functionalized nanofibers, nanofibers with embedded metal nanoparticles, nucleophilic polymers) for the detection and decontamination-integrated protective suits are currently under development.[41–43]

The concentration of chemical warfare agents in the environment is reduced by combined effects of sunlight, temperature, moisture, aeration, and microbial activity.[44] Whereas highly volatile agents are rapidly diluted with air or washed by water, extremely reactive compounds immediately react with other available compounds in the environment.[10] Availability of a proper decontamination method is therefore crucial, mainly in the case of contamination by persistent and highly toxic agents. Decontamination can be defined as a method essentially involving the conversion of toxic chemicals into harmless products by degradation, or as a method for reduction or removal of chemical agents in such a way that they are no longer hazardous.[34,45] A selected decontamination technique should meet required parameters regarding: safety, efficiency, cost, waste production, possible automation, and universality.[45] The requirements for harsh methods and formulations for stockpile destruction (such as the two-stage incineration or hydrolysis using sodium hydroxide) naturally differ from demands for methods for large-scale decontamination of the environment and/or sensitive surfaces of contaminated equipment.[46] The formulations for the latter should be compatible with the treated material, be easy to prepare–apply–remove, be able to adhere to vertical surfaces, evince no or low toxicity, be non-flammable, and not interfere with a monitoring equipment.[34,47]

Personal decontamination is even more delicate and is based on the stepwise use of adsorbents, washing, active ingredients (*e.g.*, chloramines), and antidotes or supportive medical treatment.[34] The exposure to chemical warfare agents can be proven by biomonitoring methods mainly based on mass spectrometry and immunochemistry.[48] Using these approaches, resulting adducts with DNA or proteins, urine and plasma metabolites, and/or enzymes' inhibition can be detected.[48,49] The relevance of oximes, antinicotinics, (bio)scavengers, and antioxidants as antidotes in the treatment of poisoning has been intensively studied.[35,38,39]

Generally, decontamination technologies can be divided into mechanical, physical, chemical, and biological approaches.[34,45] Mechanical decontamination is based on the removal of the bulk agent or covering the contaminated surface.[34] Physical decontamination covers dilution and washing, adsorption (*e.g.*, by Fuller's earth), evaporation (*e.g.*, thermal treatment), and reverse osmosis and ultra-filtration. Both mechanical and physical decontamination approaches are universal; however, they do not lead to neutralization of the treated chemicals.[45]

On the contrary, chemical decontamination refers to neutralization of the agent to a non-toxic product by chemical reactions – mainly by hydrolysis, oxidation or reduction.[34,44–46] Proper reaction conditions during decontamination processes are of the highest importance as the degradation of chemical warfare agents can lead to toxic byproducts or products that can adversely affect the environment and potentially harm the user.[9,44] For example, hydrolysis of VX at neutral to alkaline pH leads to the formation of

Table 18.1 Comparison of decontaminants of chemical warfare agents.

Decontaminant	Advantages	Disadvantages
Chemical	Broad range effectiveness Good stability Easy preparation	Toxic, sometimes toxic byproducts Aggressive, corrosive Stoichiometric Non-compatible with other agents
Enzymatic	Non-toxic Non-corrosive Environmentally acceptable Highly efficient Operating at mild conditions Catalytic Compatible with other agents Broadly applicable	Less stable Laborious isolation Highest activity in water Specificity, selectivity Narrow operation parameters Sometimes cofactor requirement Prone to inhibition Allergenic

toxic and environmentally stable *S*-(2-diisopropylaminoethyl) methylphosphonothioic acid (EA 2192).[44] Moreover, universal bleaches and strong oxidants may be unsuitable under certain conditions as they corrode, etch or erode the treated surface (Table 18.1).[34] Biological decontamination means the application of environmentally benign enzymes, which is a concept with truly exciting potential.[45]

18.3 Enzymes in Decontamination of Warfare Chemicals

Enzymes that catalyze chemical reactions at or near the diffusion limit are highly efficient biological catalysts that increase the reaction rate up to 10^7–10^{10} times, compared to the un-catalyzed reaction.[50] Such incredible catalytic efficiency together with biodegradability, non-toxicity, and non-flammability makes enzymes an interesting choice for decontamination of chemical warfare agents (Table 18.1). Decontamination formulations employing enzymes are currently developed for the personal treatment of skin, wounds, and eyes and/or decontamination of sensitive surfaces. The possible use of enzymes in prophylaxis and medical treatment as bioscavengers is a very attractive topic nowadays.[51–54]

18.3.1 Enzymes Converting Nerve Agents

The most well known enzyme associated with organophosphorus compounds is acetylcholine esterase (AChE). This serine hydrolase, like butyrylcholine esterase (BuChE), stoichiometrically inactivates nerve chemical agents in a suicide reaction that irreversibly inhibits the enzyme.[55] Over the years, a number of enzymes that convert toxic nerve agents by hydrolysis (phosphotriesterases) or oxidation (chloroperoxidase) have been isolated, characterized, and described in the literature.[33,56,57] Although various phosphotriesterases (EC 3.1.8) differ in their primary and tertiary structure, and the catalytic mechanism, they are all metal-dependent hydrolases containing a hydrophobic active site with three detached binding pockets.[33] The nomenclature of phosphotriesterases is inconsistent and confusing as they are often named according to their first identified substrate.[9,54] Phosphotriesterases can be subdivided into two groups based on preference for the bond cleavage: organophosphate hydrolases preferring cleavage of a P–O bond (EC 3.1.8.1, *e.g.*, bacterial phosphotriesterase, paraoxonase) and diisopropylfluorophosphatases preferring cleavage of P–F or P–CN bonds (EC 3.1.8.2, *e.g.*, squid DFPase).[54] As chloroperoxidase can oxidize both VX and sulfur mustard, it will be described in the next subsection as an enzyme converting blister agents (Table 18.2).[57]

The functional human acetylcholine esterase (hAChE, EC 3.1.1.7) at cholinergic synapses is made up of four 60 kDa α/β-hydrolase subunits associated with either the collagenous protein ColQ, or with a transmembrane proline-rich membrane anchor protein.[58,59] The essential catalytic triad, formed by a nucleophile (Ser203), a catalytic acid (Glu334), and a catalytic base (His477), is hidden in an active site connected with the bulk solvent by a tunnel.[58,60,61] The enzyme catalyzes the rapid hydrolysis of the neurotransmitter acetylcholine into acetic acid and choline *via* an acyl–enzyme intermediate in a diffusion-controlled reaction.[62,63] AChE serves as a biomarker of the exposure to organophosphate nerve agents and as a biorecognition component of biosensors.[48,49,64] During unsuccessful development of second-generation bioscavengers, a number of mutants were prepared by protein engineering.[52] The first variant of hAChE which showed organophosphate hydrolase activity was mutated in three positions Gly122His/Tyr124Gln/Ser125Thr.[55] The double variants Tyr124His/Tyr72Asp and Phe338Ala/Tyr337Ala exhibit improved reactivation rate and slower aging after inhibition, respectively.[52,55] Goldenzweig *et al.* designed and constructed hAChE variant bearing 51 mutations with 2000-fold higher expression in *Escherichia coli* and 20 °C higher thermostability.[59]

Human butyrylcholine esterase (BuChE, EC 3.1.1.8) is a serine hydrolase which is present in human plasma.[53] The size of this tetrameric enzyme is 340 kDa. The subunits with α/β-hydrolase fold interact with each other *via* a four-helix bundle at the C-termini (Figure 18.1).[53,62] This tetramerization domain surrounds a polyproline-rich peptide. The active site of BuChE is located at the bottom of a deep gorge lined by several aromatic residues and connected with the surface by a tunnel.[61,65] The catalysis by BuChE is similar to that of AChE, and requires an essential catalytic triad Ser198, Glu325, His438, and an oxyanion hole formed by Gly116, Gly117, Ala199.[66] Ser198 acts as a nucleophile to attack the carbonyl carbon of the substrate.[65] The acylation leads to an acyl–enzyme intermediate which is subsequently hydrolyzed by a water molecule. The Glu325–His438 pair is crucial for activation of the nucleophile and for proton transfer.

Ser198 is irreversibly covalently modified by the reaction with organophosphorus esters, which stoichiometrically inactivates BuChE and the organophosphorus agent (Scheme 18.3).[55] A single amino acid substitution

Table 18.2 Enzymatic decontaminants of chemical warfare agents.

	Enzymatic decontaminant			
	Accepted name (EC number)	Alternative name[128]	Reaction catalyzed	k_{cat}/K_m level/M^{-1} min^{-1}
Nerve agents	Cholinesterases (EC 3.1.1.8)	Acylcholine acylhydrolase, Benzoylcholinesterase, Butyrylcholine esterase, Choline esterase II, Pseudocholinesterase	Acetylcholine + H_2O → choline + carboxylate	10^2 (sarin),[55] 10^3 (VX)[55]
	Phosphotriesterases (EC 3.1.8.1)	A-esterase, Aryldialkylphosphatase, Aryltriphosphatase, Paraoxon hydrolase, Paraoxonase	Aryl dialkyl phosphate + H_2O → dialkyl phosphate + aryl alcohol	10^6 (sarin),[55] 10^4 (VX)[55]
	Diisopropylfluoro-phosphatases (EC 3.1.8.2)	Dialkylfluorophosphatase, DFPase, Organophosphorus acid anhydrolase, Somanase, Tabunase	Diisopropyl fluorophosphate + H_2O → diisopropyl phosphate + fluoride	10^6 (sarin),[55] 0 (VX)[55]
Blister agents	Haloalkane dehalogenases (EC 3.8.1.5)	1-Chlorohexane halidohydrolase	1-Haloalkane + H_2O → primary alcohol + halide	ND[a]
	Chloroperoxidases (EC 1.11.1.10)	CPOase, Vanadium haloperoxidase	$RH + Cl^- + H_2O_2 \rightarrow RCl + 2H_2O$	10^1 (SM)[96]

[a]ND – not determined.

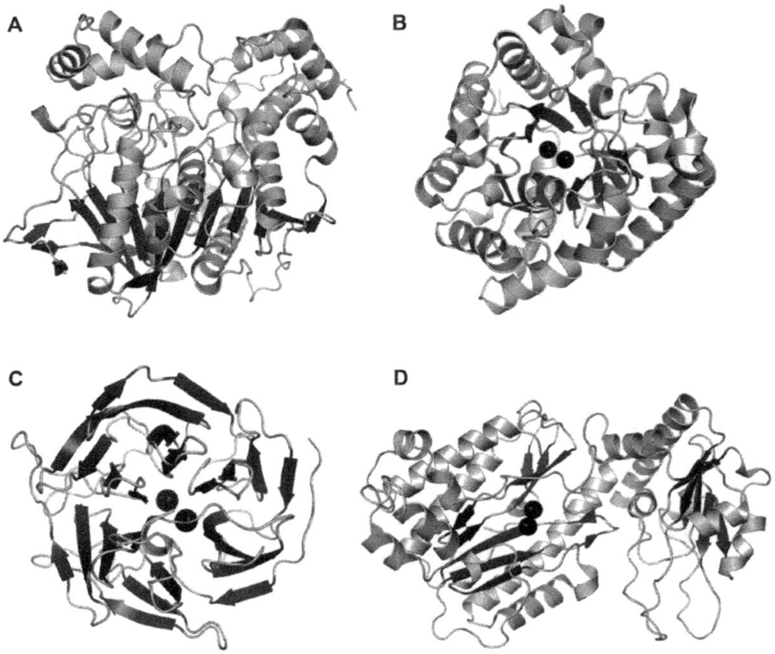

Figure 18.1 Structures of enzymes degrading nerve agents. (A) A subunit of butyrylcholinesterase with α/β-hydrolase fold (PDB ID: 1P0I), (B) a zinc containing subunit of a phosphotriesterase with a TIM barrel fold (PDB ID: 1HZY), (C) a calcium containing dialkylfluorophosphatase with a six-bladed β-propeller fold (PDB ID: 1E1A), and (D) a manganese containing subunit of a prolidase with a "pita bread" fold (PDB ID: 3L7G).[66,75,93,129]

Gly117His, which is close to both the nucleophile and the oxyanion hole, is essential for the acquirement of phosphotriesterase activity by the enzyme.[67] The mutant enzyme can hydrolyze sarin and VX. Although the physiological role of BuChE remains unclear, it is generally accepted that BuChE has a role in detoxification of poisons.[53] Plasmatic BuChE has been used as a biomarker of exposure to organophosphate nerve agents. Wild-type and mutant enzymes have been tested as stoichiometric and catalytic bioscavengers, respectively.[53,55,68,69]

The bacterial phosphotriesterase (PTE, EC 3.1.8.1) is one of the most studied organophosphate-hydrolyzing enzymes.[10,33] The enzyme was initially identified in bacterium *Brevundimonas diminuta* (formerly *Pseudomonas diminuta*, the American isolate) and *Flavobacterium* sp. ATCC 27551 in soil contaminated by insecticides parathion and diazinon, respectively.[10,33,70–72] Two other enzymes with sequence identity greater than 90% and similar activity profile were found in the soil bacteria *Agrobacterium radiobacter* P230 and *Chryseobacterium balustinum*.[33,73] *B. diminuta* PTE is a 72 kDa dimeric enzyme.[72,74] Each subunit adopts a TIM barrel motif with a binuclear zinc

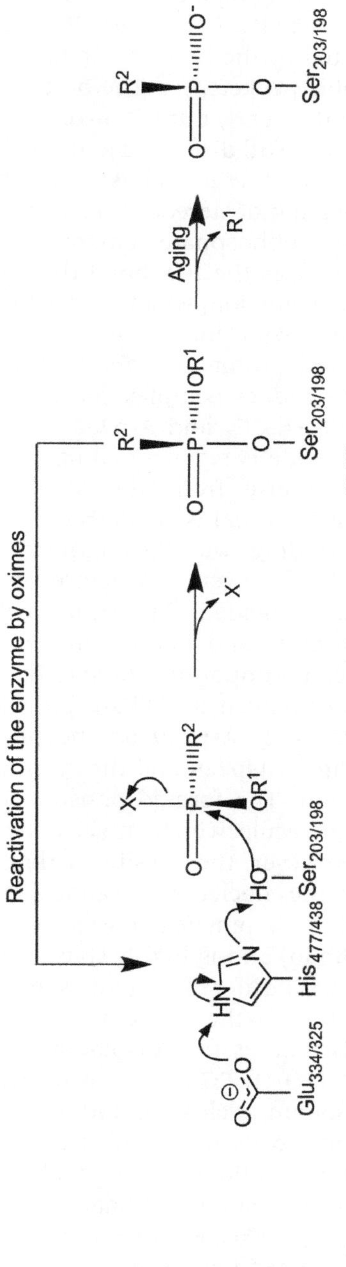

Scheme 18.3 Inhibition of acetylcholinesterase and butyrylcholinesterase.[55,130]

center situated in the active site at the C-terminal end of the barrel (Figure 18.1).[75] The zinc ions are bridged together by carboxylated Lys169 and hydroxide ion, and coordinated to the protein by the side chains of His55, His57, His201, His230, carboxylated Lys169, and Asp301. The naturally-bound zinc ions can be replaced with cadmium or manganese ions without loss of enzymatic activity. *B. diminuta* PTE has a remarkably broad substrate specificity granted by the nonspecific nature of substrate binding site with three hydrophobic pockets.[33,76] The best-known substrate is paraoxon that is hydrolyzed at the nearly diffusion-controlled rate.[77] The catalytic mechanism of this enzyme is still discussed and not fully understood.[72] The mechanism proposed by Aubert *et al.* is based on a S_N2-like simple substitution which leads to inversion of stereochemistry at the phosphorus center (Scheme 18.4).[33,78] An organophosphate compound binds to the active site in an orientation that polarizes the P–X bond through coordination of the phosphoryl oxygen to the β-zinc ion, which may weaken the binding of the bridging hydroxide to the β-zinc ion.[78] Upon the nucleophilic attack on the phosphorus center and proton transfer to Asn301, the P–X bond is broken, and an enzyme–product complex formed as a phosphate anion bridges the two zinc ions. His254 and Asp233 help to shuttle the proton away. The active-site hydroxide is regenerated upon product release.

The dialkylfluorophosphatase from the Mediterranean squid *Loligo vulgaris* (squid DFPase, EC 3.1.8.2) was isolated and characterized on the grounds of F.C. Hoskin's findings with the enzyme of the Northwest Atlantic squid *Loligo pealei*.[10,79–81] The overall structure of 35 kDa DFPase from *L. vulgaris* resembles a six-bladed β-propeller with two calcium ions (Figure 18.1).[82,83] High-affinity and low-affinity calcium ions bound in a central water-filled tunnel are important for stability and catalysis, respectively.[10,83,84] A reaction mechanism of DFPase proposed by Blum *et al.* involves a nucleophilic attack by Asp229 on the phosphorus atom of the organophosphate substrate.[85] Asp229 and the substrate are coordinated by the low-affinity calcium ion. The formed phosphoenzyme intermediate is hydrolyzed by a water molecule which attacks the carboxylate group of Asn229 (Scheme 18.5). However, the possibility that Asp229 could activate the water molecule for the nucleophilic attack is also considered.[86,87] *L. vulgaris* DFPase is of high practical relevance as it hydrolyzes sarin, soman, cyclosarin and tabun, and is highly stable over a wide temperature range and is compatible with different organic solvents.[10,33] DFPase can be produced in sufficient yield in the fermentor culture of the yeast *Pichia pastoris*.[88] DFPase was also successfully displayed on the surface of *E. coli*.[89]

Human paraoxonase 1 (PON1, EC 3.1.8.1) hydrolyzing γ- and δ-lactones, paraoxon, tabun, sarin, soman, cyclosarin and VX, is structurally similar to the squid DFPase.[10,33] Due to its broad substrate specificity, PON1 represents the most promising candidate as a prophylactic medical countermeasure against highly toxic organophosphates.[90] Paraoxonase from archea *Sulfolobus solfataricus* MT4 (SsoPox, EC 3.1.8.1), which is homologous to the *B. diminuta* PTE, exhibits incredible thermostability and resistance to urea

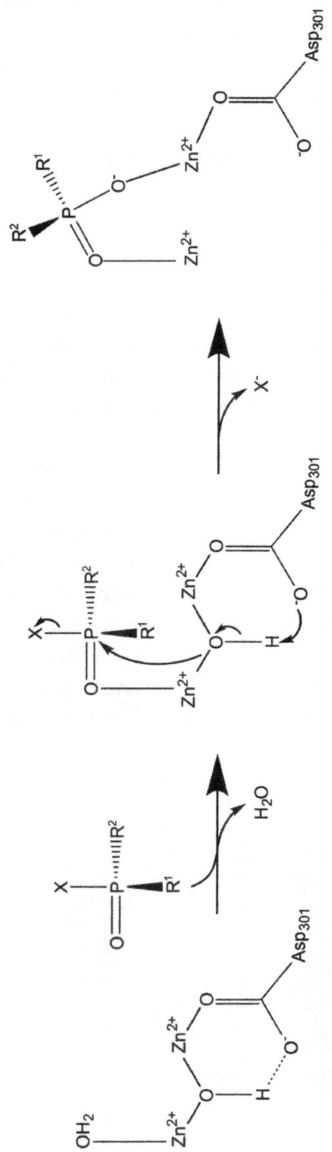

Scheme 18.4 Reaction mechanism of phosphotriesterase.[33,78]

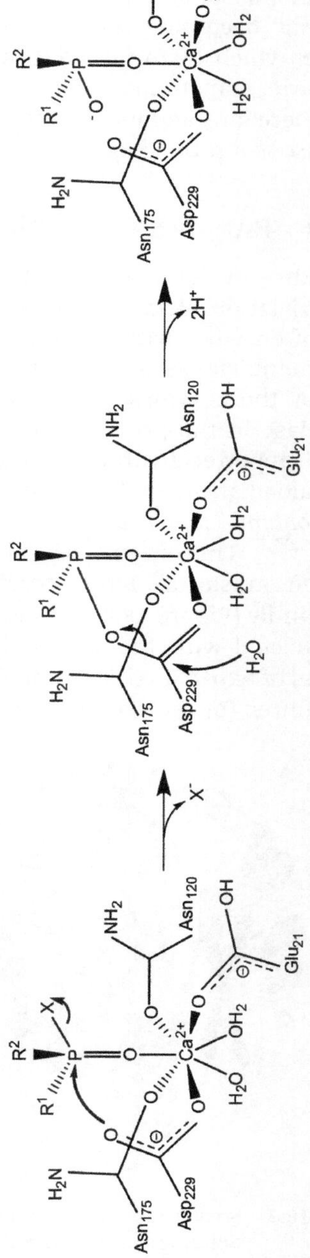

Scheme 18.5 Reaction mechanism of dialkylfluorophosphatase.[85]

thanks to its hyperthermophilic origin.[46,91] As the lactonase activity of this enzyme is two orders of magnitude higher than the promiscuous phosphotriesterase activity, the possibility to improve its phosphotriesterase activity by mutagenesis has been intensively studied.[46,92]

Organophosphate acid anhydrolase (OPAA, EC 3.1.8.1) from a moderately halophilic bacterium *Alteromonas* sp. JD6.5 is a tetramer of 60 kDa polypeptides which, apart from degrading proline-containing peptides, hydrolyses DFP, sarin, soman and cyclosarin.[33,46] The C-terminus domain contains two manganese ions and its fold resembles a "pita bread" (Figure 18.1).[93] Human prolidase is a promising enzyme for development of bioscavengers.[51]

18.3.2 Enzymes Converting Blister Agents

The hydrolytic detoxification of blister agents, particularly sulfur mustard, is not a trivial problem due to its poor solubility in water.[17,29,36] The combination of enzymes with suitable carriers in decontamination mixtures (*e.g.*, gels or emulsions) is, therefore, necessary.[10] Sulfur mustard can be hydrolyzed by the enzymes haloalkane dehalogenases, or oxidized by a chloroperoxidase in the presence of peroxide and a halide ion.[9,57,94–96] The first enzyme reported hydrolysing the carbon–halogen bond in sulfur mustard was haloalkane dehalogenase (HLD, EC 3.8.1.5), specifically LinB from *Sphingobium japonicum* UT26 (formerly *Sphingomonas paucimobilis* UT26).[9,94,97] This enzyme exhibits also a weak dehalogenase activity towards nitrogen mustard.[9] LinB structurally belongs to the α/β-hydrolase fold superfamily (Figure 18.2).[98] The active site is hidden inside the enzyme and is connected with the solvent by an access tunnel.[99,100] A catalytic triad consists of Asp108, His272, and Glu132.[101] Another two amino acid residues are required for the stabilization of a halide.[102] The residue Asp108 acts as a

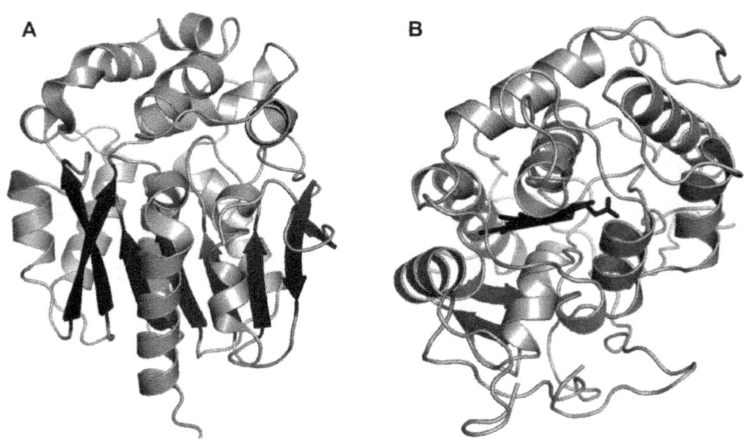

Figure 18.2 Structures of enzymes degrading blister agents. (A) A haloalkane dehalogenase with α/β-hydrolase fold (PDB ID: 1MJ5), (B) a chloroperoxidase with a heme (PDB ID: 1CPO).[100,106]

nucleophile causing the release of halide ion and formation of a covalent alkyl–enzyme intermediate. The intermediate is subsequently hydrolyzed by a water molecule that is activated by His272 acting as a base catalyst. A catalytic acid (Glu132) stabilizes the charge developed on the imidazole ring of the His272 during the reaction (Scheme 18.6).[103,104] While spontaneous hydrolysis results in the production of a potent alkylating compound (the reactive sulfonium ion), enzymatic hydrolysis proceeds *via* non-toxic intermediates.[9] Further research on haloalkane dehalogenases led to the identification of other enzymes converting sulfur mustard, *i.e.*, DhaA from *Rhodococcus rhodochrous* NCIMB 13064 and DmbA from *Mycobacterium bovis* 5033/66. Screening using a simple fluorescent activity assay on twenty haloalkane dehalogenases revealed weak activity towards sulfur mustard also in the case of DadB from halophilic bacterium *Alcanivorax dieselolei* B-5[105] (Bidmanova *et al.*, in preparation).

Another enzyme that catalyzes detoxification of sulfur mustard is chloroperoxidase isolated from the fungus *Caldariomyces fumago* (CPO, EC 1.11.1.10).[57,96] This heme peroxidase bears a cysteine-ligated heme that uses peroxides as electron acceptors.[96] It acts as peroxidase, halogenase, catalase, and exhibits also oxygen insertion activities. CPO folds into a unique tertiary structure dominated by eight helical segments (Figure 18.2).[106] Substrates probably bind into a hydrophobic patch above the heme. The catalytic base, essential for cleavage of the peroxide O–O bond, is Glu183 rather than His as in other peroxidases. Two independent studies demonstrated rapid degradation of sulfur mustard catalyzed by CPO with the addition of hydrogen peroxide and sodium chloride.[57,96] While Amitai *et al.* identified sulfoxide, sulfone, and sulfoxidivinyl as the oxidation products of the reaction, Popiel and Nawała observed only sulfur mustard sulfoxide. Amitai *et al.* demonstrated that CPO can be used also for oxidation of nerve agent VX.[57]

18.4 Practical Applications of Enzymes Converting Warfare Chemicals

Not many enzyme-based technologies for neutralization or biodegradation of warfare chemicals are fully developed and available for current use even though there is often an active research and IP-protection on-going. Most of the applications have been demonstrated at a small scale, while many others still require identification or engineering of more suitable biocatalysts. To reach commercialization or field use, large-scale production of the components needs to be achieved.[8] Recombinant DNA technology for gene cloning in a variety of expression hosts (*e.g.*, bacteria, yeasts, and fungi) and the ability to overexpress target proteins in large quantities have opened the door to bulk enzyme production.[10] Industrial scale production using recombinant bacterial strains has been developed for enzymes converting nerve gas agents and organophosphate-based pesticides. These enzymes have been confirmed to be successfully functioning in a variety of water-based scaled systems such

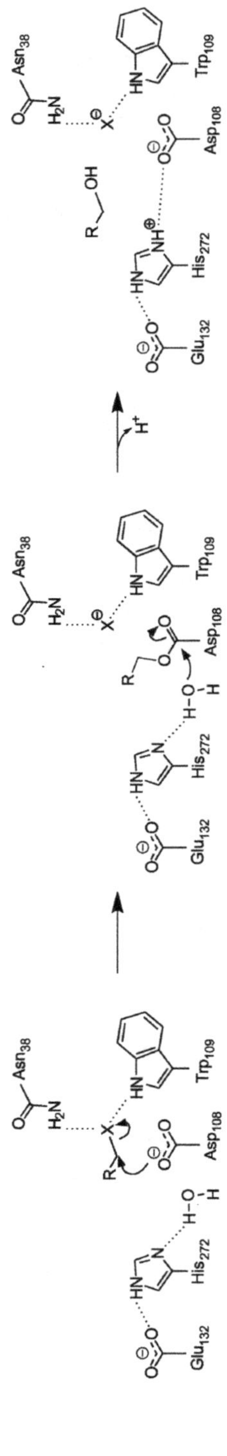

Scheme 18.6 Reaction mechanism of haloalkane dehalogenase.[103]

as fire-fighting foams and sprays, aqueous degreasers and aircraft deicers, and consequently commercially used in DEFENZ™ line of decontamination products capable of breaking down organophosphate type materials, including G-type nerve agents such as soman and sarin (Genencor–DuPont, USA).[8] Another example of a commercial decontamination kit is Yperzyme™ which contains a recombinant haloalkane dehalogenase as an active component and offers a safe and environmentally friendly alternative to chemical detoxification of sulfur mustard (Enantis, Czech Republic).

Stability is another characteristic influencing the practical applicability of enzymatic catalysts. Stability for long-term storage can be significantly increased by drying of an enzyme preparation (*e.g.*, freeze or spray drying) and further influenced by a wide variety of additives. The more critical operational stability is specifically dependent on the property of an enzyme and reaction conditions. Decontamination formulations (*e.g.*, foams and emulsions) often include co-solvents for solubilization of agents with limited water solubility and so-called thickened agents increasing the persistency of otherwise volatile chemical agents.[10] Enzymes are in this way effective and stable in conditions far away from physiological conditions for which they have evolved. This disparity between natural and operational conditions can be overcome thanks to modern advances in molecular biology, biochemistry, bioinformatics and computational chemistry. Enzyme stability at elevated temperatures and/or in the presence of organic solvents can be modulated by protein engineering or by other methods such as immobilization (see Chapter 16), chemical modification (see Chapter 16), post-translation modification (see Chapter 6), or application of new nanomaterials and nanostructures.[107–111]

Due to the inherent gentleness of enzymes and the costs of their preparation, formulations employing enzymes are currently demanded mainly for personal decontamination of skin, wounds, and eyes. Two different decontamination products based on organophosphate degrading/scavenging enzymes have been described so far: (i) a reusable sponge, and (ii) a polyurethane foam.[45] Significant advancement has also been made in the field of bioscavengers that inactivate nerve agents in the bloodstream before they can reach AChE.[52] Phase I clinical trials with a bioscavenger for the protection of humans against nerve agents have been successfully completed recently.[51,53]

Another application of enzymes converting warfare chemicals is in biosensing, providing an early alarm in case of terrorist attacks, as well as monitoring the presence of these chemicals in the environment. Detectors based on isolated enzymes or whole cells offer advantages in terms of high sensitivity, miniaturization, integration, low cost, and power requirements, comparing favorably with traditional analytical techniques, *e.g.*, chromatography and mass spectrometry.[112] Inhibition of AChE or BuChE has been mostly utilized in detectors for nerve agents.[64,113] The need to regenerate or replace the enzyme during measurement and longer incubation times have shifted the recent focus towards catalytic systems based on organophosphorus acid anhydrolase and organophosphorus hydrolase.[114] Despite an increasing number of reports on biosensor technologies capable of detecting warfare chemicals, there are apparently only very few items of this kind available commercially.[115] Practical examples of commercial detectors are colorimetric tubes ChP 71, respectively 05 (Oritest, Czech Republic), disposable dipsticks Detehit (Oritest, Czech Republic), and the M256A1 kit and M-272 Water Test Kit containing enzyme-based detector tickets (Luxfer Magtech, USA). All these systems use AChE enabling detection of nerve agents.[115] The only report of a catalytic biosensor for sulfur mustard is recently developed colorimetric test stripe based on haloalkane dehalogenase.[116] The lack of suitable biorecognition molecules and inherent instability of the respective enzymes have been the greatest impediments to the commercialization of biosensors for warfare agents.[117] Nevertheless, on-going research provides better enzyme catalysts, new methods for their immobilization, chemistries stable through the range of the working environments, and optimization of detection by application of the latest trends in nanotechnology, lab-on-chip, and functional materials.

18.5 Conclusions and Perspectives

Enzymes provide unprecedented catalytic activity and substrate specificity, which make them suitable catalysts for detoxification of dangerous compounds including warfare chemical agents. As natural biomolecules, enzymes are fully biodegradable, non-toxic, perform at mild conditions of neutral pH and moderate temperature, and are harmless to the environment and biota. These characteristics are well suited for decontamination of sensitive surfaces, application in the treatment of human victims or prophylaxis for personnel with a high risk of exposure. The majority of enzymes are compatible with each other and function under similar conditions, providing a good opportunity to use them in mixtures. Mixed enzymatic preparations provide several advantages, (i) they can be applied universally with no need for analytical steps and identification of agent identity, dramatically reducing time from an exposure to treatment, (ii) they can simplify handling, as one formulation fits to all cases, and (iii) they can reduce logistic requirements related to the application of multiple formulations.

Successful implementation of enzymes for detection and detoxification of chemical warfare agents, however, depends on a number of requirements, such as sufficient activity and stability at operational conditions, long-term storage stability, high productivity and simplicity of separation at a higher scale, which makes enzyme technologies economical. It is not generally simple to obtain an optimal catalyst with a proper combination of all required characteristics.

Natural sequence diversity is explored by using next generation sequencing techniques applied to individual organisms as well as metagenomes isolated from various environments. The number of protein sequences grows exponentially doubling every 18 months.[118] Novel systems combining experiment with data analysis and modeling need to be developed to address this "big data" problem.[119] High-throughput determination of protein structures is targeted systematically by the structural genomics projects. This approach aims to provide a three-dimensional model structure for all of the tractable macromolecules that are encoded by complete genomes.[119] Since the protein structure is closely linked to the protein function, the high-throughput determination of tertiary structures has the potential to provide the first insight into the molecular function and mechanism of thousands of natural proteins. Experimental verification and assignment of functional properties are still desirable. The recent development of lab-on-a-chip technology significantly accelerates experimental testing.

Natural enzymes have evolved to act under different environmental conditions, however, they often do not provide optimal process window and need to be engineered. Both directed evolution and rational design are broadly applied to modify valuable properties of targeted proteins. Advances in bioinformatics, molecular modeling, and genetic manipulations are extremely helpful in the engineering of enzymes. The semi-rational design is being frequently applied to focus on desired protein properties with the minimal experimental effort required for library screening (see Chapter 7).

The development of high-throughput experimental techniques, *e.g.*, high-throughput gas chromatography or mass spectrometry, allows exploration of the advanced complexity of designed libraries. Lab-on-a-chip technologies again can significantly accelerate experimental screening. Moreover, a number of software tools for protein design have been developed and provided to the non-expert community *via* user-friendly web applications, *e.g.*, HotSpot Wizard, 3DM, RosettaDesign, FoldX, Erebus/Eris, *etc.*[120–124] Significant progress has been made in the field of protein stabilization. Several recently developed methods, *e.g.*, FireProt, FRESCO, PROSS, are available for designing of proteins with increased stability under operational conditions.[59,125,126]

The technological utility of enzymes can be enhanced by their use in organic solvents rather than in their natural aqueous reaction media (see Chapter 16). Changing the solvent is not only feasible, it sometimes also grants novel catalytic properties. Enzymes can catalyze reactions impossible in water, show increased stability and exhibit so-called 'molecular memory'. Enzyme-catalyzed reactions in organic solvents, and even in supercritical fluids and the gas phase, have found numerous potential applications, some of which are already commercialized.[127]

Recent advances in nanotechnologies have led to the discovery of new materials effective for enzyme immobilization. Various nanostructured materials, *e.g.*, carbon nanotubes and nanoparticles, often based on combined organic and inorganic species, provide effective support for enzyme immobilization due to their intrinsically large surface areas and physicochemical properties such as pore size, hydrophilic/hydrophobic balance, and surface chemistry. The immobilization of enzymes is particularly valuable for enhancement of enzyme stability and for multiple uses of enzymes in the process.

Another advantage of using enzymes includes efficient enzyme production at lower costs by using recombinant DNA technologies. Traditional industrial microbiology has been greatly advanced by modern molecular methods. There is a vast catalog of expression plasmids, a great number of readily cultivable microbial cell factories and genetically stable, non-pathogenic host strains or engineered chassis. Many efficient cultivation strategies, as well as cell-free methods for *in vitro* translation, contribute to the development of improved cost-effective industrial production processes.

We live in a world where dangerous chemical weapons are being produced, stockpiled and abused. In response to this, protective and decontamination technologies mandated by the Chemical Weapons Convention need to be developed. The increasing number of newly isolated or tailor-made enzymes with improved technological characteristics and available cost-effective industrial production processes will play a prominent role in this endeavor. Molecular technologies together with novel materials provide a good potential for development of effective enzyme-based decontamination formulation.

Acknowledgements

Support from the project no. LQ1605 from the National Program of Sustainability II of the Ministry of Education of Czech Republic is acknowledged.

References

1. M. Alcalde, M. Ferrer, F. J. Plou and A. Ballesteros, *Trends Biotechnol.*, 2006, **24**, 281.
2. T. D. Sutherland, I. Horne, K. M. Weir, C. W. Coppin, M. R. Williams, M. Selleck, R. J. Russell and J. G. Oakeshott, *Clin. Exp. Pharmacol. Physiol.*, 2004, **31**, 817.
3. T. P. Ruggaber and J. W. Talley, *Pract. Period. Hazard. Toxic Radioact. Waste Manag.*, 2006, **10**, 73.
4. B. K. Singh, *Nat. Rev. Microbiol.*, 2009, **7**, 156.
5. C. S. Karigar and S. S. Rao, *Enzyme Res.*, 2011, **2011**, 805187.
6. Z. Yongliang, L. Lifang, F. Yuanyuan, G. Jianping, F. Jianping and J. Weibin, *Afr. J. Microbiol. Res.*, 2013, **7**, 2127.
7. M. A. Rao, R. Scelza, R. Scotti and L. Gianfreda, *J. Soil Sci. Plant Nutr.*, 2010, **10**, 333.
8. J. J. DeFrank, S. P. Harvey, V. K. Rastogi, T. Cheng and C. S. Penet, *Enzymatic Decontamination: from Concept to Commercialization*, Edgewood Chemical Biological Center, 2004.
9. Z. Prokop, F. Oplustil, J. DeFrank and J. Damborsky, *Biotechnol. J.*, 2006, **1**, 1370.
10. M. M. Blum and A. Richardt, in *Decontamination of Warfare Agents*, ed. A. Richardt and M. M. Blum, Wiley-VCH Verlag GmbH & Co. KGaA, 2008, pp. 135–162.
11. L. Szinicz, *Toxicology*, 2005, **214**, 167.

12. N. H. Johnson, J. C. Larsen and E. Meek, in *Handbook of Toxicology of Chemical Warfare Agents*, ed. R. C. Gupta, Academic Press, Boston, 2nd edn, 2015, pp. 7–15.
13. A. Richardt, in *Decontamination of Warfare Agents*, ed. A. Richardt and M. M. Blum, Wiley-VCH Verlag GmbH & Co. KGaA, 2008, pp. 11–19.
14. A. L. Ternay, Jr., J. K. Smart and H. Salem, in *Chemical Warfare Agents*, CRC Press, 2007, pp. 1–17.
15. K. Ilchmann and J. Revill, *Sci. Eng. Ethics*, 2014, **20**, 753.
16. D. Kimball, *Chemical weapons convention signatories states-parties*, 2015. https://www.armscontrol.org/factsheets/cwcsig.
17. S. Chauhan, S. Chauhan, R. D'Cruz, S. Faruqi, K. K. Singh, S. Varma, M. Singh and V. Karthik, *Environ. Toxicol. Pharmacol.*, 2008, **26**, 113.
18. United Nations, *United Nations News Center*, 2013, http://www.un.org/apps/news/story.asp?NewsID=45856#.
19. United Nations, *United Nations News Services Section*, 2016, http://www.un.org/apps/news/story.asp?NewsID=54795&#.
20. Anonymous, *The Guardian*, 2016, https://www.theguardian.com/world/2016/aug/25/assad-regime-isis-chemical-attacks-syria-un-investigators.
21. R. Browne and E. Labott, *CNN*, 2016, http://www.cnn.com/2016/08/25/politics/un-report-chemical-weapons-syria/index.html.
22. M. I. Greenberg, K. J. Sexton and D. Vearrier, *Clin. Toxicol.*, 2016, **54**, 79.
23. H. Sanderson, P. Fauser, M. Thomsen, P. Vanninen, M. Soderstrom, Y. Savin, I. Khalikov, A. Hirvonen, S. Niiranen, T. Missaen, A. Gress, P. Borodin, N. Medvedeva, Y. Polyak, V. Paka, V. Zhurbas and P. Feller, *Environ. Sci. Technol.*, 2010, **44**, 4389.
24. F. Welling, in *Deterring Terrorism: Theory and Practice*, ed. A. Wenger and A. Wilner, Stanford Security Studies, Stanford, CA, 2012, pp. 273–307.
25. H. Morita, N. Yanagisawa, T. Nakajima, M. Shimizu, H. Hirabayashi, H. Okudera, M. Nohara, Y. Midorikawa and S. Mimura, *Lancet*, 1995, **346**, 290.
26. K. Yokoyama, A. Yamada and N. Mimura, *Am. J. Med.*, 1996, **100**, 586.
27. R. N. Saladi, E. Smith and A. N. Persaud, *Clin. Exp. Dermatol.*, 2006, **31**, 1.
28. H. Thiermann, F. Worek and K. Kehe, *Chem. Biol. Interact.*, 2013, **206**, 435.
29. A. Grabowski, A. Richardt and M. M. Blum, in *Decontamination of Warfare Agents*, ed. A. Richardt and M. M. Blum, Wiley-VCH Verlag GmbH & Co. KGaA, 2008, pp. 55–66.
30. R. Sferopoulos, *A Review of Chemical Warfare Agent (CWA) Detector Technologies and Commercial-Off-The-Shelf Items*, Defense Technical Information Center, Ft. Belvoir, 2009.
31. A. Watson, D. Opresko, R. A. Young, V. Hauschild, J. King and K. Bakshi, in *Handbook of Toxicology of Chemical Warfare Agents*, ed. R. C. Gupta, Academic Press, Boston, 2nd edn, 2015, pp. 87–109.
32. H. P. Benschop and L. P. A. De Jong, *Acc. Chem. Res.*, 1988, **21**, 368.
33. A. N. Bigley and F. M. Raushel, *Biochim. Biophys. Acta*, 2013, **1834**, 443.

34. B. Singh, G. K. Prasad, K. S. Pandey, R. K. Danikhel and R. Vijayaraghavan, *Def. Sci. J.*, 2010, **60**, 428.
35. F. Worek, H. Thiermann and T. Wille, *Chem. Biol. Interact.*, 2016, **259**, 93.
36. R. A. Young and C. B. Bast, in *Handbook of Toxicology of Chemical Warfare Agents*, ed. R. C. Gupta, Academic Press, Boston, 2nd edn, 2015, pp. 69–86.
37. H. J. Altmann and A. Richardt, in *Decontamination of Warfare Agents*, ed. A. Richardt and M. Blum, Wiley-VCH Verlag GmbH & Co. KGaA, 2008, pp. 83–115.
38. C. S. McElroy and B. J. Day, *Biochem. Pharmacol.*, 2016, **100**, 1.
39. M. Pohanka, *Mini Rev. Med. Chem.*, 2012, **12**, 742.
40. S. W. Borron, F. J. Baud, B. Megarbane and C. Bismuth, *Am. J. Emerg. Med.*, 2007, **25**, 551.
41. K. Ganesan, S. K. Raza and R. Vijayaraghavan, *J. Pharm. Bioallied Sci.*, 2010, **2**, 166.
42. M. Boopathi, B. Singh and R. Vijayaraghavan, *Open Text. J.*, 2008, **1**, 1.
43. L. Bromberg, W. R. Creasy, D. J. McGarvey, E. Wilusz and T. A. Hatton, *ACS Appl. Mater. Interfaces*, 2015, **7**, 22001.
44. S. S. Talmage, A. P. Watson, V. Hauschild, N. B. Munro and J. King, *Curr. Org. Chem.*, 2007, **11**, 285.
45. A. W. Khan, S. Kotta, S. H. Ansari, J. Ali and R. K. Sharma, *Def. Sci. J.*, 2013, **63**, 487.
46. P. Jacquet, D. Daude, J. Bzdrenga, P. Masson, M. Elias and E. Chabriere, *Environ. Sci. Pollut. Res.*, 2016, **23**, 8200.
47. A. J. Russell, J. A. Berberich, G. F. Drevon and R. R. Koepsel, *Annu. Rev. Biomed. Eng.*, 2003, **5**, 1.
48. D. Noort, H. P. Benschop and R. M. Black, *Toxicol. Appl. Pharmacol.*, 2002, **184**, 116.
49. M. Moshiri, E. Darchini-Maragheh and M. Balali-Mood, *Daru-J. Pharm. Sci.*, 2012, **20**, 81.
50. M. E. Stroppolo, M. Falconi, A. M. Caccuri and A. Desideri, *Cell. Mol. Life Sci. CMLS*, 2001, **58**, 1451.
51. R. C. diTargiani, L. Chandrasekaran, T. Belinskaya and A. Saxena, *Chem. Biol. Interact.*, 2010, **187**, 349.
52. F. Nachon, X. Brazzolotto, M. Trovaslet and P. Masson, *Chem. Biol. Interact.*, 2013, **206**, 536.
53. O. Lockridge, *Pharmacol. Ther.*, 2015, **148**, 34.
54. F. Worek, H. Thiermann and T. Wille, *Toxicol. Lett.*, 2016, **244**, 143.
55. P. Masson, F. Nachon, C. A. Broomfield, D. E. Lenz, L. Verdier, L. M. Schopfer and O. Lockridge, *Chem. Biol. Interact.*, 2008, **175**, 273.
56. J. J. DeFrank and W. E. White, in *Organofluorines*, ed. A. H. Neilson, Springer, Berlin, Heidelberg, 2002, pp. 295–343.
57. G. Amitai, R. Adani, M. Hershkovitz, P. Bel, I. Rabinovitz and H. Meshulam, *J. Appl. Toxicol.*, 2003, **23**, 225.
58. H. Dvir, I. Silman, M. Harel, T. L. Rosenberry and J. L. Sussman, *Chem. Biol. Interact.*, 2010, **187**, 10.

59. A. Goldenzweig, M. Goldsmith, S. E. Hill, O. Gertman, P. Laurino, Y. Ashani, O. Dym, T. Unger, S. Albeck, J. Prilusky, R. L. Lieberman, A. Aharoni, I. Silman, J. L. Sussman, D. S. Tawfik and S. J. Fleishman, *Mol. Cell*, 2016, **63**, 337.
60. M. C. Franklin, M. J. Rudolph, C. Ginter, M. S. Cassidy and J. Cheung, *Proteins*, 2016, **84**, 1246.
61. A. Gora, J. Brezovsky and J. Damborsky, *Chem. Rev.*, 2013, **113**, 5871.
62. M. Holmquist, *Curr. Protein Pept. Sci.*, 2000, **1**, 209.
63. D. M. Quinn, *Chem. Rev.*, 1987, **87**, 955.
64. E. A. Songa and J. O. Okonkwo, *Talanta*, 2016, **155**, 289.
65. X. Chen, L. Fang, J. Liu and C. G. Zhan, *Biochemistry*, 2012, **51**, 1297.
66. Y. Nicolet, O. Lockridge, P. Masson, J. C. Fontecilla-Camps and F. Nachon, *J. Biol. Chem.*, 2003, **278**, 41141.
67. C. B. Millard, O. Lockridge and C. A. Broomfield, *Biochemistry*, 1995, **34**, 15925.
68. P. Masson, E. Carletti and F. Nachon, *Protein Pept. Lett.*, 2009, **16**, 1215.
69. P. Masson and O. Lockridge, *Arch. Biochem. Biophys.*, 2010, **494**, 107.
70. C. M. Serdar, D. T. Gibson, D. M. Munnecke and J. H. Lancaster, *Appl. Environ. Microbiol.*, 1982, **44**, 246.
71. W. W. Mulbry, J. S. Karns, P. C. Kearney, J. O. Nelson, C. S. McDaniel and J. R. Wild, *Appl. Environ. Microbiol.*, 1986, **51**, 926.
72. P. Masson, in *Handbook of Toxicology of Chemical Warfare Agents*, ed. R. C. Gupta, Academic Press, Boston, 2nd edn, 2015, pp. 1107–1123.
73. I. Horne, T. D. Sutherland, R. L. Harcourt, R. J. Russell and J. G. Oakeshott, *Appl. Environ. Microbiol.*, 2002, **68**, 3371.
74. E. Carletti, L. Lacquamet, M. Loiodice, D. Rochu, P. Masson and F. Nachon, *J. Enzyme Inhib. Med. Chem.*, 2009, **24**, 1045.
75. M. M. Benning, H. Shim, F. M. Raushel and H. M. Holden, *Biochemistry*, 2001, **40**, 2712.
76. I. Cherny, P. Greisen, Y. Ashani, S. D. Khare, G. Oberdorfer, H. Leader, D. Baker and D. S. Tawfik, *ACS Chem. Biol.*, 2013, **8**, 2394.
77. S. Caldwell, J. Newcomb, K. Schlecht and F. Raushel, *Biochemistry*, 1991, **30**, 7438.
78. S. D. Aubert, Y. C. Li and F. M. Raushel, *Biochemistry*, 2004, **43**, 5707.
79. F. C. Hoskin, *Biol. Bull.*, 1969, **137**, 389.
80. F. C. Hoskin, *Science*, 1971, **172**, 1243.
81. F. C. Hoskin and A. H. Roush, *Science*, 1982, **215**, 1255.
82. J. Hartleib and H. Rüterjans, *Protein Expr. Purif.*, 2001, **21**, 210.
83. J. Koepke, E. I. Scharff, C. Lucke, H. Ruterjans and G. Fritzsch, *Acta Crystallogr. Sect. -Biol. Crystallogr.*, 2003, **59**, 1744.
84. J. Hartleib, S. Geschwindner, E. I. Scharff and H. Ruterjans, *Biochem. J.*, 2001, **353**, 579.
85. M. M. Blum, F. Löhr, A. Richardt, H. Rüterjans and J. C. H. Chen, *J. Am. Chem. Soc.*, 2006, **128**, 12750.
86. S. Elias, S. Saphier, I. Columbus and Y. Zafrani, *Environ. Sci. Technol.*, 2014, **48**, 2893.

87. T. Wymore, M. J. Field, P. Langan, J. C. Smith and J. M. Parks, *J. Phys. Chem. B*, 2014, **118**, 4479.
88. A. Richardt, M. M. Blum and S. Mitchell, *Chem. Unserer Zeit*, 2006, **40**, 252.
89. A. M. Latifi, A. Karami and S. Khodi, *Appl. Biochem. Biotechnol.*, 2015, **177**, 624.
90. M. Valiyaveettil, Y. Alamneh, L. Biggemann, I. Soojhawon, B. P. Doctor and M. P. Nambiar, *Biochem. Biophys. Res. Commun.*, 2010, **403**, 97.
91. L. Merone, L. Mandrich, M. Rossi and G. Manco, *Extremophiles*, 2005, **9**, 297.
92. R. D. Gupta, M. Goldsmith, Y. Ashani, Y. Simo, G. Mullokandov, H. Bar, M. Ben-David, H. Leader, R. Margalit, I. Silman, J. L. Sussman and D. S. Tawfik, *Nat. Chem. Biol.*, 2011, 7, 120.
93. N. K. Vyas, A. Nickitenko, V. K. Rastogi, S. S. Shah and F. A. Quiocho, *Biochemistry*, 2010, **49**, 547.
94. Z. Prokop, J. Damborsky, F. Oplustil, A. Jesenska and Y. Nagata, Patent WO 2006/128390 A1, US 7,888,103, 2005.
95. T. Koudelakova, S. Bidmanova, P. Dvorak, A. Pavelka, R. Chaloupkova, Z. Prokop and J. Damborsky, *Biotechnol. J.*, 2013, **8**, 32.
96. S. Popiel and J. Nawala, *Enzyme Microb. Technol.*, 2013, **53**, 295.
97. Y. Nagata, T. Nariya, R. Ohtomo, M. Fukuda, K. Yano and M. Takagi, *J. Bacteriol.*, 1993, **175**, 6403.
98. D. L. Ollis, E. Cheah, M. Cygler, B. Dijkstra, F. Frolow, S. M. Franken, M. Harel, S. J. Remington, I. Silman and J. Schrag, *Protein Eng.*, 1992, **5**, 197.
99. J. Marek, J. Vevodova, I. K. Smatanova, Y. Nagata, L. A. Svensson, J. Newman, M. Takagi and J. Damborsky, *Biochemistry*, 2000, **39**, 14082.
100. A. J. Oakley, M. Klvana, M. Otyepka, Y. Nagata, M. C. J. Wilce and J. Damborsky, *Biochemistry*, 2004, **43**, 870.
101. K. Hynkova, Y. Nagata, M. Takagi and J. Damborsky, *FEBS Lett.*, 1999, **446**, 177.
102. E. Chovancova, J. Kosinski, J. M. Bujnicki and J. Damborsky, *Proteins*, 2007, **67**, 305.
103. K. H. Verschueren, F. Seljée, H. J. Rozeboom, K. H. Kalk and B. W. Dijkstra, *Nature*, 1993, **363**, 693.
104. Z. Prokop, M. Monincova, R. Chaloupkova, M. Klvana, Y. Nagata, D. B. Janssen and J. Damborsky, *J. Biol. Chem.*, 2003, **278**, 45094.
105. A. Li and Z. Shao, *PLoS One*, 2014, **9**, e89144.
106. M. Sundaramoorthy, J. Terner and T. L. Poulos, *Structure*, 1995, **3**, 1367.
107. A. K. Kanugula, E. R. Repalle, J. P. Pandey, G. Sripad, C. K. Mitra, D. K. Dubey and D. Siddavattam, *Indian J. Biochem. Biophys.*, 2011, **48**, 29.
108. W. Wei, J. Du, J. Li, M. Yan, Q. Zhu, X. Jin, X. Zhu, Z. Hu, Y. Tang and Y. Lu, *Adv. Mater.*, 2013, **25**, 2212.
109. T. Ohmori, K. Kawahara, K. Nakayama, A. Shioda, S. Ishikawa, M. Kanamori-Kataoka, S. Kishi, A. Komano and Y. Seto, *Forensic Toxicol.*, 2013, **31**, 37.

110. V. Stepankova, S. Bidmanova, T. Koudelakova, Z. Prokop, R. Chaloupkova and J. Damborsky, *ACS Catal.*, 2013, **3**, 2823.
111. N. Suthiwangcharoen and R. Nagarajan, *Biomacromolecules*, 2014, **15**, 1142.
112. F. Arduini, V. Scognamiglio, D. Moscone and G. Palleschi, in *Biosensors for Security and Bioterrorism Applications*, ed. D. P. Nikolelis and G.-P. Nikoleli, Springer International Publishing, 2016, pp. 115–139.
113. M. Burnworth, S. J. Rowan and C. Weder, *Chem. – Eur. J.*, 2007, **13**, 7828.
114. S. Paliwal, in *Development of Enzyme-based Biosensors for the Detection of Organophosphate Neurotoxins*, Dissertation Thesis, 2008.
115. M. Pohanka, V. Vlcek, J. Zdarova-Karasova, K. Kuca, J. Cabal and J. Fusek, *Adv. Mil. Technol.*, 2012, **7**, 83.
116. S. Bidmanova, M. S. Steiner, M. Stepan, K. Vymazalova, M. A. Gruber, A. Duerkop, J. Damborsky, Z. Prokop and O. S. Wolfbeis, *Anal. Chem.*, 2016, **88**, 6044.
117. B. M. Paddle, *Biosens. Bioelectron.*, 1996, **11**, 1079.
118. UniProt Consortium, *Nucleic Acids Res.*, 2015, **43**, D204.
119. P. Vanacek, E. Sebestova, P. Babkova, S. Bidmanova, L. Daniel, P. Dvorak, V. Stepankova, R. Chaloupkova, J. Brezovsky, Z. Prokop and J. Damborsky, *ACS Catal.*, 2018, **8**, 2402.
120. J. Bendl, J. Stourac, E. Sebestova, O. Vavra, M. Musil, J. Brezovsky and J. Damborsky, *Nucleic Acids Res.*, 2016, **44**, W479.
121. R. K. Kuipers, H.-J. Joosten, W. J. H. van Berkel, N. G. H. Leferink, E. Rooijen, E. Ittmann, F. van Zimmeren, H. Jochens, U. Bornscheuer, G. Vriend, V. A. P. M. dos Santos and P. J. Schaap, *Proteins*, 2010, **78**, 2101.
122. G. Dantas, B. Kuhlman, D. Callender, M. Wong and D. Baker, *J. Mol. Biol.*, 2003, **332**, 449.
123. J. Schymkowitz, J. Borg, F. Stricher, R. Nys, F. Rousseau and L. Serrano, *Nucleic Acids Res.*, 2005, **33**, W382.
124. C. Zhu, D. D. Mowrey and N. V. Dokholyan, *Methods Mol. Biol. Clifton NJ*, 2017, **1529**, 227.
125. M. Musil, J. Stourac, J. Bendl, J. Brezovsky, Z. Prokop, J. Zendulka, T. Martinek, D. Bednar and J. Damborsky, *Nucleic Acids Res.*, 2017, **45**, W393.
126. H. J. Wijma, R. J. Floor, P. A. Jekel, D. Baker, S. J. Marrink and D. B. Janssen, *Protein Eng. Des. Sel.*, 2014, **27**, 49.
127. A. M. Klibanov, *Nature*, 2001, **409**, 241.
128. SIB Swiss Institute of Bioinformatics, *ExPASy – Bioinformatics Resource Portal*, 2016, http://enzyme.expasy.org/.
129. E. I. Scharff, J. Koepke, G. Fritzsch, C. Lücke and H. Rüterjans, *Structure*, 2001, **9**, 493.
130. M. B. Colovic, D. Z. Krstic, T. D. Lazarević-Pašti, A. M. Bondzic and V. M. Vasic, *Curr. Neuropharmacol.*, 2013, **11**, 315.

Subject Index

Chemistry